D0776772

Benzimidazoles and Congeneric Tricyclic Compounds

IN TWO PARTS

PART TWO

This is the fortieth volume in the series

THE CHEMISTRY OF HETEROCYCLIC COMPOUNDS

THE CHEMISTRY OF HETEROCYCLIC COMPOUNDS

A SERIES OF MONOGRAPHS

ARNOLD WEISSBERGER and EDWARD C. TAYLOR

Editors

BENZIMIDAZOLES AND CONGENERIC TRICYCLIC COMPOUNDS

PART 2

Edited by

P. N. PRESTON

DEPARTMENT OF CHEMISTRY,
HERIOT-WATT UNIVERSITY,
EDINBURGH, SCOTLAND

With contributions by

M. F. G. STEVENS
DEPARTMENT OF PHARMACY,
UNIVERSITY OF ASTON,
BIRMINGHAM,
ENGLAND

G. TENNANT
DEPARTMENT OF CHEMISTRY,
UNIVERSITY OF EDINBURGH,
EDINBURGH,
SCOTLAND

AN INTERSCIENCE ® PUBLICATION

JOHN WILEY & SONS

New York . Chichester . Brisbane . Toronto

An Interscience ® Publication

Library of Congress Cataloging in Publication Data:
Main entry under title:

Benzimidazoles and congeneric tricyclic compounds.

(The Chemistry of heterocyclic compounds;
-v. 40, pt. 1 ISSN 0069-3154)
"An Interscience publication."
Includes index
1. Benzimidazoles. I. Preston, P. N.
QD401.B46 547'.593 80-17383
ISBN 0-471-03792-3 (v. 1)
ISBN 0-471-08189-2 (v. 2)

The Chemistry of Heterocyclic Compounds

The chemistry of heterocyclic compounds is one of the most complex branches of organic chemistry. It is equally interesting for its theoretical implications, for the diversity of its synthetic procedures, and for the physiological and industrial significance of heterocyclic compounds.

A field of such importance and intrinsic difficulty should be made as readily accessible as possible, and the lack of a modern detailed and comprehensive presentation of heterocyclic chemistry is therefore keenly felt. It is the intention of the present series to fill this gap by expert presentations of the various branches of heterocyclic chemistry. The subdivisions have been designed to cover the field in its entirety by monographs which reflect the importance and the interrelations of the various compounds, and accommodate the specific interests of the authors.

In order to continue to make heterocyclic chemistry as readily accessible as possible, new editions are planned for those areas where the respective volumes in the first edition have become obsolete by overwhelming progress. If, however, the changes are not too great so that the first editions can be brought up-to-date by supplementary volumes, supplements to the respective volumes will be published in the first edition.

ARNOLD WEISSBERGER

Research Laboratories
Eastman Kodak Company
Rochester, New York

EDWARD C. TAYLOR

Princeton University
Princeton, New Jersey

Preface to Part 2

More than 25 years have elapsed since the publication in this series of *Imidazole and Its Derivatives* by Klaus Hofmann. In updating this work, Leroy Townsend has undertaken the task of editing a volume on monocyclic imidazoles, and the present book covers the chemistry of benzimidazole and its dihydro derivatives, as well as congeneric tricyclic compounds that contain a condensed benzimidazole moiety. Because many ring systems are covered, it has proved necessary to divide the volume into Part 1 (Chapters 1 to 5) and Part 2 (Chapters 6 to 10).

Chapters 1 to 3 on benzimidazoles, benzimidazole *N*-oxides, and dihydro derivatives update the book of Hofmann through Volume 87 of *Chemical Abstracts.* The chemistry of tricyclic compounds containing a condensed benzimidazole moiety is covered comprehensively from early literature through the same Volume 87 of *Chemical Abstracts.*

Chapters 4 to 9 on the condensed ring systems are organized in terms of the position and size of the ring fused to the benzimidazole skeleton (denoted "6-5"). Thus Chapters 4 through 8 are concerned with compounds in which fusion of the third ring is at the benzo and imidazole rings respectively.

Chapter 9 deals with the chemistry of tricyclic compounds in which a benzimidazole moiety may be considered to be formally annulated from N-1 to C-7.

The growth of benzimidazole chemistry in the past 25 years has paralleled that of purines and stems from the determination of the partial structures of nucleic acids in the early 1950s. Benzimidazoles and congeneric compounds are substrates that might act as inhibitors in nucleic acid biosynthesis, and their relative ease of preparation and low cost make them attractive as potential pharmacological agents. The variety of marketed products described in chapter 10 bears witness to the large commitment to benzimidazole chemistry. I hope that this book will stimulate further research, particularly on the synthesis of new tricyclic derivatives and related condensed analogs.

I am indebted to a number of friends and colleagues who have contributed to this book. It has been a pleasure to collaborate with David Smith and with Malcolm Stevens and George Tennant, and I thank them for their large collective contribution. Information on commercially marketed products is difficult to obtain, but my task was simplified with the generous assistance of Colin C. Beard, Gerald Farrow, Janet M. Shether, Brian K. Snell, and Ian S. Swanson. I also thank my wife, Veronica, who carried out an initial estimate of the magnitude of literature on benzimidazoles and congeneric tricyclic compounds. Thanks are due also to Susan Bobby who typed part of the manuscript, Anthony F. Fell who translated a number of documents from Russian, and my former research students Alex Davidson and Ian E. P. Murray who helped to check the manuscript. Finally, I express my appreciation of the help and enthusiasm of the Series Editors, Edward C. Taylor and Arnold Weissberger, of Stanley F. Kudzin, and of the staff of John Wiley and Sons, Inc.

P. N. Preston

Edinburgh, Scotland
January 1981

Contents

PART TWO

PART ONE

CHAPTER 6

Condensed Benzimidazoles of Type 6-5-5

G. TENNANT

6.1. TRICYCLIC 6-5-5 FUSED BENZIMIDAZOLES WITH NO ADDITIONAL HETEROATOMS

Union of a five-membered carbocyclic ring in 6-6-5 fashion with benzimidazole involves fusion across the *N*(1)–*C*(2) bond in the latter and gives rise to a single structural type corresponding to the pyrrolo[1,2-*a*]benzimidazole ring system (Scheme 6.1). The latter is encountered in 1*H* (**6.1**), 3*H*

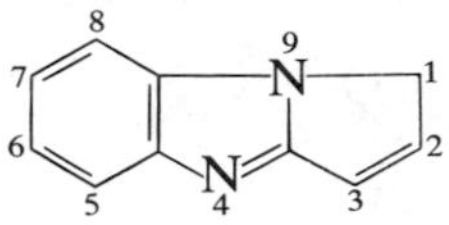

(6.1) 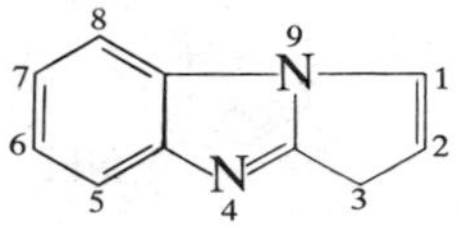 **(6.2)**

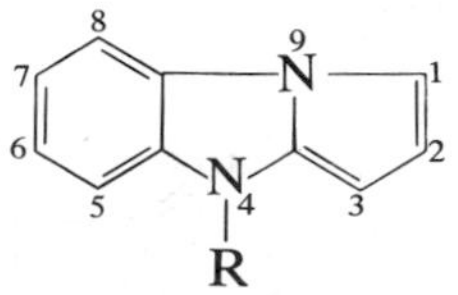

(6.3) 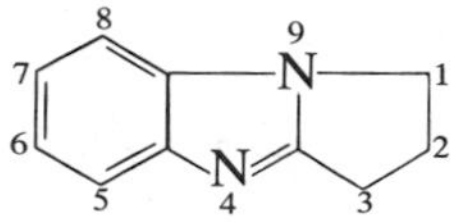 **(6.4)**

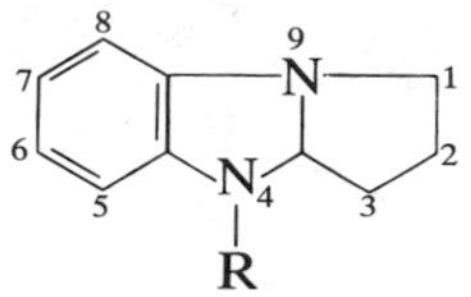

(6.5)

Scheme 6.1

(**6.2**), and 4*H* (**6.3**) tautomeric forms as well as in the guise of 1*H*-2,3-dihydro (**6.4**) and 1*H*-2,3,3*a*,4-tetrahydro (**6.5**) structures (Scheme 6.1 and Table 6.1). Of these the 4*H* system (**6.3**) has attracted most attention because of its potentially aromatic character.

TABLE 6.1. TRICYCLIC 6-5-5 FUSED BENZIMIDAZOLE RING SYSTEMS WITH NO ADDITIONAL HETEROATOMS

Structure[a]	Name[b]
(6.1)	1*H*-Pyrrolo[1,2-*a*]benzimidazole
(6.2)	3*H*-Pyrrolo[1,2-*a*]benzimidazole
(6.3)	4*H*-Pyrrolo[1,2-*a*]benzimidazole
(6.4)	2,3-Dihydro-1*H*-pyrrolo[1,2-*a*]benzimidazole
(6.5)	2,3,3*a*,4-Tetrahydro-1*H*-pyrrolo[1,2-*a*]benzimidazole

[a] Cf. Scheme 6.1.
[b] Based on the Ring Index.

6.1.1. Synthesis

Ring-closure Reactions of Benzimidazole Derivatives

The condensation[1–3] of *ortho*-phenylenediamine (**6.6**) and its derivatives with maleic anhydrides (**6.7**) to give pyrrolo[1,2-*a*]benzimidazol-1-ones (**6.10**) (Scheme 6.2 and Table 6.2) is plausibly explained in terms of the

Scheme 6.2

TABLE 6.2. SYNTHESIS OF 1*H*-PYRROLO[1,2-*a*]BENZIMIDAZOLES BY RING-CLOSURE REACTIONS OF BENZIMIDAZOLE DERIVATIVES AND RELATED PROCESSES.

Starting materials	Reaction conditions[a]	Product	Yield (%)	m.p. (°C)	Solvent of crystallization	Crystal form	Ref.
(**6.6**) + (**6.7**; $R^1 = R^2 = Ph$)	*A*	(**6.10**; $R^1 = R^2 = Ph$)	—[b]	186	—[c]	Brown needles	1
(**6.8**; $R^1 = R^2 = Ph$)	*B*	(**6.10**; $R^1 = R^2 = Ph$)	—[b]	—	—	—	1
(**6.9**; $R^1 = R^2 = Ph$)	*C*	(**6.10**; $R^1 = R^2 = Ph$)	85	—	—	—	1
(**6.6**) + (**6.7**; $R^1 = R^2 = H$)	—[d]	(**6.10**; $R^1 = R^2 = Me$)	—[b]	—[e]	—[c]	—[f]	3
(**6.6**) + (**6.7**; $R^1 = OMe$, $R^2 = Ph$)	*D*	(**6.10** $R^1 = OMe$, $R^2 = Ph$)	7	169	Acetone–chloroform	Orange-red crystals	2
(**6.6**) + (**6.14**; $R = H$, $R^1 = Ph$)	*E*	(**6.10**; $R^1 = R^2 = Ph$)	2	184	Ethanol	Red solid	4
(**6.6**) + (**6.14**; $R = H$, $R^1 = OMe$)	*F*	(**6.10**; $R^1 = OH$, $R^2 = Ph$)	76	258 (decomp.)	Dimethylformamide–water	Red needles	2
(**6.6**) + (**6.14**; $R = NO_2$, $R^1 = OH$)	*G*	(**6.10**; $R^1 = OH$, $R^2 = p\text{-}NO_2C_6H_4$)	57	316 (decomp.)	Dimethylformamide–acetic acid	Red powder	5

[a] *A* = heat in the melt; *B* = heat in a high b.p. solvent or in the melt; *C* = heat at 250°; *D* = heat at 180–190°/0.5 hr; *E* = AcOH/(reflux)(40 min) *F* = AcOH/(room temp)(12 hr); *G* = AcOH/(room temp)(2 hr).

[b] Yield not quoted.

[c] Solvent not specified.

[d] Reaction conditions not specified.

[e] Melting point not quoted.

[f] Crystal form not specified.

formation and thermal cyclization of 2-benzimidazolylacrylic acid intermediates (**6.8**), thus justifying the inclusion of such reactions under the present heading. Indeed, the thermal cyclization (Scheme 6.2) of the acid (**6.8**; $R^1 = R^2 = Ph$) to the pyrrolo[1,2-*a*]benzimidazolone (**6.10**; $R^1 = R^2 = Ph$) (albeit in unspecified yield) has been demonstrated.[1] Equally, however, ring-closure reactions of the type [Scheme 6.2; (**6.6**) + (**6.7**) → (**6.10**)] may involve the corresponding *N*-(2-aminophenyl)maleimides (**6.9**) as intermediates, since it has also been shown[1] that the compound (**6.9**; $R^1 = R^2 = Ph$) undergoes thermal cyclization at 250° to afford the pyrrolo[1,2-*a*]benzimidazol-1-one (**6.10**; $R^1 = R^2 = Ph$) in high yield (Table 6.2). Information on the general scope and efficiency of 1*H*-pyrrolo[1,2-*a*]benzimidazole syntheses based on the condensation of *ortho*-phenylenediamines with maleic anhydrides is lacking, and in view of their simple character such reactions merit more detailed study. Of particular interest is the possibility of isomer formation when unsymmetrically substituted maleic anhydrides are employed as substrates. Thus, irrespective of whether a benzimidazole derivative or an *N*-(2-aminophenyl)maleimide is involved as intermediate ring-closure using an unsymmetrically substituted maleic anhydride, (**6.7**; $R^1 \neq R^2$) should lead to two possible isomeric pyrrolo[1,2-*a*]benzimidazol-1-ones. In the only extant example[2] of this situation the condensation of *ortho*-phenylenediamine (**6.6**) with the methoxy-substituted anhydride (**6.7**; $R^1 = OMe$, $R^2 = Ph$) led exclusively to the 3-methoxypyrrolobenzimidazolone (**6.10**; $R^1 = OMe$, $R^2 = Ph$), whose formation is consistent with either preferential initial [Scheme 6.2; (**6.6**) + (**6.7**; $R^1 = OMe$, $R^2 = Ph$) → (**6.8**; $R^1 - OMc$, $R^2 = Ph$)] or final [Scheme 6.2; (**6.9**; $R^1 = OMe$, $R^2 = Ph$) → (**6.10**; $R^1 = OMe$, $R^2 = Ph$)] condensation between an amino group and the carbonyl group not deactivated toward nucleophilic attack by the methoxyl substituent. The nature of *ortho*-phenylenediamine–maleic anhydride condensations is such that the products are of necessity 1*H*-pyrrolo[1,2-*a*]-benzimidazol-1-ones (**6.10**) and not the isomeric 3*H*-pyrrolo[1,2-*a*]-benzimidazol-3-ones (**6.11**). 1*H*-Pyrrolo[1,2-*a*]benzimidazol-1-ones are also the end-products of the reactions of *ortho*-phenylenediamines with cyclobutene-3,4-diones in acetic acid [Scheme 6.2; (**6.6**) + (**6.14**) → (**6.10**)].[2,4,5] This type of condensation gives very poor yields (Table 6.2) when 1,2-diphenylcyclobutene-3,4-dione (**6.14**; $R = H$, $R^1 = Ph$) is used as substrate,[4] whereas employing 2-aryl-1-hydroxycyclobutene-3,4-diones (**6.14**; $R^1 = OH$) leads[2,5] to the corresponding 2-aryl-3-hydroxy-1*H*-pyrrolo-[1,2-*a*]benzimidazol-1-ones (**6.10**; $R^1 = OH$, R^2 = phenyl or *p*-nitrophenyl) in good yield (Table 6.2). The latter reactions are suggested[2] to follow a course (Scheme 6.2) involving the formation and ring expansion–ring contraction of a quinoxaline intermediate [Scheme 6.2; (**6.6**) + (**6.14**; $R^1 = OH$) → (**6.15**) → (**6.13**) → (**6.12**) → (**6.10**)]. The reactions (Scheme 6.3) of 2-azido-1-methylbenzimidazole (**6.16**) with acetylenic esters (methyl propiolate, dimethyl acetylenedicarboxylate) in acetonitrile under reflux as well as resulting in the expected cycloaddition to the azido group, are

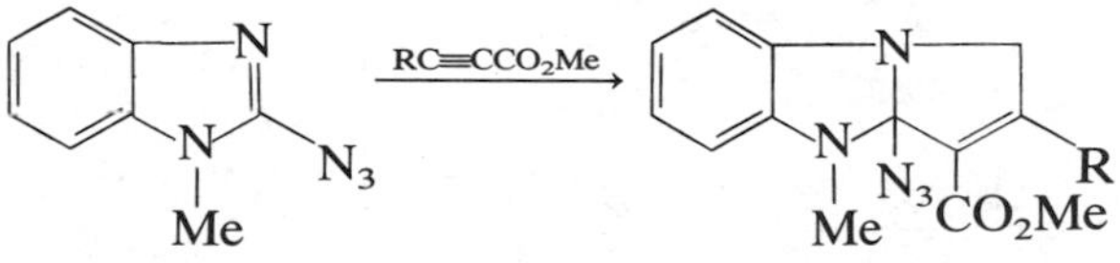

$RC{\equiv}CCO_2Me$

(**6.16**) (**6.17**)

Scheme 6.3

reported[6] to afford moderate to high yields of products formulated as the 1*H*-pyrrolo[1,2-*a*]benzimidazole derivatives (**6.17**; R = H or CO_2Me), though probably inadvertently, since the combustion analysis and mass spectral properties[6] of the supposed diester product (**6.17**; R = CO_2Me) are consistent with a C_{14} rather than a C_{15} structure. Moreover, the ^1H NMR spectra reported[6] for these products lack signals attributable to the *C*(1) methylene protons in the structures (**6.17**; R = H or CO_2Me). Closer scrutiny of the structures of these compounds is in any case warranted in view of their unorthodox mode of formation (i.e., annelation of the imidazole ring in preference to the anticipated exclusive cycloaddition to the azido group).

3*H*-Pyrrolo[1,2-*a*]benzimidazoles are readily accessible, usually in high yield (Table 6.3), by the thermal condensation of *ortho*-phenylenediamine and its derivatives with γ-ketocarboxylic acids [Scheme 6.4; (**6.6**) + (**6.18**) → → (**6.20**)].[7–9] The probable intermediacy of the corresponding 2-benzimidazolylethyl ketones in these reactions is supported by the ready thermal cyclization of 4-(2-benzimidazolyl)-2-butanone (**6.19**; R^1 = Me, $R^2 = R^3 = R^4$ = H) to 1-methyl-3*H*-pyrrolo[1,2-*a*]benzimidazole (**6.20**; R^1 = Me, $R^2 = R^3 = R^4$ = H).[7] In some instances the γ-keto acid can be replaced by a suitable γ-ketonitrile, in which case condensation is conducted under acidic conditions (Table 6.3).[9] Reaction (Scheme 6.4) of *ortho*-phenylenediamine (**6.6**) with 1,2-diaroyl-1,2-diphenylethylenes (**6.23**) in refluxing methanolic acetic acid affords high yields (Table 6.3) of 1,2,3,3-tetraaryl-3*H*-pyrrolo-[1,2-*a*]benzimidazoles (**6.20**; $R^1 = R^3$ = Ar, $R^2 = R^4$ = phenyl).[10] These reactions are readily explained[10] in terms of initial condensation to give benzimidazole derivatives convertible by cyclization and subsequent vinylogous Wagner–Meerwein rearrangement into the observed products [Scheme 6.4; (**6.6**) + (**6.23**) → (**6.22**) → (**6.21**) → (**6.20**)].

1-Substituted 2-alkylbenzimidazoles [Scheme 6.5; (**6.24**)] are quaternized by α-halogeno ketones to give benzimidazolium salts (**6.25**), which are smoothly cyclized by base treatment to afford the corresponding 4*H*-pyrrolo[1,2-*a*]benzimidazoles (**6.27**) in high yield (Table 6.4).[11–19] This highly versatile synthetic method has been exploited[11–19] for the synthesis of a wide variety of 4*H*-pyrrolo[1,2-*a*]benzimidazoles bearing alkyl or aryl substituents at all three possible sites in the pyrrole nucleus (Table 6.4). The cyclization step [(**6.25**) → → (**6.27**)] in these syntheses is most commonly effected by simply heating the isolated benzimidazolium salt (**6.25**) under

TABLE 6.3. SYNTHESIS OF 3*H*-PYRROLO[1,2-*a*]BENZIMIDAZOLES BY RING CLOSURE REACTIONS OF ORTHOPHENYLENEDIAMINE

Starting materials	Reaction conditions[a]	Product	Yield (%)	m.p. (°C)	Ref.
(**6.6**) + (**6.18**; $R^1 = Me$, $R^2 = R^3 = R^4 = H$)	*A*	(**6.20**; $R^1 = Me$, $R^2 = R^3 = R^4 = H$)[b]	88	205–207	7, 8
(**6.6**) + (**6.18**; $R^1 = R^3 = Me$, $R^2 = R^4 = H$)	*A*	(**6.20**; $R^1 = R^3 = Me$, $R^2 = R^4 = H$)[c]	87	210–212	7, 8
(**6.6**) + (**6.18**; $R^1 = Me$, $R^2 = R^4 = H$, $R^3 = Et$)	*A*	(**6.20**; $R^1 = Me$, $R^2 = R^4 = H$, $R^3 = Et$)[d]	73	154–156	7, 8
(**6.6**) + (**6.18**; $R^1 = Me$, $R^2 = R^4 = H$, $R^3 = Pr^n$)	*A*	(**6.20**; $R^1 = Me$, $R^2 = R^4 = H$, $R^3 = Pr^n$)[f]	—[e]	116–118	7, 8
(**6.6**) + (**6.18**; $R^1 = Me$, $R^2 = R^4 = H$, $R^3 = Bu^n$)	*A*	(**6.20**; $R^1 = Me$, $R^2 = R^4 = H$, $R^3 = Bu^n$)[g]	—[e]	110–111	7, 8
(**6.6**) + (**6.18**; $R^1 = R^3 = R^4 = Me$, $R^2 = H$, CN for CO_2H)	*B*	(**6.20**; $R^1 = R^3 = R^4 = Me$, $R^2 = H$)	71	180–189	9
(**6.6**) + (**6.23**; Ar = Ph)	*C*	(**6.20**; $R^1 = R^2 = R^3 = R^4 = Ph$)	80	206[h]	10
(**6.6**) + (**6.23**; Ar = *p*-$MeOC_6H_4$)	*C*	(**6.20**; $R^1 = R^3 =$ *p*-$MeOC_6H_4$, $R^2 = R^4 = Ph$)	80	246[h]	10

[a] *A* = 80–200°; *B* = HCl/reflux; *C* = AcOH, MeOH/(reflux)/(2 hr).
[b] Forms a hydrochloride, m.p. 281–283°.
[c] Forms a hydrochloride, m.p. 260–263°.
[d] Forms a hydrochloride, m.p. 147°.
[e] Yield not quoted.
[f] Forms a hydrochloride, m.p. 204–206°.
[g] Forms a hydrochloride, m.p. 210–213°.
[h] Crystallized from methanol–acetic acid.

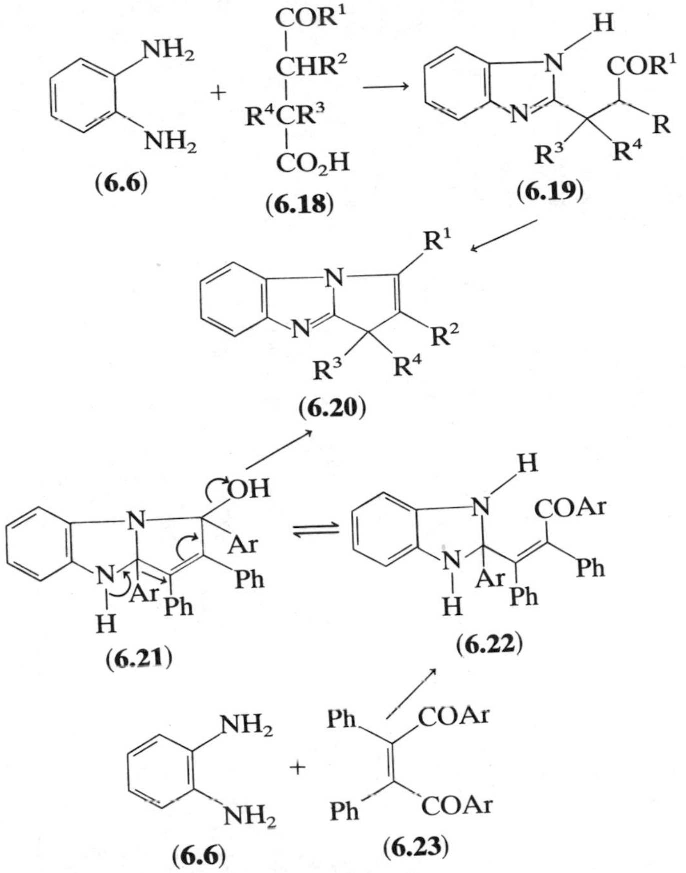

NH_2 NH_2 (6.6) + COR^1 CHR^2 R^4CR^3 CO_2H (6.18) → H N N COR^1 R R^3 R^4 (6.19)

N N R^1 R^2 R^3 R^4 (6.20)

OH N N Ar Ph Ar Ph H (6.21) ⇌ H N COAr N Ar Ph Ph H (6.22)

NH_2 NH_2 (6.6) + Ph COAr Ph COAr (6.23)

Scheme 6.4

N N CH_2R^2 R^1 (6.24) —$R^3COCHBr$ (R^4)→ R^4 $CHCOR^3$ $\overset{+}{N}$ N CH_2R^2 R^1 Br^- (6.25)

R^4 $^-CCOR^3$ $\overset{+}{N}$ N CH_2R^2 R^1 (6.26) → N N R^4 R^3 R^1 R^2 (6.27)

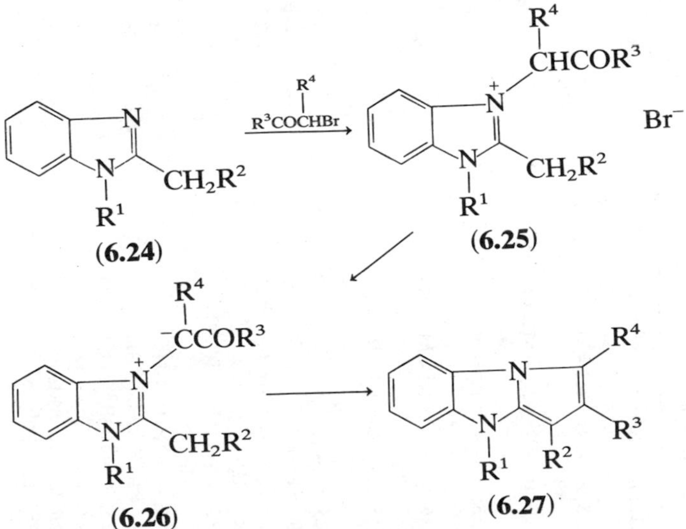

Scheme 6.5

TABLE 6.4. SYNTHESIS OF 4*H*-PYRROLO[1,2-*a*]BENZIMIDAZOLES BY RING-CLOSURE REACTIONS OF BENZIMIDAZOLE DERIVATIVES

Starting material	Reaction conditions[a]	Product	Yield (%)	m.p. (°C)	Solvent of crystallisation	Crystal form	Ref.
(**6.25**; $R^1 = R^3 = Me$, $R^2 = R^4 = H$)	*A*	(**6.27**; $R^1 = R^3 = Me$, $R^2 = R^4 = H$)[b]	46	90	Ethanol	Colorless plates	12
(**6.25**; $R^1 = R^2 = R^3 = Me$, $R^4 = H$)	*B*	(**6.27**; $R^1 = R^2 = R^3 = Me$, $R^4 = H$)[c]	50	99	Methanol	Colorless plates[d]	12
(**6.25**; $R^1 = R^3 = R^4 = Me$, $R^2 = H$)	*C*	(**6.27**; $R^1 = R^3 = R^4 = Me$, $R^2 = H$)[e]	69	96	Methanol	Colorless plates[f]	12
(**6.25**; $R^1 = R^3 = R^4 = Me$, $R^2 = H$)	*D*	(**6.27**; $R^1 = R^3 = R^4 = Me$, $R^2 = H$)	65	96	Isopropanol	Colorless plates[f]	13
(**6.25**; $R^1 = R^3 = R^4 = Me$, $R^2 = H$)	*E*	(**6.27**; $R^1 = R^3 = R^4 = Me$, $R^2 = H$)[g]	74	114–116	Ethanol–ether	—[h]	15
(**6.25**; $R^1 = R^2 = R^3 = R^4 = Me$)	*F*	(**6.27**; $R^1 = R^2 = R^3 = R^4 = Me$)[i]	39	165–166	Ethanol	Colorless crystals	14
(**6.25**; $R^1 = R^3 = Me$, $R^2 = H$, $R^4 = Pr^n$)	*E*	(**6.27**; $R^1 = R^3 = Me$, $R^2 = H$, $R^4 = Pr^n$)	70	136–138	Acetone	—[h]	15
(**6.25**; $R^1 = R^3 = Me$, $R^2 = Ph$, $R^4 = H$)	*E*	(**6.27**; $R^1 = R^3 = Me$, $R^2 = Ph$, $R^4 = H$)[j]	86	152–153	Ethanol	—[h]	15
(**6.25**; $R^1 = Et$, $R^2 = R^4 = H$, $R^3 = Me$)	*E*	(**6.27**; $R^1 = Et$, $R^2 = R^4 = H$, $R^3 = Me$)[j]	—[k]	177–178	Ethanol	—[h]	11
(**6.25**; $R^1 = Et$, $R^2 = H$, $R^3 = R^4 = Me$)	*E*	(**6.27**; $R^1 = Et$, $R^2 = H$, $R^3 = R^4 = Me$)[j]	78	138–140	Water	—[h]	15
(**6.25**; $R^1 = Et$, $R^2 = H$, $R^3 = Me$, $R^4 = Ph$)	*E*	(**6.27**; $R^1 = Et$, $R^2 = H$, $R^3 = Me$, $R^4 = Et$)[j]	82	139–140	Acetone–water	—[h]	15
(**6.25**; $R^1 = R^3 = Me$, $R^2 = R^4 = H$)[l]	*E*	(**6.27**; $R^1 = R^3 = Me$, $R^2 = R^4 = H$)[i,m]	90	207–208	Dimethylformamide–water	—[h]	15
(**6.25**; $R^1 = CH_2Ac$, $R^2 = R^4 = H$, $R^3 = Me$)	*E*	(**6.27**; $R^1 = CH_2Ac$, $R^2 = R^4 = H$, $R^3 = Me$)	74	169–171 (decomp.)	Ethanol	—[h]	19
(**6.25**; $R^1 = Me$, $R^2 = R^4 = H$, $R^3 = Ph$)	*E*	(**6.27**; $R^1 = Me$, $R^2 = R^4 = H$, $R^3 = Ph$)[n]	95	109–111	Ethanol	Colorless solid	11

(**6.25**; $R^1 = Me, R^2 = R^4 = H, R^3 = Ph$)	*A*	(**6.27**; $R^1 = Me, R^2 = R^4 = H, R^3 = Ph$)[c]	—[k]	114	Ethanol	Colorless needles	12
(**6.25**; $R^1 = Et, R^2 = R^4 = H, R^3 = Ph$)	*E*	(**6.27**; $R^1 = Et, R^2 = R^4 = H, R^3 = Ph$)[p]	97	119–120	Ethanol	Colorless solid	11
(**6.25**; $R^1 = CH_2Ph, R^2 = R^4 = H, R^3 = Ph$)	*E*	(**6.27**; $R^1 = CH_2Ph, R^2 = R^4 = H, R^3 = Ph$)	92	123–124 (decomp.)	Ethanol	—[h]	17
(**6.25**; $R^1 = Me, R^2 = R^4 = H, R^3 = p\text{-}BrC_6H_4$)	*E*	(**6.27**; $R^1 = Me, R^2 = R^4 = H, R^3 = p\text{-}BrC_6H_4$)[q]	86	155–156	Ethanol	Colorless solid	11
(**6.25**; $R^1 = Me, R^2 = R^4 = H, R^3 = p\text{-}NO_2C_6H_4$)	*E*	(**6.27**; $R^1 = Me, R^2 = R^4 = H, R^3 = p\text{-}NO_2C_6H_4$)	90	180–182	Ethanol–dimethylformamide	Red crystals	11
(**6.26**; $R^1 = Me, R^2 = R^4 = H, R^3 = p\text{-}NO_2C_6H_4$)	*G*	(**6.27**; $R^1 = Me, R^2 = R^4 = H, R^3 = p\text{-}NO_2C_6H_4$)	96–98	—	—	—	16
(**6.25**; $R^1 = Me, R^2 = R^4 = H, R^3 = m\text{-}NO_2C_6H_4$)	*E*	(**6.27**; $R^1 = Me, R^2 = R^4 = H, R^3 = m\text{-}NO_2C_6H_4$)	92	170.5–171.5	Dimethylformamide	Red crystals	11
(**6.26**; $R^1 = Me, R^2 = R^4 = H, R^3 = m\text{-}NO_2C_6H_4$)	*G*	(**6.27**; $R^1 = Me, R^2 = R^4 = H, R^3 = m\text{-}NO_2C_6H_4$)	96–98	—	—	—	16
(**6.25**; $R^1 = Me, R^2 = R^4 = H, R^3 = p\text{-}MeC_6H_4$)	*E*	(**6.27**; $R^1 = Me, R^2 = R^4 = H, R^3 = p\text{-}MeC_6H_4$)	67	131–132	Ethanol	—[h]	13
(**6.25**; $R^1 = Me, R^2 = R^4 = H, R^3 = p\text{-}MeOC_6H_4$)	*E*	(**6.27**; $R^1 = Me, R^2 = R^4 = H, R^3 = p\text{-}MeOC_6H_4$)	84	141–142	Ethanol	—[h]	15
(**6.25**; $R^1 = Et, R^2 = R^4 = H, R^3 = p\text{-}BrC_6H_4$)	*E*	(**6.27**; $R^1 = Et, R^2 = R^4 = H, R^3 = p\text{-}BrC_6H_4$)[r]	95	123–124	Ethanol	Colorless solid	11
(**6.25**; $R^1 = Et, R^2 = R^4 = H, R^3 = p\text{-}NO_2C_6H_4$)	*E*	(**6.27**; $R^1 = Et, R^2 = R^4 = H, R^3 = p\text{-}BrC_6H_4$)[s]	81	118–121	Acetone–light petroleum	Red crystals	11
(**6.25**; $R^1 = Et, R^2 = R^4 = H, R^3 = 2\text{-thienyl}$)	*E*	(**6.27**; $R^1 = Et, R^2 = R^4 = H, R^3 = 2\text{-thienyl}$)[j]	79	178–179	Water	—[h]	15
(**6.25**; $R^1 = Me, R^2 = R^4 = H, R^3 = Ph$)[l]	*E*	(**6.27**; $R^1 = Me, R^2 = R^4 = H, R^3 = Ph$)[m]	98	149–150	Dimethylformamide	—[h]	15
(**6.25**; $R^1 = Me, R^2 = R^4 = H, R^3 = p\text{-}BrC_6H_4$)[l]	*E*	(**6.27**; $R^1 = Me, R^2 = R^4 = H, R^3 = p\text{-}BrC_6H_4$)[m]	82	203–204	Dimethylformamide	—[h]	15
(**6.25**; $R^1 = CH_2Ph, R^2 = R^4 = H, R^3 = p\text{-}MeOC_6H_4$)	*E*	(**6.27**; $R^1 = CH_2Ph, R^2 = R^4 = H, R^3 = p\text{-}MeOC_6H_4$)	95	159–160 (decomp.)	Dimethylformamide	—[h]	17

TABLE 6.4 (*Continued*)

Starting materials	Reaction conditions[a]	Product	Yield (%)	m.p. (°C)	Solvent of crystallisation	Crystal form	Ref.
(**6.25**; $R^1 = Me, R^2 = R^3 = Ph, R^4 = H$)	*E*	(**6.27**; $R^1 = Me, R^2 = R^3 = Ph, R^4 = H$)	96	157–159	Dimethylformamide	Colorless solid	11
(**6.25**; $R^1 = CH_2Ph, R^2 = R^3 = Ph, R^4 = H$)	*E*	(**6.27**; $R^1 = CH_2Ph, R^2 = R^3 = Ph, R^4 = H$)	94	157–158	Ethanol–dimethylformamide	—[h]	17
(**6.25**; $R^1 = Me, R^2 = Ph, R^3 = p\text{-}BrC_6H_4, R^4 = H$)	*E*	(**6.27**; $R^1 = Me, R^2 = Ph, R^3 = p\text{-}BrC_6H_4, R^4 = H$)	99	156–158	Ethanol–dimethylformamide	Colorless solid	11
(**6.25**; $R^1 = Me, R^2 = Ph, R^3 = p\text{-}NO_2C_6H_4, R^4 = H$)	*E*	(**6.27**; $R^1 = Me, R^2 = Ph, R^3 = p\text{-}NO_2C_6H_4, R^4 = H$)	92	184–185	Ethanol–dimethylformamide	Red crystals	11
(**6.25**; $R^1 = Me, R^2 = Ph, R^3 = m\text{-}NO_2C_6H_4, R^4 = H$)	*E*	(**6.27**; $R^1 = Me, R^2 = Ph, R^3 = m\text{-}NO_2C_6H_4, R^4 = H$)	97	170–172	Dimethylformamide	—[h]	15
(**6.25**; $R^1 = R^2 = Me, R^3 = Ph, R^4 = H$)	*D*	(**6.27**; $R^1 = R^2 = Me, R^3 = Ph, R^4 = H$)	66	36[t]	Ethanol	—[h]	12
(**6.25**; $R^1 = R^2 = Me, R^3 = Ph, R^4 = H$)	*E*	(**6.27** $R^1 = R^2 = Me, R^3 = Ph, R^4 = H$)[u]	97	136–137	Dimethylformamide	—[h]	15
(**6.25**; $R^1 = R^2 = Me, R^3 = p\text{-}BrC_6H_4, R^4 = H$)	*E*	(**6.27**; $R^1 = R^2 = Me, R^3 = p\text{-}BrC_6H_4, R^4 = H$)	91	162–163	Dimethylformamide	—[h]	15
(**6.25**; $R^1 = R^2 = Me, R^3 = p\text{-}NO_2C_6H_4, R^4 = H$)	*E*	(**6.27**; $R^1 = R^2 = Me, R^3 = p\text{-}NO_2C_6H_4, R^4 = H$)	83	143–144	Dimethylformamide	—[h]	15
(**6.25**; $R^1 = R^4 = Me, R^2 = H, R^3 = Ph$)	*H*	(**6.27**; $R^1 = R^4 = Me, R^2 = H, R^3 = Ph$)[v]	85	145–146	Methanol	Yellow plates	12
(**6.25**; $R^1 = R^4 = Me, R^2 = H, R^3 = Ph$)	*E*	(**6.27**; $R^1 = R^4 = Me, R^2 = H, R^3 = Ph$)	84	141–142	Ethanol	—[h]	15
(**6.25**; $R^1 = R^4 = Me, R^2 = H, R^3 = p\text{-}PhC_6H_4$)	*E*	(**6.27**; $R^1 = R^4 = Me, R^2 = H, R^3 = p\text{-}PhC_6H_4$)	91	191–192	Dimethylformamide	—[h]	15
(**6.25**; $R^1 = R^2 = R^4 = Me, R^3 = Ph$)	*E*	(**6.27**; $R^1 = R^2 = R^4 = Me, R^3 = Ph$)[j]	87	172–173	Acetone–water	—[h]	15

(**6.25**; $R^1 = CH_2COPh$, $R^2 = R^4 = H$, $R^3 = Ph$)	*E*	(**6.27**; $R^1 = CH_2COPh$, $R^2 = R^4 = H$, $R^3 = Ph$)	—[k]	160–161 (decomp.)	Ethanol	—[h]	18
(**6.25**; $R^1 = CH_2COPh$, $R^2 = R^4 = H$, $R^3 = Ph$)[l]	*E*	(**6.27**; $R^1 = CH_2COPh$, $R^2 = R^4 = H$, $R^3 = Ph$)[m]	47	164–166	Methanol–dimethylformamide	—[h]	19
(**6.25**; $R^1 = p\text{-}BrC_6H_4COCH_2$, $R^2 = R^4 = H$, $R^3 = p\text{-}BrC_6H_4$)	*E*	(**6.27**; $R^1 = p\text{-}BrC_6H_4COCH_2$, $R^2 = R^4 = H$, $R^3 = p\text{-}BrC_6H_4$)	68	180–181	Ethanol–dimethylformamide	—[h]	19
(**6.25**; $R^1 = p\text{-}NO_2C_6H_4COCH_2$, $R^2 = R^4 = H$, $R^3 = p\text{-}NO_2C_6H_4$)	*E*	(**6.27**; $R^1 = p\text{-}NO_2C_6H_4COCH_2$, $R^2 = R^4 = H$, $R^3 = p\text{-}NO_2C_6H_4$)	98	238–240	Dimethylformamide	—[h]	19
(**6.25**; $R^1 = p\text{-}BrC_6H_4COCH_2$, $R^2 = Ph$, $R^3 = p\text{-}BrC_6H_4$, $R^4 = H$)	*E*	(**6.27**; $R^1 = p\text{-}BrC_6H_4COCH_2$, $R^2 = Ph$, $R^3 = p\text{-}BrC_6H_4$, $R^4 = H$)	—[k]	196–197	Ethanol–dimethylformamide	—[h]	18
(**6.25**; $R^1 = CH_2COPh$, $R^2 = R^3 = Ph$, $R^4 = H$)[l]	*E*	(**6.27**; $R^1 = CH_2COPh$, $R^2 = R^3 = Ph$, $R^4 = H$)[m]	67	188–189	Methanol–dimethylformamide	—[h]	19
(**6.25**; $R^1 = p\text{-}BrC_6H_4COCH_2$, $R^2 = Me$, $R^3 = p\text{-}BrC_6H_4$, $R^4 = H$)	*E*	(**6.27**; $R^1 = p\text{-}BrC_6H_4COCH_2$, $R^2 = Me$, $R^3 = p\text{-}BrC_6H_4$; $R^4 = H$)	—[k]	167–169	Ethanol–dimethyl formamide	—[h]	18
(**6.24**; $R^1 = Me$, $R^2 = CO_2Et$)	*I*	(**6.27**; $R^1 = Me$, $R^2 = CO_2Et$, $R^3 = Ph$, $R^4 = H$)	41	94–95	Ethanol	—[h]	20
(**6.24**; $R^1 = Me$, $R^2 = CN$)	*J*	(**6.27**; $R^1 = Me$, $R^2 = CN$, $R^3 = Ph$, $R^4 = H$)	45	160–161	Methanol–dimethylformamide	—[h]	20
(**6.28**; $R^1 = Me$, $R^2 = CO_2Et$)	*K*	(**6.29**; $R^1 = R^3 = Me$, $R^2 = CO_2Et$)	67	153–154	Isopropanol	—[h]	21
(**6.28**; $R^1 = Me$, $R^2 = CO_2Et$)	*K*	(**6.29**; $R^1 = Me$, $R^2 = CO_2Et$, $R^3 = Et$)	40	110–111	Ethanol	—[h]	21
(**6.28**; $R^1 = Me$, $R^2 = CO_2Et$)	*K*	(**6.29**; $R^1 = Me$, $R^2 = CO_2Et$, $R^3 = Ph$)	32	156	Ethanol	—[h]	21
(**6.28**; $R^1 = Et$, $R^2 = CO_2Et$)	*K*	(**6.29**; $R^1 = Et$, $R^2 = CO_2Et$, $R^3 = Me$)	63	162–163	Ethanol	—[h]	21
(**6.28**; $R^1 = Me$, $R^2 = CN$)	*K*	(**6.29**; $R^1 = R^3 = Me$, $R^2 = CN$)	74	259	Ethanol	—[h]	21
(**6.28**; $R^1 = Me$, $R^2 = CN$)	*K*	(**6.29**; $R^1 = Me$, $R^2 = CN$, $R^3 = Et$)	58	158–159	Isopropanol	—[h]	21

TABLE 6.4 (*Continued*)

Starting materials	Reaction conditions[a]	Product	Yield (%)	m.p. (°C)	Solvent of crystallisation	Crystal form	Ref.
(**6.28**; $R^1 = Et, R^2 = CN$)	*K*	(**6.29**; $R^1 = Et, R^2 = CN, R^3 = Me$)	70	187–188	Ethanol	—[h]	21
(**6.28**; $R^1 = Me, R^2 = Ph$)	*K*	(**6.29**; $R^1 = R^3 = Me, R^2 = Ph$)	55	246	Isopropanol	—[h]	21
(**6.28**; $R^1 = Me, R^2 = Ph$)	*K*	(**6.29**; $R^1 = Me, R^2 = Ph, R^3 = Et$)	41	207–208	Isopropanol	—[h]	21
(**6.30**; $R^1 = Me, R^2 = H$)	*L*	(**6.34**; $R^1 = Me, R^2 = H$)	32	88–89	—	—	22
(**6.30**; $R^1 = Et, R^2 = H$)	*L*	(**6.34**; $R^1 = Et, R^2 = H$)	30	90–100 (decomp.)	—	—	22
(**6.30**; $R^1 = R^2 = Me$)	*L*	(**6.34**; $R^1 = R^2 = Me$)	41	178–179 (decomp.)	—	—	22
(**6.30**; $R^1 = Me, R^2 = Ph$)	*M*	(**6.34**; $R^1 = Me, R^2 = Ph$)[j]	76	151–152	—	—	22
(**6.30**; $R^1 = Et, R^2 = Ph$)	*M*	(**6.34**; $R^1 = Et, R^2 = Ph$)[j]	73	132–133 (decomp.)	—	—	22
(**6.35**)	*M*	(**6.34**; $R^1 = H, R^2 = Ph$)[w]	81	64–65	Ethanol–water	Yellow prisms	22

[a] *A* = 0.7% Na_2CO_3, Na_2SO_3, H_2O/(80–90°)(2 hr); *B* = 0.7% Na_2CO_3, Na_2SO_3, H_2O/(90–95°)(40 min); *C* = 0.7% Na_2CO_3, H_2O/(80°)(1.5 hr); *D* = NaOEt, Na_2SO_3, EtOH/(reflux)(20 min); *E* = $NaHCO_3$, H_2O/(reflux)(2–8 hr); *F* = Na_2CO_3, Na_2SO_3, H_2O/(100°)(2.5 hr); *G* = H_2O/(reflux)(15–30 min); *H* = 0.8% Na_2CO_3, H_2O/(90°)(1 hr); *I* = $PhCOCH_2Br$, acetone/(reflux)(96 hr); *J* = $PhCOCH_2Br$, acetone/(reflux)(4 hr); *K* = $(R^3CO)_2O$, Et_3N/(130–140°)(1 hr); *L* = $NaHCO_3$, $NaHSO_3$, H_2O/(reflux)(4 hr); *M* = KOH, THF/room temp.)(4–14 hr).

[b] Forms a perchlorate, colorless needles, m.p. 199–200°.

[c] Forms a picrate, yellow plates, m.p. 164° (from ethanol) and a perchlorate, m.p. 212° (from water).

[d] Turn green in air.

[e] Forms a perchlorate, colorless needles, m.p. 178° (from ethanol).

[f] Turn red in air.

[g] Forms a picrate, m.p. 151–153° (from water).

[h] Crystal form not specified.

[i] Hydrochloride; free base forms colorless crystals which rapidly turn red in air; forms a perchlorate, colorless crystals, m.p. 194–195° (from acetic acid then ethanol).

[j] Picrate.

[k] Yield not quoted.
[l] 5,6-Dimethyl derivative.
[m] 6,7-Dimethyl derivative.
[n] Forms a picrate, m.p. 202° (from acetone).
[o] Forms a hydrochloride, m.p. 273° (from water) and a hydriodide, yellow needles, m.p. 276°.
[p] Forms a picrate, m.p. 192–194°.
[q] Forms a picrate, m.p. 200–201° (from acetic acid).
[r] Forms a picrate, m.p. 205–207° (from acetic acid).
[s] Forms a picrate, m.p. 185–186° (from acetone).
[t] This m.p. differs widely from that cited in Ref. 15.
[u] Forms a picrate, m.p. 155–156° (decomp.) (from water).
[v] Forms a picrate, yellow needles , m.p. 185° (from ethanol), a perchlorate, yellow crystals, m.p. 208° (from aqueous hydrochloric acid),and a hydriodide, yellow needles, m.p. 245° (from aqueous hydriodic acid).
[w] Forms a picrate, yellow prisms, m.p. 251–253° (from ethanol).

TABLE 6.5. THE EFFECT OF VARYING THE BASIC CATALYST ON THE EFFICIENCY OF THE CYCLIZATION OF 1,2-DIMETHYL-3-PHENACYLBENZIMIDAZOLIUM BROMIDE (**6.25**; $R^1 = Me$, $R^2 = R^4 = H$, $R^3 = Ph$) TO 4-METHYL-2-PHENYL-4*H*-PYRROLO[1,2-*a*]-BENZIMIDAZOLE (**6.27**; $R^1 = Me$, $R^2 = R^4 = H$, $R^3 = Ph$)

Basic catalyst[a]	Yield (%)	Basic catalyst[a]	Yield (%)
Sodium ethoxide	93	Calcium carbonate	18
Sodium methoxide	86	Sodium phosphate	70
Sodium hydroxide	91	Sodium acetate[b]	7
Calcium hydroxide	95	Triethylamine	50
Potassium carbonate	86	Ammonia	91
Sodium carbonate	91	Di-*n*-butylamine	36
Sodium hydrogen carbonate	95	*n*-Butylamine	70
Ammonium carbonate	35	Pyridine[b]	19

[a] Reaction conditions: H_2O or EtOH/(reflux)(5 hr).
[b] Reaction conditions: H_2O or EtOH/(reflux)(20 hr).

reflux for a few hours with aqueous sodium hydrogen carbonate[11,15–17,19] or aqueous sodium carbonate[12,14,16] with or without the addition of sodium sulfite[12–14] to inhibit the subsequent oxidation of the 4*H*-pyrrolo[1,2-*a*]-benzimidazole products, which tends to occur making purification difficult. Other bases that have been used successfully to catalyze the cyclization of benzimidazolium salts of the type (**6.25**) to 4*H*-pyrrolo[1,2-*a*]benzimid-azoles (**6.27**) include alkali metal hydroxides[11,16,19] and alkoxides,[12,13,16] ammonia,[16] benzyltrimethylammonium hydroxide,[16] and amines (primary, secondary, and tertiary).[12,16] A detailed study[16] of the variation in the efficiency of the cyclization of 1,2-dimethyl-3-phenacylbenzimidazolium bromide (**6.25**; $R^1 = Me$, $R^2 = R^4 = H$, $R^3 = Ph$) to 4-methyl-2-phenyl-4*H*-pyrrolo[1,2-*a*]benzimidazole (**6.27**; $R^1 = Me$, $R^2 = R^4 = H$, $R^3 = Ph$), using different catalysts, reveals (Table 6.5) that ammonium and alkaline earth metal carbonates, sodium acetate, and certain amines (e.g., di-*n*-butylamine, pyridine) are inefficient catalysts for transformations of this type. The probable intermediacy of benzimidazolium betaines [Scheme 6.5; (**6.26**)] in the cyclizations of the benzimidazolium salts (**6.25**) is demonstrated[16] by their isolation under suitable conditions and their ready transformation (Table 6.4) into the corresponding 4*H*-pyrrolo[1,2-*a*]benzimidazoles (**6.27**) merely on warming with water or on attempted crystallization from organic solvents. Where the 2-alkyl group in the original benzimidazole (**6.24**) is activated by a substituent such as ethoxycarbonyl or cyano, simply warming in acetone solution with the α-halogeno ketone is sufficient to accomplish direct conversion into the 3-ethoxycarbonyl or 3-cyano-4*H*-pyrrolo[1,2-*a*]-benzimidazole (**6.27**; $R^2 = CO_2Et$ or CN) thus opening up routes to the otherwise difficultly accessible 3-carboxylic acids of the series.[20] In a further synthetically useful variant, 1-substituted 2-methylbenzimidazolium salts bearing a benzyl, ethoxycarbonylmethyl, or cyanomethyl substituent at *N*(3)

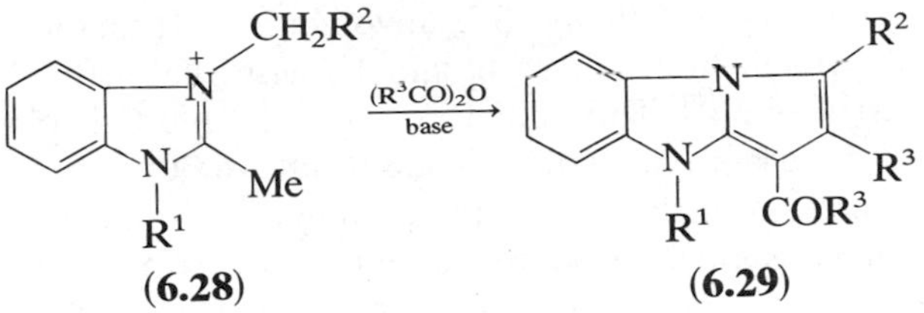

(R^{3}CO)$_2$O, base

(6.28) (6.29)

Scheme 6.6

[Scheme 6.6; (**6.28**; $R^2 = Ph$, CO_2Et, or CN)] have been shown[21] to condense with acid anhydrides in the presence of a base such as triethylamine to afford in a single step, moderate to good yields (Table 6.4) of 4*H*-pyrrolo[1,2-*a*]benzimidazoles with an acyl substituent at *C*(3) and a phenyl, ethoxycarbonyl, or cyano substituent at *C*(1) (**6.29**; $R^2 = Ph$, CO_2Et, or CN).

2-Methylated 4*H*-pyrrolo[1,2-*a*]benzimidazoles are also the end-products of the sodium hydrogen carbonate–sodium hydrogen sulfite, or potassium hydroxide-mediated cyclizations of 1-substituted 2-alkyl-3-(2-propynyl)-benzimidazolium bromides [Scheme 6.7; (**6.30**)].[22] These transformations are reported[22] to proceed in moderate to excellent yield (Table 6.4) and are

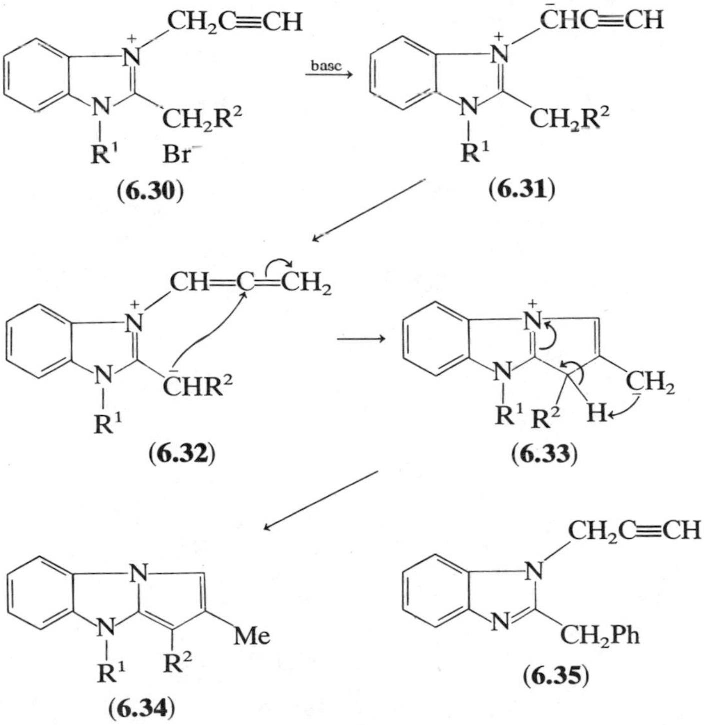

(6.30) (6.31) (6.32) (6.33) (6.34) (6.35)

Scheme 6.7

rationalized by a course (Scheme 6.7) involving the formation and cyclization of an allenylbenzimidazolium betaine intermediate [(**6.30**) → (**6.31**) → (**6.32**) → (**6.33**) → (**6.34**)]. The enhanced yields (Table 6.4) observed in the cyclizations of the 2-benzylbenzimidazolium salts (**6.30**; $R^2 = Ph$) are consistent with stabilization of the proposed carbanion intermediate (**6.32**) by the phenyl substituent. Similar carbanion stabilization also accounts for the high yield (Table 6.4) base-catalyzed cyclization of 2-benzyl-1-(2-propynyl)benzimidazole (**6.35**) to 2-methyl-3-phenyl-4*H*-pyrrolo[1,2-*a*]benzimidazole (**6.34**; $R^1 = H$, $R^2 = Ph$).[22]

4*H*-Pyrrolo[1,2-*a*]benzimidazoles are isolated in very low yield (Table 6.6) from the reactions (Scheme 6.8) of 1-substituted 3-acylmethylbenzimidazolium bromides (**6.36**) with acetylenic esters under basic conditions.[23,24] These transformations are readily explicable[23,24] in terms of the *in situ* formation of benzimidazolium ylid intermediates and their 1,3-dipolar cycloaddition to the acetylenic ester to afford dihydropyrrolo[1,2-*a*]benzimidazoles convertible by oxidation in the reaction medium into the observed products [Scheme 6.8; (**6.37**) → (**6.38**) → (**6.39**)]. 4*H*-Pyrrolo[1,2-*a*]benzimidazoles are also formed in low yield (Table 6.6) in the cycloaddition

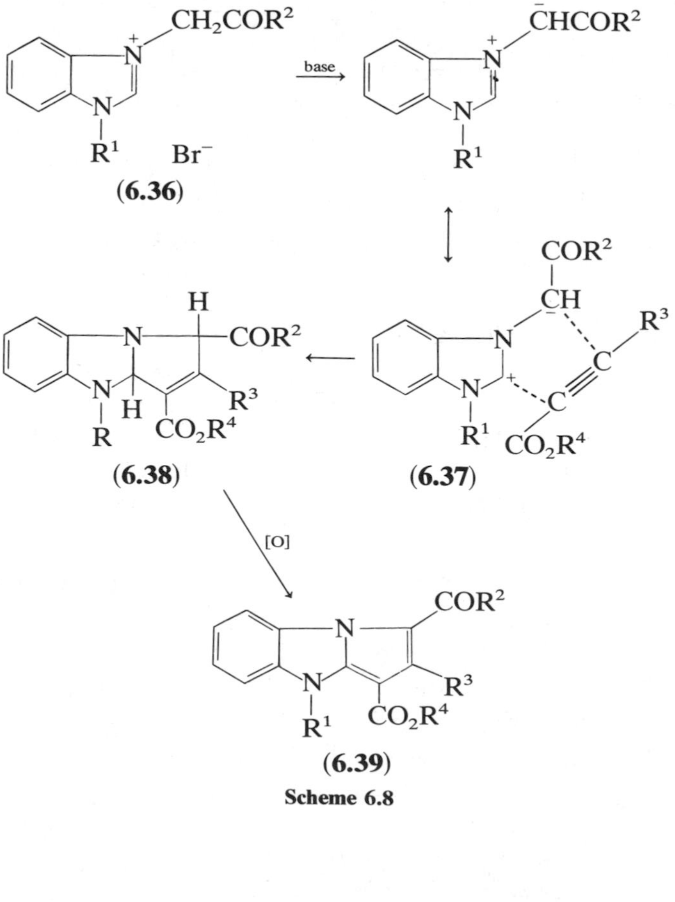

Scheme 6.8

(**6.40**) (**6.41**)

Scheme 6.9

reactions of simple benzimidazole derivatives with acetylenic esters.[25–29] Thus, prolonged heating of benzimidazole with methyl propiolate in acetonitrile gives the adduct [Scheme 6.9; (**6.41**)] in 10% yield.[25] In contrast, 1,2-disubstituted benzimidazoles react with dimethyl acetylenedicarboxylate to give low yields of 4*H*-pyrrolo[1,2-*a*]benzimidazoles [Scheme 6.10; (**6.45**)] derived by loss of the *C*(2) substituent.[26–28] These deep-seated transformations are accounted for (Scheme 6.10) by reaction of the benzimidazole derivative with two molecules of the acetylenic ester to give a zwitterionic intermediate convertible by cyclization and subsequent fragmentation into the 4*H*-pyrrolo[1,2-*a*]benzimidazole products [Scheme 6.10; (**6.42**) → (**6.43**) → (**6.44**) → (**6.45**)]. On the other hand, the reaction[29] of ethyl 2-(*N*-methylbenzimidazolyl)acetate with dimethyl acetylenedicarboxylate to give the 4*H*-pyrrolo[1,2-*a*]benzimidazole tricarboxylic ester [Scheme 6.11; (**6.50**)] is best accommodated by a course (Scheme 6.11)[29]

(**6.42**) (**6.43**) (**6.44**) (**6.45**)

Scheme 6.10

TABLE 6.6. SYNTHESIS OF 4*H*-PYRROLO[1,2-*a*]BENZIMIDAZOLES BY CYCLOADDITION REACTIONS OF BENZIMIDAZOLE DERIVATIVES WITH ACETYLENIC ESTERS

Starting material	Reaction conditions[a]	Product	Yield (%)	m.p. (°C)	Solvent of crystallization	Crystal form	Ref.
(**6.36**; $R^1 = Me$, $R^2 = Ph$)	*A*	(**6.39**; $R^1 = Me$, $R^2 = Ph$, $R^3 = H$, $R^4 = Et$)	4	164	Ethanol	Colorless needles	23
(**6.36**; $R^1 = Me$, $R^2 = p\text{-}BrC_6H_4$)	*A*	(**6.39**; $R^1 = Me$, $R^2 = p\text{-}BrC_6H_4$, $R^3 = H$, $R^4 = Et$)	5	224	Ethanol	Colorless needles	23
(**6.36**; $R^1 = Et$, $R^2 = Ph$)	*A*	(**6.39**; $R^1 = R^4 = Et$, $R^2 = Ph$, $R^3 = H$)	4	155–157	Ethanol	Colorless needles	23
(**6.36**; $R^1 = Et$, $R^2 = p\text{-}BrC_6H_4$)	*A*	(**6.39**; $R^1 = R^4 = Et$, $R^2 = p\text{-}BrC_6H_4$, $R^3 = H$)	4	186	Ethanol	Colorless needles	23
(**6.36**; $R^1 = CH_2Ph$, $R^2 = Ph$)	*B*	(**6.39**; $R^1 = CH_2Ph$, $R^2 = Ph$, $R^3 = H$, $R^4 = Et$)	17	175–178	Methanol	Colorless crystals	24
(**6.36**; $R^1 = Me$, $R^2 = OMe$)	*A*	(**6.39**; $R^1 = Me$, $R^2 = OMe$, $R^3 = H$, $R^4 = Et$)	5	185	Ethanol	Colorless needles	23
(**6.36**; $R^1 = Et$, $R^2 = OMe$)	*A*	(**6.39**; $R^1 = Et$, $R^2 = OMe$, $R^3 = H$, $R^4 = Et$)	2	160	Ethanol	Colorless needles	23
(**6.36**; $R^1 = Me$, $R^2 = Ph$)	*C*	(**6.39**; $R^1 = R^4 = Me$, $R^2 = Ph$, $R^3 = CO_2Me$)	11	130–131	Ethanol	Colorless needles	23

(**6.36**; $R^1 = CH_2Ph$, $R^2 = Ph$)	*D*	(**6.39**; $R^1 = CH_2Ph$, $R^2 = Ph$, $R^3 = CO_2Me$, $R^4 = Me$)	8	212–213	Dichloroethane	Colorless crystals	24
(**6.36**; $R^1 = Me$, $R^2 = OMe$)	*C*	(**6.39**; $R^1 = R^4 = Me$, $R^2 = OMe$, $R^3 = CO_2Me$)	7	177–178	Ethanol	Colorless prisms	23
(**6.36**; $R^1 = Et$, $R^2 = OMe$)	*C*	(**6.39**; $R^1 = Et$, $R^2 = OMe$, $R^3 = CO_2Me$, $R^4 = Me$)	2	134–135	Ethanol	Colorless prisms	23
(**6.40**)	*E*	(**6.41**)	10	198	Methanol	Colorless needles	25
(**6.42**; $R = R^1 = Me$, $R^2 = H$)	*F*	(**6.45**; $R = Me$)	8	182	Methanol	Colorless needles	26
(**6.42**; $R = Me$, $R^1 = Ph$, $R^2 = H$)	*G*	(**6.45**; $R = Me$)	2	183	—	—	27
(**6.42**; $R = CH\overset{t}{=}CHCO_2Me$, $R^1 = Ph$, $R^2 = H$)	*H*	(**6.45**; $R = CH\overset{t}{=}CHCO_2Me$)	36	203	Methanol	Rods	28
(**6.42**; $R = CH\overset{t}{=}CHCO_2Me$, $R^1 = R^2 = Ph$)	*I*	(**6.45**; $R = CH\overset{t}{=}CHCO_2Me$)	10	—	—	—	28
(**6.42**; $R = MeO_2C\overset{\mid}{C}{=}CHCO_2Me$)	*I*	(**6.45**; $R = MeO_2C\overset{\mid}{C}{=}CHCO_2Me$)	0.2	237–239	Methanol	Rods	28
(**6.46**)	*I*	(**6.50**)	3	140	Methanol	Plates	29

[a] *A* = $HC{\equiv}CCO_2Et$, K_2CO_3, dimethylformamide/(room temp.)(2 days); *B* = $HC{\equiv}CCO_2Et$, Et_3N, benzene/(reflux)(3 hr); *C* = $MeO_2CC{\equiv}CCO_2Me$, K_2CO_3, dimethylformamide/(room temp.)(3 days); *D* = $MeO_2CC{\equiv}CCO_2Me$, Et_3N, benzene/(room temp.)(24 hr); *E* = $HC{\equiv}CCO_2Me$, MeCN/(reflux)(5–7 days); *F* = $MeO_2CC{\equiv}CCO_2Me$, THF or ether/(room temp.)(7 days); *G* = $MeO_2CC{\equiv}CCO_2Me$, MeCN/(reflux)(5 hr); *H* = $MeO_2CC{\equiv}CCO_2Me$, MeCN/(reflux)(12 hr); I = $MeO_2CC{\equiv}CCO_2Me$, MeCN/(reflux)(5–10 days).

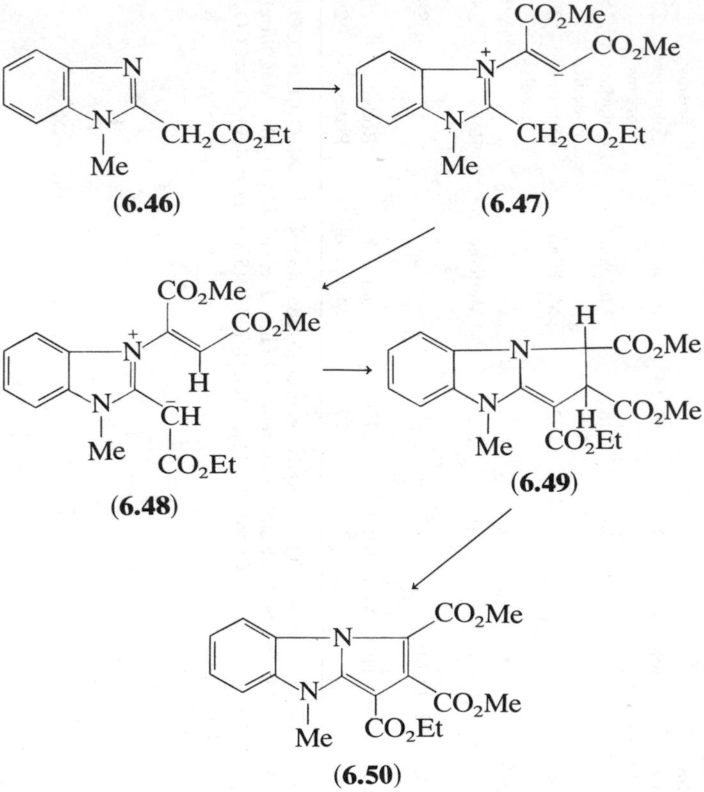

(**6.46**) (**6.47**)

(**6.48**) (**6.49**)

(**6.50**)

Scheme 6.11

initiated by the formation of the zwitterionic intermediate (**6.48**) cyclization of which affords a dihydropyrrolo[1,2-*a*]benzimidazole (**6.49**) readily oxidizable to the observed product (**6.50**).

2,3-Dihydro-1*H*-pyrrolo[1,2-*a*]benzimidazoles [Scheme 6.12; (**6.54**)] are readily synthesized in moderate to high yield (Table 6.7) by the base-catalyzed cyclization of 2-(γ-halogenopropyl)benzimidazoles (**6.55**; X = Cl or Br).[30,31] The ready formation of 2-(γ-chloropropyl)benzimidazoles (**6.55**; X = Cl) by chlorination of the corresponding 2-(γ-hydroxypropyl)benzimidazoles (**6.53**) provides the basis for an improved version[32] of such cyclizations, in which the hydroxy compound (**6.53**) is heated with thionyl chloride in dimethylformamide to give directly moderate to high yields of the requisite 2,3-dihydro-1*H*-pyrrolo[1,2-*a*]benzimidazoles (**6.54**). Products of the latter type are also formed even more directly, if less efficiently (Table 6.7), by the thermal cyclization [Scheme 6.12; (**6.53**) → (**6.54**)] of 2-(γ-hydroxypropyl)benzimidazoles, which can either be preformed[33] or generated *in situ* by the condensation of *ortho*-phenylenediamine derivatives with γ-lactones [Scheme 6.12; (**6.51**) + (**6.52**) → (**6.53**)].[33–35] It is interesting to

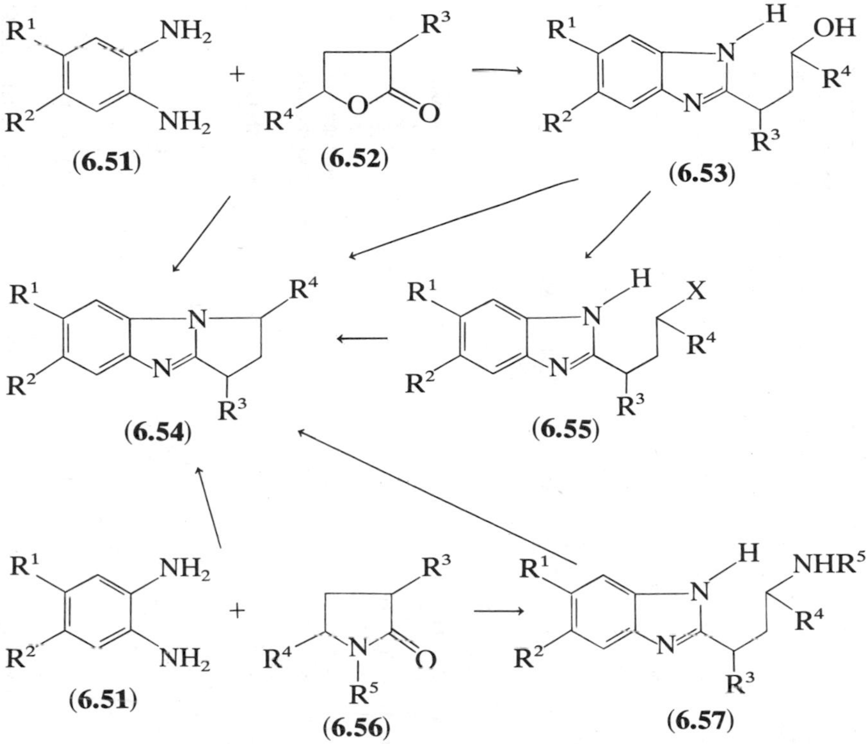

Scheme 6.12

compare the low yields of 2,3-dihydro-1*H*-pyrrolo[1,2-*a*]benzimidazoles (Table 6.7) obtained in the cyclizations of 2-(γ-hydroxypropyl)benzimidazoles (**6.53**) with the uniformly high yields of such products (Table 6.7) claimed[36] for the acid-catalyzed ring-closures of 2-(γ-aminopropyl)benzimidazoles accessible in isolated form or by the *in situ* reaction of *ortho*-phenylenediamine derivatives with pyrrolidin-2(1*H*)-ones [Scheme 6.12; (**6.51**) + (**6.56**) → (**6.57**)]. Formation of 2,3-dihydro-1*H*-pyrrolo[1,2-*a*]benzimidazoles by cyclization of γ-functionalized 2-propylbenzimidazoles (cf. Scheme 6.12) unlike the alternative synthetic approach of ring-closure of *N*-(*ortho*-substituted phenyl)pyrrolidines (see later) is unambiguous in relation to the site of substituents in the pyrrolidine nucleus of the final products. However, the use of nuclear-substituted benzimidazole precursors in such cyclizations can lead, by ring-closure in two possible senses, to isomer mixtures (Table 6.7),[32,36] which in some cases can be successfully separated.[32] In other instances, single isomers, often of unestablished orientation,[33] are produced. In the context of the latter situation, Freedman, Payne, and Day have shown[32] that, irrespective of the electronic bias of the substituent, ring-closure of 5(6)-substituted 2-(γ-chloropropyl)benzimidazoles occurs preferentially at the nitrogen atom *para* to the occupied site in

TABLE 6.7. SYNTHESIS OF 2,3-DIHYDRO-1*H*-PYRROLO[1,2-*a*]BENZIMIDAZOLES BY RING-CLOSURE REACTIONS OF BENZIMIDAZOLE DERIVATIVES AND RELATED PROCESSES

Starting materials	Reaction conditions[a]	Product	Yield (%)	m.p. (°C)	Solvent of crystallization	Ref.
(**6.55**; $R^1 = R^2 = R^3 = R^4 = H$, $X = Cl$)[b]	*A*	(**6.54**; $R^1 = R^2 = R^3 = R^4 = H$)[c]	70	115.5–116.5	Ethyl acetate	30
(**6.55**; $R^1 = R^2 = R^3 = R^4 = H$, $X = Br$)	*B*	(**6.54**; $R^1 = R^2 = R^3 = R^4 = H$)[d,e]	43	115–117	*n*-Heptane	31
(**6.55**; $R^1 = R^2 = R^3 = R^4 = H$, $X = Br$)	*C*	(**6.54**; $R^1 = R^2 = R^3 = R^4 = H$)[d,e]	quant.	—	—	31
(**6.55**; $R^1 = X = Cl$, $R^2 = R^3 = R^4 = H$)	*A*	(**6.54**; R^1 or $R^2 = Cl$ or H, $R^3 = R^4 = H$)[b,f]	72	236–237	—	30
(**6.55**; $R^1 = R^2 = R^4 = H$, $R^3 = Me$, $X = Cl$)[b]	*A*	(**6.54**; $R^1 = R^2 = R^4 = H$, $R^3 = Me$)[b,g]	85	188–189	—	30
(**6.55**; $R^1 = X = Cl$, $R^2 = R^4 = H$, $R^3 = Me$)[b]	*A*	(**6.54**; R^1 or $R^2 = Cl$ or H, $R^3 = Me$, $R^4 = H$)[b,f]	40	236–238	—	30
(**6.55**; $R^1 = R^2 = R^4 = H$, $R^3 = n\text{-}C_7H_{15}$, $X = Cl$)[b]	*D*	(**6.54**; $R^1 = R^2 = R^4 = H$, $R^3 = n\text{-}C_7H_{15}$)[b]	38	93–95	Ether–benzene	30
(**6.55**; $R^1 = X = Cl$, $R^2 = R^4 = H$, $R^3 = n\text{-}C_7H_{15}$)[b]	*D*	(**6.54**; R^1 or $R^2 = Cl$ or H, $R^3 = n\text{-}C_7H_{15}$, $R^4 = H$)[b,f]	40	176–177	—	30
(**6.53**; $R^1 = R^2 = R^3 = R^4 = H$)	*E*	(**6.54**; $R^1 = R^2 = R^3 = R^4 = H$)	44	115	Ethyl acetate	32
(**6.53**; $R^1 = Me$, $R^2 = R^3 = R^4 = H$)	*E*	(**6.54**; $R^1 = R^3 = R^4 = H$, $R^2 = Me$)	43	144–145	Ether–hexane	32
(**6.53**; $R^1 = NO_2$, $R^2 = R^3 = R^4 = H$)	*E*	(**6.54**; $R^1 = R^3 = R^4 = H$, $R^2 = NO_2$)	60	209–210	Ethyl acetate	32
(**6.53**; $R^1 = Cl$, $R^2 = R^3 = R^4 = H$)	*E*	{(**6.54**; $R^1 = Cl$, $R^2 = R^3 = R^4 = H$) + (**6.54**; $R^1 = R^3 = R^4 = H$, $R^2 = Cl$)}	49; 48	133–134; 96–98	—; —	32
(**6.53**; $R^1 = R^2 = Me$, $R^3 = R^4 = H$)	*E*	(**6.54**; $R^1 = R^2 = Me$, $R^3 = R^4 = H$)	61	177–179	Ethyl acetate	32
(**6.53**; $R^1 = R^2 = MeO$, $R^3 = R^4 = H$)	*E*	(**6.54**; $R^1 = R^2 = MeO$, $R^3 = R^4 = H$)	50	194–196	Ethyl acetate	32
(**6.53**; $R^1 = R^2 = R^3 = R^4 = H$)[h]	*E*	(**6.54**; $R^1 = R^2 = R^3 = R^4 = H$)[i]	45	133–135	Benzene	32

(**6.53**; $R^1 = R^2 = R^4 = H$, $R^3 = Me$)	*F*	(**6.54**; $R^1 = R^2 = R^4 = H$, $R^3 = Me$)[j]	43	35–40	—[k]	33
(**6.53**; $R^1 = R^2 = Cl$, $R^3 = Me$, $R^4 = H$)	*G*	(**6.54**; $R^1 = R^2 = Cl$, $R^3 = Me$, $R^4 = H$)	18	147–149	*n*-Heptane	33
(**6.51**; $R^1 = R^2 = H$) + (**6.52**; $R^3 = R^4 = H$)	*H*	(**6.54**; $R^1 = R^2 = R^3 = R^4 = H$)	29	115	Ethyl acetate[l]	35
(**6.51**; $R^1 = R^2 = H$) + (**6.52**; $R^3 = H$, $R^4 = Me$)	*I*	(**6.54**; $R^1 = R^2 = R^3 = H$, $R^4 = Me$)	10	—[m]	—	34
(**6.51**; $R^1 = Cl$, $R^2 = H$) + (**6.52**; $R^3 = Me$, $R^4 = H$)	*J*	(**6.54**; R^1, $R^2 = Cl$ or H, $R^3 = Me$, $R^4 = H$)[j]	16	101–103	*n*-Heptane	33
(**6.51**; $R^1 = R^2 = Cl$) + (**6.52**; $R^3 = Me$, $R^4 = H$)	*J*	(**6.54**; $R^1 = R^2 = Cl$, $R^3 = Me$, $R^4 = H$)	15	149	*n*-Heptane	33
(**6.57**; $R^1 = R^2 = R^3 = R^4 = R^5 = H$)	*K*	(**6.54**; $R^1 = R^2 = R^3 = R^4 = H$)	76	118[n]	Ethyl acetate	36
(**6.57**; $R^1 = R^2 = R^3 = R^4 = R^5 = H$)	*L*	(**6.54**; $R^1 = R^2 = R^3 = R^4 = H$)	85	—	—	36
(**6.57**; $R^1 = R^5 = Me$, $R^2 = R^3 = R^4 = H$)	*M*	{(**6.54**; $R^1 = R^3 = R^4 = H$, $R^2 = Me$) + (**6.54**; $R^1 = Me$, $R^2 = R^3 = R^4 = H$)}[o]	73	—[p]	—	36
(**6.51**; $R^1 = R^2 = H$) + (**6.56**; $R^3 = R^4 = R^5 = H$)	*N*	(**6.54**; $R^1 = R^2 = R^3 = R^4 = H$)	74	118	Ethyl acetate	36
(**6.51**; $R^1 = R^2 = H$) + (**6.56**; $R^3 = R^4 = H$, $R^5 = Me$)	*N*	(**6.54**; $R^1 = R^2 = R^3 = R^4 = H$)	72	—	—	36
(**6.51**; $R^1 = Me$, $R^2 = H$) + (**6.56**; $R^3 = R^4 = R^5 = H$)	*N*	{(**6.54**; $R^1 = Me$, $R^2 = R^3 = R^4 = H$) + (**6.54**; $R^1 = R^3 = R^4 = H$, $R^2 = Me$)}[o]	71	—[p]	—	36
(**6.58**; R = H)	*O*	(**6.60**; R = H)[j]	43	170–171	Ethanol or nitrobenzene	37, 38, 39

TABLE 6.7 (*Continued*)

Starting Materials	Reaction condition[a]	Product	Yield (%)	m.p. (°C)	Solvent of crystallization	Ref.
(**6.59**)	*O*	(**6.60**; R = H)[j]	62	171–172	—[q]	38, 43
(**6.58**; R = H)	*P*	(**6.60**; R = H)[j]	—[r]	172–175	Ether or nitrobenzene	39, 40
(**6.58**; R = H)	*Q*	(**6.60**; R = H)[j]	78	—	—	41
(**6.58**; R = H)	*R*	(**6.61**)[j]	68	200–201	Chloroform	42
(**6.58**; R = OH)	*P*	(**6.60**; R = OAc)[j]	—[r]	121–122 (decomp.)	Ether	40
(**6.62**; R = H)	*S*	(**6.63**; R = H)	85	195	Dichloroethane	44
(**6.62**; R = OMe)	*S*	(**6.63**; R = OMe)	85	250	Methanol	44
(**6.62**; R = NMe_2)	*S*	(**6.63**; R = NMe_2)	85	252	Ethanol	44

[a] *A* = NaOEt, EtOH/(reflux)(2 hr); *B* = KOH, EtOH/(room temp.)(20 min); *C* = NaOEt, EtOH, high dilution/(room temp.)(20 min.); *D* = NaOEt, EtOH/(room temp.)(2 hr); *E* = $SOCl_2$, dimethylformamide/(reflux)(45 min); *F* = α-methylbutyrolactone/(125°)(5 hr); *G* = α-methylbutyrolactone/(reflux)(8 hr); *H* = 270°, autoclave/7 hr; *I* = 270°/3 hr; *J* = reflux/5.5–8 hr; *K* = Montmorillonite catalyst/(290–310°)(10 hr); *L* = 85% H_3PO_4/(300°)(12 hr); *M* = methanesulfonic acid/(300–320°)(10 hr); *N* = 85% H_3PO_4/(300°)(2–4 hr); *O* = sublimation at 230–240°; *P* = 95–130°/30 min; *Q* = dicyclohexylcarbodiimide, pyridine, dimethylformamide/(5–10°)(10–12 hr); *R* = $POCl_3$, dimethylformamide/(room temp.)(24 hr); *S* = crystallization from acetic acid.

[b] Hydrochloride.

[c] Forms a hydrochloride, colorless, hygroscopic crystals, m.p. 235–237° (from ethanol–ether).

[d] Colorless prisms.

[e] Forms a picrate, yellow needles, m.p. 222–222.5 (from 2-methoxyethanol).

[f] Position of the chlorine substituent not determined.

[g] Free base has m.p. 68.5–70°.

[h] 4,7-Dimethoxy derivative.

[i] 5,8-Dimethoxy derivative.

[j] Colorless crystals.

[k] Purified by distillation.

[l] Purified initially by distillation, b.p. 130°/0.2 mm Hg.

[m] Brown oil, characterized as the picrate, yellow prisms, m.p. 222–224° (from ethanol).

[n] b.p. 148–150°/0.2 mm Hg.

[o] Isomer mixture not separated.

[p] Oil, b.p. 143–145°/0.2 mm Hg.

[q] Purified by sublimation.

[r] Yield not quoted.

the nucleus. For example,[32] cyclization of the benzimidazole derivatives (**6.55**; $R^1 = Me$ or NO_2, $R^2 = R^3 = R^4 = H$, $X = Cl$) leads to the single 2,3-dihydro-1*H*-pyrrolo[1,2-*a*]benzimidazoles (**6.54**; $R^1 = R^3 = R^4 = H$, $R^2 = Me$ or NO_2), albeit in only moderate yield (Table 6.7). In contrast, ring-closure of the chloro derivative (**6.55**; $R^1 = X = Cl$, $R^2 = R^3 = R^4 = H$) affords a separable mixture of the 6-chloro- and 7-chloropyrrolo[1,2-*a*]benzimidazoles (**6.54**; R^1 or $R^2 = Cl$, R^2 or $R^1 = R^3 = R^4 = H$).[32] The aforementioned preferential cyclization[32] of the methyl-substituted benzimidazole (**6.54**; $R^1 = Me$, $R^2 = R^3 = R^4 = H$) to the single pyrrolo[1,2-*a*]benzimidazole isomer (**6.54**; $R^1 = R^3 = R^4 = H$, $R^2 = Me$) stands in contrast to the isomer mixture obtained[36] (cf. Table 6.7) in the analogous cyclization of the 2-(γ-aminopropyl)benzimidazole (**6.57**; $R^1 = R^5 = Me$, $R^2 = R^3 = R^4 = H$). This latter result casts some doubt on the specificity of the cyclization of electronically biased 5(6)-substituted benzimidazoles by preferential ring-closure at the nitrogen atom *para* to the substituent.

2,3-Dihydro-1*H*-pyrrolo[1,2-*a*]benzimidazol-1-ones [Scheme 6.13; (**6.60**)] are readily available as products of the dehydrative cyclization[37–42] of 3-(2-benzimidazolyl)propionic acids [Scheme 6.13; (**6.58**) → (**6.60**)] or, less orthodoxly, in the parent case by the thermal ring-contraction of a benzodiazepinedione [Scheme 6.13; (**6.59**) → (**6.60**; R = H)].[38,43] The former cyclizations proceed in moderate to good yield (Table 6.7) and can be effected

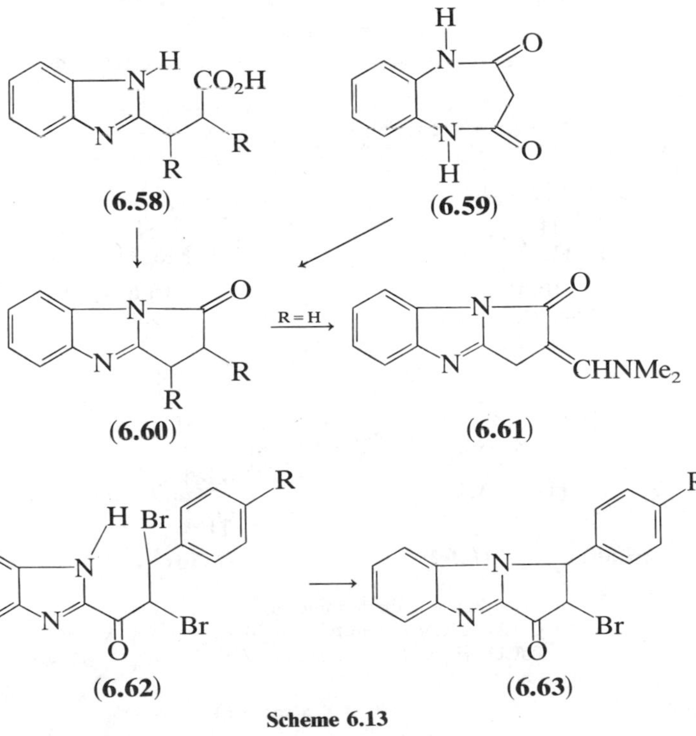

(**6.58**) (**6.59**)

(**6.60**) R = H (**6.61**)

(**6.62**) (**6.63**)

Scheme 6.13

thermally[37–39] or in the presence of a dehydrating agent such as acetic anhydride[39,40] or dicyclohexylcarbodiimide.[41] The use of phosphorus oxychloride in conjunction with dimethylformamide as the dehydrating agent results in Vilsmeier–Haack condensation (see later) subsequent to cyclization, the product then being the corresponding dimethylaminomethylene derivative [Scheme 6.13; (**6.58**; R = H) → (**6.60**; R = H) → (**6.61**)].[42] 1-Aryl-2-bromo-1,2-dihydro-3*H*-pyrrolo[1,2-*a*]benzimidazol-3-ones result (Table 6.7) on attempted crystallization of the dibromo adducts of 2-cinnamoylbenzimidazoles [Scheme 6.13; (**6.62**) → (**6.63**)].[44]

2,3,3*a*,4-Tetrahydro-1*H*-pyrrolo[1,2-*a*]benzimidazoles, more commonly encountered as the reduction products of 2,3-dihydro-1*H*-pyrrolo[1,2-*a*]-bemzimidazoles (see later) are also formed in low yield by the *in situ* cycloaddition reactions of *N*-acylmethylenebenzimidazolium ylids with electron-deficient alkenes [Scheme 6.14; (**6.64**; R^1 = Me or CH_2Ph, R^2 = Ph

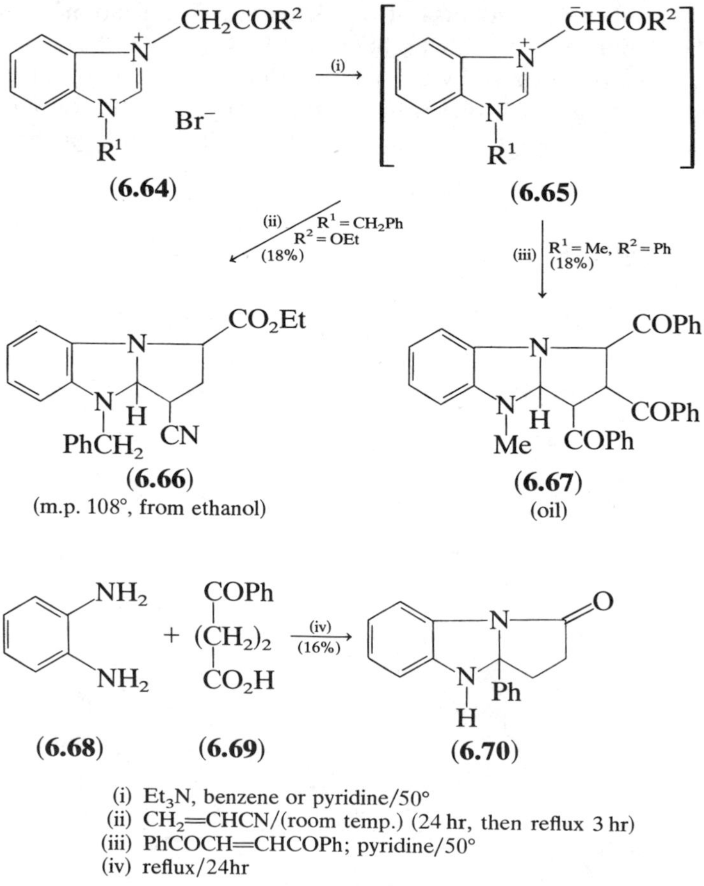
CH2COR2
Br−
R1
(6.64)
(i)
CHCOR2
R1
(6.65)
(ii)
R1 = CH2Ph
R2 = OEt
(18%)
(iii)
R1 = Me, R2 = Ph
(18%)
CO2Et
PhCH2
CN
(6.66)
(m.p. 108°, from ethanol)
COPh
COPh
Me
COPh
(6.67)
(oil)
NH2
NH2
COPh
(CH2)2
CO2H
(iv)
(16%)
O
Ph
(6.68)
(6.69)
(6.70)
(i) Et3N, benzene or pyridine/50°
(ii) CH2=CHCN/(room temp.) (24 hr, then reflux 3 hr)
(iii) PhCOCH=CHCOPh; pyridine/50°
(iv) reflux/24hr

Scheme 6.14

or OEt) → (**6.65**; $R^1 = Me$ or CH_2Ph, $R^2 = Ph$ or OEt) → (**6.66**) or (**6.67**)][45,46] and in a single instance[38] by the thermal condensation of *ortho*-phenylenediamine with a γ-keto carboxylic acid [Scheme 6.14; (**6.68**) + (**6.69**) → (**6.70**)].

Ring-closure Reactions of Other Heterocycles

Substituent interaction[47] in *N*-(*ortho*-substituted phenyl)pyrrolidines is a rich source of 2,3-dihydro-1*H*-pyrrolo[1,2-*a*]benzimidazoles. Thus, *N*-(2-aminophenyl)pyrrolidines[47–49] or their *N*-acyl derivatives[47,49–51] are smoothly cyclized by peroxytrifluoroacetic acid[48,49] or performic acid[47,49–51] (in both cases generated *in situ* from the corresponding carboxylic acid and hydrogen peroxide) to afford uniformly high yields (Table 6.8) of the corresponding 2,3-dihydro-1*H*-pyrrolo[1,2-*a*]benzimidazoles [Scheme 6.15; (**6.71**; $R^3 = H$ or COR) → (**6.74**)]. The mechanism involved in these ring-closures has been the subject of some controversy.[47,49,52,53] However, support for the proposal[47,49,52] that cyclization involves the initial formation and subsequent Polonovski rearrangement of a pyrrolidine *N*-oxide intermediate [Scheme 6.16; (**6.79**) → (**6.80**) → (**6.81**) → (**6.82**) → (**6.83**) → (**6.84**)] is provided by the demonstration[52] that *N*-(2-benzamidophenyl)pyrrolidine *N*-oxide (**6.79**; R = NHCOPh) cyclizes readily under acidic conditions to afford 2,3-dihydro-1*H*-pyrrolo[1,2-*a*]benzimidazole (**6.84**) in good yield (Table 6.8). Less compelling evidence is afforded by the analogous reductive ring-closure[53] of *N*-(2-nitrophenyl)pyrrolidine *N*-oxide [Scheme 6.16; (**6.79**; $R = NO_2$) → → → (**6.84**)], since this transformation could be the result of deoxygenation followed by reductive cyclization (see later) of the *N*-(2-nitrophenyl)pyrrolidine produced. The utility of the Hg(II)–EDTA complex as an oxidizing agent for effecting the high yield (Table 6.8) cyclization of *N*-(2-aminophenyl)pyrrolidine and its *N*-acyl derivatives to 2,3-dihydro-1*H*-pyrrolo[1,2-*a*]benzimidazoles has recently been emphasized.[54,55] The acid-catalyzed cyclization of *N*-(2-benzamidophenyl)-pyrrolidin-2-one is reported[56] to afford 2,3-dihydro-1*H*-pyrrolo[1,2-*a*]-benzimidazole, albeit in unspecified yield.

2,3-Dihydro-1*H*-pyrrolo[1,2-*a*]benzimidazoles are also the end-products of the thermolysis of *N*-(2-azidophenyl)pyrrolidines in high-boiling solvents such as nitrobenzene or diethyleneglycol dimethyl ether [Scheme 6.15; (**6.72**) → (**6.74**)].[57–59] Ring-closure of this type is generally less efficient (Table 6.8) than peracid-mediated cyclization of the corresponding amines (see before) and is believed to occur by cyclizative insertion in a nitrene intermediate followed by oxidation of the 2,3,3*a*,4-tetrahydro-1*H*-pyrrolo-[1,2-*a*]benzimidazole produced [Scheme 6.17; (**6.72**) → (**6.85**) → (**6.86**) → (**6.74**)]. This pathway to product is supported by the demonstration[59] that thermolysis of *N*-(2-azidophenyl)pyrrolidin-2-one leads to 2,3,3*a*,4-tetrahydro-1*H*-pyrrolo[1,2-*a*]benzimidazol-1-one [Scheme 6.18; (**6.87**) → (**6.88**)]

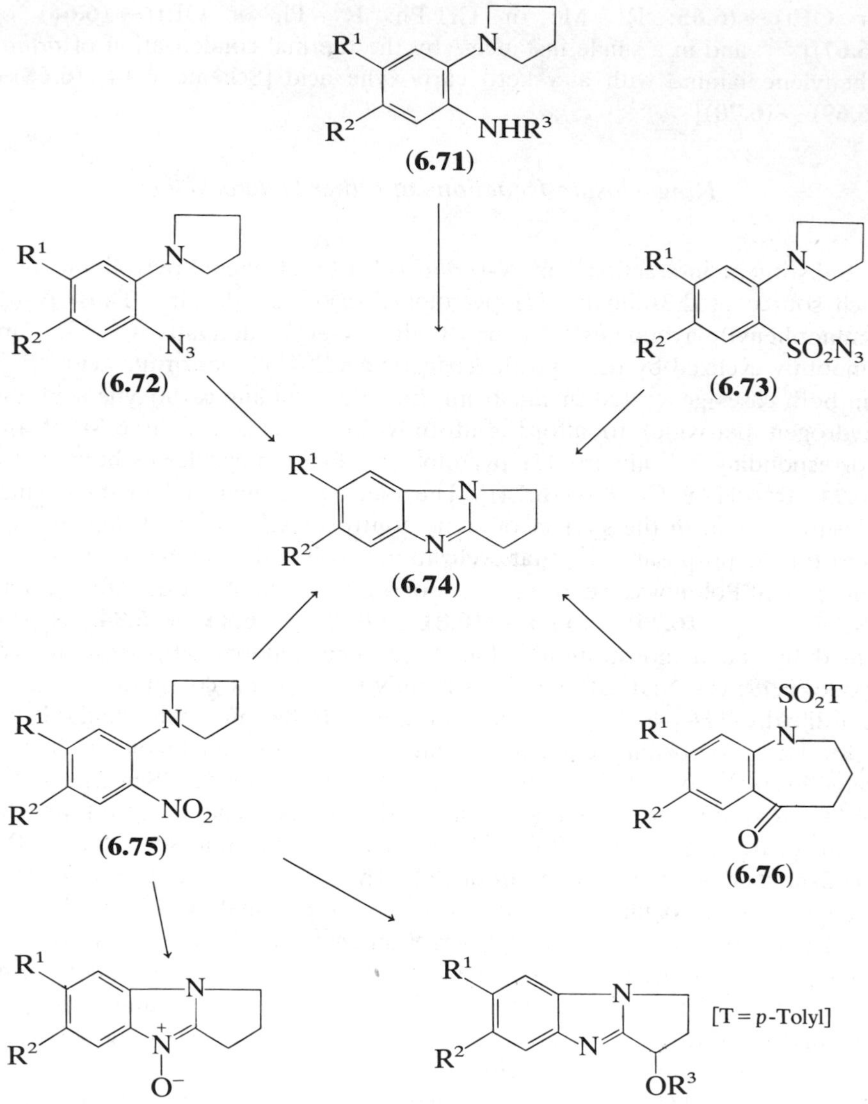

(**6.71**) (**6.72**) (**6.73**) (**6.74**) (**6.75**) (**6.76**) (**6.77**) (**6.78**)

[T = *p*-Tolyl]

Scheme 6.15

in high yield (Table 6.9). The thermal conversion of the sulfonyl azide [Scheme 6.15; (**6.73**; $R^1 = H$, $R^2 = NO_2$)] into the pyrrolobenzimidazole derivative (**6.74**; $R^1 = H$, $R^2 = NO_2$) is considered[60] to be initiated by Curtius rearrangement in a sulfonyl nitrene intermediate.

Interaction between an aromatic nitro substituent and the *C*(2) methylene center in an ortho-situated pyrrolidine ring is observed under a variety of conditions and almost invariably leads to 2,3-dihydro-1*H*-pyrrolo[1,2-*a*]-benzimidazoles as the major products.[47] In the context of such ring-closures, the reductive cyclization of *N*-(2-nitrophenyl)pyrrolidines [Scheme 6.15;

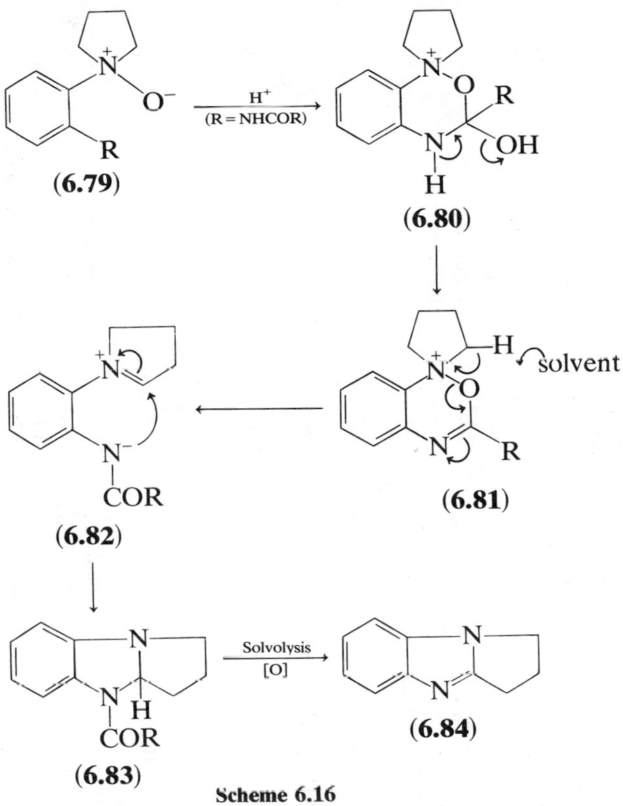

(6.79) $\xrightarrow[(R = NHCOR)]{H^+}$ (6.80)

(6.80) → (6.81) (solvent)

(6.81) → (6.82)

(6.82) → (6.83)

(6.83) $\xrightarrow[[O]]{Solvolysis}$ (6.84)

Scheme 6.16

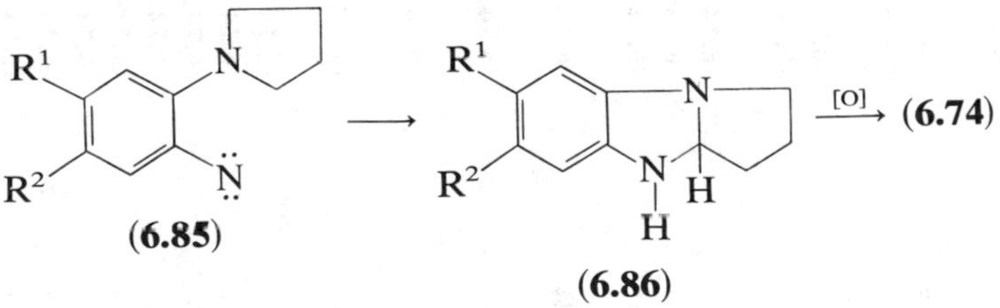

(6.85) → (6.86) $\xrightarrow{[O]}$ (6.74)

Scheme 6.17

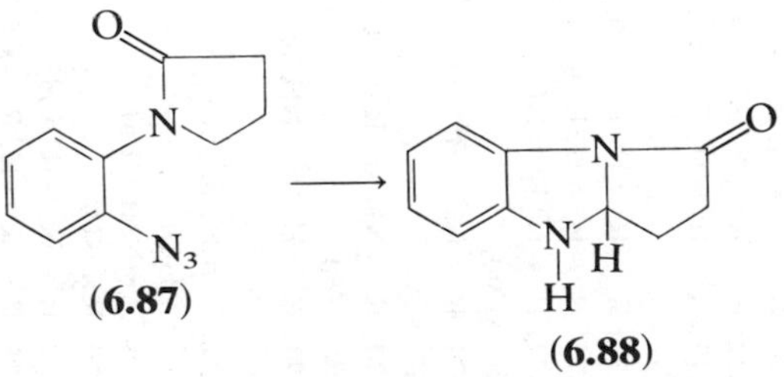

(6.87) → (6.88)

Scheme 6.18

TABLE 6.8. SYNTHESIS OF 2,3-DIHYDRO-1*H*-PYRROLO[1,2-*a*]BENZIMIDAZOLES BY RING-CLOSURE REACTIONS OF *N*-(2-SUBSTITUTED PHENYL)PYRROLIDINES AND RELATED PROCESSES

Starting material	Reaction conditions[a]	Product	Yield (%)	m.p. (°C)	Solvent of crystallization	Ref.
(**6.71**; $R^1 = R^2 = R^3 = H$)	*A*	(**6.74**; $R^1 = R^2 = H$)	81	114–115	Benzene–light petroleum	48
(**6.71**; $R^1 = R^2 = R^3 = H$)	*B*	(**6.74**; $R^1 = R^2 = H$)	85–95	115	—	49
(**6.71**; $R^1 = R^2 = R^3 = H$)	*C*	(**6.74**; $R^1 = R^2 = H$)	92	115	—	47
(**6.71**; $R^1 = R^2 = H$, $R^3 = Ac$)	*B*	(**6.74**; $R^1 = R^2 = H$)	74	115	—	49
(**6.71**; $R^1 = R^2 = H$, $R^3 = COPh$)	*B*	(**6.74**; $R^1 = R^2 = H$)	85	115	—	49
(**6.71**; $R^1 = R^3 = H$, $R^2 = Cl$)	*A*	(**6.74**; $R^1 = H$, $R^2 = Cl$)	75	133–134	Benzene–light petroleum	48
(**6.71**; $R^1 = R^3 = H$, $R^2 = Me$)	*A*	(**6.74**; $R^1 = H$, $R^2 = Me$)	86	144	2-Butanone	48
(**6.71**; $R^1 = R^3 = H$, $R^2 = NO_2$)	*A*	(**6.74**; $R^1 = H$, $R^2 = NO_2$)	72	209–210	2-Butanone	48
(**6.71**; $R^1 = H$, $R^2 = NO_2$, $R^3 = Ac$)	*B*	(**6.74**; $R^1 = H$, $R^2 = NO_2$)	80–90	208	—	50
(**6.71**; $R^1 = NO_2$, $R^2 = H$, $R^3 = Ac$)	*B*	(**6.74**; $R^1 = NO_2$, $R^2 = H$)	80–90	205	—[b]	47
(**6.71**; $R^1 = NHAc$, $R^2 = H$, $R^3 = Ac$)	*B*	(**6.74**; $R^1 = NHAc$, $R^2 = H$)	91	236	—[b]	51
(**6.71**; $R^1 = H$, $R^2 = NHAc$, $R^3 = Ac$)	*B*	(**6.74**; $R^1 = H$, $R^2 = NHAc$)	43	256	—[b]	51
(**6.71**; $R^1 = H$, $R^2 = CO_2Et$, $R^3 = CHO$)	*B*	(**6.74**; $R^1 = H$, $R^2 = CO_2Et$)	62	139	—[b]	47
(**6.79**; $R = NHCOPh$)	*D*	(**6.74**; $R^1 = R^2 = H$)	71	114–115	—	52
(**6.79**; $R = NO_2$)	*E*	(**6.74**; $R^1 = R^2 = H$)	50	—	—	53
(**6.71**; $R^1 = R^2 = R^3 = H$)	*F*	(**6.74**; $R^1 = R^2 = H$)	quant.	115	Cyclohexane	54
(**6.71**; $R^1 = R^2 = H$, $R^3 = CONH_2$)	*F*	(**6.74**; $R^1 = R^2 = H$)	85	114	Ether	54
(**6.72**; $R^1 = R^2 = H$)	*G*	(**6.74**; $R^1 = R^2 = H$)[c]	68	115	—[d]	57
(**6.72**; $R^1 = H$, $R^2 = Cl$)	*G*	(**6.74**; $R^1 = H$, $R^2 = Cl$)	46	137	Benzene-*n*-hexane	58
(**6.72**; $R^1 = H$, $R^2 = Br$)	*G*	(**6.74**; $R^1 = H$, $R^2 = Br$)	—[e]	150	—[b]	58
(**6.72**; $R^1 = H$, $R^2 = F$)	*G*	(**6.74**; $R^1 = H$, $R^2 = F$)	—[e]	128	—[b]	58
(**6.72**; $R^1 = R^2 = H$)[f]	*G*	(**6.74**; $R^1 = R^2 = H$)[g]	—[e]	122	—[b]	58
(**6.72**; $R^1 = R^2 = Cl$)	*G*	(**6.74**; $R^1 = R^2 = Cl$)	—[e]	215	—[b]	58
(**6.72**; $R^1 = H$, $R^2 = Me$)	*G*	(**6.74**; $R^1 = H$, $R^2 = Me$)	—[e]	146	—[b]	58

(**6.72**; $R^1 = H$, $R^2 = SO_2N$(pyrrolidinyl))	*G*	(**6.74**; $R^1 = H$, $R^2 = SO_2N$(pyrrolidinyl))	—[e]	239	—[b]	58
(**6.72**; $R^1 = H$, $R^2 = CO_2Et$)	*G*	(**6.74**; $R^1 = H$, $R^2 = CO_2Et$)	—[e]	134	—[b]	58
(**6.72**; $R^1 = R^2 = H$)[h]	*G*	(**6.74**; $R^1 = R^2 = H$)[i]	—[e]	96	—[b]	58
(**6.72**; $R^1 = Cl$, $R^2 = CO_2Et$)	*G*	(**6.74**; $R^1 = Cl$, $R^2 = CO_2Et$)	—[e]	138	—[b]	58
(**6.72**; $R^1 = Br$, $R^2 = CO_2Et$)	*G*	(**6.74**; $R^1 = Br$, $R^2 = CO_2Et$)	—[e]	114	—[b]	58
(**6.75**; $R^1 = R^2 = H$)	*H*	(**6.74**; $R^1 = R^2 = H$)	quant.	115	—	61
(**6.75**; $R^1 = H$, $R^2 = CF_3$)	*H*	(**6.74**; $R^1 = H$, $R^2 = CF_3$)	quant.	148	—[b]	61
(**6.75**; $R^1 = R^2 = H$)	*I*	(**6.77**; $R^1 = R^2 = H$)[j]	51	224	—[b]	62
(**6.75**; $R^1 = R^2 = H$)	*J*	(**6.77**; $R^1 = R^2 = H$)[j]	78	—	—	62
(**6.75**; $R^1 = H$, $R^2 = Cl$)	*I*	(**6.77**; $R^1 = H$, $R^2 = Cl$)	70	182	—[b]	62
(**6.75**; $R^1 = H$, $R^2 = Cl$)	*J*	(**6.77**; $R^1 = H$, $R^2 = Cl$)	13	—	—	62
(**6.75**; $R^1 = H$, $R^2 = NO_2$)	*I*	(**6.77**; $R^1 = H$, $R^2 = NO_2$)[j]	32	212	—[b]	62
(**6.75**; $R^1 = H$, $R^2 = NO_2$)	*K*	(**6.77**; $R^1 = H$, $R^2 = NO_2$)[k]	16	180 (decomp.)	—[b]	63
(**6.75**; $R^1 = H$, $R^2 = CO_2H$)	*I*	(**6.77**; $R^1 = H$, $R^2 = CO_2H$)	61	255–260	—[b]	62
(**6.75**; $R^1 = H$, $R^2 = CF_3$)	*I*	(**6.77**; $R^1 = H$, $R^2 = CF_3$)	33	196–198	—[b]	62
(**6.75**; $R^1 = R^2 = H$)	*L*	(**6.78**; $R^1 = R^2 = R^3 = H$)	67	190	Ethyl acetate	65
(**6.75**; $R^1 = H$, $R^2 = NO_2$)	*L*	(**6.78**; $R^1 = H$, $R^2 = NO_2$, $R^3 = Ac$)[c]	68	148	Ethyl acetate–light petroleum (b. p. 60–80°)	65

[a] $A = CF_3CO_2H$, 30% H_2O_2, CH_2Cl_2/(reflux)(15–30 min); B = 98% HCO_2H, 30% H_2O_2/(100°)(10–15 min); $C = CH_3CO_2H$, 30% H_2O_2 (conditions not specified); D = 2 M HCl/(reflux)(15 min); E = Sn, HCO_2H/(reflux)(3 hr); F = HgO, EDTA, EtOH–H_2O (1:1)/(room temp.)(few min); $G = PhNO_2$/(170°)(0.5 hr); $H = TiCl_2$, conc. HCl/(80°)(1 hr); I = conc. HCl/(reflux)(1–20 hr); $J = h\nu$, 1 M HCl, MeOH, H_2O/(room temp.)(48–54 hr); $K = h\nu$, AcOH, H_2O/(room temp.)(several hr); $L = ZnCl_2$, Ac_2O/(reflux)(3–4 hr).

[b] Solvent of crystallization not specified.

[c] Colorless needles.

[d] Purified by sublimation.

[e] Yield not quoted.

[f] 6-Chloro derivative.

[g] 8-Chloro derivative.

[h] 6-Ethoxycarbonyl derivative.

[i] 8-Ethoxycarbonyl derivative.

[j] Hydrochloride.

[k] Buff needles.

(**6.75**) → (**6.74**)] promoted by titanous chloride in acidic solution has perhaps the greatest synthetic potential. At least in the cases reported[47,61] these reactions afford the corresponding 2,3-dihydro-1*H*-pyrrolo[1,2-*a*]-benzimidazoles in essentially quantitative yield (Table 6.8) and are believed[61] to owe their efficiency to the intermediacy of an organometallic complex, in which the metal plays the dual role of reducing and chelating agent. The first-formed products in such cyclizations are probably the corresponding 2,3-dihydro-1*H*-pyrrolo[1,2-*a*]benzimidazole 4-*N*-oxides, which undergo subsequent deoxygenation to the parent 2,3-dihydro-1*H*-pyrrolo[1,2-*a*]benzimidazoles isolated. In accord with this view, 2,3-dihydro-1*H*-pyrrolo[1,2-*a*]benzimidazole *N*-oxides are formed, admittedly in variable yield (Table 6.8), in the absence of titanous chloride, when *N*-(2-nitrophenyl)pyrrolidines are heated with concentrated hydrochloric acid [Scheme 6.15; (**6.75**) → (**6.77**)].[47,62] 2,3-Dihydro-1*H*-pyrrolo[1,2-*a*]benzimidazole 4-*N*-oxides (**6.77**) are also the end-products (Table 6.8) of the photolysis of *N*-(2-nitrophenyl)pyrrolidines (**6.75**) in acetic acid[63] or methanolic hydrochloric acid.[47,62] The mechanistic details of the acid-catalyzed cyclizations of *N*-(2-nitrophenyl)pyrrolidines to 2,3-dihydro-1*H*-pyrrolo[1,2-*a*]benzimidazole 4-*N*-oxides have not been established, though for the purely thermal processes involvement of an aci-nitro intermediate has been proposed.[47,62] However, it is conceivable that, as in the case of other acid-catalyzed aromatic nitro group ortho side-chain interactions,[64] cyclization is the end result of *direct* aldol-type condensation between the nitro group and an α-methylene center in the pyrrolidine ring activated by protonation of the adjacent nitrogen atom. 2,3-Dihydro-1*H*-pyrrolo[1,2-*a*]-benzimidazole 4-*N*-oxides are also plausible intermediates in the zinc chloride–acetic anhydride mediated ring-closure of *N*-(2-nitrophenyl)pyrrolidines (**6.75**), which leads, depending on the nature of the workup, to 3-hydroxy- or 3-acetoxy-2,3-dihydro-1*H*-pyrrolo[1,2-*a*]benzimidazoles [Scheme 6.15; (**6.78**; $R^3 = H$ or Ac)],[47,65] in good yield (Table 6.8).

Isolated instances of ring-closure reactions leading to 2,3-dihydro-1*H*-pyrrolo[1,2-*a*]benzimidazoles include the transformation (Scheme 6.19)[66] of the azo-sulfonate (**6.89**) into the interesting betaine structure (**6.90**) and the transannular cyclization (Scheme 6.15) via Schmidt rearrangement of the diazepinone (**6.76**; $R^1 = R^2 = H$, T = toluene-*p*-sulfonyl) to 2,3-dihydro-1*H*-pyrrolo[1,2-*a*]benzimidazole (**6.74**; $R^1 = R^2 = H$).[67]

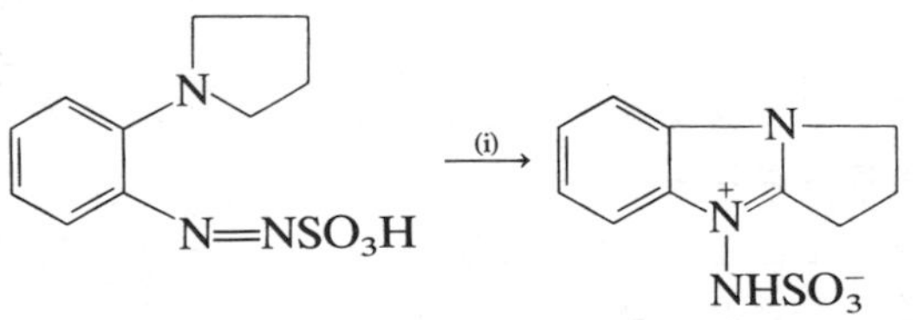

N
N=NSO$_3$H
(i)
N
$\overset{+}{N}$
NHSO$_3^-$

(**6.89**) (**6.90**)

(i) SO$_2$,H$_2$O/(70°)(0.5 hr) (m.p. 304–306°)

Scheme 6.19

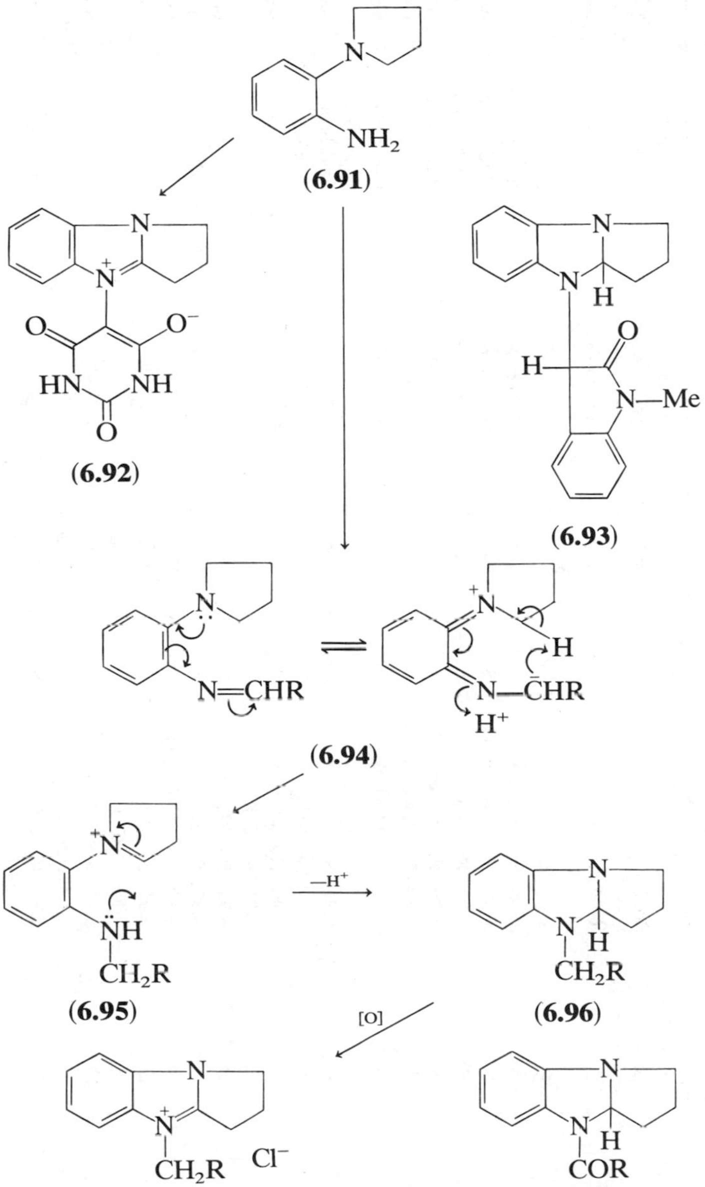

Scheme 6.20

2,3,3*a*,4-Tetrahydro-1*H*-pyrrolo[1,2-*a*]benzimidazole derivatives are readily accessible[68] by the acid-catalyzed cyclization of *N*-(2-alkylideneamino or 2-arylideneamino)pyrrolidines either preformed or prepared *in situ* by the condensation of *N*-(2-aminophenyl)pyrrolidine with aliphatic, aromatic, or heteroaromatic aldehydes [Scheme 6.20; (**6.91**) → (**6.94**) → → (**6.96**)]. Ring-closure reactions of this type proceed in high yield (Table 6.9) and on

TABLE 6.9. SYNTHESIS OF 2,3,3*a*,4-TETRAHYDRO-1*H*-PYRROLO[1,2-*a*]BENZIMIDAZOLES BY RING-CLOSURE REACTIONS OF *N*-(2-AMINOPHENYL)PYRROLIDINE DERIVATIVES

Starting material	Reaction conditions[a]	Product	Yield (%)	m.p. (°C)	Solvent of crystallization	Crystal form	Ref.
(**6.91**)[b]	*A*	(**6.96**; R = Ph)[b]	89	135	Ethanol–ether	Colorless needles	68
(**6.94**; R = Ph)	*B*	(**6.96**; R = Ph)	89	—[c]	—[d]	—[e]	68
(**6.91**)[b]	*A*	(**6.96**; R = *o*-$O_2NC_6H_4$)[b]	89	155	Ethanol	Yellow prisms	68
(**6.94**; R = *o*-$O_2NC_6H_4$)	*B*	(**6.96**; R = *o*-$O_2NC_6H_4$)	86	64	Ethanol	Red prisms	68
(**6.94**; R = *m*-$O_2NC_6H_4$)	*B*	(**6.96**; R = *m*-$O_2NC_6H_4$)	82	92	Light petroleum	Red prisms	68
(**6.94**; R = *p*-$O_2NC_6H_4$)	*B*	(**6.96**; R = *p*-$O_2NC_6H_4$)	85	91	Light petroleum	Red prisms	68
(**6.91**)[b]	*A*	(**6.96**; R = 3,4-$Cl_2C_6H_3$)[b]	84	125–130	Ethanol	—[f]	68
(**6.91**)[b]	*A*	(**6.96**; R = *m*-HOC_6H_4)[b]	81	156	Ethanol	Colorless needles	68
(**6.94**; R = 2-Cl, 5-$O_2NC_6H_3$)	*B*	(**6.96**; R = 2-Cl, 5-$O_2NC_6H_3$)[b]	96	125	Ethanol	Red-brown prisms	68
(**6.91**)	*C*	(**6.96**; R = 4-pyridyl)	87	—[g]	—[d]	—[e]	68
(**6.91**)	*C*	(**6.96**; R = 2-thienyl)	85	—[h]	—[d]	—[e]	68
(**6.91**)	*D*	(**6.97**; R = Me)[i]	79	140	—[j]	—[f]	68
(**6.91**)	*D*	(**6.97**; R = Ph)[k]	86	219–220	—[j]	—[f]	68
(**6.91**)	*D*	(**6.97**; R = *p*-HOC_6H_4)[l]	80	252	—[j]	—[f]	68
(**6.91**)	*D*	(**6.97**; R = 3,4-$(MeO)_2C_6H_3$)[l]	82	240 (decomp.)	—[j]	—[f]	68
(**6.91**)	*D*	(**6.97**; R = 3-HO, 4-$MeOC_6H_3$)[l]	81	240 (decomp.)	—[j]	—[f]	68
(**6.91**)	*D*	(**6.97**; R = 2-Cl, 5-$O_2NC_6H_3$)[i]	84	290 (decomp.)	—[j]	—[f]	68
(**6.91**)	*D*	(**6.97**; R = CH═CHPh)	86	258–260	—[j]	—[f]	68
(**6.91**)	*D*	(**6.97**; R = 2-pyridyl)	75	163	—[j]	—[f]	68
(**6.91**)	*D*	(**6.97**; R = 3-pyridyl)	75	169 (decomp.)	—[j]	—[f]	68

(**6.91**)	*E*	(**6.92**)[m]	—[n]	>300 (decomp.)	Acetic acid–water	—[f]	68
(**6.91**)	*E*	(**6.92**)[i]	52	>360 (decomp.)	Water	Colorless prisms	69
(**6.91**)	*F*	(**6.93**)	—[n]	145–150 (decomp.)	—[j]	—[f]	68
(**6.72**; $R^1 = R^2 = H$)	*F*	(**6.98**; R = Me)	32	85–86	—[j]	Colorless prisms	57
(**6.72**; $R^1 = R^2 = H$)	*F*	(**6.98**; R = Ph)	30	90	—[j]	Colorless prisms	57
(**6.91**; NHCOPh for NH_2)	*G*	(**6.98**; R = Ph)	84	88–89.5	Ethanol–water	—[f]	54
(**6.87**)	*H*	(**6.88**)	41	122–124	Carbon tetrachloride	Colorless needles	59
(**6.87**)	*I*	(**6.88**)	96	—	—	—	59

[a] *A* = RCHO, EtOH/(room temp.)(14 hr); *B* = conc. HCl, EtOH/(room temp.)(14 hr); *C* = RCHO, conc. HCl, EtOH/(room temp.)(14 hr); *D* = RCHO, CF_3CO_2H, CCl_4/(reflux)(1 hr); *E* = Alloxan, conc. HCl, EtOH/(room temp.)(14 hr or 3 days); *F* = Ac_2O or $(EtCO)_2O$/heat. *G* = HgO, EDTA, EtOH–H_2O (1 : 1)/(room temp.)(few min); *H* = $PhNO_2$/(reflux)(3 hr); *I* = Diethyleneglycol dimethyl ether/(reflux)(3 hr).
[b] Hydrochloride.
[c] b.p. 160°/0.6 mm Hg.
[d] Purified by distillation.
[e] Yellow liquid.
[f] Crystal form not specified.
[g] b.p. 165–170°/1 mm Hg.
[h] b.p. 160–165°/1 mm Hg.
[i] Hydrate.
[j] Solvent of crystallization not specified.
[k] Dihydrate.
[l] Hemihydrate.
[m] Acetic acid solvate.
[n] Yield not quoted.

the basis of deuterium labeling studies[68] are suggested to follow a course (Scheme 6.20) initiated by intramolecular proton transfer [(**6.94**)→(**6.95**)→(**6.96**)]. In contrast, the condensation of the amine (**6.91**) with aliphatic or aromatic aldehydes in carbon tetrachloride in the presence of trifluoroacetic acid leads to high yields (Table 6.9) of the corresponding 2,3-dihydro-1*H*-pyrrolo[1,2-*a*]benzimidazolium salts [Scheme 6.20; (**6.97**)] as a result of oxidation of the first-formed tetrahydro compounds (**6.96**) by the solvent. The formation (Scheme 6.20 and Table 6.9) of the betaine (**6.92**)[68,69] by condensation of the amine (**6.91**) in ethanolic hydrochloric acid can likewise be rationalized[68,69] in terms of initial anil formation followed by cyclization to and *in situ* oxidation of a tetrahydro-1*H*-pyrrolo[1,2-*a*]benzimidazole intermediate [cf. Scheme 6.20; (**6.91**)→(**6.94**)→→(**6.96**)→(**6.97**)]. In the similar condensation[68] of the amine (**6.91**) with *N*-methylisatin, the tetrahydro intermediate (**6.93**) is stable enough to be isolated (Table 6.9). The isolation in moderate yield (Table 6.9) of 4-acyl-2,3,3*a*,4-tetrahydro-1*H*-pyrrolo[1,2-*a*]benzimidazoles [Scheme 6.20; (**6.98**)] when *N*-(2-azidophenyl)pyrrolidine [Scheme 6.15; (**6.72**; $R^1 = R^2 = H$)] is thermolyzed[57] in carboxylic anhydrides and the high yield thermolysis (Table 6.9) of *N*-(2-azidophenyl)pyrrolidin-2-one to 2,3,3*a*,4-tetrahydro-1*H*-pyrrolo[1,2-*a*]-benzimidazol-1-one [Scheme 6.18; (**6.87**)→(**6.88**)][59] provide evidence for the intermediacy of 2,3,3*a*,4-tetrahydro-1*H*-pyrrolo[1,2-*a*]benzimidazoles in the presumed nitrene-initiated cyclizations of *N*-(2-azidophenyl)pyrrolidines leading to 2,3-dihydro-1*H*-pyrrolo[1,2-*a*]benzimidazole derivatives (see before). 4-Benzoyl-2,3,3*a*,4-tetrahydro-1*H*-pyrrolo[1,2-*a*]benzimidazole (**6.98**; R = Ph) is also formed in high yield (Table 6.9) by the oxidative cyclization of *N*-(2-benzamidophenyl)pyrrolidine [Scheme 6.20; (**6.91**; NHCOPh for NH_2)].[54]

6.1.2. Physicochemical Properties

Spectroscopic Studies

Infrared Spectra. *N*(4)-Unsubstituted 4*H*-pyrrolo[1,2-*a*]benzimidazoles exhibit[22] IR *NH* absorption at ca. 3500 cm^{-1}, whereas the NH group in *N*(4)-unsubstituted 2,3,3*a*,4-tetrahydro-1*H*-pyrrolo[1,2-*a*]benzimidazol-1-ones [Table 6.11; (**6.103**; R = H or Ph)] gives rise to IR absorption at somewhat lower frequencies (3390–3280 cm^{-1}).[38,59] Broad absorption at 3100–2200 cm^{-1} in the IR spectra[2,5] of 3-hydroxy-1*H*-pyrrolo[1,2-*a*]benzimidazol-1-ones [Table 6.11; (**6.101**; $R^1 = OH$)] demonstrates the strongly chelated nature of the hydroxyl group in such molecules. The presence of a band at 3400 cm^{-1} (OH) and the absence of absorption at 1470 cm^{-1} (N═O) in the IR spectra[70] of the acid salts of 1-nitroso-4*H*-pyrrolo[1,2-*a*]-benzimidazoles show that these compounds exist in the oximino tautomeric form [Scheme 6.21; (**6.105**)] rather than the nitroso form (**6.104**).

TABLE 6.10. INFRARED SPECTRA OF CARBONYL DERIVATIVES OF 4*H*-PYRROLO[1,2-*a*]BENZIMIDAZOLES (**6.99**) AND 2,3,3*a*,4-TETRAHYDRO-1*H*-PYRROLO[1,2-*a*]BENZIMIDAZOLES (**6.100**)

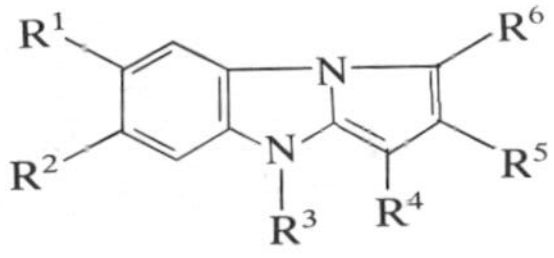

(**6.99**) 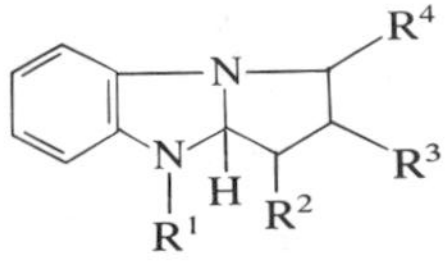 (**6.100**)

Compound	R^1	R^2	R^3	R^4	R^5	R^6	Medium	$\nu_{max}(cm^{-1})$	Ref.
(**6.99**)	H	H	Me	Me	Me	COMe	—[a]	1640–1620	12, 13
(**6.99**)	H	H	Me	COMe	Me	Me	—[a]	1640–1620	12, 13
(**6.99**)	H	H	Me	CO_2Et	H	COPh	—[b]	1690, 1620	23
(**6.99**)	H	H	Me	CO_2Et	H	p-BrC_6H_4CO	—[b]	1685, 1610	23
(**6.99**)	H	H	Et	CO_2Et	H	COPh	—[b]	1695, 1620	23
(**6.99**)	H	H	Et	CO_2Et	H	p-BrC_6H_4CO	—[b]	1680, 1610	23
(**6.99**)	H	H	CH_2Ph	CO_2Et	H	COPh	—[a]	1702, 1630	24
(**6.99**)	H	H	Me	CO_2Et	H	CO_2Me	—[b]	1705, 1650	23
(**6.99**)	H	H	Et	CO_2Et	H	CO_2Me	—[b]	1675, 1650	23
(**6.99**)	H	H	Me	COMe	H	CO_2Et	—[a]	1680, 1640	21
(**6.99**)	H	H	Me	COMe	H	CN	—[a]	2200, 1640	21
(**6.99**)	H	H	Me	CO_2Me	CO_2Me	COPh	—[b]	1735, 1685	23
(**6.99**)	H	H	CH_2Ph	CO_2Me	CO_2Me	COPh	—[a]	1722, 1695, 1618	24
(**6.99**)	H	H	Me	CO_2Me	CO_2Me	CO_2Me	—[b]	1742, 1690, 1655	23
(**6.99**)	H	H	Me	CO_2Et	CO_2Me	CO_2Me	—	1740, 1726, 1694	29
(**6.99**)	H	H	Et	CO_2Me	CO_2Me	CO_2Me	—[b]	1740, 1710, 1690	23
(**6.99**)	H	H	$CH{=}CHCO_2Me$	CH_2CO_2Me	CO_2Me	H	—[a]	1735, 1706, 1691	25
(**6.99**)	H	H	CH_2COMe	H	Me	H	—[c]	1740	19
(**6.99**)	H	H	p-$BrC_6H_4COCH_2$	H	p-BrC_6H_4	H	—[c]	1710	19
(**6.99**)	H	H	p-$NO_2C_6H_4COCH_2$	H	p-$NO_2C_6H_4$	H	—[c]	1692	19
(**6.99**)	Me	Me	CH_2COPh	H	Ph	H	—[c]	1696	19
(**6.99**)	Me	Me	CH_2COPh	Ph	Ph	H	—[c]	1688	19
(**6.100**)	Me	COPh	COPh	COPh	—	—	—[d]	1680	46
(**6.100**)	CH_2Ph	CN	H	CO_2Et	—	—	—[a]	2140, 1745	45

[a] Medium not specified.
[b] KBr.
[c] Nujol.
[d] $CHCl_3^-$.

TABLE 6.11. INFRARED SPECTRA OF 1*H*-PYRROLO[1,2-*a*]BENZIMIDAZOL-1-ONES (**6.101**), 2,3-DIHYDRO-1*H*-PYRROLO[1,2-*a*]BENZIMIDAZOL-1-ONE (**6.102**) AND 2,3,3*a*,4-TETRAHYDRO-1*H*-PYRROLO[1,2-*a*]BENZIMIDAZOL-1-ONES (**6.103**)

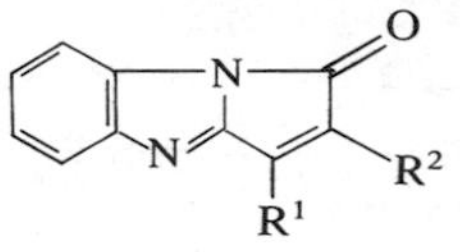

(**6.101**) 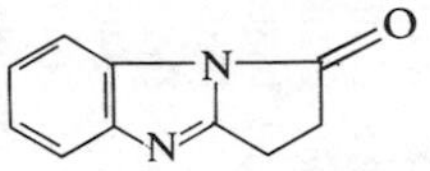 (**6.102**) 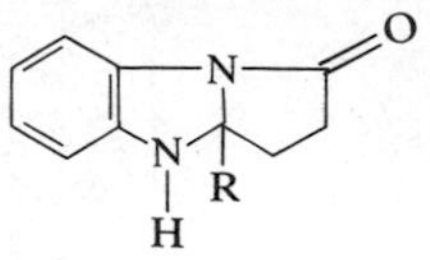 (**6.103**)

Compound	R	R^1	R^2	Medium	$\nu_{max}(cm^{-1})$	Ref.
(**6.101**)	—	Ph	Ph	—[a]	1745	4
(**6.101**)	—	OH	Ph	—[b]	3100–2200[c]	2
(**6.101**)	—	OH	$p\text{-}NO_2C_6H_4$	—[a]	3100–2200, 1740	5
(**6.102**)	—	—	—	—[b]	1760	41
(**6.102**)	—	—	—	—[d]	1755	38
(**6.102**)	—	—	—	—[e]	1755	38
(**6.103**)	H	—	—	—[e]	3280, 1690	59
(**6.103**)	Ph	—	—	—[a]	3390, 1705	38

[a] Medium not specified.
[b] KBr.
[c] C=O absorption not quoted.
[d] $CHCl_3$.
[e] Nujol.

The IR spectra[71] of 6- and 7-azido-2,3-dihydro-1*H*-pyrrolo[1,2-*a*]benzimidazoles contain orthodox IR azide absorption at 2120 cm^{-1}, which serves to characterize these otherwise relatively unstable molecules. The similar IR stretching frequencies (Table 6.10) for the cyano groups in the molecules (**6.99**; $R^1 = R^2 = R^5 = H$, $R^3 = Me$, $R^4 = Ac$, $R^6 = CN$)[21] and (**6.100**; $R^1 = CH_2Ph$, $R^2 = CN$, $R^3 = H$, $R^4 = CO_2Et$)[45] implies a surprising lack of sensitivity to conjugate effects on the part of the cyano substituent in the former case. In contrast, the marked variation in carbonyl stretching frequency (Table 6.10) with the site of substitution, observed for carbonyl derivatives of 4*H*-pyrrolo[1,2-*a*]benzimidazoles is directly attributable to the presence

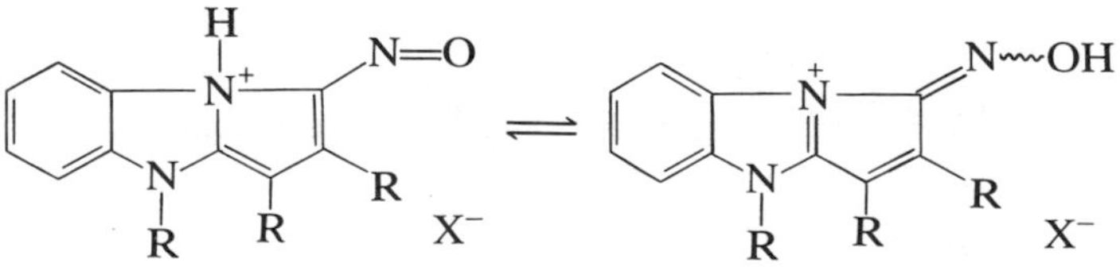

(**6.104**) ⇌ (**6.105**)

[X = Cl or ClO_4]
Scheme 6.21

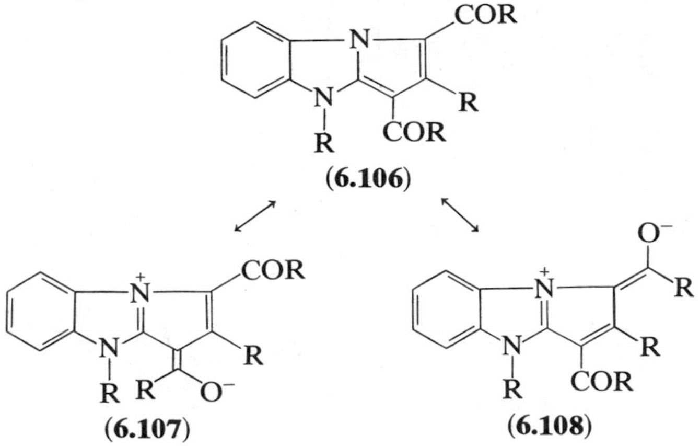

(6.106)

(6.107)

(6.108)

Scheme 6.22

or absence of conjugative interaction with the bridgehead nitrogen atom.[12,13] Thus, in accord with their vinylogous amide character [Scheme 6.22; (**6.106**) ↔ (**6.107**) ↔ (**6.108**)] ester and ketonic substituents at the *C*(1) and *C*(3) positions in 4*H*-pyrrolo[1,2-*a*]benzimidazoles absorb at significantly lower frequencies (cf. Table 6.10) than their *C*(2) counterparts. Conversely, the much higher carbonyl frequencies (Table 6.11) observed for 1*H*-pyrrolo[1,2-*a*]benzimidazol-1-ones (**6.101**) and their 2,3-dihydro analogs (**6.102**) compared with normal amides may be attributed[38] to the absence of the usual nitrogen–carbonyl interaction in the latter as a result of preferential conjugation between the bridgehead nitrogen atom in (**6.101**) and (**6.102**) with the other nitrogen center and the benzene ring. In accord with this interpretation are the markedly lower carbonyl frequencies (Table 6.11) found for 2,3,3*a*,4-tetrahydro-1*H*-pyrrolo[1,2-*a*]benzimidazol-1-ones (**6.103**). The IR spectra[10] of simple 3*H*-pyrrolo[1,2-*a*]benzimidazoles contain characteristic bands at 1630–1625 cm^{-1}.

Ultraviolet Spectra. The UV spectra (Table 6.12) of 2,3-dihydro-1*H*-pyrrolo[1,2-*a*]benzimidazoles (**6.111**), not unexpectedly in view of the saturated nature of the fused five-membered ring present, resemble that of 1-methylbenzimidazole. Correspondingly, the UV absorption[69] of the 2,3-dihydro-1*H*-pyrrolo[1,2-*a*]benzimidazolium betaine [Scheme 6.20; (**6.92**)] is akin to that of the benzimidazolium cation. The essential lack of conjugation involving the *C*(1)–*C*(2) double bond in 3*H*-pyrrolo[1,2-*a*]benzimidazoles [Table 6.12; (**6.109**)], and the carbonyl group in 2,3-dihydro-1*H*-pyrrolo[1,2-*a*]benzimidazol-1-one [Table 6.12; (**6.112**)] likewise results in UV spectra (Table 6.12) not too far removed from those of simple benzene derivatives. In contrast, the UV spectra[2] of the 1*H*-pyrrolo[1,2-*a*]benzimidazole derivatives [Scheme 6.23; (**6.113**) and (**6.114**)] exhibit the marked bathochromic shift in UV maxima anticipated from the extensive conjugation present.

TABLE 6.12. ULTRAVIOLET SPECTRA OF 3*H*-PYRROLO[1,2-*a*]BENZIMIDAZOLES (**6.109**), 4*H*-PYRROLO[1,2-*a*]BENZIMIDAZOLES (**6.110**), AND 2,3-DIHYDRO-1*H*-PYRROLO[1,2-*a*]BENZIMIDAZOLES (**6.111**) AND (**6.112**)

(**6.109**) (**6.110**) (**6.111**) (**6.112**)

Compound	R	R^1	R^2	R^3	R^4	Solvent[a]	λ_{max} (nm) (log ϵ)	
(**6.109**	—	Ph	Ph	—	—	*A*	272(4.26), 310(3.8)	10
(**6.109**)	—	Ph	*p*-$MeOC_6H_4$	—	—	*A*	272(4.43), 310(3.94)	10
(**6.110**)	—	Me	H	*p*-MeC_6H_4	H	*B*	263(4.20), 283(4.20), 330(3.80)	15
(**6.110**)	—	Me	H	*p*-PhC_6H_4	Me	*B*	256(4.22), 369(4.60)	15
(**6.110**)	—	Me	Ph	*m*-$NO_2C_6H_4$	H	*B*	268(4.35)	15
(**6.110**)	—	Me	CO_2Et	H	CO_2Me	*B*	246(4.73), 313(4.07), 326(4.02)	23
(**6.110**)	—	Et	CO_2Et	H	CO_2Me	*B*	246(4.63), 294(4.18), 313(4.10), 327(4.17)	23
(**6.110**)	—	Me	CO_2Et	H	*p*-BrC_6H_4CO	*B*	227(4.41), 333(3.77)	23
(**6.110**)	—	Et	CO_2Et	H	*p*-BrC_6H_4CO	*B*	228(4.42), 332 (3.81)	23
(**6.110**)	—	Me	CO_2Me	CO_2Me	CO_2Me	*B*	247(4.49), 292(4.29), 331(4.22)	23
(**6.110**)	—	Me	CO_2Me	CO_2Me	CO_2Me	*A*	214(4.39), 247(4.45), 291(4.36), 332(4.27)	26, 27

(6.110)	—	Me	CO_2Me	CO_2Me	CO_2Me	*C*	214(4.39), 247(4.45), 291(4.36), 332(4.27)	26, 27
(6.110)	—	Me	CO_2Et	CO_2Et	CO_2Et	*A*	214(4.66), 247(4.74), 291(4.64), 332(4.56)	27
(6.110)	—	Me	CO_2Et	CO_2Et	CO_2Et	*C*	214(4.65), 247(4.71), 291(4.63), 332(4.54)	27
(6.110)	—	Me	CO_2Et	CO_2Me	CO_2Me	*A*	216(4.45), 248(4.51), 292(4.41), 332.5(4.31)	29
(6.110)	—	Et	CO_2Me	CO_2Me	CO_2Me	*B*	215(4.43), 247(4.48), 292(4.38), 331(4.31)	23
(6.110)	—	Me	CO_2Me	CO_2Me	COPh	*B*	234(4.51), 330(3.83)	23
(6.110)	—	$CH\overset{t}{=}CHCO_2Et$	CH_2CO_2Me	CO_2Me	H	*A*	215(4.29), 240(4.50), 262 inf (4.35), 315(4.41)	25
(6.110)	—	$CH\overset{t}{=}CHCO_2Et$	CH_2CO_2Me	CO_2Me	H	*C*	211(4.29), 238(4.46), 260 inf (4.23)	25
(6.111)	H	—	—	—	—	*B*	255(3.69), 284(3.66), 292(3.65)	48
(6.111)	H	—	—	—	—	*D*	243(3.72), 248(2.95), 266(3.58), 271(3.71), 278(3.72)	67
(6.111)	Cl	—	—	—	—	*B*	255(3.69), 284(3.66), 292(3.65)	48
(6.111)	Me	—	—	—	—	*B*	255(3.69), 284(3.66), 292(3.65)	48
(6.112)	—	—	—	—	—	*B*	233(4.22), 275 inf (3.20)	38

[a] *A* = methanol; *B* = ethanol; *C* = methanol + 3 drops of 72% $HClO_4$aq; *D* = phosphate buffer (pH 8).

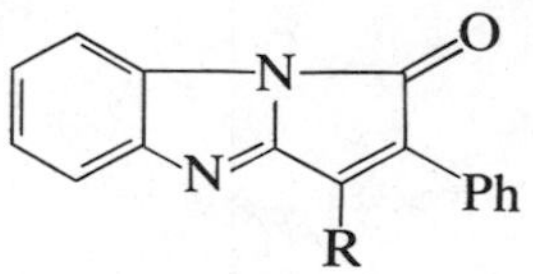

(**6.113**) 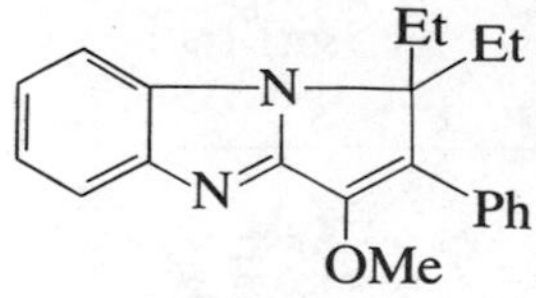 (**6.114**)

Scheme 6.23

The UV spectra (Table 6.12) of 4*H*-pyrrolo[1,2-*a*]benzimidazoles (**6.110**) are typified by the presence of two or more intense absorption bands at relatively long wavelength consistent with the high degree of delocalization and aromatic character (see later) associated with these molecules. These spectral properties have been put to practical use in the formulation of cyanine dyes based on 4*H*-pyrrolo[1,2-*a*]benzimidazole frameworks which absorb strongly in the visible region at 520–680 nm.[12] The somewhat longer wavelength absorption of 1-acyl-4*H*-pyrrolo[1,2-*a*]benzimidazoles in comparison with the less extensively conjugated 3-acyl isomers has been used as a criterion for assigning the position of acylation in 4*H*-pyrrolo[1,2-*a*]benzimidazoles (see later).[13,21] The low basicity of polyacylated 4*H*-pyrrolo[1,2-*a*]benzimidazoles is demonstrated by the invariability of their UV absorption (Table 6.12) in changing from neutral to acidic solution.

Nuclear Magnetic Resonance Spectra. Representative examples of the ^{1}H NMR spectra of pyrrolo[1,2-*a*]benzimidazole derivatives of the different structural types are collected in Tables 6.13–6.16.

The *C*(1) protons in the 3*H*-pyrrolo[1,2-*a*]benzimidazole derivatives [Table 6.13; (**6.115**)] absorb in the range δ 6.80–8.00, whereas the protons at *C*(3) give rise to signals at higher field (δ 5.15–5.38).[17] The relatively high-field position for the latter in comparison with the low-field absorption of the *C*(3) proton in 4*H*-pyrrolo[1,2-*a*]benzimidazoles provides a clear cut distinction between the 3*H* and 4*H* tautomeric forms of pyrrolo[1,2-*a*]-benzimidazoles of utility in assigning the structures of the debenzylation products of 4-benzyl-4*H*-pyrrolo[1,2-*a*]benzimidazoles.[17]

^{1}H NMR signals associated with protons at the *C*(1) position in 4*H*-pyrrolo[1,2-*a*]benzimidazoles [Table 6.14; (**6.116**)] appear at significantly lower field (δ 6.45–6.55) than those of their *C*(3)*H* counterparts (δ 4.95–5.83) with which they are weakly coupled ($J \approx 1$ hz) (Table 6.14).[13,17] Additional splitting ($J \approx 1$ Hz) of the *C*(1) proton resonances, which arises as a result of coupling with *C*(2) alkyl substituents, is absent in the signals due to the *C*(3) protons. The ready differentiation of unoccupied *C*(1) and *C*(3) positions in 4*H*-pyrrolo[1,2-*a*]benzimidazoles on the basis of these differing ^{1}H NMR characteristics of the *C*(1) and *C*(3) protons allows the determination of the site of electrophilic substitution in such molecules (see later).[13,21,70]

TABLE 6.13. ¹H NMR SPECTRA[a,b] OF 1*H*-PYRROLO[1,2-*a*]BENZIMIDAZOLE AND 3*H*-PYRROLO[1,2-*a*]BENZIMIDAZOLE DERIVATIVES (**6.114**)–(**6.115**)

(**6.114**) (**6.115**)

Compound	R	Solvent[c]	*H*(1)	*H*(3)	ArH	Ref.
(**6.114**)	—	*A*	0.40t[d], 2.20oct[e]	4.30[f]	7.10–7.90m	2
(**6.115**)	Ph	*B*	6.80–8.00m[g]	5.15	6.80–8.00m[g]	17
(**6.115**)	*p*-$MeOC_6H_4$	*B*	6.80–8.00m[g]	5.38	6.80–8.00m[g]	17

[a] δ values in ppm measured from TMS.
[b] Signals are sharp singlets unless otherwise specified as t = triplet; oct = octet; m = multiplet.
[c] *A* = $CDCl_3$; *B* = Me_2SO.
[d] Me of Et.
[e] CH_2 of Et.
[f] MeO.
[g] ArH + *H*(1).

Comparison of the ¹H NMR spectra of 4*H*-pyrrolo[1,2-*a*]benzimidazoles in neutral (carbon tetrachloride) and acidic (trifluoroacetic acid) solution (Table 6.14) demonstrates preferential protonation at *C*(1) in the absence of a *C*(1) substituent and competing protonation at *C*(1) and *C*(3) when a *C*(1) alkyl substituent is present.[14,72,74,77] A feature of the ¹H NMR absorption (Table 6.14) of the resulting 4*H*-pyrrolo[1,2-*a*]benzimidazolium cations (**6.117**) and (**6.118**) is the predictable downfield shift (compared with the parent bases) in the resonances of protons at the alternative *C*(3) and *C*(1) vinyl sites and in attached alkyl substituents.[14,72] Additionally the protons of *N*(4) alkyl substituents show a shift to lower field roughly double that experienced by a *C*(3) methyl substituent (Table 6.14) implying considerable delocalization of the positive charge in the 4*H*-pyrrolo[1,2-*a*]benzimidazolium cations (**6.117**) and (**6.118**) to the *N*(4) and *N*(9) centers.[72] A further feature of the ¹H NMR absorption (Table 6.14) of cations of the types (**6.117**) and (**6.118**) is homoallyl coupling (J = 1.5–2.5 Hz) between *C*(1) and *C*(3) vinyl methyl groups and protons at the alternative *C*(3) and *C*(1) positions.[72] Studies[77] of the effect of structure on the ¹H NMR absorption of 4*H*-pyrrolo[1,2-*a*]benzimidazoles in trifluoroacetic acid also provide useful information on the equilibration and relative stability of 4*H*-pyrrolo[1,2-*a*]-benzimidazolium cations.

C(1) Acyl substituents in 4*H*-pyrrolo[1,2-*a*]benzimidazoles exert a paramagnetic anisotropic effect at *C*(8), resulting in the specific deshielding of *H*(8), which resonates in the range δ 8.13–9.80.[23] This effect permits the

TABLE 6.14. ^{1}H NMR SPECTRA[a,b] OF 4*H*-PYRROLO[1,2-*a*]BENZIMIDAZOLE DERIVATIVES (**6.116**)–(**6.119**)

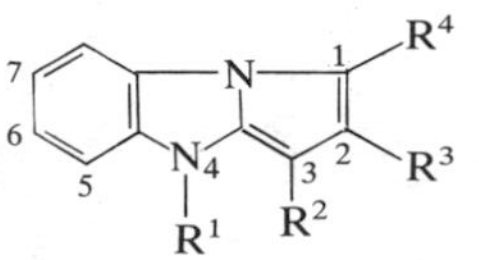

(**6.116**) 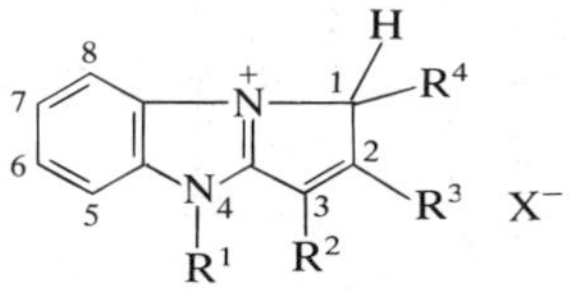 (**6.117**) 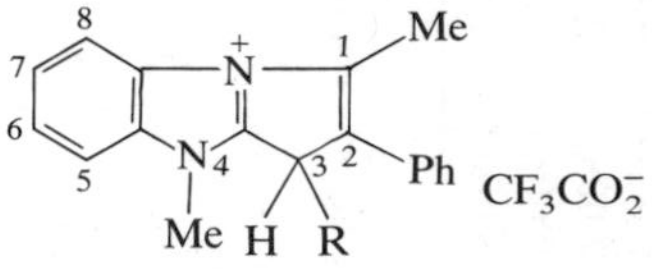 (**6.118**) 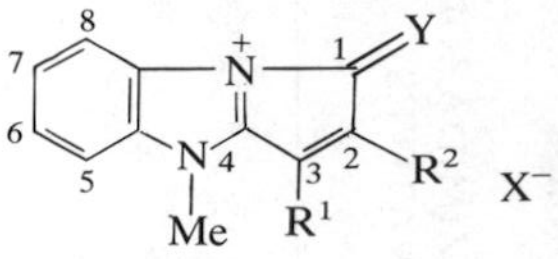 (**6.119**)

Compound	Solvent[c]	*H*(1)	*H*(2)	*H*(3)	Me	*N*(4)CH	ArH	Others	Ref.
(**6.116**; $R^1 = R^3 = Me$, $R^2 = R^4 = H$)	*A*	6.45m	—	4.95d[d]	1.97d[d,e]	—[f]	6.75	—	13
(**6.116**; $R^1 = R^3 = Me$, $R^2 = R^4 = H$)	*B*	6.55	—	5.15	2.23[e]	3.51	6.95–7.35m	—	21
(**6.116**; $R^1 = R^3 = R^4 = Me$, $R^2 = H$)	*A*	—	—	4.94	2.30[g]	—[f]	6.82	—	13
(**6.116**; $R^1 = R^2 = R^3 = Me$, $R^4 = H$)	*A*	6.50q[h]	—	—	1.96[e] 1.95d[e,h] 1.99[i]	—[f]	6.77	—	13
(**6.116**; $R^1 = Me$, $R^2 = R^4 = H$, $R^3 = Ph$)	*B*	—[j]	—	5.55	—	3.44	6.90–8.00m[l]	—	17
(**6.116**; $R^1 = Me$, $R^2 = R^4 = H$, $R^3 = Ph$)	*C*	—[j]	—	5.57d[k]	—	3.51	6.50–7.00m	—	72
(**6.116**; $R^1 = CH_2Ph$, $R^2 = R^4 = H$, $R^3 = Ph$)	*D*	—[j]	—	5.83	—	5.25	7.00–8.00m[l]	—	17
(**6.116**; $R^1 = CH_2Ph$, $R^2 = R^4 = H$, $R^3 = Ph$)	*C*	—[j]	—	5.53d[k]	—	5.05	6.50—7.00[l]	—	72
(**6.116**; $R^1 = CH_2Ph$, $R^2 = R^4 = H$, $R^3 = p\text{-}MeOC_6H_4$)	*D*	—[j]	—	5.69	3.79[m]	5.24	7.08–8.00[l]	—	17
(**6.116**; $R^1 = R^2 = Me$, $R^3 = p\text{-}BrC_6H_4$, $R^4 = H$)	*E*	—[l]	—	—	2.38[i]	3.72	—[n]	—	17
(**6.116**; $R^1 = R^4 = Me$, $R^2 = H$, $R^3 = p\text{-}PhC_6H_4$)	*E*	—	—	5.59	2.80[g]	3.60	—[n]	—	17
(**6.116**; $R^1 = CH_2Ph$, $R^2 = R^3 = Ph$, $R^4 = H$)	*C*	—[j]	—	—	—	5.04	7.00–7.60m[l]	—	17, 72
(**6.116**; $R^1 = R^2 = Me$, $R^3 = Ph$, $R^4 = H$)	*C*	—[j]	—	—	2.36[i]	3.69	6.50–7.00m	—	72

(**6.116**; $R^1 = R^4 = Me$, $R^2 = H$, $R^3 = Ph$)	*C*	—	—	5.35	2.69[g]	3.50	6.50–7.00m	—	72
(**6.116**; $R^1 = R^2 = R^4 = Me$, $R^3 = Ph$)	*C*	—	—	—	2.57[g] 2.25[i]	3.70	6.50–7.00m	—	72
(**6.116**; $R^1 = R^4 = Me$, $R^2 = R^3 = Ph$)	*C*	—	—	—	2.65[g]	3.46	6.50–7.00m	—	72
(**6.116**; $R^1 = R^3 = Me$, $R^2 = H$, $R^4 = CHO$)	*A*	—	—	5.28	2.15[e]	—[n]	6.88	—[n]	13
(**6.116**; $R^1 = R^3 = Me$, $R^2 = H$, $R^4 = COMe$)	*A*	—	—	5.30	2.13[e]	—[n]	6.89	—[n]	13
(**6.116**; $R^1 = R^3 = Me$, $R^2 = H$, $R^4 = COPh$)	*A*	—	—	5.36	—[n]	—[o]	6.98	—	13
(**6.116**; $R^1 = R^3 = Me$, $R^2 = H$, $R^4 = N{=}NPh$)	*F*	—	—	6.29	3.20[e]	4.20	7.60–8.50m	—	70
(**6.116**; $R^1 = R^3 = Me$, $R^2 = COMe$, $R^4 = CO_2Et$)	*B*	—	—	—	2.33[e] 2.59[o] 1.42t[p,q]	4.01	7.05–7.25m	4.35q[p,q]	21
(**6.116**; $R^1 = Me$, $R^2 = COPh$, $R^3 = Ph$, $R^4 = CO_2Et$)	*B*	—	—	—	0.97t[p,q]	3.91	6.75–7.45m	3.97q[r,q]	21
(**6.116**; $R^1 = R^3 = Me$, $R^2 = COMe$, $R^4 = CN$)	*G*	—	—	—	2.28[i] 2.46[o]	3.87	6.95–7.55m	—	21
(**6.116**; $R^1 = CH_2Ph$, $R^2 = CO_2Et$, $R^3 = H$, $R^4 = COPh$)	*C*	—	8.95m	—	1.27t[p]	6.10	7.22–8.00m	4.09–4.45q[r]	24
(**6.116**; $R^1 = Me$, $R^2 = R^3 = CO_2Me$, $R^4 = COPh$)	*C*	—	—	—	3.76[m] 3.83[m]	2.90	7.23m 7.73	—	23
(**6.116**; $R^1 = CH_2Ph$, $R^2 = R^3 = CO_2Me$, $R^4 = COPh$)	*C*	—	—	—	3.30[m] 3.72[m]	5.72	6.86–7.63m 8.08–8.36m[s]	—	24
(**6.116**; $R^1 = Me$, $R^2 = R^3 = R^4 = CO_2Me$)	*C*	—	—	—	3.86[m] 3.93[m] 4.00[m]	4.23	7.36m 8.40[t]	—	23
(**6.116**; $R^1 = Me$, $R^2 = R^3 = R^4 = CO_2Me$)	*C*	—	—	—	3.93[m] 3.87[m] 3.80[m]	4.12	7.00–7.40m 8.64[t]	—	26, 27

TABLE 6.14 (*Continued*)

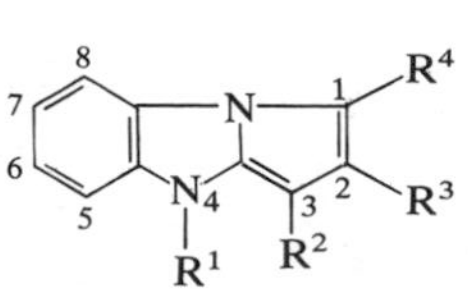

(**6.116**) 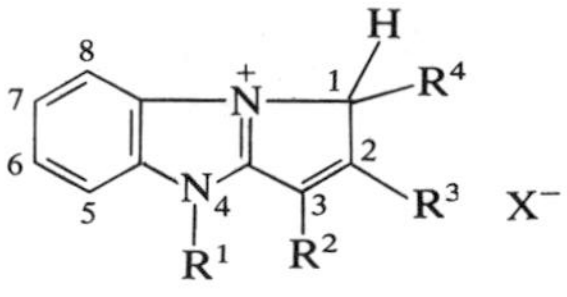 (**6.117**) 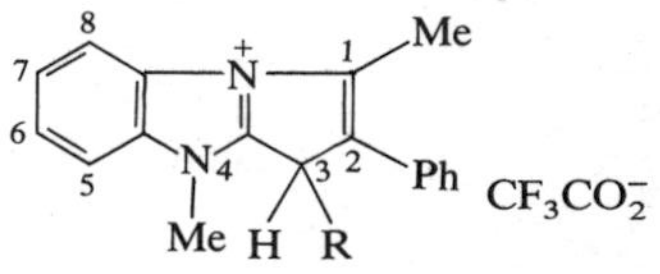 (**6.118**) 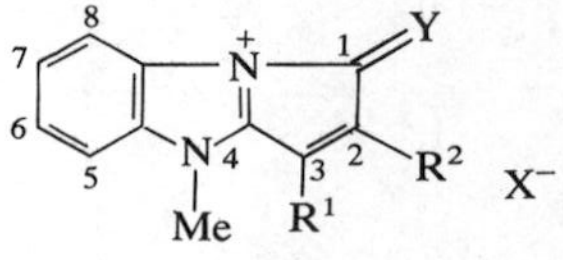 (**6.119**)

Compound	Solvent[c]	*H*(1)	*H*(2)	*H*(3)	Me	*N*(4)CH	ArH	Others	Ref.
(**6.116**; $R^1 = Et$, $R^2 = R^3 = R^4 = CO_2Me$)	*C*	—	—	—	3.90[m] 3.95[m] 4.02[m]	1.43t[p] 4.76q[r]	7.33m 8.40[t]	—	23
(**6.116**; $R^1 = Me$, $R^2 = R^3 = R^4 = CO_2Et$)	*C*	—	—	—	1.41t[p] 1.36t[p] 1.34t[p]	4.18	7.00–7.40m 8.70m[t]	4.01–4.56m[r]	27
(**6.116**; $R^1 = Me$, $R^2 = CO_2Et$, $R^3 = R^4 = CO_2Me$)	*C*	—	—	—	3.93[m] 3.99[m] 1.36t[p,q]	4.22[s]	7.20–7.60m 8.80m[t]	4.33q[r,q]	29
(**6.116**; $R^1 = CH \overset{t}{=} CHCO_2Me$, $R^2 = R^3 = R^4 = CO_2Me$)	*C*	—	—	—	3.90[m] 3.94[m] 3.96[m] 4.00[m]	9.62d[v,w] 6.37d[x,w]	7.80m 7.30–7.60m 8.90m[s]	—	28
(**6.116**; $R^1 = MeO_2C\overset{\vert}{C}{=}CHCO_2Me$, $R^2 = R^3 = R^4 = CO_2Me$)	*C*	—	—	—	3.82[m] 3.79[m] 3.76[m] 3.60[m] 3.53[m]	6.75[x]	7.45 8.87m[s]	—	28
(**6.116**; $R^1 = CH \overset{t}{=} CHCO_2Me$, $R^2 = CH_2CO_2Me$, $R^3 = CO_2Me$, $R^4 = H$)	*C*	7.74	—	—	3.90[m] 3.85[m] 3.75[m]	7.98d[v,w] 5.36d[x,w]	7.05–7.60m	4.93[y]	25
(**6.117**; $R^1 = R^2 = R^3 = Me$, $R^4 = H$, $X = Cl$)	*G*	4.72	—	—	2.25[e] 2.19[i]	3.95	7.40m	—	14

(**6.117**; $R^1 = R^3 = R^4 = Me$, $R^2 = H$, $X = Cl$)	*G*	5.00q[q]	—	6.60q[z]	2.22d[e,d] 1.60[g]	3.85	7.45m	—	14
(**6.117**; $R^1 = R^2 = R^3 = R^4 = Me$, $X = Cl$)	*G*	4.95q[q]	—	—	2.19[i] 2.31[e] 1.68[g,q]	4.05	7.50	—	14
(**6.117**; $R^1 = Me$, $R^2 = R^4 = H$, $R^3 = Ph$, $X = CF_3CO_2$)	*G*	5.57	—	7.38	—	4.17	6.50–7.00m	—	72
(**6.117**; $R^1 = CH_2Ph$, $R^2 = R^4 = H$, $R^3 = Ph$, $X = CF_3CO_2$)	*G*	5.54	—	7.31	—	5.67	6.50–7.00m	—	72
(**6.117**; $R^1 = R^2 = Me$, $R^3 = Ph$, $R^4 = H$, $X = CF_3CO_2$)	*G*	5.33q[aa]	—	—	266d[i,aa]	4.28	6.50–7.00m	—	72
(**6.117**; $R^1 = CH_2Ph$, $R^2 = R^3 = Ph$, $R^4 = H$, $X = CF_3CO_2$)	*G*	5.38	—	—	—	5.69	6.50–7.00m	—	72
(**6.117**; $R^1 = R^4 = Me$, $R^2 = H$, $R^3 = Ph$, $X = CF_3CO_2$)	*G*	5.93q[q]	—	7.31	1.90d[g,q]	4.15	6.50–7.00m	—	72
(**6.117**; $R^1 = R^2 = R^4 = Me$, $R^3 = Ph$, $X = CF_3CO_2$)	*G*	—[n]	—	—	1.72d[g,q] 2.63[i,aa]	4.25	6.50–7.00m	—	72
(**6.117**; $R^1 = R^4 = Me$, $R^2 = R^3 = Ph$, $X = CF_3CO_2$)	*G*	5.82q[q]	—	—	1.89d[g,q]	3.57	6.50–7.00m	—	72
(**6.118**; $R = H$)	*G*	—	—	4.38q[aa]	2.87d[g,aa]	4.20	6.50–7.00m	—	72
(**6.118**; $R = Me$)	*G*	—	—	4.68q[aa]	2.84[g,aa] 1.64d[i,q]	4.25	6.50–7.00m	—	72
(**6.118**; $R = Ph$)	*G*	—	—	5.68q[aa]	2.92d[g,aa]	3.90	6.50–7.00m	—	72
(**6.119**; $R^1 = H$, $R^2 = Me$, $Y = NNHPh$, $X = CF_3CO_2$)	*G*	—	—	6.80q[aa]	2.73d[e,d]	3.92	7.30–7.70m	—	70
(**6.119**; $R^1 = H$, $R^2 = Me$, $Y = NOH$, $X = ClO_4$)	*G*	—	—	6.76q[d]	2.15d[e,d]	3.77	7.35	—	70
(**6.117**; $R^1 = R^3 = R^4 = Me$, $R^2 = CONHCOPh$, $X = CF_3CO_2$)	*G*	4.93q[q]	—	—	1.52d[g,q] 2.18[e]	3.70	—[n]	—[n]	74
(**6.117**; $R^1 = R^2 = R^4 = Me$, $R^2 = CSNHCOPh$, $X = CF_3CO_2$)	*G*	4.96q[q]	—	—	1.53d[g,q] 2.02[e]	3.71	—[n]	—[n]	74
(**6.119**; $R^1 = H$, $R^2 = Me$, $Y = C(OH)NHCOPh$, $X = CF_3CO_2$)	*G*	—	—	6.46	1.98[e]	3.57	—[n]	—	74
(**6.119**; $R^1 = H$, $R^2 = Me$, $Y = C(SH)NHCOPh$, $X = CF_3CO_2$)	*G*	—	—	6.39	1.98[e]	3.57	—[n]	—	74

TABLE 6.14 (*Continued*)

(**6.116**) (**6.117**) (**6.118**) (**6.119**)

Compound	Solvent[c]	$H(1)$	$H(2)$	$H(3)$	Me	$N(4)CH$	ArH	Others	Ref.
(**6.119**; $R^1 = H$, $R^2 = Ph$, $Y = C(OH)NHCOPh$, $X = CF_3CO_2$)	*G*	—	—	6.38	—	3.82	—[n]	—	74
(**6.119**; $R^1 = H$, $R^2 = Ph$, $Y = C(SH)NHCOPh$, $X = CF_3CO_2$)	*G*	—	—	6.53	—	3.73	—[n]	—	74
(**6.119**; $R^1 = R^2 = Me$, $Y = C(OH)NHCOPh$, $X = CF_3CO_2$)	*G*	—	—	—	2.05[e] 2.05[i]	3.82	—[n]	—	74
(**6.119**; $R^1 = R^2 = Me$, $Y = C(SH)NHCOPh$, $X = CF_3CO_2$)	*G*	—	—	—	2.06[e] 2.13[i]	3.95	—[n]	—	74
(**6.119**; $R^1 = Me$, $R^2 = Ph$, $Y = C(OH)NHCOPh$, $X = CF_3CO_2$)	*G*	—	—	—	2.15[i]	3.92	—[n]	—	74
(**6.119**; $R^1 = Me$ $R^2 = Ph$, $Y = C(SH)NHCOPh$, $X = CF_3CO_2$)	*G*	—	—	—	2.10[i]	3.93	—[n]	—	74

[a] δ in ppm measured from TMS.
[b] Signals are sharp singlets unless otherwise specified as: d = doublet; t = triplet; q = quartet; m = multiplet.
[c] *A* = tetrahydrofurane; *B* = CCl_4; *C* = $CDCl_3$; *D* = Me_2SO; *E* = $CHCl_3$; *F* = CS_2; *G* = CF_3CO_2H.
[d] $J_{Me(2)-H(3)} = 1.0–1.5$ Hz.
[e] *Me*(2).
[f] *Me*(4) not quoted.
[g] *Me*(1).
[h] $J_{Me(2)-H(1)} = 1.0$ Hz.
[i] *Me*(3).

[j] $H(1)$ masked by ArH.
[k] $J_{H(1)-H(3)} = 1.0$ Hz.
[l] ArH + $H(1)$.
[m] MeO.
[n] not quoted.
[o] COMe.
[p] Me of Et group.
[q] $J = 7.0–7.5$ Hz.
[r] CH_2 of Et group.
[s] $H(5)$.
[t] $H(8)$.
[u] This signal may be interchanged with MeO.
[v] Vinyl $H(1)$.
[w] $J = 13.6–14.5$ Hz.
[x] Vinyl $H(2)$.
[y] CH_2.
[z] Ill-defined.
[aa] Homoallylic coupling: $J_{Me(3)-H(1)}$ or $J_{Me(1)-H(3)} = 1.5$ Hz; $J_{Me(3)-H(1)}$ or $J_{Me(1)-H(3)} = 2.5$ Hz.

TABLE 6.15. ^{1}H NMR SPECTRA[a,b] OF 2,3-DIHYDRO-1*H*-PYRROLO[1,2-*a*]BENZIMIDAZOLE DERIVATIVES (**6.120**)–(**6.124**)

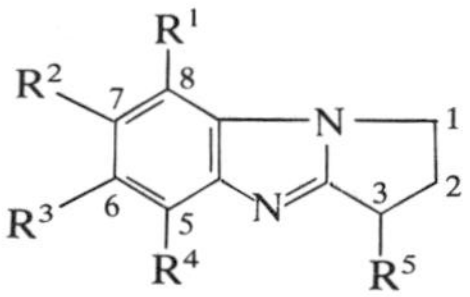

(**6.120**) 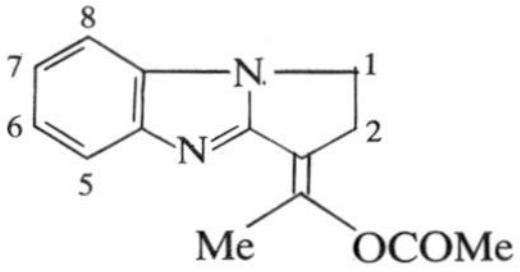 (**6.121**) 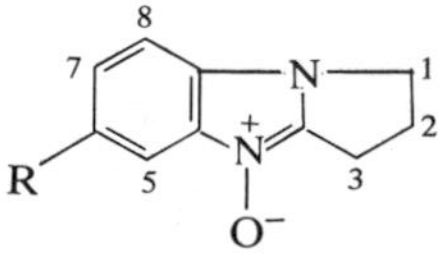 (**6.122**) 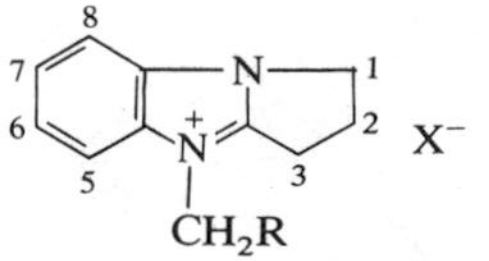 (**6.123**) 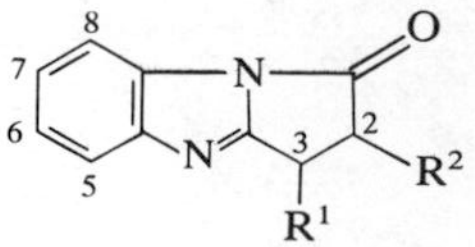 (**6.124**)

Compound	Solvent[c]	*H*(1)	*H*(2)	*H*(3)	*H*(5)	*H*(6)	*H*(7)	*H*(8)	Others	Ref.
(**6.120**; $R^1 = R^2 = R^3 = R^4 = R^5 = H$)	*A*	3.60t[d]	2.25m	2.75t[d]	← 7.00–7.70m →				—	30
(**6.120**; $R^1 = R^2 = R^3 = R^4 = R^5 = H$)	*B*	3.91t[e]	← 2.00–3.10m →		7.60m	← 7.10m →			—	67
(**6.120**; $R^1 = R^3 = R^4 = R^5 = H$, $R^2 = Cl$)	*B*	4.01t[e]	← 2.48–3.22m →		7.62d[f]	7.18dd[g]	—	7.26d[h]	—	73
(**6.120**; $R^1 = R^2 = R^3 = R^5 = H$, R^4 = 1-pyrrolidinyl)	*B*	4.00t[e]	← 2.44–3.21m →		—	6.32dd[i]	7.12t[j]	6.64dd[l]	3.79t[e,k] 1.82–2.13m[l]	73
(**6.120**; $R^1 = R^4 = R^5 = H$, $R^2 = R^3 = Cl$)	*B*	4.06t[e]	← 2.53–3.27m →		7.78	—	—	7.36	—	73
(**6.120**; $R^1 = R^4 = R^5 = H$, $R^2 = NHAc$, $R^3 = NO_2$)	*B*	4.58–4.84t[m]	← 2.90–3.90m →		9.13	—	—	8.88	2.55[n]	71
(**6.120**; $R^1 = R^4 = R^5 = H$, $R^2 = NH_2$, $R^3 = NO_2$)	*B*	4.55–4.80t[m]	← 2.90–3.85m →		8.91	—	—	7.97	—	71
(**6.120**; $R^1 = R^4 = R^5 = H$, $R^2 = N_3$, $R^3 = NO_2$)	*B*	4.55–4.93t[m]	← 2.90–3.85m →		8.53	—	—	7.77	—	71
(**6.120**; $R^1 = R^2 = R^5 = H$, $R^3 = NO_2$, $R^4 = Cl$)	*B*	4.22t[e]	← 2.64–3.32m →		—	—	7.80d[f]	7.23d[f]	—	73
(**6.120**; $R^1 = R^2 = R^5 = H$ $R^3 = NO_2$, R^4 = 1-pyrrolidinyl)	*B*	4.08t[e]	← 2.54–3.18m →		—	—	7.73d[f]	6.55d[f]	3.79t[e,k] 1.81–2.11m[l]	73
(**6.120**; $R^1 = R^2 = R^5 = H$, $R^3 = NO_2$, R^4 = 1-piperidinyl)	*B*	4.12t[e]	← 2.64–3.27m →		—	—	7.68d[f]	6.79d[f]	3.38–3.60m[k] 1.63–1.86m[l]	73
(**6.120**; $R^1 = R^2 = R^5 = H$, $R^3 = NO_2$, R^4 = 1-perhydroazepinyl)	*B*	4.03t[e]	← 2.48–3.21m →		—	—	7.55d[f]	6.70d[f]	3.42–3.69m[k] 1.48–1.93m[l]	73
(**6.120**; $R^1 = R^2 = R^3 = R^4 = H$, $R^5 = OH$)	*B*	3.80–4.60m	4.57q[o]	7.80m	← 7.10–7.40m →				7.8br[p]	65
(**6.120**; $R^1 = R^2 = R^4 = H$, $R^3 = NO_2$, $R^5 = OAc$)	*B*	4.37m	2.50–3.60m	6.28q[o]	8.60d[h]	—	8.19d[f]	7.48d[f]	2.17q	65

(**6.121**)	*B*	4.03t[e]	3.17m	—	7.80m	←— 7.00–7.45m —→			2.18q 2.58[r,h]	65
(**6.122**; R = H)[s]	*C*	4.13t[e]	2.62quin.	3.18t[e]	←— 7.14–7.38m —→				—	62
(**6.122**; R = Cl)	*B*	4.26t[e]	2.85q[e]	3.37t[e]	7.90d[t]	←— 7.27–7.36m —→			—	62
(**6.122**; R = NO_2)[s]	*D*	4.58t[e]	2.80q[e]	3.60t[e]	8.69d[h]	—	8.45dd[g]	7.92d[f]	—	62
(**6.123**; R = Me)[u]	*D*	4.49t[j]	2.90quin.[j]	3.54t[j]	←— 7.00–7.50m —→				1.57t[j] 4.47q[j]	68
(**6.123**; R = Ph)[u]	*D*	4.34t[j]	2.85quin.[j]	3.35t[j]	←— 7.20–7.50m —→				5.25[v] 7.40[w]	68
(**6.123**; R = *p*-HOC_6H_4)[u]	*D*	4.40t[j]	2.88m	3.43t[j]	←— 7.40–7.70m —→				6.88d[w,f] 7.31d[w,f] 5.43[v]	68
(**6.123**; R = 3,4-$(MeO)_2C_6H_3$)[u]	*D*	4.40t[j]	2.60m	3.40t[j]	←— 6.90–7.80m —→				6.20[x] 6.25[x] 5.42[v]	68
(**6.123**; R = CH═CHPh)[u]	*D*	4.55t[j]	3.00m	3.50t[j]	←— 7.00–7.70m —→				5.85–6.60m[y] 4.88[v]	68
(**6.124**; $R^1 = R^2$ = H)	*E*	—	←— 3.17 —→		←— 7.20–7.90 —→				—	38
(**6.124**: $R^1 = R^2$ = H)	*F*	—	←— 3.50–4.00 —→		←— 7.80m —→			8.20m	—	38
(**6.124**; R^1 = OH, R^2 = Ph)	*G*	—	4.26d	5.20d(br)	←— 7.10–7.90m —→				6.40(br)[p]	2

[a] δ in ppm measured from TMS.
[b] Signals are sharp singlets unless otherwise specified as: d = doublet; t = triplet; q = quartet; quin. = quintet; dd = double doublet; m = multiplet.
[c] A = pyridine; *B* = $CdCl_3$; *C* = CF_3CO_2D; *D* = D_2O; *E* = $(CD_3)_2SO$; *F* = CF_3CO_2H; *G* = $CDCl_3$-$(CD_3)_2SO$.
[d] J = 6.7 Hz.
[e] J = 7 Hz.
[f] J = 9 Hz.
[g] J = 2 and 9 Hz.
[h] J = 2 Hz.
[i] J = 7.5 and 1.5 Hz.
[j] J = 7.5 Hz.
[k] $N(CH_2)_2$ of C(5) substituent.
[l] $(CH_2)_2$ of C(5) substituent.
[m] J value not quoted.
[n] COMe.
[o] J = 7 and 4 Hz.
[p] OH.
[q] OCOMe.
[r] CH_3C═C.
[s] Hydrochloride.
[t] J = 2.5 Hz.
[u] Chemical shift in ppm relative to HOD.
[v] ═$\overset{+}{N}CH_2$—.
[w] ArH.
[x] MeO.
[y] —CH═CH—.

TABLE 6.16. ^{1}H NMR SPECTRA[a,b] OF 2,3,3a,4-TETRAHYDRO-1H-PYRROLO[1,2-a]BENZIMIDAZOLE DERIVATIVES (**6.125**)–(**6.127**)

(**6.125**) (**6.126**) (**6.127**)

Compound	Solvent[c]	H(1)	H(2)	H(3)	H(3a)	H(5)	H(6)	H(7)	H(8)	Others	Ref.
(**6.125**; R^1 = Me, R^2 = R^4 = H, R^3 = Ph)	A	←3.32→		2.20	4.92	← —[a] →				2.69[e]	75
(**6.125**; R^1 = CH_2Ph, R^2 = R^3 = R^4 = H)	B	2.80–3.80m	←1.70m→		5.03m	←6.10–7.50m→				4.22[f] 7.23[g]	68
(**6.125**; R^1 = CH_2Ph, R^2 = CN, R^3 = H, R^4 = CO_2Et)	B	4.06t[h]	2.20–2.40m	2.91–3.20q[h]	5.40–5.49d[h]	←6.86→				7.52m[g] 4.61[f] 4.30q[i,h] 1.34t[j,h]	45
(**6.126**)	B	4.30–5.20m	←1.50–3.00m→		6.15br	←6.25–8.20m→				3.78[f]	76
(**6.127**; R = H)	B	—	←2.30–2.80m→		5.83t[m]	←6.50–7.10m→			7.42	4.50br[n]	59
(**6.127**; R = Ph)	B	—	←2.60m→		—	←6.50–7.10m→			7.25–7.60	7.25–7.60m[g]	38
(**6.127**; R = Ph)	C	—	←2.36–2.84m→		—	6.60m	←6.85m→		7.55m	7.25–7.45m[g]	38

[a] δ values in ppm measured from TMS.
[b] Signals are sharp singlets unless otherwise specified as: d = doublet; t = triplet; q = quartet; m = multiplet.
[c] A = CCl_4; B = $CDCl_3$; C = $(CD_3)_2SO$.
[d] Not quoted.
[e] NMe.
[f] CH_2Ph.
[g] Ph.
[h] J value not quoted.
[i] CH_2 of Et.
[j] Me of Et.
[k] $\overset{+}{N}CH_2Ph$.
[l] J = 13 Hz.
[m] J = 7 Hz.
[n] NH.

orientation of the 4*H*-pyrrolo[1,2-*a*]benzimidazole products formed in the cycloaddition reactions of benzimidazolium ylides with acetylenic esters.[23]

The protons at *C*(1) in 2,3-dihydro-1*H*-pyrrolo[1,2-*a*]benzimidazole derivatives [Table 6.15; (**6.120**)] resonate as a triplet in the range δ 3.6–4.6 clearly distinguishable from the *C*(2) and *C*(3) protons, which collectively give rise to a multiplet at δ 2.0–3.9. The effect of a quaternary center at *N*(4), as in the salts (**6.123**) and the *N*-oxides (**6.122**), is to produce an overall downfield shift in the pyrrolidine ring proton resonances with specific deshielding of the protons at *C*(3). As a result, the signals due to the protons at each of the sites in the pyrrolidine ring become clearly discernable (Table 6.15). The deshielding effect of a quaternary nitrogen center is also seen in the comparatively low-field position of the *C*(1) proton resonances in the salt [Table 6.16; (**6.126**)]. In general, however, the pyrrolidine ring protons in 2,3,3*a*,4-tetrahydro-1*H*-pyrrolo[1,2-*a*]benzimidazoles [Table 6.16; (**6.125**)] absorb at higher field than their counterparts in 2,3-dihydro-1*H*-pyrrolo[1,2-*a*]benzimidazoles [Table 6.15; (**6.120**)]. In contrast, protons at the bridgehead [*C*(3a)] position in 2,3,3*a*,4-tetrahydro-1*H*-pyrrolo[1,2-*a*]-benzimidazoles absorb uniformly at relatively low field in the range δ 4.9–6.2 (Table 6.16) making them clearly distinguishable from protons at the other ring positions. The relative deshielding of the *C*(8) proton in 2,3-dihydro-1*H*-pyrrolo[1,2-*a*]benzimidazol-1-ones [Table 6.15; (**6.124**)] and their 2,3,3*a*,4 tetrahydro analogs [Tables 6.16; (**6.127**)] may be attributed to the paramagnetic anisotropic effect at the *C*(8) position, of the *C*(1) carbonyl group in these molecules.

Mass Spectra. Relatively little information is available concerning the mass spectral fragmentation pathways available to pyrrolo[1,2-*a*]benzimidazoles of the various structural types. The most comprehensive study to date appears to be that of Anisimova and his colleagues,[78] who have examined in detail the mass spectral fragmentation undergone by 4*H*-pyrrolo[1,2-*a*]benzimidazoles [Table 6.17; (**6.128**)]. Apart from the primary loss of HCN observed in all cases, the electron-impact induced fragmentation (Scheme 6.24) of 4*H*-pyrrolo[1,2-*a*]benzimidazoles lacking a *C*(3) methyl substituent (**6.128**; $R^2 \neq Me$) is initiated by loss of the *N*(4) substituent from the molecular ion, the driving force being the formation of stabilized cations of the type (**6.129**). Subsequent breakdown of the latter can then be rationalized[78] in terms of scission of the pyrrole ring with loss of *C*(2) and *C*(3) and the attached substituents as a discrete unit (i.e., an alkyne) and concomitant formation of the common ion (**6.131**), which fragments further in orthodox fashion by loss of HCN [Scheme 6.24; (**6.131**) $\rightarrow$ (**6.132**) $\rightarrow$ etc]. This fragmentation pattern is akin to that observed for thiazolo[3,2-*a*]benzimidazoles and imidazo[1,2-*a*]benzimidazoles (see later). In the case of 3-methyl-substituted 4*H*-pyrrolo[1,2-*a*]benzimidazoles, mass spectral fragmentation occurs by initial hydrogen atom loss from the molecular ion to give cations of the type (**6.130**) (Scheme 6.24).[78]

TABLE 6.17. MASS SPECTRA[a,b] OF 4*H*-PYRROLO[1,2-*a*]BENZIMIDAZOLE DERIVATIVES (**6.128**)[78]

(**6.128**)

R^1	R^2	R^3	*m/e* (rel. abundance, %)
Me	H	Ph	248(2.1), 247(20.9), 246(100), 245(7.5), 244(2.4), 232(5.5), 231(14.9), 230(3.7), 229(3.6), 204(1.3), 203(1.3), 202(1.5), 200(1), 149(1.1), 129(7.3), 128(1.7), 127(1.2), 122(1.2), 118(1.2), 115(2.1), 103(2.7), 102(6.6), 91(1.3), 81(1), 78(3.4), 77(4.5), 71(1), 57(2.1), 56(8.1), 55(2.1), 51(1.4), 43(7.6), 42(3.6), 41(6.7), 40(5.1).
Me	Me	Ph	262(2.2), 261(20), 260(100), 259(49.4), 258(5.8), 257(2.5), 246(3.3), 245(15.6), 244(6.8), 243(8.9), 242(3), 218(1.2), 183(2.4), 169(1.2), 168(1.3), 157(1.1), 149(1.3), 140(1), 117(2.4), 116(2), 115(3.2), 109(1), 103(1), 102(2.3), 92(1), 91(1), 89(1), 77(3.5), 76(2.4), 75(1.1), 74(1), 64(1), 62(1), 50(1.2), 49(1).
CH_2Ph	H	Ph	324(3.2), 323(21.2), 322(76.3), 321(18.2), 233(3.4), 232(28.8), 231(100), 204(1.0), 129(6.3), 128(1.2), 115(1), 103(1.9), 102(2.8), 92(2), 91(1), 77(1.4), 65(1.4).

[a] Measured at 50 eV and an emission current of 75 mA at 125°.
[b] Only peaks with $m/e > 39$ are indicated; peaks with intensities <1% are not shown.

Primary loss of the *N*(4)-pyrimidyl substituent is also a feature of the mass spectral fragmentation[69] of the 2,3-dihydro-1*H*-pyrrolo[1,2-*a*]-benzimidazole betaine (**6.92**) (cf. Scheme 6.20), while the most intense peaks in the mass spectra[30] of simple 2,3-dihydro-1*H*-pyrrolo[1,2-*a*]-benzimidazoles correspond to the formation of cations of the type (**6.133**) (cf. Scheme 6.24).

General Studies

The typically basic character of 3*H* and 4*H*-pyrrolo[1,2-*a*]benzimidazoles, 2,3-dihydro-1*H*-pyrrolo[1,2-*a*]benzimidazoles, and 2,3,3*a*,-4-tetrahydro-1*H*-pyrrolo[1,2-*a*]benzimidazoles is qualitatively demonstrated[12,14,22,31,37,48,79] by the tendency for such molecules to form well-defined, stable, acid salts (e.g., hydrochlorides, perchlorates, and picrates, cf. Tables 6.3, 6.4, and 6.7–6.9). Surprisingly, however, only a few quantitative studies[72,77] of the basicity of pyrrolo[1,2-*a*]benzimidazoles of the various structural types appear to have been carried out. Basicity

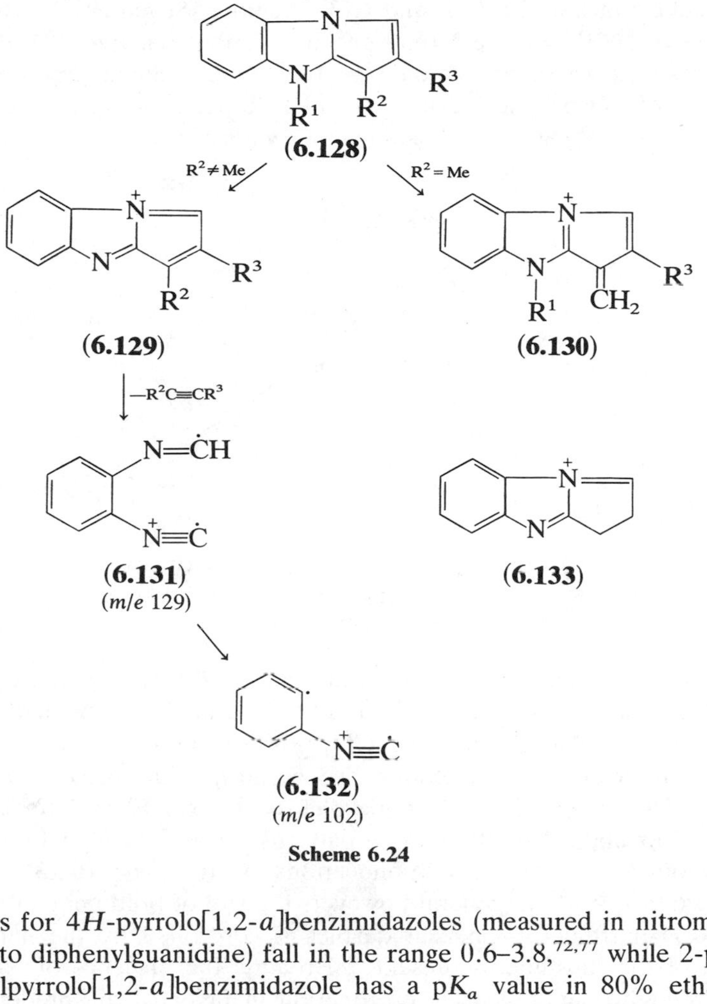

(**6.128**)

$R^2 \neq Me$ $R^2 = Me$

(**6.129**) (**6.130**)

$-R^2C{\equiv}CR^3$

(**6.131**) (*m/e* 129) (**6.133**)

(**6.132**) (*m/e* 102)

Scheme 6.24

constants for 4*H*-pyrrolo[1,2-*a*]benzimidazoles (measured in nitromethane relative to diphenylguanidine) fall in the range 0.6–3.8,[72,77] while 2-phenyl-4-methylpyrrolo[1,2-*a*]benzimidazole has a pK_a value in 80% ethanol of 6.86. These data demonstrate 4*H*-pyrrolo[1,2-*a*]benzimidazoles to be more basic than pyrrolo[1,2-*a*]pyridines (indolizines) (pK_a = 3.5 to 5.2 in 60% ethanol) but less basic than pyrrolo[1,2-*a*]imidazoles (pK_a ca. 8.6 in 80% ethanol).[72] The introduction of a substituent into the *C*(1) position in 4*H*-pyrrolo[1,2-*a*]benzimidazoles leads to a considerable decrease in basicity.[72]

6.1.3. Reactions

Reactions with Electrophiles

Molecular orbital calculations[80] imply a high degree of aromatic character[12] for 4*H*-pyrrolo[1,2-*a*]benzimidazoles and also their susceptibility to electrophilic attack at the *C*(1) and *C*(3) positions [Scheme 6.25; (**6.134**) ↔ (**6.135**) ↔ (**6.136**)] with the former site being more reactive than the latter in this respect. These predictions are fully vindicated in practice by the propensity of 4*H*-pyrrolo[1,2-*a*]benzimidazole derivatives to undergo substitution at the *C*(1) and *C*(3) positions by a variety of electrophilic reagents.

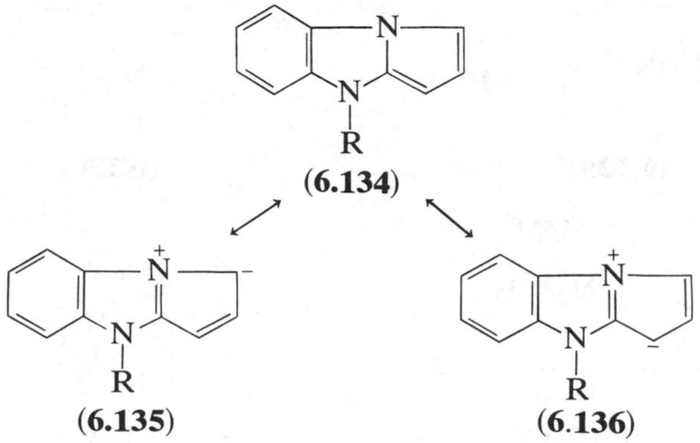

(**6.134**)

(**6.135**) (**6.136**)

Scheme 6.25

Protonation. The simplest manifestation of the susceptibility of 4*H*-pyrrolo[1,2-*a*]benzimidazoles to electrophilic attack is provided by their ready protonation at the *C*(1) and *C*(3) positions in acidic media. Thus, in accord with theoretical predictions (see before), solutions of *C*(1)-unsubstituted 4*H*-pyrrolo[1,2-*a*]benzimidazoles or their acid salts in trifluoroacetic acid exhibit ^{1}H NMR splitting patterns consistent only with exclusive protonation at *C*(1) to give monocations of the type (**6.137**; $R^4 = H$) (Scheme 6.26).[14,72,77] In general however, the site of predominant protonation in 4*H*-pyrrolo[1,2-*a*]benzimidazoles is controlled by the nature and position of the substituents present.[72] Broadly, the presence of *C*(1) alkyl groups tends to suppress *C*(1) protonation in favor of the predominating formation of *C*(3) protonated monocations [Scheme 6.26; (**6.138**; $R^4 =$ alkyl)], whereas the reverse is true for *C*(3) alkyl groups. For example,[72] protonation of 1-methyl-4*H*-pyrrolo[1,2-*a*]benzimidazoles affords a mixture of the *C*(1) and *C*(3) monocations (**6.137**; $R^4 = Me$) and (**6.138**; $R^4 = Me$) with the latter predominating, whereas the *C*(1)-protonated species is largely in excess in the mixture of monocations (**6.137**; $R^2 = R^4 = Me$) and (**6.138**; $R^2 = R^4 = Me$) produced from 1,3-dimethyl-4*H*-pyrrolo[1,2-*a*]benzimidazoles. On the other hand, a *C*(3) phenyl substituent has a facilitating effect on *C*(3) protonation.[72] The temperature and acid strength of the medium also play an important role as demonstrated[77] by the fact that at fixed acidity lowering of the temperature results in a decrease in protonation at *C*(1) in favor of increased formation of the *C*(3) monocation. This effect has been interpreted[77] in terms of initial kinetic control of monocation formation and a low rate of attainment of the equilibrium [Scheme 6.26; (**6.137**) ⇌ (**6.138**)] and hence predominance of the thermodynamically more stable *C*(1) monocation at low temperature. This interpretation has the consequence that, under conditions of kinetic control, protonation of 4*H*-pyrrolo[1,2-*a*]benzimidazoles occurs fastest at the less basic *C*(3) center.[77] The electrophilic deuterium exchange (in $CDCl_3/CD_3OD$) observed for the *C*(1) and *C*(3) positions in 4*H*-pyrrolo[1,2-*a*]benzimidazoles has been shown[77] to follow first-order kinetics.

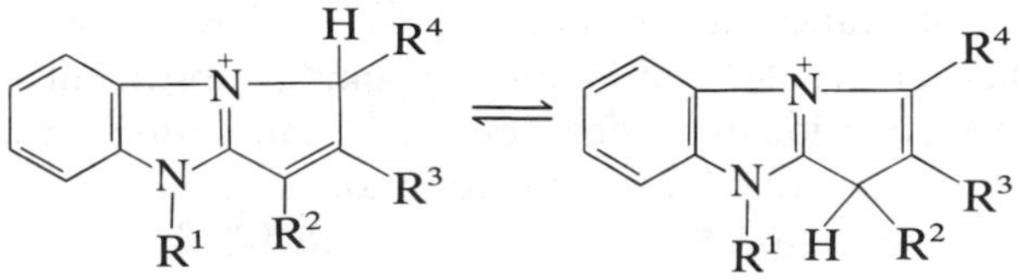

(**6.137**) (**6.138**)

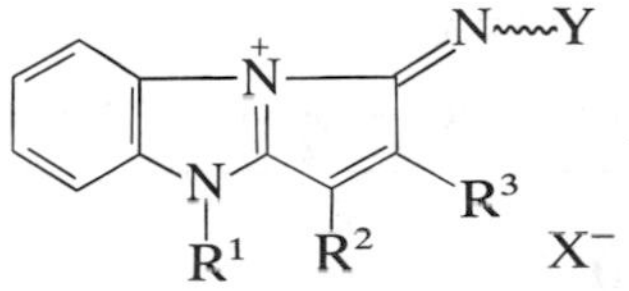

(**6.139**)
(X = Cl or ClO_4)

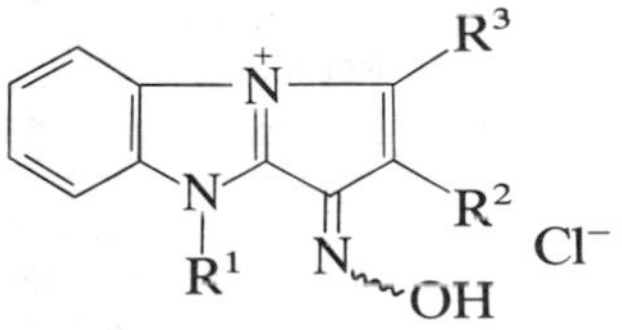

(**6.140**)

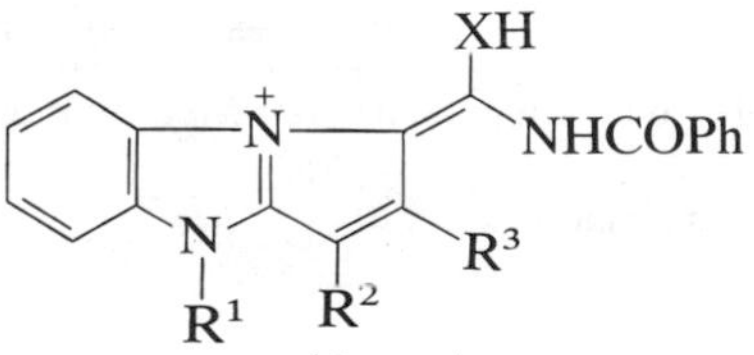

(**6.141**)

Scheme 6.26

4*H*-Pyrrolo[1,2-*a*]benzimidazoles with heteroatom substituents tend to protonate on the side chain rather than the pyrrole ring. This situation is exemplified[70] by the protonation of *C*(1) and *C*(3) arylazo and nitroso derivatives of 4*H*-pyrrolo[1,2-*a*]benzimidazoles to give stable salts (hydrochlorides and perchlorates), whose spectral properties are consistent with the hydrazone and oxime structures (**6.139**; Y = NHAr or OH) and (**6.140**) (Scheme 6.26). The formation[74] of cations of the type (**6.141**; X = O or S) (Scheme 6.26) on dissolution of the parent *C*(1) acyl or thioacyl 4*H*-pyrrolo-[1,2-*a*]benzimidazoles in trifluoroacetic acid contrasts with the more orthodox *C*(1) protonation undergone by the corresponding *C*(3) acyl and thioacyl 4*H*-pyrrolo[1,2-*a*]benzimidazoles and is rationalized[74] in terms of conjugative stabilization of the structures (**6.141**; X = O or S) as opposed to the alternative *C*(1)-protonated forms.

ALKYLATION. The reactions of 4*H*-pyrrolo[1,2-*a*]benzimidazoles with alkylating agents are less well documented than their reactions with acylating agents (see later). *N*(4)-Unsubstituted 4*H*-pyrrolo[1,2-*a*]benzimidazoles react with alkyl halides under alkaline conditions in orthodox fashion to afford good yields (Table 6.18) of the corresponding *N*(4)-alkyl derivatives, e.g. [Scheme 6.27; (**6.142**) → (**6.143**)].[22] In reactions (Table 6.18) akin to those of related substrates with protic acids (see before), *N*(4)-methyl-4*H*-pyrrolo[1,2-*a*]benzimidazoles containing thioamide substituents at *C*(1) or *C*(3) react with methyl iodide under neutral ("sealed tube") conditions to give quaternary salts formed by alkylation of the side chain at sulfur [Scheme 6.27; (**6.144**) → (**6.145**) and (**6.146**) → (**6.147**)].[81]

TABLE 6.18. ALKYLATION REACTIONS OF 4*H*-PYRROLO[1,2-*a*]BENZIMIDAZOLE DERIVATIVES (**6.142**), (**6.144**), AND (**6.146**).

Starting materials	R^1	R^2	Reaction conditions[a]	Product	R^1	R^2	Yield (%)	m.p. (°C)	Ref.
(**6.142**)	—	—	*A*	(**6.143**)	—	—	68	—[b]	22
(**6.144**)	H	Me	*B*	(**6.145**)[c]	H	Me	48	172	81
(**6.144**)	H	Ph	*B*	(**6.145**)[c]	H	Ph	62	169	81
(**6.144**)	Me	Me	*B*	(**6.145**)[c]	Me	Me	45	196	81
(**6.144**)	Me	Ph	*B*	(**6.145**)[c]	Me	Ph	43	198	81
(**6.146**)	—	—	*B*	(**6.147**)[c]	—	—	62	173–174	81

[a] *A* = MeI, KOH, EtOH/(room temp.)(1 hr); *B* = MeI/(100°, sealed tube)(1.5 hr).
[b] Picrate has m.p. 132–133°.
[c] Crystallized from nitromethane.

2,3-Dihydro-1*H*-pyrrolo[1,2-*a*]benzimidazoles are alkylated preferentially at *N*(4) [as opposed to the bridgehead center *N*(9)] under effectively neutral conditions by a wide variety of reagents including chlorides,[59,68] bromides,[58] iodides,[58] and tosylates,[58,62] to afford moderate to high yields (Table 6.19) of quaternary salts [Scheme 6.28; (**6.148**)] of utility as starting materials for the synthesis of cyanine dyes.[58,82] *N*(4)-Arylmethyl-2,3,3*a*,4-tetrahydro-1*H*-pyrrolo[1,2-*a*]benzimidazoles, on the other hand, are reported[76] to undergo benzylation at the bridgehead position to give quaternary salts of the type (**6.149**) (Scheme 6.28) in unspecified yield. *C*(3)-Hydroxy substituents in 1*H*-pyrrolo[1,2-*a*]benzimidazoles behave in an orthodox manner toward methylation by diazomethane in tetrahydrofurane giving moderate to good yields (Table 6.19) of enol ethers of the types (**6.150**) and (**6.151**) (Scheme 6.28).

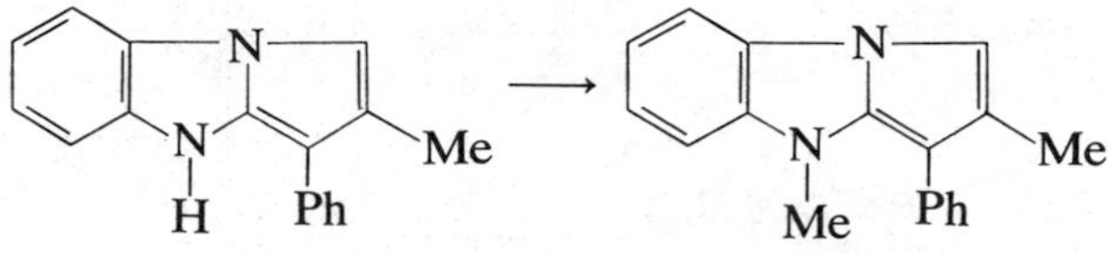

N N H Ph Me → N N Me Ph Me

(6.142) (6.143)

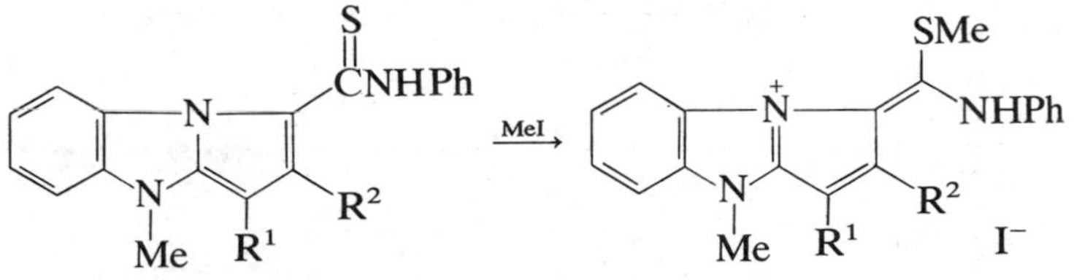

S CNHPh N N Me R[1] R[2] $\xrightarrow{MeI}$ SMe NHPh N N Me R[1] R[2] I^-

(6.144) (6.145)

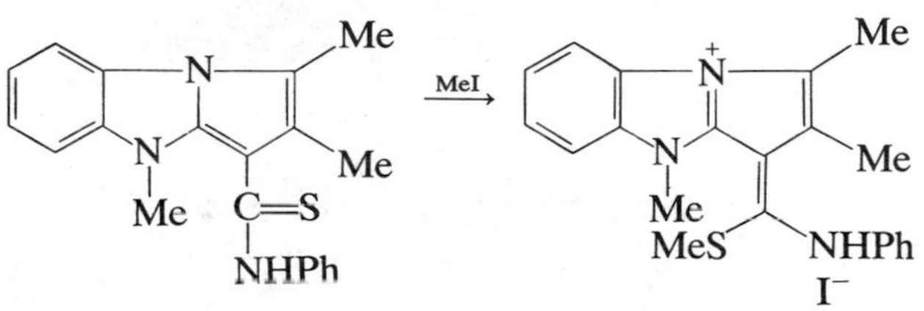

Me N N Me Me C=S NHPh $\xrightarrow{MeI}$ Me N N Me Me MeS NHPh I^-

(6.146) (6.147)

Scheme 6.27

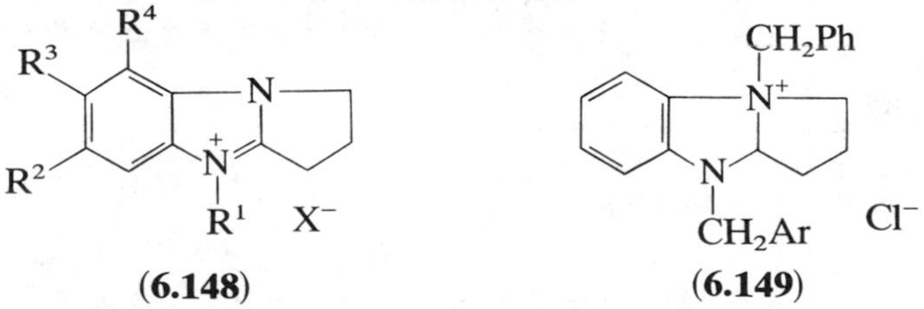

R[4] R[3] R[2] N N R[1] X^-

(6.148)

CH_2Ph N N CH_2Ar Cl^-

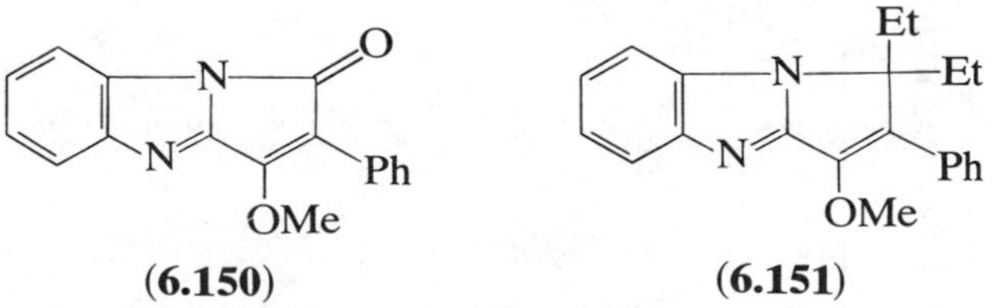

(6.149)

O N N Ph OMe

(6.150)

Et Et N N Ph OMe

(6.151)

Scheme 6.28

TABLE 6.19. ALKYLATION REACTIONS OF 2,3-DIHYDRO-1*H*-PYRROLO[2,3-*a*]BENZIMIDAZOLES, 2,3,3*a*,4-TETRAHYDRO-1*H*-PYRROLO[1,2-*a*]BENZIMIDAZOLES, AND 1*H*-PYRROLO[1,2-*a*]BENZIMIDAZOLES

Alkylating agent[a]	Reaction time (hr)	Reaction temp. (°C)	Solvent	Product	R^1	R^2	R^3	R^4	X	Yield (%)	m.p. (°C)	Ref.
A	2	100	—[b]	(**6.148**)	Me	H	H	H	Cl	79	140	68
B	0.5	—[c]	—[d]	(**6.148**)	Me	H	H	H	I	83	220	58
C	15	110	—[e]	(**6.148**)	Et	H	H	H	I	87	198	58
D	36	100	—[b]	(**6.148**)	Et	H	H	H	TSO_3^f	92	133–134	82
E	0.25	100	—[b]	(**6.148**)	CH_2Ph	H	H	H	Cl	quant.	—[g]	59
F	6	105	—[e]	(**6.148**)	$(CH_2)_2OH$	H	H	H	Br	—[h]	180	58
C	15	110	—[e]	(**6.148**)	Et	Cl	H	H	I	—[h]	242	58
C	3.5	105–110	—[e]	(**6.148**)	Et	H	H	H	Cl	—[h]	238	58
C	5.5	110	—[e]	(**6.148**)	Et	H	Cl	H	I	—[h]	238	58
C	16	110	—[e]	(**6.148**)	Et	F	H	H	I	—[h]	237	58
C	15	110	—[e]	(**6.148**)	Et	H	F	H	I	—[h]	210	58
C	15	110	—[e]	(**6.148**)	Et	Br	H	H	I	—[h]	250	58
B	15	100	—[e]	(**6.148**)	Me	CO_2H	H	H	I	—[h]	304	58
B	16	125	—[e]	(**6.148**)	Me	H	H	CO_2H	I	—[h]	265	58
B	3	90	—[e]	(**6.148**)	Me	CO_2Et	H	H	I	—[h]	238	58
B	3.5	90	—[e]	(**6.148**)	Me	H	H	CO_2Et	I	—[h]	190	58
C	16	105	—[e]	(**6.148**)	Et	CN	H	H	I	—[h]	>250	58
F	15	125	—[i]	(**6.148**)	$(CH_2)_2OH$	CN	H	H	Br	—[h]	207–209	58
C	8	110	—[e]	(**6.148**)	Et	H	CN	H	I	—[h]	242	58
F	3	105	—[e]	(**6.148**)	$(CH_2)_2OH$	H	CN	H	Br	—[h]	246	58
C	8	100	—[i]	(**6.148**)	Et	Me	H	H	I	—[h]	202	58
B	0.25	—[c]	—[j]	(**6.148**)	Me	H	NH_2	H	I	—[h]	282	58
C	16	110	—[e]	(**6.148**)	Et	H	NHAc	H	I	—[h]	230	58

C	16	110	—[e]	(**6.148**)	Et	Cl	Cl	H	I	—[h]	>250	58
F	4	110	—[e]	(**6.148**)	$(CH_2)_2OH$	Cl	Cl	H	I	—[h]	>250	58
B	2	95	—[e]	(**6.148**)	Me	CO_2Et	Cl	H	I	—[h]	250	58
C	16	110	—[e]	(**6.148**)	Et	CO_2H	Cl	H	I	—[h]	230	58
C	16	110	—[e]	(**6.148**)	Et	Br	CN	H	I	—[h]	>250	58
C	16	110	—[e]	(**6.148**)	Et	Cl	CN	H	I	—[h]	>250	58
C	16	110	—[e]	(**6.148**)	Et	F	CN	H	I	—[h]	280	58
E	2	—[c]	—[k]	(**6.149**; Ar = Ph)	—	—	—	—	—	—[h]	132–133[l]	76
E	2	—[c]	—[k]	(**6.149**; Ar = *o*-$NO_2C_6H_4$)	—	—	—	—	—	—[h]	110–122[l] (decomp.)	76
G	10	—[m]	—[n]	(**6.150**)	—	—	—	—	—	44	169[o]	2
G	5	—[m]	—[n]	(**6.151**)	—	—	—	—	—	63	177[p]	2

[a] A = MeCl; *B* = MeI; *C* = EtI; *D* = TSO_3Et; *E* = $PhCH_2Cl$; *F* = $HO(CH_2)_2Br$; *G* = CH_2N_2.
[b] Without cosolvent.
[c] Under reflux.
[d] Acetone.
[e] Sealed tube conditions.
[f] T = toluene-*p*-sulfonyl.
[g] Melting point not specified.
[h] Yield not specified.
[i] Nitromethane.
[j] Methanol.
[k] Ethanol.
[l] Needles from ethanol.
[m] Room temp.
[n] Tetrahydrofurane.
[o] Orange-red crystals from acetone–chloroform.
[p] Crystallized from ethanol.

ACYLATION. The high degree of aromatic character and susceptibility to electrophilic attack of the 4*H*-pyrrolo[1,2-*a*]benzimidazole ring system (see before) is demonstrated by the ability of its simple alkyl and aryl derivatives to undergo Friedel–Crafts type reactions with acylating agents.[12,13] Acylation occurs preferentially at the more reactive *C*(1) position and only takes place at *C*(3) when the former site is occupied by a substituent. Formylation[13] is readily achieved in moderate to high yield (Table 6.20) under Vilsmeier–Haack conditions ($POCl_3$–DMF) and provides a useful general method for the synthesis of 4*H*-pyrrolo[1,2-*a*]benzimidazole-1- and 3-carboxaldehydes [Scheme 6.29; (**6.152**; $R^1 = H$) and (**6.153**; $R^1 = H$)]. Acetylation[12,13] occurs even more simply on heating with acetic anhydride allowing synthetic access, again usually in good yield (Table 6.20), to 1- and 3-acetyl-4*H*-pyrrolo[1,2-*a*]benzimidazoles [Scheme 6.29; (**6.152**; $R^1 = Me$) and (**6.153**; $R^1 = Me$)]. Correspondingly, benzoyl chloride in the presence of pyridine effects the smooth transformation of suitable 4*H*-pyrrolo[1,2-*a*]-benzimidazole substrates in good yield (Table 6.20) into the *C*(1) and *C*(3) benzoyl derivatives [Scheme 6.29; (**6.152**; $R^1 = Ph$) and (**6.153**; $R^1 = Ph$)].[13]

4*H*-Pyrrolo[1,2-*a*]benzimidazoles are also acylated at the *C*(1) and *C*(3) positions in good yield (Table 6.20) by aryl and acyl isocyanates and isothiocyanates, the products of these synthetically useful reactions being the corresponding carboxamide and thiocarboxamide derivatives [Scheme 6.29; (**6.152**; $R^1 = NHR$), (**6.153**; $R^1 = NHR$), (**6.154**), and (**6.155**)].[74,81] The protic salts (e.g., perchlorates) of 4*H*-pyrrolo[1,2-*a*]benzimidazoles are activated at *C*(1) and *C*(3) toward aldol-type condensation with aldehydes. Reactions of this type have been utilized[12] for the synthesis of methine dyestuffs containing a 4*H*-pyrrolo[1,2-*a*]benzimidazole chromophore, as

TABLE 6.20. ACYLATION REACTIONS OF 4*H*-PYRROLO[1,2-*a*]BENZIMIDAZOLE DERIVATIVES

Reaction conditions[a]	Product	R^1	R^2	R^3	Yield (%)	m.p. (°C)	Solvent of crystallization	Ref.
A	(**6.152**)	H	Me	H	35	140	Cyclohexane	13
A	(**6.152**)	H	Ph	H	64	142	Cyclohexane	13
A	(**6.152**)	H	Me	Me	61	222	2-Propanol	13
A	(**6.153**)	H	Me	Me	60	184	Ethanol	13
B	(**6.152**)	Me	Me	H	78	157	Ethanol	13
B	(**6.152**)	Me	Et	H	53	132	Ethanol	13
B	(**6.152**)	Me	Ph	H	75	147	Ethanol	13
B	(**6.152**)	Me	Me	Me	52	199[b]	Ethanol	12
B	(**6.152**)	Me	Ph	Me	77	188	Ethanol	13
B	(**6.153**)	Me	Me	Me	61	173[b]	Ethanol	12, 13
C	(**6.152**)	Ph	Me	H	53	145	Ethanol	13
C	(**6.152**)	Ph	Ph	H	50	211	Ethanol	13
C	(**6.152**)	Ph	Me	Me	47	192	Ethanol	13
C	(**6.152**)	Ph	Ph	Me	71	219	Ethanol	13

TABLE 6.20 (*Continued*)

Reaction conditions[a]	Product	R^1	R^2	R^3	Yield (%)	m.p. (°C)	Solvent of crystallization	Ref.
C	(**6.153**)	Ph	Me	Me	87	168	Ethanol	13
D	(**6.152**)	NHPh	Me	H	31	171–172	Ethanol or dimethyl-formamide	81
D	(**6.152**)	NHPh	Ph	H	67	182–183	Ethanol or dimethyl-formamide	81
D	(**6.152**)	NHPh	Me	Me	66	218	Ethanol or dimethyl-formamide	81
D	(**6.153**)	NHPh	Me	Me	80	220	Ethanol or dimethyl-formamide	81
E	(**6.154**)	Ph	Me	H	49	162–163[b]	Ethanol or dimethyl-formamide	81
E	(**6.154**)	Ph	Ph	H	61	177–178[c]	Ethanol or dimethyl-formamide	81
E	(**6.154**)	Ph	Me	Me	58	170[c]	Ethanol or dimethyl-formamide	81
E	(**6.155**)	Ph	Me	Me	70	176–177[c]	Ethanol or dimethyl-formamide	81
F	(**6.152**)	NHCOPh	Me	H	76	193–194[c]	Ethanol or nitromethane	74
F	(**6.152**)	NHCOPh	Ph	H	81	191–192	—[d]	74
F	(**6.152**)	NHCOPh	Me	Me	80	196–197	—[d]	74
F	(**6.152**)	NHCOPh	Ph	Me	93	192–193	—[d]	74
F	(**6.153**)	NHCOPh	Me	Me	83	169–170	—[d]	74
G	(**6.154**)	COPh	Me	H	75	168–169[e]	Ethanol or nitromethane	74
G	(**6.154**)	COPh	Ph	H	80	165–166	—[d]	74
G	(**6.154**)	COPh	Me	Me	79	165–166	—[d]	74
G	(**6.154**)	COPh	Ph	Me	82	181–182	—[d]	74
G	(**6.155**)	COPh	Me	Me	80	183–184	—[d]	74
H	(**6.157**)	—	—	—	69	232	Ethanol	12
H	(**6.158**)	—	—	—	41	236	Ethanol	12

[a] *A* = $POCl_3$, dimethylformamide/(room temp.)(0.5 hr), then warm briefly at 50–60°; *B* = Ac_2O/(100°)(5 min); *C* = PhCOCl, pyridine/(100°)(5 min); *D* = PhN=C=O/(100°, sealed tube)(15 min); *E* = PhN=C=S/(100°, sealed tube)(15 min); *F* = PhCON=C=O, ether/(room temp.)(few min); *G* = PhCON=C=S, ether/(room temp.)(few min); *H* = p-$Me_2NC_6H_4CHO$, Ac_2O/(reflux)(few min).

[b] Colorless needles.

[c] Yellow solid.

[d] Solvent of crystallization not specified.

[e] Red crystals.

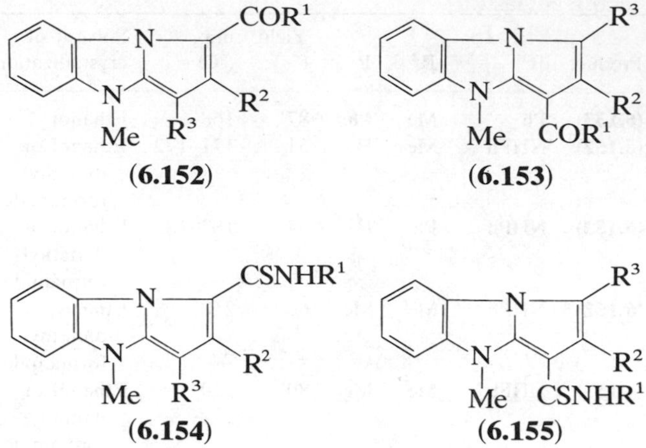

(**6.152**) (**6.153**)

(**6.154**) (**6.155**)

Scheme 6.29

exemplified by the transformations [Scheme 6.30; (**6.156**)→(**6.157**) or (**6.158**)] (Table 6.20).[12]

The *C*(3) center in 2,3-dihydro-1*H*-pyrrolo[1,2-*a*]benzimidazole [Scheme 6.31; (**6.159**)] is sufficiently activated by the adjacent azomethine center to undergo acylation under relatively mild conditions. For example, treatment of 2,3-dihydro-1*H*-pyrrolo[1,2-*a*]benzimidazole (**6.159**) with acetic anhydride in the presence of zinc chloride[65] or with benzoyl chloride in pyridine[56]

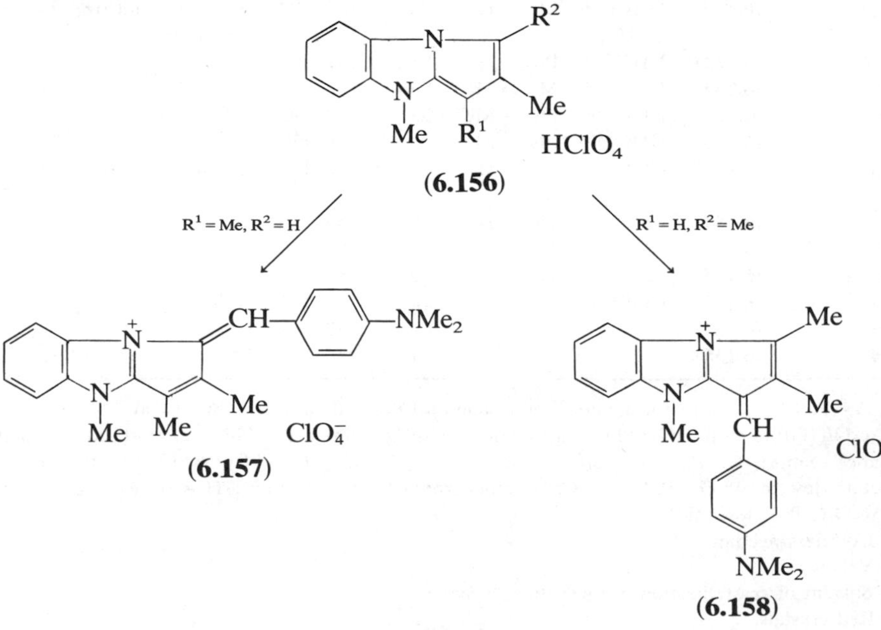

(**6.156**)

$R^1 = Me, R^2 = H$

$R^1 = H, R^2 = Me$

(**6.157**) (**6.158**)

Scheme 6.30

results in specific acylation at *C*(3) with formation of enol esters of the type (**6.160**; R = Me or Ph). In the benzoylation of (**6.159**) using benzoyl chloride in alkaline solution,[56] enol ester formation is accompanied by ring-opening to the product (**6.161**) (Scheme 6.31). Not unexpectedly, quaternization of the *N*(4) center in 2,3-dihydro-1*H*-pyrrolo[1,2-*a*]benzimidazole (**6.159**) enhances the reactivity of the *C*(3) position toward acylation. As a consequence, *C*(3) acylative condensation reactions of *N*(4)-alkyl-2,3-dihydro-1*H*-pyrrolo[1,2-*a*]benzimidazolium salts, e.g. [Scheme 6.32; (**6.162**) → (**6.163**)] have found widespread use[58,82] for the construction of methine dyestuffs incorporating a 2,3-dihydro-1*H*-pyrrolo[1,2-*a*]benzimidazole nucleus. The lack of reactivity toward base-catalyzed aldol-type condensation reported[41] for 2,3-dihydro-1*H*-pyrrolo[1,2-*a*]benzimidazol-1-one (**6.164**) (Scheme 6.32) is surprising and contrasts with enhanced reactivity of the sulfur analog [Scheme 6.32; (**6.165**)] in such processes (see later).

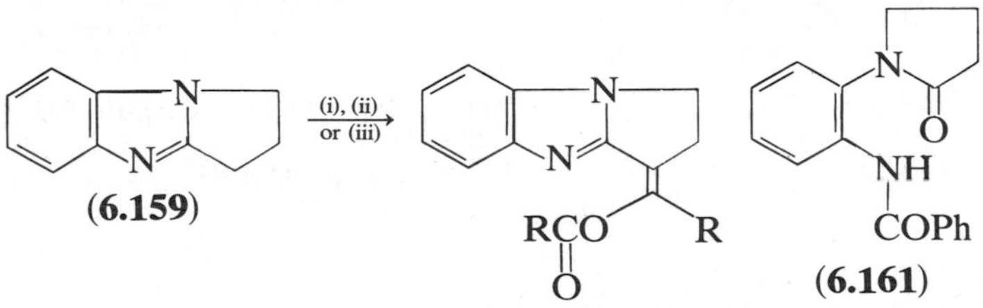

(**6.159**) —(i), (ii) or (iii)→ (**6.160**), (**6.161**) (m.p. 139°)

R	m.p. (°C)
Me	126.5
Ph	231

(i) Ac_2O, $ZnCl_2$/(reflux)/(3 hr)
(ii) PhCOCl, pyridine/(100°)/(15 min)
(iii) PhCOCl, 15% NaOH/(room temp.)(10 min)

Scheme 6.31

The orthodox behavior of amino substituents in the aromatic nucleus of 2,3-dihydro-1*H*-pyrrolo[1,2-*a*]benzimidazoles toward acylation[58] has been exploited for the synthesis of fused systems as illustrated by the transformation[83] shown in Scheme 6.33.

Halogenation, Nitration, Nitrosation, Diazotization, and Diazo Coupling. With the exception of the reported[84] chlorination of the aromatic nucleus in 2,3-dihydro-1*H*-pyrrolo[1,2-*a*]benzimidazole by sulfuryl chloride, examples of electrophilic halogenation reactions of pyrrolo[1,2-*a*]-benzimidazoles of the various structural types are surprisingly lacking. In contrast, the behavior of both 4*H*-pyrrolo[1,2-*a*]benzimidazoles and 2,3-dihydro-1*H*-pyrrolo[1,2-*a*]benzimidazoles toward nitration is reasonably well documented.

The attempted nitration[70] of alkyl- and aryl-substituted 4*H*-pyrrolo[1,2-*a*]benzimidazoles with fuming nitric acid results in the formation of nitrates,

(6.162) → (i) → (6.163)

R	X	Yield (%)	m.p. (°C)
H	TSO_3	15	250–251 (decomp.)
Cl	I	20	317 (decomp.)

	X
(6.164)	CH_2
(6.165)	S

(i) $Me_2NC_6H_4CHO$, piperidine, EtOH/(reflux)/(0.5–2 hr)
[T = toluene-*p*-sulfonyl]

Scheme 6.32

which resist further nitration. On the other hand, treatment of 1- and 3-acetyl-4*H*-pyrrolo[1,2-*a*]benzimidazoles with fuming nitric acid in acetic acid[70] results in the nitrative replacement of the acyl substituent to give the corresponding 1- and 3-nitro-4*H*-pyrrolo[1,2-*a*]benzimidazoles [Scheme 6.34; (**6.166**; $X = NO_2$) and (**6.168**; $X = NO_2$)], albeit in only low yield (Table 6.21). Subsequent nitration of the benzene nucleus is not observed, presumably as a result of conjugative deactivation by the nitro group already present in the *C*(1) or *C*(3) position [cf. Scheme 6.34; (**6.166**; $X = NO_2$) ↔ (**6.167**) and (**6.168**; $X = NO_2$) ↔ (**6.169**)]. The presence of such conjugative

H_2N + COMe–CHMe–CO_2Et → (85%) (i) → (m.p. 260°)

(i) ethyl polyphosphate/(165°)/(0.75 hr)

Scheme 6.33

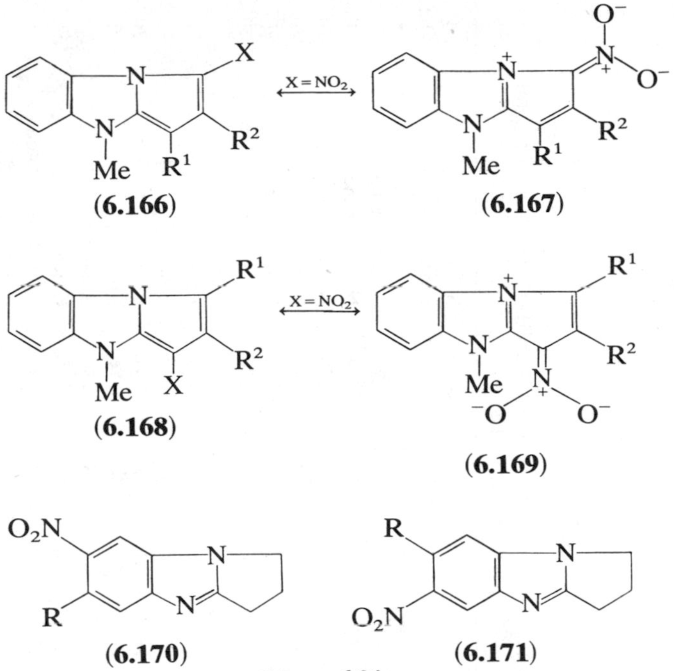

(6.166) (6.167)

(6.168) (6.169)

(6.170) (6.171)

Scheme 6.34

interaction is supported by the exceptionally low frequency[70] of the IR bands due to the nitro groups in these molecules. In contrast, 2,3-dihydro-1*H*-pyrrolo[1,2-*a*]benzimidazoles containing a *C*(6) or *C*(7) amino substituent or a *C*(6) halogeno substituent are smoothly nitrated by "mixed acid" at low temperature to give the corresponding *C*(7) or *C*(6) nitro derivatives [Scheme 6.34; (**6.170**) or (**6.171**)] in high yield (Table 6.21).[58,71]

In contrast to their resistance to nitration (see before) simple *C*(1) and *C*(3)-unsubstituted 4*H*-pyrrolo[1,2-*a*]benzimidazoles are rapidly nitrosated[70] at room temperature by sodium nitrite in acetic acid to give the *C*(1) or *C*(3) nitroso derivative [Scheme 6.34; (**6.166**; X = NO) or (**6.168**; X = NO)] in excellent yield (Table 6.21). As in the case of acylation the nitrosation of 4*H*-pyrrolo[1,2-*a*]benzimidazoles takes place preferentially at *C*(1) and only occurs at the *C*(3) position when the *C*(1) position is substituted. 4*H*-Pyrrolo[1,2-*a*]benzimidazoles also couple readily at the *C*(1) and *C*(3) positions with arenediazonium cations giving the respective azo derivatives [Scheme 6.34; (**6.166**; X = N=NAr) or (**6.168**; X = N=NAr)] in very high yield (Table 6.21).[70]

Amino substituents in the benzene nucleus of 2,3-dihydro-1*H*-pyrrolo-[1,2-*a*]benzimidzoles can be diazotized under standard conditions[32,51] to give diazonium salts which undergo the usual displacement reactions (see later).

TABLE 6.21. NITRATION, NITROSATION, AND DIAZO-COUPLING REACTIONS OF 4*H*-PYRROLO[1,2-*a*]BENZIMIDAZOLES AND 2,3-DIHYDRO-1*H*-PYRROLO[1,2-*a*]BENZIMIDAZOLES

Reaction conditions[a]	Product	X	R	R^1	R^2	Yield (%)	m.p. (°C)	Solvent of crystallization	Ref.
A	**(6.166)**	NO_2	—	H	Me	32	229	Nitromethane	70
A	**(6.166)**	NO_2	—	H	Ph	30	304	Nitromethane	70
A	**(6.166)**	NO_2	—	Me	Me	36	276	Nitromethane	70
A	**(6.166)**	NO_2	—	Me	Ph	40	274	Nitromethane	70
B	**(6.168)**	NO_2	—	Me	Me	17	254	Nitromethane	70
C	**(6.166)**	NO	—	H	Me[b]	93	276	Ethanol–perchloric acid	70
C	**(6.166)**	NO	—	Me	Me[c]	74	207	Ethanol[d]	70
C	**(6.168)**	NO	—	Me	Me[e]	87	245	Ethanol[d]	70
D	**(6.166)**	N═NPh	—	H	Me	58	141	Ethanol[f]	70
D	**(6.166)**	N═NPh	—	H	Ph	60	176	Ethanol[f]	70
D	**(6.166)**	N═NPh	—	Me	Me[g]	65	180	Ethanol[f]	70
D	**(6.168)**	N═NPh	—	Me	Me	51	155	Ethanol[f]	70
D	**(6.166)**	*p*-$Et_2NC_6H_4N$═N	—	H	Me	62	161	Dimethylformamide–water[f]	70
D	**(6.166)**	*p*-$Et_2NC_6H_4N$═N	—	H	Ph	82	191	Dimethylformamide–water[f]	70
D	**(6.166)**	*p*-$Et_2NC_6H_4N$═N	—	Me	Me	56	180	Dimethylformamide–water[f]	70

D	**(6.166)**	$p\text{-}HO_2CC_6H_4N{=}N$	—	H	Me	86	241	Ethanol[f]	70
D	**(6.166)**	$p\text{-}HO_2CC_6H_4N{=}N$	—	H	Ph	78	208	Ethanol[f]	70
D	**(6.166)**	$p\text{-}O_2NC_6H_4N{=}N$	—	H	Ph	75	253	Dimethylformamide–water[f]	70
D	**(6.166)**	$p\text{-}O_2NC_6H_4N{=}N$	—	Me	Me	46	262	Dimethylformamide–water[f]	70
E	**(6.170)**	—	Br	—	—	74	201	Ethanol	58
E	**(6.170)**	—	Cl	—	—	—[h]	203	—[i]	58
E	**(6.170)**	—	F	—	—	—[h]	236	—[i]	58
F	**(6.170)**	—	NHAc	—	—	75	212	Ethanol[j]	71
F	**(6.171)**	—	NHAc	—	—	75	208	Ethanol[j]	71

[a] $A = HNO_3$(S.G. 1.52), AcOH/(room temp.)(short period); $B = HNO_3$(S.G. 1.52), AcOH/(100°)(30 min); $C = NaNO_2$, AcOH/(room temp.)(10 min); $D = Ar\overset{+}{N}_2$, NaOAc, AcOH/(5–10°)(short time); $E = HNO_3$(S.G. 1.42), conc. H_2SO_4/(0–5°)(short time); $F = HNO_3$(S.G. 1.5), conc. H_2SO_4/(0°)(10 min, then room temp.)(1 hr).

[b] Perchlorate, yellow needles.

[c] Forms a hydrochloride, yellow-orange needles, m.p. 260° (from water).

[d] Forms green crystals.

[e] Forms a perchlorate, orange-yellow needles, m.p. 244° (from acetic acid).

[f] Forms red needles.

[g] Forms a perchlorate, m.p. 281° (from nitromethane).

[h] Yield not quoted.

[i] Solvent of crystallization not specified.

[j] Forms yellow needles.

Reactions with Nucleophiles

HYDROXYLATION. The lability, toward hydrolytic cleavage, of the pyrroline ring in simple 3*H*-pyrrolo[1,2-*a*]benzimidazoles is illustrated (Scheme 6.35) by the reported[9] ring-opening of 1,3,3-trimethyl-3*H*-pyrrolo[1,2-*a*]benzimidazole (**6.172**) to the *N*-(2-aminophenyl)pyrrolinone (**6.173**) merely on attempted crystallization from hydroxylic solvents. On the other hand, the

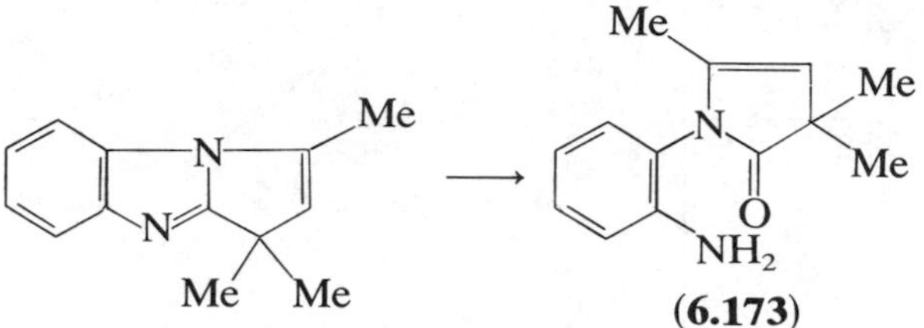

(**6.172**) → (**6.173**)

Scheme 6.35

inertness toward hydroxylation, and consequent stability to hydrolysis of the pyrrole ring in 4*H*-pyrrolo[1,2-*a*]benzimidazoles, is indicated by the demonstration[12,13] that *C*(1) and *C*(3) acetyl substituents in such molecules are removable by forcing acidic hydrolysis, leaving the parent ring system intact. The similar stability to hydrolysis of the pyrrolidine ring in 2,3-dihydro-1*H*-pyrrolo[1,2-*a*]benzimidazoles is illustrated (Table 6.22) by the survival of the ring system under acidic and basic conditions which serve to demethylate

TABLE 6.22. HYDROLYTIC REACTIONS OF 2,3-DIHYDRO-1*H*-PYRROLO[1,2-*a*]-BENZIMIDAZOLE DERIVATIVES

Hydrolysis conditions[a]	2,3-Dihydro-1*H*-pyrrolo[1,2-*a*]-benzimidazole: Substrate	Product	Yield (%)	m.p. (°C)	Solvent of crystallization	Ref.
A	5,6-Dimethoxy-	5,8-Dihydroxy-	50	305–306	Ethanol	32
B	6-Acetamido-7-nitro-	6-Amino-7-nitro-	—[b]	275 (decomp.)	2-Ethoxyethanol	71
B	7-Acetamido-6-nitro-	7-Amino-6-nitro-	—[b]	320 (decomp.)	2-Ethoxyethanol	71
C	6-Ethoxycarbonyl-	6-Carboxy-	quant.	300	Acetic acid–water	58
C	8-Ethoxycarbonyl-	8-Carboxy-	—[b]	310–311	—[d]	58
C	6-Ethoxycarbonyl-7-chloro-	6-Carboxy-7-chloro-	—[b]	>270	—[d]	58

[a] *A* = conc. HCl/100°, pressure; *B* = conc. HCl/(reflux)(1 hr); *C* = 2.5 M NaOH, EtOH/(reflux)(5 min).
[b] Yield not quoted.
[c] Forms red prisms or needles.
[d] Solvent not specified.

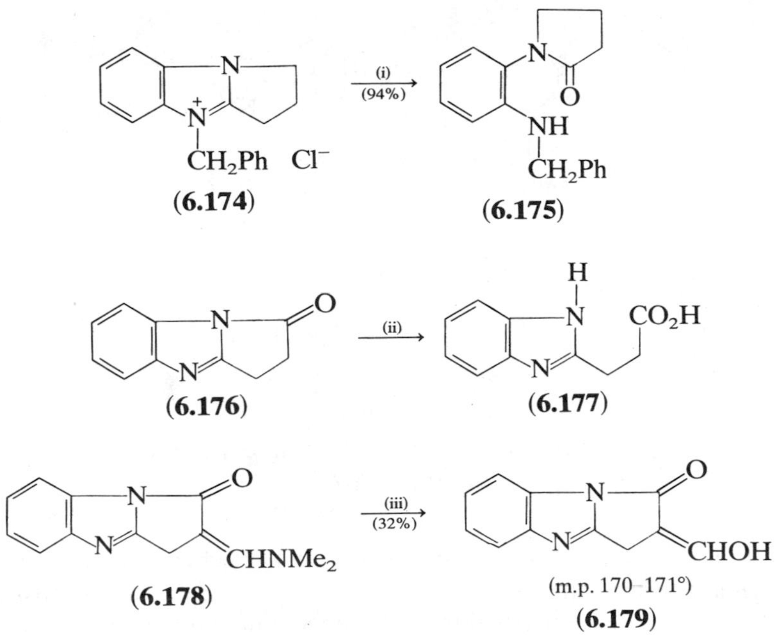

(6.174) → (6.175) [(i), (94%)]

(6.176) → (6.177) [(ii)]

(6.178) → (6.179) (m.p. 170–171°) [(iii), (32%)]

(i) 4 *M* NaOH/(heat)/(1 min)
(ii) Na_2CO_3 aq./heat
(iii) 3% HCl, EtOH/(room temp.)(3 hr)

Scheme 6.36

methoxy substituents,[32] deacetylate acetamido groups,[71] and hydrolyze ethoxycarbonyl substituents,[58] giving the corresponding phenols, amines, and carboxylic acids, usually in good yield (cf. Table 6.22). *C*(3)-Acyl substituents in 2,3-dihydro-1*H*-pyrrolo[1,2-*a*]benzimidazoles are also hydrolytically removed under alkaline conditions without disruption of the ring system.[56] On the other hand, the presence of a quaternary or a carbonyl center at *N*(4) and *C*(1), respectively, confers hydrolytic instability on the 2,3-dihydro-1*H*-pyrrolo[1,2-*a*]benzimidazole ring system as demonstrated by the ring-opening reactions [(**6.174**) → (**6.175**)][59] and [(**6.176**) → (**6.177**)][37] portrayed in Scheme 6.36. However, the successful acid-catalyzed hydrolysis (Scheme 6.36)[42] of the dimethylaminomethylene derivative (**6.178**) to the hydroxymethylene product (**6.179**) shows that 2,3-dihydro-1*H*-pyrrolo[1,2-*a*]benzimidazoles can be manipulated in protic media provided the conditions are mild enough.

As in the case of 2,3-dihydro-1*H*-pyrrolo[1,2-*a*]benzimidazole (see before), the hydroxide-catalyzed ring-opening of 1*H*-pyrrolo[1,2-*a*]benzimidazol-1-ones [Scheme 6.37; (**6.180**)] as initiated by attack at the carbonyl group, and affords 2-benzimidazolylacrylic acids of the type (**6.181**; R = Ph)[1] or [in the case of 3-hydroxy substrates, e.g. (**6.180**; R = OH)] by subsequent decarboxylation, the 2-benzimidazolyl ketone [e.g. (**6.182**)].[2] The presence

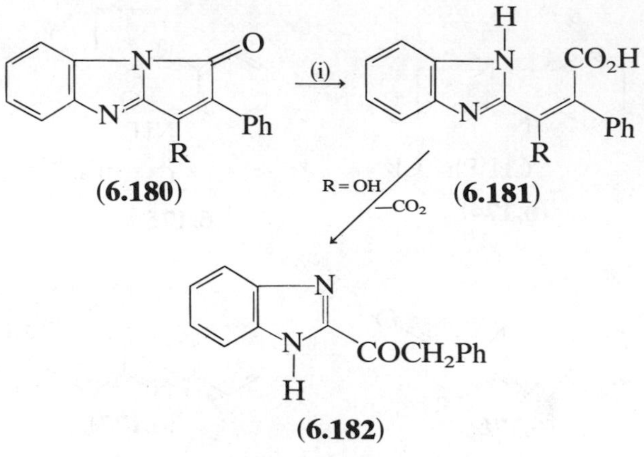

N O N Ph R (i) H N CO₂H N Ph R

(**6.180**) R = OH —CO_2 (**6.181**)

N N $COCH_2Ph$ H

(**6.182**)

(i) 30% KOH aq., EtOH/(room temp.)/(10 min)

Scheme 6.37

of the bridgehead quaternary center in 2,3,3*a*,4-tetrahydro-1*H*-pyrrolo[1,2-*a*]benzimidazolium salts such as (**6.183**) (Scheme 6.38), promotes hydroxide ion attack at the *C*(3*a*) position with concomitant ring-opening to eight-membered carbinolamines [e.g. (**6.184**)], which coexist in equilibrium with the open-chain aminoaldehydes [e.g. (**6.185**)].[76]

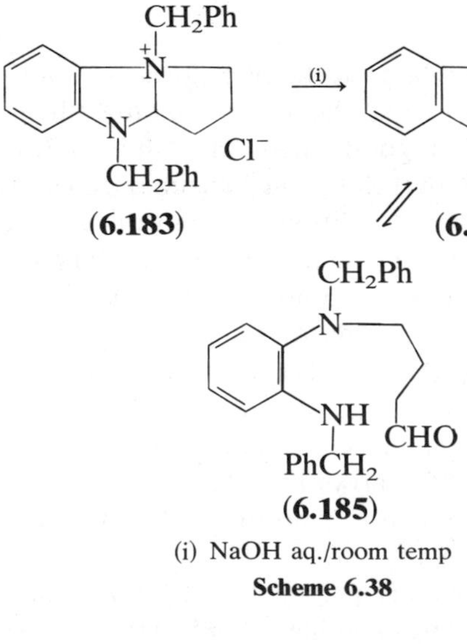

CH_2Ph + N N Cl^- CH_2Ph (i) CH_2Ph N N OH CH_2Ph

(**6.183**) (**6.184**)

CH_2Ph N NH CHO $PhCH_2$

(**6.185**)

(i) NaOH aq./room temp

Scheme 6.38

Amination. The nucleophilic addition of amines to the *C*(1)–*C*(2) double bond in 1,3,3-trimethyl-1*H*-pyrrolo[1,2-*a*]benzimidazole [cf. Scheme 6.35; (**6.172**)] to give adducts of the type (**6.186**) (Scheme 6.39) is briefly reported in a patent.[85]

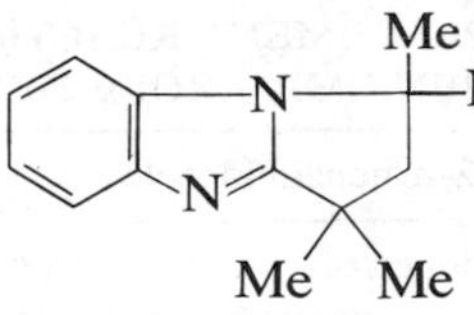

(6.186)

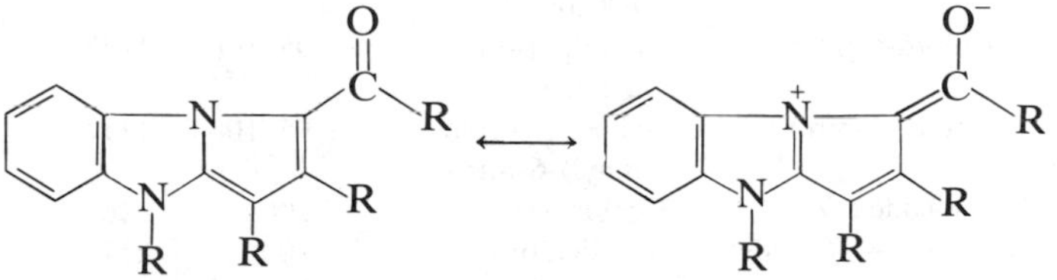

(6.187) ⟷ **(6.188)**

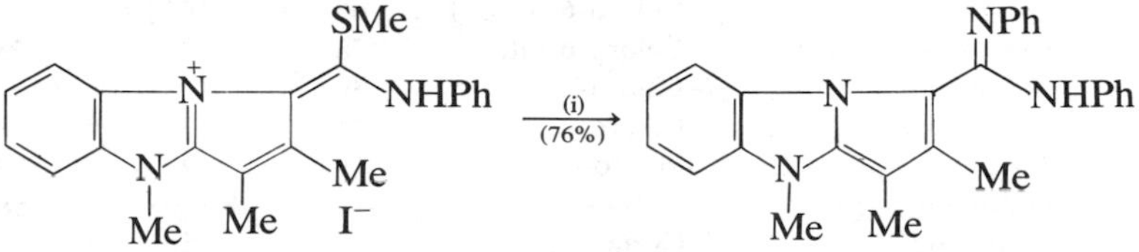

(6.189) $\xrightarrow[(76\%)]{(i)}$ **(6.190)**

(m.p. 108–109°)

(i) $PhNH_2$/(130–140°, sealed tube) (30 min), then NH_3

Scheme 6.39

The failure of $C(1)$ and $C(3)$ aldehydic and ketonic substituents in 4*H*-pyrrolo[1,2-*a*]benzimidazoles to undergo the usual condensation reactions with amino reagents (amines, hydrazines, hydroxylamine) is attributed[13] to the amide-like character imparted by resonance interaction with the bridgehead nitrogen atom [Scheme 6.39; (**6.187**) ↔ (**6.188**)]. *S*-Methylthioimidate salts of the 4*H*-pyrrolo[1,2-*a*]benzimidazole series undergo orthodox aminative replacement reactions as illustrated by the transformation[81] [(**6.189**) → (**6.190**)] (Scheme 6.39).

The activation of the chlorine atom in 5-chloro-6-nitro-2,3-dihydro-1*H*-pyrrolo[1,2-*a*]benzimidazole to nucleophilic attack is demonstrated by its smooth replacement by cyclic secondary amines to give the corresponding 5-amino-6-nitro-2,3-dihydro-1*H*-pyrrolo[1,2-*a*]benzimidazoles in excellent yield (Table 6.23).[73]

Halogenation. As in the case of electrophilic halogenation (see before) the nucleophilic halogenation of pyrrolo[1,2-*a*]benzimidazoles is only rarely observed. Noteworthy examples are essentially confined to the 2,3-dihydro-1*H*-pyrrolo[1,2-*a*]benzimidazole ring system and in particular to 2,3-dihydro-1*H*-pyrrolo[1,2-*a*]benzimidazole 4-*N*-oxides [Scheme 6.40; (**6.191**)]. Analogy with simpler α-alkylated heterocyclic *N*-oxides[86] suggests

TABLE 6.23. NUCLEOPHILIC DISPLACEMENT REACTIONS OF 2,3-DIHYDRO-1*H*-PYRROLO[1,2-*a*]BENZIMIDAZOLE DERIVATIVES

Reaction conditions[a]	2,3-Dihydro-1*H*-pyrrolo[1,2-*a*]benzimidazole		Yield (%)	m.p. (°C)	Ref.
	Substrate	Product			
A	5-Chloro-6-nitro	5-(1-Pyrrolidinyl)-6-Nitro-	95–100	155[b]	73
A	5-Chloro-6-nitro-	5-(1-Piperidinyl)-6-nitro-	95–100	193[b]	73
A	5-Chloro-6-nitro	5-(1-Perhydroazepinyl)-6-nitro-	95–100	131[b]	73
B	4-*N*-Oxide	7-chloro-	100	136	73
B	6-Chloro 4-*N*-oxide	6,7-Dichloro-	100	211	73
B	6-Nitro 4-*N*-oxide	{5-Chloro-6-nitro- + 8-Chloro-6-nitro-}	32 68	184 162	73
C	6-Nitro 4-*N*-oxide	5-Chloro-6-nitro-	78	—	73
D	6-Amino	6-Chloro-	30	133–134	32
E	7-Amino	7-Chloro-	21	136[c]	58
F	7-Amino	7-Fluoro-	19	124[e]	58
G	6-Bromo-	6-Cyano-	—[d]	190[b]	58
H	7-Amino-	7-Cyano-	—[d]	155	58
H	7-Amino-6-bromo-	6-Bromo-7-cyano-	48	224[f]	58
H	7-Amino-6-chloro-	6-Chloro-7-cyano-	—[d]	215	58
H	7-Amino-6-fluoro-	6-Fluoro-7-cyano-[f]	—[d]	210	58
I	7-Amino-	7-Azido-	—[d]	150 (decomp.)	51
I	6-Amino-	6-Azido-	—[d]	136	51
I	7-Amino-6-nitro-	7-Azido-6-nitro-[f]	—[d]	190 (decomp.)	71
I	6-Amino-7-nitro-	6-Azido-7-nitro-	—[d]	186 (decomp.)	71

[a] *A* = amine, EtOH/(reflux)(12 hr); *B* = $POCl_3$, $CHCl_3$/(reflux)(4 hr); *C* = conc. HCl/(110°)(144 hr); *D* = $NaNO_2$, HCl/(0°), then CuCl, conc. HCl/(room temp.); *E* = $NaNO_2$, HCl/(0°), then CuCl, conc. HCl/(50–60°); *F* = $NaNO_2$, HBF_4/(0°), then tetralin/(reflux); *G* = CuCN, $PhNO_2$/(reflux 1.5 h), then NaCN, H_2O/(100°)/(few min); *H* = $NaNO_2$, HCl/(0°), then CuCN, KCN/(room temp.)(0.5 hr), then 50–60° 15 min); *I* = $NaNO_2$, HCl/(0°), then NaN_3, NaOAc, H_2O.

[b] Crystallized from ethanol.

[c] Crystallized from benzene.

[d] Yield not specified.

[e] Boiling point 166°/3 mm Hg.

[f] Crystallized from benzene-hexane.

that these molecules should be susceptible to nucleophilic halogenation at the *C*(3) position. In practice, reaction of 2,3-dihydro-1*H*-pyrrolo[1,2-*a*]-benzimidazole 4-*N*-oxide and its 6-chloro derivative (**6.191**; R = H or Cl) with phosphorus oxychloride in refluxing chloroform results in specific chlorination at the *C*(7) position in the benzene ring giving 7-chloro- and 6,7-dichloro-2,3-dihydro-1*H*-pyrrolo[1,2-*a*]benzimidazoles (**6.193**; R = H

N
R
N
O
(**6.191**)

Cl
a
N
R
N
OX
(**6.192**)

a

$R = NO_2$

$R = NO_2$

H
Cl
R
N
N

H
Cl
N
O_2N
N

N
O_2N
N
Cl
H

Cl
N
R
N
(**6.193**)

Cl
N
O_2N
N
(**6.194**)

N
O_2N
N
Cl
(**6.195**)

(**Scheme 6.40**)

or Cl) in essentially quantitative yield (Table 6.23).[73] The analogous reaction of 6-nitro-2,3-dihydro-1*H*-pyrrolo[1,2-*a*]benzimidazole 4-*N*-oxide (**6.191**; R = NO_2), on the other hand, leads to competing attack at *C*(5) and *C*(8), affording a mixture of 5- and 8-chloro-2,3-dihydro-1*H*-pyrrolo[1,2-*a*]benzimidazoles (**6.195**) and (**6.194**) again in good overall yield (Table 6.23) with the latter product predominating.[73] The exclusive formation of the 5-chloro product (**6.195**) in high yield (Table 6.23) by simply heating the nitro *N*-oxide (**6.191**; R = NO_2) in concentrated hydrochloric acid[73] is indicative of the enhanced reactivity of this substrate to halogenation of this type. The chlorination reactions of 2,3-dihydro-1*H*-pyrrolo[1,2-*a*]benzimidazole 4-*N*-oxides may be rationalized mechanistically (Scheme 6.40) in terms of the initial coordination of the chlorinating agent at the *N*-oxide oxygen atom. The resulting species (**6.192**; X = H or $POCl_2$) produced would then be susceptible to attack by chloride ion at all three available sites in the benzene ring (cf. Scheme 6.40) and hence is a plausible common intermediate for the subsequent alternative *C*(5), *C*(7), and *C*(8) chlorination observed. However, this mechanistic rationale does not readily account for the contrasting chlorination reactions of the parent and 6-chloro *N*-oxides (**6.191**; R = H or Cl), on the one hand, and the nitro *N*-oxide (**6.191**; R = NO_2), on the other, in terms of the site of substitution. Nor does it explain the apparent dichotomy in the behavior of the nitro *N*-oxide (**6.191**; R = NO_2) toward chlorination by phosphorus oxychloride as opposed to chlorination in concentrated hydrochloric acid. Substitution reactions of this type obviously warrant further study in order to clarify these details.

Halogen substituents are also introduced into the benzene nucleus of 2,3-dihydro-1*H*-pyrrolo[1,2-*a*]benzimidazoles via Sandmeyer reactions of the corresponding amines. Halogenation of this type[32,58] proceeds in moderate yield (Table 6.23) and, though not strictly nucleophilic in a mechanistic sense, is included for convenience under the present heading.

Miscellaneous Reactions. The transformation[2] [(**6.196**) → (**6.197**)] (Scheme 6.41) exemplifies what appears to be the sole reported example of carbanion attack on a pyrrolo[1,2-*a*]benzimidazole derivative. On the other hand, the behavior of pyrrolo[1,2-*a*]benzimidazoles toward nucleophilic attack by cyanide ion has been reported in a number of instances, as

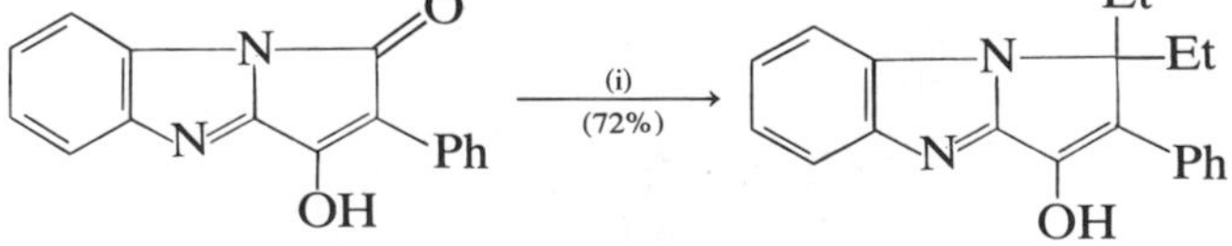

(**6.196**) —(i) (72%)→ (**6.197**)

(i) EtMgBr, ether/(reflux)/(8 hr)

Scheme 6.41

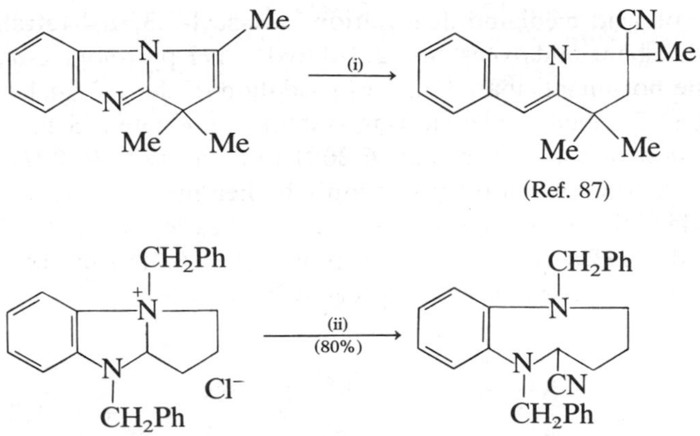

(Ref. 87)

(Ref. 76)

(i) HCN liq./80°
(ii) NaCN, H_2O/room temp.

Scheme 6.42

illustrated by the reactions[76,87] shown in Scheme 6.42. 2,3-Dihydro-1*H*-pyrrolo[1,2-*a*]benzimidazoles containing cyano substituents in the benzene nucleus are synthesized in orthodox fashion by reaction of the corresponding halogeno compounds under forcing conditions with cuprous cyanide[58] (Table 6.23) or by the Sandmeyer reactions[32,58] of appropriate amines (Table 6.23). Reactions[51,71] related to the latter type also allow synthetic access to azido-substituted 2,3-dihydro-1*H*-pyrrolo[1,2-*a*]benzimidazoles (Table 6.23).

Oxidation

Relatively little information is available regarding the behavior of pyrrolo-[1,2-*a*]benzimidazoles toward oxidizing agents. However, the relative stability of the 2,3-dihydro-1*H*-pyrrolo[1,2-*a*]benzimidazole ring system to oxidation is indicated by the successful chromic acid oxidation[32] of the hydroquinone (**6.198**) to the quinone (**6.199**) (Scheme 6.43) in moderate yield,

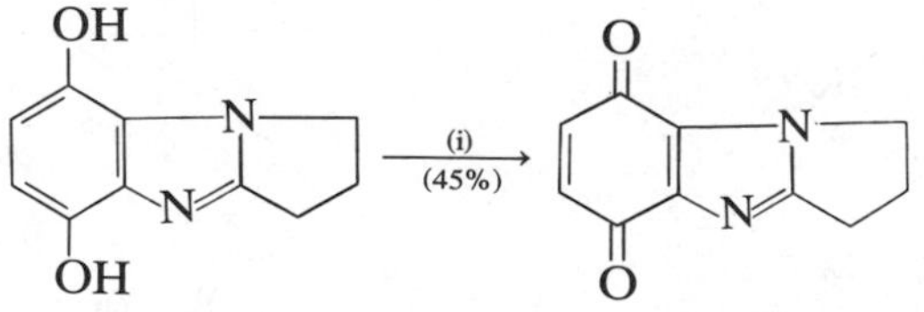

(**6.198**) (**6.199**)

(i) CrO_3,H_2O/60°

Scheme 6.43

and by the peracid-mediated deacylation of 4-acyl-2,3,3*a*,4-tetrahydro-1*H*-pyrrolo[1,2-*a*]benzimidazoles to 2,3-dihydro-1*H*-pyrrolo[1,2-*a*]benzimidazole.[57] The not-unanticipated ease of oxidation of the 2,3,3*a*,4-tetrahydro-1*H*-pyrrolo[1,2-*a*]benzimidazole ring system is illustrated (Scheme 6.44) by the conversion of the derivatives (**6.200**) into the salts (**6.201**) with concomitant formation of chloroform, simply by heating under reflux in carbon tetrachloride.[68] Similar oxidation is effected by heating the hydrochlorides of the tetrahydro-1*H*-pyrrolo[1,2-*a*]benzimidazoles (**6.200**) in acetone, which is coreduced to 2-propanol in the process.[68]

(**6.200**) —(i)→ (**6.201**) + $CHCl_3$

(i) CCl_4/(reflux)/(1 hr)

Ar	Yield (%)	m.p. (°C)
Ph	86	219–220
2-Cl,5-$NO_2C_6H_3$	84	290 (decomp.)

Scheme 6.44

Reduction

The zinc and acetic acid reduction (Scheme 6.45) of readily available 1-diazoaryl-4*H*-pyrrolo[1,2-*a*]benzimidazoles (**6.202**) which provides synthetic access (albeit in low yield)[70] to the unstable 1-amino derivatives (**6.203**) is dependent on the stability of the 4*H*-pyrrolo[1,2-*a*]benzimidazole ring

(**6.202**) —(i)→ (**6.203**) picrate

(i) Zn, AcOH

R	Yield (%)	m.p. (°C)
Ph	13	221
Me	not quoted	249

Scheme 6.45

system to metal–proton donor reduction of this type. The debenzylation of the 4-benzyl-4*H*-pyrrolo[1,2-*a*]benzimidazoles (**6.204**) and (**6.205**) (Scheme 6.46) using sodium in liquid ammonia,[17] is accompanied by an unprecedented hydrogen shift, the products, formed in good yield, being the corresponding 3*H*-pyrrolo[1,2-*a*]benzimidazoles (**6.209**). However, the stability of the 4*H*- and 3*H*-pyrrolo[1,2-*a*]benzimidazole ring systems toward metal–proton donor reduction under basic conditions, implicit in these transformations, is not extended to the diphenyl derivative (**6.206**). The debenzylation and hydrogen shift induced in this substrate by treatment[17] with sodium in liquid ammonia is accompanied by further reduction, the product being the 2,3-dihydro-1*H*-pyrrolo[1,2-*a*]benzimidazole derivative

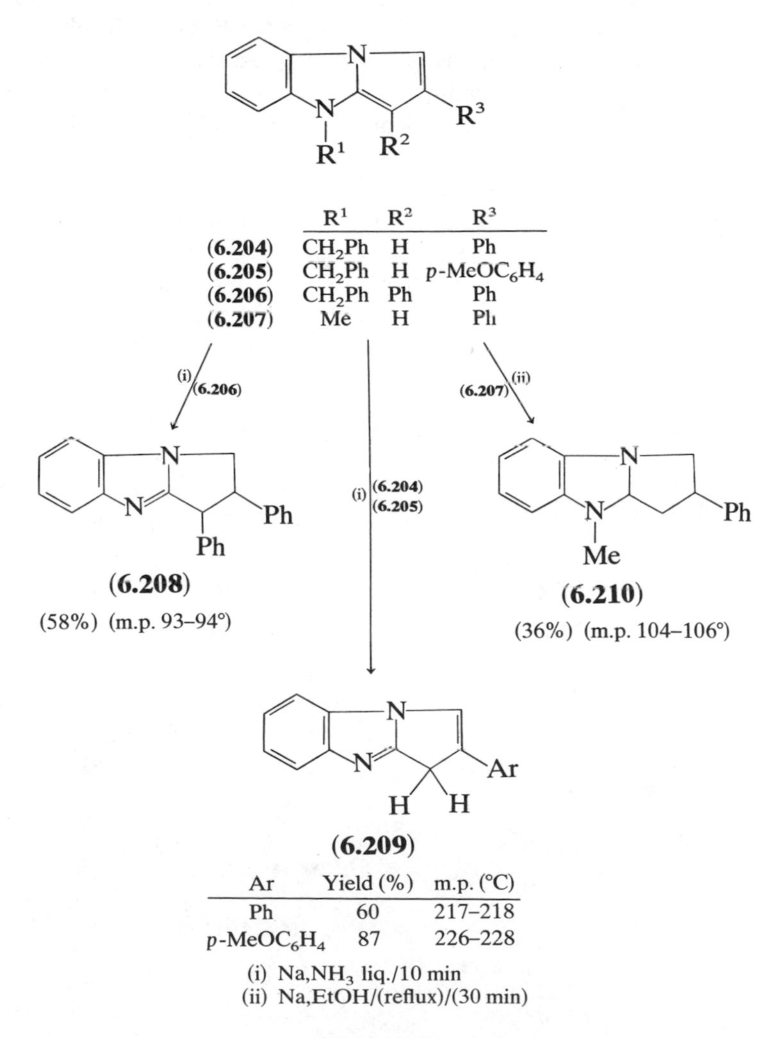

	R^1	R^2	R^3
(**6.204**)	CH_2Ph	H	Ph
(**6.205**)	CH_2Ph	H	*p*-$MeOC_6H_4$
(**6.206**)	CH_2Ph	Ph	Ph
(**6.207**)	Me	H	Ph

Ar	Yield (%)	m.p. (°C)
Ph	60	217–218
p-$MeOC_6H_4$	87	226–228

(i) Na,NH_3 liq./10 min
(ii) Na,EtOH/(reflux)/(30 min)

Scheme 6.46

(**6.208**). Sodium and ethanol reduction[75] of the *N*-methylpyrrolobenzimidazole (**6.207**) (Scheme 6.46) proceeds a stage further and affords the 2,3,3*a*,4-tetrahydro-1*H*-pyrrolo[1,2-*a*]benzimidazole (**6.210**).

Zinc in acetic acid effects the specific reduction [Scheme 6.47; (**6.211**)→(**6.212**)][2] of the carbon–carbon double bond in the enone (**6.211**) which somewhat surprisingly is reported[2] to be inert to complex metal hydride reduction under a variety of conditions. This inertness contrasts with the ready reductive ring-opening induced in 2,3-dihydro-1*H*-pyrrolo[1,2-*a*]-benzimidazol-1-one by treatment[38] with lithium aluminum hydride [Scheme 6.47; (**6.213**)→(**6.214**)]. *N*(9)-Quaternary salts derived from 2,3,3*a*,4-tetrahydro-1*H*-pyrrolo[1,2-*a*]benzimidazoles are also susceptible to sodium borohydride-promoted ring-opening [Scheme 6.48; (**6.215**)→(**6.216**)],[76] whereas in the analogous reduction of *N*(4)-benzyl-2,3-dihydro-1*H*-pyrrolo[1,2-*a*]benzimidazolium salts the ring system remains intact, the product being the corresponding *N*(4)-benzyl-2,3,3*a*,4-tetrahydro-1*H*-pyrrolo[1,2-*a*]benzimidazole, e.g. [Scheme 6.48; (**6.217**)→(**6.218**)].[68] Debenzylation of compounds of the latter type can be accomplished by catalytic hydrogenation.[59] However, the products are not the anticipated parent 2,3,3*a*,4-tetrahydro-1*H*-pyrrolo[1,2-*a*]benzimidazoles, but the 2,3-dihydro-1*H*-pyrrolo[1,2-*a*]benzimidazoles produced by their ready *in situ* oxidation, e.g. [Scheme 6.48; (**6.218**)→(**6.219**)→(**6.220**)].[59] The catalytic hydrogenation of nitro-substituted 2,3-dihydro-1*H*-pyrrolo[1,2-*a*]benzimidazoles proceeds in standard fashion to give the corresponding amines (Table 6.24).[32,36,58]

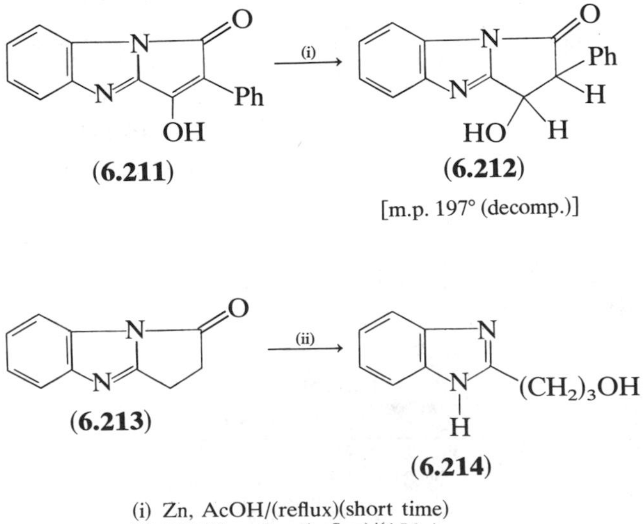

(**6.211**) —(i)→ (**6.212**) [m.p. 197° (decomp.)]

(**6.213**) —(ii)→ (**6.214**)

(i) Zn, AcOH/(reflux)(short time)
(ii) $LiAlH_4$, ether/(reflux)/(15 hr)

Scheme 6.47

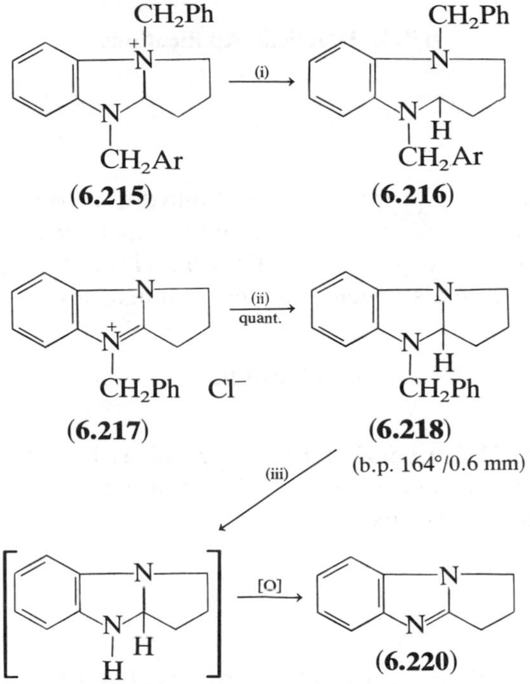

(6.215) (6.216)

(6.217) (6.218) (b.p. 164°/0.6 mm)

(6.219) (6.220)

(i) $NaBH_4$,H_2O/(room temp.)(short time)
(ii) $NaBH_4$,H_2O/(0°)(short time)
(iii) 10%, Pd-C, EtOH/atm. press., room temp.

Scheme 6.48

TABLE 6.24. CATALYTIC HYDROGENATION OF NITRO-2,3-DIHYDRO-1*H*-PYRROLO[1,2-*a*]BENZIMIDAZOLES

Hydrogenation conditions[a]	2,3-Dihydro-1*H*-pyrrolo[1,2-*a*]benzimidazole Substrate	Product	Yield (%)	m.p. (°C)	Solvent of crystallization	Ref.
A	6-Nitro-	6-Amino-[b]	quant.	294–297	Ethanol–hexane	32
B	6-Bromo-7-nitro-	7-Amino-6-bromo-	75	264	Ethanol	58
B	6-Chloro-7-nitro-	7-Amino-6-chloro-	—[c]	264	—[d]	58
B	6-Fluoro-7-nitro-	7-Amino-6-fluoro-	—[c]	230	—[d]	58

[a] *A* = H_2, PtO_2, EtOH/room temp., atm. press; *B* = H_2, Raney–Ni, methyl glycol/room temp., atm. press.
[b] Dihydrochloride.
[c] Yield not quoted.
[d] Solvent of crystallization not specified.

6.1.4. Practical Applications

Biological Properties

The pharmacological properties of 2,3-dihydro-1*H*-pyrrolo[1,2-*a*]benzimidazoles have attracted attention[30] because of their structural similarity to alkaloids such as deoxypeganine. 2,3-Dihydro-1*H*-pyrrolo[1,2-*a*]benzimidazole derivatives have also been patented as fungicides.[85]

Dyestuffs

Various pyrrolo[1,2-*a*]benzimidazole derivatives have found application in the formulation of cyanine dyes[12,58,88] and also as photographic sensitizing[82] and nucleating[89] agents.

Polymers

Two patents describe the use of 3*H*-pyrrolo[1,2-*a*]benzimidazoles[3] and 2,3-dihydro-1*H*-pyrrolo[1,2-*a*]benzimidazoles[90] as monomers in polymer synthesis.

6.2. TRICYCLIC 6-5-5 FUSED BENZIMIDAZOLES WITH ONE ADDITIONAL HETEROATOM

Six ring systems having a tricyclic 6-5-5 fused benzimidazole structure with one additional heteroatom (oxygen, sulfur, or nitrogen) have been documented to date (cf. Scheme 6.49 and Table 6.25). Of these, the oxazolo[3,2-*a*]benzimidazole ring system, so far described only in the 2,3-dihydro form (**6.221**), represents the sole example of an oxygen-containing framework of this type. Sulfur-containing 6-5-5 fused benzimidazoles with one additional heteroatom include the 1*H*,3*H*-thiazolo[3,4-*a*]benzimidazole (**6.224**) and thiazolo[3,2-*a*]benzimidazole (**6.222**) ring systems, the latter also being encountered in the 2,3-dihydro form (**6.223**).

Fully nitrogen-containing structures of the type under consideration are represented by the pyrazolo[2,3-*a*]benzimidazole framework, known in 1*H*- (**6.225**), 4*H*- (**6.226**), and tetrahydro (**6.227**) forms, the 1*H*- and 9*H*-imidazo[1,2-*a*]benzimidazole ring systems (**6.228**) and (**6.229**) and their 2,3-dihydro derivatives (**6.230**) and (**6.231**), and the respective 4*H*-imidazo-[3,4-*a*]benzimidazole and 1*H*,3*H*-imidazo[1,5-*a*]benzimidazole isosteres (**6.232**) and (**6.233**).

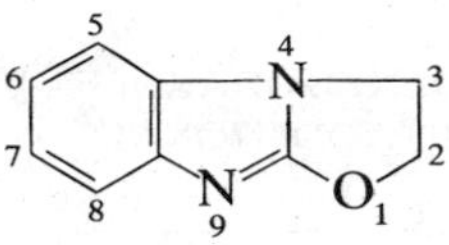

(**6.221**) 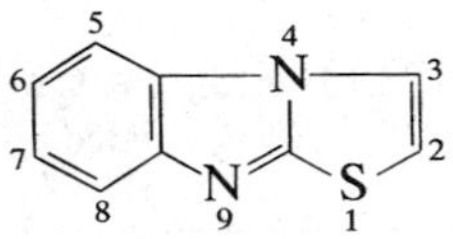 (**6.222**)

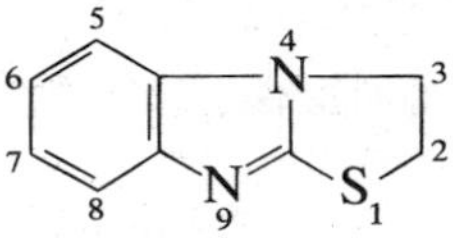

(**6.223**) 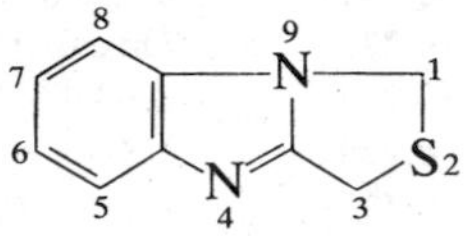 (**6.224**)

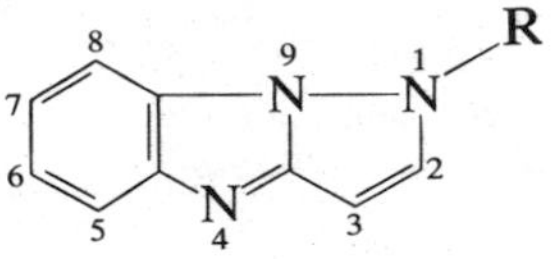

(**6.225**) 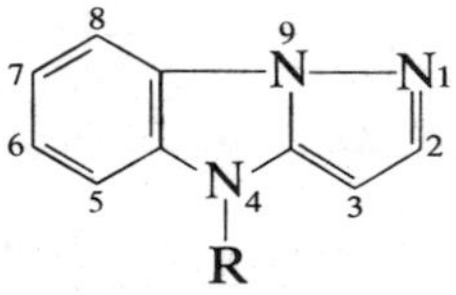 (**6.226**)

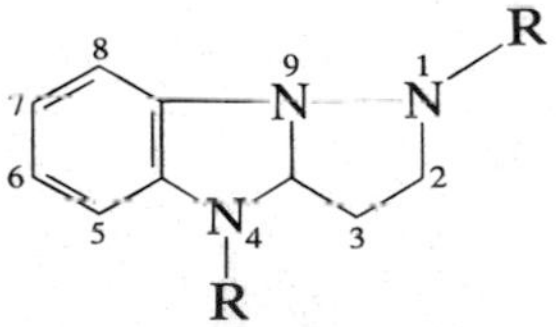

(**6.227**) 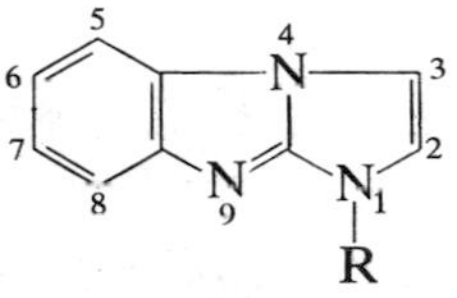 (**6.228**)

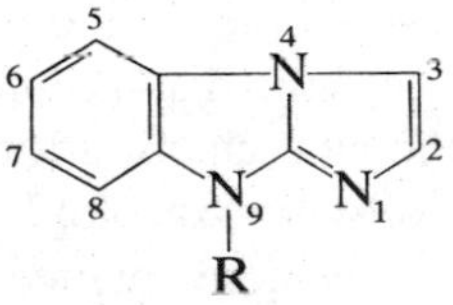

(**6.229**) 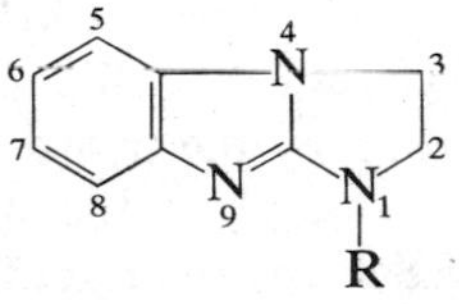 (**6.230**)

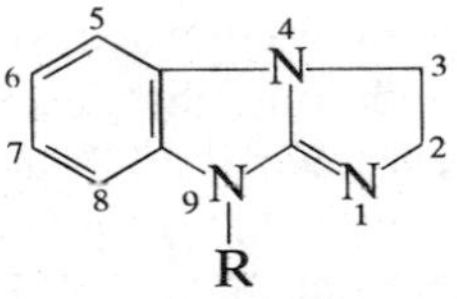

(**6.231**) 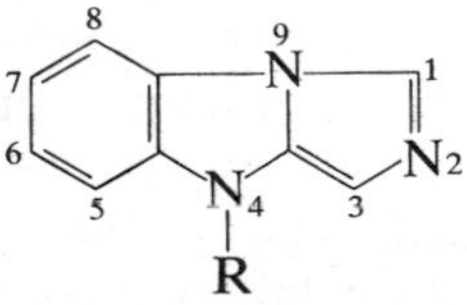 (**6.232**)

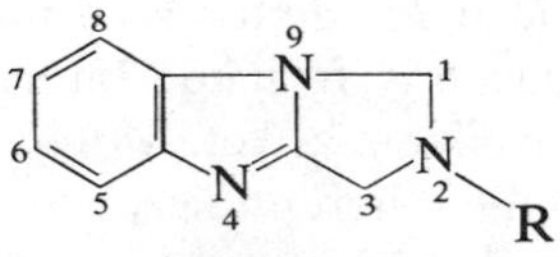

(**6.233**)

Scheme 6.49

TABLE 6.25 TRICYCLIC 6-5-5 FUSED BENZIMIDAZOLE RING SYSTEMS WITH ONE ADDITIONAL HETEROATOM

Structure[a]	Name[b]
(**6.221**)	2,3-Dihydro-oxazolo[3,2-*a*]benzimidazole
(**6.222**)	Thiazolo[3,2-*a*]benzimidazole
(**6.223**)	2,3-Dihydrothiazolo[3,2-*a*]benzimidazole
(**6.224**)	1*H*,3*H*-Thiazolo[3,4-*a*]benzimidazole
(**6.225**)	1*H*-Pyrazolo[2,3-*a*]benzimidazole
(**6.226**)	4*H*-Pyrazolo[2,3-*a*]benzimidazole
(**6.227**)	2,3,3*a*,4-Tetrahydro-1*H*-pyrazolo[2,3-*a*]benzimidazole
(**6.228**)	1*H*-Imidazo[1,2-*a*]benzimidazole
(**6.229**)	9*H*-Imidazo[1,2-*a*]benzimidazole
(**6.230**)	2,3-Dihydro-1*H*-imidazo[1,2-*a*]benzimidazole
(**6.231**)	2,3-Dihydro-9*H*-imidazo[1,2-*a*]benzimidazole
(**6.232**)	4*H*-Imidazo[3,4-*a*]benzimidazole
(**6.233**)	1*H*,3*H*-Imidazo[1,5-*a*]benzimidazole

[a] Cf. Scheme 6.49.
[b] Based on the Ring Index.

6.2.1. Synthesis

Ring-closure Reactions of Benzimidazole Derivatives

2-Chloro-1-(2-hydroxyethyl)benzimidazoles undergo smooth base-catalyzed cyclization by the intramolecular nucleophilic displacement of the 2-chloro group by the hydroxy substituent in the side chain, affording high yields (Table 6.26) of the corresponding 2,3-dihydrooxazolo[3,2-*a*]benzimidazoles [Scheme 6.50; (**6.236**) → (**6.237**)].[91,92] The convenience and flexibility of this useful general synthetic method are enhanced by the fact that the requisite 2-chloro-1-(2-hydroxyethyl)benzimidazoles are readily synthesized by the condensation of the sodium salts of 2-chlorobenzimidazoles with variously substituted epoxides under conditions that also tend to promote spontaneous cyclization to the 2,3-dihydrooxazolo[3,2-*a*]benzimidazole [Scheme 6.50; (**6.234**) + (**6.235**) → (**6.236**) → (**6.237**)] (Table 6.26).[91,92] However, the nature of these ring-closures is such that the use of unsymmetrically substituted epoxides with marginal electronic bias in the direction of ring-opening will tend to lead to mixtures of products, or structural ambiguity when a single product is formed. Information on these possible synthetic limitations is not available as yet. Moreover such "*in situ*" condensation reactions suffer from the inherent disadvantage of lack of control over the two directions of ring-closure possible when 2-chlorobenzimidazoles substituted in the benzene ring (**6.234**; $R^3 \neq H$) are used as substrates. This situation is demonstrated by the reaction[91] of 2,5(6)-dichlorobenzimidazole

TABLE 6.26. SYNTHESIS OF 2,3-DIHYDRO-OXAZOLO[3,2-*a*]BENZIMIDAZOLES BY RING-CLOSURE REACTIONS OF 2-CHLORO-1-(2-HYDROXYETHYL)BENZIMIDAZOLE DERIVATIVES.

Starting materials ($R^1 \rightarrow R^3$ unspecified = H)	Reaction conditions[a]	Product ($R^1 \rightarrow R^3$ unspecified = H)	Yield (%)	m.p. (°C)	Solvent of crystallization	Ref.
(**6.236**)	*A*	(**6.237**)	—[b]	104–105	Benzene	92
(**6.236**; $R^1 = Et$)	*A*	(**6.237**; $R^1 = Et$)	—[b]	105–107	Benzene	92
(**6.236**; $R^1 = CONH_2$)	*B*	(**6.237**; $R^1 = CONH_2$)	75	236–237	MeOH	91
(**6.234**) + (**6.235**; $R^1 = CH_2Cl$)	*C*	(**6.237**; $R^1 = CH_2Cl$)	—[b]	136–138	Benzene	91, 92
(**6.234**) + (**6.235**; $R^1 = CH_2OH$)	*D*	(**6.237**; $R^1 = CH_2OH$)	52	192–193	Ethanol	91, 92
(**6.234**) + (**6.235**; $R^1 = CH_2OPh$)	*E*	(**6.237**; $R^1 = CH_2OPh$)	70	160–161	Ethanol	91
(**6.234**) + (**6.235**; $R^1 = CH_2N$ O)	*F*	(**6.237**; $R^1 = CH_2N$ O)	64	147–148	Benzene	91, 92
(**6.234**) + (**6.235**; $R^1 = CH_2N$ O O)	*F*	(**6.237**; $R^1 = CH_2N$ O O)	63	246–248	*n*-Butanol	91
(**6.234**) + (**6.235**; $R^2 = Me$, $R^3 = CO_2Et$)	*C*	(**6.237**; $R^1 = CO_2Et$, $R^2 = Me$)	—[b]	147–148	Benzene	92
(**6.234**; $R^3 = 5(6)\text{-Cl}$) + (**6.235**; $R^1 = CH_2Cl$)	*C*	(**6.237**; $R^1 = CH_2Cl$, $R^3 = 6(7)\text{-Cl}$)	66	127–128	Benzene	91

[a] *A* = NaOH, H_2O, EtOH/(reflux)(5–14 hr); *B* = NaOMe, MeOH/(reflux)(8 hr); *C* = no solvent/(room temp.)(14–72 hr); *D* = benzene/(room temp.)(14 hr); *E* = benzene/(reflux)(4–16 hr); *F* = benzene/(room temp.)(3 days).

[b] Yield not specified.

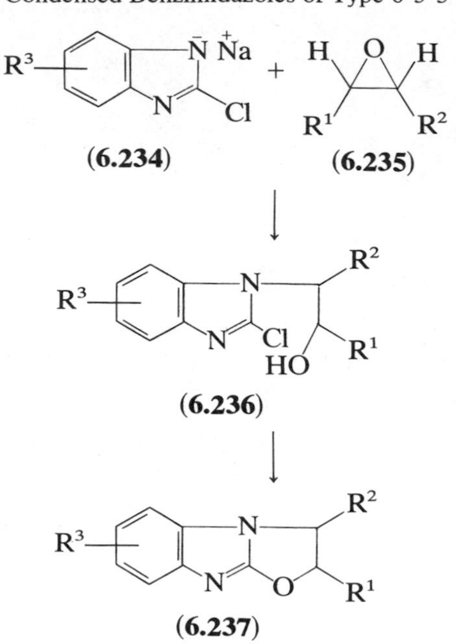

(6.234) (6.235)

(6.236)

(6.237)

Scheme 6.50

sodium salt (**6.234**; R^3 = 5-chloro) with epichlorohydrin (**6.235**; $R^1 = CH_2Cl$, R^2 = H) to give an apparently single 2,3-dihydrooxazolo[3,2-*a*]benzimidazole product (**6.237**; $R^1 = CH_2Cl$, R^2 = H, R^3 = Cl) in good yield (Table 6.26) whose 6- or 7-chloro orientation was not established. The synthesis of 2,3-dihydrooxazolo[3,2-*a*]benzimidazole (**6.237**; $R^1 \rightarrow R^3$ = H) by the sodium hydride catalyzed condensation of 2-benzimidazolone with 1,2-dibromoethane has recently been reported[93] but without experimental details.

Thiazolo[3,2-*a*]benzimidazoles containing alkyl or aryl substituents in the *C*(2) and/or *C*(3) positions [Scheme 6.51; (**6.242**; R^1, R^2 = alkyl or aryl)] are generally accessible, usually in high yield (Table 6.27) by the dehydrative cyclization of 2-(β-oxoalkylthio)benzimidazoles (**6.240**) or the related acetals [**6.240**; $CH(OR)_2$ for COR^1] using a variety of acid catalysts, the most common of which is polyphosphoric acid,[94–99] though hydrochloric acid,[100–102] hydrobromic acid,[100,103] phosphoric acid,[100] sulfuric acid,[100] and even acetic acid, [100,104] have also been employed successfully. The acid-catalyzed nature of the ring-closures [(**6.240**) → (**6.242**)] is demonstrated[100] by the fact that 2-(2-benzimidazolylthio)propanone (**6.240**; R^1 = Me, $R^2 \rightarrow R^6$ = H) is unaffected by heating in ethanol, whereas its hydrochloride on similar treatment is efficiently converted (Table 6.27) into 3-methylthiazolo[3,2-*a*]benzimidazole (**6.242**; R^1 = Me, $R^2 \rightarrow R^6$ = H). Cyclization reactions of the type [(**6.240**) → (**6.242**)] probably involve the intermediate formation of the corresponding 3-hydroxy-2,3-dihydrothia-

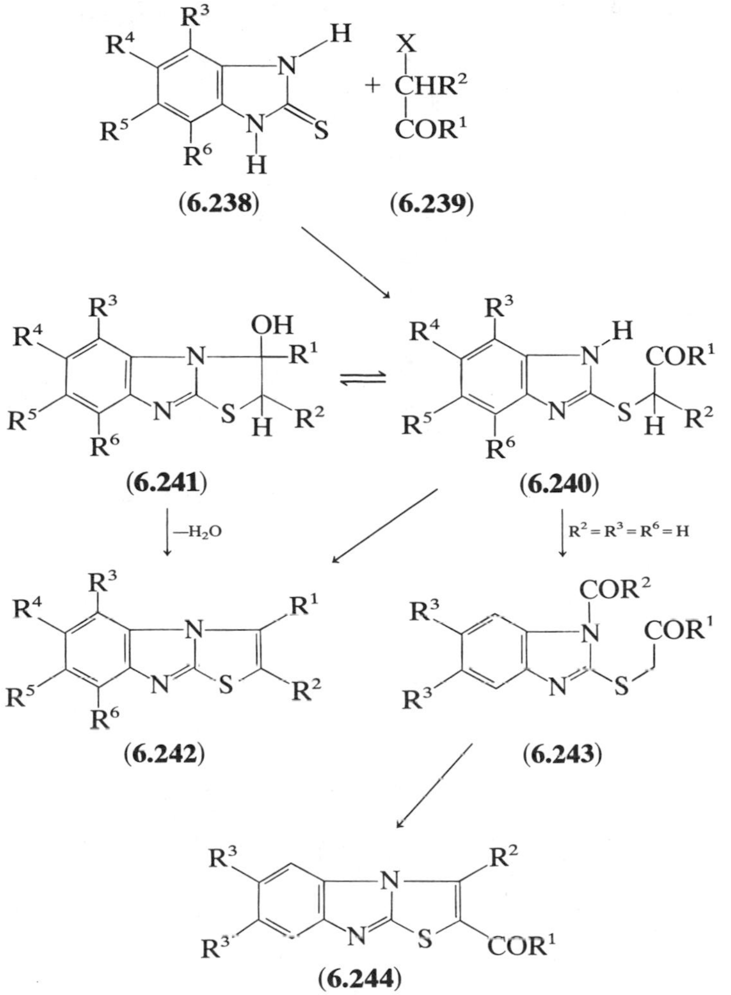

(**6.238**) (**6.239**)

(**6.241**) (**6.240**)

$-H_2O$ $R^2 = R^3 = R^6 = H$

(**6.242**) (**6.243**)

(**6.244**)

Scheme 6.51

zolo[3,2-*a*]benzimidazoles (**6.241**) isolable in certain instances (see later) as the stable tautomeric forms in the ring-chain equilibria [(**6.240**) ⇌ (**6.241**)][94,102,103,105–109] and convertible in high yield (Table 6.27) into thiazolo[3,2-*a*]benzimidazoles (**6.242**) by dehydration with reagents such as phosphorus oxychloride in combination with pyridine,[103,105] polyphosphoric acid,[94,107] hydrochloric acid,[102] or sulfuric acid.[105] The 2-(β-oxoalkylthio)benzimidazoles (**6.240**) or tautomeric 3-hydroxy-2,3-dihydro-thiazolo[3,2-*a*]benzimidazoles (**6.241**) required as substrates for the synthesis of thiazolo[3,2-*a*]benzimidazoles (**6.242**) are readily available by the uncatalyzed condensation of 2-benzimidazolethiones [Scheme 6.51; (**6.238**)] with free α-chloro- or α-bromoaldehydes (**6.239**; $R^1 = H$, R^2 = alkyl or aryl,

TABLE 6.27. SYNTHESIS OF ALKYL AND ARYLTHIAZOLO[3,2-*a*]BENZIMIDAZOLES BY RING-CLOSURE REACTIONS OF BENZIMIDAZOLE DERIVATIVES

Starting materials ($R^1 \rightarrow R^6$ unspecified = H)	Reaction conditions[a]	Product ($R^1 \rightarrow R^6$ unspecified = H)	Yield (%)	m.p. (°C)	Solvent of crystallization	Ref.
(**6.241**)	*A*	(**6.242**)	77	135.5–136.5	Hexane	103
(**6.241**)	*B*	(**6.242**)	94	140	Methanol–water	94
(**6.240**; $CH(OEt)_2$ for COR^1)	*C*	(**6.242**)[b]	—[c]	141.5 142.5	Ethanol–water	106
(**6.241**; $R^4 = R^5 = Me$)	*A*	(**6.242**; $R^4 = R^5 = Me$)	78	164–165	Benzene	103
(**6.241**; $R^3 = R^6 = Me$)	*A*	(**6.242**; $R^3 = R^6 = Me$)	93	134–135	Hexane	103
(**6.238**) + (**6.239**; $R^1 = Me$, $X = Cl$)	*D*	(**6.242**; $R^1 = Me$)[d]	73	161–162	Ethanol–water	100
(**6.238**) + (**6.239**; $R^1 = Me$, $X = Cl$)	*E*	(**6.242**; $R^1 = Me$)	89	164–165	Ethanol	110
(**6.238**) + (**6.239**; $R^1 = Me$, $X = Cl$)	*F*	(**6.242**; $R^1 = Me$)	—[c]	165	Methanol	104
(**6.241**; $R^1 = Me$)	*G*	(**6.242**; $R^1 = Me$)	78–96	161–162	Ethanol–water	100
(**6.241**; $R^1 = Me$)[e]	*H*	(**6.242**; $R^1 = Me$)	95	—	—	100
(**6.241**; $R^1 = Me$)	*I*	(**6.242**; $R^1 = Me$)	quant.	160–161	Hexane	103
(**6.238**)	*J*	(**6.242**; $R^1 = Me$)	35	162–163	Ethanol	101
(**6.247**)	*K*	(**6.249**)	68	165–166	Benzene	120
(**6.247**)[m]	*L*	(**6.249**)	78	161–162	Ethanol	119
(**6.238**; $R^5 = MeO$) + (**6.239**; $R^1 = Me$, $X = Cl$)	*D*	(**6.242**; $R^1 = Me$, $R^4 = MeO$)	66	145	Ethanol	97
(**6.247**; $R^2 = R^3 = Me$)	*K*	(**6.249**; $R^2 = R^3 = Me$)	75	206–207	Benzene	120
(**6.247**; $R^3 = Cl$)	*K*	(**6.249**; R^2 or $R^3 = Cl$)[g]	80	138–139	Benzene	120
(**6.238**; $R^4 = R^5 = Me$) + (**6.239**; $R^1 = Me$, $X = Cl$)	*D*	(**6.242**; $R^1 = R^4 = R^5 = Me$)[h]	98	205–206	Ethanol–water	100

(**6.241**; $R^2 = Me$)	*A*	(**6.242**; $R^2 = Me$)	93	158–159	Hexane	103
(**6.241**; $R^2 = Me$)	*B*	(**6.242**; $R^2 = Me$)	90	158–159	Hexane	115
(**6.245**; $R = Me$)	*M*	(**6.246**; $R = Me$)	78	157–158	Ethanol–water	118
(**6.247**)	*N*	(**6.250**)	—[c]	161–162	—[i]	120
(**6.238**) + (**6.239**; $R^1 = Et$, $X = Br$)	*D*	(**6.242**; $R^1 = Et$)[j]	72	—[k]	—	100
(**6.247**; $R = Me$)	*N*	(**6.250**; $R = Me$)	—[c]	103–105	—[l]	120
(**6.238**) + (**6.239**; $R^1 = Pr^n$, $X = Br$)	*D*	(**6.242**; $R^1 = Pr^n$)[m]	65	76–77	Ethanol–water	100
(**6.245**; $R = Bu^t$)	*M*	(**6.246**; $R = Bu^t$)	99	131–132	Ethanol–water	118
(**6.247**; $R^1 = Ph$)	*L*	(**6.249**; $R^1 = Ph$)	80	80–81	Methanol	120
(**6.247**; $R = H$, $R^1 = Ph$, $R^2 = R^3 = Me$)	*L*	(**6.249**; $R^1 = Ph$, $R^2 = R^3 = Me$)	82	201–202	Methanol	120
(**6.247**; $R^1 = Ph$, $R^3 = Cl$)	*L*	(**6.249**; $R^1 = Ph$, R^2 or $R^3 = Cl$)[g]	85	124–125	Methanol	120
(**6.238**) + (**6.239**; $R^1 = Ph$, $X = Br$)	*O*	(**6.242**; $R^1 = Ph$)	40	139.5 140.5	Dioxane–water	100
(**6.240**; $R^1 = Ph$)	*P*	(**6.242**; $R^1 = Ph$)[n]	94	—	—	100
(**6.240**; $R^1 = Ph$)	*Q*	(**6.242**; $R^1 = Ph$)	43	140–142	Ethanol	103
(**6.240**; $R^1 = Ph$)	*B*	(**6.242**; $R^1 = Ph$)	48	140	Ethanol	95
(**6.240**; $R^1 = Ph$)	*B*	(**6.242**; $R^1 = Ph$)	85	140	Ethanol	94
(**6.238**)	*R*	(**6.242**; $R^1 = Ph$)	50–60	139	Benzene–hexane	122
(**6.240**; $R^1 = p\text{-}ClC_6H_4$)	*B*	(**6.242**; $R^1 = p\text{-}ClC_6H_4$)	77	203	Ethanol	94
(**6.240**; $R^1 = p\text{-}ClC_6H_4$)	*B*	(**6.242**; $R^1 = p\text{-}ClC_6H_4$)	45	200	Ethanol–water	95
(**6.240**; $R^1 = p\text{-}BrC_6H_4$)	*B*	(**6.242**; $R^1 = p\text{-}BrC_5H_4$)	72	197	Ethanol	94
(**6.240**; $R^1 = p\text{-}BrC_6H_4$)	*B*	(**6.242**; $R^1 = p\text{-}BrC_5H_4$)	46	200	Ethanol–water	95
(**6.240**; $R^1 = p\text{-}BrC_6H_4$)	*P*	(**6.242**; $R^1 = p\text{-}BrC_6H_4$)[o]	97	200.5–201	1-Butanol	100
(**6.240**; $R^1 = p\text{-}FC_6H_4$)	*B*	(**6.242**; $R^1 = p\text{-}FC_6H_4$)	44	274	Chloroform–light petroleum	99
(**6.240**; $R^1 = p\text{-}MeC_6H_4$)	*B*	(**6.242**; $R^1 = p\text{-}MeC_6H_4$)	46	150	Ethanol–water	95
(**6.240**; $R^1 = p\text{-}MeOC_6H_4$)	*B*	(**6.242**; $R^1 = p\text{-}MeOC_6H_4$)	44	155	Ethanol–water	95
(**6.240**; $R^1 = m\text{-}NO_2C_6H_4$)	*P*	(**6.242**; $R^1 = m\text{-}NO_2C_6H_4$)[p]	87	196.5–197	Dimethylformamide–water	100

TABLE 6.27 (*Continued*)

Starting materials ($R^1 \rightarrow R^6$ unspecified = H)	Reaction conditions[a]	Product ($R^1 \rightarrow R^6$ unspecified = H)	Yield (%)	m.p. (°C)	Solvent of crystallisation	Ref.
(**6.240**; $R^1 = p\text{-}NO_2C_6H_4$)	*P*	(**6.242**; $R^1 = p\text{-}NO_2C_6H_4$)[q]	89	230–231	1-Butanol	100
(**6.240**; $R^1 = 4\text{-F, 3-}MeC_6H_3$)	*B*	(**6.242**; $R^1 = 4\text{-F, 3-}MeC_6H_3$)	42	170	Chloroform–light petroleum	99
(**6.240**; $R^1 = 3\text{-Cl, 4-}FC_6H_4$)	*B*	(**6.242**; $R^1 = 3\text{-Cl, 4-}FC_6H_4$)	54	230–234	Chloroform–light petroleum	99
(**6.240**; $R^1 = Ph$, $R^5 = MeO$)	*B*	(**6.242**; $R^1 = Ph$, $R^4 = MeO$)	52	155	—[i]	97
(**6.240**; $R^1 = p\text{-}MeC_6H_4$, $R^5 = MeO$)	*B*	(**6.242**; $R^1 = p\text{-}MeC_6H_4$, $R^4 = MeO$)	56	140	Ethanol	97
(**6.240**; $R^1 = p\text{-}PhC_6H_4$, $R^5 = MeO$)	*B*	(**6.242**; $R^1 = p\text{-}PhC_6H_4$, $R^4 = MeO$)	46	185	—[i]	97
(**6.240**; $R^1 = p\text{-}BrC_6H_4$, $R^5 = MeO$)	*B*	(**6.242**; $R^1 = p\text{-}BrC_6H_4$, $R^4 = MeO$)	42	175	—[i]	97
(**6.240**; $R^1 = Ph$, $R^5 = Cl$)	*B*	(**6.242**; $R^1 = Ph$, $R^4 = Cl$)	51	212	Ethanol	98
(**6.240**; $R^1 = p\text{-}MeC_6H_4$, $R^5 = Cl$)	*B*	(**6.242**; $R^1 = P\text{-}MeC_6H_4$, $R^4 = Cl$)	—[c]	184	—[i]	98
(**6.240**; $R^1 = p\text{-}BrC_6H_4$, $R^5 = Cl$)	*B*	(**6.242**; $R^1 = BrC_6H_4$, $R^4 = Cl$)	—[c]	210	—[i]	98
(**6.240**; $R^1 = p\text{-}PhC_6H_4$, $R^5 = Cl$)	*B*	(**6.242**; $R^1 = p\text{-}PhC_6H_4$, $R^4 = Cl$)	—[c]	150	—[i]	98
(**6.240**; $R^1 = p\text{-}BrC_6H_4$, $R^5 = Me$)	*B*	(**6.242**; $R^1 = p\text{-}BrC_6H_4$, R^4 or $R^5 = Me$)[g]	64	240	Ethanol	96
(**6.240**; $R^1 = p\text{-}MeC_6H_4$, $R^6 = Me$)	*B*	(**6.242**; $R^1 = p\text{-}BrC_6H_4$, R^3 or $R^6 = Me$)[g]	53	231–232	Ethanol	96
(**6.238**; $R^4 = R^5 = Me$ + (**6.239**; $R^1 = Ph$, $X = Br$)	*O*	(**6.242**; $R^1 = Ph$, $R^4 = R^5 = Me$)[r]	40	192–193	Ethanol–water	100
(**6.240**; $R^1 = Ph$, $R^4 = R^5 = Me$)	*B*	(**6.242**; $R^1 = Ph$, $R^4 = R^5 = Me$)	66	189–190	Ethanol	96
(**6.238**; $R^4 = R^5 = Me$)	*R*	(**6.242**; $R^1 = Ph$, $R^4 = R^5 = Me$)	72	192	Acetone–light petroleum (b.p. 40–80°)	122
(**6.240**; $R^1 = p\text{-}BrC_6H_4$, $R^4 = R^5 = Me$)	*B*	(**6.242**; $R^1 = p\text{-}BrC_6H_4$, $R^4 = R^5 = Me$)	71	276	Chloroform–light petroleum (b.p. 40–60°)	96
(**6.240**; $R^1 = m\text{-}BrC_6H_4$, $R^4 = R^5 = Me$)	*P*	(**6.242**; $R^1 = m\text{-}BrC_6H_4$, $R^4 = R^5 = Me$)	98	265–266	Dimethylformamide	100
(**6.240**; $R^1 = m\text{-}NO_2C_6H_4$, $R^4 = R^5 = Me$)	*P*	(**6.242**; $R^1 = m\text{-}NO_2C_6H_4$, $R^4 = R^5 = Me$)	60	305–306	Dimethylformamide	100
(**6.241**; $R^2 = Ph$)	*S*	(**6.242**; $R^2 = Ph$)	90	168–170	Ethanol–water	107
(**6.245**; $R = Ph$)	*M*	(**6.246**; $R = Ph$)	95	166–167	Ethanol	118

(**6.245**; R = Ph)	*I*	(**6.246**; R = Ph)	58	—	—	118
(**6.245**; R = Ph)	*Q*	(**6.246**; R = Ph)	99	—	—	118
(**6.245**; R = p-ClC_6H_4)	*M*	(**6.246**; R = p-ClC_6H_4)	93	225–226	Ethanol	118
(**6.245**; R = p-BrC_6H_4)	*M*	(**6.246**; R = p-BrC_6H_4)	98	222–223	Ethanol	118
(**6.245**; R = p-MeC_6H_4)	*M*	(**6.246**; R = p-MeC_6H_4)	99	178–179	Ethanol	118
(**6.245**; R = p-$MeOC_6H_4$)	*M*	(**6.246**; R = p-$MeOC_6H_4$)	89	162–163	Ethanol	118
(**6.245**; R = p-PhC_6H_4)	*M*	(**6.246**; R = p-PhC_6H_4)	99	205–206	Ethanol	118
(**6.245**; R = 2-naphthyl)	*M*	(**6.246**; R = 2-naphthyl)	93	228–229	Ethanol	118
(**6.245**; R = 2-thienyl)	*M*	(**6.246**; R = 2-thienyl)	97	150–151	Ethanol	118
(**6.238**) + (**6.239**; $R^1 = R^2$ = Me, X = Br)	*D*	(**6.242**; $R^1 = R^2$ = Me)[f]	64	273–275	—[i]	111
(**6.238**) + (**6.239**; $R^1 = R^2$ = Me, X = Br)	*D*	(**6.242**; $R^1 = R^2$ = Me)[s]	42	151–152	Ethanol–water	100
(**6.241**; $R^1 = R^2$ = Me)	*T*	(**6.242**; $R^1 = R^2$ = Me)	89	—	—	100
(**6.241**; $R^1 = R^2$ = Me)	*U*	(**6.242**; $R^1 = R^2$ = Me)	99	—	—	100
(**6.241**; $R^1 = R^2$ = Me)	*V*	(**6.242**; $R^1 = R^2$ = Me)	97	—	—	100
(**6.241**; $R^1 = R^2$ = Me)	*I*	(**6.242**; $R^1 = R^2$ = Me)	95	154	Ethanol	101
(**6.238**; $R^4 = R^5$ = Me) + (**6.239**; $R^5 = R^6$ = Me, X = Br)	*D*	(**6.242**; $R^1 = R^2 = R^4 = R^5$ = Me)[t]	88	162–164	Ethanol–water	100
(**6.241**; R^1 = Et, R^2 = Me)	*P*	(**6.242**; R^1 = Et, R^2 = Me)[u]	93	127–128	Ethanol–water	100
(**6.238**) + (**6.239**; R^1 = Me, R^2 = Et, X = Br)	*D*	(**6.242**; R^1 = Me, R^2 = Et)[v]	93	107–108	Ethanol–water	100
(**6.238**) + (**6.239**; R^1 = Me, R^2 = Pr^n, X = Cl)	*D*	(**6.242**; R^1 = Me, R^2 = Pr^n)[w]	76	—[x]	—	100
(**6.241**; R^1 = Me, R^2 = CH_2Ph)	*P*	(**6.242**; R^1 = Me, R^2 = CH_2Ph)	79	122–123	Ethanol–water	100
(**6.240**; R^1 = Me, R^2 = Ph)	*P*	(**6.242**; R^1 = Me, R^2 = Ph)[y]	91	131–132	Ethanol–water	100
(**6.238**) + (**6.239**; R^1 = Ph, R^2 = Me, X = Br)	*W*	(**6.242**; R^1 = Ph, R^2 = Me)	20	179–180	Ethanol–water	100

TABLE 6.27 (*Continued*)

Starting materials ($R^1 \rightarrow R^6$ unspecified = H)	Reaction conditions[a]	Product ($R^1 \rightarrow R^6$ unspecified = H)	Yield (%)	m.p. (°C)	Solvent of crystallization	Ref.
(**6.241**; $R^1 = Ph$, $R^2 = Me$)	*P*	(**6.242**; $R^1 = Ph$, $R^2 = Me$)	92	—	—	100
(**6.241**; $R^1 = R^2 = Ph$)	*P*	(**6.242**; $R^1 = R^2 = Ph$)[z]	98	193–194	1-Butanol	100
(**6.238**; $R^4 = R^5 = Me$) + (**6.239**; $R^1 = R^2 = Ph$, X = Cl)	*O*	(**6.242**; $R^1 = R^2 = Ph$, $R^4 = R^5 = Me$)	51	225–226	Methanol	100
(**6.241**; $R^1 = p\text{-}ClC_6H_4$, $R^2 = CH_2CO_2H$)	*S*	(**6.242**; $R^1 = p\text{-}ClC_6H_4$, $R^2 = CH_2CO_2H$)	23	242–243	Dimethylformamide	102
(**6.238**) + (**6.239**; $R^1 = CH_2P(O)(OEt)_2$)	*D*	(**6.242**; $R^1 = CH_2P(O)(OEt)_2$)[f]	85	134	—[i]	113
(**6.240**; CN for COR^1)	*X*	(**6.242**; $R^1 = R^2 = CO_2Me$)	30	166–167	Benzene–hexane	121

[a] *A* = $POCl_3$, pyridine/(room temp.)(1 hr); *B* = polyphosphoric acid/(150–170°)(2–4 hr); *C* = conc. H_2SO_4/(30°)(15 min, then 18–20°)(3 hr); *D* = EtOH/(reflux)(3–6 hr); *E* = melt/3–5 min; *F* = NaOEt, EtOH/(50–60°)(1 hr, then AcOH)(reflux)(4 hr); *G* = conc. HCl, EtOH/(reflux)(4 hr); *H* = H_2O/reflux (reaction time not specified); *I* = 4–6% HCl/(reflux)(4–15 hr); *J* = acetone, I_2/(28 hr)(reflux); *K* = $Hg(OAc)_2$, AcOH, conc. H_2SO_4/(reflux)(5 hr); *L* = NaOEt, EtOH/(reflux)(3 hr); *M* = $POCl_3$/(reflux)(0.5–1 hr); *N* = HMPT/(reflux)(15–40 min); *O* = hexanol/(reflux)(1–22 hr); *P* = $POCl_3$/(reflux)(1–22 hr); *Q* = conc. HBr/(reflux)(2–5 hr); *R* = $Hg(C{\equiv}CPh)_2$, PhN=C=S, pyridine/(reflux)(6 hr); *S* = 4 M HCl/(reflux)(13–20 hr); *T* = conc. H_2SO_4/(18–22°)(32 hr); *U* = 85–89% H_3PO_4/(95–100°)(1 hr); *V* = AcOH/(reflux)(4 hr); *W* = Dimethylformamide/(reflux)(5 hr); X = $MeO_2CC{\equiv}CCO_2Me$, tetrahydrofurane/(reflux)(30 hr).

[b] Forms a hydrochloride, m.p. 193–194° (decomp.) (from ethanol), and a picrate, yellow needles, m.p. 235–236° (decomp.) (from ethanol).

[c] Yield not quoted.

[d] Forms a picrate, m.p. 241–242°.

[e] Hydrochloride.

[f] Hydrobromide.

[g] Single isomer formed; *C*(6) or *C*(7) orientation for the substituent not determined.

[h] Forms a picrate, m.p. 285–286°, and a hydrochloride, m.p. 263–266°.

[i] Solvent of crystallization not specified.

[j] Forms a hydrochloride, m.p. 221–222°.

[k] Oil, b.p. 227–228/760 mm Hg.

[l] Purified by tlc.

[m] Forms a picrate, m.p. 248–250°, and a hydrochloride, m.p. 226–228°.

[n] Forms a picrate, m.p. 242–243°.
[o] Forms a picrate, m.p. 264–266°.
[p] Forms a picrate, m.p. 244–245°.
[q] Forms a picrate, m.p. 268–270°.
[r] Forms a picrate, m.p. 263–264°.
[s] Forms a picrate, m.p. 233–234°, and a hydrochloride, m.p. 267–268°.
[t] Forms a picrate, m.p. 259–260°, and a hydrochloride, m.p. 308–309°.
[u] Forms a picrate, m.p. 127–128°.
[v] Forms a picrate, m.p. 224–226°, and a hydrochloride, m.p. 227–229°.
[w] Forms a picrate, m.p. 212–213°.
[x] Oil, b.p. 237–238°/760 mm Hg.
[y] Forms a picrate, m.p. 232–232°.
[z] Forms a picrate, m.p. 215–216°.

X = Cl or Br)[103,106] or their acetals [**6.239**; R^2 = alkyl or aryl, X = Cl or Br, $CH(OAlk)_2$ for COR^1][94,103,105–107] or α-chloro or α-bromo alkyl or aryl ketones (**6.239**; R^1, R^2 = alkyl or aryl, X = Cl or Br),[100,102,103,108,109] in aqueous (or in the case of acetals, aqueous acidic) media[102,105,106,109] or in solvents such as alcohols (methanol, ethanol, hexanol)[100,105,106,109] 2-butanone,[103,108] dimethoxyethane,[109] dimethylformamide,[105,106] or acetic acid,[109] with or without the application of heat. Preferential condensation of the α-halogeno carbonyl compound at sulfur rather than at nitrogen in the benzimidazolethione to give the *S*-alkyl derivative (**6.240**) and not an *N*-alkylated isomer is demonstrated by the structures of the derived thiazolo[3,2-*a*]benzimidazoles whose orientation is established unequivocally by mass spectroscopy.[107] It follows that ring-closure reactions of the types [(**6.240**) → (**6.241**) or (**6.242**)] are unambiguous in relation to the position of substituents in the thiazole ring of the products. However, this is not the case for substituents in the benzene ring. Cyclization of a 2-(β-oxoalkylthio)benzimidazole (**6.240**) unsymmetrically substituted in the benzene nucleus can, depending on which of the nonequivalent benzimidazole nitrogen atoms is involved in the ring-closure, lead to two possible isomeric products. In all such cyclizations[96,98] reported to date one isomer tends to predominate to the virtual exclusion of the other. However, there has been no systematic study of the orientational preference shown by substituents in cyclizations of this type, largely because of the difficulty of establishing with certainty the site of substituents in the cyclic product. More success in this direction has been achieved with 2,3-dihydrothiazolo[3,2-*a*]benzimidazol-3-one derivatives (see later) and the orientation of similarly unsymmetrically substituted thiazolo[3,2-*a*]benzimidazoles has tended to be assigned, somewhat unwisely, by analogy.[97,98] Within this limitation, it has been shown[97,98] that 5(6)-chloro- or methoxy-substituted 2′-(2-benzimidazolylthio)acetophenones (**6.240**; R^1 = aryl, $R^2 = R^3 = R^6$ = H, R^4, R^5 = Cl or MeO) tend to ring-close preferentially at the benzimidazole nitrogen atom *meta* (rather than *para*) to the substituent, the products being the corresponding 3-aryl-6-chloro- or methoxythiazolo[3,2-*a*]benzimidazoles (**6.242**; R^1 = aryl, $R^2 = R^3 = R^5 = R^6$ = H, R^4 = Cl or MeO). This orientational preference is the opposite to that observed in closely analogous alkylative ring-closures leading to 2,3-dihydro-1*H*-pyrrolo[1,2-*a*]benzimidazoles and if correct has the surprising consequence that the acylative ring-closure involved takes place at the predictably less basic of the two available benzimidazole nitrogen centers. In view of these apparent anomalies, the orientational preference involved in cyclizations of the type [Scheme 6.51; (**6.240**) → (**6.241**) → (**6.242**)] warrants more detailed scrutiny.

In some instances, thiazolo[3,2-*a*]benzimidazoles are reported to be formed in high yield (Table 6.27) without the need to isolate intermediates of the types (**6.240**) and (**6.241**) by the direct condensation of a 2-benzimidazolethione (**6.238**) with an α-halogeno carbonyl compound (**6.239**) in the melt[110] or in hot ethanolic solution.[97,100,104,111–114] Thiazolo[3,2-*a*]benzimidazole formation can be achieved even more directly (albeit in only moderate yield, cf. Table 6.27) by condensation of a 2-benzimidazolethione with an α-iodo ketone generated *in situ* by reaction of the corresponding alkyl ketone with iodine (the Ortoleva–King reaction).[101] In this way 2-benzimidazolethione (**6.238**; $R^3 \rightarrow R^6 = H$) reacts with acetone in the presence of iodine to give 3-methylthiazolo[3,2-*a*]benzimidazole (**6.242**; $R^1 = Me$, $R^2 \rightarrow R^6 = H$) in 35% yield.[101] The use of β-dicarbonyl compounds (e.g., ethyl acetoacetate) in this modified procedure provides synthetic access to thiazolo[3,2-*a*]benzimidazoles having an acyl substituent at *C*(2) [e.g. (**6.242**; $R^2 = CO_2Et$)] though again only in low yield (Table 6.28).[101] On the other hand, 2-acetyl and 2-ethoxycarbonylthiazolo-[3,2-*a*]benzimidazoles (**6.242**; $R^2 = COMe$ or CO_2Et) are formed in high yield (Table 6.28) by the more orthodox cyclization of 3-(2-benzimidazolylthio)pentane-2,4-diones (**6.240**; $R^1 = Me$, $R^2 = COMe$)[100,101,103,115] or ethyl 2-(2-benzimidazolylthio)acetoacetates (**6.240**; $R^1 = Me$, $R^2 = CO_2Et$)[115] catalysed by hydrochloric acid or acetic anhydride in the presence of pyridine. The facility of this type of cyclization is highlighted by the observation[116] that 2-(2-benzimidazolylthio)-1-phenylbutane-1,3-dione (**6.240**; $R^1 = Me$, $R^2 = COPh$) is partly converted into the isomeric 2-acylthiazolo[3,2-*a*]benzimidazoles (**6.242**; $R^1 = Me$, $R^2 = COPh$) and (**6.242**; $R^1 = Ph$, $R^2 = COMe$) merely on standing in chloroform. However, undoubtedly the most convenient general method for the synthesis of 2-acylated thiazolo[3,2-*a*]benzimidazoles is provided by the thermal cyclization of *N*(1)-acyl-2-(β-oxoalkylthio)benzimidazoles either preformed or generated *in situ* by the reaction of suitable 2-(β-oxoalkylthio)benzimidazoles with the sodium salts of carboxylic acids (formic, acetic, propionic) in the presence of acetic or propionic anhydride [Scheme 6.51; (**6.240**; $R^2 = R^3 = R^6 = H$) $\rightarrow$ (**6.243**) $\rightarrow$ (**6.244**)].[117] Ring-closure of this type gives uniformly high yields (Table 6.28) of both 2-alkanoyl and 2-aroylthiazolo[3,2-*a*]benzimidazoles and is made flexible by the ready availability of the 2-(β-oxoalkylthio)benzimidazoles (**6.240**; $R^2 = R^3 = R^6 = H$) required as starting materials. The conversion[116] of 2′-(2-benzimidazolylthio)acetophenone (**6.240**; $R^1 = Ph$, $R^2 \rightarrow R^6 = H$) into 2-benzoyl-3-phenylthiazolo[3,2-*a*]benzimidazole (**6.244**; $R^1 = R^2 = Ph$, $R^3 = H$) is explicable in terms of the intermediate formation and dehydrative cyclization of the *N*-benzoyl intermediate (**6.243**; $R^1 = R^2 = Ph$, $R^3 = H$) and hence exemplifies a further version of ring-closure of the general type [Scheme 6.52; (**6.243**) $\rightarrow$ (**6.244**)].

An alternative synthetic route to thiazolo[3,2-*a*]benzimidazoles, which complements that through 2-(β-oxoalkylthio)benzimidazoles (see before) in giving high yields (Table 6.27) of 2-alkyl and most notably 2-aryl derivatives, involves the ring-closure of *N*(1)-(β-oxoalkyl)-2-benzimidazolethiones [Scheme 6.52; (**6.245**) $\rightarrow$ (**6.246**)].[118] Cyclization of this type is readily effected by heating with phosphorus oxychloride, or with hydrobromic or sulfuric acids, and derives its synthetic utility from the fact that it provides synthetic access to 2-arylthiazolo[3,2-*a*]benzimidazoles, products not readily available through 2-(β-oxoalkylthio)benzimidazoles because of the relatively inaccessible nature of α-halogeno arylacetaldehydes. On the other

TABLE 6.28. SYNTHESIS OF ACYLTHIAZOLO[3,2-*a*]BENZIMIDAZOLES BY RING-CLOSURE REACTIONS OF BENZIMIDAZOLE DERIVATIVES

Starting material ($R^1 \rightarrow R^6$ unspecified = H)	Reaction conditions[a]	Product ($R^1 \rightarrow R^3$ unspecified = H)	Yield (%)	m.p. (°C)	Solvent of crystallization	Ref.
(**6.240**; $R^1 = Me$, $R^2 = CO_2Et$)	*A*	(**6.244**; $R^1 = OEt$, $R^2 = Me$)	97	122–123	Ethanol	115
(**6.238**)	*B*	(**6.244**; $R^1 = OEt$, $R^2 = Me$)	15	123	Ethanol	101
(**6.240**; $R^1 = Me$, $R^2 = COMe$)	*A*	(**6.244**; $R^1 = R^2 = Me$)	99	167–168	Ethanol	115
(**6.240**; $R^1 = Me$, $R^2 = COMe$)	*A*	(**6.244**; $R^1 = R^2 = Me$)	quant.	163–165	Ethanol	103
(**6.240**; $R^1 = Me$, $R^2 = COMe$)	*C*	(**6.244**; $R^1 = R^2 = Me$)	89	—	—	103
(**6.240**; $R^1 = Me$, $R^2 = COMe$)	*C*	(**6.244**; $R^1 = R^2 = Me$)	95	165–166	Ethanol	101
(**6.240**; $R^1 = Me$, $R^2 = COMe$)	*D*	(**6.244**; $R^1 = R^2 = Me$)	95	163–164	Ethanol	100
(**6.240**; $R^1 = Ph$)	*E, F*	(**6.244**; $R^1 = Ph$)	90–93	163–164	Ethanol–water	117
(**6.243**; $R^1 = Ph$)	*G*	(**6.244**; $R^1 = Ph$)	90–93	—	—	117
(**6.240**; $R^1 = p\text{-}BrC_6H_4$)	*E, F*	(**6.244**; $R^1 = p\text{-}BrC_6H_4$)	93–98	227–228	Dioxane–water	117
(**6.243**; $R^1 = p\text{-}BrC_6H_4$)	*G*	(**6.244**; $R^1 = p\text{-}BrC_6H_4$)	93–98	—	—	117
(**6.240**; $R^1 = p\text{-}NO_2C_6H_4$)	*E*	(**6.244**; $R^1 = p\text{-}NO_2C_6H_4$)	90–96	277–278	Dioxane	117
(**6.243**; $R^1 = p\text{-}NO_2C_6H_4$)	*G*	(**6.244**; $R^1 = p\text{-}NO_2C_6H_4$)	90–96	—	—	117
(**6.240**; $R^1 = Ph$, $R^4 = R^5 = Me$)	*E*	(**6.244**; $R^1 = Ph$, $R^3 = Me$)	90	186–187	Ethanol	117
(**6.243**; $R^1 = Ph$, $R^3 = Me$)	*G*	(**6.244**; $R^1 = Ph$, $R^3 = Me$)	90	—	—	117
(**6.240**; $R^1 = p\text{-}BrC_6H_4$, $R^4 = R^5 = Me$)	*E*	(**6.244**; $R^1 = p\text{-}BrC_6H_4$, $R^3 = Me$)	91–97	255–256	1-Butanol	117
(**6.243**; $R^1 = p\text{-}BrC_6H_4$, $R^3 = Me$)	*G*	(**6.244**; $R^1 = p\text{-}BrC_6H_4$, $R^3 = Me$)	91–97	—	—	117
(**6.240**; $R^1 = p\text{-}NO_2C_6H_4$, $R^4 = R^5 = Me$)	*E*	(**6.244**; $R^1 = p\text{-}NO_2C_6H_4$, $R^3 = Me$)	91	279–280	1-Butanol	117

(**6.243**; $R^1 = p\text{-}NO_2C_6H_4$, $R^1 = R^3 = Me$)	*G*	(**6.244**; $R^1 = p\text{-}NO_2C_6H_4$, $R^3 = Me$)	91	—	—	117
(**6.240**)	*H*	(**6.244**; $R^2 = Me$)[b]	81	227–228	Ethanol	117
(**6.243**; $R^2 = Me$)	*G*	(**6.244**, $R^2 = Me$)	81	—	—	117
(**6.240**; $R^1 = Me$)	*H*	(**6.244**; $R^1 = R^2 = Me$)	96	163–164	Ethanol–water	117
(**6.243**; $R^1 = R^2 = Me$)	*G*	(**6.244**; $R^1 = R^2 = Me$)	96	—	—	117
(**6.240**; $R^1 = p\text{-}BrC_6H_4$)	*H*	(**6.244**; $R^1 = p\text{-}BrC_6H_4$, $R^2 = Me$)	86	217–218	Ethanol-1-butanol	117
(**6.243**; $R^1 = p\text{-}BrC_6H_4$, $R^2 = Me$)	*G*	(**6.244**; $R^1 = p\text{-}BrC_6H_4$, $R^2 = Me$)	86	—	—	117
(**6.240**; $R^1 = R^4 = R^5 = Me$)	*H*	(**6.244**; $R^1 = R^2 = R^3 = Me$)	85	229–230	Ethanol	117
(**6.243**; $R^1 = R^2 = R^3 = Me$)	*G*	(**6.244**; $R^1 = R^2 = R^3 = Me$)	85	—	—	117
(**6.240**; $R^1 = p\text{-}BrC_6H_4$)	*H*	(**6.244**; $R^1 = p\text{-}C_6H_4$, $R^2 = Et$)	83–85	160–161	Methanol	117
(**6.243**; $R^1 = p\text{-}BrC_6H_4$, $R^2 = Et$)	*G*	(**6.244**; $R^1 = p\text{-}BrC_6H_4$, $R^2 = Et$)	83–85	—	—	117
(**6.240**; $R^1 = R^4 = R^5 = Me$)	*H*	(**6.244**; $R^1 = R^3 = Me$, $R^2 = Et$)	73	132–133	Methanol	117
(**6.243**; $R^1 = R^3 = Me$, $R^2 = Et$)	*G*	(**6.244**; $R^1 = R^3 = Me$, $R^2 = Et$)	73	—	—	117
(**6.240**; $R^1 = Bu^t$, $R^4 = R^5 = Me$)	*H*	(**6.244**; $R^1 = Bu^t$, $R^2 = Et$, $R^3 = Me$)	80	161–162	Ethanol	117
(**6.243**; $R^1 = Bu^t$, $R^2 = Et$, $R^3 = Me$)	*G*	(**6.244**; $R^1 = Bu^t$, $R^2 = Et$, $R^3 = Me$)	80	—	—	117
(**6.240**; $R^1 = Ph$, $R^4 = R^5 = Me$)	*H*	(**6.244**; $R^1 = Ph$, $R^2 = Et$, $R^3 = Me$)	90	187–188	Ethanol	117
(**6.243**; $R^1 = Ph$, $R^2 = Et$, $R^3 = Me$)	*G*	(**6.244**; $R^1 = Ph$, $R^2 = Et$, $R^3 = Me$)	90	—	—	117

[a] *A* = Ac_2O, pyridine/(90–100°)(3 hr); *B* = $CH_3COCH_2CO_2Et$, I_2/(reflux)(28 hr); *C* = 5–6% HCl/(reflux)(3–5 hr); *D* = $POCl_3$/reflux (time not specified); *E* = HCO_2Na, 85% HCO_2H, Ac_2O/(reflux)(1–2 hr); *F* = 85% HCO_2H, Ac_2O with or without dimethylformamide/(150–155°, autoclave)(2–6 hr); *G* = sodium salt of the corresponding acid, Ac_2O or $(EtCO)_2O$/(reflux)(0.5–2 hr); *H* = NaOAc or NaOCOEt, Ac_2O or $(EtCO)_2O$/(reflux or 120–160°)(0.5–2 hr).

[b] Forms a *p*-nitrophenylhydrazone, m.p. 250–251° (decomp.) (from acetic acid).

hand, the *N*(1)-(β-oxoalkyl)-2-benzimidazolethiones required as substrates for ring-closure of the type [(**6.245**)→(**6.246**)] are readily prepared from 2-chlorobenzimidazoles by *N*-alkylation with an α-halogeno carbonyl compound followed by treatment with thiourea.[118]

(**6.245**) (**6.246**)

Scheme 6.52

2- and 3-Alkylthiazolo[3,2-*a*]benzimidazoles are also the end-products of cyclization reactions undergone by 2-(2-propynylthio)benzimidazoles (Scheme 6.53). Thiazolo[3,2-*a*]benzimidazole formation of this type proceeds in uniformly high yield (Table 6.27) and is accomplished catalytically using ethanolic sodium ethoxide[119,120] or mercuric acetate in acetic acid,[120] or purely thermally by heating under reflux in hexamethylphosphoric triamide (HMPT).[121] However, the course followed in the purely thermal process differs markedly from that involved in the catalytically mediated ring-closures. In the latter, cyclization is the result of direct interaction between a benzimidazole nitrogen atom and the triple bond in the propynylthio side chain, the product being the corresponding 3-alkylthiazolo[3,2-*a*]benzimidazole [Scheme 6.53; (**6.247**; R = H)→(**6.249**)].[119,120a] In the thermally induced process, on the other hand, ring formation is preceded by [3,3] sigmatropic shift, the resulting *N*(1)-allenyl-2-benzimidazolethione intermediate then cyclizing to a 2-alkylthiazolo[3,2-*a*]benzimidazole [Scheme 6.53; (**6.247**; $R^1 = R^2 = R^3 = H$, R = H or Me)→(**6.248**; R = H or Me)→(**6.250**; R = H or Me)].[120b] The thiazolo[3,2-*a*]benzimidazole diester (**6.242**; $R^1 = R^2 = CO_2Me$, $R^3 \rightarrow R^6 = H$) is the product formed in low yield

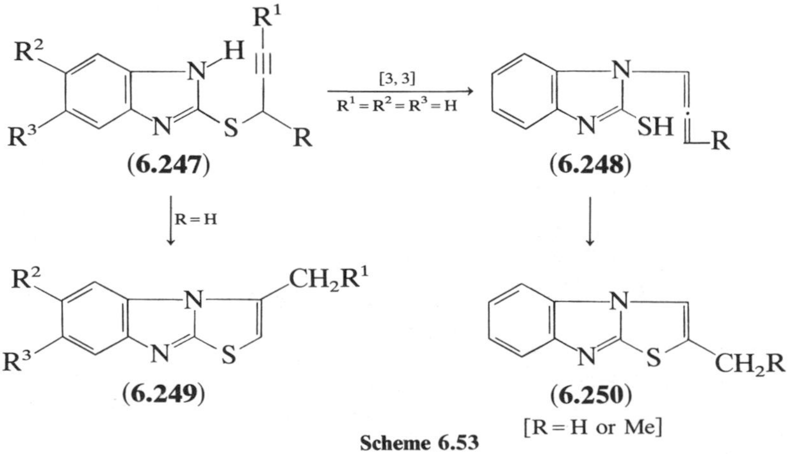

(**6.247**) (**6.248**) (**6.249**) (**6.250**)

[R = H or Me]

Scheme 6.53

(Table 6.27) by the somewhat unusual reaction of (2-benzimidazolylthio)acetonitrile (**6.241**; $R^2 \rightarrow R^6 = H$, CN for COR^1) with dimethyl acetylenedicarboxylate.[121] The course followed in this transformation is not clear. 3-Phenylthiazolo[3,2-*a*]benzimidazoles are also formed in unorthodox, but nonetheless efficient, fashion (Table 6.27) by the reaction of 2-benzimidazolethiones with mercury bisphenylacetylide in the presence of phenylisothiocyanate.[122] These deep-seated transformations are accounted for[122] by a course (Scheme 6.54) involving the formation of, and extrusion of mercury from an eight-membered cyclic intermediate (**6.251**) followed by ring-contraction of the fused dithiazepine (**6.252**) produced.

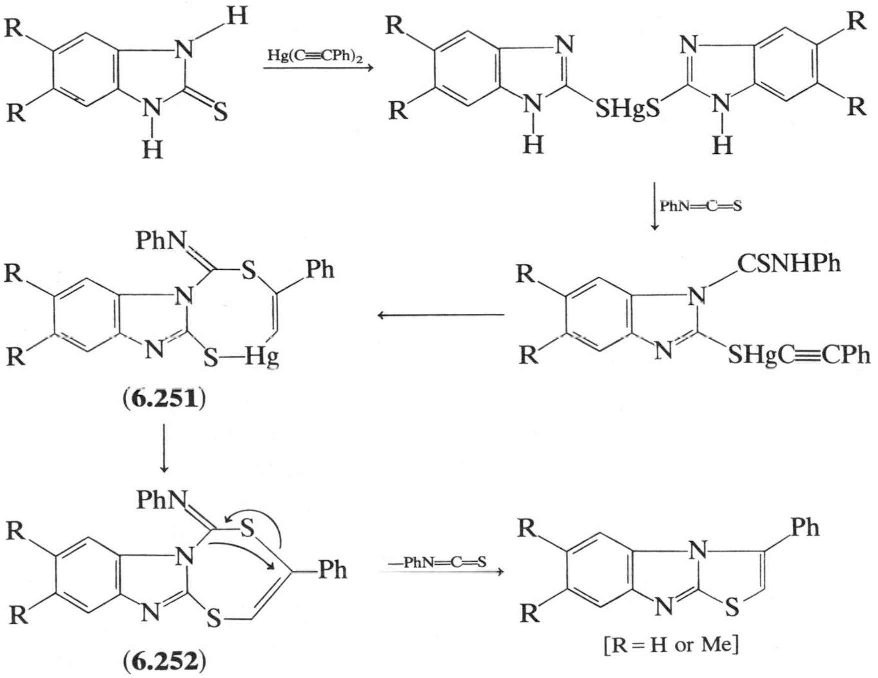

[R = H or Me]

Scheme 6.54

2,3-Dihydrothiazolo[3,2-*a*]benzimidazole derivatives are most simply constructed by the thermal or base-catalyzed dehydrohalogenative cyclization of 2-(β-halogenoalkylthio)benzimidazoles [Scheme 6.55; (**6.254**; X = Cl or Br) → (**6.256**)], which can either be preformed[123] or prepared *in situ* by reaction of a 2-benzimidazolethione with a 1,2-dihalogenoalkane such as dichloro-,[94,123,124] dibromo-,[93] or bromochloroethane,[125] and cyclized without isolation. To judge from the limited amount of information available,[123] ring-closure of this type proceeds in good yield (Table 6.29) only with the preformed 2-(β-halogenoalkylthio)benzimidazole and in conjunction with

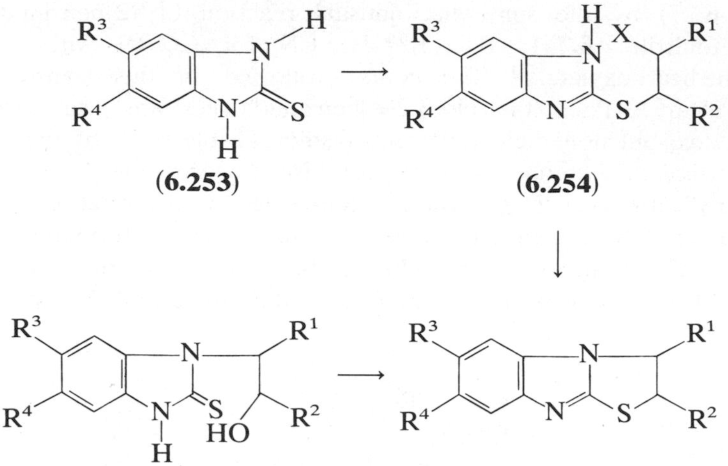

(6.253) → (6.254) ↓ (6.256) ← (6.255)

Scheme 6.55

catalysis by a relatively strong base, such as an alkali metal hydroxide[94,123–125] or sodium hydride,[93] or with heating under reflux in a suitable solvent such as toluene.[123] In the context of the base-promoted process it is noteworthy that the reportedly[126] successful condensation of 2-benzimidazolethione with 1,2-dibromoethane in the presence of potassium carbonate to give 2,3-dihydrothiazolo[3,2-*a*]benzimidazole has been suggested[123] to be in error. The melting point associated with the latter compound (cf. Table 6.29) is also a source of some confusion, having been assigned values as widely differing as 108–109°,[123] 110°,[94] 141–142°,[124] and last but not least, 239–240°.[125] A value of ca. 110° may be assumed on statistical grounds if nothing else!

As in the case of related cyclizations leading to thiazolo[3,2-*a*]benzimidazole derivatives (see before), ring formation of the type [(**6.254**) → (**6.256**)] using substrates unsymmetrically substituted in the benzene ring can lead to two possible, isomeric products, depending on which of the benzimidazole nitrogen atoms is involved in the ring-closure step. The single 2,3-dihydrothiazolo[3,2-*a*]benzimidazole products obtained in the cyclization of the *C*5(7)-chloro- and nitro-2-(β-halogenoalkylthio)benzimidazoles [Scheme 6.55; (**6.254**; $R^1 = R^2 = H$, R^3 or $R^4 = Cl$ or NO_2, $X = Cl$)] have been assigned the *C*(7) orientation (**6.256**; $R^1 = R^2 = R^3 = H$, $R^4 = Cl$ or NO_2) but without compelling evidence to exclude the alternative *C*(6) formulations (**6.256**; $R^1 = R^2 = R^4 = H$, $R^3 = Cl$ or NO_2). If correct, the orientation assigned to the nitro product in particular requires, contrary to expectation, that alkylative cyclization of the type [(**6.254**) → (**6.256**)] occurs preferentially through ring-closure at the less basic of the two benzimidazole nitrogen atoms. For this reason firmer evidence for the orientation of the

TABLE 6.29. SYNTHESIS OF 2,3-DIHYDROTHIAZOLO[3,2-*a*]BENZIMIDAZOLES BY RING-CLOSURE REACTIONS OF 2-BENZIMIDAZOLETHIONE DERIVATIVES

Starting materials ($R^1 \rightarrow R^6$ unspecified = H)	Reaction conditions[a]	Product ($R^1 \rightarrow R^6$ unspecified = H)	Yield (%)	m.p. (°C)	Solvent of crystallization	Ref.
(**6.253**)	*A*	(**6.256**)[b]	—[c]	141–142	—[d]	124
(**6.253**)	*A*	(**6.256**)	—[c]	110	Methanol–water	94
(**6.253**)	*B*	(**6.256**)[e]	—[c]	239–240	—[d]	125
(**6.253**)	*C*	(**6.256**)[f]	42	108–109	Methanol–water	123
(**6.254**; X = Cl)	*D*	(**6.256**)	70	—	—	123
(**6.254**; X = Cl)	*E*	(**6.256**)	45	—	—	123
(**6.255**)	*F*	(**6.256**)	79	—	—	123
(**6.253**; R^4 = Cl)	*B*	(**6.256**; R^4 = Cl)[g]	—[c]	176–178	—[d]	125
(**6.253**; $R^4 = NO_2$)	*B*	(**6.256**; $R^4 = NO_2$)[g]	—[c]	180–182	—[d]	125
(**6.255**; $R^4 = NO_2$)	*F*	(**6.256**; $R^4 = NO_2$)	90	229–230	Dimethylformamide	128
(**6.255**; R^4 = NHAc)	*F*	(**6.256**; R^4 = NHAc)[h]	82	>300	Ethanol–water	128
(**6.254**; $R^3 = R^4$ = Me)	*D*	(**6.256**; $R^3 = R^4$ = Me)	81	167–168	Ethanol–water	123
(**6.254**; $R^3 = R^4$ = Me)	*E*	(**6.256**; $R^3 = R^4$ = Me)	92	—	—	123
(**6.238**) + (**6.239**; X = Cl)	*G*	(**6.241**)[i]	quant.	180–180.5	Ethanol	103
(**6.238**) + (**6.239**; X = Cl)	*H*	(**6.241**)[j]	95	194–196 (decomp.)	Ethanol	106
(**6.238**) + (**6.239**; X = Cl)	*I*	(**6.241**)	89–93	—	—	106
(**6.238**) + (**6.239**; X = Cl)	*J*	(**6.241**)	89–93	—	—	106
(**6.238**) + (**6.239**; X = Br, $CH(OEt)_2$ for COR^1)	*H*	(**6.241**)	71	—	—	106

TABLE 6.29 (*Continued*)

Starting materials ($R^1 \rightarrow R^6$ unspecified = H)	Reaction conditions[a]	Product ($R^1 \rightarrow R^6$ unspecified = H)	Yield (%)	m.p. (°C)	Solvent of crystallization	Ref.
(**6.240**; $CH(OEt)_2$ for COR^1)	*K*	(**6.241**)[g]	82	—	—	106
(**6.238**) + (**6.239**; X = Cl, $CH(OEt)_2$ for COR^1)	*L*	(**6.241**)[k]	89	200–202	Methanol	94
(**6.238**; $R^4 = R^5$ = Me) + (**6.239**; X = Cl)	*G*	(**6.241**; $R^4 = R^5$ = Me)	88	185–202 (decomp.)	Tetrahydrofurane	103
(**6.238**; $R^4 = R^5$ = Me) + (**6.239**; X = Cl)	*M*	(**6.241**; $R^4 = R^5$ = Me)[l]	92	202–205 (decomp.)	Ethanol	106
(**6.238**; $R^3 = R^6$ = Me)[l] + (**6.239**; X = Cl)	*G*	(**6.241**; $R^3 = R^6$ = Me)	89	191.5–193.5 (decomp.)	Tetrahydrofurane	103
(**6.238**) + (**6.239**; R^1 = Me, X = Cl)	*G*	(**6.241**; R^1 = Me)	97	111–112	Diethyl ether	103
(**6.238**) + (**6.239**; R^1 = Et, X = Br)	*G*	(**6.241**; R^1 = Et)	—[c]	94–96	Acetonitrile, acetone, or ethanol–water	108
(**6.238**) + (**6.239**; $R^1 = Pr^i$, X = Cl)	*G*	(**6.241**; $R^1 = Pr^i$)	—[c]	109–111	Acetonitrile, acetone, or ethanol–water	108
(**6.238**) + (**6.239**; $R^1 = CF_3$, X = Br)	*G*	(**6.241**; $R^1 = CF_3$)	—[c]	138–139	Acetonitrile, acetone or ethanol–water	108
(**6.238**) + (**6.239**; $R^1 = CH_2Cl$, X = Cl)	*G*	(**6.241**; $R^1 = CH_2Cl$)	—[c]	181–184	Acetonitrile, acetone or ethanol–water	108

(**6.238**) + (**6.239**; $R^1 = CH_2SPh$, X = Cl)	*G*	(**6.241**; $R^1 = CH_2SPh$)	—[c]	138–139	Acetonitrile, acetone or ethanol–water	108
(**6.238**) + (**6.239**; R^1 = cyclohexyl, X = Cl)	*G*	(**6.241**; R^1 = cyclohexyl)	—[c]	155–156	Acetonitrile, acetone or ethanol–water	108
(**6.238**; $R^5 = NO_2$) + (**6.239**; R^1 = Me, X = Cl)	*G*	(**6.241**; R^1 = Me, $R^5 = NO_2$)	—[c]	122–124	Acetonitrile, acetone or ethanol–water	108
(**6.238**; $R^5 = NO_2$) + (**6.239**; $R^1 = CF_3$, X = Br)	*G*	(**6.241**; $R^1 = CF_3$, $R^5 = NO_2$)	—[c]	206.5–208.5	Acetonitrile, acetone or ethanol–water	108
(**6.238**) + (**6.239**; R^2 = Me, X = Br, $CH(OEt)_2$ for COR^1)	*N*	(**6.241**; R^2 = Me)[m]	quant.	201–203	Tetrahydrofurane	103
(**6.238**) + (**6.239**; R^2 = Me, X = Br, $CH(OMe)_2$ for COR^1)	*O*	(**6.241**; R^2 = Me)	90	198–200	Methanol	107
(**6.238**) + (**6.239**; R^2 = Me, X = Br, $CH(OEt)_2$ for COR^1)	*H*	(**6.241**; R^2 = Me)[n]	96	196–197	Ethanol–water	106
(**6.238**; $R^4 = R^5$ = Me) + (**6.239**; R^2 = Me, X = Br, $CH(OEt)_2$ for COR^1)	*H*	(**6.241**; $R^2 = R^4 = R^5$ = Me)	90	230–231	Dimethylformamide	106
(**6.238**) + (**6.239**; R^2 = Ph, X = Br, $CH(OMe)_2$ for COR^1)	*O*	(**6.241**; R^2 = Ph)	82	195.5–196.5	Methanol	107
(**6.240**; R^1 = Ph, $R^2 = CH_2CO_2Et$)[o]	*P*	(**6.241**; R^1 = Ph, $R^2 = CH_2CO_2Et$)[o]	67	129–131	Acetone	102, 109

TABLE 6.29 (*Continued*)

Starting materials ($R^1 \rightarrow R^6$ unspecified = H)	Reaction conditions[a]	Product ($R^1 \rightarrow R^6$ unspecified = H)	Yield (%)	m.p. (°C)	Solvent of crystallization	Ref.
(**6.238**) + (**6.239**; $R^1 = Ph$, $R^2 = CH_2CO_2Me$, X = Br)	*Q*	(**6.241**; $R^1 = Ph$, $R^2 = CH_2CO_2Me$)[o]	99	—[r]	Acetonitrile	109
(**6.238**) + (**6.239**; $R^1 = Ph$, $R^2 = CH_2CO_2Me$, X = Br)	*R*	(**6.241**; $R^1 = Ph$, $R^2 = CH_2CO_2Me$)[o]	44	145–147	Acetone	109
(**6.240**; $R^1 = Ph$, $R^2 = CH_2CO_2Me$)[o]	*S*	(**6.241**; $R^1 = Ph$, $R^2 = CH_2CO_2Me$)[o]	53	145–147	Acetone	102
(**6.240**; $R^1 = p\text{-}ClC_6H_4$, $R^2 = CH_2CO_2Et$)	*S*	(**6.241**; $R^1 = p\text{-}ClC_6H_4$, $R^2 = CH_2CO_2Et$)	83	140–142	Acetone	102, 109
(**6.240**; $R^1 = p\text{-}ClC_6H_4$, $R^2 = CH_2CO_2Me$)	*P*	(**6.241**; $R^1 = p\text{-}ClC_6H_4$, $R^2 = CH_2CO_2Me$)	55	132–134	Acetone	102, 109
(**6.240**; $R^1 = p\text{-}ClC_6H_4$, $R^2 = CH_2CO_2Et$, R^4 or $R^5 = NO_2$)	*P*	(**6.241**; $R^1 = p\text{-}ClC_6H_4$, $R^2 = CH_2CO_2Et$, R^4 or $R^5 = NO_2$)[p]	95	99–101	Benzene–acetonitrile	102, 109
(**6.238**) + (**6.239**; $R^1 = p\text{-}ClC_6H_4$, $R^2 = CH_2CO_2H$)	*R*	(**6.241**; $R^1 = p\text{-}ClC_6H_4$, $R^2 = CH_2CO_2H$)[q]	75	163–165	Dioxane–acetonitrile	102, 109
(**6.240**; $CH(OEt)_2$ for COR^1)	*T*	(**6.241**; OEt for OH)[s]	78	102.5–103.5	Ethanol–water	106

[a] *A* = $Cl(CH_2)_2Cl$, $NaHCO_3$, 20% NaOH, Pr^iOH/(reflux)(3 hr); *B* = $Br(CH_2)_2Cl$, NaOH, EtOH/(reflux)(3 hr); *C* = $Cl(CH_2)_2Cl$, $NaHCO_3$, KOH, Pr^iOH/(reflux)(3 hr); *D* = KOH, MeOH, H_2O/(reflux)(3 hr); *E* = toluene/(reflux)(30 min); *F* = $SOCl_2$/(reflux)(7–10 min); *G* = 2-butanone/(reflux)(2–8 hr); *H* = H_2O/(reflux)(3–4 hr); *I* = EtOH/(reflux)(4 hr); *J* = dimethylformamide/(60–65°)(1 hr, then 100°)(10 min); *K* = 36% HCl/(reflux)(1.5 hr); *L* = 5*M* HCl/(reflux)(3 hr); *M* = 50% EtOH-H_2O/(reflux)(1 hr); *N* = 20% HCl/(reflux)(30 min); *O* = 4 M HCl/(reflux)(7–13 hr); *P* = $NaHCO_3$, H_2O, benzene or $CHCl_3$/(room temp.)(few min); *Q* = MeOH/(reflux)(5 hr); *R* = AcOH/(100°)(1–3 hr); *S* = H_2O/(room temp.)(5 hr); *T* = $POCl_3$/(reflux)(1.5 hr).

[b] Forms a methiodide, m.p. 185–187° (decomp.).

[c] Yield not quoted.

[d] Solvent of crystallization not specified.
[e] Forms a hydrochloride m.p. 217–219°.
[f] Forms a hydrobromide, m.p. 219–220° (decomp.) (from aqueous hydrobromic acid), and a picrate, m.p. 230–231° (from acetic acid).
[g] Hydrochloride.
[h] Hydrochloride monohydrate; free base has m.p. 243–245° (from water).
[i] Hydrochloride, free base has m.p. 180–205° (decomp.) (from dioxane).
[j] Forms a hydrochloride, m.p. 184–186° (decomp.) (from dioxane-water) and a picrate, yellow needles, m.p. 225–226° (decomp.) (from ethanol).
[k] Forms a hydrochloride, m.p. 198°.
[l] Forms a picrate, m.p. 275–280° (decomp.) from ethanol.
[m] *Cis/trans* isomer mixture.
[n] Forms a picrate, yellow crystals, m.p. 179–180° (decomp.) (from ethanol-water).
[o] Hydrobromide.
[p] Position of nitro substituent not established.
[q] Forms a hydrobromide m.p. 203–205°.
[r] Melting point not quoted.
[s] Forms a picrate, yellow needles from ethanol, m.p. 209–211° with resolidification and remelting at 235–236°.

products of such cyclizations would appear desirable. Heating in polyphosphoric acid is reported[95] to effect the cyclization of 2-(2-allylthio)benzimidazole in low yield (35%) to 3-methyl-2,3-dihydrothiazolo[3,2-*a*]benzimidazole (m.p. 168–170°) by a process which is presumably related mechanistically to the foregoing 2-(β-halogenoalkylthio)benzimidazole cyclizations.

General synthetic access to 2,3-dihydrothiazolo[3,2-*a*]benzimidazole derivatives is also provided by the halogenative–dehydrohalogenative ring-closure induced in *N*(1)-(β-hydroxyalkyl)-2-benzimidazolethiones by treatment with thionyl chloride [Scheme 6.55; (**6.255**) → (**6.256**)]. The excellent yields achievable (Table 6.29) coupled with the ready availability of the *N*-(β-hydroxyalkyl)benzimidazolethione starting materials (**6.255**) from 2-chlorobenzimidazoles (by *N*-alkylation with a β-halogeno alcohol followed by treatment with thiourea) makes this type of 2,3-dihydrothiazolo[3,2-*a*]-benzimidazole synthesis an attractive alternative to that through 2-(β-halogenoalkylthio)benzimidazoles (see before). Moreover, unlike ring-closure of the latter type, the nature of the cyclization step [(**6.255**) → (**6.256**)] in *N*-(β-hydroxyalkyl)benzimidazolethione cyclizations is such that, provided the orientation of the starting material is secure, ring-formation is unambiguous in terms of the ultimate site of substituents in the benzene nucleus.[128]

As already briefly discussed, 2-(β-oxoalkylthio)benzimidazoles are capable of coexisting in ring-chain tautomeric equilibrium with the corresponding 3-hydroxy-2,3-dihydrothiazolo[3,2-*a*]benzimidazoles [cf. Scheme 6.51; (**6.240**) ⇌ (**6.241**)]. Careful IR and ^1H NMR studies[103,108] of such equilibria reveal that 2-(β-oxoalkylthio)benzimidazoles containing bulky or electron-withdrawing substituents, which can conjugate with the carbonyl group, exist preferentially in the open-chain form (**6.240**) both in the solid state and in solution. Conversely, the presence of sterically undemanding, electron-donating substituents incapable of conjugating with the carbonyl group favor the ring-closed form to the extent that, in the solid state at least, the molecule exists essentially as the 3-hydroxy-2,3-dihydrothiazolo[3,2-*a*]-benzimidazole (**6.241**). Simple, unsubstituted, and *C*(2) or *C*(3) alkyl or *C*(2) aryl-substituted 3-hydroxy-2,3-dihydrothiazolo[3,2-*a*]benzimidazoles (**6.241**; $R^1 = H$ or alkyl, $R^2 = H$, alkyl or aryl) are therefore synthetically readily accessible (Scheme 6.51) in high yield (Table 6.29) by the condensation of 2-benzimidazolethiones (**6.238**) with α-halogeno ketones[102,103,109] and aldehydes[103,106] under neutral conditions or with α-halogeno aldehyde acetals[94,103,106,107] in acidic media. 3-Aryl-3-hydroxy-2,3-dihydrothiazolo-[3,2-*a*]benzimidazoles (**6.241**; $R^1 =$ aryl) are generally inaccessible by this route being usually unstable relative to the open-chain 2-(2-benzimidazolylthio)acetophenone tautomers (**6.240**; $R^1 =$ aryl), though the presence of certain α-substituents in the latter [e.g. (**6.240**; $R^1 =$ aryl, $R^2 = CH_2CO_2R$)], somewhat unexpectedly, favors formation[102,109] of the ring-closed isomer (**6.241**; $R^1 =$ aryl, $R^2 = CH_2CO_2R$) in high yield (Table 6.29). Acid-catalyzed cyclization[106] of the preformed acetal [Scheme 6.51; (**6.240**; $R^1 \rightarrow R^6 = H$, $CH(OEt)_2$ for COR^1)] gives 3-hydroxy-2,3-dihydrothiazolo[3,2-*a*]benzimidazole (**6.241**; $R^1 \rightarrow R^6 = H$) in good yield (Table 6.29) through the presumed intermediacy of the corresponding 2-(β-oxoalkylthio)benzimidazole (**6.240**; $R^1 \rightarrow R^6 = H$). Conversely, phosphorus oxychloride-mediated ring-closure of the acetal (**6.240**; $R^1 \rightarrow R^6 = H$, $CH(OEt)_2$ for COR^1) leads to 3-ethoxy-2,3-dihydrothiazolo[3,2-*a*]benzimidazole (**6.241**; $R^1 \rightarrow R^6 = H$, OEt for OH) also in good yield (Table 6.29).[106]

The preferential alkylation of benzimidazolethiones by α-halogeno carbonyl compounds and their acetals at sulfur rather than nitrogen, and the consequent 3-hydroxy structure [Scheme 6.51; (**6.241**)] for the resulting ring tautomeric products, rather than the alternative 2-hydroxy formulation (**6.241**; $R^2 = OH$, H for OH), follow from the oxidation[103] of the parent system (**6.241**; $R^1 \rightarrow R^6 = H$) to 2,3-dihydrothiazolo[3,2-*a*]benzimidazol-3-one whose orientation has been firmly established (see later). On the other hand, the orientational preference governing the ring-closure reactions of 2-(β-oxoalkylthio)benzimidazoles to 3-hydroxy-2,3-dihydrothiazolo[3,2-*a*]-benzimidazoles substituted in the benzene ring has not been investigated beyond the qualitative demonstration of single isomer formation.[102,108]

2,3-Dihydrothiazolo[3,2-*a*]benzimidazol-3-ones are generally accessible by a logical extension of the foregoing 2-(β-oxoalkylthio)benzimidazole cyclizations involving the acylative ring closure of 2-benzimidazolylthioacetic acids and esters [Scheme 6.56; (**6.257**; $R^6 = H$ or alkyl) → (**6.259**)].

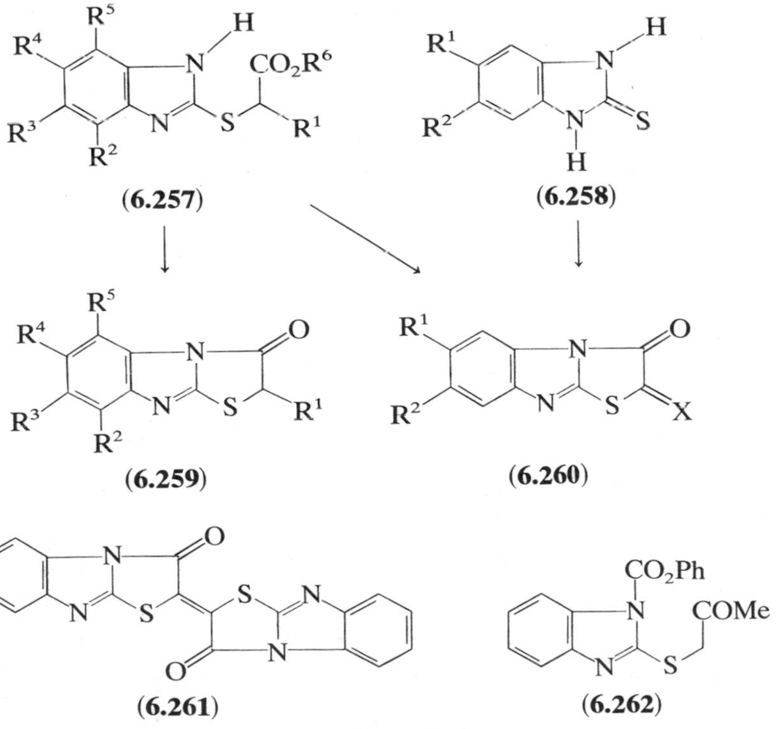

(**6.257**) (**6.258**)

(**6.259**) (**6.260**)

(**6.261**) (**6.262**)

Scheme 6.56

TABLE 6.30. SYNTHESIS OF 2,3-DIHYDROTHIAZOLO[3,2-*a*]BENZIMIDAZOL-3-ONES BY RING-CLOSURE REACTIONS OF 2-BENZIMIDAZOLETHIONE DERIVATIVES

Starting material ($R^1 \rightarrow R^6$ unspecified = H)	Reaction conditions[a]	Product ($R^1 \rightarrow R^6$ unspecified = H)	Yield (%)	m.p. (°C)	Solvent of crystallization	Ref.
(**6.257**)	*A*	(**6.259**)	64	181	Ethanol	129
(**6.257**)	*A*	(**6.259**)	55	180	—[b]	131
(**6.257**)	*A*	(**6.259**)	67–69	181	Ethanol–benzene	132
(**6.257**)	*A*	(**6.259**)	66	180	—[b]	101
(**6.257**)	*B*	(**6.259**)	65	178	Ethanol	135
(**6.257**)	*C*	(**6.259**)	98	181	Ethanol	137
(**6.257**; $R^6 = Et$)	*D*	(**6.259**)	93	181	Methanol	94
(**6.257**; $R^6 = Et$)	*E*	(**6.259**)	40	178–181	*o*-dichlorobenzene	138
(**6.257**; $R^6 = Et$)	*E*	(**6.259**)	25	179–180	Ethanol	134
(**6.257**; $R^6 = Me$ or Et)	*F*	(**6.259**)	25	181	Ethanol	140
(**6.257**)	*G*	(**6.261**)	40	> 300	Nitrobenzene	130
(**6.257**; R^3 or $R^4 = Me$)	*A*	(**6.259**; $R^4 = Me$)	70	183–184 (decomp.)	Ethanol	133
(**6.257**; R^2 or $R^5 = Me$)	*A*	(**6.259**; $R^2 = Me$)	65	191–192	Ethanol	133
(**6.257**; R^3 or $R^4 = MeO$)	*A*	(**6.259**; $R^4 = OMe$)	48	154	Ethanol	97
(**6.257**; R^3 or $R^4 = Cl$)	*A*	(**6.259**; $R^4 = Cl$)	64	183–184	Benzene	98
(**6.257**; R^3 or $R^4 = NO_2$)	*A*	(**6.259**; $R^4 = NO_2$)	70	220	Benzene–ethanol	136
(**6.257**; $R^3 = R^4 = Me$)	*A*	(**6.259**; $R^3 = R^4 = Me$)	76	177–178	Ethanol	133
(**6.257**; $R^3 = R^4 = Me$)	*H*	(**6.259**; $R^3 = R^4 = Me$)	80–82	175–176 (decomp.)	Ethanol	134
(**6.257**; $R^2 = R^5 = Me$)	*A*	(**6.259**; $R^2 = R^5 = Me$)	74	172–173	Ethanol	133
(**6.257**; $R^1 = Me$)	*B*	(**6.259**; $R^1 = Me$)	79	101	Methanol	94
(**6.257**; $R^1 = Me$)	*I*	(**6.259**; $R^1 = Me$)	43	204 (decomp.)	—[b]	101
(**6.257**; $R^1 = Et$)	*B*	(**6.259**; $R^1 = Et$)	82	—[c]	—	94
(**6.257**; $R^1 = Pr^n$)	*B*	(**6.259**; $R^1 = Pr^n$)	78	58	Methanol	94
(**6.257**; $R^1 = Bu^n$)	*B*	(**6.259**; $R^1 = Bu^n$)	75	90	Methanol	94
(**6.257**; $R^1 = R^3 = R^4 = Me$)	*H*	(**6.259**; $R^1 = R^3 = R^4 = Me$)	99	123–124	Ligroin	134
(**6.258**)	*J*	(**6.259**; $R^1 = CH_2CO_2H$)	59	—	—[b]	144
(**6.258**)	*K*	(**6.259**; $R^1 = CH_2CO_2H$)	28	—	—[b]	147
(**6.263**; $R^1 = Me$)	*A*	(**6.264**; $R^1 = Me$)	78	255	—[b]	129

(**6.263**; R^1 = Me, R^2 = NO_2)	*A*	(**6.264**; R^1 = Me, R^2 = NO_2)	81	238 (decomp.)	—[b]	129
(**6.263**; R^1 = Ph)	*A*	(**6.264**; R^1 = Ph)	44	222	Xylene	129
(**6.263**; R^1 = Ph, R^2 = NO_2)	*A*	(**6.264**; R^1 = Ph, R^2 = NO_2)	62	221	—[b]	129
(**6.257**; R^1 = COMe, R^6 = Et)	*L*	(**6.260**; X = C(Me)OAc)	—[d]	192–194	Benzene	141
(**6.258**)	*M*	(**6.260**; X = CHPh)	18	214	—[e]	149
(**6.258**)	*M*	(**6.260**; X = *o*-HOC_6H_4CH)	34	178	—[e]	149
(**6.258**)	*M*	(**6.260**; X = CHCH = CHPh)	55	234	—[e]	149
(**6.258**)	*M*	(**6.260**; X = $AcNHC_6H_4CH$)[f]	27	172	—[e]	149
(**6.258**)	*M*	(**6.260**; X = *m*-$O_2NC_6H_4CH$)	62	230	—[e]	149
(**6.258**)	*M*	(**6.260**; X = 2-$HOC_{10}H_6CH$)	58	236	—[e]	149
(**6.258**)	*M*	(**6.260**; X = *p*-$MeOC_6H_4CH$)	52	238	—[e]	149
(**6.258**)	*M*	(**6.260**; X = *p*-$NO_2C_6H_4CH$)	37	280	—[e]	149
(**6.257**)	*N*	(**6.260**; X = CHPh)	82	226	Acetic acid	148
(**6.257**)	*O*	(**6.260**; X = CHPh)	90	—	—	148
(**6.257**)	*N*	(**6.260**; X = *p*-ClC_6H_4CH)[g]	84	286–287	Dimethylformamide	148
(**6.257**)	*O*	(**6.260**; X = *p*-ClC_6H_4CH)	95	—	—	148
(**6.257**)	*N*	(**6.260**; X = *p*-$NO_2C_6H_4CH$)[g]	85	315	Dimethylformamide	148
(**6.257**)	*O*	(**6.260**; X = *p*-$NO_2C_6H_4CH$)	98	—	—	148
(**6.257**)	*N*	(**6.260**; X = *p*-MeC_6H_4CH)[h]	81	250.5–251	Acetic acid	148
(**6.257**)	*O*	(**6.260**; X = *p*-MeC_6H_4CH)	92	—	—	148
(**6.257**)	*N*	(**6.260**; X = *p*-$MeOC_6H_4CH$)[h]	83	238–239	Acetic acid	148
(**6.257**)	*O*	(**6.260**; X = *p*-$MeOC_6H_4CH$)	94	—	—	148
(**6.257**)	*N*	(**6.260**; X = *p*-BrC_6H_4CH)	80	269–272	Dimethylformamide	148
(**6.257**)	*O*	(**6.260**; X = *p*-BrC_6H_4CH)	93	—	—	148
(**6.257**)	*N*	(**6.260**; X = *p*-IC_6H_4CH)	78	294–295	Dimethylformamide	148
(**6.257**)	*O*	(**6.260**; X = *p*-IC_6H_4CH)	90	—	—	148
(**6.257**)	*O*	(**6.260**; X = *p*-$H_2NC_6H_4CH$)	85	193–195	Chloroform	148
(**6.257**)	*N*	(**6.260**; X = *p*-NCC_6H_4CH)	81	269	Dimethylformamide	148
(**6.257**)	*O*	(**6.260**; X = *p*-NCC_6H_4CH)	94	—	—	148
(**6.257**)	*N*	(**6.260**; X = *p*-$Me_2NC_6H_4CH$)	68	169	Ethanol	148
(**6.257**)	*O*	(**6.260**; X = *p*-$Me_2NC_6H_4CH$)	81	—	—	148
(**6.257**)	*N*	(**6.260**; X = 2-thienyl CH)	85	169.5	Acetic acid	148
(**6.257**)	*O*	(**6.260**; X = 2-thienyl CH)	96	—	—	148
(**6.257**)	*N*	(**6.260**; X = 2-furyl CH)	89	235.5–236	Dimethylformamide	148
(**6.257**)	*O*	(**6.260**; X = 2-furyl CH)	96	—	—	148
(**6.257**)	*N*	(**6.260**; X = CHCH=CHPh)	81	242–243	Acetic acid	148
(**6.257**)	*O*	(**6.260**; X = CHCH=CHPh)	89	—	—	148

TABLE 6.30 (*Continued*)

Starting material ($R^1 \rightarrow R^6$ unspecified = H)	Reaction conditions[a]	Product ($R^1 \rightarrow R^6$ unspecified = H)	Yield (%)	m.p. (°C)	Solvent of crystallization	Ref.
(**6.257**)	*P*	(**6.265**; X = O)	45	245	Ethanol	148
(**6.257**)	*P*	(**6.265**; X = NH)	40	283	Ethanol	148
(**6.257**; R^6 = Et)	*Q*	(**6.265**; X = *p*-$(ClCH_2CH_2)_2NC_6H_4CH$)	69	218	Acetic acid	132
(**6.258**)	*R*	(**6.260**; $CHCO_2Me$)	96	175–176	Acetone	145
(**6.258**)	*R*	(**6.260**; $CHCO_2Me$)	—[d]	192–193	Acetone	147
(**6.258**)	*R*	(**6.260**; $CHCO_2Et$)	26	145–150	Methanol	145
(**6.258**; R^1 or R^2 = Me)	*R*	(**6.260**; R^1 or R^2 = Me, X = $CHCO_2Me$)[i]	73	180–183	Benzene	145
(**6.258**; R^1 or R^2 = NO_2)	*R*	(**6.260**; R^1 or R^2 = NO_2, X = $CHCO_2Me$)[i]	66	243–245	Benzene	145
(**6.258**; R^1 or R^2 = Cl)	*R*	(**6.260**; R^1 or R^2 = Cl, X = $CHCO_2Me$)[i]	73	190–194	Acetone	145
(**6.258**; R^1 or R^2 = CO_2H)	*R*	(**6.260**; R^1 or R^2 = CO_2H, X = $CHCO_2Me$)[i]	66	238–240	Acetone	145
(**6.258**)	*S*	(**6.260**; X = NPh)	71	163–165	Ethanol–water	151
(**6.258**)	*T*	(**6.260**; X = *p*-$O_2NC_6H_4NHN$)	47	249–250	Dioxane	151
(**6.258**)	*T*	(**6.260**; X = *p*-$MeOC_6H_4NHN$)	52	202–204	Dioxane	151
(**6.258**)	*T*	(**6.260**; X = *p*-MeC_6H_4NHN)	53	263	Dioxane	151

[a] *A* = Ac_2O, pyridine/(100°)(5–15 min); *B* = Ac_2O, pyridine/(100°)(3–48 hr); *C* = dicyclohexylcarbodiimide, pyridine/(5–10°)(12 hr); *D* = ethyl polyphosphate/(100°)(2 hr); *E* = *ortho*-dichlorobenzene/(reflux)(1 hr); *F* = sodium sand, benzene/(reflux)(0.5 hr); *G* = $SOCl_2$, pyridine, benzene/(warm)(10 min); *H* = Ac_2O/(reflux)(4–8 min); *I* = Ac_2O, pyridine/(room temp.)(48 hr); *J* = maleic anhydride, dioxane/(reflux)(24 hr); *K* = maleic anhydride, glyme/(160°)(reaction time not specified); *L* = Ac_2O/(reflux)(3 hr); *M* = NaOAc, AcOH/(reflux)(2–3 hr); *N* = ArCHO, KOAc, Ac_2O/(130–140°)(15 min); *O* = ArCHO, pyridine/(120°)(10 min); *P* = phthalic anhydride or phthalimide, Ac_2O/(140°)(15 min); *Q* = piperidine, EtOH/(reflux)(4 hr); *R* = $MeO_2CC{\equiv}CCO_2Me$, or $EtO_2CC{\equiv}CCO_2Et$, MeOH/(reflux)(1–2 hr); *S* = PhN=C(Cl)COCl, Et_3N, benzene/(100°)(3 hr); *T* = ArNHN=C(Cl)COCl, Et_3N, benzene-dioxane/(reflux)(3 hr).

[b] Solvent of crystallization not specified.

[c] Oil.

[d] Yield not quoted.

[e] Crystallized from ethanol, dioxane, or acetic acid.

[f] Position of the AcNH substituent not specified.

[g] For details of the ortho and meta isomers cf. Ref. 148.

[h] For details of the ortho isomer cf. Ref. 148.

[i] Single isomer obtained; site of substituent not established.

This mode of thiazolo[3,2-*a*]benzimidazole ring formation normally proceeds in high yield (Table 6.30) and is most efficiently achieved by the action of dehydrating agents, such as acetic anhydride (usually in conjunction with pyridine)[94,97,98,101,129–136] or dicyclohexylcarbodiimide,[137] on 2-benzimidazolylthioacetic acids (**6.257**; $R^6 = H$) readily accessible by the condensation of 2-benzimidazolethiones with α-halogeno carboxylic acids. The cyclization of 2-benzimidazolylthioacetic acids by heating with thionyl chloride in pyridine tends to afford dimeric products of the type (**6.261**) (Scheme 6.56) rather than simple 2,3-dihydrothiazolo[3,2-*a*]benzimidazol-3-ones.[130] The ring-closure of ethyl 2-benzimidazolylthioacetate (**6.257**; $R^1 \rightarrow R^5 = H$, $R^6 =$ Et), which occurs only in low yield by heating in 1,2-dichlorobenzene[138,139] or with sodium in benzene,[140] gives 2,3-dihydrothiazolo[3,2-*a*]benzimidazol-3-one (**6.259**; $R^1 \rightarrow R^5 = H$) in high yield (Table 6.30) when catalyzed by hot ethyl polyphosphate.[94] The extension of the acetic anhydride–pyridine mediated cyclization of 2-benzimidazolylthioacetic acids to α-alkyl derivatives (**6.257**; $R^1 =$ alkyl) affords[94,101,134] moderate to high yields (Table 6.30) of 2-alkyl-2,3-dihydrothiazolo[3,2-*a*]benzimidazol-3-ones (**6.259**; $R^1 =$ alkyl), molecules previously reported[129] to be inaccessible by ring-closure of this type. Treatment[141] of ethyl 2-(2-benzimidazolylthio)acetoacetates [Scheme 6.56; (**6.257**; $R^1 =$ COMe, $R^6 =$ Et)] with acetic anhydride in the absence of pyridine leads to ring-closure through the ester group (and not the acetyl substituent as observed in the presence of pyridine—see before) the products being the enol acetates [**6.260**; X = C(OAc)R] which are presumably derived by subsequent acetylation of initially formed 2-acetyl-2,3-dihydrothiazolo[3,2-*a*]benzimidazol-3-ones (**6.259**; $R^1 =$ COMe). The interesting sodium hydride catalyzed transformation[142] of the benzimidazolylthiopropanone derivative [Scheme 6.56; (**6.262**)] into 2-acetyl-2,3-dihydrothiazolo[3,2-*a*]benzimidazol-3-one (**6.259**; $R^2 \rightarrow R^5 = H$, $R^1 =$ COMe) is rationalized in terms of the intermediacy of phenyl 2-(2-benzimidazolylthio)acetoacetate (**6.257**; $R^1 =$ COMe, $R^2 \rightarrow R^5 = H$, $R^6 =$ Ph), whose formation by facile $N \rightarrow C$ acyl shift provides a model for enzyme-mediated transcarboxylation involving carboxybiotin. The acetic anhydride–pyridine promoted cyclization of *N*(1)-substituted 2-benzimidazolylthioacetic acids [Scheme 6.57, (**6.263**; $R^1 =$ Me or Ph)] leads to anhydro products having interesting tricyclic mesoionic structures (**6.264**; $R^1 =$ Me or Ph).[129] The chemistry of these molecules has not been investigated to any extent,[129]

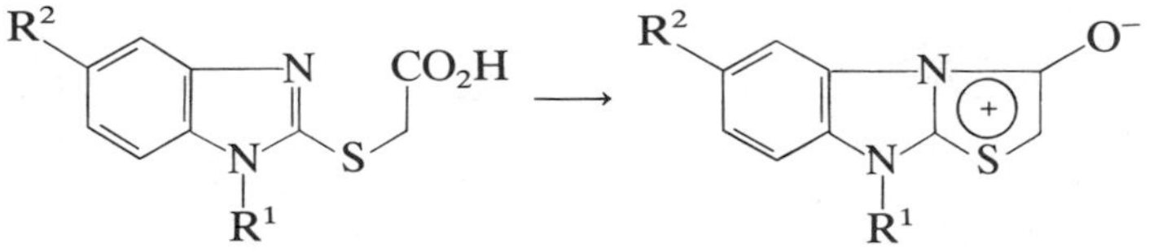

(**6.263**) → (**6.264**)

Scheme 6.57

despite current widespread interest in mesoionic compounds in general.[143]

As for other cyclizations leading to thiazolo[3,2-*a*]benzimidazole derivatives (see before), ring-closure of 2-benzimidazolylthioacetic acids and esters unsymmetrically substituted in the benzene ring can give rise to two possible isomeric 2,3-dihydrothiazolo[3,2-*a*]benzimidazol-3-one products by alternative ring-formation at the two nonequivalent benzimidazole nitrogen atoms. The fact that this situation leads to structural ambiguity when single products are formed has been ignored in some instances,[136,141] while in others,[144,145] though recognized, has not been resolved. In other studies[97,98,133] a choice between possible $C(6)$ and $C(7)$ orientations for monosubstituted 2,3-dihydrothiazolo[3,2-*a*]benzimidazol-3-ones has been made on the basis of the ^1H NMR splitting patterns of the protons at $C(5)$ and $C(8)$, which can be differentiated (with allowance for the perturbing effect of the substituent) by the enhanced deshielding of the former as a result of the anisotropic effect of the proximate $C(3)$ carbonyl group. The $C(6)$ orientations assigned[97,98,133] on this basis indicate that ring-closure takes place preferentially at the nitrogen atom *meta* rather than *para* to the substituent, irrespective of whether this is chloro, methyl, or methoxy. The orientational preference exhibited by the latter two substituents is particularly surprising, implying as it does that a presumed electrophilic (acylative) cyclization process involves ring-closure at the predictably less basic of the two available nitrogen centers. In view of this unexpected result, further detailed studies of the factors governing orientational preference in 2-benzimidazolylthioacetic acid cyclizations would appear to be warranted.

The formation of 2,3-dihydrothiazolo[3,2-*a*]benzimidazol-3-ones by the direct condensation of 2-benzimidazolethiones with α-halogeno carboxylic acids does not appear to have been described, the reported reaction[146] of 2-benzimidazolethione itself with chloroacetic acid in the presence of sodium acetate to give 2,3-dihydrothiazolo[3,2-*a*]benzimidazol-3-one having been shown[132] to be incorrect. On the other hand, the uncatalyzed condensation of 2-benzimidazolethione with maleic anhydride to give the acetic acid derivative [Scheme 6.56; (**6.259**; $R^1 = CH_2CO_2H$, $R^2 \rightarrow R^5 = H$)], albeit in low yield (Table 6.30) appears to be well enough authenticated.[144,147] In closely related processes,[148] 2-benzimidazolylthioacetic acid reacts with phthalic anhydride or phthalimide in acetic anhydride in the presence of potassium acetate to afford moderate yields (Table 6.30) of the 2,3-dihydrothiazolo[3,2-*a*]benzimidazol-3-one derivatives [Scheme 6.58; (**6.265**; X = O or NH)]. The fact that 2,3-dihydrothiazolo[3,2-*a*]benzimidazol-3-ones condense with aromatic aldehydes under a variety of conditions (see later) to give the corresponding 2-arylidene derivatives [Scheme 6.56; (**6.260**; X = CHAr)] makes them plausible intermediates in the formation (Table 6.30) of analogous products when 2-benzimidazolethione is allowed to react with aromatic aldehydes and chloroacetic acid in acetic acid containing sodium acetate.[149] However, the possibility that the aldol condensation involved in these ring-closures occurs prior to and not subsequent to ring-formation is

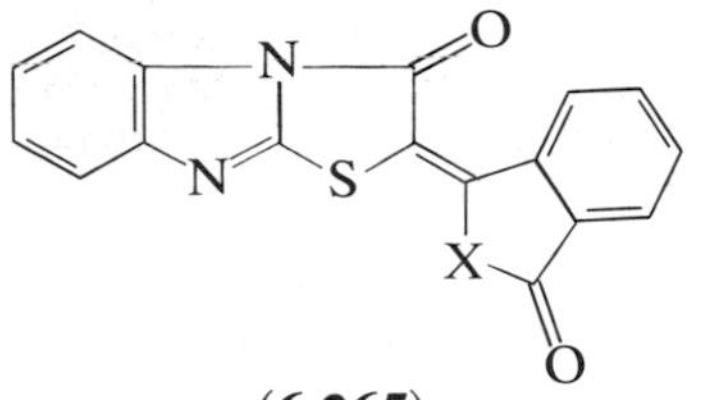

Scheme 6.58

suggested by the reported[138] cyclization of a 2-arylidene derivative of 2-benzimidazolylthioacetic acid to the corresponding 2-arylidene-2,3-dihydrothiazolo[3,2-*a*]benzimidazol-3-one [Scheme 6.56; (**6.260**; $R^1 = R^2 = H$, X = CHAr)] simply on attempted crystallization. That 2-benzimidazolylthioacetic acid and its ethyl ester can function as substrates for the synthesis of 2-arylidene-2,3-dihydrothiazolo[3,2-*a*]benzimidazol-3-ones is readily demonstrated[132,148] by their smooth acetic anhydride or pyridine promoted condensation with aromatic aldehydes to give the anticipated arylidene products (**6.260**; $R^1 = R^2 = H$, X = CHAr).

2-Alkylidene-2,3-dihydrothiazolo[3,2-*a*]benzimidazol-3-ones of the type (**6.260**; X = $CHCO_2R$) are also formed in moderate to high yield (Table 6.30) as the end-products of the uncatalyzed condensation reactions of 2-benzimidazolethiones with acetylenic esters.[121,145,150] A recent study[147] is concerned with the knotty problem of product orientation in cycloaddition of this type and has confirmed by chemical and crystallographic means the 3-oxo structure (**6.260**; $R^1 = R^2 = H$, X = CH_2CO_2Me) (as opposed to a 2-oxo or six-membered formulation) originally assigned[150] to one of the major products of the reaction of 2-benzimidazolethione with dimethylacetylenedicarboxylate in acetic acid or methanol. Similar problems of product orientation are encountered in the base-catalyzed condensation reactions of 2-benzimidazolethiones with 2-arylimino and 2-arylhydrazono chloroacetyl chlorides which afford moderate yields (Table 6.30) of monoarylimines and monoarylhydrazones of 2,3-dihydrothiazolo[3,2-*a*]-benzimidazole-2,3-diones.[151] These condensates appear to have been assigned[151] the *C*(3)-oxo orientations [Scheme 6.56; (**6.260**; X = NAr or NNHAr)] (as opposed to the alternative 2-oxo structures) purely on the basis of analogy with similar products derived by coupling reactions at the reactive *C*(2) position in preformed 2,3-dihydrothiazolo[3,2-*a*]benzimidazol-3-ones (see later) yet apparently without the demonstration of common product formation. More rigorous evidence in support of the 3-oxo formulations (**6.260**; X = NAr or NNHAr) for the benzimidazolethione derived products would therefore be desirable.

Synthetic access to the rare 1*H*,3*H*-thiazolo[3,2-*a*]benzimidazole ring system is dependent on the spontaneous thermal cyclization (Scheme 6.59)

TABLE 6.31 SYNTHESIS OF 1-IMINO-1*H*,3*H*-THIAZOLO[3,4-*a*]BENZIMIDAZOLE DERIVATIVES BY RING-CLOSURE REACTIONS OF 2-(*a*-CHLOROALKYL)- AND 2-(α-THIOCYANOALKYL)BENZIMIDAZOLES[a]

Starting material ($R^1 \rightarrow R^3$ unspecified = H)	Reaction conditions[b]	Product ($R^1 \rightarrow R^3$ unspecified = H)	Yield (%)	m.p. (°C)	Solvent of crystallization
(**6.266**; Cl for SCN)	*A*	(**6.267**)	42	169–170	Methanol
(**6.266**)	*B*	(**6.267**)	40	—	—
(**6.266**; R^3 = Me, Cl for SCN)	*A*	(**6.267**; R^3 = Me)	23	117–118	Light petroleum
(**6.266**; R^1 or R^2 = Cl, Cl for SCN)	*C*	(**6.267**; R^2 = Cl) +	59	156–158	Diethyl ether
		(**6.267**; R^1 = Cl)	—[c]	161–162	Diethyl ether
(**6.266**; R^1 or R^2 = Me, Cl for SCN)	*C*	(**6.267**; R^2 = Me)	19	152–153	Diethyl ether

[a] From Refs. 152 and 153.
[b] *A* = NH_4SCN, MeOH/(reflux)(1 hr); *B* = MeOH/(reflux)(1 hr); *C* = NH_4SCN, dimethylformamide/(50°)(3.5 hr).
[c] Yield not specified.

of 2-(α-thiocyanoalkyl)benzimidazoles (**6.266**) either preformed or generated *in situ* by the action of ammonium thiocyanate on 2-(α-chloroalkyl)-benzimidazoles.[152,153] The only moderate yields (Table 6.31) of 1-imino-1*H*,3*H*-thiazolo[3,4-*a*]benzimidazoles (**6.267**) so obtained are offset by the convenience in practice of such cyclizations based as they are on 2-(α-chloroalkyl)benzimidazole starting materials readily available by the reaction of *ortho*-phenylenediamine derivatives with α-chlorocarboxylic acids. Ring-closure fails[153] for 2-(α-thiocyanoalkyl)benzimidazole substrates having strongly electron-withdrawing substituents (e.g., nitro) in the *C*5(6) position presumably due to the diminished nucleophilicity of the imidazole ring nitrogen atoms. Cyclization of 2-(α-thiocyanoalkyl)benzimidazoles with less electronically demanding substituents (e.g., chloro, methyl) at the *C*5(6) position proceeds readily but affords mixtures of both possible ring-closed products [i.e. (**6.267**; R^1 = Cl or Me) and (**6.267**; R^2 = Cl or Me)], which at least in the examples described,[152,153] are amenable to separation (Table 6.31) and characterization on the basis of their respective ^{1}H NMR absorption. The ring-closure of 2-(α-mercaptomethyl)benzimidazole to 1*H*,3*H*-thiazolo[3,4-*a*]benzimidazol-1-one (m.p. 212–214°) [Scheme 6.59; (**6.266**;

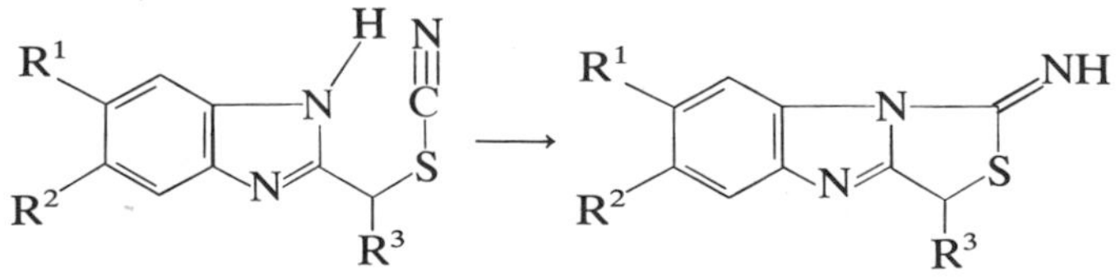

(**6.266**) → (**6.267**)

Scheme 6.59

$R^1 \rightarrow R^3$ = H, H for CN) → (**6.267**; $R^1 \rightarrow R^3$ = H, O for NH)] using phosgene in the presence of pyridine has been briefly described[153] but without experimental details.

The cycloaddition reactions[154] (Scheme 6.60) of the 1-methylbenzimidazolium 3-imine (**6.268**) with activated alkenes (e.g., methyl acrylate, acrylonitrile) afford, in unspecified yield, oily intermediates assigned 2,3,3*a*,4-tetrahydro-1*H*-pyrazolo[2,3-*a*]benzimidazole structures of the type (**6.269**; R = CO_2Me or CN). These transformations appear to be the only examples of pyrazolo[2,3-*a*]benzimidazole synthesis involving ring-closure in benzimidazole derivatives.

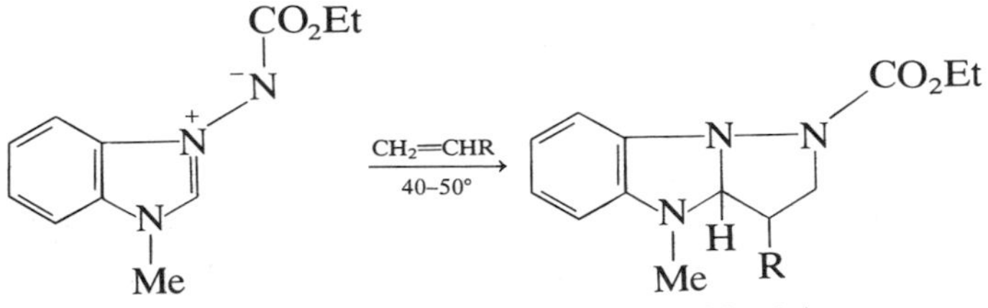

(**6.268**) (**6.269**)
[R = CN or CO_2Me]

Scheme 6.60

Synthetic routes to imidazo[1,2-*a*]benzimidazoles of the various structural types are based almost exclusively on cyclization reactions of 2-aminobenzimidazole precursors (cf. Tables 6.32, 6.33, 6.34, 6.35, and 6.36). For example ring-closure in *N*(1)-(β-oxoalkyl)- or 2-(β-oxoalkylamino)-benzimidazoles affords general synthetic access to variously substituted 1*H*- and 9*H*-imidazo[1,2-*a*]benzimidazoles (Scheme 6.61). 2-Alkyl- and 2-aryl *N*-unsubstituted 1(9)*H*-imidazo[1,2-*a*]benzimidazoles (**6.275**; R^1 = alkyl or aryl, R^2 = H) or (**6.276**; R^1 = H, R^2 = alkyl or aryl) in particular are readily prepared in high yield (Table 6.33) by the smooth cyclization of 2-amino-*N*(1)-(β-oxoalkyl)benzimidazoles [(**6.272**) ⇌ (**6.273**); R^1 or R^2 = alkyl or aryl, R^2 or R^1 = H] achieved thermally in alcoholic solvents[155,156] or in hydrochloric acid[157] or, in the form of their hydrobromides, using methanolic alkali.[155,156] Moreover, the 2-amino-*N*(1)-(β-oxoalkyl)benzimidazoles required as substrates need not be preformed but can be prepared and ring-closed *in situ* by the reaction of 2-chloro-*N*(1)-(β-oxoalkyl)-benzimidazoles with ammonia [Scheme 6.61; (**6.270**) → (**6.272**; R^2 = H) → (**6.275**; R^2 = H)],[118a,158] or more conveniently, but in lower yield (Table 6.33) by heating 2-aminobenzimidazoles with α-halogeno carbonyl compounds in alcoholic[155,156] or dimethylformamide[159] solution [Scheme 6.61; (**6.271**; $R^1 = R^2$ = H) → (**6.273**; R^1 = H) → (**6.276**; R^1 = H)]. Correspondingly the *in situ* condensation of 2-aminobenzimidazole with bromoacetaldehyde

TABLE 6.32 SYNTHESIS OF 1*H*-IMIDAZO[1,2-*a*]BENZIMIDAZOLES (**6.275**) BY RING-CLOSURE REACTIONS OF 2-CHLORO-1-(β-OXOALKYL)BENZIMIDAZOLES (**6.270**).[a]

Reaction conditions[b]	Product (**6.275**)			Yield (%)	m.p. (°C)	Solvent of crystallization
	R	R^1	R^2			
A	H	Me	$(CH_2)_2OH$	87	167–169	Water
A	H	Me	$CH_2CH{=}CH_2$[c]	67	224–226	Dioxane
A	H	Me	CH_2Ph	93	128–130	Acetone–water
A	H	Me	Ph	57	143–145	Acetone–water
B	H	Me	*m*-MeC_6H_4[c]	54	204–206	Methanol
A	H	Me	*p*-MeC_6H_4[d]	64	111–113	Acetone–water
B	H	Me	*p*-$MeOC_6H_4$[d]	85	113–115	Acetone–water
A	H	Me	*p*-$EtOC_6H_4$[d]	53	122–124	Acetone–water
B	H	Me	*a*-naphthyl[c]	47	180–182	Methanol
A	H	Ph	Me[e,f]	88	127	Ethanol
A	H	Ph	$(CH_2)_2OH$	83	166–168	Methanol–water
A	H	Ph	$(CH_2)_2NEt_2$[g]	70	192–194	Acetic acid
A	H	Ph	Bu^i	82	207–209	Methanol
A	H	Ph	C_6H_{11}[c]	66	203–205	Methanol
A	H	Ph	CH_2Ph	92	130–132	Methanol–water
A	H	Ph	Ph	71	204–206	Methanol–water
A	H	Ph	*m*-MeC_6H_4	71	188–190	Acetone–water
A	H	Ph	*p*-MeC_6H_4	65	188–190	Acetone–water
A	H	Ph	*p*-HOC_6H_4	50	348–350 (decomp.)	Dimethylformamide–water
A	H	Ph	*p*-$MeOC_6H_4$	88	205–207	Methanol
A	H	Ph	*p*-$EtOC_6H_4$[d]	87	149–151	Acetone–water
A	H	Ph	*a*-naphthyl	45	212–214	Acetone–water
A	H	*p*-$MeOC_6H_4$	H	83	295–297 (decomp.)	Dimethylformamide–water
A	H	*p*-$MeOC_6H_4$	$(CH_2)_2OH$	80	188–190	Acetone–water
B	H	*p*-$MeOC_6H_4$	Ph[d]	82	125–127	Acetone–water
A	H	*p*-BrC_6H_4	H	64	316–318 (decomp.)	Dimethylformamide–water
A	H	*p*-BrC_6H_4	$(CH_2)_2OH$	87	184–186	Acetone–water
A	H	*p*-BrC_6H_4	Bu^i[c]	77	225–227	Dioxane
B	H	*p*-BrC_6H_4	Ph	72	182–184	Acetone–water
B	H	*p*-BrC_6H_4	*p*-MeC_6H_4[d]	85	188–190	Acetone–water
A	H	2-thienyl	H	34	281–283 (decomp.)	Dimethylformamide–water
A	Me	Me	CH_2Ph	83	192–194	Acetone–water
A	Me	Me	Ph	75	198–200	Acetone–water
A	Me	Ph	$(CH_2)_2OH$	59	163–165	Acetone–water
A	Me	Ph	Bu^n[c]	54	211–213	Methanol
A	Me	Ph	Ph	51	154–156	Acetone–water

[a] From Refs. 118a, 157, and 158.

[b] *A* = R^2NH_2, MeOH/(140–180°, autoclave)(6 hr); *B* = R^2NH_2, dimethylformamide/(reflux)(4 hr);

[c] Picrate; free base is an oil.

[d] Monohydrate.

[e] Also obtained in quantitative yield from (**6.272**; R = H, R^1 = Ph, R^2 = Me) by treating with conc. HCl under reflux for 2 hr.

[f] Forms a hydrochloride, m.p. 243° (decomp.) (from ethanol–ether or acetone–methanol).

[g] Dipicrate.

TABLE 6.33. SYNTHESIS OF ALKYL AND ARYL 9*H*-IMIDAZO[1,2-*a*]BENZIMIDAZOLES BY RING-CLOSURE REACTIONS OF 2-AMINOBENZIMIDAZOLE DERIVATIVES

Starting material	Reaction conditions[a]	Product	Yield (%)	m.p. (°C)	Solvent of crystallization	Ref.
(**6.273**; $R = R^1 = H$, $CH(OEt)_2$ for COR^2)	*A*	(**6.276**; $R = R^1 = R^2 = H$)	95	190 (decomp.)	Methanol	155
(**6.271**; $R = R^1 = R^2 = H$)	*B*	(**6.276**; $R = R^2 = H$, $R^1 = CH_2COMe$)[b]	16	208–210	Methanol	156
(**6.271**; $R = R^1 = R^2 = H$)	*B*	(**6.276**; $R = R^1 = H$, $R^2 = Me$)	30	251–253 (decomp.)	—[c]	156
(**6.279**; $R = H$)	*C*	(**6.281**; $R = H$)	83	195–196	Benzene	170
(**6.280**; $R = H$)	*C*	(**6.281**; $R = H$)	76	—	—	170
(**6.273**; $R = H$, $R^1 = R^2 = Me$)	*D*	(**6.276**; $R = H$, $R^1 = R^2 = Me$)[d]	88	200	—[c]	165
(**6.279**; $R = Me$)	*C*	(**6.281**; $R = Me$)	95	94	—[c]	169
(**6.279**; $R = Me$)	*E*	(**6.281**; $R = Me$)	85–88	—	—	169
(**6.274**; $R^1 = Me$, $R^2 = COMe$)	*F*	(**6.277**; $R = Me$)	52	85	Light petroleum	171
(**6.273**; $R = H$, $R^1 = Et$, $R^2 = Me$)	*G*	(**6.276**; $R = H$, $R^1 = Et$, $R^2 = Me$)[e]	—[f]	236–238	—[c]	161
(**6.279**; $R = Et$)	*C*	(**6.281**; $R = Et$)	93	236	—[c]	169
(**6.279**; $R = Et$)	*E*	(**6.281**; $R = Et$)	85–88	—	—	169
(**6.278**; $R = Et$, $R^1 = H$, $R^2 = Br$)	*E*	(**6.281**; $R = Et$)	85–90	—	—	169
(**6.271**; $R = R^2 = H$, $R^1 = CH_2Ph$)	*H*	(**6.267**; $R = H$, $R^1 = CH_2Ph$, $R^2 = Me$)[g]	87	111	Ethanol–water	166
(**6.273**; $R = H$, $R^1 = CH_2Ph$, $R^2 = Me$)	*D*	(**6.276**; $R = H$, $R^1 = CH_2Ph$, $R^2 = Me$)	95	—	—	166
(**6.279**; $R = CH_2Ph$)	*C*	(**6.281**; $R = CH_2Ph$)	94	111–112	—[c]	169
(**6.278**; $R^1 = H$, $R^2 = Br$)	*E*	(**6.281**; $R = CH_2Ph$)	85–90	—	—	169
(**6.278**; $R^1 = Br$, $R^2 = H$)	*E*	(**6.281**; $R = CH_2Ph$)	85–90	—	—	169
(**6.273**; $R = H$, $R^1 = (CH_2)_2NEt_2$, $R^2 = Me$)	*D*	(**6.276**; $R = H$, $R^1 = (CH_2)_2NEt_2$, $R^2 = Me$)[h]	90	—[i]	—	162b
(**6.279**; $R = CH_2C{\equiv}CH$)	*E*	(**6.281**; $R = CH{=}C{=}CH_2$)[j]	50	67–69	Ethanol–water	170
(**6.280**; $R = CH{=}C{=}CH_2$)	*E*	(**6.281**; $R = CH{=}C{=}CH_2$)	72	—	—	170
(**6.273**; $R = R^1 = H$, $R^2 = Ph$)[b]	*I*	(**6.276**; $R = R^1 = H$, $R^2 = Ph$)	85	269–270 (decomp.)	Methanol	155, 156

TABLE 6.33 (*Continued*)

Starting material	Reaction conditions[a]	Product	Yield (%)	m.p. (°C)	Solvent of crystallization	Ref.
(**6.273**; $R = R^1 = H$, $R^2 = Ph$)	*J*	(**6.276**; $R = R^1 = H$, $R^2 = Ph$)	—[f]	—	—	156
(**6.273**; $R = R^1 = H$, $R^2 = Ph$)	*D*	(**6.276**; $R = R^1 = H$, $R^2 = Ph$)	91	310	Dimethylformamide	157
(**6.270**; $R = H$, $R^1 = Ph$)	*K*	(**6.276**; $R = R^1 = H$, $R^2 = Ph$)	87	285–287 (decomp.)	Acetic acid	118a, 158
(**6.273**; $R = R^1 = H$, $R^2 = p\text{-}ClC_6H_4$)[b]	*I*	(**6.276**; $R = R^1 = H$, $R^2 = p\text{-}ClC_6H_4$)	92	275–276 (decomp.)	Dimethylformamide	155, 156
(**6.273**; $R = R^1 = H$, $R^2 = p\text{-}ClC_6H_4$)	*J*	(**6.276**; $R = R^1 = H$, $R^2 = p\text{-}ClC_6H_4$)	90	—	—	155
(**6.273**; $R = R^1 = H$, $R^2 = p\text{-}BrC_6H_4$)[b]	*I*	(**6.276**; $R = R^1 = H$, $R^2 = p\text{-}BrC_6H_4$)	96	279–280 (decomp.)	Dimethylformamide	155, 156
(**6.273**; $R = R^1 = H$, $R^2 = p\text{-}BrC_6H_4$)	*J*	(**6.276**; $R = R^1 = H$, $R^2 = p\text{-}BrC_6H_4$)	90	—	—	155
(**6.271**; $R = R^1 = R^2 = H$)	*L*	(**6.276**; $R = R^1 = H$, $R^2 =$ 2-(5-nitrothienyl))	24	300	Methanol–chloroform	159
(**6.273**; $R = H$, $R^1 = Me$, $R^2 = Ph$)[b]	*M*	(**6.276**; $R = H$, $R^1 = Me$, $R^2 = Ph$)[b]	44	301–304 (decomp.)	Methanol	167
(**6.273**; $R = H$, $R^1 = Me$, $R^2 = Ph$)	*D*, *G*	(**6.276**; $R = H$, $R^1 = Me$, $R^2 = Ph$)[k]	92	120	Ethanol–water	162a
(**6.273**; $R = H$, $R^1 = Me$, $R^2 = p\text{-}BrC_6H_4$)	*D*, *G*	(**6.276**; $R = H$, $R^1 = Me$, $R^2 = p\text{-}BrC_6H_4$)	66	153	Methanol	162a
(**6.273**; $R = H$, $R^1 = Me$, $R^2 = o\text{-}O_2NC_6H_4$)	*G*	(**6.276**; $R = H$, $R^1 = Me$, $R^2 = o\text{-}O_2NC_6H_4$)	52	118	Methanol	163
(**6.273**; $R = H$, $R^1 = Me$, $R^2 = m\text{-}O_2NC_6H_4$)	*D*	(**6.276**; $R = H$, $R^1 = Me$, $R^2 = m\text{-}O_2NC_6H_4$)	70	225	Ethanol–dimethyl-formamide	163
(**6.273**; $R = H$, $R^1 = Me$, $R^2 = p\text{-}O_2NC_6H_4$)	*N*	(**6.276**; $R = H$, $R^1 = Me$, $R^2 = p\text{-}O_2NC_6H_4$)	82	192	Ethanol	163
(**6.273**; $R = H$, $R^1 = Me$, $R^2 =$ 2-naphthyl)	*N*	(**6.276**; $R = H$, $R^1 = Me$, $R^2 =$ 2-naphthyl)[l]	83	—[i]	—	164
(**6.274**; $R^1 = Ph$, $R^2 = COMe$)	*D*	(**6.277**; $R = Ph$)	51	117–118	Octane	171
(**6.273**; $R = H$, $R^1 = Et$, $R^2 = Ph$)	*G*	(**6.276**; $R = H$, $R^1 = Et$, $R^2 = Ph$)[m]	87	93–93.5	Ethanol–water	161

(**6.273**; $R = H$, $R^1 = Et$, $R^2 = p\text{-}BrC_6H_4$)	*O*	(**6.276**; $R = H$, $R^1 = Et$, $R^2 = p\text{-}BrC_6H_4)^{n}$	84	112–113	Ethanol	161
(**6.273**; $R = H$, $R^1 = Et$, $R^2 = m\text{-}O_2NC_6H_4$)	*O*	(**6.276**; $R = H$, $R^1 = Et$, $R^2 = m\text{-}O_2NC_6H_4)^{o}$	77	212–213	Ethanol	161
(**6.273**; $R = H$, $R^1 = Et$, $R^2 = p\text{-}O_2NC_6H_4$)	*O*	(**6.276**; $R = H$, $R^1 = Et$, $R^2 = p\text{-}O_2NC_6H_4)^{p}$	98	156–157	Ethanol	161
(**6.273**; $R = Me$, $R^1 = Et$, $R^2 = p\text{-}O_2NC_6H_4$)	*G*	(**6.276**; $R = Me$, $R^1 = Et$, $R^2 = p\text{-}O_2NC_6H_4$)	68	163	Ethanol	163
(**6.273**; $R = H$, $R^1 = CH_2Ph$, $R^2 = Ph$)	*D*, *G*	(**6.276**; $R = H$, $R^1 = CH_2Ph$, $R^2 = Ph$)	93	147	Methanol	162a
(**6.273**; $R = H$, $R^1 = CH_2COPh$, $R^2 = Ph)^{b}$	*J*	(**6.276**; $R = H$, $R^1 = CH_2COPh$, $R^2 = Ph)^{q}$	96	268–269	—^{c}	155, 156
(**6.273**; $R = H$, $R^1 = CH_2COPh$, $R^2 = Ph$)	*D*	(**6.276**; $R = H$, $R^1 = CH_2COPh$, $R^2 = Ph)^{r}$	quant.	265	Ethanol–water	157
(**6.273**; $R = H$, $R^1 = p\text{-}ClC_6H_4COCH_2$, $R^2 = p\text{-}ClC_6H_4$)	*J*	(**6.276**; $R = H$, $R^1 = p\text{-}ClC_6H_4COCH_2$, $R^2 = p\text{-}ClC_6H_4)^{r}$	97	321–323	—^{c}	155,156
(**6.273**; $R = H$, $R^1 = p\text{-}BrC_6H_4COCH_2$, $R^2 = p\text{-}BrC_6H_4$)	*J*	(**6.276**; $R = H$, $R^1 = p\text{-}BrC_6H_4COCH_2$, $R^2 = p\text{-}BrC_6H_4)^{t}$	98	239–240	—^{c}	155, 156
(**6.273**; $R = H$, $R^1 = (CH_2)_2NEt_2$, $R^2 = Ph$)	*D*	(**6.276**; $R = H$, $R^1 = (CH_2)_2NEt_2$, $R^2 = Ph)^{u}$	95	—^{i}	—	162b
(**6.273**; $R = H$, $R^1 = (CH_2)_2NEt_2$, $R^2 = p\text{-}ClC_6H_4)^{b}$	*P*	(**6.276**; $R = H$, $R^1 = (CH_2)_2NEt_2$, $R^2 = p\text{-}ClC_6H_4)^{b}$	75	197–199	Ethanol	155, 160
(**6.273**; $R = H$, $R^1 = (CH_2)_2NEt_2$, $R^2 = p\text{-}ClC_6H_4)^{b}$	*J*	(**6.276**; $R = H$, $R^1 = (CH_2)_2NEt_2$, $R^2 = p\text{-}ClC_6H_4)^{b}$	68	—	—	160
(**6.273**; $R = H$, $R^1 = (CH_2)_2NEt_2$, $R^2 = p\text{-}BrC_6H_4)^{b}$	*P*	(**6.276**; $R = H$, $R^1 = (CH_2)_2NEt_2$, $R^2 = p\text{-}BrC_6H_4)^{b}$	78	187–189	Ethanol	155, 160
(**6.273**; $R = H$, $R^1 = (CH_2)_2NEt_2$, $R^2 = p\text{-}BrC_6H_4)^{b}$	*J*	(**6.276**; $R = H$, $R^1 = (CH_2)_2NEt_2$, $R^2 = p\text{-}BrC_6H_4)^{b}$	78	—	—	160
(**6.273**; $R = H$, $R^1 = (CH_2)_2NMe_2$, $R^2 = p\text{-}ClC_6H_4)^{b}$	*P*	(**6.276**; $R = H$, $R^1 = (CH_2)_2NMe_2$, $R^2 = p\text{-}ClC_6H_4)^{b}$	70	190–192	Ethanol	155, 160
(**6.273**; $R = H$, $R^1 = (CH_2)_2NMe_2$, $R^2 = p\text{-}BrC_6H_4)^{b}$	*P*	(**6.276**; $R = H$, $R^1 = (CH_2)_2NMe_2$, $R^2 = p\text{-}BrC_6H_4)^{b}$	65	218–219	Ethanol	155, 160
(**6.273**; $R = H$, $R^1 = (CH_2)_2N\bigcirc$, $R^2 = Ph$)	*D*	(**6.276**; $R = H$, $R^1 = (CH_2)_2N\bigcirc$, $R^2 = Ph)^{b}$	95	80–82	Ethanol–water	162b

(*Footnotes overleaf*)

(*Footnotes to Table 6.33*)

[a] *A* = 2M HCl/(room temp.)(14 hr); *B* = $MeCOCH_2Br$, Bu^nOH/(130–135°)(30 min); *C* = NaOEt, EtOH/(reflux)(1–5 hr); *D* = conc. HCl/(reflux)(2–7 hr); *E* = KOH, tetrahydrofurane/(room temp.)(2–24 hr); *F* = 10% HCl/(reflux)(1 hr); *G* = $POCl_3$/(reflux)(4–12 hr); *H* = I_2, MeCOMe/(reflux)(1 hr, then conc. HCl)(reflux)(1.5 hr); *I* = 10% NaOH, MeOH/(room temp.)(few min); *J* = EtOH or MeOH/(reflux)(10–24 hr); *K* = NH_3/(160–180°, autoclave)(6 hr); *L* = 5-nitro-2-thienyl bromomethyl ketone (O_2N–thiophene–$COCH_2Br$), dimethylformamide/(reflux)(10 min); *M* = 48% HBr/(reflux)(6 hr); *N* = $POCl_3$, conc. HCl/(reflux)(8–20 hr); *O* = 85% HCO_2H/(reflux)(5 hr); *P* = no solvent/(190–210°)(10 min).

[b] Hydrobromide.

[c] Solvent of crystallization not specified.

[d] Hydrochloride; free base has m.p. 94° (from ethanol); hydrates on standing.

[e] Picrate.

[f] Yield not quoted.

[g] Forms a hydrate, m.p. 80°.

[h] Forms a dihydrochloride, m.p. 256–258° (from ethanol), and a dipicrate, m.p. 160–162° (from ethanol).

[i] Oil.

[j] Forms a picrate, yellow prisms, m.p. 178–179° (from ethanol).

[k] Decomposes on storage at room temp.

[l] Forms a hydrochloride, m.p. 262° (decomp.) (from ethanol–water), and a picrate, m.p. 248° (decomp.) from ethanol–dimethylformamide).

[m] Forms a picrate, m.p. 238–240° (decomp.) (from ethanol).

[n] Forms a picrate, m.p. 228–230°.

[o] Forms a picrate, m.p. 224–226°.

[p] Forms a picrate, m.p. 238–240°.

[q] Hydrobromide; free base has m.p. 218° (from ethanol).

[r] Hydrobromide; free base has m.p. 207–208° (from ethanol–water).

[s] Hydrobromide; free base has m.p. 218–219° (from ethanol).

[t] Hydrobromide; free base has m.p. 240–242°.

[u] Forms a dihydrochloride, m.p. 247° (decomp.) (from ethanol) and a dipicrate, m.p. 205–206° (from ethanol).

[v] Forms a dihydrochloride, m.p. 269° (decomp.) (from ethanol–diethyl ether) and a dipicrate, m.p. 225° (from acetic acid).

TABLE 6.34 SYNTHESIS OF ACYL-9*H*-IMIDAZO[1,2-*a*]BENZIMIDAZOLES BY RING-CLOSURE REACTIONS OF 2-AMINOBENZIMIDAZOLE DERIVATIVES

Starting material	Reaction conditions[a]	Product	Yield (%)	m.p. (°C)	Solvent of crystallization	Ref.
(**6.274**; $R^1 = Ph$, $R^2 = H$)	*A*	(**6.284**; $R^1 = R^2 = Me$) + (**6.284**; $R^1 = Ph$, $R^2 = Me$)	40; 32	177; 140–142	Acetone; Acetone	171
(**6.283**; $R^1 = Ph$, $R^2 = Me$)	*B*	(**6.285**; $R^1 = Ph$, $R^2 = Me$)	50	—	—	171
(**6.282**; $R = H$, $R^1 = Me$, $R^2 = COPh$)	*C*	(**6.285**; $R = H$, $R^1 = Me$, $R^2 = Ph$)	58	152	Octane	171
(**6.282**; $R = H$, $R^1 = Me$, $R^2 = COPh$)	*D*	(**6.285**; $R = H$, $R^1 = R^2 = Ph$)	61	156	Octane	171
(**6.282**; $R = H$, $R^1 = Me$, $R^2 = CH_2CO_2H$)	*E*	(**6.286**; $R = H$, $R^1 = R^2 = Me$)	50	178	Ethanol–water	173
(**6.282**; $R = H$, $R^1 = Me$, $R^2 = CH_2COMe$)[b]	*F*	(**6.286**; $R = H$, $R^1 = R^2 = Me$)	100	178	Ethanol	172
(**6.282**; $R = H$, $R^1 = Me$, $R^2 = CH_2COPh$)	*F*	(**6.286**; $R = H$, $R^1 = R^2 = Me$) + (**6.286**; $R = H$, $R^1 = R^2 = Me$, Ph for Me)	42; 56	178; 212	Ethanol; Ethanol	172
(**6.282**; $R = H$, $R^1 = CH_2Ph$, $R^2 = CH_2COMe$)[b]	*F*	(**6.286**; $R = H$, $R^1 = CH_2Ph$, $R^2 = Me$)[c]	95	170	Ethanol	172
(**6.282**; $R = Me$, $R^1 = Et$, $R^2 = CH_2COMe$)[b]	*F*	(**6.286**; $R = R^2 = Me$, $R^1 = Et$)[d]	96	176	Ethanol	172
(**6.284**; $R = H$, $R^1 = Me$, $R^2 = Ph$)[b]	*G*	(**6.286**; $R = H$, $R^1 = Me$, $R^2 = Ph$)	48	156	Benzene	172
(**6.282**; $R = H$, $R^1 = Me$, $R^2 = CH_2CO_2Me$)[e]	*H*	(**6.286**; $R = H$, $R^1 = Me$, $R^2 = OMe$)	89	138	Ethanol–water	173

TABLE 6.34 (*Continued*)

Starting material	Reaction conditions[a]	Product	Yield (%)	m.p. (°C)	Solvent of crystallization	Ref.
(**6.282**; R = H, R^1 = Me, $R^2 = CH_2CO_2C_7H_{15}$)[e]	*H*	(**6.286**; R = H, R^1 = Me, $R^2 = OC_7H_{15}$)	76	79–80	Light petroleum	173
(**6.282**; R = H, $R^1 = CH_2Ph$, $R^2 = CH_2CO_2Me$)[e]	*H*	(**6.286**; R = H, $R^1 = CH_2Ph$, R^2 = OMe)	85	112	Light petroleum	173

[a] *A* = Ac_2O, NaOAc/(room temp.)(few min); *B* = dimethylformamide/(80–90°)(20 hr); *C* = $MeCOCH_2Br$, dimethylformamide/(80°)(20 hr); *D* = $PhCOCH_2Br$, dimethylformamide/(80°)(16 hr); *E* = Ac_2O/(reflux)(3 hr); *F* = Ac_2O, NaOAC/(reflux)(30–60 min); *G* = Et_3N, dimethylformamide/(reflux)(5 hr); *H* = Ac_2O, NaOAc/(reflux)(5 hr).

[b] Hydrobromide.

[c] Forms a dinitrophenylhydrazone, m.p. 218° (from ethanol–dimethylformamide).

[d] Forms a dinitrophenylhydrazone, m.p. 216° (from ethanol).

[e] Hydrochloride.

TABLE 6.35 SYNTHESIS OF AMINO-9*H*-IMIDAZO[1,2-*a*]BENZIMIDAZOLES BY RING-CLOSURE REACTIONS OF 2-AMINOBENZIMIDAZOLE DERIVATIVES

Starting material	Reaction conditions[a]	Product	Yield (%)	m.p. (°C)	Solvent of crystallization	Ref.
(**6.273**; R = H, R^2 = Me, R^2 = NHPh)	*A*	(**6.276**; R = H, R^1 = Me, R^2 = NHPh)[b]	90–95	226	Ethanol–dimethyl-formamide	174
(**6.273**; R = H, R^1 = Me, R^2 = N(Me)Ph)	*A*	(**6.276**; R = H, R^1 = Me, R^2 = N(Me)Ph)[b]	90–95	162	Ethanol	174
(**6.273**; R = H, R^1 = Me, R^2 = *p*-$O_2NC_6H_4NH$)	*A*	(**6.276**; R = H, R^1 = Me, R^2 = *p*-$O_2NC_6H_4NH$)[b]	90–95	189	Acetic acid	174
(**6.273**; R = H, R^1 = Me, R^2 = *p*-ClC_6H_4NH)	*A*	(**6.276**; R = H, R^1 = Me, R^2 = *p*-ClC_6H_4NH)[b]	90–95	234	Dimethylformamide	174
(**6.273**; R = H, R^1 = Me, R^2 = *p*-$EtO_2CC_5H_4NH$)	*A*	(**6.276**; R = H, R^1 = Me, R^2 = *p*-$EtO_2CC_6H_4NH$)	90–95	230	Dimethylformamide	174
(**6.287**; R = Me)	*B*	(**6.288**; R = Me, Ar = *p*-$O_2NC_6H_4$)[c]	—[d]	232	Ethanol	175
(**6.287**; R = Ph)	*C*	(**6.288**; R = Ar = Ph)[c]	97	137–138	Ethanol	175
(**6.287**; R = Ph)	*D*	(**6.289**)[e]	91	235 (decomp.)	Ethanol	175

[a] *A* = $POCl_3$/(reflux)(3 hr); *B* = *p*-$O_2NC_6H_4CHO$, EtOH/reflux (reaction time not specified); *C* = PhCHO/(130°, melt)(5–7 min); *D* = Ac_2O/reflux (reaction time not specified).

[b] Picrate; free bases and their hydrochlorides are unstable.

[c] Orange needles.

[d] Yield not quoted.

[e] Colorless plates.

TABLE 6.36 SYNTHESIS OF 2,3-DIHYDRO-1*H*- AND 9*H*-IMIDAZO[1,2-*a*]BENZIMIDAZOLES BY RING-CLOSURE REACTIONS OF 2-AMINOBENZIMIDAZOLE DERIVATIVES

Starting material (R, $R^2 \rightarrow R^5$ unspecified = H)	Reaction conditions[a]	Product (R, $R^1 \rightarrow R^4$ unspecified = H)	Yield (%)	m.p. (°C)	Solvent of crystallization	Ref.
(**6.290**)	*A*	(**6.291**)	17	97.5–98	Diethylether-light petroleum	179
(**6.290**; $R = NO_2$)	*A*	(**6.291**)[b]	92	161–163	Ethanol	178
(**6.292**)	*B*	(**6.293**)	68	204–206	Dioxane	176
(**6.292**; $R^2 = NO_2$)	*B*	(**6.293**; $R^2 = NO_2$)[c]	75	262–263	Ethanol	176
(**6.292**; $R^1 = Et$)	—[d]	(**6.293**; $R^1 = Et$)[e]	—[f]	267–268 (decomp.)	Acetic acid	161
(**6.292**; $R^1 = CH_2Ph$)[g]	*C*	(**6.293**; $R^1 = CH_2Ph$)[h]	22	—[i]	—	176
(**6.271**; $R^1 = Ph$)	*D*	(**6.293**; $R^1 = Ph$)[j]	14	225–228	Acetone	177
(**6.292**; $R^1 = Ph$)	*E*	(**6.293**; $R^1 = Ph$)[j]	37	—	—	177
(**6.292**; $R^1 = p\text{-}ClC_6H_4$, $R^2 = Cl$)	*C*	(**6.293**; $R^1 = p\text{-}ClC_6H_4$, $R^2 = Cl$)	—[f]	—[k]	—	177
(**6.294**; $R^1 = NH_2$)	*F*	(**6.296**)	78	214	Ethanol	180
(**6.295**)	*G*	(**6.296**)	23	—	—	180
(**6.294**; $R = CH_2CO_2H$)	*H*	(**6.296**; $R = CH_2CON$[piperazine ring]NMe)	60	227–229	Acetone	181
(**6.297**; $R = Me$)	*F*	(**6.298**; $R = Me$)	85	265–266	Dimethylformamide	173
(**6.297**; $R = CH_2Ph$)	*F*	(**6.298**; $R = CH_2Ph$)	90	198	Ethanol	173
(**6.299**; $R^3 = Cl$)	*I*	(**6.301**; $R^3 = Cl$)	78	268–270	Xylene	182
(**6.299**; $R^2 = Cl$)	*I*	(**6.301**; $R^2 = Cl$) + (**6.301**); $R^4 = Cl$)	42 22	291–293 335–337	Xylene Xylene	182
(**6.302**)	*I*	(**6.301**; $R^3 = Cl$) + (**6.301**; $R^2 = Cl$)	23 18	— —	— —	182
(**6.299**; $R^1 = Cl$)	*I*	(**6.301**; $R^1 = Cl$)	80	296–298	Nitromethane	182
(**6.299**; $R^3 = Br$)	*I*	(**6.301**; $R^3 = Br$)	55	272–276	Xylene	182

(**6.299**; $R^3 = NO_2$)	*I*[l]	(**6.301**; $R^3 = NO_2$)	52	280–283	Xylene	182
(**6.299**; $R^3 = CN$)	*I*	(**6.301**; $R^3 = CN$)	—[f]	255–259	—[m]	183
(**6.299**; $R^3 = F$)	*I*	(**6.301**; $R^3 = F$)	—[f]	265–267	—[m]	183
(**6.299**; $R^3 = COPh$)	*I*	(**6.301**; $R^3 = COPh$)	—[f]	239–241	—[m]	183
(**6.299**; $R^1 = R^3 = Cl$)	*I*	(**6.301**; $R^1 = R^3 = Cl$)	80	285–287	Xylene	182
(**6.299**; $R^1 = R^5 = Cl$)	*J*	(**6.301**; $R^1 = R^4 = Cl$)	17	326–330	Nitrobenzene	184
(**6.299**; $R^1 = R^4 = Cl$)	*I*	(**6.301**; $R^1 = R^4 = Cl$)	80	—	—	184
(**6.299**; $R^1 = R^3 = R^5 = Br$)	*K*	(**6.301**; $R^1 = R^3 = R^4 = Br$)	20	300–304	Xylene	184
(**6.303**)	*L*	(**6.305**)	94	237–238	Benzene-acetone	6
(**6.306**; $R^1 = CH_2Ph$)	*M*	(**6.307**; $R = CH_2Ph$, $X = O$)	58	110	Octane	185
(**6.306**)	*N*	(**6.307**; $X = NCF_3$)	70	175 (decomp.)	—[m]	187
(**6.306**; $R^2 = CO_2Me$)	*N*	(**6.307**; $R = CO_2Me$, $X = NCF_3$)	74	152–153 (decomp.)	Acetone	187
(**6.306**; $R^2 = CO_2Et$)	*N*	(**6.307**; $R = CO_2Et$, $X = NCF_3$)	76	108 (decomp.)	—[m]	187
(**6.306**)	*M*	(**6.308**; $X = Y = O$)	66	324–325	Acetic acid	185
(**6.306**; $R^1 = Me$)	*M*	(**6.308**; $R = Me$, $X = Y = O$)	90	295	Ethanol	185
(**6.306**; $R^1 = Me$, $R^2 = COCO_2Me$)	*O*	(**6.308**; $R = Me$, $X = Y = O$)	25	—	—	185
(**6.306**; $R^1 = Et$)	*M*	(**6.308**; $R = Et$, $X = Y = O$)	67	269	Ethanol–water	185
(**6.306**; $R^1 = Et$, $R^2 = COCO_2Me$)	*P*	(**6.308**; $R = Et$, $X = Y = O$)	27	—	—	185
(**6.306**; $R^1 = CH_2Ph$)	*M*	(**6.308**; $R = CH_2Ph$, $X = Y = O$)	66	172–173	Ethanol	185
(**6.306**; $R^1 = CH_2Ph$, $R^2 = COCO_2Me$)	*P*	(**6.308**; $R = CH_2Ph$, $X = Y = O$)	20	—	—	185
(**6.306**)	*Q*	(**6.308**; $X = O$, $Y = NPh$)	69	218–219	Ethanol	151
(**6.306**)	*R*	(**6.308**; $X = O$, $Y = p\text{-}ClC_6H_4NHN$)	56	180–182	Ethanol	151, 186
(**6.306**)	*R*	(**6.308**; $X = O$ $Y = o\text{-}ClC_6H_4NHN$)	49	225–227	Ethanol	151, 186

(*Footnotes overleaf*)

(*Footnotes to Table 6.36*)

[a] $A = 160–190°$ (no solvent)/(2 hr); $B = SOCl_2$, dimethylformamide/(0–2°)(1 hr, then reflux 3–8 hr); $C = SOCl_2$, $CHCl_3$/(reflux)(2–3 hr, then NaOH or KOH, H_2O, MeOH/(reflux)(2–4 hr); $D = Br(CH_2)_2Br$, toluene/(reflux)(24 hr); $E = SOCl_2$, $CHCl_3$/(reflux)(2 hr); $F =$ AcO/(reflux)(5–20 min); $G = ClCH_2CO_2Me$/(reflux)(1 hr); $H = SOCl_2$/(reflux)(10 min) then treatment with HN◯NMe (piperazine ring) (room temp.)(30 min); $I =$ xylene/(reflux)(15 min); $J =$ bromobenzene/(reflux)(5 min); $K =$ xylene/(reflux)(45 min); $L = Ph_2C{=}C{=}O$, dioxane/(room temp.)(14 hr); $M = (COCl)_2$, dioxane/(15–20°)(1 hr, then Et_3N)(reflux)(30 min); $N = (FC{=}NCF_3)_2$, NaF, MeCN/(−30° then 0°)(2 hr); $O =$ melt (240–250°)/8 hr; $P =$ melt (180–200°)/5 hr; $Q = PhN{=}C(Cl)COCl$, Et_3N, benzene–dioxane/(reflux)(3 hr); $R =$ ArNHN=C(Cl)COCl, Et_3N, benzene–dioxane/(100°)(3 hr).

[b] Forms a hydrochloride, m.p. 245–248°.

[c] Forms a hydrochloride, m.p. 262–263°.

[d] Reaction conditions unspecified.

[e] Picrate.

[f] Yield not quoted.

[g] Hydrochloride.

[h] Forms a picrate, m.p. 204–206° (ethanol).

[i] Yellow oil, b.p. 192–194°/0.15 mm Hg.

[j] Hydrobromide.

[k] Oil.

[l] Time of reflux, 2 hr.

[m] **Solvent of crystallization unspecified.**

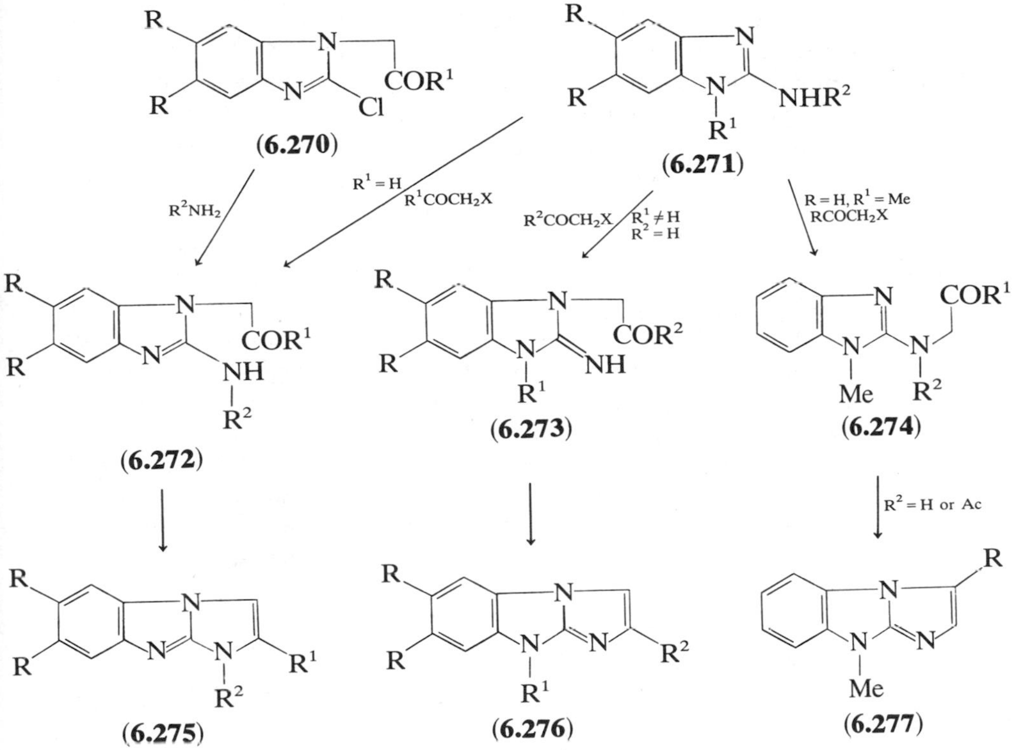

Scheme 6.61

diethylacetal in the presence of hydrochloric acid[155] affords the parent 9*H*-imidazo[1,2-*a*]benzimidazole (**6.276**; $R = R^1 = R^2 = H$) in excellent yield (Table 6.33). General synthetic access to 2-alkyl and 2-aryl *N*(1)-substituted 1*H*-imidazo[1,2-*a*]benzimidazoles (**6.275**; R^1 = alkyl or aryl, $R^2 \neq H$) (Table 6.32) is provided by the ring-closure of 2-(substituted amino)-*N*(1)-(β-oxoalkyl)benzimidazoles (**6.272**; R^1 = alkyl or aryl, $R^2 \neq H$) catalyzed by hydrochloric acid[157] or promoted spontaneously in the course of the reactions of 2-chloro-*N*(1)-(β-oxoalkyl)benzimidazoles (**6.270**) with aliphatic and aromatic amines.[157,158] Conversely, 2-alkyl and 2-aryl *N*(9)-substituted 9*H*-imidazo[1,2-*a*]benzimidazoles (**6.276**; $R \neq H$, R^2 = alkyl or aryl) are synthesized usually in excellent yield (Table 6.33) by the cyclization of preformed *N*(3)-substituted *N*(1)-(β-oxoalkyl)-2-iminobenzimidazolines (**6.273**; $R^1 \neq H$, R^2 = alkyl or aryl) or their hydrobromides, made readily available by the *N*(1)-alkylation of *N*(3)-substituted 2-aminobenzimidazoles with α-halogeno carbonyl compounds. Ring-closure of this type can be induced in a purely thermal fashion in the presence[155,156,160] or absence[155,160] of an alcoholic solvent or under the influence of a variety of acidic catalysts including phosphorus oxychloride,[161–164] hydrochloric acid,[157,162,166] hydrobromic acid,[167] and formic acid.[161] A potentially useful, yet little exploited "one pot" version of such 9*H*-imidazo[1,2-*a*]benzimidazole synthesis is based on the *in situ* formation of the *N*(1)-(β-oxoalkyl)-2-iminobenzimidazoline precursor by *N*-β-oxoalkylation of *N*(1)-substituted 2-aminobenzimidazoles with acetone in the presence of iodine (the Ortoleva–King reaction[168]). This modification is exemplified[166] by the reaction of 2-amino-1-benzylbenzimidazole with acetone and iodine to give, via the presumed intermediacy of the iminobenzimidazoline (**6.273**; R = H, $R^1 = CH_2Ph$, $R^2 = Me$), 9-benzyl-2-methyl-9*H*-imidazo[1,2-*a*]benzimidazole (**6.276**; R = H, $R^1 = CH_2Ph$, $R^2 = Me$) in high yield (Table 6.33). 9-Substituted-2-methyl-9*H*-imidazo[1,2-*a*]benzimidazoles are also the end-products (Table 6.33) of the sodium ethoxide or potassium hydroxide mediated ring-closure of *N*(3)-substituted *N*(1)-(2-propynyl)-2-iminobenzimidazolines or of their vinyl bromide precursors [Scheme 6.62; (**6.278**; R^1 or $R^2 = Br$) → (**6.279**) → → (**6.281**)].[169,170] A course for these reactions involving base-catalyzed rearrangement [Scheme 6.62; (**6.279**) → (**6.280**)] prior to cyclization is demonstrated[170] by the isolation of allene intermediates of the type (**6.280**) and their smooth transformation into the corresponding 2-methyl-9*H*-imidazo[1,2-*a*]benzimidazoles (**6.281**) on treatment with base. Derivatives of the 9*H*-imidazo[1,2-*a*]benzimidazole ring system are also accessible in high yield (Table 6.33) by the hydrochloric acid-catalyzed ring-closure of *N*(1)-substituted 2-(β-oxoalkylamino) benzimidazoles [Scheme 6.61; (**6.274**) → (**6.277**)].[171] Cyclization of this type complements that through *N*(1)-(β-oxoalkyl)-2-iminobenzimidazolines [e.g. (**6.273**)] in providing synthetic access to 3-substituted (as opposed to 2-substituted) 9*H*-imidazo[1,2-*a*]benzimidazoles.

Scheme 6.62

Ring-closure in *N*(1)-(β-oxoalkyl)-2-iminobenzimidazoline derivatives also provides the basis for the efficient synthesis (Scheme 6.63) of 9*H*-imidazo[1,2-*a*]benzimidazoles containing an ester or ketonic substituent at the $C(3)$ position. For example, reaction with acetic anhydride in the presence of sodium acetate affords acetylimino intermediates (**6.284**; R^2 = alkyl, aryl, or *O*-alkyl), which undergo spontaneous intramolecular aldol-type condensation, affording the corresponding 3-acyl- or 3-alkoxycarbonyl-9*H*-imidazo[1,2-*a*]benzimidazoles (**6.286**; R^2 = alkyl, aryl, or *O*-alkyl) in moderate yield (Table 6.34).[172,173] One disadvantage[172] of this approach is the tendency for "acyl scrambling" to occur giving mixtures of the 3-acyl and 3-acetyl derivatives, the latter as a result of deacylation of the former and reacetylation in Friedel–Crafts fashion (see later). However, this difficulty can be circumvented by cyclization of the preformed acetyliminobenzimidazoline (**6.284**) using triethylamine as catalyst.[172] 2-Acyl-9*H*-imidazo-[1,2-*a*]benzimidazoles are synthesized[171] in moderate yield (Table 6.34) by the *in situ* formation and dehydrative cyclization of *N*(1)-acyl-2-(β-oxoalkylimino)benzimidazolines [Scheme 6.63; (**6.283**) → (**6.285**)].[171] This otherwise convenient synthetic method is complicated by the tendency for the *N*(1)-acyl-2-iminobenzimidazoline (**6.282**; R^2 = acyl) used as starting

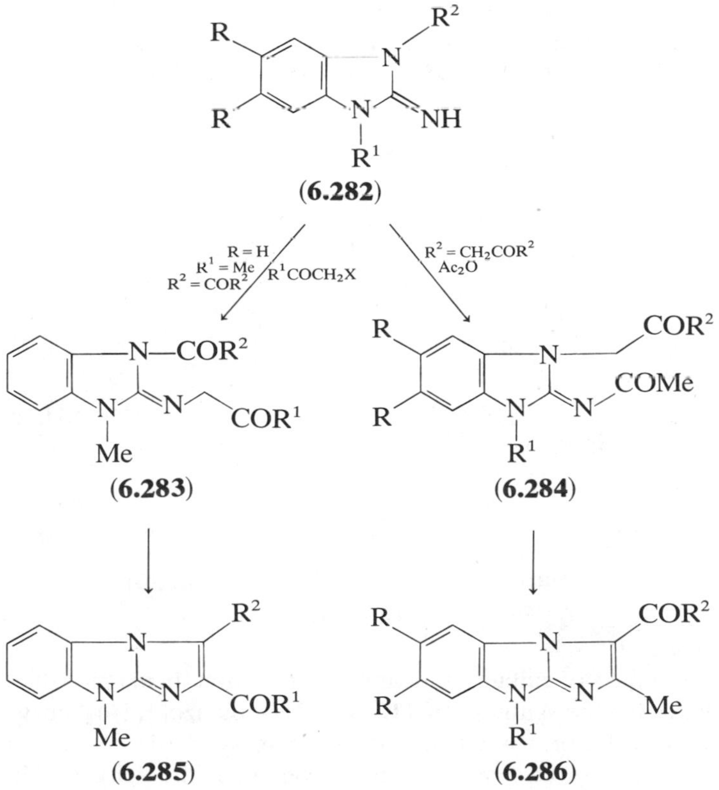

(**6.282**)

R = H, R^1 = Me, R^2 = COR^2; R^1COCH_2X

R^2 = CH_2COR^2; Ac_2O

(**6.283**) (**6.284**)

(**6.285**) (**6.286**)

Scheme 6.63

material to undergo initial thermal rearrangement to the corresponding 2-acylaminobenzimidazole with consequent ultimate contamination of the 2-acyl-9*H*-imidazo[1,2-*a*]benzimidazole product by the 3-acyl isomer. However, the thermal cyclization of the preformed *N*(1)-acyl-2-(β-oxoalkylimino)benzimidazoline (**6.283**) (as the hydrobromide) in dimethylformamide gets round this difficulty and affords the required 2-acyl-9*H*-imidazo[1,2-*a*]-benzimidazole in reasonable yield (Table 6.34).[171] The alternative approach[171] to 2-acyl-9*H*-imidazo[1,2-*a*]benzimidazoles, of *in situ* acylation–cyclization of 2-(β-oxoalkylamino)benzimidazoles [Scheme 6.61; (**6.274**; $R^2 = H$)] is only moderately successful and also tends to result in acyl scrambling as a consequence of deacylation of the initial product followed by reacylation by the reagent.

The phosphorus oxychloride-catalyzed cyclization[174] of *N*(1)-(β-carbamoylmethyl)-2-iminobenzimidazolines [Scheme 6.63; (**6.282**; $R^2 = CH_2CONHR$)] provides a high yield (Table 6.35) route to the otherwise difficultly accessible and highly unstable 2-amino derivatives of 9*H*-imidazo-[1,2-*a*]benzimidazoles. The equally elusive 3-amino-9*H*-imidazo[1,2-*a*]-benzimidazoles can be isolated[175] in the form of acetyl and benzylidene derivatives produced by the reaction of their open-chain 2-(α-cyanoalkyl)-benzimidazole tautomers with acetic anhydride and aromatic aldehydes, respectively [Scheme 6.64; (**6.287**) → (**6.288**), or (**6.289**)].

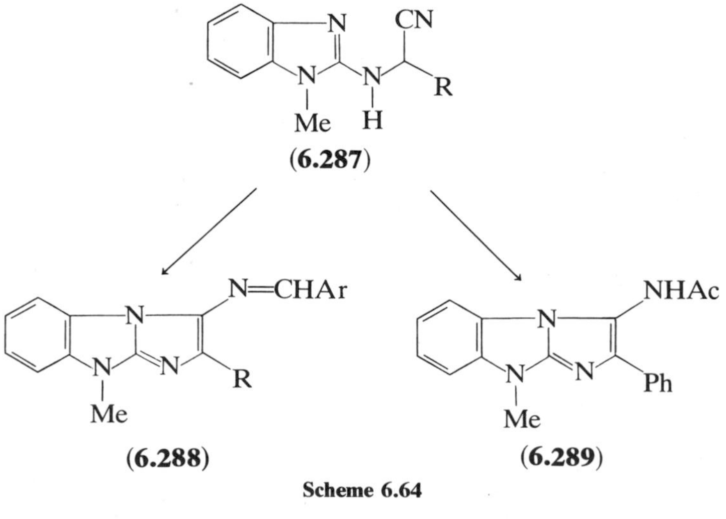

Scheme 6.64

The importance of suitably functionalized 2-aminobenzimidazoles as key intermediates for the synthesis of 1*H*- and 9*H*-imidazo[1,2-*a*]benzimidazole derivatives (see before) is equally apparent in synthetic routes leading to their 2,3-dihydro counterparts. A simple example is the dehydrochlorinative ring-closure of *N*(1)-(β-chloroalkyl)-2-aminobenzimidazoles [prepared by the *in situ* reaction of the corresponding (β-hydroxyalkyl)benzimidazoles with

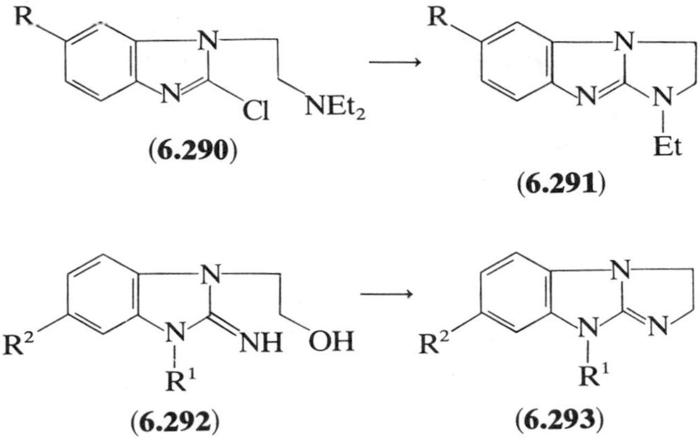

(6.290) → (6.291)

(6.292) → (6.293)

Scheme 6.65

thionyl chloride] which provides an efficient (Table 6.36) and potentially general method for the synthesis of tautomeric 2,3-dihydro-1(9)*H*-imidazo[1,2-*a*]benzimidazole derivatives [Scheme 6.65; (**6.292**; $R^1 = H$)→→(**6.293**; $R^1 = H$)].[176] Extension[161,176,177] of this methodology to *N*(3)-substituted *N*(1)-(β-hydroxyalkyl)-2-iminobenzimidazolines (**6.292**; $R^1 \neq H$) affords high yield (Table 6.36) synthetic access to *N*(9)-alkyl- and aryl-2,3-dihydro-9*H*-imidazo[1,2-*a*]benzimidazoles (**6.293**; R^1 = alkyl or aryl). Products of the latter type are also formed, though in poor yield (Table 6.36) by the direct annelation of *N*(1)-substituted-2-aminobenzimidazoles with 1,2-dibromoethane.[177] The corresponding chlorinative–dehydrochlorinative cyclization of 2-(*N*-substituted amino)-*N*(1)-(β-hydroxyalkyl)benzimidazoles does not appear to have been reported, the anticipated *N*(1)-substituted 2,3-dihydro-1*H*-imidazo[1,2-*a*]benzimidazole products of such ring-closure being available, though in variable yield (Table 6.36) by the alternative dealkylative thermal cyclization of *N*(1)-(β-dialkylaminoalkyl)-2-chlorobenzimidazoles, e.g. [Scheme 6.65; (**6.290**) → (**6.291**)].[178,179]

The dehydrative cyclization (Scheme 6.66) of [2-(*N*-substituted amino)-1-benzimidazolyl]acetic acid derivatives, e.g. (**6.294**; R^1 = OH, OMe, or NH_2), which can be induced thermally[180] or by heating with acetic anhydride[180] or thionyl chloride,[181] represents a simple and potentially general method for the synthesis of *N*(1)-substituted 2,3-dihydro-1*H*-imidazo[1,2-*a*]-benzimidazol-2-ones, e.g. (**6.296**). *N*(9)-Substituted 2,3-dihydro-9*H*-imidazo[1,2-*a*]benzimidazol-2-ones are formed in excellent yield (Table 6.36) by the analogous, acetic anhydride-catalyzed cyclization of *N*(3)-substituted *N*(1)-(2-iminobenzimidazolyl)acetic acids [Scheme 6.67; (**6.297**)→(**6.298**)].[173] (2-Benzimidazolyl) aminoacetic esters, on the other hand, cyclize thermally to 2,3-dihydro-1(9)*H*-imidazo[1,2-*a*]benzimidazol-3-ones only in low yield (Table 6.36).[182] Moreover, the efficiency of 2,3-dihydro-1(9)*H*-imidazo[1,2-*a*]benzimidazole synthesis of this type is further

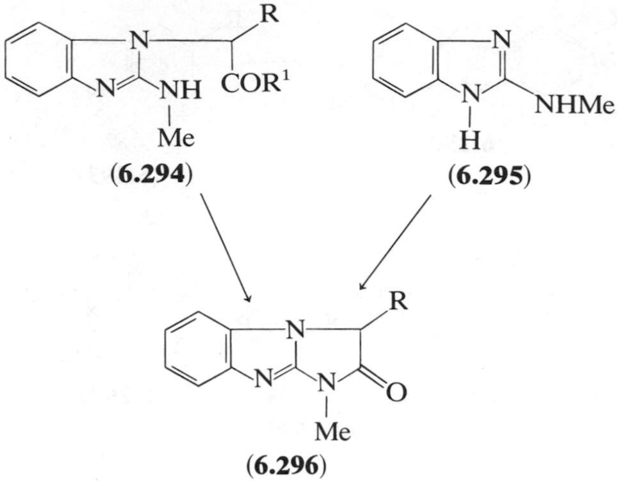

(**6.294**) (**6.295**)

(**6.296**)

Scheme 6.66

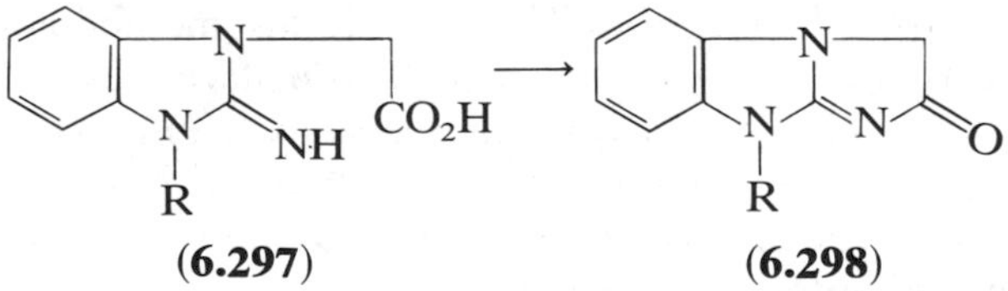

(**6.297**) (**6.298**)

Scheme 6.67

diminished by the tendency for substrates unsymmetrically substituted in the benzene ring to give rise to isomer mixtures as a result of alternative ring-closure at the nonequivalent benzimidazole nitrogen atoms, e.g. [Scheme 6.68; (**6.302**)→(**6.301**; $R^1 = R^3 = R^4 = H$, $R^2 = Cl$) + (**6.301**; $R^1 = R^2 = R^4 = H$, $R^3 = Cl$)].[182] 2,3-Dihydro-1(9)*H*-imidazo[1,2-*a*]benzimidazol-3-ones are also formed less orthodoxly, but nonetheless in high yield (Table 6.36) by the smooth thermal rearrangement of adducts (1-cyano-2-aryl-4,4-diphenylazetidin-3-ones) derived by the cycloaddition of diphenylketene to arenediazo cyanides [Scheme 6.68; (**6.299**) → → (**6.301**)].[182–184] Applied to diazetidinones (**6.299**) containing *ortho*- or *para*-substituted phenyl groups, transformations of this type allow the unambiguous synthesis of 5- and 7-substituted 2,3-dihydro-1(9)*H*-imidazo[1,2-*a*]benzimidazol-3-ones (**6.301**) in high yield (Table 6.36).[182,183] On the other hand, mixtures of 6- and 8-substituted products (**6.301**) result from the thermolysis[182] of 2-(*meta*-substituted phenyl)diazetidinones (**6.299**), and the rearrangement of substrates (**6.299**), in which both of the ortho positions of the phenyl ring are occupied by halogen substitutents, leads to nuclear substituted

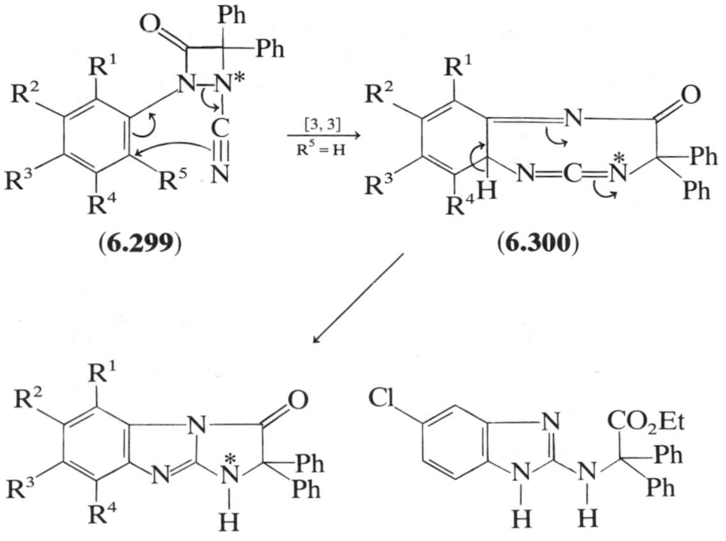

(6.299) (6.300) (6.301) (6.302)

Scheme 6.68

dihydroimidazo[1,2-*a*]benzimidazol-3-ones (**6.301**) derived by unprecedented halogen migration.[184] The activation parameters observed[183] for the deep-seated thermal transformations of 1-cyano-2-aryldiazetidinones (**6.299**) into 2,3-dihydro-1(9)*H*-imidazo[1,2-*a*]benzimidazol-3-ones (**6.301**) are consistent with a course (Scheme 6.68) initiated by Cope rearrangement and concluded by transannular cyclization in the eight-membered carbodiimide (**6.300**) formed. This course is further substantiated by the results of ^{15}N-labeling studies[184] which demonstrate, in accordance with initial Cope rearrangement, that *N*(1) in the substrate (**6.299**) becomes *N*(1) in the final product (**6.301**). A mechanism (Scheme 6.69) involving the initial formation and subsequent cyclization of a zwitterionic intermediate (**6.304**) has been proposed[6] to account for the reaction of 2-azido-1-methylbenzimidazole (**6.305**) with diphenylketene to give the 2,3-dihydro-9*H*-imidazo-[1,2-*a*]benzimidazol-3-one (**6.305**) in high yield (Table 6.36).

Cyclization of intermediate *N*-chlorooxalyl derivatives accounts for the smooth reaction[185] (Scheme 6.70) of 2-(*N*-substituted amino)benzimidazoles (**6.306**; $R^1 = H$, $R^2 \neq H$) and *N*(1)-substituted 2-aminobenzimidazoles (**6.306**; $R^1 \neq H$, $R^2 = H$) with oxalyl chloride followed by triethylamine, to afford high yields (Table 6.36) of 2,3-dihydro-1*H*- and 9*H*-imidazo[1,2-*a*]benzimidazole-2,3-diones (**6.307**; X = O), and (**6.308**; X = O), respectively. Products of the latter type are also obtained,[185] though less efficiently by the thermal ring-closure of *N*(1)-alkyl 2-(*N*-methoxalylamino)benzimidazoles [Scheme 6.70; (**6.306**; R^1 = alkyl, R^2 = $COCO_2Me$)→(**6.308**; R = alkyl, X = O)]. The products, obtained in good

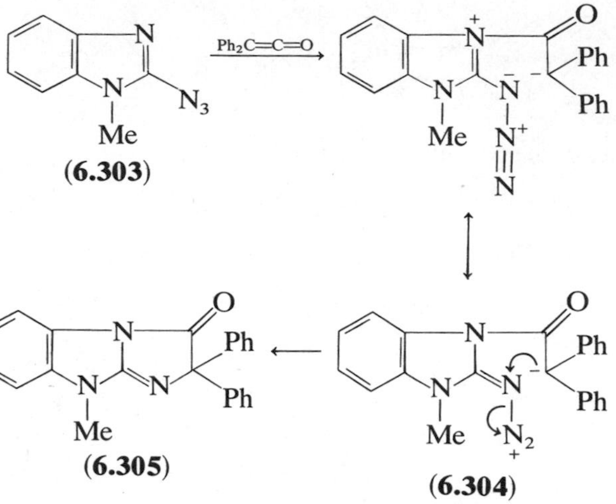

N, N, Me, N_3 —($Ph_2C{=}C{=}O$)→ N⁺, O, Ph, N, N⁻, Ph, Me, N⁺, N

(6.303)

↕

N, O, Ph, N, N, Ph, Me ← N, O, Ph, N, N⁻, Ph, Me, N_2⁺

(6.305) (6.304)

Scheme 6.69

yield (Table 6.36) by the triethylamine-catalyzed condensation of 2-aminobenzimidazole with the mono-*N*-phenylimine and mono-*N*-arylhydrazones of oxalyl chloride are formulated[151,186] as the 3-*N*-phenylimine and 3-*N*-arylhydrazones of 2,3-dihydro-1(9)*H*-imidazo[1,2-*a*]benzimidazole-2,3-dione [Scheme 6.70; (**6.308**; X = O, X = NPh or NNHAr)] but without any compelling evidence to exclude the alternative 2-imino-3-oxo formulations (**6.308**; X = NPh or NNHPh, Y = O). Further investigation of the orientation of these products would therefore be desirable. 2,3-Dihydro-1*H*-imidazo[1,2-*a*]benzimidazole-2,3-di(trifluoromethyl)-imines (**6.308**; X = Y = NCF_3) are the somewhat unusual end-products (Table 6.36) of the annelation reactions of 2-aminobenzimidazole derivatives with perfluoro-2,5-diazahexa-2,4-diene [$F_3CN{=}C(F)C(F){=}NCF_3$].[187]

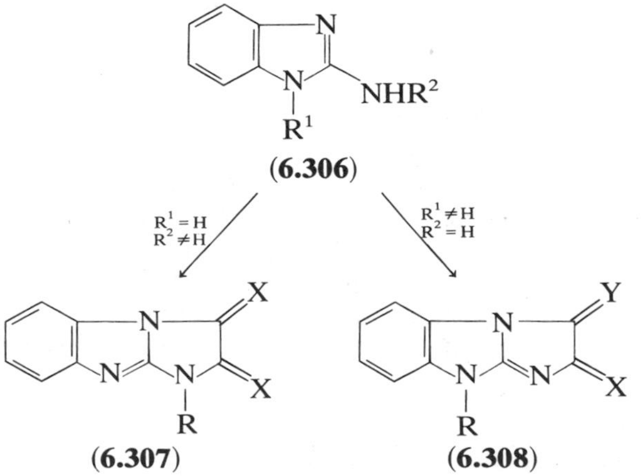

N, N, NHR^2, R^1

(6.306)

$R^1 = H$, $R^2 \neq H$ ↙ ↘ $R^1 \neq H$, $R^2 = H$

N, X, N, N, X, R N, Y, N, N, X, R

(6.307) (6.308)

Scheme 6.70

Until recently imidazo[1,5-*a*]benzimidazoles of the different structural types (cf. Scheme 6.49) were unknown due to the lack of suitable methods for their synthesis. However, such 6-5-5 fused benzimidazoles are now generally accessible in high yield (Tables 6.37 and 6.38) by a variety of procedures sharing the common feature of ring-closure in suitable 2-(α-aminoalkyl)benzimidazole derivatives (Scheme 6.71). A simple example[188] is provided by the condensation of the amine (**6.309**; $R = R^3 = H$, $R^1 = R^2 = Ph$) with ethyl orthoformate under reflux to afford the imidazo[1,5-*a*]benzimidazole derivative (**6.310**) in good yield. The closely related cyclodehydration of 2-(α-acylaminoalkyl)benzimidazoles (**6.309**; R^3 = acyl), promoted by heating with phosphorus oxychloride, proceeds in excellent yield (Table 6.37) and permits flexible synthetic access to 4*H*-imidazo[1,5-*a*]benzimidazoles (**6.311**) unsubstituted at *C*(1) and *C*(3) or containing alkyl or aryl substituents at one or both of these positions.[189–193] The thermal transformation of the thiourea derivative (**6.309**; $R = Me$, R^1 = 4-pyridyl, $R^2 = H$, R^3 = CSNHPh), in moderate yield (Table 6.37), into the 1-mercapto-4*H*-imidazo[1,5-*a*]benzimidazole (**6.311**; $R = Me$, R^1 = 4-pyridyl, R^2 = SH) illustrates a further synthetic aspect of cyclization of this type.[189]

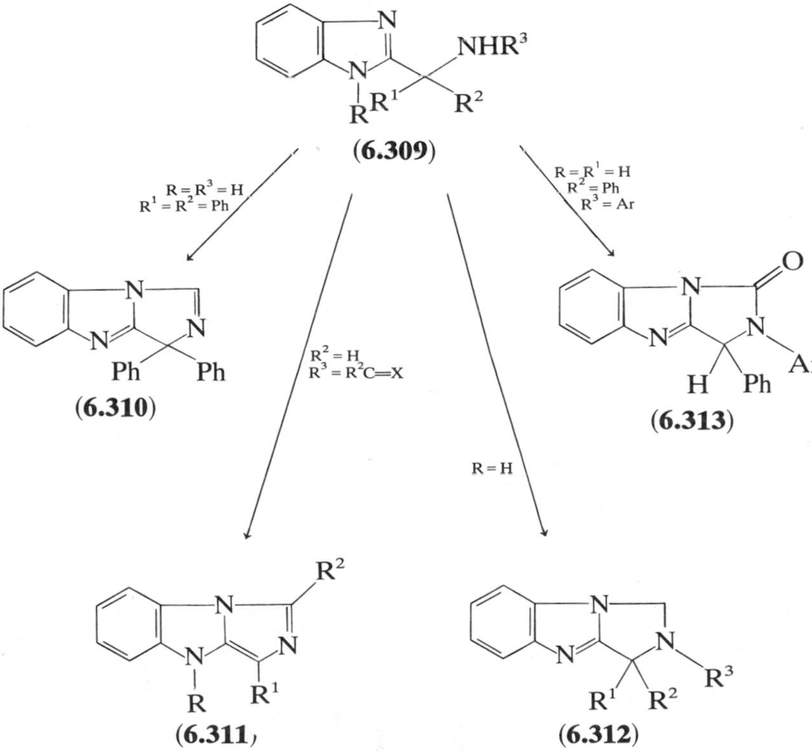

Scheme 6.71

TABLE 6.37. SYNTHESIS OF 4*H*-IMIDAZO[1,5-A]BENZIMIDAZOLE DERIVATIVES (**6.311**) BY RING-CLOSURE REACTIONS OF 3-(α-AMINOALKYL)BENZIMIDAZOLES (**6.309**)

Starting material (**6.309**)				Reaction	Product (**6.311**)			Yield	m.p.	Solvent of	
R	R^1	R^2	R^3	conditions[a]	R	R^1	R^2	(%)	(°C)	crystallization	Ref.
CH_2Ph	H	H	CHO	*A*	CH_2Ph	H	H[b]	42	107–108	—[c]	191
Me	Me	H	CHO	*B*	Me	Me	H[d]	55	99–101.5	—[e]	190
Me	Ph	H	CHO	*C*	Me	Ph	H	89	130.5–131	Benzene	189
Me	4-Pyridyl	H	CHO	*C*	Me	4-Pyridyl	H[f]	94	204.5–206	Ethanol–water	189
Me	H	H	COMe	*A*	Me	H	Me[g,h]	62	92–94	—[i]	191
Me	Me	H	COMe	*B*	Me	Me	Me[j]	75	119–121.5	Benzene–hexane	190
Me	Ph	H	Me	*C*	Me	Ph	Me	99	133–135	Ethyl acetate	189
Me	H	H	Ph	*A*	Me	H	Ph[k]	41	117–118.5	—[l]	192
CH_2Ph	H	H	Ph	*D*	CH_2Ph	H	Ph	93	147.5–150	Ethanol	193
Me	4-Pyridyl	H	CSNHPh	*E*	Me	4-Pyridyl	SH[m]	49–51	215.5–216	Dimethylformamide–water	189
Me	4-Pyridyl	H	CSNHPh	*F*	Me	4-Pyridyl	SH	49–51	—	—	189

[a] *A* = $POCl_3$, toluene/reflux—until evolution of HCl ceases; *B* = $POCl_3$ benzene/reflux—until evolution of HCl ceases, *C* = $POCl_3$, benzene/(reflux)(4 hr); *D* = $POCl_3$, benzene/(reflux)(35 hr); *E* = phenetole/(reflux)(5 min); *F* = melt (175–185°)/5–7 min.

[b] Forms a picrate, yellow crystals m.p. 189–190.5° (from ethanol).

[c] Purified by distillation, b.p. 207–208°/0.3 mm Hg.

[d] Deliquesces in air, turning green in color; forms a picrate, yellow crystals, mp. 197.5–199.5° (from acetone).

[e] Purified by distillation.

[f] Forms a picrate, yellow crystals, m.p. 264–264.5° (decomp.).

[g] Unstable in air, turning violet in color.

[h] Forms a picrate, yellow crystals, m.p. 214° (decomp.) (from acetone).

[i] Purified by distillation, b.p. 135–136°/0.25 mm Hg.

[k] Forms a picrate, yellow needles, m.p. 220–221.5° (decomp.) (from ethanol).

[l] Purified by distillation, b.p. 219–221°/0.25 mm Hg.

[m] Forms a picrate, red crystals, m.p. 235.5–237 (decomp.).

TABLE 6.38. SYNTHESIS OF 1*H*, 3*H*-IMIDAZO[1, 5-*a*]BENZIMIDAZOLE DERIVATIVES (**6.312**) AND (**6.313**) BY RING-CLOSURE REACTIONS OF 2-(α-AMINOALKYL)BENZIMIDAZOLES (**6.309**).

Starting material ($R,R^1 \rightarrow R^3$ unspecified = H)	Reaction conditions[a]	Product ($R^1 \rightarrow R^3$ unspecified = H)	Yield (%)	m.p. (°C)	Solvent of crystallization	Ref.
(**6.309**; $R^1 = Ph, R^3 = C_6H_{11}$)	*A*	(**6.312**; $R^1 = Ph, R^3 = C_6H_{11}$)	72	125–126	Toluene–light petroleum	188
(**6.309**; $R^1 = Ph, R^3 = CH_2Ph$)	*A*	(**6.312**; $R^1 = Ph, R^3 = CH_2Ph$)	56	128–130	Acetone	188
(**6.309**; $R^1 = Ph, R^3 = p\text{-}MeOC_6H_4$)	*A*	(**6.312**; $R^1 = Ph, R^3 = p\text{-}MeOC_6H_4$)	78	167–170	Acetone–benzene –light petroleum	188
(**6.309**; $R^1 = Ph, R^3 = p\text{-}MeC_6H_4$)	*A*	(**6.312**; $R^1 = Ph, R^3 = p\text{-}MeC_6H_4$)	74	195–196	Ethanol	188
(**6.309**; $R^1 = Ph, R^3 = p\text{-}ClC_6H_4$)	*A*	(**6.312**; $R^1 = Ph, R^3 = p\text{-}ClC_6H_4$)	—[b]	218–220	—[c]	194
(**6.309**; $R^1 = R^2 = Ph$)	*A*	(**6.312**; $R^1 = R^2 = Ph$)	81	195–196	Benzene-hexane	188
(**6.309** $R^1 = R^2 = Ph, R^3 = C_6H_{11}$)	*A*	(**6.312**; $R^1 = R^2 = Ph, R^3 = C_5H_{11}$)	86	222–224	Ethanol	188
(**6.309** $R^1 = R^2 = Ph, R^3 = p\text{-}ClC_6H_4$)	*A*	(**6.312**; $R^1 = R^2 = Ph, R^3 = p\text{-}ClC_6H_4$)	41	274–276	Dimethylformamide –water	188
(**6.309**; $R^1 = R^2 = Ph$, $R^3 = p\text{-}O_2NC_6H_4CO$)	*A*	(**6.312**; $R^1 = R^2 = Ph$, $R^3 = p\text{-}O_2NC_6H_4CO$)	89	193–195	Toluene–light petroleum	188
(**6.309** $R^1 = Ph, R^3 = p\text{-}MeOC_6H_4$)	*B*	(**6.313**; $Ar = p\text{-}MeOC_6H_4$)	83	169–171	Ethanol–water or acetone–cyclohexane	188
(**6.309**; $R^1 = Ph, R^3 = p\text{-}MeC_6H_4$)	*B*	(**6.313**; $Ar = p\text{-}MeC_6H_4$)	87	203–205	Ethanol	188

[a] *A* = 30% HCHO aq., ethanol or dioxane (reflux/4 hr); *B* = $COCl_2$, Et_3N, tetrahydrofurane/(140°)(30 min).
[b] Yield not quoted.
[c] Solvent of crystallization not specified.

Mannich-like condensation reactions of 2-(α-aminoalkyl)benzimidazoles (**6.309**) with formaldehyde provide the principal method for the synthesis of derivatives (**6.312**) of the 1*H*,3*H*-imidazo[1,5-*a*]benzimidazole ring system.[188,194] This mode of ring-closure generally occurs in good yield (Table 6.38), though not unexpectedly it fails completely for substrates (**6.309**) containing electron-withdrawing amino substituents (e.g., *p*-nitrophenyl). A further synthetic limitation is revealed by the reported[188] failure of aldehydes (other than formaldehyde) and ketones, to participate in 1*H*,3*H*-imidazo[1,5-*a*]benzimidazole formation [(**6.309**)→(**6.312**)]. In contrast, treatment with phosgene in the presence of triethylamine effects the smooth ring-closure of 2-(α-aminoalkyl)benzimidazoles (**6.309**) to 1*H*,3*H*-imidazo[1,5-*a*]benzimidazol-1-ones (**6.313**) in high yield (Table 6.38).[188]

Ring-closure Reactions of Other Heterocycles

In addition to their more commonplace synthesis by cyclization of 2-benzimidazolethione derivatives (see before), thiazolo[3,2-*a*]benzimidazoles of various structural types are also the end-products of a miscellany of transformations undergone by thiazole derivatives (Scheme 6.72). These include the photolysis of *N*(1)-(2-thiazolyl)benzo-1,2,3-triazole (**6.314**) which represents a convenient, if low yield method for the synthesis of the parent thiazolo[3,2-*a*]benzimidazole (**6.317**),[195] and the condensation of 1,4-benzoquinones with 2-aminothiazole to afford nuclear hydroxythiazolo[3,2-*a*]benzimidazoles of unestablished orientation, in unspecified yield [e.g. (**6.315**)+(**6.316**)→(**6.318**)].[196] Correspondingly, access to perhydrothiazolo[3,2-*a*]benzimidazoles [e.g. (**6.320**) and (**6.322**)] in variable yield (Scheme 6.72) is provided by the condensation of 2-chlorocyclohexanone (**6.319**) with 2-aminothiazole (**6.316**),[103] and by the brominative cyclization of substrates of the type (**6.321**).[197]

Derivatives of the 4*H*-pyrazolo[2,3-*a*]benzimidazole ring system [Scheme 6.73; (**6.325**)] are generally accessible in moderate to excellent yield (Table 6.39) by the dehydrative or deaminative ring-closure of *N*(1)-(2-aminoaryl)-5-pyrazolones (**6.323**) or *N*(1)-(2-aminoaryl)-5-aminopyrazoles (**6.324**; $R^4 = NH_2$). Cyclizations of these types are variously effected by heating with a high-boiling amine (e.g., aniline or *m*-toluidine) in the absence or presence of the amine hydrochloride,[198–200] or with hydrochloric,[201] sulfuric,[202–204] or acetic[205,206] acids, and afford 4*H*-pyrazolo[2,3-*a*]benzimidazoles with a wide choice of substituents (e.g., alkyl,[198,199,202–204] aryl,[198–200] amino,[201,205,206] carboxyl[198–200]) at the *C*(2) position. *C*(2)- and *C*(3)-substituted 4*H*-pyrazolo[2,3-*a*]benzimidazoles are also formed in moderate yield (Table 6.39) by means of intramolecular nucleophilic displacement of the *ortho*-chloro group by the aminopyrazole substituent in *N*(1)-(2-chlorophenyl)-5-aminopyrazoles (**6.324**; $R^2 = R^3 = H$, $R^4 = Cl$). These ring-closure reactions are promoted by treatment with potassium in liquid ammonia[207] or by

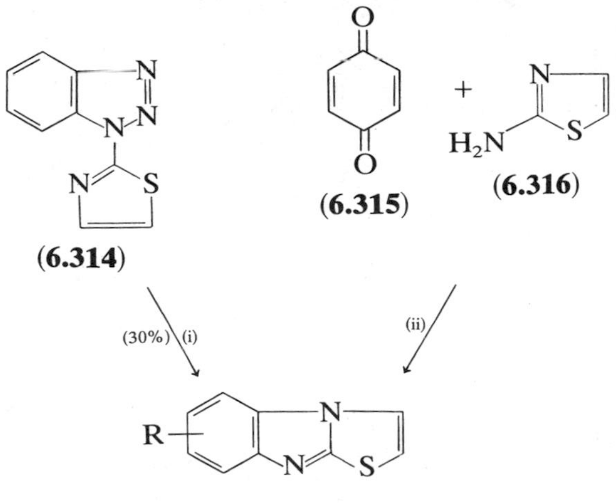

N N N N S (**6.314**)

O O (**6.315**) + N S H_2N (**6.316**)

(30%) (i) (ii)

R N N S

	R	m.p. (°C)
(**6.317**)	H	140
(**6.318**)	OH	>150

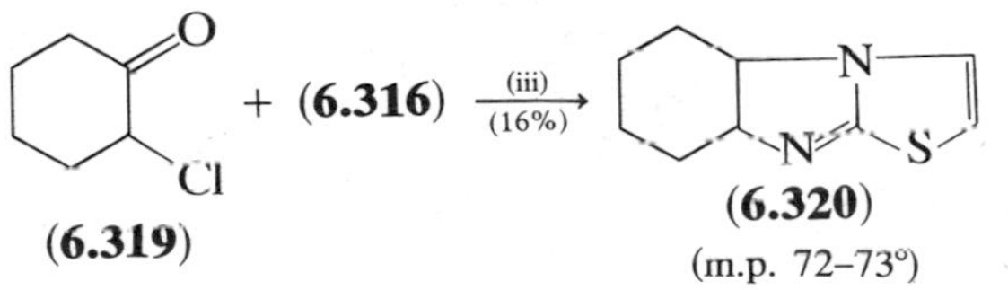

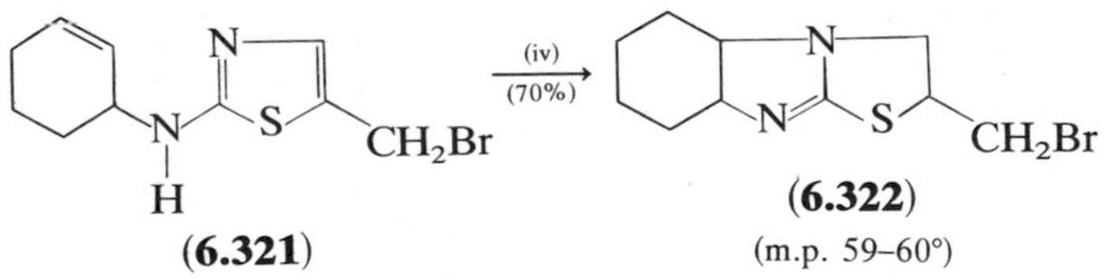

O Cl (**6.319**) + (**6.316**) —(iii) (16%)→ N N S (**6.320**) (m.p. 72–73°)

N N S CH_2Br H (**6.321**) —(iv) (70%)→ N N S CH_2Br (**6.322**) (m.p. 59–60°)

(i) $h\nu$ (λ_{max} 360 nm), benzene or ethanol
(ii) AcOH, 90–100°
(iii) reflux, 15 min
(iv) Br_2, CCl_4

Scheme 6.72

heating with Cu(II) oxide in dimethylformamide,[208] and, under the former conditions at least, may follow a benzyne pathway.

The annelation of *ortho*-phenylenediamine by 2-chlorooxazoles [Scheme 6.74; (**6.326**)+(**6.327**)→(**6.328**) or (**6.329**)][209,210] exemplifies a little used alternative to ring-closure in 2-aminobenzimidazole derivatives (see before) as a means for the construction of the 9*H*-imidazo[1,2-*a*]benzimidazole ring system.

TABLE 6.39. SYNTHESIS OF 4*H*-PYRAZOLO[2,3-*a*]BENZIMIDAZOLES (**6.325**) BY RING-CLOSURE REACTIONS OF PYRAZOLE DERIVATIVES

Starting material (R, $R^1 \rightarrow R^3$ unspecified = H)[a]	Reaction conditions[b]	Product (**6.325**) ($R^1 \rightarrow R^3$ unspecified = H)[a]	Yield (%)	m.p. (°C)	Solvent of crystallization	Ref.
(**6.323**; R = Me)	*A*	(R = Me)	40	246–248	Methanol	199
(**6.324**; R = Me, R^4 = Cl)	*B*	(R = Me)	60	240	—[c]	207
(**6.324**; R = *n*-$C_{17}H_{35}$, R^4 = NH_2)	*C*	(R = *n*-$C_{17}H_{35}$)	—[d]	119–120	—[c]	202
(**6.323**; R = *n*-$C_{17}H_{35}$, R^3 = Cl)	*C*	(R = *n*-$C_{17}H_{35}$, R^3 = Cl)	75	147	Ethyl acetate	203
(**6.324**; R = *n*-$C_{17}H_{35}$, R^3 = Cl, R^4 = NH_2)	*C*	(R = *n*-$C_{17}H_{35}$, R^3 = Cl)	—[d]	140–146	—[e]	202
(**6.323**; R = *n*-$C_{17}H_{35}$, R^3 = Br)	*C*	(R = *n*-$C_{17}H_{35}$, R^3 = Br)	—[d]	150	—[e]	203
(**6.323**; R = *n*-$C_{17}H_{35}$, R^3 = F)	*C*	(R = *n*-$C_{17}H_{35}$, R^3 = F)	—[d]	72	—[e]	203
(**6.323**; R = *n*-$C_{17}H_{35}$, R^3 = OEt)	*C*	(R = *n*-$C_{17}H_{35}$, R^3 = OEt)	—[d]	99–100	—[e]	203
(**6.323**; R = *n*-$C_{17}H_{35}$, R^3 = CO_2H)	*C*	(R = *n*-$C_{17}H_{35}$, R = CO_2H)	—[d]	125	Ethyl acetate	203
(**6.323**; R = *n*-$C_{17}H_{35}$, R^3 = CO_2Pr^i)	*C*	(R = *n*-$C_{17}H_{35}$, R^3 = CO_2Pr^i)	—[d]	115	—[e]	203
(**6.323**; R = *n*-$C_{17}H_{35}$, R^2 = R^3 = Cl)	*C*	(R = *n*-$C_{17}H_{35}$, R^2 = R^3 = Cl)	—[d]	144	Methanol	203
(**6.324**; R = *n*-$C_{17}H_{35}$, R^2 = Me, R^4 = NH_2)	*C*	(R = *n*-$C_{17}H_{35}$, R^2 = Me)	—[d]	123–125	—[e]	202
(**6.323**; R = *n*-$C_{17}H_{35}$, R^2 = CF_3)	*C*	(R = *n*-$C_{17}H_{35}$, R^2 = CF_3)	74	152–154	Methanol	204
(**6.324**; R = *n*-$C_{17}H_{35}$, R^2 = CF_3, R^4 = NH_2)	*C*	(R = *n*-$C_{17}H_{35}$, R^2 = CF_3)	62	158–159	—[d,e]	202
(**6.324**; R = *n*-$C_{17}H_{35}$, R^2 = OMe, R^4 = NH_2)	*C*	(R = *n*-$C_{17}H_{35}$, R^2 = OMe)	—[d]	114–116	—[e]	202
(**6.324**; R = *n*-$C_{17}H_{35}$, R^2 = CO_2H, R^4 = NH_2)	*C*	(R = *n*-$C_{17}H_{35}$, R^2 = CO_2H)	—[d]	>260 (decomp.)	—[e]	202
(**6.323**; R = *n*-$C_{17}H_{35}$, R^2 = SO_3H)	*D*	(R = *n*-$C_{17}H_{35}$, R^2 = SO_3H)	76	—[f]	—[e]	198
(**6.324**; R = *n*-$C_{17}H_{35}$ R^2 = SO_3H, R^4 = NH_2)	*C*	(R = *n*-$C_{17}H_{35}$, R^2 = SO_3H)	—[d]	>260 (decomp.)	—[e]	202
(**6.324**; R = *n*-$C_{17}H_{35}$, R^2 = SO_2NMe_2, R^4 = NH_2)	*C*	(R = *n*-$C_{17}H_{35}$, R^2 = SO_2NMe_2)	83	182–183	—[e]	202
(**6.324**; R = *n*-$C_{11}H_{23}$, R^2 = CF_3, R^4 = NH_2)	*C*	(R = *n*-$C_{11}H_{23}$, R^2 = CF_3)	—[d]	172–173	—[e]	202

(**6.324**; R = n-$C_{13}H_{27}$, $R^2 = CF_3$, $R^4 = NH_2$)	*C*	(R = n-$C_{13}H_{27}$, $R^2 = CF_3$)	—[d]	165–167	—[e]	202
(**6.324**; R = n-$C_{15}H_{31}$, $R^2 = CF_3$, $R^4 = NH_2$)	*C*	(R = n-$C_{15}H_{31}$, $R^2 = CF_3$)	—[d]	156–158	—[e]	202
(**6.323**; R = Ph)	*D*	(R = Ph)	71	260	Acetone–water	198
(**6.324**; R = Ph, $R^4 = NH_2$)	*C*	(R = Ph)	—[d]	>260 (decomp.)	—[e]	202
(**6.324**; R = Ph, $R^2 = CF_3$, $R^4 = NH_2$)	*C*	(R = Ph, $R^2 = CF_3$)	—[d]	244–247	—[e]	202
(**6.323**; R = Ph, $R^2 = SO_3H$)	*E*	(R = Ph, $R^2 = SO_3H$)	97	>310 (decomp.)	—[e]	198, 199
(**6.323**; R = Ph, $R^2 = SO_3H$)	*F*	(R = Ph, $R^2 = SO_3H$)	88	280 (decomp.)	—[e]	200
(**6.323**; R = CO_2H)	*D*	(R = CO_2H)	45	—[f]	—[e]	198
(**6.323**; R = CO_2H)	*F*	(R = CO_2H)	50	270	Methanol	199
(**6.323**; R = $R^2 = CO_2H$)	*F*	(R = $R^2 = CO_2H$)	82	>270 (decomp.)	—[e]	199
(**6.323**; R = CO_2H, $R^2 = SO_3H$)	*D*	(R = CO_2H, $R^2 = SO_3H$)	60	>320	—[e]	199
(**6.323**; R = NHPh)	*G*	(R = NHPh)[g]	quant.	232–235	Methanol	201
(**6.323**; R = p-$O_2NC_6H_4NH$)	*G*	(R = p-$O_2NC_6H_4NH$)	46	>260	—[e]	201
(**6.323**; R = NHCOMe)	*H*	(R = NHCOMe)	80	—[f]	—[e]	205
(**6.324**; $R^1 = CO_2Et$, $R^4 = Cl$)	*I*	($R^1 = CO_2Et$)	15	171–172	—[e]	208
(**6.324**; $R^1 = CN$, $R^4 = Cl$)	*I*	($R^1 = CN$)	42	281–282	—[e]	208

[a] n-$C_{17}H_{35}$ = stearyl.
[b] *A* = HCl, AcOH, H_2O/(reflux)(6 hr); *B* = K, NH_3 liq./30 min; *C* = 20% H_2SO_4/(100–120°)(2 hr); *D* = $PhNH_2$/(140–170°)(2–3 hr); *E* = m-toluidine/(160°)(2–3 hr); *F* = $PhNH_2$, $PhNH_3Cl^-$/(140–170°)(1–2 hr); *G* = conc. HCl, Pr^nOH/(reflux)(time not specified); *H* = AcOH/(reflux)(2 hr); *I* = Cu(II)O, dimethylformamide/heat (precise conditions not specified).
[c] Purified by sublimation.
[d] Yield not quoted.
[e] Solvent of crystallization not specified.
[f] Melting point not quoted.
[g] Hydrochloride.

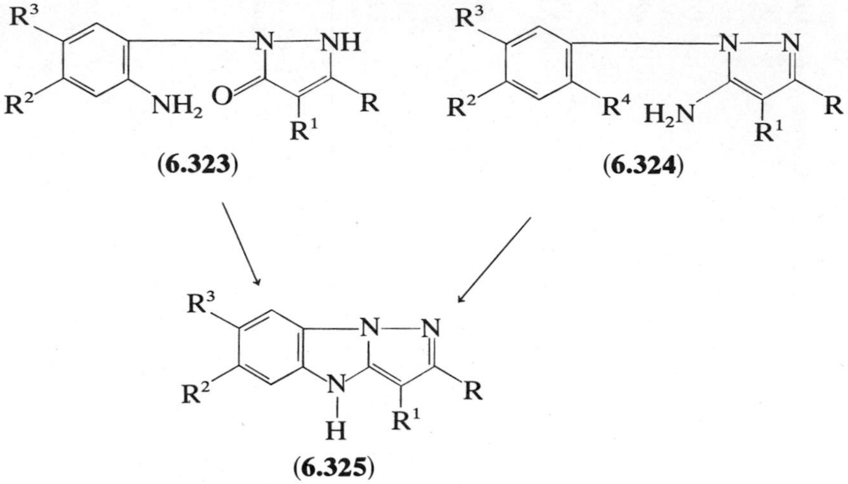

(6.323) (6.324)

(6.325)

Scheme 6.73

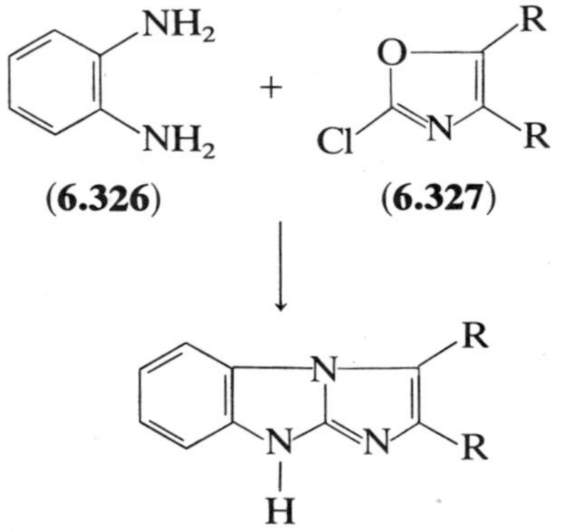

(6.326) (6.327)

	R	Yield (%)	m.p. (°C)	Solvent of crystallization
(6.328)	Ph	65	297–298	Ethyl acetate
(6.329)	Pr^i	30	272–273	Benzene

Scheme 6.74

6.2.2. Physicochemical Properties

Spectroscopic Studies

INFRARED SPECTRA. The infrared spectra of derivatives of several of the 6-5-5 fused benzimidazole ring systems with one additional heteroatom (cf. Tables 6.40–6.45) display a number of structurally revealing and diagnostically useful features. The I.R. spectra[91,92] of 2,3-dihydrooxazolo[3,2-

TABLE 6.40. INFRARED SPECTRA[a] OF 2,3-DIHYDROOXAZOLO[3,2-*a*]BENZIMIDAZOLES (**6.330**)

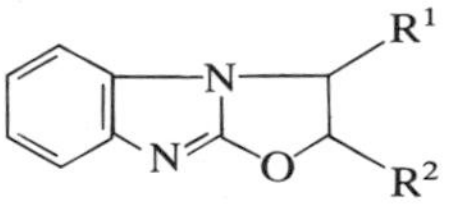

(**6.330**)

Substituents		ν_{max} cm^{-1}			
R^1	R^2	OH, NH	C=O	Others	Ref.
H	H	—	—	1635, 1590	92
H	Et	—	—	1635, 1590	92
H	CH_2Cl	—	—	1635, 1585, 1550	91, 92
H	CH_2OH	3280	—	1630, 1590, 1550	91, 92
H	CH_2OAc	—	1745	1640, 1590, 1560	91
H	CH_2OPh	—	—	1630, 1600, 1590, 1550	91
H	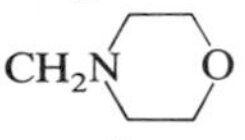 CH_2N (morpholino)	—	—	1630, 1590, 1550	91, 92
H	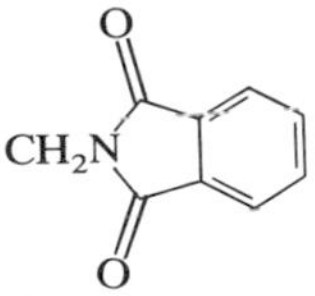 CH_2N (phthalimido)	—	1770, 1720	1635, 1610, 1590, 1560	91
H	$CONH_2$	3300, 3150	1685	1640, 1590, 1565	91
Me	CO_2Et	—	1755	1635, 1590	91

[a] Measured for suspensions in Nujol.

a]benzimidazole derivatives (Table 6.40) in particular are typified by a series of bands at 1640–1630, 1590–1585, and 1560–1550 cm^{-1}, which appear to be characteristic of this particular ring system. An IR band in the range 1490–1460 cm^{-1} is likewise a feature of the spectra of simple thiazolo[3,2-*a*]benzimidazoles (Table 6.41),[103,115,128,195] while 3-arylthiazolo[3,2-*a*]benzimidazoles are specifically associated[95] with IR absorption in the ranges 1620–1610 and 1585–1580 cm^{-1}. Resonance interaction with the N(4) bridgehead nitrogen atom [Scheme 6.75; (**6.339**)↔(**6.340**)] imparts "vinylogous amide" character to *C*(2) carbonyl derivatives (esters, aldehydes, ketones, carboxylic acids, amides) of thiazolo[3,2-*a*]benzimidazoles thus accounting for the exceptionally low IR carbonyl stretching frequencies observed for such compounds (Table 6.41).[103,115,117] On the other hand, the adjacent bridgehead nitrogen atom appears to have minimal effect on the *C*(3) carbonyl substitutent in 2,3-dihydrothiazolo[3,2-*a*]benzimidazol-3-ones, which absorbs uniformly at high frequencies (1740–1730 cm^{-1}) (Table 6.42)[97,98,101,103,133,134,136,151] diminished only by conjugation with arylidene[101,133,134,136,151] or imine[134,151,211] unsaturation at the *C*(2)

TABLE 6.41. INFRARED SPECTRA OF THIAZOLO[3,2-*a*]BENZIMIDAZOLE DERIVATIVES (**6.331**) AND (**6.332**)

(**6.331**) (**6.332**)

Compound	R^1	R^2	R^3	Medium	ν_{max} (cm^{-1}) OH, NH	C=O	Others	Ref.
(**6.331**)	H	H	—	KBr	—	—	1459	103, 195
(**6.331**)	H	H[a]	—	KBr	—	—	1462	103
(**6.331**)	H	H[b]	—	KBr	—	—	1471	103
(**6.331**)	H	Me	—	KBr	—	—	1461	103
(**6.331**)	Me	H	—	KBr	—	—	1463	103
(**6.331**)	Ph	H	—	KBr	—	—	1479	103
(**6.331**)	H	CHOHMe	—	KBr	3228	—	—	103
(**6.331**)	H	CH(OAc)Me	—	KBr	—	1734	—	103
(**6.331**)	Me	CH_2OH	—	KBr	3300–3000 br	—	—	103
(**6.331**)	Me	CHOHMe	—	Nujol	3200	—	1490	115
(**6.331**)	Me	CO_2Et	—	KBr	—	1711	—	103
(**6.331**)	Me	CO_2Et	—	$CHCl_3$	—	1708	1490	115
(**6.331**)	Me	CHO	—	Nujol	—	1678	—	117
(**6.331**)	H	COMe	—	KBr	—	1660	—	103
(**6.331**)	Me	COMe	—	$CHCl_3$	—	1674	1490	115
(**6.331**)	Me	COMe	—	KBr	—	1646	1483	103
(**6.331**)	Me	COMe	—	Nujol	—	1650	—	117
(**6.331**)	Me	C(Me)=NOH	—	Nujol	2800–2600 br	—	1490	115
(**6.331**)	OCO_2Et	COMe	—	KBr	—	1770, 1680	—	142
(**6.331**)	Me	COMe	—	Nujol	—	1658	—	117
(**6.331**)	Et	COMe[a]	—	Nujol	—	1650	—	117
(**6.331**)	Et	$COBu^t$[a]	—	Nujol	—	1660	—	117
(**6.331**)	H	COPh	—	Nujol	—	1635	—	117
(**6.331**)	H	COPh[a]	—	Nujol	—	1630	—	117
(**6.331**)	H	*p*-BrC_6H_4CO	—	Nujol	—	1657	—	117
(**6.331**)	H	*p*-$O_2NC_6H_4CO$	—	Nujol	—	1640	—	117

(6.331)	H	p-BrC_6H_4CO[a]	—	Nujol	—	1648	—	117
(6.331)	H	p-$O_2NC_6H_4CO$[a]	—	Nujol	—	1651	—	117
(6.331)	Me	p-BrC_6H_4CO	—	Nujol	—	1623	—	117
(6.331)	Et	COPh[a]	—	Nujol	—	1639	—	117
(6.331)	Et	p-BrC_6H_4CO	—	Nujol	—	1648	—	117
(6.331)	H	CO_2H	—	Nujol	2353	1675	1490	115
(6.331)	H	CONHPh	—	$CHCl_3$	3300	1670	1490	115
(6.332)	H	H	H[c]	Nujol or KBr	—	—	1635–1595, 1480–1438	128
(6.332)	H	H	H[d]	Nujol or KBr	3245, 3198	1698	1635–1595, 1480–1438	128
(6.332)	OH	H	H	KBr	3200–2500 br[e]	—	—	103
(6.332)	OH	H	H	Nujol	3060–2680 br[e]	—	—	106
(6.332)	OH	H	H	$CHCl_3$, MeCN, or dimethyl sulfoxide	3670–3660[f] 3561–3545[e] 3480–3420[g] 3320–3150[h]	1720–1685	—	108
(6.332)	OH	H	H[a]	KBr	3200–2600 br[e]	—	—	103
(6.332)	OH	H	H[b]	KBr	3200–2600 br[e]	—	—	103
(6.332)	OAc	H	H	KBr	—	1732	—	103
(6.332)	OAc	H	H	$CHCl_3$	—	1750	—	103
(6.332)	OH	Me	H	KBr	3200–2600[e]	—	—	103
(6.332)	OH	Me	H	$CHCl_3$	3650[f] 3554[e] 3448[g] 3250[h]	1715	—	103
(6.332)	OH	Et	H	KBr	3200–2700[e]	—	—	108
(6.332)	OH	Pr[i]	H	KBr	3300–2700[e]	—	—	108
(6.332)	OH	H	Me	KBr	3200–2600 br[e]	—	—	103
(6.332)	OH	H	Me	$CHCl_3$	3605[f] 3448[g]	1710	—	212
(6.332)	OH	H	Me	Nujol	3060–2680 br[e]	—	—	106
(6.332)	OAc	H	Me	KBr	—	1748	—	103

[a] 6,7-Dimethyl derivative.
[b] 5,8-Dimethyl derivative.
[c] 7-Nitro derivative.
[d] 7-Acetamido derivative.
[e] Bonded OH.
[f] Free OH.
[g] Free NH.
[h] Bonded NH.

TABLE 6.42. INFRARED CARBONYL STRETCHING FREQUENCIES OF 2,3-DIHYDROTHIAZOLO[3,2-*a*]BENZIMIDAZOL-3-ONE DERIVATIVES (**6.333**)

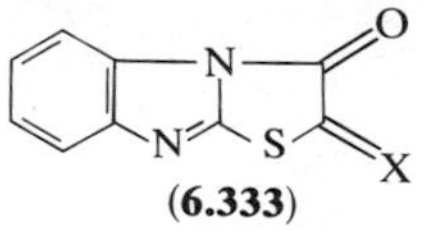

(**6.333**)

Substituent (X)	Medium	ν_{max}(C=O) (cm^{-1})	Ref.
H, H	KBr	1742	103
H, H	KBr	1728	101
H, H	KBr	1748	137
H, H[a]	KCl	1745	133
H, H[b]	KCl	1750	133
H, H[c]	KBr	1734	97
H, H[d]	Nujol	1745	98
H, H[e]	Nujol	1740	136
H, H[f]	KCl	1735	133
H, H[f]	—[g]	1736	134
H, H[h]	KCl	1735	133
H, Me	Nujol	1704	101
H, Me[f]	—[g]	1738	134
CHPh	Nujol	1725	134
CHPh[f]	KBr	1680	133
CHPh[e]	Nujol	1730	136
p-$MeOC_6H_4CH$	Nujol	1713	134
p-$NO_2C_6H_4CH$	Nujol	1726	134
p-$Me_2NC_6H_4CH$	KBr	1708	101
p-$MeOC_6H_4CH$[e]	Nujol	1735	136
p-ClC_6H_4CH[e]	Nujol	1725	136
$CHCO_2H$	KBr	1685	145
PhN	—[g]	1770–1710	151
p-$Me_2NC_6H_4N$	Nujol	1735	134
p-$MeOC_6H_4NHN$	Nujol	1732	134
o-$MeOC_6H_4NHN$	KBr	1730	151
p-BrC_6H_4NHN	Nujol	1740	134

[a] 6-Methyl derivative.
[b] 8-Methyl derivative.
[c] 6-Methoxy derivative.
[d] 6-Chloro derivative.
[e] 6-Nitro derivative,
[f] 6,7-Dimethyl derivative.
[g] Medium not specified.
[h] 5,8-Dimethyl derivative.

TABLE 6.43. INFRARED SPECTRA OF 1*H*- AND 9*H*-IMIDAZO[1,2-*a*]BENZIMIDAZOLE DERIVATIVES (**6.334**) AND (**6.335**)

(**6.334**) (**6.335**)

Compound	R^1	R^2	R^3	Medium	ν_{max} (cm^{-1})					Ref.
					OH, NH	C≡C	C=O	C=N	Others	
(**6.334**)	Me	Ph	—	$CHCl_3$	—	—	—	1643	—	166
(**6.334**)	Me	Ph	—[a]	$CHCl_3$	—	—	1665	1638	—	157
(**6.335**)	H	H	H	KBr	—[b]	—	—	1613		155
(**6.335**)	H	Me	H	—[c]	—[b]	—	—	1660	—	156
(**6.335**)	H	Me	H	$CHCl_3$	3480	—	—	1650	1615	166
(**6.335**)	CH=C=CH_2	Me	H	$CHCl_3$	—	1980[d]	—	—	—	170
(**6.335**)	H	Pr^i	Pr^i	$CHCl_3$	3485	—	—	—	—	210
(**6.335**)	H	Ph	H	—[c]	—[b]	—	—	1619	—	156
(**6.335**)	H	Ph	H	Nujol	3225	—	—	1683	—	166
(**6.335**)	H	Ph	H	$CHCl_3$	3435	—	—	1643	—	166
(**6.335**)	H	*p*-ClC_6H_4	H	—[c]	—[b]	—	—	1611	—	156
(**6.335**)	H	*p*-BrC_6H_4	H	—[c]	—[b]	—	—	1622	—	156
(**6.335**)	Me	Me	NH_2	Nujol	3487, 3380	2143[e]	—	1643[f]	—	175
(**6.335**)	Me	Ph	N=CHPh	Nujol	—	—	—	1632	—	175
(**6.335**)	Me	Me	C≡CH	—[c]	—	2210	—	—	—	213
(**6.335**)	Me	Ph	C≡CH	—[c]	—	2210	—	—	—	213
(**6.335**)	CH_2Ph	Me	C≡CH	—[c]	—	2210	—	—	—	213
(**6.335**)	Me	Me	CH(OH)C≡CPh	Nujol or $CHCl_3$	3595, 3300–3200	2240 w	—	—	—	164
(**6.335**)	Me	Ph	CH(OH)C≡CPh	Nujol or $CHCl_3$	3595, 3300–3200	2240 w	—	—	—	164
(**6.335**)	Me	Me	CO_2Me	$CHCl_3$	—	—	1710	—	—	173
(**6.335**)	Me	Me	$CO_2C_7H_{15}$	Nujol	—	—	1710	—	—	173
(**6.335**)	CH_2Ph	Me	CO_2Me	$CHCl_3$	—	—	1710	—	—	173

TABLE 6.43 (*Continued*)

(**6.334**) (**6.335**)

Compound	R^1	R^2	R^3	Medium	ν_{max} (cm^{-1}) OH, NH	C≡C	C═O	C═N	Others	Ref.
(**6.335**)	Me	COMe	Me	Nujol	—	—	1624	—	—	171
(**6.335**)	Me	COMe	Ph	Nujol	—	—	1619	—	—	171
(**6.335**)	Me	COPh	Me	Nujol	—	—	1710	—	—	171
(**6.335**)	Me	Me	CHO	—[c]	—	—	1640	—	—	165
(**6.335**)	Me	Ph	CHO	—[c]	—	—	1638	—	—	165
(**6.335**)	Me	Me	COMe	Nujol	—	—	1630	—	—	173
(**6.335**)	Me	Me	COMe	$CHCl_3$	—	—	1640–1620	—	—	172
(**6.335**)	Me	Ph	COMe	$CHCl_3$	—	—	1640–1620	—	—	172
(**6.335**)	Et	Me	COMe	$CHCl_3$	—	—	1640–1620	—	—	172
(**6.335**)	CH_2Ph	Me	COMe	$CHCl_3$	—	—	1640–1620	—	—	172
(**6.335**)	Me	Me	COPh	$CHCl_3$	—	—	1640–1620	—	—	172
(**6.335**)	Me	Ph	COPh	$CHCl_3$	—	—	1640–1620	—	—	172
(**6.335**)	Me	Me	COCH═CHPh	Nujol or $CHCl_3$	—	—	1645–1635	—	—	214
(**6.335**)	Me	Ph	COCH═CHPh	Nujol or $CHCl_3$	—	—	1645–1635	—	—	214
(**6.335**)	Me	Me	COC≡CPh	Nujol or $CHCl_3$	—	2215–2210	1575–1570	—	—	164
(**6.335**)	Me	Ph	COC≡CPh	Nujol or $CHCl_3$	—	2215–2210	1575–1570	—	—	164
(**6.335**)	Me	Me	NO_2	—[c]	—	—	—	—	1526, 1345	163

[a] Formyl derivative; orientation not established.
[b] ν_{max} (N—H) not quoted.
[c] Medium not specified.
[d] ν_{max} (CH═C═CH_2).
[e] ν_{max} (C≡N) of open-chain tautomer.
[f] May be due to NH def.

TABLE 6.44. INFRARED SPECTRA OF 2,3-DIHYDRO-1*H*- AND 9*H*-IMIDAZO[1,2-*a*]BENZIMIDAZOLE DERIVATIVES (**6.336**) AND (**6.337**)

(**6.336**) (**6.337**)

Compound	R	X	Y	Medium	$\nu_{max}(cm^{-1})$ OH, NH	C=O	C=N	Ref.
(**6.336**)	H	H, H	H, H	—[a]	—	—	1640	176
(**6.336**)	CH_2Ph	H, H	H, H	—[a]	—	—	1630	176
(**6.337**)	CH_2Ph	H, H	H, H	—[a]	—	—	1660	176
(**6.337**)	Me	H, OH	H, OH	Nujol	3490–3130	—	—	185
(**6.337**)	CH_2Ph	H, OH	H, OH	Nujol	3370–3100	—	—	185
(**6.337**)	CH_2Ph	Et, OH	Et, OH	Nujol	3380–3300	—	—	185
(**6.336**)	Me	H, H	O	Nujol	—	1760	—	180
(**6.336**)	Me	H, CH_2CON(piperazine ring)NMe	O	CH_2Cl_2	—	1750, 1655–1640	—	181
(**6.337**)	Me	H, H	O	Nujol	—	1725–1720	—	173
(**6.337**)	CH_2Ph	H, H	O	Nujol	—	1725–1720	—	173
(**6.336**)	H	O	Ph, Ph[b]	Nujol	3100	1760	1660	182
(**6.336**)[c]	H	O	Ph, Ph	Nujol	—[d]	1770	1660	184
(**6.337**)[e]	H	O	Ph, Ph	Nujol	—[d]	1780	1650	184
(**6.336**)	H	O	O	Nujol	3329	1690	—	185
(**6.336**)	CH_2Ph	O	O	Nujol	—	1799, 1764	—	185
(**6.337**)	Me	O	O	Nujol	—	1748, 1694	1632	185
(**6.337**)	Et	O	O	Nujol	—	1760, 1700	—	185
(**6.337**)	CH_2Ph	O	O	Nujol	—	1758, 1705	1624	185
(**6.337**)	$(CH_2)_2NEt_2$	O	O	Nujol	—	1767, 1703	—	185
(**6.337**)	H	NPh	O	—[a]	—[d]	1770–1710	1660	151
(**6.337**)	H	*p*-ClC_6H_4NHN	O	KBr	3320–3250	1750–1730	1610–1585	151
(**6.337**)	H	*o*-ClC_6H_4NHN	O	KBr	3320–3250	1750–1730	1610–1585	151

[a] Medium not specified.
[b] 7-Chloro derivative.
[c] 5,8-Dichloro derivative.
[d] ν_{max} (N–H) not quoted.
[e] 5,7,8-Tribromo derivative.

TABLE 6.45. INFRARED SPECTRA[a] OF 4*H*-IMIDAZO[1,5*a*]BENZIMIDAZOLE DERIVATIVES (**6.338**)

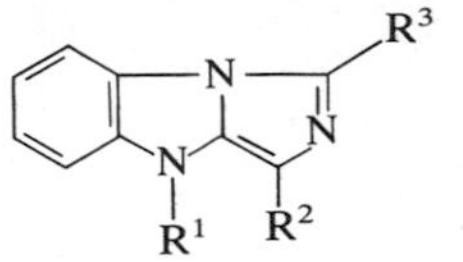

(**6.338**)

Substituents			ν_{max}(cm^{-1})			
R¹	R²	R³	OH, NH	C≡N	C=O	Ref.
Me	Ph	CH_2OH	3120	—	—	215
Me	Me	CN	—	2210	—	216
Me	Ph	CN	—	2215	—	216
CH_2Ph	CN	Ph	—	2200	—	193
Me	Ph	CHO	—	—	1640	215
Me	Ph	COMe	—	—	1645	215
CH_2Ph	CHO	Ph	—	—	1662, 1650	193
Me	COMe	Me	—	—	1634	192
Me	COMe	Ph	—	—	1642	192
CH_2Ph	COMe	Ph	—	—	1640	193

[a] Measured for suspensions in Nujol.

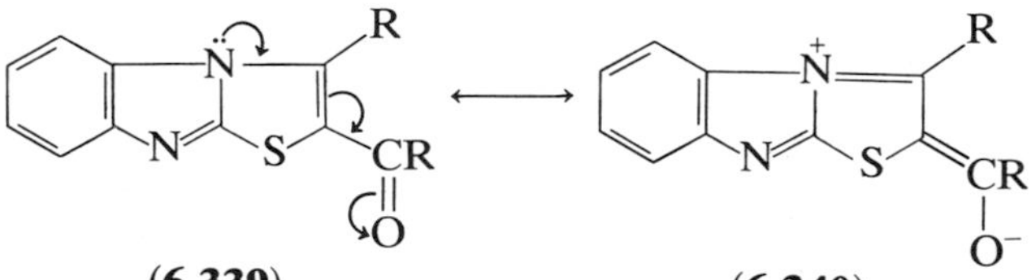

(**6.339**) ⟷ (**6.340**)

Scheme 6.75

position (Table 6.42). Conversely, the conjugative lowering of the *C*(3) carbonyl frequency in 2,3-dihydrothiazolo[3,2-*a*]benzimidazol-3-ones by a *C*(2) arylhydrazono substituent,[134,151] together with the presence of IR NH absorption,[134] implies the existence of the latter at least partially in the hydrazone tautomeric form rather than completely in the azo form as inferred from the claimed[211] absence of NH absorption and the presence of IR azo bands at 1580–1550 cm^{-1}. Bands at 3300–2600 cm^{-1} attributable to hydrogen-bonded hydroxyl groups and the lack of carbonyl absorption in the solid state IR spectra (Table 6.41) of certain 2-(β-oxoalkylthio)benzimidazoles [cf. page 108, Scheme 6.51; (**6.240**)] demonstrate the existence of these molecules (at least in the solid phase) entirely in the ring tautomeric 3-hydroxy-2,3-dihydrothiazolo[3,2-*a*]benzimidazole form.[102,103,106,108,109,212] On the other hand, IR spectra run in chloroform, acetonitrile, or dimethyl sulfoxide (Table 6.41)[108,212] contain bands due to both NH and OH groups as well as carbonyl absorption, thereby revealing the existence in solution of the ring-chain tautomeric equilibrium [Scheme 6.51; (**6.240**)⇌(**6.241**)].[108] The presence of well-defined NH absorption and the total absence of bands due to cyano groups in the IR spectra[153] of 1-imino-1*H*,3*H*-thiazolo[3,4-*a*]benzimidazoles [cf. page 116, Scheme 6.59; (**6.267**)] excludes the possible open-chain tautomeric 2-(α-thiocyano-alkyl)benzimidazole structure (**6.266**) for these molecules.

The IR spectrum of 1(4)*H*-pyrazolo[2,3-*a*]benzimidazole[208] exhibits characteristically broad NH absorption at 3240–2500 cm^{-1}, which is shifted to higher frequencies in the range 3485–3225 cm^{-1} and becomes sharper in the IR spectra of *N*-unsubstituted 1(9)*H*-imidazo[1,2-*a*]benzimidazoles (Table 6.43).[166,210] The presence of IR cyano absorption in addition to the anticipated NH bands at 3487 and 3380 cm^{-1} in the spectrum of 3-amino-2,9-dimethyl-9*H*-imidazo[1,2-*a*]benzimidazole (Table 6.43)[175] is indicative of the existence of this molecule to a significant extent in the open-chain 2-(α-cyanoalkylamino)benzimidazole form [cf. page 132, Scheme 6.64; (**6.287**)].

Acetylenic[213] and nitro[163] substituents at the *C*(3)-position in 9*H*-imidazo[1,2-*a*]benzimidazoles show the expected IR absorption characteristics. On the other hand, carbonyl substituents at this site and at the *C*(2) position in 9*H*-imidazo[1,2-*a*]benzimidazoles absorb at anomalously low frequencies (Table 6.43),[164,165,171–174,214] presumably due to "vinylogous amide" character imparted by resonance interaction with either the *N*(1) or *N*(4) (bridgehead) nitrogen atom, a phenomenon already encountered in the *C*(2) carbonyl derivatives of thiazolo[3,2-*a*]benzimidazoles (see before). The lower frequency (1725–1720 cm^{-1}) for the carbonyl absorption in *N*(9)-substituted 2,3-dihydro-9*H*-imidazo[1,2-*a*]benzimidazol-2-ones (Table 6.44)[173] compared with that (1760–1750 cm^{-1}) in their *N*(1)-substituted 2,3-dihydro-1*H*-imidazo[1,2-*a*]benzimidazol-2-one counterparts (Table 6.44)[180,181] is consistent with the conjugated nature of the carbonyl substituent in the former structures and is diagnostic for distinguishing the 1*H*- and 9*H*-tautomeric forms of 2,3-dihydroimidazo[1,2-*a*]benzimidazol-2-one derivatives. High-frequency IR carbonyl absorption at 1780–1760 cm^{-1} and bands at 1660–1650 cm^{-1} attributable to the C=N group are characteristic features of the IR spectra of 2,3-dihydro-1(9)*H*-imidazo[1,2-*a*]benzimidazol-3-one derivatives (Table 6.44).[182,194] Correspondingly, the IR spectra of *N*(1)-substituted 2,3-dihydro-1*H*-imidazo[1,2-*a*]benzimidazole-2,3-diones (Table 6.44)[185] contain intense carbonyl absorption at 1799 and 1764 cm^{-1} assignable to the *C*(3) and *C*(2) carbonyl groups, respectively. The lower frequencies (1770–1740 and 1705–1690 cm^{-1}) observed (Table 6.44)[185] for the C(3) and *C*(2) carbonyl substituents in *N*(9)-substituted 2,3-dihydro-9*H*-imidazo[1,2-*a*]benzimidazole-2,3-diones are consistent with the effect of increased conjugation and again

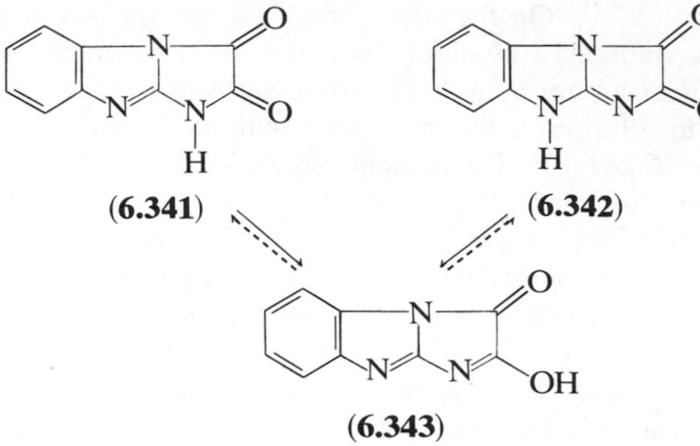

(**6.341**) (**6.342**)

(**6.343**)

Scheme 6.76

serve to distinguish these molecules from their *N*(1)-substituted 2,3-dihydro-1*H*-imidazo[1,2-*a*]benzimidazole-2,3-dione isomers. Comparison of the IR spectra of both structural types with that of the parent 2,3-dihydro-1(9)*H*-imidazo[1,2-*a*]benzimidazole-2,3-dione (Table 6.44)[185] demonstrates the existence of the latter molecule predominantly in the enol form [Scheme 6.76; (**6.343**)] as opposed to either of the two possible NH tautomeric forms (**6.341**) or (**6.342**). Interestingly, the IR absorption of the *C*(3) arylhydrazones of 2,3-dihydro-1(9)*H*-imidazo[1,2-*a*]benzimidazole-2,3-dione (Table 6.44)[151] is more in accord with fully ketonized structures [(**6.341**) or (**6.342**); *C*(3)O replaced by ArNHN] rather than the alternative enol formulations [**6.343**; ArNHN for *C*(3)O].

Simple, unfunctionalized, 1(9)*H*-imidazo[1,2-*a*]benzimidazoles give rise to characteristic IR C═N absorption in the range 1660–1610 cm^{-1} (Table 6.43).[155,156,166] The ready differentiation of *N*(9)- and *N*(1)-substituted 2,3-dihydro-1(9)*H*-imidazo[1,2-*a*]benzimidazoles on the basis of the conjugated or unconjugated character, respectively, of the IR C═N absorption exhibited by such molecules (Table 6.44)[176] permits the demonstration of the preferred site of alkylation in 2,3-dihydro-1(9)*H*-imidazo[1,2-*a*]benzimidazole derivatives.

Carbonyl[192,193,215] (but not cyano)[216] substituents at the *C*(1) and *C*(3) positions in 4*H*-imidazo[1,5-*a*]benzimidazoles are associated with exceptionally low-frequency IR absorption (Table 6.45),[192,193,215] which, as in the case of carbonyl substituted 1(9)*H*-imidazo[1,2-*a*]benzimidazoles (see before) can be attributed to the effect of resonance interaction with the bridgehead nitrogen atom.

Ultraviolet Spectra. The UV absorption characteristics of thiazolo[3,2-*a*]benzimidazoles vary significantly with the nature of the substituents in the thiazole ring. The UV spectra of thiazolo[3,2-*a*]benzimidazole and its *C*(2) and *C*(3) methyl derivatives (Table 6.46) are typified by the presence of an intense imidazole band at 240–250 nm and a somewhat less intense $\pi \rightarrow \pi^*$

TABLE 6.46. ULTRAVIOLET SPECTRA[a] OF THIAZOLO[3,2-*a*]BENZIMIDAZOLE DERIVATIVES (**6.344**), (**6.345**), (**6.346**), AND (**6.347**)

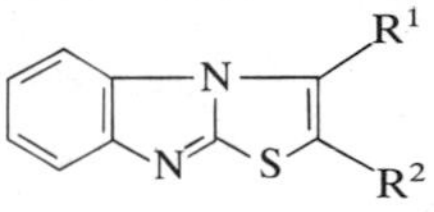

(**6.344**)

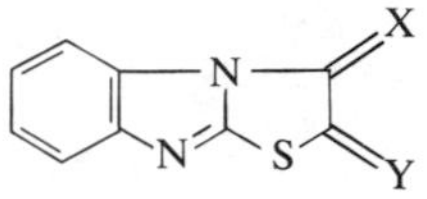

(**6.345**)

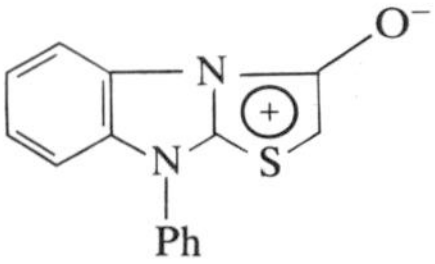

(**6.346**)

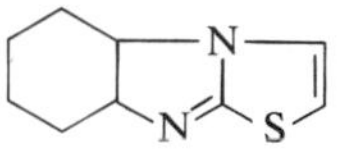

(**6.347**)

Compound	R^1	R^2	X	Y	λ_{max}(nm)(log ε)	Ref.
(**6.344**)	H	H	—	—	250 (3.76), 286 (3.84), 293 (3.94), 306 sh (3.75)	94, 107
(**6.344**)	H	H	—	—	215 (4.48), 242 (4.18), 247 (4.16), 275 (3.99), 283 sh (3.90)	103, 195
(**6.344**)[b]	H	H	—	—	215 (4.59), 245 (4.28), 277 (4.05)	103
(**6.344**)[c]	H	H	—	—	221 (4.57), 241 (4.16), 277 (4.04)	103
(**6.344**)	H	Me	—	—	215 (4.63), 242 (4.36), 248 (4.40), 275 (4.12), 286 (3.95)	103
(**6.344**)	H	Me	—	—	244 (3.89), 250 (3.95), 277 (4.09)	107
(**6.344**)	Me	H	—	—	213 (4.58), 238.5 (4.40), 245.5 (4.40), 277 (4.15), 288 sh	103
(**6.344**)	Me	H	—	—	240 (4.17), 247 (4.46), 280 (4.91)	107
(**6.344**)	H	Ph	—	—	218 (4.19), 277 (4.14), 311 (4.11)	107
(**6.344**)	Ph	H	—	—	235 (4.26), 248 (4.10), 268 (4.10), 283 (4.01)	103
(**6.344**)	Ph	H	—	—	235 (4.30), 251 (4.13), 270 (4.15), 285 sh (4.05)	94, 107
(**6.344**)	*p*-ClC_6H_4	H	—	—	235 (4.29), 275 (4.12)	94
(**6.344**)	*p*-BrC_6H_4	H	—	—	236 (4.37), 275 (4.18)	94
(**6.344**)	H	COMe	—	—	212 (4.67), 271 (4.62)	103
(**6.344**)	Me	COMe	—	—	213 (4.62), 273 (4.64)	103
(**6.344**)	Me	CO_2Et	—	—	212 (4.49), 266 (4.49)	103
(**6.345**)	—	—	H,H	H,H	251 (3.93), 286 (4.08), 293 (4.10)	94
(**6.345**)	—	—	H, OH	H, H	250 (3.96), 282 (4.02), 282 (4.06)	94, 107

TABLE 6.46 (*Continued*)

(**6.344**)

(**6.345**)

(**6.346**)

(**6.347**)

Compound	R^1	R^2	X	Y	λ_{max}(nm)(log ε)	Ref.
(**6.345**)	—	—	Me, OH	H, H	249 (3.99), 284 (4.12), 291 (4.15)	107
(**6.345**)	—	—	H, OH	Me, H	250 (3.98), 284 (4.04), 291 (4.08)	107
(**6.345**)	—	—	H, OH	Ph, H	250 (4.08), 284 (4.07), 291 (4.10)	107
(**6.345**)	—	—	O	H, H	238 (4.28), 282 (4.01), 291 (3.94)	129
(**6.345**)	—	—	O	H, Me	239 (4.49), 283 (3.94), 292 (3.92)	94
(**6.345**)	—	—	O	H, Et	240 (4.10), 283 (4.06), 292 (3.84)	94
(**6.345**)	—	—	O	H, Pr^n	240 (4.62), 283 (4.08), 292 (4.01)	94
(**6.345**)	—	—	O	H, Bu^n	240 (4.46), 283 (3.89), 292 (3.78)	94
(**6.345**)	—	—	O	$CHCO_2Me$	217 (3.43), 257 (3.52)[d]	145
(**6.345**)	—	—	O	$CHCO_2Et$	214 (2.23), 257 (2.36)[d]	145
(**6.346**)	—	—	—	—	255 (4.08), 286 (3.90), 293 (3.89), 350 (3.99)	129
(**6.346**)	—	—	—	—	270 (4.06), 287 (3.88), 294 (3.88), 360 (3.88)[e]	129
(**6.346**)	—	—	—	—	288 (3.90), 295 (3.91), 365.5 (3.84)[f]	129
(**6.346**)	—	—	—	—	292 (3.83), 352 (4.02)[g]	129
(**6.347**)	—	—	—	—	213 (4.35), 260 (3.98)	103

[a] Measured for solutions in ethanol, unless otherwise specified.
[b] 6,7-Dimethyl derivative.
[c] 5,8-Dimethyl derivative.
[d] Spectrum measured in methanol.
[e] Spectrum measured in dioxane.
[f] Spectrum measured in benzene.
[g] Spectrum measured in pyridine–water.

band at longer wavelength (270–290 nm).[94,103,107,195,217] The introduction of a *C*(3)-phenyl-substituent results in a hypsochromic shift in both bands,[107] whereas the presence of a *C*(2)-phenyl group causes the disappearance of the imidazole band, an increase in the intensity of the $\pi \rightarrow \pi^*$ band, and the appearance of a charge-transfer band at 308–314 nm.[94,103,107] Curiously, other studies[217] report the reverse effects for *C*(2) and *C*(3) phenyl substituents in thiazolo[3,2-*a*]benzimidazoles. The UV spectra[103] of thiazolo[3,2-*a*]benzimidazoles containing a *C*(2) carbonyl substituent (acetyl, ethoxycarbonyl) (Table 6.46), while retaining a $\pi \rightarrow \pi^*$ band at ca. 270 nm, also lack imidazole absorption at ca. 240 nm. Not unexpectedly, in view of their lesser unsaturation, 2,3-dihydrothiazolo[3,2-*a*]benzimidazole derivatives exhibit lower intensity imidazole and $\pi \rightarrow \pi^*$ UV absorption (Table 6.46)[107] in comparison with their thiazolo[3,2-*a*]benzimidazole parents. Interestingly, the intensity of the lower wavelength imidazole band but not that of the $\pi \rightarrow \pi^*$ band in 2,3-dihydrothiazolo[3,2-*a*]benzimidazoles is increased by the introduction of a *C*(3) carbonyl moiety as evidenced by the UV spectra of 2,3-dihydrothiazolo[3,2-*a*]benzimidazol-3-ones (Table 6.46).[94,129] The effect of increased conjugation in such structures is demonstrated by the UV absorption (Table 6.46) of the corresponding *C*(2) arylidene derivatives (**6.345**; X = O, Y = CHAr) which consists of three intense bands at 257–273, 281–296, and 319–390 nm, believed[149] to originate in amido, benzimidazole, and K-band transitions respectively. In contrast, *C*(2)-alkylidene-2,3-dihydrothiazolo[3,2-*a*]benzimidazol-3-ones of the type (**6.345**; X = O, Y = $CHCO_2Me$ or $CHCO_2Et$) exhibit relatively short-wavelength UV absorption at 210–240 and 260–320 nm (Table 6.46).[145] The high intensity and relatively long-wavelength bands in the UV spectrum of the mesoionic compound (**6.346**) (Table 6.46)[129] further demonstrate the dramatic effect of increased delocalization on the UV absorption of the thiazolo[3,2-*a*]benzimidazole ring system. The presence of intense bands at 382 and 400 nm in the UV spectra[211] of 2-arylhydrazono-2,3-dihydrothiazolo[3,2-*a*]benzimidazol-3-ones (**6.345**; X = O, Y = NNHAr) has been cited as evidence for the azo as opposed to hydrazone structures of these molecules. The UV absorption [λ_{max} nm (log ε), 226 (4.36), 275 (3.41), and 285 (3.73)][153] of 1-imino-1*H*,3*H*-thiazolo[3,4-*a*]benzimidazole [cf. page 116. Scheme 6.59; (**6.267**; $R^1 \rightarrow R^3 = H$)] is significantly different from that of the structurally related 2,3-dihydrothiazolo[3,2-*a*]benzimidazol-3-one (**6.345**; X = O, Y = H_2) (cf. Table 6.46).

The 4*H*-pyrazolo[2,3-*a*]benzimidazole nucleus[207] absorbs strongly in the UV at 229 and 306 nm (log ε 4.42 and 4.00) and consequently has been widely used as a chromophoric unit in dyestuffs, which show intense absorption in the visible region at ca. 525–560 nm.[200,218,219]

Simple alkyl derivatives of 1(9)*H*-imidazo[1,2-*a*]benzimidazoles give rise to intense UV absorption at ca. 240 nm (log ε 4.46),[158,210,220] which not unexpectedly is shifted to longer wavelength (ca. 280–300 nm) by phenyl or styryl substitution at the *C*(2) position,[165] and into the visible region

(370–420 nm) by acyl or alkenyl substitution at *C*(3).[214] The increased delocalization afforded by metal complexation or incorporation in a cyanine framework is illustrated by the strong absorption in the visible region associated with 1(9)*H*-imidazo[1,2-*a*]benzimidazole transition metal complexes[210,220] and cyanine dyestuffs,[164] respectively. The UV spectra[151] of compounds formulated as 3-arylhydrazono-1(9)*H*-imidazo[1,2-*a*]benzimidazol-2-ones are characterized by the presence of three strong absorption bands at 206, 291–301, and 394 nm. The presence of the intact benzimidazole unit in *N*(1)-substituted 2,3-dihydro-1*H*-imidazo[1,2-*a*]-benzimidazoles confers UV absorption at 242–247 and 278–300 nm[176,193] attributable to imidazole and benzenoid transitions, respectively. Conversely, the absence of a discrete benzimidazole nucleus in *N*(9)-substituted 2,3-dihydro-9*H*-imidazo[1,2-*a*]benzimidazoles accounts for their lack of imidazole UV absorption at 242–247 nm and allows the clear structural differentiation of *N*(1)- and *N*(9)-substituted 2,3-dihydro-1(9)*H*-imidazo-[1,2-*a*]benzimidazole derivatives.[176,193] Moreover, the presence of a band at 242–247 nm in the UV spectra[176,193] of *N*-unsubstituted 2,3-dihydro-1(9)*H*-imidazo[1,2-*a*]benzimidazoles implies their preferential existence in the 1*H* as opposed to the 9*H* tautomeric form.

The UV absorption of 4*H*-imidazo[1,5-*a*]benzimidazole derivatives is markedly influenced by both the nature and the site of substituents in the imidazole ring, those at the *C*(3) position having a particularly strong effect on band position and intensity. The UV spectra of simple *C*(1) and *C*(3) alkyl derivatives[190,191] are typified by the presence of three equally intense (log ε ca. 3.8) bands at 240–250, 270–280, and 320–330 nm. The introduction of a *C*(3) aryl substituent results in a marked increase in the intensity of all three bands accompanied by a bathochromic shift of ca. 20–30 nm in the two longer wavelength absorptions at 270–280 and 320–330 nm. In contrast, the presence of a *C*(1) aryl substituent, though resulting in a similar overall intensification of the UV absorption (Table 6.47),[192] produces a shift to longer wavelength only in the 320–330 nm band. Interestingly the disparity in their influence on the UV absorption of 4*H*-imidazo[1,5-*a*]benzimidazoles is apparently reversed for *C*(1) and *C*(3) acyl, nitroso, and nitro substituents (Table 6.47). Whereas substituents of these types at the *C*(1) position[215,221] produce both an intensification and a marked shift to longer wavelength in all three bands, those at the *C*(3) position,[192,193] while causing an increase in the intensity and complexity of the UV absorption have little effect on its wavelength.

Nuclear Magnetic Resonance Spectra. Representative examples of the ^{1}H N.M.R. spectra of 6-5-5 fused benzimidazole derivatives having one additional heteroatom are collected in Tables 6.48–6.55.

In accord with their closer proximity to the oxazole ring oxygen atom, the *C*(2) protons in 2,3-dihydrooxazolo[3,2-*a*]benzimidazole derivatives absorb at lower field than those at the *C*(3) position (Table 6.48).[91,92]

TABLE 6.47. ULTRAVIOLET SPECTRA[a] OF 4*H*-IMIDAZO[1,5-*a*]BENZIMIDAZOLE DERIVATIVES (**6.348**)

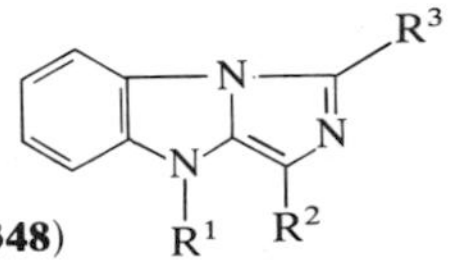

(**6.348**)

(**6.348**)	R^1	R^2	R^3	λ_{max}(nm)(log ε)	Ref.
	Me	H	Ph	230 inf. (4.17), 250 inf. (4.02), 280(4.00), 352(4.16)	192
	CH_2Ph	H	Ph	230 inf. (4.22), 251 inf. (4.02), 281(4.05), 351(4.18)	192
	Me	Ph	CH_2OH	232(4.41), 255(3.90), 300(4.33)	215
	Me	Ph	CHO	261(4.28), 298(4.32), 386(4.36)	215
	Me	Ph	COMe	257(4.22), 300(4.29), 382(3.35)	215
	Me	Ph	NO	265(4.22), 286(4.06), 322(3.90), 427(4.27)	215
	Me	Ph	NO_2	218(4.44), 269(4.34), 304(4.11), 452(4.32)	221
	$PhCH_2$	CN	Ph	232(4.30), 280(4.34), 327(4.12)	193
	Me	COMe	Me	214(4.34), 224(4.33), 242(3.95), 303(4.12), 332(4.44)	192
	Me	COMe	Ph	225(4.28), 329(4.54)	192
	CH_2Ph	COMe	Ph	225(4.36), 328(4.53)	193
	Me	NO	Me	228(4.04), 294(4.21)	192
	Me	NO	Ph	238(4.30), 259(4.29), 283(4.29)	192

[a] Measured for solutions in ethanol.

TABLE 6.48. ^{1}H NMR SPECTRA[a,b] OF 2,3-DIHYDROOXAZOLO[3,2-*a*]BENZIMIDAZOLE DERIVATIVES (**6.349**)

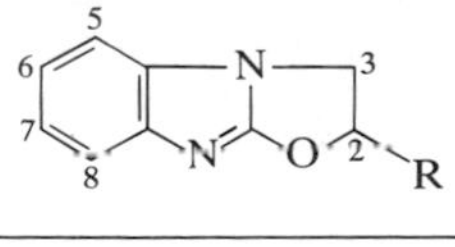

(**6.349**)

(**6.349**)	R	Solvent[c]	H(2)	H(3)	H(5)	H(6)	H(7)	H(8)	Others	Ref.
	Et	*A*	5.25q[d]	3.90–4.25t[d]	←—— 7.10m and 7.50m ——→				1.10t[d,e] 1.90q[d,f]	92
	CH_2Cl	*A*	5.40q[d]	3.90–4.24m[g]	←—— 7.10m and 7.50m ——→				—	92
	$CONH_2$	*B*	5.85q[d]	4.40m	←—— 7.10m ——→				7.70[h] 8.00[h]	91

[a] δ in ppm measured from TMS.
[b] Signals are sharp singlets unless denoted as: t = triplet; q = quartet; m = multiplet.
[c] *A* = $CDCl_3$; *B* = $(CH_3)_2SO$.
[d] Coupling constant not quoted.
[e] Me of Et group.
[f] CH_2 of Et group.
[g] Overlapping NCH_2 and CH_2Cl.
[h] NH.

TABLE 6.49. ^{1}H NMR SPECTRA[a,b] OF THIAZOLO[3,2-*a*]BENZIMIDAZOLE DERIVATIVES (**6.350**)

(**6.350**)

(**6.350**)	R^1	R^2	Solvent[c]	H(2)	H(3)	H(5)	H(6)	H(7)	H(8)	Me	Others	Ref.
	H	H	*A*	6.70d[d]	7.60d[d]	← 7.00—7.83m →				—	—	103
	H	H	*B*	7.07d[d]	8.10d[d]	7.79m	7.40m	7.40m	7.79m	—	—	195
	H	H	*A*	6.73d[d]	7.61d[d]	7.58	← 7.19 →		7.70	—	—	222
	H	H	*C*	6.83d[d]	7.72d[d]	7.70	← —[e] →		7.70	—	—	222
	H	H[f]	*D*	7.52d[d]	8.19d[d]	7.96	← —[e] →		7.86	—	—	222
	H	H[f]	*E*	7.57d[d]	8.28d[d]	8.00	← —[e] →		7.89	—	—	222
	H	H[g]	*D*	7.51d[d]	8.18d[d]	7.98	← —[e] →		7.98	—	—	222
	H	H[h]	*D*	7.68d[d]	8.40d[d]	8.12	← —[e] →		7.80	—	—	222
	H	H[i]	*A*	6.63d[d]	7.46d[d]	7.52[j]	—	—	7.26[j]	2.30	—	103
	H	H[i,f]	*D*	7.42d[d]	8.07d[d]	7.69	—	—	7.63	—[e]	—	222
	H	H[i,f]	*E*	7.52d[d]	8.20d[d]	7.82	—	—	7.63	—[e]	—	222
	H	H[i,h]	*D*	7.61d[d]	8.28d[d]	7.85	—	—	7.52	—[e]	—	222
	H	H[k]	*A*	6.68d[l]	7.70d[d]	—	← 6.92q[l] →		—	2.55 2.64	—	103
	H	Me	*A*	—	7.30q[m]	← 7.15–7.90m →				2.40d[m]	—	103
	H	Me	*A*	—	7.00– 7.80m	← —[e] →				2.53	—	107
	H	Me	*A*	—	7.26q[m]	7.51	← 7.23 →		7.73	2.34	—	222
	H	Me[i]	*A*	—	7.18q[m]	7.19	—	—	7.49	2.30 2.35	—	222
	Me	H	*A*	6.11q[m]	—	← 7.00–7.90m →				2.59d[m]	—	103
	Me	H	*A*	6.11	—	← —[e] →				2.59	—	107
	Me	H	*A*	6.27d[m]	—	← 7.72m →				2.65	—	222
	Me	H[i]	*A*	6.18d[m]	—	7.40	—	—	7.47	2.25 2.50d[m]	—	222

Me	H	*C*	6.28d[m]	—	7.54	—	—	7.47	—[e]	—	222
Me	H[f]	*D*	6.93d[m]	—	7.78	—	—	7.64	—[e]	—	222
Me	H[f]	*E*	7.04d[m]	—	7.92	—	—	7.64	—[e]	—	222
Me	H[g]	*D*	6.94d[m]	—	7.77	—	—	7.69	—[e]	—	222
Me	H[h]	*D*	7.13d[m]	—	7.83	—	—	7.55	—[e]	—	222
Me	Me	*A*	—	—	← 7.18, 7.66 →				2.21[n] 2.45[o]	—	222
Me	Me[i]	*A*	—	—	7.36	—	—	7.46	2.24[n] 2.46[o] 2.33	—	222
Ph	H	*A*	6.60	—	← 7.00–8.10m →				—	7.50[p]	103
Ph	H	*A*	6.82	—	← 7.10–7.60m →				—	—	122
Ph	H	*A*	6.55	—	← —[e] →				—	—	107
p-MeC_6H_4	H	*A*	6.51	—	← 7.00–7.90m →				2.49	—	95
p-BrC_6H_4	H	*A*	6.64	—	← 7.10–7.90m →				—	—	96
Ph	H[p]	*A*	6.69	—	7.07–7.82m	—	← 7.07–7.82m →		—	7.07–7.82m[p]	98
p-BrC_6H_4	H[q]	*A*	6.70	—	← 7.02–7.88m →				2.50	—	96
p-BrC_6H_4	H[r]	*A*	6.64	—	← 7.08–7.82m →				2.65	—	96
p-BrC_6H_4	H[h]	*A*	6.76	—	7.61	—	—	7.02	2.30 2.39	7.65q[p]	96
H	CH(OH)Me	*A*	—	7.00–7.90[t]	← 7.00–7.90m →				1.59d[u]	5.00q[u,v] 5.70br[w]	103
H	CH(OAc)Me	*A*	—	7.75	← 7.10–7.90m →				1.67d[u] 2.08[x]	6.13q[u]	103
Me	CH_2OH	*A*	—	—	← 7.00–7.90m →				2.52	4.55 5.50br[w]	103
H	COMe	*F*	—	9.3	← 7.10–8.00m →				2.45[x]	—	103
Me	COMe	*A*	—	—	← 7.00–7.80m →				2.47[x] 2.95	—	103
Me	CO_2Et	*A*	—	—	← 7.00–7.90m →				1.40t[y,z] 3.00	4.38q[z]	103
OCO_2Et	COMe	*A*	—	—	← 7.50–8.50m →				1.63t[y,aa] 2.60[x]	4.70q[aa]	142

(Footnotes overleaf)

(*Footnotes to Table* 6.49)

[a] δ in ppm measured from TMS.
[b] Signals are sharp singlets unless denoted as br = broad; d = doublet; t = triplet; q = quartet; m = multiplet.
[c] $A = CDCl_3$; $B = CD_3OD$; $C = CH_2Cl_2$; $D = CH_2Cl_2\text{-}CF_3CO_2H$; $E = CF_3CO_2H$; $F = (CD_3)_2SO$.
[d] $J = 4.5$–4.8 Hz.
[e] δ values not quoted.
[f] Trifluoroacetate.
[g] Hydrochloride.
[h] Methiodide.
[i] 6,7-Dimethyl derivative.
[j] These signal assignments may be interchanged.
[k] 5,8-Dimethyl derivative.
[l] $J = 11.1$ and 7.5 Hz.
[m] $J = 1.5$–2.5 Hz.
[n] Me(2).
[o] Me(3).
[p] Protons of the aryl substituent.
[q] 6-Chloro derivative.
[r] 6(7)-Methyl derivative; orientation not established.
[s] 5(8)-Methyl derivative; orientation not established.
[t] *H*(3) obscured by ArH absorption.
[u] $J = 6.5$ Hz.
[v] C*H*OH.
[w] OH.
[x] Me of Ac group.
[y] Me of Et group.
[z] $J = 7.2$ Hz.
[aa] *J* value not quoted.

TABLE 6.50. ^{1}H NMR SPECTRA[a,b] OF 2,3-DIHYDROTHIAZOLO[3,2-*a*]BENZIMIDAZOLE DERIVATIVES (**6.351**)

(**6.351**)

(**6.351**)	R¹	R²	R³	Solvent[c]	H(2)	H(3)	H(5)	H(6)	H(7)	H(8)	Me	Others	Ref.
	H	H	H	—[d]	←— 3.97m —→		←— —[e] —→				—	—	94
	H	H	H	*A*	←— 3.93 —→		←— 7.60 —→				—	—	125
	Me	H	H	*B*	3.53[f]	3.97 sex[f]	←— 7.20—7.50m —→				1.29d[f]	—	95
	OH	H	H	*C*[g]	3.92q[h] 6.62q[i]	—[e]	←— 7.30—7.90m —→				—	—	103
	OH	H	H	*D*	3.77q[g] 4.45q[i]	6.38q[h]	←— —[e] —→				—	—	212
	OAc	H	H	*B*	3.70d[k] 4.29q[l]	—[m]	←— 7.00–7.70m —→				2.05	—	103
	OH	H	H[n]	*C*[g]	3.64q[h] 4.24q[i]	6.30q[j]	7.43[o]	—	—	7.50[o]	2.28	—	103
	OH	H	H[p]	*C*[g]	3.65q[h] 4.25q[i]	6.32q[j]	—	←— 6.80 —→		—	2.42 2.53	—	103
	OH	H	Me[q]	*C*[g]	4.12oct[q] 4.65m[t]	5.82d[r,s] 5.98d[t,u]	←— 6.90–7.60m —→				1.40		103
	OH	H	Me	*D*[v]	4.30[r] 4.75[t]	6.01[r,s] 6.17[t,u]	←— —[e] —→				1.54	—	212
	OAc	H	Me	*B*	4.20q[r] 4.78oct[t]	6.75[r] 7.0d[t,w]	←— 7.10–7.80m —→				1.63 1.69 2.10	—	103
	OH	Me	H	*C*[g]	3.94q[x]	—	←— 6.95–7.72 —→				1.94 2.32[y]	4.35[y]	103
	OH	Me	H	*C*	3.78d[h] 3.92d[i]	—	←— —[e] —→				1.84 2.30[y]	4.23[y]	212

TABLE 6.50 (*Continued*)

(6.351)	R^1	R^2	R^3	Solvent[c]	H(2)	H(3)	H(5)	H(6)	H(7)	H(8)	Me	Others	Ref.
	OH	Me	Me	*E*	4.26[r]	—	←	—	—[e]	→	1.07[t,aa]	4.76[y]	212
					4.46[t]	—	←	—	—[e]	→	1.13[y]		
											1.18[r,aa]		
											1.31[t,z]		
											2.09[r,z]		
											2.31[y]		
	OH	H	Ph	—[d]	5.95[o]	6.85[o]	←	—	—[e]	→	—	—	107

[a] δ in ppm measured from TMS.
[b] Signals are sharp singlets unless denoted as: d = doublet; q = quartet; sex = sextet; oct = octet; m = multiplet.
[c] $A = D_2O$; $B = CDCl_3$; $C = (CD_3)_2SO$; $D = (CD_3)_2NCDO$; $E = C_5D_5N$.
[d] Solvent not specified.
[e] δ values not quoted.
[f] $J = 6.5$ Hz.
[g] After exchange with CF_3CO_2H.
[h] H(2) trans to OH; $J = 11.9$–12.5 and 1–2.5 Hz.
[i] H(2) cis to OH; $J = 11.9$–12.5 and 5.0–6.0 Hz.
[j] $J = 5.0$–6.0 and 1.5–2.5 Hz.
[k] H(2) cis to OAc; $J = 13.5$ Hz.
[l] H(2) trans to OAc; $J = 13.5$ and 5.0 Hz.
[m] H(3) obscured by ArH.
[n] 6,7-Dimethyl derivative.
[o] These signal assignment may be interchanged.
[p] 5,8-Dimethyl derivative.
[q] Mixture of cis/trans isomers in the ratio 4:5.
[r] Trans isomer.
[s] $J = 2.0$–2.5 Hz.
[t] Cis isomer.
[u] $J = 5.2$ Hz.
[v] Mixture of cis/trans isomers in ratio 3:4.
[w] $J = 7.0$–7.5 Hz.
[x] $J = 11.5$ and 8.8 Hz.
[y] Me and CH_2 of open-chain tautomer.
[z] Me (2).
[aa] Me (3).

TABLE 6.51. ^{1}H NMR SPECTRA[a,b] OF 2,3-DIHYDROTHIAZOLO[3,2-*a*]BENZIMIDAZOL-3-ONE DERIVATIVES (**6.352**)

(**6.352**)

(**6.352**)	R	Solvent[c]	H(2)	H(5)	H(6)	H(7)	H(8)	Me	Others	Ref.
	H	*A*	4.5	←———	7.00–7.00m	———	———→	—	—	103
	H	*A*	4.34	7.89	7.39	7.39	7.69	—	—	133
	Me	*A*	4.65q[d]	←———	—[e]	———	———→	1.85d[d]	—	94
	Et	*A*	4.68q[d]	←———	—[e]	———	———→	1.12t[d]	2.20	94
	Pr^{n}	*A*	4.55q[d]	←———	—[e]	———	———→	1.01t[d]	—	94
	Bu^{n}	*A*	4.60q[d]	←———	—[e]	———	———→	1.00t[d]	—	94
	H[f]	*B*	4.66	7.47d[g]	—	7.05q[g,h]	7.57d[h]	3.86	—	97
	H[i]	*A*	4.43	7.92q[j,k]	—	7.34q[h,j]	7.51q[h,k]		—	98
	H[l]	*A*	4.35	7.76	—	7.18	7.48	2.47	—	133
	H[m]	*A*	4.33	7.66	7.18	7.18	—	2.57	—	133
	H[n]	*A*	4.30	7.64	—	—	7.34	2.30	—	133
	H[o]	*A*	4.32	—	6.93d[p]	7.17d[p]	—	2.52 2.77	—	133

[a] δ in ppm measured from TMS.
[b] Signals are sharp singlets unless denoted as: d = doublet; t = triplet; q = quartet.
[c] *A* = $CDCl_3$; *B* = $(CD_3)_2SO$.
[d] J – 7.0–9.0 Hz.
[e] δ values not quoted.
[f] 6-Methoxy derivative.
[g] $J_{5,7}$ = 3.0 Hz.
[h] $J_{7,8}$ = 8.5 Hz.
[i] 6-Chloro derivative.
[j] $J_{5,7}$ = 1.8 Hz.
[k] $J_{5,8}$ = 0.6 Hz.
[l] 6-Methyl derivative.
[m] 8-Methyl derivative.
[n] 6,7-Dimethyl derivative.
[o] 5,8-Dimethyl derivative.
[p] J = 8.0 Hz.

TABLE 6.52. ^{1}H NMR SPECTRA[a,b] OF 1*H*,3*H*-THIAZOLO[3,4-*a*]BENZIMIDAZOLE DERIVATIVES (**6.353**)[c]

(**6.353**)

(**6.353**)	R^{1}	R^{2}	X	H(3)	H(5)	H(6)	H(7)	H(8)	Others
	H	H	NH	4.70	←———	7.30–7.93m	——→	8.00–8.24m	9.74[d]
	H	Cl	NH	4.68	7.75[e]	—	7.38q[e,f]	8.04d[f]	9.74[d]
	Cl	H	NH	4.68	7.71d[f]	7.38q[e,f]	—	8.04[e]	9.71[d]
	H	H	O	4.82	←———	———	7.53–8.00m	———→	—

[a] δ in ppm measured in $(CD_3)_2SO$ solution from TMS.
[b] Signals are sharp singlets unless denoted as: d = doublet; q = quartet; m = multiplet.
[c] From Ref. 153.
[d] NH.
[e] J_{meta} = 2.0 Hz.
[f] J_{ortho} = 8.5 Hz.

TABLE 6.53. ^{1}H NMR SPECTRA[a,b] OF PYRAZOLO[2,3-*a*]BENZIMIDAZOLE DERIVATIVES (**6.354**) AND (**6.355**)

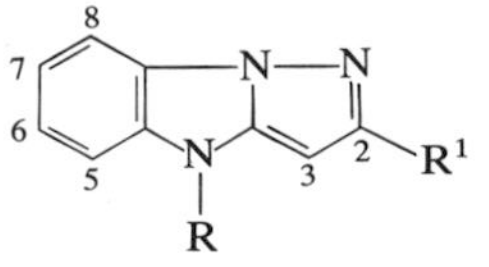

(**6.354**) 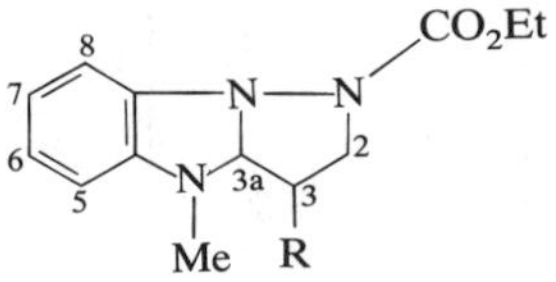   (**6.355**)

Compound	R	R[1]	Solvent[c]	H(2)	H(3)	H(3a)	H(5)	H(6)	H(7)	H(8)	Me	Others	Ref.
(**6.354**)	H	H	*A*	7.62d[d]	5.72d[d]	—	←6.90–7.50m→			7.55–7.80m	—	11.30[e]	208
(**6.354**)	H	Me	*B*	—	5.58	—	←7.10–7.90m→				2.46	—	207
(**6.354**)	H	Me	*A*	—	5.67	—	7.43[f,g,h]	7.26[i,j,k]	7.17[g,i,k]	7.70[h,j,k]	2.37	11.37[e]	223
(**6.354**)	H	Me	*C*	—	6.28q[l]	—	←[m]→				2.68d[l]	10.39[e]	223
(**6.354**)	COMe	Me	*B*	—	5.73	—	8.25	7.22	7.29	7.61	2.43	—	223
(**6.355**)	CO_2Me	—	*B*	3.18–3.45m	5.10–5.21m	4.05–4.45m[n]	←6.45–7.10m→				1.38t[o,p] 2.79[q] 2.87[q] 3.76[r]	4.05–4.55m[n]	154
(**6.355**)	CN	—	*B*	3.16–3.66m	5.06–5.21m	4.02–4.55m[n]	←6.45–7.40 m→				1.37t[o,p] 2.90[q] 2.98[q]	4.02–4.45m[n]	154

[a] δ in ppm measured from TMS.
[b] Signals are sharp singlets unless denoted as: d = doublet; t = triplet; q = quartet; m = multiplet.
[c] $A = (CD_3)_2SO$; $B = CDCl_3$; $C = CF_3CO_2H$.
[d] $J = 2.25$ Hz.
[e] NH.
[f] $J_{5,6} = 8.08$ Hz.
[g] $J_{5,7} = 1.14$ Hz.
[h] $J_{5,8} = 0.62$ Hz.
[i] $J_{6,7} = 7.43$ Hz.
[j] $J_{6,8} = 1.26$ Hz.
[k] $J_{7,8} = 8.02$ Hz.
[l] $J = 0.75$ Hz.
[m] Poorly resolved.
[n] Overlapping H(3a) and CH_2 of Et group.
[o] Me of Et group.
[p] $J = 7.0$ Hz.
[q] NMe.
[r] OMe.

TABLE 6.54. ^{1}H NMR SPECTRA[a,b] OF IMIDAZO[1,2-*a*]BENZIMIDAZOLE DERIVATIVES (**6.356**)–(**6.358**)

(**6.356**) (**6.357**) (**6.358**)

Compound	R	R[1]	R[2]	X	Solvent[c]	H(2)	H(3)	H(5)	H(6)	H(7)	H(8)	N(Alk)	Others	Ref.
(**6.356**)	H	H	H	—	*A*	←			7.50–8.10m		→	—	—	155
(**6.356**)	H	H	Me	—	*A*	←			6.90–7.10m		→	—	2.05	166
(**6.356**)	Me	Me	H	—	*B*	←			—[d]		→	3.60	2.20	171
(**6.356**)	Me	Ph	H	—	*B*	←			—[d]		→	3.65	—	171
(**6.356**)	Me	Ph	COMe	—	*C*	←			—[d]		→	3.77	2.10	171
(**6.356**)	Me	Ph	COPh	—	*B*	←			—[d]		→	3.70	—	171
(**6.357**)	CH_2Ph	—	—	H, H	*D*	← —[d]	→	←	7.10m	→	7.40m	4.65	—	176
(**6.358**)	CH_2Ph	—	—	H, H	*D*	← —[d]	→	←		6.75m	→	4.93	—	176
(**6.357**)	Me	—	—	O	*A*	—	4.73	←		—[d]	→	3.22	—	180
(**6.358**)	Me	—	—	O	*A*	—	4.75	←		—[d]		→	—	180

[a] δ in ppm measured from TMS.
[b] Signals are sharp singlets unless denoted as: m = multiplet.
[c] *A* = CF_3CO_2H; *B* = CCl_4; *C* = $CHCl_3$; *D* = $CDCl_3$.
[d] δ values not quoted.

TABLE 6.55. ^{1}H NMR SPECTRA[a] OF 4*H*-IMIDAZO[1,5-*a*]BENZIMIDAZOLE DERIVATIVES (**6.359**)

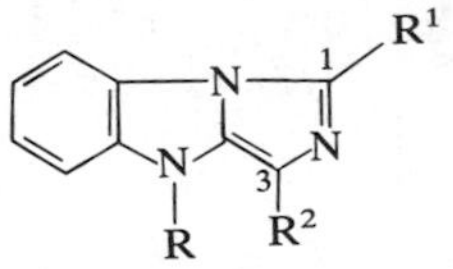

(**6.359**)

(**6.359**)	R	R[1]	R[2]	Solvent[b]	H(1)	H(3)	N(Alk)	Me	Others	Ref.
	CH_2Ph	H	H	*A*	8.02	6.26	—[d]	—	—	224
	Me	H	Me	*A*	7.53	—	3.17	—[d]	—	224
	Me	H	Ph	*A*	7.84	—	3.35	—	—	224
	Me	Me	H	*A*	—	5.96	3.04	—[d]	—	224
	Me	Ph	H	*B*	—	6.51	3.47	—	—	192, 224
	CH_2Ph	Ph	H	*B*	—	6.42	5.09	—	—	193
	Me	CHO	Ph	*B*	—	—	3.78	—	—[d]	224
	CH_2Ph	Ph	CHO	*B*	—	—	—[d]	—	9.84[c]	193
	Me	Me	COMe	*B*	—	—	4.19	2.52[e] 2.71[f]	—	192
	Me	Ph	COMe	*B*	—	—	4.30	2.67[e]	—	192
	CH_2Ph	Ph	COMe	*B*	—	—	6.13	2.70[e]	—	193
	Me	Me	NO	*A*	—	—	4.14	2.71[f]	—	192
	Me	Ph	NO	*B*	—	—	4.18	—	—	192

[a] δ in ppm measured from TMS.
[b] *A* = CD_3OD; *B* = $CDCl_3$.
[c] *CHO*.
[d] δ values not quoted.
[e] Me of Ac group.
[f] Me (1).

^{1}H NMR signals due to the *C*(2) and *C*(3) protons in *C*(2)/*C*(3)-unsubstituted thiazolo[3,2-*a*]benzimidazoles are found at uniformly low field in the ranges δ 6.6–7.1 and δ 7.4–8.1 and appear as doublets, due to *H*(2)/*H*(3) spin interaction, with a characteristic coupling constant, $J_{2,3}$ = 4.5–4.8 Hz (Table 6.49).[94,103,195,222] The observed order of shielding for the various protons in thiazolo[3,2-*a*]benzimidazole, namely *H*(8) < *H*(3) ≈ *H*(5) < *H*(6) ≈ *H*(7) < *H*(2) is in accord with the calculated[222] π-electron densities for the molecule. The latter also account for the greater shielding of the methyl substituent in 2-methylthiazolo[3,2-*a*]benzimidazoles compared with that in 3-methylthiazolo[3,2-*a*]benzimidazoles (Table 6.49).[97,103,107,119,222] Shielding by the adjacent methyl groups produces the anticipated upfield shift in the resonances of the thiazole protons in 2- and 3-methylthiazolo[3,2-*a*]benzimidazoles, which appear as well-defined quartets due to allylic coupling ($J_{Me,H}$ = 1.5 Hz) (Table 6.49).[97,103,222] In contrast, the aryl substituent in 2- and 3-arylthiazolo[3,2-*a*]-

benzimidazoles has little effect on the chemical shift of the adjacent thiazole proton (Table 6.49).[95,96,98,99,102,107,122] A *C*(2)-acyl substituent, on the other hand, has the expected marked deshielding effect on protons and methyl substituents at the *C*(3)-position in thiazolo[3,2-*a*]benzimidazoles (Table 6.49).[103] Not unexpectedly, protonation of thiazolo[3,2-*a*]benzimidazoles results in a shift of all of the ring proton resonances to lower field, *H*(2) and *H*(3) being deshielded to the greatest extent (δ 0.65–0.69 and δ 0.47, respectively) (Table 6.49).[222] The enhanced deshielding of *H*(2) and *H*(3), which results from the protonation of thiazolo[3,2-*a*]benzimidazoles, implies extensive delocalization of the positive charge to the heteroatoms of the thiazole ring, in the cation produced.

The methylene protons of 2,3-dihydrothiazolo[3,2-*a*]benzimidazoles give rise to ^{1}H NMR absorption in the range δ 3.5–4.0 (Table 6.50).[94,95,125] ^{1}H NMR spectroscopy provides a valuable alternative method to IR spectroscopy (see before) for the study[102,103,107–109,212] of ring-chain tautomerism in 3-hydroxy-2,3-dihydrothiazolo[3,2-*a*]benzimidazole derivatives [cf. page 108, Scheme 6.51; (**6.241**) ⇌ (**6.240**)]. For example, the ^{1}H NMR spectra (Table 6.50)[103,212] of 3-hydroxy-2,3-dihydrothiazolo[3,2-*a*]benzimidazole and derivatives substituted in the benzene ring are typified by the presence of discrete signals in the ranges δ 3.66–3.77, 4.24–4.35, and 6.30–6.38. The attribution[103,212] of these signals, on the basis of splitting pattern and associated coupling constants (Table 6.50) to the *C*(2) protons trans and cis to the hydroxyl group and the proton at *C*(3), respectively, is clearly consistent only with cyclic (3-hydroxy-2,3-dihydrothiazolo[3,2-*a*]benzimidazole) as opposed to open-chain structures. Correspondingly, on the basis of its ^{1}H NMR absorption (Table 6.50),[103,212] 3-hydroxy-2-methyl-2,3-dihydrothiazolo[3,2-*a*]benzimidazole (**6.351**; R^1 = OH, R^2 = H, R^3 = Me) exists entirely as a mixture of cis and trans isomeric ring-closed forms. On the other hand, the ^{1}H NMR spectrum (Table 6.50)[103,212] of the isomeric 3-hydroxy-3-methyl-2,3-dihydrothiazolo[3,2-*a*]benzimidazole (**6.351**; R^1 = OH, R^2 = Me, R^3 = H) contains, in addition to proton resonances due to the cyclic form, signals at δ 2.30–2.32 and 4.23–4.35 which reveal[103,212] the presence of the open-chain keto form and hence the establishment in solution of ring-chain tautomeric equilibrium. The ^{1}H NMR absorption (Table 6.50)[212] of 2,3-dimethyl-3-hydroxy-2,3-dihydrothiazolo[3,2-*a*]benzimidazole (**6.351**; R^1 = OH, R^2 = R^3 = Me) likewise demonstrates this molecule to coexist in solution with the corresponding open-chain keto tautomer. The results of a thorough ^{1}H NMR study[108] of the effect of structure on the ring-chain tautomerism of 3-hydroxy-2,3-dihydrothiazolo-[3,2-*a*]benzimidazoles indicate that, as the bulk, or ability to participate in conjugation, of the *C*(3) substituent increases, the greater is the tendency for the molecule to exist in solution in the open-chain form.

The deshielding influence of the *C*(3) carbonyl group is apparent in the lower field resonance of the *C*(2) methylene protons of 2,3-dihydrothiazolo-[3,2-*a*]benzimidazol-3-ones (Table 6.51)[94,97,98,103,133] compared with that of the *C*(2) methylene protons in simple 2,3-dihydrothiazolo[3,2-*a*]benzimidazoles (Table 6.50). The *C*(5) proton in 2,3-dihydrothiazolo[3,2-*a*]-benzimidazol-3-ones is readily differentiated from the other benzenoid protons on the basis of its lower field ^{1}H NMR absorption (Table 6.51) as a result of the anisotropic effect of the proximate *C*(3) carbonyl group. This effect (with due allowance for perturbation by the substituent) has been utilized to establish the orientation of nuclear substituted 2,3-dihydrothiazolo[3,2-*a*]benzimidazol-3-ones formed in ring-closure reactions (see before) of 2-benzimidazolythioacetic acids.[97,98,133] Analogous deshielding (Table 6.52) of the *C*(8) proton by the *C*(1) imino substituent in 1-imino-1*H*,3*H*-thiazolo[3,2-*a*]benzimidazoles (**6.353**) likewise permits the demonstration[153] of the substitution pattern of unsymmetrically substituted derivatives formed by the ring-closure of 2-(α-thiocyanoalkyl)benzimidazoles (see page 116).

The ^{1}H NMR signal (Table 6.53) of the *C*(3) proton in 4*H*-pyrazolo[2,3-*a*]benzimidazole (**6.354**; R = R^{1} = H) appears as a doublet at δ 5.72 due to coupling with *H*(2) ($J_{2,3}$ = 2.25 Hz), which resonates at considerably lower field (δ 7.62).[208] The *C*(2) methyl substituent in 2-methyl-4*H*-pyrazolo[2,3-*a*]benzimidazole exerts the expected shielding effect on the *C*(3) proton (Table 6.53),[207,223] whereas protonation has a more dramatic deshielding effect (Table 6.53).[223] Application of the LAOCN program in conjunction with the identification of *H*(8) on the basis of its long-range coupling with the NH proton, allows the assignment of the chemical shifts and associated coupling constants of the benzenoid protons in 2-methyl-4*H*-pyrazolo[2,3-*a*]benzimidazole (**6.354**; R = H, R^{1} = Me) (Table 6.53).[223] Conversely, spin interaction between *H*(8) and the NH group requires the latter to be sited at *N*(4), thus demonstrating[223] the preferential existence of 2-methyl-4*H*-pyrazolo[2,3-*a*]benzimidazole in the 4*H* as opposed to the alternative 1*H*-tautomeric form. The 4*H*-tautomeric structure of 2-methyl-4*H*-pyrazolo[2,3-*a*]benzimidazole is further supported by the lack of allylic coupling between *H*(3) and the protons of the *C*(2) methyl substituent.[223] The bridgehead proton in 2,3,3*a*,4-tetrahydro-1*H*-pyrazolo[2,3-*a*]benzimidazole derivatives (**6.355**) resonates in the range δ 5.06–5.21 (Table 6.53).[154]

Despite their accessibility, relatively little information on the ^{1}H NMR absorption (Table 6.54) of the various types of imidazo[1,2-*a*]benzimidazole derivative is available. The resonances of *H*(2) and *H*(3) in 1(4)*H*-imidazo-[1,2-*a*]benzimidazole are not distinguishable from those of the benzenoid protons (Table 6.54).[155] On the other hand, *C*(2) and *C*(3) methyl substituents in 1(4)*H*-imidazo[1,2-*a*]benzimidazoles are distinguishable on the basis of the lower field ^{1}H NMR absorption of the latter (Table 6.54).[166,171] The protons of *N*(4)-methyl groups in 4*H*-imidazo[1,2-*a*]benzimidazoles absorb uniformly in the range δ 3.60–3.80.[171] The ^{1}H NMR absorption (δ ca. 4.9)[166] of the *C*(3) methylene protons in 2,3-dihydro-1*H*-imidazo[1,2-*a*]benzimidazoles undergoes the expected shift to lower field (Table 6.54) in 2,3-dihydro-1*H*-imidazo[1,2-*a*]benzimidazol-2-ones [e.g. (**6.357**; R = Me,

X = O)][180] due to the deshielding effect of the adjacent carbonyl group. The isomeric *N*(1)- and *N*(9)-benzyl derivatives of 2,3-dihydro-1(9)*H*-imidazo-[1,2-*a*]benzimidazole (**6.357**; R = CH_2Ph, X = H_2) and (**6.358**; R = CH_2Ph, X = CH_2) are readily distinguished by the lower field resonance (Table 6.54) of the benzyl protons in the latter resulting from combined deshielding by the cyclic azomethine group and the benzene ring.[176]

Protons at the *C*(1) position in 4*H*-imidazo[1,5-*a*]benzimidazoles (**6.359**) resonate at consistently lower field in the range δ 7.5–8.0, compared with those at the *C*(3) position which absorb at δ 5.9–6.5 (Table 6.55).[224] The enhanced shielding of *H*(3) in comparison with *H*(1) in 4*H*-imidazo[1,5-*a*]-benzimidazoles is in accord with greater electron localization at the *C*(3) position than at the *C*(1) position as predicted by molecular orbital calculations.[224] The protons of *N*(4) methyl substituents in 4*H*-imidazo[1,5-*a*]-benzimidazoles are deshielded to a significant extent (Table 6.55) by the presence of electron-withdrawing groups in the imidazole ring, this effect being most marked in the case of *C*(3) substituents of this type.[224] In accord with their 1-dialkylaminoethylbenzimidazole-like structure, 1*H*,3*H*-imidazo[1,5-*a*]benzimidazoles give rise to singlet ^{1}H NMR absorption at δ 5.03 attributable to the *C*(1) methylene protons.[188] However, the latter become nonequivalent when an asymmetric center is present at the *C*(3) position, and then absorb at δ 4.96 as an AB quartet (J_{gem} = 6 Hz).[188]

Mass Spectra. The mass spectral fragmentation of thiazolo[3,2-*a*]benzimidazoles (Tables 6.56, 6.57, and 6.58)[107,225] follows several pathways (Schemes 6.77 and 6.78), which are principally initiated by cleavage of one or other of the bonds in the thiazole ring. The (M–SH) fragmentation [(**6.360**)→ (**6.361**)] which is a feature[107,225] of 2- and 3-methyl and ethyl thiazolo[3,2-*a*]benzimidazoles is not observed for the 2- and 3-phenyl derivatives, indicating its origin in initial hydrogen abstraction from the alkyl substituent.[225] Ions (**6.361**) derived by extrusion of sulfur, on the other hand, figure prominently in the mass spectra (Tables 6.56 and 6.57)[107,225] of 2- and 3-phenylthiazolo[3,2-*a*]benzimidazoles. $(M-1)^+$ fragment ions derived by hydrogen abstraction from the molecular ions of 2- and 3-methyl-thiazolo[3,2-*a*]benzimidazoles have been variously formulated as having intact thiazolo[3,2-*a*]benzimidazole[225] or ring-contracted[107] structures, (**6.363**) and (**6.367**), or (**6.364**) and (**6.368**), respectively. Prior ring-contraction by preferential C—S bond cleavage in the molecular ion is also invoked[107] to account for the major mass spectral fragmentation pathway of thiazolo[3,2-*a*]benzimidazoles (Scheme 6.77),[107,225] which leads to the ions (**6.365**) and (**6.366**), the former subsequently undergoing further cleavage and fragmentation to the species (**6.370**) and (**6.371**). Since the sulfur-containing ions (**6.366**) are constituted from the *C*(2) carbon atom of the thiazole ring with its attached substituent their mass numbers are diagnostic of the *C*(2)/*C*(3) substitution pattern in the original thiazolo[3,2-*a*]benzimidazole.[107,225] Similar structural information is provided by a consideration of the mass numbers of the fragment ions [$(R^1C{\equiv}CR^2)^+$] and the thiirenyl

TABLE 6.56. MASS SPECTRA[a] OF THIAZOLO[3,2-*a*]BENZIMIDAZOLE DERIVATIVES (**6.360**)[b]

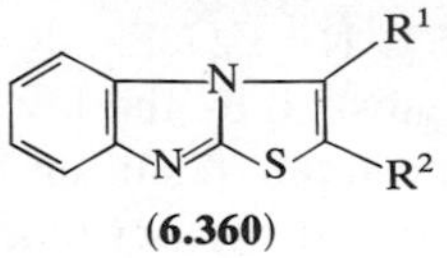

(**6.360**)

R^1	R^2	*m/e* (rel. abundance, %)
H	H	176 (5.6), 175 (13.7), 174 (100), 173 (2), 134 (3), 130 (2.4), 129 (24.9), 103 (6.1), 102 (21.2), 90 (6.8), 77 (2.3), 76 (8), 59 (9), 74 (2.7), 70 (3.8), 69 (2.3), 64 (5.6), 63 (7.6), 62 (2.7), 58 (7.3), 57 (3.3), 52 (2.3), 51 (7.1), 50 (6.7), 46 (2.6), 45 (6.8), 39 (6.5).
H	Me	190 (5.9), 189 (14), 188 (100), 187 (13.8), 160 (2), 155 (4.2), 143 (21.6), 134 (2.7), 118 (2.7), 117 (3), 116 (2.6), 103 (4.6), 102 (19), 90 (7), 77 (4.2), 76 (7.6), 75 (8.5), 74 (2), 73 (6.8), 71 (4.7), 70 (2.8), 69 (2.8), 64 (4.5), 63 (6), 52 (2.7), 51 (7), 50 (5.4), 45 (9), 44 (3), 40 (5.7), 39 (10.5).
Me	H	190 (5.9), 189 (14), 188 (100), 187 (13.8), 160 (2), 155 (4.2), 143 (21.6), 134 (2.7), 118 (2.7), 117 (3), 116 (2.6), 103 (4.6), 102 (19), 90 (7), 77 (4.2), 76 (7.6), 75 (8.5), 74 (2), 73 (6.8), 71 (4.7), 70 (2.8), 69 (2.8), 64 (4.5), 63 (6), 52 (2.7), 51 (7), 50 (5.4), 45 (9), 44 (3), 40 (5.7), 39 (10.5).
Me	Me	203 (9.7), 202 (100), 201 (27), 200 (2.5), 188 (2.8), 187 (19.7), 175 (4.9), 169 (5.7), 168 (3.7), 161 (2), 144 (3.4), 143 (13.7), 134 (7.4), 129 (2), 116 (2.3), 103 (2.3), 102 (13.5), 90 (7.7), 77 (3.1), 76 (4.9), 75 (6.3), 71 (2), 69 (2), 64 (3.4), 63 (4.9), 59 (6.3), 58 (2), 53 (6.3), 52 (2.5), 51 (6.8), 50 (4.3), 45 (3.4), 44 (3.4), 42 (3.1), 41 (4), 40 (7.7), 39 (7.4).
Et	Me	218 (5.1), 217 (13.9), 216 (100), 215 (5.6), 202 (9.9), 201 (54.5), 200 (4.5), 188 (2), 187 (6.1), 168 (2.3), 161 (2.1), 156 (2.1), 143 (13.2), 142 (2.1), 134 (4.1), 129 (3.3), 122 (2), 102 (4.3), 90 (3.8), 59 (2.8), 41 (2.5).
Me	Et	218 (5.1), 217 (13.9), 216 (80.3), 215 (4.9), 203 (5.7), 202 (16.4), 201 (100), 200 (5.7), 199 (2.5), 187 (3.1), 175 (2.5), 168 (3), 161 (2), 156 (2), 144 (2.6), 143 (21.3), 142 (3.2), 134 (11.5), 131 (3), 129 (3), 116 (3.1), 115 (3.1), 108 (3.8), 107 (2.1), 103 (2.3), 102 (13.1), 101 (2), 90 (7.4), 77 (2), 76 (2.6), 75 (3), 69 (2), 63 (2.1), 53 (2), 51 (2.3), 45 (2.1), 41 (3.6), 39 (3.7).
H	Ph	252 (6.5), 251 (20.5), 250 (100), 249 (5.1), 248 (3.2), 218 (4), 217 (2), 190 (2), 134 (1), 129 (4.1), 122 (3.2), 121 (14.4), 116 (3.6), 102 (4.4), 101 (2), 90 (3.6), 89 (2.9), 86 (10), 58 (2.1), 41 (2).
Ph	H	252 (7.4), 251 (22.3), 250 (100), 249 (5.5), 248 (7.3), 218 (2), 217 (1), 212 (2), 205 (5.8), 190 (1), 121 (1.7), 116 (1.3), 102 (3.9), 90 (2.1), 77 (3.6), 45 (1.5).
Ph	Me	266 (6.3), 265 (21.4), 264 (100), 263 (16.1), 262 (4.8), 261 (2), 237 (5.2), 231 (2.7), 212 (2), 205 (3.5), 204 (3.4), 187 (2.6), 134 (2), 116 (2), 115 (5.4), 104 (2), 103 (3), 102 (2), 90 (1), 77 (2.6), 45 (2.1).

TABLE 6.56 (*Continued*)

(**6.360**)

R^1	R^2	*m/e* (rel. abundance, %)
Ph	Ph	328 (8.1), 327 (27.1), 326 (100), 325 (2.7), 266 (2), 250 (2.7), 205 (6.3), 190 (2), 179 (2), 178 (8.9), 177 (2), 166 (2.8), 165 (15.3), 163 (4.2), 149 (2.7), 121 (2.7), 103 (2.5), 102 (4.4), 101 (2.8), 91 (2.9), 86 (12.2), 77 (5.5), 58 (3), 53 (2), 44 (2.2), 43 (2.2), 42 (2).
Ph	H[c]	280 (6.9), 279 (3.1), 278 (100), 277 (27.9), 264 (3), 263 (14.9), 262 (2.1), 261 (2.2), 176 (2), 139 (3.7), 134 (2.2), 117 (2), 116 (2), 103 (2.2), 102 (2), 91 (2.4), 77 (2.8), 45 (2).
Ph	Ph[c]	356 (6.1), 355 (29), 354 (100), 353 (21.2), 352 (4.4), 341 (2.1), 340 (7.2), 178 (4.4), 177 (3.4), 165 (3.8), 77 (1.5).
Me	COMe	236 (19), 230 (100), 216 (10), 215 (67.5), 202 (1.4), 201 (3.6), 190 (2.7), 189 (13), 188 (14.6), 155 (2.4), 144 (9.4), 143 (65.1), 142 (2.4), 134 (5.2), 129 (2.8), 128 (2), 118 (2.2), 116 (3.5), 107 (2), 103 (4.8), 102 (31.3), 101 (2.2), 91 (2.2), 90 (10.4), 89 (2.7), 77 (3.6), 76 (8.6), 75 (9.5), 71 (3.6), 70 (5.1), 69 (4.8), 64 (4.9), 63 (6.6), 58 (3.7), 52 (2.9), 51 (8), 52 (5.1), 45 (7), 43 (2.8), 42 (2.2), 39 (11).
H	COPh	308 (7.3), 307 (20), 306 (100), 305 (20), 304 (3.6), 292 (2.1), 291 (8.8), 277 (2), 229 (2.2), 201 (3.9), 157 (4), 106 (2.5), 105 (27), 78 (2), 77 (18), 44 (2.4).

[a] Measured at 50 eV.
[b] From Ref. 225.
[c] 6,7-Dimethyl derivative.

cations (**6.375**) formed by the alternative fragmentation modes of thiazolo-[3,2-*a*]benzimidazoles involving cleavage of the *S*(1)–*C*(2) and *C*(3)–*C*(4) bonds[225] or the *S*(1)–*C*(9a) and *C*(3)–*C*(4) bonds (Scheme 6.78),[107] respectively, in the thiazole ring. In the case of the methyl-substituted thiirenyl cations (**6.375**; R^1 or $R^2 = Me$) subsequent rearrangement to the thietenyl ion (**6.379**) is observed.[107] The tendency for thiazolo[3,2-*a*]benzimidazoles to undergo mass spectral fragmentation with extrusion of neutral sulfur species [Scheme 6.78; (**6.360**)→(**6.373**)+(**6.376**) and (**6.360**)→(**6.374**)→(**6.377**)] is symptomatic of the controlling influence on the mode of bond cleavage, exerted by charge localization on nitrogen.[225] Metastable transitions in the mass spectra of thiazolo[3,2-*a*]benzimidazoles (Table 6.58)[107] support the different modes of fragmentation postulated for these molecules.

TABLE 6.57. MASS SPECTRA OF THIAZOLO[3,2-*a*]BENZIMIDAZOLE DERIVATIVES (**6.360**)[a]

(**6.360**)

R^1	R^2	M^+	*m/e* (rel. abundance, %)							
			(**6.364**)/ (**6.368**)	(**6.361**)	(**6.365**)	(**6.371**)	(**6.375**)	(**6.366**)	(**6.370**)	(**6.372**)
H	H	174 (100)		142 (0.5)	129 (12.2)	102 (3.1)	58 (3.1)	45 (2.8)	102 (2.3)	90 (3.6)
H	Me	188 (100)	187 (36.0)	156 (1.4) 155 (4.3)	129 (14.6)	103 (2.7)	72 (2.0)	59 (8.6)	102 (14.0)	90 (6.6)
Me	H	188 (100)	187 (13.3)	156 (1.3) 155 (4.3)	143 (40.0)	117 (2.7)	72 (1.6)	45 (7.0)	102 (20.0)	90 (5.9)
H	Ph	250 (100)	249 (17.6)	218 (7.8) 217 (4.3)	129 (9.3) [205 (2.2)]	103 (5.4)	134 (4.3)	121 (30.2)	102 (14.1)	90 (12.8)
Ph	H	250 (100)	249 (19.2)	218 (3.2) 217 (1.5)	205 (9.2) [129 (0.7)]	179 (1.1)	134 (1.6)	45 (1.4)	102 (9.6)	90 (4.5)

[a] From Ref. 107.

TABLE 6.58. METASTABLE TRANSITIONS IN THE MASS SPECTRA OF THIAZOLO[3,2-*a*]BENZIMIDAZOLE DERIVATIVES (**6.360**)[a]

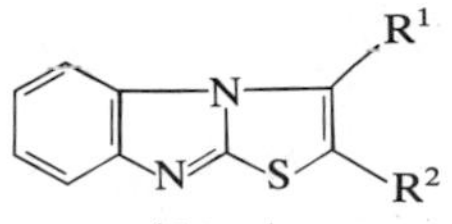

(**6.360**)

R^1	R^2	Transition
H	Me	$188^+ \rightarrow 187^+$, $188^+ \rightarrow 155^+$, $188^+ \rightarrow 129^+$, $129^+ \rightarrow 103^+$, $129^+ \rightarrow 102^+$, $188^+ \rightarrow 90^+$, $188^+ \rightarrow 72^+$
Me	H	$188^+ \rightarrow 155^+$, $188^+ \rightarrow 143^+$, $143^+ \rightarrow 102^+$, $188^+ \rightarrow 90^+$
Ph	H	$250^+ \rightarrow 249^+$, $250^+ \rightarrow 218^+$, $250^+ \rightarrow 134^+$, $205^+ \rightarrow 179^+$, $205^+ \rightarrow 103^+$, $205^+ \rightarrow 102^+$

[a] From Ref. 107.

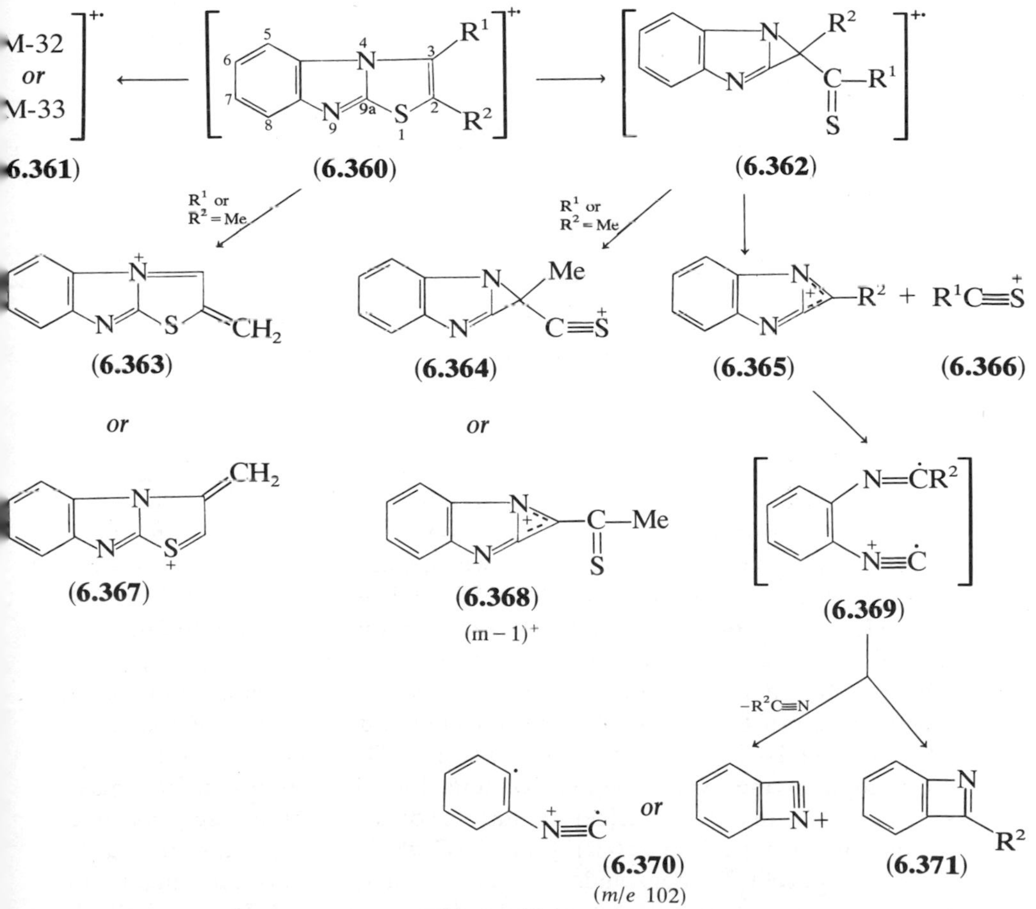

Scheme 6.77

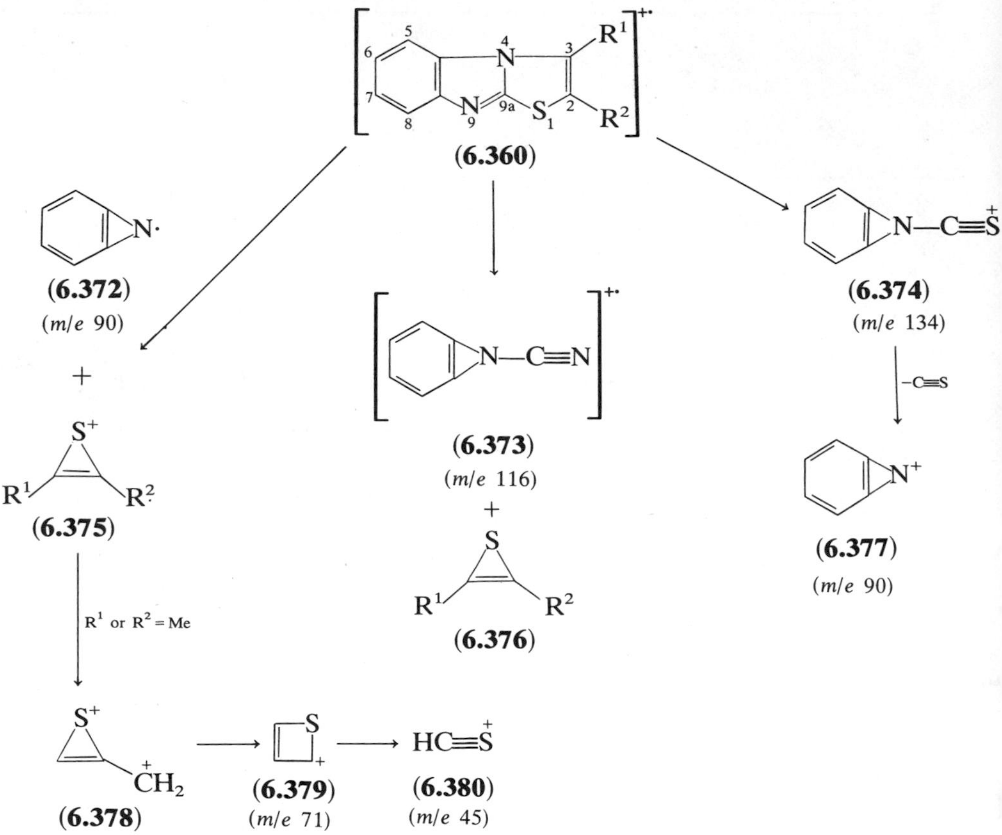

(6.360)

(6.372) (m/e 90) + (6.375)

R^1 or R^2 = Me

(6.373) (m/e 116) + (6.376)

(6.374) (m/e 134) $\xrightarrow{-C\equiv S}$ (6.377) (m/e 90)

(6.378) ⟶ (6.379) (m/e 71) ⟶ $HC\equiv\overset{+}{S}$ (6.380) (m/e 45)

Scheme 6.78

Charge localization on nitrogen apparently also dictates the course followed in the mass–spectral fragmentation of 1*H*-imidazo[1,2-*a*]benzimidazoles (Scheme 6.79; Table 6.59)[78] which closely resembles that of 4*H*-pyrrolo[1,2-*a*]benzimidazoles (cf. page 55) and proceeds by initial loss of the *N*(1) substituent to give relatively stable cations of the type (**6.382**). As well as decomposing further in standard fashion by loss of HCN, these eliminate *N*(1) and *C*(2) with its attached substituent as an intact unit (i.e., a nitrile) to afford the ion (**6.383**) also encountered in the fragmentation of 4*H*-pyrrolo[1,2-*a*]benzimidazoles (see before). The opportunity for charge localization at the *N*(1) position accounts for the presence in the mass spectra of 1*H*-imidazo[1,2-*a*]benzimidazoles of relatively intense peaks due to ions of the type [$(RC\equiv NR)^+$] and [$RC\equiv N)^+$].[78] The mass spectra of 1*H*,3*H*-imidazo[1,5-*a*]benzimidazoles contain base peaks attributable to ions produced by the fragmentation process [Scheme 6.80; (**6.385**) → (**6.386**)].

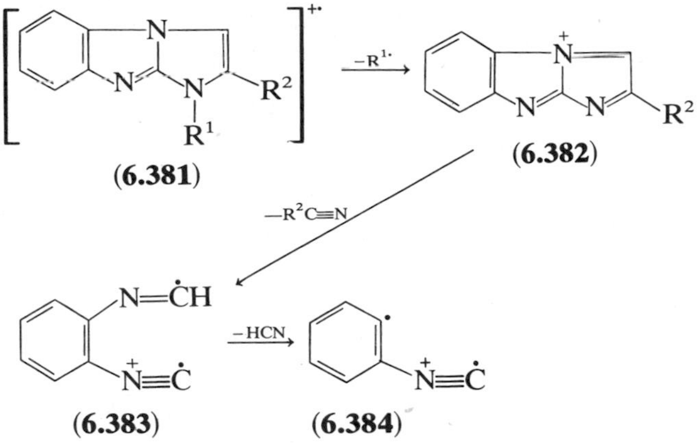

(6.381) (6.382)

(6.383) (6.384)

Scheme 6.79

TABLE 6.59. MASS SPECTRA[a,b] OF 1*H*-IMIDAZO[1,2-*a*]BENZIMIDAZOLE DERIVATIVES (**6.381**)[c]

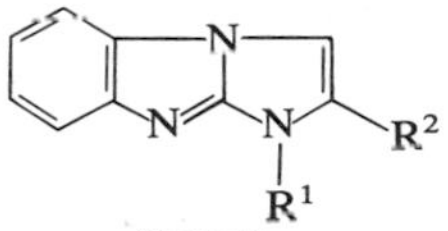

(**6.381**)

R^1	R^2	*m/e* (rel. abundance, %)
H	Ph	235 (3), 234 (25.5), 233 (100), 232 (12.3), 231 (3.6), 230 (1.1), 207 (1.3), 206 (2.3), 205 (4.7), 179 (1.7), 130 (1.3), 129 (5.1), 104 (3.8), 103 (10.4), 102 (5.1), 91 (1.2), 90 (2.8), 89 (2.3), 77 (3.2), 76 (2.1), 53 (1), 45 (1), 43 (1.1).
Me	Ph	249 (1.9), 248 (19), 247 (100), 246 (6.3), 245 (1.7), 233 (2.5), 232 (12.5), 231 (1.5), 206 (1), 205 (2), 170 (1), 130 (1.2), 129 (9.2), 118 (2.7), 117 (1.7), 104 (1.8), 103 (1.2), 102 (4.8), 77 (1.9), 45 (1).
Ph	Me	249 (1.8), 248 (8.9), 247 (100), 246 (63.6), 245 (9.1), 232 (1.1), 220 (1.4), 219 (2.7), 208 (1.2), 207 (2.7), 206 (2.1), 205 (1.8), 179 (1.5), 178 (1.1), 171 (1.4), 170 (9.1), 144 (6.4), 143 (5.2), 142 (1.0), 132 (1.1), 130 (2), 129 (5), 128 (1.4), 118 (5.3), 117 (3.1), 116 (2), 115 (2.5), 104 (2.2), 103 (4.3), 102 (4.8), 92 (1.1), 91 (1.1), 90 (2.3), 78 (1.2), 77 (9.1), 76 (2), 51 (1.2).
Ph	Ph	311 (3.1), 310 (24.2), 309 (100), 308 (66.3), 307 (18.4), 281 (1), 232 (1.5), 207 (1), 206 (1.8), 205 (4.1), 180 (2.6), 179 (1.3), 178 (1.5), 165 (1.3), 129 (1.8), 103 (2.3), 102 (1.3), 72 (1.8).

[a] Measured at 50 eV and an emission current of 75 mA at 125°.
[b] Only peaks with *m/e* >39 are indicated; peaks with intensities <1% are not shown.
[c] From Ref. 78.

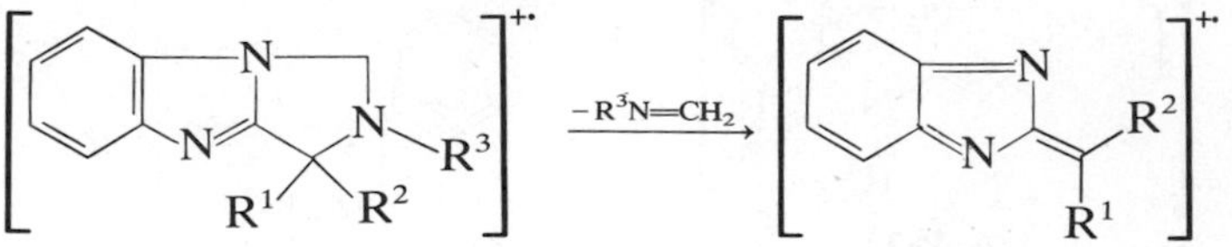

(**6.385**) (**6.386**)

Scheme 6.80

General Studies

Crystallography. X-ray analysis of the dibromo derivative [Scheme 6.81; (**6.387**)] shows[122] the thiazolo[3,2-*a*]benzimidazole ring in this molecule to be nonplanar. A single crystal study[147] of one of the products of the reaction of 2-benzimidazolethione with dimethyl acetylenedicarboxylate confirms the structure (**6.388**) tentatively assigned[147,150] on the basis of chemical evidence.

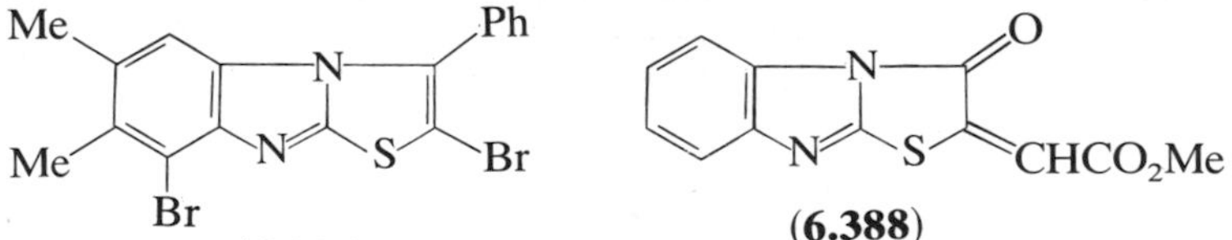

(**6.387**) (**6.388**)

Scheme 6.81

Dipole Moments. Comparison of the dipole moments of 2-methyl-1*H*-pyrazolo[2,3-*a*]benzimidazole ($\mu = 6.92$ D) and 2-methyl-4*H*-pyrazolo[2,3-*a*]benzimidazole ($\mu = 3.97$ D) calculated by CNDO/2, with the experimental value ($\mu = 3.75$ D in dioxane at 25°), demonstrates the predominance of the 4*H*-tautomeric form in dioxane at room temperature.[226]

Ionization Constants. The measured ionization constants for 2,3-dihydro-1*H*-imidazo[1,2-*a*]benzimidazole ($pK_a = 6.20–6.23 \pm 0.03–0.04$) and 9-benzyl-2,3-dihydro-9*H*-imidazo[1,2-*a*]benzimidazole ($pK_a = 8.48–8.49 \pm 0.03–0.04$) have been used to evaluate the equilibrium constant for the tautomeric equilibrium between 2,3-dihydro-1*H*- and 9*H*-imidazo[1,2-*a*]benzimidazoles.[176] The value obtained ($K_{taut.} = 1.87 \times 10^2$) indicates the predominance of the 1*H*-tautomer at equilibrium.

The pK_a data for 3,4-dimethyl-4*H*-imidazo[1,5-*a*]benzimidazole ($pK_a = 6.01$) and 4-methyl-3-phenyl-4*H*-imidazo[1,5-*a*]benzimidazole ($pK_a = 4.75$) are consistent with preferential protonation of 4*H*-imidazo[1,5-*a*]benzimidazoles in general at the *N*(2)-position as predicted on the basis of molecular orbital calculations.[215,224]

6.2.3. Reactions

Reactions with Electrophiles

The high degree of aromatic character manifest in the spectroscopic properties of the fully unsaturated thiazolo[3,2-*a*]benzimidazole, pyrazolo-[2,3-*a*]benzimidazole, 9*H*-imidazo[1,2-*a*]benzimidazole, and 4*H*-imidazo-[1,5-*a*]benzimidazole ring systems is also readily apparent in the propensity of their derivatives to undergo electrophilic substitution at available sites in the thiazole, pyrazole, and imidazole nuclei, respectively. This reactivity is consistent with the results of molecular orbital calculations which demonstrate the *C*(2) and *C*(3) positions in thiazolo[3,2-*a*]benzimidazoles[222] and 9*H*-imidazo[1,2-*a*]benzimidazoles[157] and the *C*(1) and *C*(3) positions in 4*H*-imidazo[1,5-*a*]benzimidazoles[224] to be the sites of highest π-electron localization and hence the most prone to electrophilic attack. Conversely, the significantly lower electron localization calculated[157] for the *C*(2) and *C*(3) positions in the 1*H*-imidazo[1,2-*a*]benzimidazole ring system accounts for the lower reactivity of its derivatives toward electrophilic attack (see later).

PROTONATION. The ease of protonation of simple thiazolo[3,2-*a*]benzimidazoles and 2,3-dihydrothiazolo[3,2-*a*]benzimidazoles is demonstrated by their ready solubility in dilute mineral acids[118] and by their tendency to form well-defined acid salts (e.g., hydrochlorides and picrates—see Tables 6.27 and 6.29). Changes in the ^{1}H NMR absorption of thiazolo[3,2-*a*]-benzimidazoles produced by dissolution in trifluoroacetic acid (see Table 6.49[222]) are consistent with monocation formation and, moreover, demonstrate preferential protonation at *N*(9). Proton uptake at this site is also favored on theoretical grounds, formation of the *N*(9) cation resulting in a significant overall stabilization of the π-system in thiazolo[3,2-*a*]-benzimidazoles as revealed by molecular orbital calculations.[222] The involvement of the sulfur atom and the *N*(4) nitrogen atom in the aromatic π-framework of thiazolo[3,2-*a*]benzimidazoles precludes protonation at these sites.

The formation of stable acid salts (hydrochlorides, hydrobromides, picrates—see Tables 6.32–6.34 and 6.36) by 1*H*- and 9*H*-imidazo[1,2-*a*]-benzimidazoles and their 2,3-dihydro derivatives is illustrative of the generally basic character of such molecules. Paradoxically, the hydrochlorides of 2-amino-9*H*-imidazo[1,2-*a*]benzimidazoles like the parent bases are too unstable to be isolated.[174] The ready hydrolysis[163,227] of hydrochlorides derived from 3-nitroso- and 3-nitro-9*H*-imidazo[1,2-*a*]benzimidazoles demonstrates the base-weakening effect of electron-withdrawing substituents at the *C*(3) position in the 9*H*-imidazo[1,2-*a*]benzimidazole ring system.

TABLE 6.60. ALKYLATION REACTIONS OF THIAZOLO[3,2-*a*]BENZIMIDAZOLES AND 2,3-DIHYDROTHIAZOLO[3,2-*a*]BENZIMIDAZOLES

Reaction conditions[a]	Thiazolo[3,2-*a*]benzimidazole		Yield (%)	m. p. (°C)	Solvent of crystallization	Ref.
	Substrate	Product				
A	Unsubstituted	9-Me-, iodide	—[b]	237–239	Ethanol–ether	106
A	Unsubstituted	9-Et-, iodide	—[b]	214–216	Ethanol–ether	106
B	3-Me-	9-Et-3-Me-, bromide	—[b]	284	—[c]	229
B	3-Me-	9-(n-$C_{16}H_{33}$)-3-Me-, bromide	—[b]	160–161	—[c]	229
B	3-Me-	9-$PhCH_2$-3-Me-, bromide	—[b]	229–231	—[c]	229
C	3-Me-	9-$PhCH_2$-3-Me-, bromide	—[b]	227–229	—[c]	228
B	3-Me-	9-(4-$ClC_6H_4CH_2$)-3-Me-, chloride	—[b]	227–229	—[c]	229
B	3-Me-	9-(4-$NCC_6H_4CH_2$)-3-Me-, bromide	—[b]	314–315	—[c]	229
B	3-Me-	9-(3-$F_3CC_6H_4CH_2$)-3-Me-, chloride	—[b]	234	—[c]	229
B	3-Me-	9-(3-$ClC_6H_4CH_2$)-3-Me-, bromide	—[b]	243–245	—[c]	229
B	3-Me-	9-(3-$FC_6H_4CH_2$)-3-Me-, chloride	—[b]	233–235	—[c]	229
B	3-Me-	9-(3-$MeC_6H_4CH_2$)-3-Me-, chloride	—[b]	218–219	—[c]	229
B	3-Me-	9-(3-$NCC_6H_4CH_2$)-3-Me-, bromide	—[b]	252	—[c]	229
B	3-Me-	9-$Ph(CH_2)_2$-3-Me-, bromide	—[b]	197–198	—[c]	229
B	3-Me-	9-$CH_2{=}CHCH_2$-3-Me-, bromide	—[b]	256–257	—[c]	229
A	3-Ph	9-Me-3-Ph-, iodide	86	196	Ethanol	122
A	6,7-di-Me-3-Ph-	6,7,9-tri-Me-3-Ph-, iodide	86	269	Ethanol	122

D	2,3-di-Me-	2,3,9-tri-Me-, iodide	quant.	303–304	—[c]	111
D	2,3-di-Me-	2,3-di-Me-9-[$HO_2C(CH_2)_2$]-, iodide	—[b]	250–253	—[c]	111
—[d]	2,3-Dihydro-	9-Me-2,3-dihydro-	—[b]	239–240	Ethanol–ether	123
—[d]	6,7-di-Me-2,3-dihydro-	6,7,9-tri-Me-2,3-dihydro-	—[b]	245–246	Ethanol–ether	123
E	2,3-dihydro-3-one	2-(Et_2NCH_2)-2,3-dihydro-3-one	42	210	Ethanol	163
E	2,3-dihydro-3-one	2-(Me_2NCH_2)-2,3-dihydro-3-one	45	190	Ethanol	163
E	2,3-dihydro-3-one	2-Ph_2NCH_2-2,3-dihydro-3-one	50	175	Ethanol	163
E	2,3-dihydro-3-one	2-[Ph(Et)NCH_2]-2,3-dihydro-3-one	40	250	Ethanol	163
E	2,3-dihydro-3-one	2-(*n*-Pr_2NCH_2)-2,3-dihydro-3-one	45	240	Ethanol	163
E	2,3-dihydro-3-one	2-(*n*-Bu_2NCH_2)-2,3-dihydro-3-one	50	250	Ethanol	163
E	2,3-dihydro-3-one	2-(*sec*-Bu_2NCH_2)-2,3-dihydro-3-one	45	210	Ethanol	163
E	2,3-dihydro-3-one	2-(NCH_2)-2,3-dihydro-3-one	40	215	Ethanol	163
E	2,3-dihydro-3-one	2-(O NCH_2)-2,3-dihydro-3-one	41	230	Ethanol	163

[a] *A* = alkyl halide, acetone/(reflux)(1 hr); *B* = alkyl halide, dimethylformamide/(100°)(20 min.); *C* = alkyl halide, acetonitrile/(room temp.)(16 hr); *D* = alkyl halide, ethanol/(reflux)(18 hr); *E* = Et_2NH, HCHO, AcOH/(100°)(7 hr).

[b] Yield not quoted.

[c] Solvent of crystallization not specified.

[d] Reaction conditions not specified.

Ready solubility in dilute hydrochloric acid[190,191] and stable hydrochloride[224] and picrate (Table 6.37) formation are symptomatic of the marked basicity of 4*H*-imidazo[1,5-*a*]benzimidazole derivatives. The available pK_a data as well as the results of molecular orbital calculations pinpoint *N*(2) as the preferred site of protonation in these molecules.[224]

ALKYLATION. Thiazolo[3,2-*a*]benzimidazoles are readily alkylated by reaction with alkyl bromides or alkyl iodides in acetonitrile at room temperature[228] or in ethanol,[111] acetone,[106,122] or dimethylformamide[229] at elevated temperature, to afford high yields (Table 6.60) of the corresponding *N*(9) quaternary salts. Specific alkylation at *N*(9) in thiazolo[3,2-*a*]benzimidazoles is supported by ^{1}H NMR evidence[222] and is consistent with the high electron localization at this center as indicated by the results of molecular orbital circulations.[222] 2,3-Dihydrothiazolo[3,2-*a*]benzimidazoles are also reported[123] to form quaternary salts, albeit in unspecified yield (Table 6.60), by exclusive alkylation at the *N*(9) position. The *O*-ethylation, which is reported[103] to occur when 3-hydroxy-2,3-dihydrothiazolo[3,2-*a*]benzimidazoles are treated with ethanolic hydrochloric acid, can be rationalized by prior ring-opening to the aldehyde tautomer followed by acetal formation and recyclization [Scheme 6.82; (**6.389**)→(**6.390**)→(**6.391**)→(**6.392**)]. The enhanced nucleophilic reactivity of the *C*(2) methylene group in 2,3-dihydrothiazolo[3,2-*a*]benzimidazol-3-one is illustrated by its ability to participate in Mannich condensations with formaldehyde and a variety of secondary amines giving moderate yields (Table 6.60) of the corresponding 2-dialkylaminomethyl-2,3-dihydrothiazolo[3,2-*a*]benzimidazol-3-ones.[134]

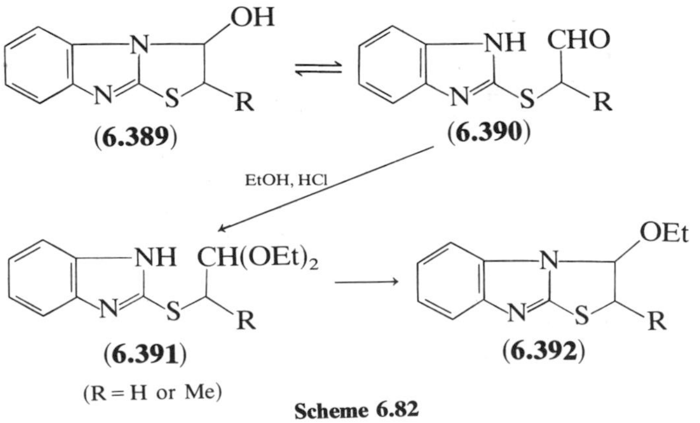

(**6.389**) ⇌ (**6.390**) —EtOH, HCl→ (**6.391**) → (**6.392**)

(R = H or Me)

Scheme 6.82

Dyestuff intermediates produced by heating 4*H*-pyrazolo[2,3-*a*]benzimidazoles with dimethyl sulfate in a high-boiling solvent (e.g., chlorobenzene or 1,2-dichlorobenzene) are formulated[230] as *N*(1)-methyl derivatives, but without any apparent evidence to exclude the alternative *N*(4)-alkyl structures. Conversely, the reaction of 3-alkylidene-3*H*-pyrazolo[2,3-*a*]-benzimidazoles with methylating agents such as methyl tosylate is reported to result in exclusive quaternization at the *N*(4)-position, e.g. [Scheme 6.83; (**6.393**)→(**6.394**)],[219] but again in the absence of rigorous structure proof for the products.

N—N N Me CHR
(**6.393**)

N—N N⁺ Me Me CHR I^-
(**6.394**)

R	Yield (%)	m.p. (°C)
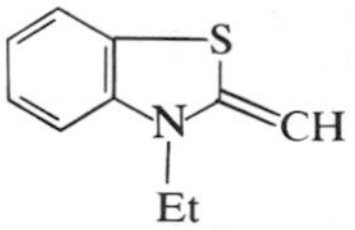 S N CH Et	51	274

(i) p-$MeC_6H_4SO_3Me$/(heat)(1 min), then treat with KI.

Scheme 6.83

Methylation of the tautomeric 2-methyl-1(9)*H*-imidazo[1,2-*a*]benzimidazole under basic conditions involves the intermediacy of the ambident anion [Scheme 6.84; (**6.395**)↔(**6.396**), R = Me] and consequently results in the formation of a difficultly separable mixture of the *N*(1)- and *N*(9)-methyl derivatives.[166] In contrast, the base-catalyzed alkylation of 2-aryl-1(9)*H*-

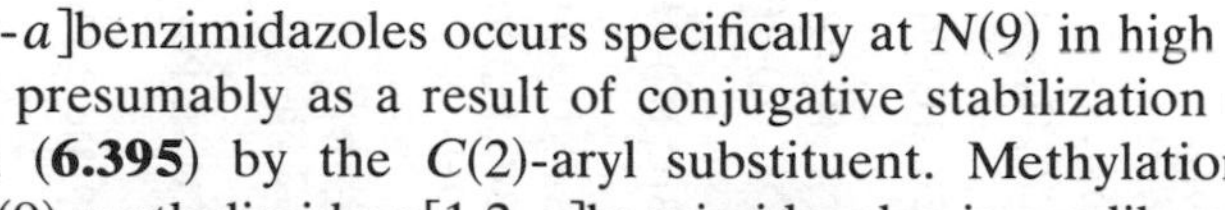

N N N R ⟷ N N N R
(**6.395**) (**6.396**)

Scheme 6.84

imidazo[1,2-*a*]benzimidazoles occurs specifically at *N*(9) in high yield (Table 6.61),[155,160] presumably as a result of conjugative stabilization of the resonance form (**6.395**) by the *C*(2)-aryl substituent. Methylation of simple *N*(1)- or *N*(9)-methylimidazo[1,2-*a*]benzimidazoles is readily accomplished by heating with methyl iodide in ethanol[157,162a,165,166,227] or by fusion with methyl tosylate[157] and produces uniformly good yields (Table 6.61) of the corresponding *N*(1),*N*(9)-dimethyl quaternary salts. In accord with the base-weakening effect of the nitro- substituent 9-methyl-3-nitroimidazo[1,2-*a*]benzimidazole is alkylated only with difficulty to give quaternary salts, which are readily dealkylated to the parent bases in alkaline media.[163] The superior conjugative stabilization of the resonance form [Scheme 6.85;

TABLE 6.61. ALKYLATION REACTIONS OF 1(9)*H*-IMIDAZO[1,2-*a*]BENZIMIDAZOLES AND 2,3-DIHYDRO-1(9)*H*-IMIDAZO[1,2-*a*]BENZIMIDAZOLES

Reaction conditions[a]	1(9)*H*-Imidazo[1,2-*a*]benzimidazole		Yield (%)	m. p. (°C)	Solvent of crystallization	Ref.
	Substrate	Product				
A	2-Ph-9*H*-	9-[$Me_2N(CH_2)_2$]-2-Ph-[b]	65	260–263	—[c]	155
A	2-Ph-9*H*-	9-[$Et_2N(CH_2)_2$]-2-Ph-[b]	70	268–169	—[c]	155
A	2-Ph-9*H*-	9-[$Me_2N(CH_2)_3$]-2-Ph-[b]	92	266–267	—[c]	155
A	2-(4-ClC_6H_4)-9*H*-	9-[$Me_2N(CH_2)_2$]-2-(4-ClC_6H_4)-[b]	71	271–272 (decomp.)	—[c]	155, 160
B	2-(4-ClC_6H_4)-9*H*-	9-[$Me_2N(CH_2)_2$]-2-(4-ClC_6H_4)-[b]	55	271–272 (decomp.)	—[c]	160
A	2-(4-ClC_6H_4)-9*H*-	9-[$Et_2N(CH_2)_2$]-2-(4-ClC_6H_4)-[d]	68	258–259	—[c]	155
A	2-(4-ClC_6H_4)-9*H*-	9-[$Me_2N(CH_2)_3$]-2-(4-ClC_6H_4)-[b]	62	261–263	—[c]	155, 160
A	2-(4-BrC_6H_4)-9*H*-	9-[$Me_2N(CH_2)_2$]-2-(4-BrC_6H_4)-[b]	62	268–270	—[c]	155, 160
A	2-(4-BrC_6H_4)-9*H*-	9-[$Et_2N(CH_2)_2$]-2-(4-BrC_6H_4)-[b]	65	245–248	—[c]	155
A	2-(4-BrC_6H_4)-9*H*-	9-[$Me_2N(CH_2)_3$]-2-(4-BrC_6H_4)-[b]	56	249–250	—[c]	155
C	2-Ph-9*H*-	9-$PhCOCH_2$-2-Ph-	80	268–269	—[c]	155, 156
D	2,9-di-Me-	1,2,9-tri-Me-, iodide	78	249	Ethanol	227
D	9-Me-2-Ph-	1,9-di-Me-2-Ph-, iodide	72	234 (decomp.)	Ethanol	162a
D	9-$PhCH_2$-2-Ph-	1-Me-9-$PhCH_2$-2-Ph-, iodide	70	257 (decomp.)	Ethanol	166
E	3-Br-9-Me-2-Ph-	3-Br-1,9-di-Me-2-Ph-, benzenesulfonate	96	227	Ethanol–ether	162a
E	3-Br-1-Me-2-Ph-	3-Br-1,9-di-Me-2-Ph-, benzenesulfonate[e]	—[f]	227	—[c]	157
F	2,9-di-Me-3-CHO-	1,2,9-tri-Me-3-CHO-, iodide	60	246–247	Ethanol–ether	165
G	9-Me-2-Ph-3-CHO-	1,9-di-Me-2-Ph-3-CHO-, iodide	40	232–233 (decomp.)	Ethanol	165

H	2,3-dihydro-1(9)*H*-	1-$PhCH_2$-2-3-dihydro-	80	115–116	Benzene–hexane	176
I	2,3-dihydro-1(9)*H*-	1-Ph_2CH-2,3-dihydro-	67	206.5–207.5	Benzene–ligroin	176
A	2,3-dihydro-1(9)*H*-	1-[$Et_2N(CH_2)_2$]-2,3-dihydro-[g]	85	—[h]	—	176
I	7-NO_2-2,3-dihydro-1(9)*H*-	1-Et-7-NO_2-2,3-dihydro-	77	171–172	Ethanol	176
H	7-NO_2-2,3-dihydro-1(9)*H*-	1-$PhCH_2$-7-NO_2-2,3-dihydro-	72	174.5–176	Dimethylformamide–water	176
I	7-NO_2-2,3-dihydro-1(9)*H*-	1-PhCH(Me)-7-NO_2-2,3-dihydro-	41	159–160	Benzene	176
I	7-NO_2-2,3-dihydro-1(9)*H*-	1-Ph_2CH-7-NO_2-2,3-dihydro-	71	217–218	Dimethylformamide–water	176
A	7-NO_2-2,3-dihydro-1(9)*H*-	1-[$Et_2N(CH_2)_2$]-7-NO_2-2,3-dihydro-	47	120–121	Benzene–hexane	176
J	2-Ph-2,3-dihydro-1(9)*H*-	1-Me-2-Ph-2,3-dihydro-[i,j]	100	112	Ethanol–water	166
K	2,3-dihydro-1(9)*H*-	9-$PhCH_2$-2,3-dihydro-[k]	80	262–264	—[c]	176
L	2,3-dihydro-1(9)*H*-	9-$PhCH_2$-2,3-dihydro-[k]	85	—	—	176
D	2-Ph-2,3-dihydro-1(9)*H*-	9-Me-2-Ph-2,3-dihydro-[l]	95	236	Ethanol	166
D	1-Me-2-Ph-2,3-dihydro-	1,9-di-Me-2-Ph-2,3-dihydro-	—[f]	234 (decomp.)	Ethanol	166
D	9-Me-2-Ph-2,3-dihydro-	1,9-di-Me-2-Ph-2,3-dihydro-	—[f]	—	—	166

[a] *A* = $NaNH_2$, NH_3liq./(1 hr), then alkyl chloride, toluene/(90°)(3 hr); *B* = NaOEt, EtOH then alkyl chloride/(90°)(3 hr); *C* = $PhCOCH_2Br$, MeOH/(reflux)(5 hr); *D* = MeI, EtOH/(reflux)(2–12 hr); *E* = $PhSO_3Me$/(80°)(15–30 min); *F* = MeI, EtOH/(reflux)(15–20 hr); *G* = MeI, EtOH/(reflux)(35–40 hr); *H* = $NaNH_2$, NH_3 liq./0.5 hr, then $PhCH_2Cl$, ether; *I* = $NaNH_2$, NH_3 liq./0.5 hr, then alkyl bromide, ether; *J* = MeI, $NaNH_2$, NH_3 liq./20 min; *K* = $PhCH_2Cl$, EtOH/(reflux)(4 hr); *L* = $PhCH_2Cl$, dimethylformamide/(reflux)(4 hr).

[b] Hydrochloride.

[c] Solvent of crystallization not specified.

[d] Dihydrochloride.

[e] Iodide has m.p. 265° (decomp.) (from ethanol–water).

[f] Yield not quoted.

[g] Forms a hydrochloride, m.p. 255–257° (from 2-propanol).

[h] Oil.

[i] Monohydrate; anhydrous product has m.p. 146°.

[j] Forms a hydrochloride, colorless needles, m.p. 234° (from ethanol–ether).

[k] Hydrochloride; free base is an oil, b.p. 192–194°/0.15 mm Hg.

[l] Hydriodide; free base has m.p. 65°; forms a hydrochloride, m.p. 258° (from ethanol–ether).

(**6.397**)] of the ambident anion [(**6.397**) ↔ (**6.398**)] provides a rationale for the observed preferential alkylation of 2,3-dihydro-1(9)*H*-imidazo[1,2-*a*]-benzimidazoles at the *N*(1)-position under strongly basic conditions (Table 6.61).[166,176] Conversely, the specific *N*(9)-alkylation of such substrates in neutral protic or aprotic media (Table 6.61) is a measure of the greater resonance stabilization of the cationic intermediate produced by attack at *N*(9) as opposed to that derived by substitution at the *N*(1)position.[166,176]

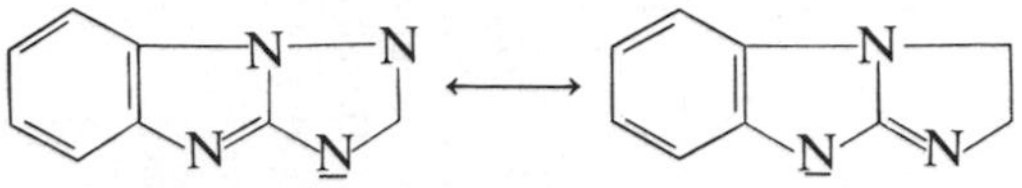

(**6.397**) (**6.398**)

Scheme 6.85

Acylation. Despite the enhanced reactivity to electrophilic attack predicted for the *C*(2) and *C*(3) positions in the thiazolo[3,2-*a*]benzimidazole ring system (see before), bona fide examples of Friedel–Crafts and related acylation reactions of thiazolo[3,2-*a*]benzimidazole derivatives do not appear to have been reported to date. The paucity of such reactions is surprising when compared with the well-documented *C*-acylation undergone by the structurally and electronically related 9*H*-imidazo[1,2-*a*]-benzimidazole and 4*H*-imidazo[1,5-*a*]benzimidazole ring systems under a variety of conditions (see later). Dehydration to and acetylation of the corresponding thiazolo[3,2-*a*]benzimidazole provides an obvious rationale for the conversion (by heating with acetic anhydride in the presence of pyridine) of 3-hydroxy-2,3-dihydrothiazolo[3,2-*a*]benzimidazoles in good to excellent yield (70–90%) into the corresponding 2-acetylthiazolo[3,2-*a*]-benzimidazoles.[103] However, the isolation[103] of *N*(1)-acetyl-2-(β-oxoalkyl)-benzimidazole derivatives from such reactions under milder conditions supports an alternative course for 2-acetylthiazolo[3,2-*a*]benzimidazole formation involving prior ring-opening to and acetylation of a 2-(β-oxoalkyl)-benzimidazole, followed by dehydrative cyclization.[103] Activation by the adjacent sulfur and *C*(3) carbonyl substitutents confers enhanced reactivity toward electrophilic attack on the *C*(2) methylene group in 2,3-dihydrothiazolo[3,2-*a*]benzimidazol-3-ones. As a result such molecules condense smoothly with aromatic aldehydes in the presence of condensation catalysts such as triethylamine,[129] piperidine,[131–134,138,231,232] sodium acetate,[101,130,133,136] or dicyclohexylcarbodiimide-pyridine,[132] to afford the corresponding 2-arylidene-2,3-dihydrothiazolo[3,2-*a*]benzimidazol-3-ones in good to excellent yields (Table 6.62). Reaction of 2,3-dihydrothiazolo[3,2-*a*]benzimidazol-3-one with functionalized aldehydes (e.g., glyoxylic acid[147]) and ketones (e.g., 2-propanone[131]) proceeds similarly (Table 6.62), while the closely related condensations with enethiol ethers[129] and enamides[129,135]

TABLE 6.62. ACYLATION AND RELATED REACTIONS OF 2,3-DIHYDROTHIAZOLO[3,2-*a*]BENZIMIDAZOLES AND 2,3-DIHYDROTHIAZOLO[3,2-*a*]BENZIMIDAZOL-3-ONES

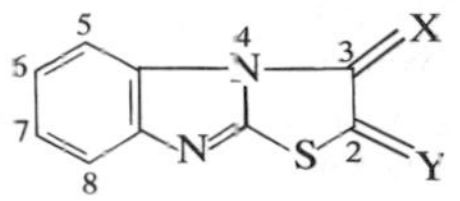

(6.399)

Reaction conditions[a]	Substrate (**6.399**)	Substrate X	Substrate Y	Product (**6.399**)	Product X	Product Y	Yield (%)	m.p. (°C)	Solvent of crystallization	Ref.
A	Unsubstituted	H, OH	H, H	Unsubstituted	H, OAc	H, H	92	102–103	Hexane	103
A	Unsubstituted	H, OH	H, Me	Unsubstituted	H, OAc	H, Me	95	—[b]	—	103
B	Unsubstituted	O	H, H	Unsubstituted	O	PhCH	25	219	Benzene	129
C	Unsubstituted	O	H, H	Unsubstituted	O	PhCH	—[c]	226	Xylene	138
D	Unsubstituted	O	H, H	Unsubstituted	O	PhCH	—[c]	216–217	Acetic acid	130
E	Unsubstituted	O	H, H	Unsubstituted	O	PhCH	78	222	—[d]	101
F	6-Me-	O	H, H	6-Me-	O	PhCH	68	221	Ethanol	133
E	6-NO_2-	O	H, H	6-NO_2-	O	PhCH	85	295	Xylene	136
E	6,7-di-Me-	O	H, H	6,7-di-Me-	O	PhCH	82	265–266	Xylene	133
F	6,7-di-Me-	O	H, H	6,7-di-Me-	O	H, H	80	254–255	Acetic acid	134
E	Unsubstituted	O	H, H	Unsubstituted	O	2-ClC_6H_4CH	55	294	—[d]	101
E	Unsubstituted	O	H, H	Unsubstituted	O	2-ClC_6H_4CH	74	186	Acetic acid	231
E	Unsubstituted	O	H, H	Unsubstituted	O	4-ClC_6H_4CH	65	280	—[d]	101
E	Unsubstituted	O	H, H	Unsubstituted	O	4-ClC_6H_4CH	80	272	Chlorobenzene	231
F	6-Me-	O	H, H	6-Me	O	4-ClC_6H_4CH	62	182	Ethanol	133
E	6,7-di-Me-	O	H, H	6,7-di-Me-	O	4-ClC_6H_4CH	73	>320	Xylene	133
E	6-NO_2-	O	H, H	6-NO_2-	O	4-ClC_6H_4CH	82	255	Xylene	136
E	6-NO_2-	O	H, H	6-NO_2-	O	4-MeC_6H_4CH	84	267	Xylene	136
D	Unsubstituted	O	H, H	Unsubstituted	O	2-HOC_6H_4CH	—[c]	214–215	Ethanol	130
E	Unsubstituted	O	H, H	Unsubstituted	O	4-HOC_6H_4CH	55	>300	—[d]	101
C	Unsubstituted	O	H, H	Unsubstituted	O	2-$MeOC_6H_4CH$	80–90	210	Xylene	138
E	Unsubstituted	O	H, H	Unsubstituted	O	2-$EtOC_6H_4CH$	75	167	Acetic acid	231

TABLE 6.62 (*Continued*)

(6.399)

Reaction conditions[a]	Substrate (**6.399**)	Substrate X	Substrate Y	Product (**6.399**)	Product X	Product Y	Yield (%)	m.p. (°C)	Solvent of crystallization	Ref.
E	Unsubstituted	O	H, H	Unsubstituted	O	$4\text{-}MeOC_6H_4CH$	40	236	—[d]	101
F	6-Me-	O	H, H	6-Me-	O	$4\text{-}MeOC_6H_4CH$	70	195	Ethanol	133
E	6,7-di-Me-	O	H, H	6,7-di-Me-	O	$4\text{-}MeOC_6H_4CH$	74	257–258	Xylene	133
F	6,7-di-Me-	O	H, H	6,7-di-Me-	O	$4\text{-}MeOC_6H_4CH$	49	238–239	Pr^nOH	134
E	Unsubstituted	O	H, H	Unsubstituted	O	$3\text{-}O_2NC_6H_4CH$	80	254	—[d]	101
D	Unsubstituted	O	H, H	Unsubstituted	O	$4\text{-}O_2NC_6H_4CH$	—[c]	>300	—[d]	130
G	6,7-di-Me-	O	H, H	6,7-di-Me-	O	$2\text{-}O_2NC_6H_4CH$	70	246–247	Acetic acid	134
G	6,7-di-Me-	O	H, H	6,7-di-Me-	O	$3\text{-}O_2NC_6H_4CH$	72	259–260	Acetic acid	134
G	6,7-di-Me-	O	H, H	6,7-di-Me-	O	$4\text{-}O_2NC_6H_4CH$	60	318–319	Dimethylform-amide	134
B	Unsubstituted	O	H, H	Unsubstituted	O	$4\text{-}Me_2NC_6H_4CH$	45	269	Ethanol	129
D	Unsubstituted	O	H, H	Unsubstituted	O	$4\text{-}Me_2NC_6H_4CH$	—[c]	267–268 (decomp.)	—[d]	130
H	Unsubstituted	O	H, H	Unsubstituted	O	$4\text{-}Me_2NC_6H_4CH$	90	269	1-Propanol	131
E	Unsubstituted	O	H, H	Unsubstituted	O	$4\text{-}Me_2NC_6H_4CH$	40	270	Acetic acid	101
F	6-Me-	O	H, H	6-Me-	O	$4\text{-}Me_2NC_6H_4CH$	81	273–274	Ethanol	133
E	6,7-di-Me-	O	H, H	6,7-di-Me-	O	$4\text{-}Me_2NC_6H_4CH$	86	>320	Xylene	133
D	Unsubstituted	O	H, H	Unsubstituted	O	PhCH=CHCH	—[c]	234–235	—[d]	130
E	Unsubstituted	O	H, H	Unsubstituted	O	MeCH=CHCH	45	212	—[d]	101
D	Unsubstituted	O	H, H	Unsubstituted	O	2-Furfurylidene	—[c]	231–232	—[d]	130
G	6,7-di-Me-	O	H, H	6,7-di-Me-	O	2-Furfurylidene	76	260–261	Acetic acid	134
I	Unsubstituted	O	H, H	Unsubstituted	O	Me_2C	50	181.5–182	Methanol	131
J	Unsubstituted	O	H, H	Unsubstituted	O	HO_2CCH	—	—[c]	—[d]	147

K	Unsubstituted	O	H, H	Unsubstituted	O	EtOCH	46	167–169	Ethanol	129
L	Unsubstituted	O	H, H	Unsubstituted	O	Me_2NCH	77	255–256	Acetic acid	234
M	Unsubstituted	O	H, H	Unsubstituted	O	MeCON(Ph)CH	55	195	Ethanol	135
N	Unsubstituted	O	H, H	Unsubstituted	O	$4\text{-}Me_2NC_6H_4N$	93	256	Chloroform	131
O	6,7-di-Me-	O	H, H	6,7-di-Me-	O	$4\text{-}Me_2NC_6H_4N$	39	300–302	Ethanol–dichloroethane	134
P	$7\text{-}NH_2$	H, H	H, H	7-AcNH	H, H	H, H	100	243–245	Water	128
Q	Unsubstituted	O	H, NH_2	Unsubstituted	O	PhNHCSNH, H[e]	60	220–221	—[d]	235
Q	Unsubstituted	O	H, NH_2	Unsubstituted	O	$2\text{-}MeC_6H_4NHCSNH$, H[e]	65	195–197	—[d]	235
Q	Unsubstituted	O	H, NH_2	Unsubstituted	O	$3\text{-}MeC_6H_4NHCSNH$, H[e]	58	198–200	—[d]	235
Q	Unsubstituted	O	H, NH_2	Unsubstituted	O	$4\text{-}MeC_6H_4NHCSNH$, H[e]	59	175–177	—[d]	235
Q	Unsubstituted	O	H, NH_2	Unsubstituted	O	$2\text{-}ClC_6H_4NHCSNH$, H[e]	62	202–204	—[d]	235
Q	Unsubstituted	O	H, NH_2	Unsubstituted	O	$3\text{-}ClC_6H_4NHCSNH$, H[e]	61	215	—[d]	235
Q	Unsubstituted	O	H, NH_2	Unsubstituted	O	$4\text{-}ClC_6H_4NHCSNH$, H[e]	59	211	—[d]	235
Q	Unsubstituted	O	H, NH_2	Unsubstituted	O	$2\text{-}BrC_6H_4NHCSNH$, H[e]	60	189	—[d]	235
Q	Unsubstituted	O	H, NH_2	Unsubstituted	O	$3\text{-}BrC_6H_4NHCSNH$, H[e]	58	189	—[d]	235
Q	Unsubstituted	O	H, NH_2	Unsubstituted	O	$4\text{-}BrC_6H_4NHCSNH$, H[e]	65	165–166	—[d]	235

[a] A = Ac_2O/(room temp.)(14 hr); *B* = ArCHO, Et_3N, EtOH/(reflux)(10 min); *C* = ArCHO, piperidine, Bu^nOH/(reflux)(2 hr); *D* = ArCHO, NaOAc, AcOH/(reflux)(reaction time unspecified); *E* = ArCHO, NaOAc, AcOH/(reflux)(2–4 hr); *F* = ArCHO, piperidine, EtOH/(100°)(1–2 hr); *G* = ArCHO, AcOH/(reflux)(20 min–1 hr); *H* = ArCHO, piperidine, EtOH/(room temp.)(1 hr); *I* = $Me_2C{=}O$, piperidine/(reflux)(6 hr); *J* = HO_2CCHO, AcOH (reaction conditions not specified); *K* = $(EtO)_3CH$, Ac_2O/(reflux)(20 min); *L* = $POCl_3$, dimethylformamide/(0°)(30 min), then (50°)(5 hr); *M* = PhNHCH=NPh, Ac_2O/(reflux)(30 min); *N* = $4\text{-}Me_2NC_6H_4NO$, EtOH/(reflux)(20 min); *O* = $4\text{-}Me_2NC_6H_4NO$, piperidine, EtOH/(reflux)(2 hr); *P* = Ac_2O/(reflux)(1.5 hr); *Q* = ArN=C=S, EtOH/(reflux)(5 hr).

[b] Colorless oil.

[c] Yield not quoted.

[d] Solvent of crystallization not specified.

[e] Hydrochloride.

provide the means for the construction of methine dyestuffs[129,228,233] incorporating the 2,3-dihydrothiazolo[3,2-*a*]benzimidazol-3-one nucleus. The methylene reactivity of 2,3-dihydrothiazolo[3,2-*a*]benzimidazol-3-one toward carbon-based electrophiles is further exemplified by its reaction[129] with ethyl orthoformate in the presence of acetic anhydride to afford the *C*(2)-ethoxymethylene derivative, albeit in only moderate yield (Table 6.62) and by its ready aminoalkenylation at the *C*(2)-position by dimethylformamide in the presence of phosphorus oxychloride (the Vilsmeier–Haack condensation).[234] In reactions closely related to those with aromatic aldehydes (see before), 2,3-dihydrothiazolo[3,2-*a*]benzimidazol-3-ones undergo uncatalyzed[131] or base (sodium carbonate[132,232] or piperidine[134]) catalyzed condensation with nitrosoarenes to give moderate to excellent yields (Table 6.62) of 2-arylimino-2,3-dihydrothiazolo[3,2-*a*]benzimidazol-3-ones. Amino- and hydroxy-substituted 2,3-dihydrothiazolo[3,2-*a*]benzimidazoles behave in an orthodox manner toward acylation as illustrated by the straightforward acetylation of nuclear amino substituents[128] and by the reaction of *C*(2) amino groups with aryl isothiocyanates to afford the corresponding thioureas (Table 6.62).[235] In accord with their ring as opposed to open-chain tautomeric structures (see page 108), 3-hydroxy-2,3-dihydrothiazolo[3,2-*a*]benzimidazoles acetylate in standard fashion giving high yields (Table 6.62) of the corresponding acetoxy derivatives.[103,142]

Acylation (Table 6.63) of 1-imino-1*H*,3*H*-thiazolo[3,4-*a*]benzimidazole (**6.400**; X = NH, Y = H_2) by acetic anhydride or acid chlorides under basic conditions occurs at both *N*(4) and at the imino substituent, demonstrating the ability of this molecule to react in the 4*H*-tautomeric form (**6.401**; X = NH, Y = H).[152] In contrast, the reaction of 1-imino-1*H*,3*H*-thiazolo-[3,4-*a*]benzimidazole (**6.400**; X = NH, Y = H_2) with isocyanates and isothiocyanates is confined to the imino substituent and leads to moderate yields (Table 6.63) of the simple ureas and thioureas.[152] The mobility of the hydrogen atoms in the *C*(3)-methylene group of 1-imino-1*H*,3*H*-thiazolo-[3,4-*a*]benzimidazole manifest in the tautomeric character of this molecule (see before) is also apparent in its uncatalyzed condensation with aromatic aldehydes to give moderate yields (Table 6.63) of the corresponding 3-arylidene derivatives (**6.400**; X = NH, Y = ArCH).[152]

The tendency for 4*H*-pyrazolo[2,3-*a*]benzimidazoles to undergo electrophilic attack at the *C*(3) position is demonstrated by their smooth reaction with aromatic aldehydes under basic conditions to afford high yields (Table 6.64) of 3-arylidene-3*H*-pyrazolo[2,3-*a*]benzimidazole derivatives (**6.403**; X = ArCH).[198] Analogous condensation reactions with enamides and related substrates are widely used for the synthesis of 3*H*-pyrazolo[2,3-*a*]-benzimidazole cyanine dyestuffs.[198] The susceptibility of 4*H*-pyrazolo[2,3-*a*]benzimidazoles to electrophilic attack at the *C*(3) position is further demonstrated by their specific aminoalkenylation at this site under Vilsmeier–Haack conditions (i.e., reaction with dimethylformamide–phosphorus

TABLE 6.63. ACYLATION REACTIONS OF 1*H*,3*H*-THIAZOLO[3,4-*a*]BENZIMIDAZOLE DERIVATIVES[a]

(6.400) (6.401)

Reaction conditions[b]	Substrate (6.400) X	Substrate (6.400) Y	Product	X	Y	Yield (%)	m.p. (°C)	Solvent of crystallization
A	NH	H, H	(**6.400**)	MeCON	H, H	30	195–197	Ether–light petroleum
B	NH	H, H	(**6.401**)	MeCON	MeCO	87	258–261	Chloroform
C	NH	H, H	(**6.401**)	PhCON	PhCO	50	232–234	Ethyl acetate
D	NH	H, H	(**6.401**)	4-ClC_6H_4CON	4-ClC_6H_4CO	65	220	Pyridine
E	NH	H, H	(**6.401**)	EtO_2CN	EtO_2C	27	162–163	Benzene
E	NH	H, H	(**6.401**)	$PhCH_2O_2CN$	$PhCH_2O_2C$	25	168–170	—[c]
F	NH	H, H	(**6.401**)	4-$MeC_6H_4SO_2N$	4-$MeC_6H_4SO_2$	20	197–198	Acetone
G	NH	H, H	(**6.400**)	PhNHCON	H, H	43	160[d]	Benzene
G	NH	H, H	(**6.400**)	$Cl_3CCONHCON$	H, H	33	180–182	Dimethyl sulfoxide–water
G	NH	H, H	(**6.400**)	4-$NO_2C_6H_4NHCON$	H, H	50	260–262	Pyridine
A	PhNHCON	H, H	(**6.401**)	PhNHCON	MeCO	77	153–155	Ethyl acetate
H	PhNHCON	H, H	(**6.400**)	PhNHCON	PhCH	63	221–224	Chloroform–ether
H	PhNHCON	H, H	(**6.400**)	PhNHCON	2-Furfurylidene	33	230–234	Chloroform–ether
H	PhNHCON	H, H	(**6.400**)	PhNHCON	2-Thienylidene	42	256–258	Ether
H	O	H, H	(**6.400**)	O	2-Thienylidene	35	211–212	Chloroform

[a] From Ref. 152.

[b] $A = Ac_2O$/(100°)/(1.5–15 min); $B = Ac_2O$, NaOAc/(100°)(0.25 hr); *C* = PhCOCl, pyridine, ethyl acetate/(reflux)(0.5 hr); *D* = 4-ClC_6H_4COCl, Et_3N, ethyl acetate/(reflux)(20 min); $E = RCO_2Cl$/(reflux, neat)(1.5 hr); *F* = 4-$MeC_6H_4SO_2Cl$, pyridine, benzene/(room temp.)(48 hr); *G* = RN=C=O, ethyl acetate/(reflux)(1 hr); *H* = ArCHO/(reflux, neat)(3 min).

[c] Solvent of crystallization not specified.

[d] With resolidification at 195–197°.

TABLE 6.64. ACYLATION AND RELATED REACTIONS OF 4*H*-PYRAZOLO[2,3-*a*]BENZIMIDAZOLE DERIVATIVES

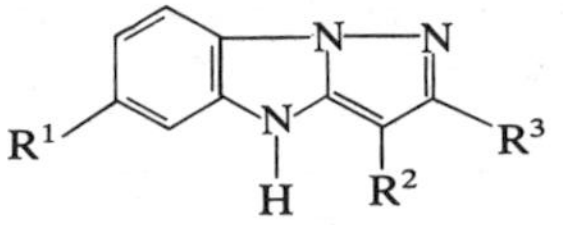

(6.402) 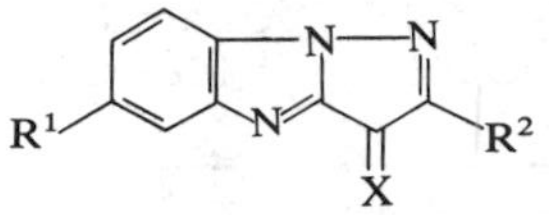 **(6.403)**

Reaction conditions[a]	Substrate (**6.402**)			Product	R¹	R²	R³	X	Yield (%)	m.p. (°C)	Solvent of crystallization	Ref.
	R¹	R²	R³									
A	H	NO	Me	(**6.402**)	H	NHAc	Me	—	68	293–293.5	—[b]	236
B	H	NH_2	Ph	(**6.402**)	H	NHAc	Ph	—	89	295–296	Methanol	236
C	SO_3H	NH_2	Ph	(**6.402**)	SO_3H	NHAc	Ph	—	61–72	<330 (decomp.)	Methanol	200, 236, 237
D	SO_3H	NH_2	Ph	(**6.402**)	SO_3H	NHCOEt	Ph	—	51–68	220 (decomp.)	Methanol–water	200, 236
E	SO_3H	NH_2	Ph	(**6.402**)	SO_3H	NHCOPh	Ph	—	61	>350	Methanol–water	236
F	SO_3H	H	$C_{17}H_{35}$	(**6.403**)	SO_3H	$C_{17}H_{35}$	—	4-$Me_2NC_6H_4CH$	90	—[c]	—[b]	198
G	H	H	Me	(**6.403**)	H	Me	—	Me_2NCH	59	216	—[b]	219
G	H	H	Ph	(**6.403**)	H	Ph	—	Me_2NCH	—[d]	235	—[b]	219
H	SO_3H	H	Ph	(**6.403**)	SO_3H	Ph	—	4-$Et_2NC_6H_4N$	43	>315 (decomp.)	—[b]	200

[a] *A* = Fe, AcOH/95–100°, then treat with Ac_2O; *B* = MeCOCl, pyridine/(80–90°)(0.5 hr); *C* = Ac_2O, pyridine/(reflux)(30 min); *D* = EtCOCl, pyridine/(80–90°)(1 hr); *E* = PhCOCl, pyridine/(80–90°)(1 hr); *F* = 4-$Me_2NC_6H_4CHO$, NaOH/(reflux)(3 hr); *G* = $POCl_3$, dimethylformamide/(warm)(1 min); *H* = 4-$Et_2NC_6H_4NH_2$, $AgNO_3$, NaCl, Na_2CO_3, H_2O/(room temp.)(30 min).

[b] Solvent of crystallization not specified.

[c] Melting point not quoted.

[d] Yield not specified.

oxychloride) (Table 6.64).[219] The oxidative condensation of 4*H*-pyrazolo-[2,3-*a*]benzimidazoles with arylamines also takes place at the *C*(3) position and provides a viable method (Table 6.64) for the synthesis of 3-arylimino-3*H*-pyrazolo[2,3-*a*]benzimidazoles (**6.403**; X = ArN).[200] The amino substituent in 3-amino-4*H*-pyrazolo[2,3-*a*]benzimidazoles undergoes acylation in orthodox fashion (Table 6.64).[200,236,237]

9*H*-imidazo[1,2-*a*]benzimidazoles are readily acylated at the *C*(2) and *C*(3) positions under a variety of conditions (Table 6.65) in Friedel–Crafts-type reactions, which substantiate theoretical predictions (see before) of the enhanced reactivity of the imidazole ring in these molecules to electrophilic attack. Acetylation is accomplished simply by heating with acetic anhydride alone[172,173] or in the presence of sodium acetate[171,174] and starting with a 2,9- or 3,9-disubstituted 9*H*-imidazo[1,2-*a*]benzimidazole affords the corresponding 3-acetyl[172–174] or 2-acetyl[171] derivative in high yield (Table 6.65). Acylation with simple[171,172] and $\alpha\beta$-unsaturated (ethylenic[214] or acetylenic[164]) acid chlorides is also readily achieved by heating in the absence of solvent[164,214] or in the presence of pyridine as catalyst.[171,172] The trichloroacetylation of 2,9-disubstituted 9*H*-imidazo[1,2-*a*]benzimidazoles followed by treatment with sodium methoxide provides an efficient method (yields > 90%) for the synthesis of the *C*(3) methoxycarbonyl derivatives.[238] The corresponding carboxylic acids are more directly accessible (yield > 90%) by lithiation of a 2,9-disubstituted 9*H*-imidazo[1,2-*a*]benzimidazole at the *C*(3) position followed by carboxylation with carbon dioxide.[238] The reactivity of the *C*(3) position in 2,9-disubstituted 9*H*-imidazo[1,2-*a*]benzimidazoles to electrophilic substitution also extends to formylation under Vilsmeier–Haack conditions (reaction with dimethylformamide–phosphorus oxychloride). Acylation of this type[164,165] proceeds in high yield (Table 6.65) and allows direct access to the synthetically useful 3-formyl derivatives, which are less conveniently prepared by lithiation of 2,9-disubstituted 3-bromo-9*H*-imidazo[1,2-*a*]benzimidazoles followed by reaction with dimethylformamide.[165] The failure[157] of 1*H*-imidazo[1,2-*a*]benzimidazoles to undergo acylation with acetic anhydride, acid chlorides, or the Vilsmeier–Haack reagent (dimethylformamide–phosphorus oxychloride) demonstrates, in accord with theoretical predictions (see before), the lower reactivity of the imidazole ring in such molecules to electrophilic attack. 2,9-Disubstituted 3-acetyl-9*H*-imidazo[1,2-*a*]benzimidazoles are converted under Vilsmeier–Haack conditions into the 3-ethynyl derivatives in high yield (Table 6.65) via isolable β-chlorovinyl aldehyde intermediates.[213] *C*(2) and *C*(3) amino substituents in 9*H*-imidazo[1,2-*a*]benzimidazoles are readily acetylated under standard conditions (Table 6.65).[174,175]

Prior lithiation at the *C*(3) position is a prerequisite of the successful reaction of 9*H*-imidazo[1,2-*a*]benzimidazoles with aldehydes, which allows access, albeit in variable yield (Table 6.65), to the corresponding *C*(3) secondary alcohol derivatives.[164] *C*(2) methyl substituents in 9*H*-imidazo-[1,2-*a*]benzimidazoles are sufficiently activated to electrophilic attack to

TABLE 6.65. ACYLATION REACTIONS OF 1(9)*H*-IMIDAZO[1,2-*a*]BENZIMIDAZOLES AND 2,3-DIHYDRO-1(9)*H*-IMIDAZO[1,2-*a*] BENZIMIDAZOLES

Reaction conditions[a]	1(9)*H*-imidazo[1,2-*a*]benzimidazole Substrate	Product	Yield (%)	m.p. (°C)	Solvent of crystallization	Ref.
A	2,9-di-Me-	2,9-di-Me-3-MeCO-[b]	87	178	Ethanol–water	173
A	2,9-di-Me-	2,9-di-Me-3-MeCO-	89	178	Ethanol	172
A	2-Ph-9-Me-	2-Ph-3-MeCO-9-Me-[c]	92	212	Ethanol–dimethyl-formamide	172
A	2-(*p*-BrC_6H_4)-9-Me-	2-(*p*-BrC_6H_4)-3-MeCO-9-Me-	85	197	Ethanol–dimethyl-formamide	172
B	2,9-di-Me-	2,9-di-Me-3-PhCO-	50	156	Benzene	172
C	2-Ph-9-Me-	2-Ph-3-PhCO-9-Me-	70	215	Ethanol	172
D	2-PhNH-9-Me-[d]	2-PhN(COMe)-3-MeCO-9-Me-	76	171	Ethanol	174
D	2-(*p*-$O_2NC_6H_4NH$)-9-Me-[d]	2-[*p*-$O_2NC_6H_4N$(COMe)]-3-MeCO-9-Me-	77	271	Dimethylformamide	174
D	2-(*p*-ClC_6H_4NH)-9-Me-[d]	2-[*p*-ClC_6H_4N(COMe)]-3-MeCO-9-Me-	84	194	Heptane	174
D	2-(*p*-$EtO_2CC_6H_4NH$)-9-Me-[d]	2-[*p*-$EtO_2C_6H_4N$(COMe)]-3-MeCO-9-Me-	88	174	Ethanol	174
D	2-PhN(Me)-9-Me-	2-PhN(Me)-3-MeCO-9-Me-	78	158	Ethanol	174
E	2,9-di-Me-	2,9-di-Me-3-(PhCH=CHCO)-	—[d]	179–180	Ethanol–dimethyl-formamide	214
E	2-Ph-9-Me-	2-Ph-3-(PhCH=CHCO)-9-Me-	—[d]	225	Ethanol–dimethyl-formamide	214
E	2-Ph-9-Me-	2-Ph-3-[2-(2-furyl)acryloyl]-9-Me-	—[d]	210 (decomp.)	Ethanol	214
F	2-Ph-9-Me-	2-Ph-3-[2-(5-nitro-2-furyl)acryloyl]-9-Me-	53	304–305 (decomp.)	Dimethylformamide	214
G	2,9-di-Me-	2,9-di-Me-3-(PhC≡CCO)-	40	203–204	Ethanol	164
G	2-Ph-9-Me-	2-Ph-3-(PhC≡CCO)-9-Me-[f]	47	195–196	Ethanol	164
D	3,9-di-Me-	3,9-di-Me-2-MeCO-	80	177	Acetone	171
H	3,9-di-Me-	3,9-di-Me-2-PhCO-	58	140–142	Acetone	171
I	2-Ph-3-NH_2-9-$PhCH_2$-	2-Ph-3-MeCONH-9-$PhCH_2$-	—[d]	212–213	Ethanol–water	175
J	2,9-di-Me-	2,9-di-Me-3-CHO-[g]	70	186	Methanol	165
J	2-Ph-9-Me-	2-Ph-3-CHO-9-Me-	88	147	Dioxane	165
J	2-(*p*-BrC_6H_4)-9-Me-	2-(*p*-BrC_6H_4)-3-CHO-9-Me-[h]	40	199–201	Ethanol	164

J	2-(2-naphthyl)-9-Me-	2-(2-naphthyl)-3-CHO-9-Me-[i]	80	254–255	Ethanol–dimethylformamide	164
J	2-Ph-9-$PhCH_2$-	2-Ph-3-CHO-9-$PhCH_2$-[j]	98	153–154	Ethanol	164
J	2-Ph-6,7-di-Me-9-Et-	2-Ph-3-CHO-6,7-di-Me-9-Et-[k]	90	215–216	Ethanol	164
K	2,9-di-Me-3-MeCO-	2,9-di-Me-3-(HC≡C)-	70–85	117–118 (decomp.)	Light petroleum	213
K	2-Ph-3-MeCO-9-Me-	2-Ph-3-(HC≡C)-9-Me-	70–85	127–128 (decomp.)	Light petroleum	213
K	2-Me-3-MeCO-9-$PhCH_2$-	2-Me-3-(HC≡C)-9-$PhCH_2$-	70–85	170	Ethanol	213
K	2-Ph-3-MeCO-9-[$Et_2N(CH_2)_2$]-	2-Ph-3-(HC≡C)-9-[$Et_2N(CH_2)_2$]-[l]	70–85	—[m]	—	213
L	2-Ph-3-MeCO-9-Me-	2-Ph-3-[HCOCH=C(Cl)]-9-Me-	—[e]	155–156	Light petroleum	213
M	2-Ph-3-Br-9-Me-	2-Ph-3-CHO-9-Me-[n]	81	147	Dioxane	165
N	2-Ph-3-Br-9-Me-	2-Ph-3-PhCHOH-9-Me-[o]	72	192 (decomp.)	Dimethylformamide	164
N	2-Ph-3-Br-9-Me-	2-Ph-3-(*p*-$Me_2NC_6H_4CHOH$)-9-Me-	55	181–182	Ethanol	164
N	2,9-di-Me-3-Br-	2,9-di-Me-3-(*p*-$Me_2NC_6H_4CHOH$)-	22	193	Ethanol	164
O	2,9-di-Me-	2-(PhCH=CH)-9-Me-	90	215	Ethanol	165
O	2,9-di-Me-	2-(*p*-$O_2NC_6H_4CH$=CH)-9-Me-	72	221 (decomp.)	Dimethylformamide	165
O	2,9-di-Me-	2-(*o*-HOC_6H_4CH=CH)-9-Me-	98	297	Dimethylformamide	165
O	2,9-di-Me-	2-(*p*-$Me_2NC_6H_4CH$=CH)-9-Me-	63	282	Ethanol	165
P	2,9-di-Me-3-MeCO-	2,9-di-Me-3-(PhCH=CHCO)-[p]	90	179–180	Ethanol–dimethylformamide	214
P	2,9-di-Me-3-MeCO-	2,9-di-Me-3-(*p*-$MeOC_6H_4CH$=CHCO)-	91	181	Ethanol	214
P	2,9-di-Me-3-MeCO-	2,9-di-Me-3-(*m*-$O_2NC_6H_4CH$=CHCO)-	94	243–244 (decomp.)	Ethanol–dimethylformamide	214
P	2,9-di-Me-3-MeCO-	2,9-di-Me-3-(*p*-$O_2NC_6H_4CH$=CHCO)-	95	292	Dimethylformamide	214
P	2-Ph-3-MeCO-9-Me-	2-Ph-3-(PhCH=CHCO)-9-Me-[q]	94	225	Ethanol–dimethylformamide	214
P	2-Ph-3-MeCO-9-Me-	2-Ph-3-(*p*-$MeOC_6H_4CH$=CHCO)-9-Me-	90	211–212	Ethanol	214
P	2-Ph-3-MeCO-9-Me-	2-Ph-3-(*m*-$O_2NC_6H_4CH$=CHCO)-9-Me-	90	323	Dimethylformamide	214
P	2-Ph-3-MeCO-9-Me-	2-Ph-3-(*p*-$O_2NC_6H_4CH$=CHCO)-9-Me-	95	288	Dimethylformamide	214
P	2-Ph-3-MeCO-9-Me-	2-Ph-3-(*p*-$Me_2NC_6H_4CH$=CHCO)-9-Me-	73	208	Ethanol	214
P	2-Ph-3-MeCO-9-[$Et_2N(CH_2)_2$]-	2-Ph-3-(*p*-$MeOC_6H_4CH$=CHCO)-9-[$Et_2N(CH_2)_2$]-	88	149–150	Ethanol	214
Q	2,9-di-Me-3-NH_2-	2,9-di-Me-3-(*p*-$O_2NC_6H_4CH$=N)-	78	230–232	Ethanol	175
Q	9-Me-2,3-dihydro-2-one	3-(*p*-$O_2NC_6H_4CH$=)-9-Me-2,3-dihydro-2-one	74	317	Dimethylformamide	180

TABLE 6.65 (*Continued*)

Reaction conditions[a]	1(9)*H*-imidazo[1,2-*a*]benzimidazole		Yield (%)	m.p. (°C)	Solvent of crystallization	Ref.
	Substrate	Product				
Q	9-Me-2,3-dihydro-2-one	3-(o-$O_2NC_6H_4CH$=)-9-Me-2,3-dihydro-2-one	62	262	Dimethylformamide	180
Q	9-$PhCH_2$-2,3-dihydro-2-one	3-(p-$O_2NC_6H_4CH$=)-9-$PhCH_2$-2,3-dihydro-2-one	70	300–301	Dimethylformamide	180
R	1-Me-2,3-dihydro-2-one	1-Me-3-(p-$O_2NC_6H_4CH$=)-2,3-dihydro-2-one	70	284	—[r]	180
S	2,3-dihydro-1(9)*H*-	1-guanyl-2,3-dihydro-	31	270–272 (decomp.)	Water	176

[a] *A* = Ac_2O/(reflux)(1–3 hr); *B* = PhCOCl, benzene/(room temp.)(1 hr); *C* = PhCOCl, pyridine/(80°)(2–3 min., then boil briefly); *D* = Ac_2O, NaOAc/(reflux)(3–3.5 hr); *E* = RCH=CHCOCl/(40–80°)(reaction time not specified); *F* = 2-(5-nitro-2-furyl)acryloyl chloride/(100–120°)(reaction time not specified); *G* = PhC≡C COCl/(40°)(20 min); *H* = PhCOCl, pyridine/(110–120°)(10 min); *I* = Ac_2O/(reflux)(brief); *J* = $POCl_3$, dimethylformamide/(room temp.)(30 min, then 100°)(20 min–2 hr); *K* = $POCl_3$, dimethylformamide/(room temp.)(30 min, then 60–70°)(2 hr, then treat with KOAc)/(60–70°)(1–1.5 hr); *L* = $POCl_3$ dimethylformamide, KOAc/(10°)(reaction time not specified); *M* = Bu^nLi, toluene–ether/(−75°)(5 hr, then dimethylformamide–ether)(−75°)(1 hr, then 20°)(3 hr); *N* = Bu^nLi, toluene/(−75°)(5 hr, then ArCHO, ether)(−75°)(1 hr, then room temp.)(14 hr); *O* = ArCHO, melt/(65–100°)(5–10 min); *P* = ArCHO, 40% NaOH aq., EtOH/(warm)(few min); *Q* = ArCHO, AcOH or Ac_2O/(reflux)(15–20 min); *R* = p-$O_2NC_6H_4CHO$, Ac_2O/(reflux)(1 hr); *S* = 2,5-dimethyl-1-guanylpyrazole, EtOH/(reflux)(4 hr).

[b] Forms a 2,4-dinitrophenylhydrazone, dark brown needles, m.p. 260–261° (from dimethylformamide), and a hydrochloride, colorless needles, m.p. 276° (from ethanol).

[c] Forms a hydrochloride, m.p. 217° (from ether–ethanol).

[d] Hydrochloride.

[e] Yield not quoted.

[f] Forms a 2,4-dinitrophenylhydrazone, red crystals, m.p. 272° (from dimethylformamide).

[g] Forms a 2,4-dinitrophenylhydrazone, black crystals, m.p. 288° (from dimethylformamide), and an oxime, m.p. 265° (from ethanol–water).

[h] Forms a 2,4-dinitrophenylhydrazone, m.p. 307–308° (from dimethylformamide).

[i] Forms a 2,4-dinitrophenylhydrazone, m.p. 330° (from dimethylformamide).

[j] Forms a 2,4-dinitrophenylhydrazone, m.p. 248–250° (from dimethylformamide).

[k] Froms a 2,4-dinitrophenylhydrazone, m.p. 288° (decomp.) (from dimethylformamide).

[l] Forms a dipicrate, m.p. 168–169°.

[m] Oil.

[n] Forms a 2,4-dinitrophenylhydrazone, red needles, m.p. 304° (from dimethylformamide).

[o] Forms a hydrochloride, m.p. 134° (from ethanol).

[p] Forms a 2,4-dinitrophenylhydrazone, red crystals, m.p. 228–229° (decomp.) (from dimethylformamide).

[q] Forms a 2,4-dinitrophenylhydrazone, red crystals, m.p. 259–260° (decomp.) (from dimethylformamide).

[r] Solvent of crystallization not specified.

undergo uncatalyzed condensation with aromatic aldehydes affording moderate to excellent yields (Table 6.65) of the *C*(2) styryl derivatives.[165] Despite the strongly electron-donating (and hence deactivating) character of the 9*H*-imidazo[1,2-*a*]benzimidazole ring system, attached *C*(3) acetyl groups participate in orthodox aldol-type condensation reactions with aromatic aldehydes giving high yields (Table 6.65) of the anticipated $\alpha\beta$-unsaturated keto derivatives.[214] The condensation of *C*(3) amino substituents in 9*H*-imidazo[1,2-*a*]benzimidazoles with aromatic aldehydes leads to Schiff base formation in good yield (Table 6.65).[175] As a result of the lower reactivity of their component *C*(3) methylene groups, 2,3-dihydro-1*H*- and 9*H*-imidazo[1,2-*a*]benzimidazol-2-ones are much less able than 2,3-dihydrothiazolo[3,2-*a*]benzidazol-3-ones (see before) to take part in aldol-type condensation reactions and only form *C*(3) arylidene derivatives (Table 6.65) with the more reactive aromatic aldehydes (e.g., *p*-nitrobenzaldehyde).[180]

4*H*-Imidazo[1,5-*a*]benzimidazoles are readily acylated at the *C*(1) and *C*(3) positions under a variety of conditions, thus substantiating theoretical predictions[224] of the susceptibility of these sites to electrophilic attack. Acetylation at the *C*(1) and *C*(3) positions in 4*H*-imidazo[1,5-*a*]benzimidazoles is achieved smoothly and in high yield (84–96%) (Table 6.66) simply by heating with acetic anhydride in the presence of sodium acetate.[192,193,215] Formylation under Vilsmeier Haack conditions likewise affords excellent yields (Table 6.66) of 4*H*-imidazo[1,5-*a*]benzimidazole 1- and 3-carboxaldehydes.[193,215,216] *C*(1)-Unsubstituted 4*H*-imidazo[1,5-*a*]-benzimidazoles also react readily with aqueous formaldehyde to give high yields (Table 6.66) of the corresponding *C*(1)-hydroxymethyl derivatives.[215] In contrast, the reaction of *C*(3)-unsubstituted 4*H*-imidazo[1,5-*a*]benzimidazoles with aqueous formaldehyde leads not to the anticipated *C*(3)-hydroxymethyl derivatives but to dimeric products (Table 6.66) formed by their self-condensation.[192] 4*H*-Imidazo[1,5-*a*]benzimidazoles are readily aminoalkylated[215,221] at the *C*(1) position under Mannich conditions (condensation with secondary amines in the presence of formaldehyde) (Table 6.66), and undergo cyanoethylation at the same site on treatment with acrylonitrile (Table 6.66).[215]

Halogenation. Reagents such as bromine in chloroform[239] or *N*-bromosuccinimide in acetone[122] effect the bromination of thiazolo[3,2-*a*]benzimidazoles at the electron-rich *C*(2) position giving moderate to high yields (Table 6.67) of the corresponding *C*(2)-bromo derivatives. Bromine readily adds to the carbon–carbon double bond in 2-arylidene-2,3-dihydrothiazolo-[3,2-*a*]benzimidazol-3-ones giving vermilion colored dibromo adducts in moderate to high yield (Table 6.67).[101,133]

The transformation shown in Scheme 6.86 represents what appears to be the sole example of the halogenation of the 4*H*-pyrazolo[2,3-*a*]benzimidazole ring system,[203] and illustrates the reactivity of the *C*(3) position in the latter to electrophilic attack.

TABLE 6.66. ACYLATION REACTIONS OF 4*H*-IMIDAZO[1,5]*a*]BENZIMIDAZOLES

Reaction conditions[a]	4*H*-Imidazo[1,5-*a*]benzimidazole		Yield (%)	m.p. (°C)	Solvent of crystallization	Ref.
	Substrate	Product				
A	3-Ph-4-Me-	1-MeCO-3-Ph-4-Me-	94	209–210.5	Ethanol	215
A	1,4-di-Me-	1,4-di-Me-3-MeCO-	84	223.5–225.5	Ethanol	192
A	1-Ph-4-Me-	1-Ph-3-MeCO-4-Me-	89	247–249	Ethanol	192
B	1-Ph-4-$PhCH_2$-	1-Ph-3-MeCO-4-$PhCH_2$-	96	182–183	Ethanol	193
C	3,4-di-Me-	1-CHO-3,4-di-Me[b]-	72	178–179	Benzene	216
D	3-Ph-4-Me-	1-CHO-3-Ph-4-Me[c]-	92	199.5–200.5	Ethanol	215, 216
E	1-Ph-4-$PhCH_2$-	1-Ph-3-CHO-4-$PhCH_2$-[d]	100	190–192	Ethanol	193
F	3-Ph-4-Me-	1-$HOCH_2$-3-Ph-4-Me-	98	159–160 (decomp.)	Ethyl acetate	215
G	1,4-di-Me-	$(1,4\text{-di-Me-3})_2CH_2$-	90	201–203	Ethanol–water	192
H	1,4-di-Me-	$(1,4\text{-di-Me-3})_2CH_2$-	76	—	—	192
I	1-Ph-4-Me-	$(1\text{-Ph-4-Me-3})_2CH_2$-	88	249.5–252	—[e]	192
H	3-Ph-4-Me-	1-(Me_2NCH_2)-3-Ph-4-Me-	99	114.5–116.5	Ether	215

J	3-Ph-4-Me-	1-(MeN NCH$_2$)-3-Ph-4-Me-[f]	96	220–223	Ether–ethanol	221
K	3-Ph-4-Me-	1-[NC(CH$_2$)$_2$]-3-Ph-4-Me-[g]	32	184–186	Ethanol	215

[a] A = Ac_2O, NaOAc/(reflux)(2–3 hr); *B* = Ac_2O, NaOAc/(90–100°)(1.5 hr), then (120°)(30 min); *C* = $POCl_3$, dimethylformamide/(room temp.)(several days); *D* = $POCl_3$, dimethylformamide/(25°)(4 hr); *E* = $POCl_3$, dimethylformamide/(room temp.)(16 hr); *F* = 30% HCHO aq., H_2O/(reflux)(4 hr); *G* = 30% HCHO aq., H_2O/(reflux)(30 min); *H* = 40% Me_2NH aq., 30% HCHO aq., EtOH, H_2O/(room temp.)(24–48 hr); *I* = 15% HCHO aq., EtOH/(room temp.)(12 hr); *J* = MeN NH, 30% HCHO aq., EtOH/(room temp.)(5 hr); *K* = CH_2═CHCN, Rodionov reagent/(reflux)(8 hr).

[b] Forms an oxime, yellow crystals, m.p. 260–261.5 (decomp.) (from ethanol), and a thiosemicarbazone, yellow crystals, m.p. 225.5–226° (decomp.) (from methanol).

[c] Forms an oxime, yellow crystals, m.p. 214–214.5° (decomp.) (from ethanol), and a thiosemicarbazone, m.p. 215–216° (decomp.) (from acetic acid).

[d] Forms an oxime, colorless crystals, m.p. 191–193° (from ethanol), and a thiosemicarbazone, yellow needles, m.p. 225.5–227.5° (decomp.) (from ethanol).

[e] Solvent of crystallization not specified.

[f] Dihydrochloride.

[g] Forms a picrate, m.p. 187–188°.

TABLE 6.67. BROMINATION REACTIONS OF THIAZOLO[3,2-*a*]BENZIMIDAZOLES AND 2,3-DIHYDROTHIAZOLO[3,2-*a*]-BENZIMIDAZOL-3-ONES

Reaction conditions[a]	Thiazolo[3,2-*a*]benzimidazole: Substrate	Product	Yield (%)	m.p. (°C)	Solvent of crystallization	Ref.
A	3-Ph-	2-Br-3-Ph-	61	208	Acetone	122
A	3-Ph-6,7-di-Me-	2-Br-3-Ph-6,7-di-Me- +	28	220	Ethanol	122
		2,8-di-Br-3-Ph-6,7-di-Me-	14	245	Benzene–ethyl acetate	
B	2-(PhCH=)-2,3-dihydro-3-one	2-Br-2-[PhCH(Br)]-2,3-dihydro-2-one	50	207	—[b]	101
B	2-(PhCH=)-6-Me-2,3-dihydro-3-one	2-Br-2-[PhCH(Br)]-6-Me-2,3-dihydro-3-one-	48	285–286 (decomp.)	—[b]	133
B	2-(PhCH=)-6,7-di-Me-2,3-dihydro-3-one	2-Br-2-[PhCH(Br)]-6,7-di-Me-2,3-dihydro-3-one	52	217–218	—[b]	133
B	2-(*p*-HOC_6H_4CH=)-2,3-dihydro-3-one	2-Br-2-[*p*-$HOC_6H_4CH(Br)$]-2,3-dihydro-3-one	80	>300	—[b]	101
B	2-(*p*-$MeOC_6H_4CH$=)-2,3-dihydro-3-one	2-Br-2-[*p*-$MeOC_6H_4CH(Br)$]-2,3-dihydro-3-one	50	240	—[b]	101
B	2-(*p*-$MeOC_6H_4CH$=)-6-Me-2,3-dihydro-3-one	2-Br-2-[*p*-$MeOC_6H_4CH(Br)$]-6-Me-2,3-dihydro-3-one	45	244–246 (decomp.)	—[b]	133
B	2-(*p*-$MeOC_6H_4CH$=)-6,7-di-Me-2,3-dihydro-3-one	2-Br-2-[*p*-$MeOC_6H_4CH(Br)$]-6,7-di-Me-2,3-dihydro-3-one	62	208–209 (decomp.)	—[b]	133
B	2-(*m*-$O_2NC_6H_4CH$=)-2,3-dihydro-3-one	2-Br-2-[*m*-$O_2NC_6H_4CH(Br)$]-2,3-dihydro-3-one	80	258	—[b]	101
B	2-(*p*-$Me_2NC_6H_4CH$=)-2,3-dihydro-3-one	2-Br-2-[*p*-$Me_2NC_6H_4CH(Br)$]-2,3-dihydro-3-one	58	190	—[b]	101
B	2-(*p*-$Me_2NC_6H_4CH$=)-6-Me-2,3-dihydro-3-one	2-Br-[*p*-$Me_2NC_6H_4CH(Br)$]-6-Me-2,3-dihydro-3-one	48	213–214 (decomp.)	—[b]	133
B	2-(*p*-$Me_2NC_6H_4CH$=)-6,7-di-Me-2,3-dihydro-3-one	2-Br-[*p*-$Me_2NC_6H_4CH(Br)$]-6,7-di-Me-2,3-dihydro-3-one	68	262–263 (decomp.)	—[b]	133
B	2-(*p*-ClC_6H_4CH=)-6-Me-2,3-dihydro-3-one	2-Br-[*p*-$ClC_6H_4CH(Br)$]-6-Me-2,3-dihydro-3-one	40	180 (decomp.)	—[b]	133
B	2-(*p*-ClC_6H_4CH=)-6,7-di-Me-2,3-dihydro-3-one	2-Br-[*p*-$ClC_6H_4CH(Br)$]-6,7-di-Me-2,3-dihydro-3-one	59	267–268 (decomp.)	—[b]	133

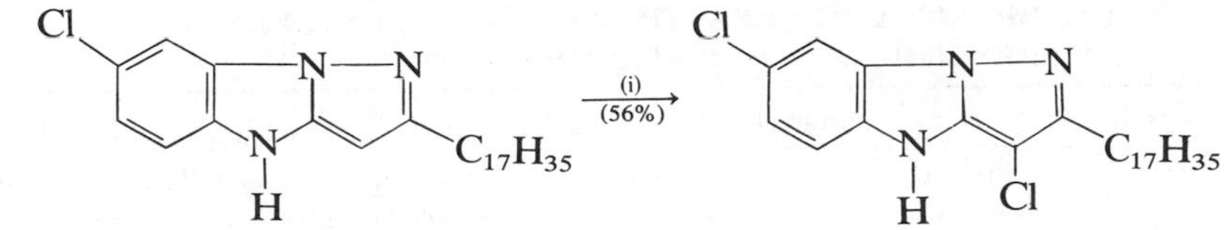

(i) SO_2Cl_2, NaOAc, AcOH/(40°)(15 min) (m.p. 115–116°)

Scheme 6.86

C(3)-Unsubstituted imidazo[1,2-*a*]benzimidazoles are readily brominated under mild conditions to give uniformly high yields (Table 6.68) of the *C*(3)-bromo derivatives.[162a,163,164] The facility of these transformations is a measure of the high-electron localization at the *C*(3) position in the 9*H*-imidazo[1,2-*a*]benzimidazole ring system (see before). The resistance of the imidazole ring in 1*H*-imidazo[1,2-*a*]benzimidazoles to electrophilic attack (see later) does not apply to bromination, which occurs readily and affords the *C*(3)-bromo derivatives in high yield (Table 6.68).[157] Not unexpectedly, thc bromination[166,180] of 2,3-dihydroimidazo[1,2-*a*]benzimidazoles occurs at

TABLE 6.68. BROMINATION REACTIONS[a] OF 1*H*- AND 9*H*-IMIDAZO[1,2-*a*] BENZIMIDAZOLES

1*H*- or 9*H*-Imidazo[1,2-*a*]benzimidazole		Yield		Solvent of	
Substrate	Product	(%)	m.p.(°C)	crystallization	Ref.
2,9-di-Me-	2,9-di-Me-3-Br-[b,c]	89	274	Ethanol–ether	164
2-Ph-9-Me-	2-Ph-3-Br-9-Me-[b,d]	98	245 (decomp.)	—[e]	162a
2-(*p*-$O_2NC_6H_4$)-9-Me-	2-(*p*-$O_2NC_6H_4$)-3-Br-9-Me-[b,f]	90	268	Dimethylformamide	163
2-(*p*-BrC_6H_4)-9-Me-	2-(*p*-BrC_6H_4)-3-Br-9-Me-	65	170	Ethanol–dimethyl formamide	163
2-Ph-6,7-di-Me-9-Et-	2-Ph-3-Br-6,7-di-Me-9-Et-[b,g]	90	>220 (decomp.)	—[e]	163
2-Ph-9-$PhCH_2$-	2-Ph-3-Br-9-CH_2Ph-[b,h]	88	244	Ethanol	164
2-Ph-9-[$Et_2N(CH_2)_2$]-	2-Ph-3-Br-9-[$Et_2N(CH_2)_2$]-[b,i]	82	167–168	Ethanol–ether	164
1-Me-2-Ph-	1-Me-2-Ph-3-Br-[b,j]	quant.	259 (decomp.)	Ethanol	157

[a] Br_2, $CHCl_3$/(20°)(60 min).
[b] Hydrobromide.
[c] Free base has m.p. 148° (from ethanol).
[d] Free base has m.p. 148° (from ethanol).
[e] Solvent of crystallization not specified.
[f] Free base has m.p. 256° (from dimethylformamide).
[g] Free base has m.p. 158° (from ethanol).
[h] Free base has m.p. 172° (from ethanol).
[i] Free base is an oil.
[j] Free base has m.p. 205° (from ethanol).

TABLE 6.69. BROMINATION REACTIONS[a] OF 4*H*-IMIDAZO[1,5-*a*]-BENZIMIDAZOLE DERIVATIVES

4*H*-Imidazo[1,5*a*]benzimidazole					
Substrate	Product	Yield (%)	m.p. (°C)	Solvent of crystallization	Ref.
1-Ph-4-Me-	1-Ph-3-Br-4-Me-	24	152–154	ether	192
3-Ph-4-Me-	1-Br-3-Ph-4-Me-	72	191–194	methanol	221

[a] *N*-Bromosuccinimide, CCl_4/(reflux)(2 hr).

unspecified sites in the benzene ring and not in the imidazoline nucleus. 4*H*-Imidazo[1,5-*a*]benzimidazoles are brominated in moderate to high yield (Table 6.69) at the electron-rich *C*(1)[221] and *C*(3)[192] positions by heating with *N*-bromosuccinimide in carbon tetrachloride.

Nitrosation and Nitration. The ready conversion of *C*(3)-unsubstituted 4*H*-pyrazolo[2,3-*a*]benzimidazoles into *C*(3)-nitroso derivatives by treatment with sodium nitrite in hydrochloric or sulfuric acids (Table 6.70)[199,200] is a further illustration of the susceptibility of the *C*(3) position in the 4*H*-pyrazolo[2,3-*a*]benzimidazole ring system to electrophilic attack.

9-Alkyl-2-aryl-9*H*-imidazo[1,2-*a*]benzimidazoles react with sodium nitrite in acetic acid to give the corresponding *C*(3)-nitroso derivatives in high yield (Table 6.71).[175,227] In contrast, the *C*(3)-nitrosation of 2,9-dimethyl-9*H*-imidazo[1,2-*a*]benzimidazole is followed by ring-opening to an imino-benzimidazoline derivative (Scheme 6.87).[227] Analogous ring-opening occurs on attempted nitrosation of 9-methyl-2,3-dihydro-9*H*-imidazo[1,2-*a*]benzimidazol-2-one (Scheme 6.87).[180] The failure of 1*H*-imidazo[1,2-*a*]benzimidazoles to undergo nitrosation[157] is a measure of the low reactivity

TABLE 6.70. NITROSATION REACTIONS OF 4*H*-PYRAZOLO[2,3-*a*] BENZIMIDAZOLE DERIVATIVES

Reaction conditions[a]	4*H*-Pyrazolo[2,3-*a*]benzimidazole Substrate	Product	Yield (%)	m.p. (°C)	Solvent of crystallization	Ref.
A	2-Me-	2-Me-3-NO-	—[b]	—[c]	—[d]	199
A	2-Ph-	2-Ph-3-NO-	—[b]	—[c]	—[d]	199
A	2-Ph-6-SO_3H-	2-Ph-3-NO-6-SO_3H-	98	260	—[d]	200
A	2-$CONH_2$-	2-$CONH_2$-3-NO-	—[b]	—[c]	—[d]	199
B	2-CO_2H-	2-CO_2H-3-NO-	—[b]	—[c]	—[d]	199
B	2-CO_2H-6-SO_3H-	2-CO_2H-6-SO_3H-	—[b]	—[c]	—[d]	199

[a] $A = 10\%\ NaNO_2$, $20\%\ H_2SO_4$ aq./−5 to +5°; $B = 10\%\ NaNO_2$, 20% HCl aq./0–10°.
[b] Yield not quoted.
[c] Melting point not quoted.
[d] Solvent of crystallization not specified.

TABLE 6.71. NITROSATION AND NITRATION REACTIONS OF 9*H*-IMIDAZO[1,2-*a*]BENZIMIDAZOLE DERIVATIVES

Reaction conditions[a]	9*H*-Imidazo[1,2-*a*]benzimidazole				Solvent of crystallization	Ref.
	Substrate	Product	Yield (%)	m.p. (°C)		
A	2-Ph-9-Me-	2-Ph-3-NO-9-Me-[b,c]	92	247	Ethanol	227
B	2-(*p*-BrC_6H_4)-9-Me-	2-(*p*-BrC_6H_4)-3-NO-9-Me-[d]	84	244 (decomp.)	Dimethyl-formamide	175
A	2-Ph-9-$PhCH_2$-	2-Ph-3-NO-9-$PhCH_2$-[e,f]	94	215	Ethanol	175
A	2-Ph-6,7-di-Me-9-Et-	2-Ph-3-NO-6,7-di-Me-9-Et-[b]	100	265	Dimethyl-formamide	175
C	2,9-di-Me-[g]	2,9-di-Me-3-NO_2-	88	248[h]	Dimethyl-formamide	163
D	2,9-di-Me-	2,9-di-Me-3-NO_2-	70	—	—	163
D	2-(*p*-BrC_6H_4)-9-Me-	2-(*p*-BrC_6H_4)-3-NO_2-9-Me-	87	271–272	Dimethyl-formamide	163
D	2-Ph-9-Me-	2-(*p*-$O_2NC_6H_4$)-9-Me-X-NO_2-[i]	90	297–298	Dimethyl-formamide	163
D	2-(*p*-$O_2NC_6H_4$)-9-Me-	2-(*p*-$O_2NC_6H_4$)-9-Me-X-NO_2-[i]	quant.	—	—	163
E	2,9-di-Me-3-NH_2-	2,9-di-Me-3-[1-(2-hydroxy-naphthyl-azo]-	75	235–236	Ethanol	175

[a] *A* = $NaNO_2$, AcOH, H_2O/(20°)(short time); *B* = $NaNO_2$, AcOH, H_2O, EtOH/(70–80°)(15–20 min); *C* = conc. H_2SO_4/(−5 to −10°)(1 hr); *D* = KNO_3, conc. H_2SO_4/(−5 to −12°)(1 hr); *E* = $NaNO_2$, H^+ then treatment with β-naphthol in NaOH aq.

[b] Green needles.

[c] Forms a hydrochloride, red needles, m.p. 208–209°.

[d] Green solid.

[e] Dark-green plates.

[f] Forms a hydrochloride, red solid, m.p. 194–195°.

[g] Nitrate.

[h] Forms a monohydrate, m.p. 221°.

[i] Position of the second nitro group not established.

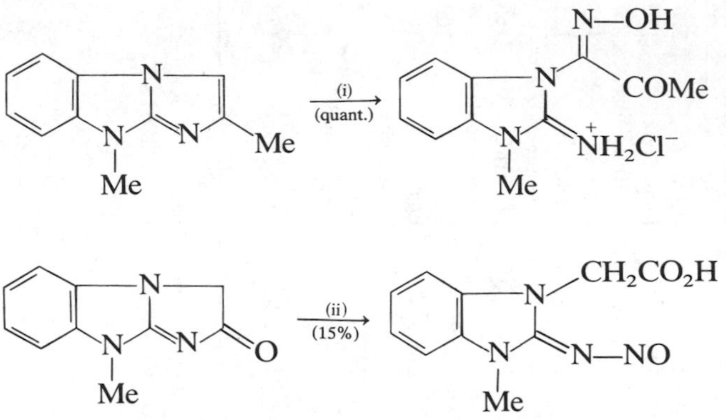

(i) $NaNO_2$, conc. HCl, EtOH/(0°)(30 min)
(ii) $NaNO_2$, AcOH/(room temp.)(48 hr)

Scheme 6.87

of the imidazole ring in such molecules to electrophilic attack. 3-Amino-9*H*-imidazo[1,2-*a*]benzimidazoles, when stable, can be diazotized to give diazonium salts which undergo orthodox diazo-coupling reactions with substrates such as β-naphthol (Table 6.71).[175]

9*H*-Imidazo[1,2-*a*]benzimidazoles can be nitrated at the *C*(3) position under standard conditions. For example, treatment of 2,9-dimethyl-9*H*-imidazo[1,2-*a*]benzimidazole with potassium nitrate-concentrated sulfuric acid affords the *C*(3)-nitro derivative in high yield (Table 6.71).[163] The same compound is obtained in lower yield (Table 6.71) by rearrangement of the nitrate salt of 2,9-dimethyl-9*H*-imidazo[1,2-*a*]benzimidazole in concentrated sulfuric acid.[163] 2-(*p*-Bromophenyl)-9-methyl-9*H*-imidazo[1,2-*a*]-benzimidazole also nitrates at the *C*(3) position in the imidazole ring giving the corresponding nitro derivative in high yield (Table 6.71).[163] In contrast, the nitration[163] of 9-methyl-2-phenyl-9*H*-imidazo[1,2-*a*]benzimidazole affords a dinitro product derived by substitution at the *para* position in the phenyl substituent and at an unestablished site in the fused benzene ring. The lack of substitution in the imidazole ring in this instance can be attributed to preferential initial nitration of the phenyl substituent with consequent deactivation of the *C*(3) position to further attack. The benzene nucleus in 9*H*-imidazo[1,2-*a*]benzimidazoles has been shown[214] to nitrate less readily than an attached furan substituent.

The reactivity of the imidazole ring in 4*H*-imidazo[1,5-*a*]benzimidazoles toward electrophilic substitution extends to nitrosation, which is readily accomplished by treatment with sodium nitrite in acetic acid at room temperature or below, and affords high yields of *C*(1)[215] and *C*(3)[192] nitroso derivatives (Table 6.72).

Diazo Coupling. The activated methylene substituent in 2,3-dihydrothiazolo[3,2-*a*]benzimidazol-3-ones couples readily with aryldiazonium salts

TABLE 6.72. NITROSATION REACTIONS OF 4*H*-IMIDAZO[1,5-*a*]BENZIMIDAZOLE DERIVATIVES

Reaction conditions[a]	4*H*-Imidazo[1,5-*a*]benzimidazole Substrate	Product	Yield (%)	m.p. (°C)	Solvent of crystallization	Ref.
A	3-Ph-4-Me-	1-NO-3-Ph-4-Me-[b]	83	222–222.5 (decomp.)	Ethanol	215
B	1,4-di-Me-	1,4-di-Me-3-NO-[c]	95	125.5–126.5	Water	192
B	1-Ph-4-Me-	1-Ph-3-NO-4-Me-[d]	75	148–151	Ligroin	192

[a] *A* = $NaNO_2$, AcOH/(5°)(15 min); *B* = $NaNO_2$, AcOH/(room temp.)(15 min).
[b] Dark-green needles.
[c] Colorless needles.
[d] Yellow crystals.

under weakly acidic[235] or basic[132,134,211,231] conditions to give high yields (Table 6.73) of *C*(2)-arylhydrazono derivatives. The spectroscopic properties of these products (see page 152) suggest their existence to a significant extent in the tautomeric azo form [Scheme 6.88; (**6.404**; $A \rightleftharpoons B$)].

4*H*-Pyrazolo[2,3-*a*]benzimidazoles couple at the electron-rich *C*(3) position with a wide range of aryl and hetaryldiazonium salts giving the corresponding azo derivatives usually in high yield. Reactions of this type

TABLE 6.73. DIAZO COUPLING REACTIONS OF 2,3-DIHYDROTHIAZOLO[3,2-*a*]-BENZIMIDAZOL-3-ONES

Reaction conditions[a]	2,3-Dihydrothiazolo[3,2-*a*]benzimidazol-3-one Substrate	Product	Yield (%)	m.p. (°C)	Solvent of crystallization	Ref.
A	Unsubstituted	2-(PhNHN=)-	70	255.5–256 (decomp.)	Methanol	131, 211
B	Unsubstituted	2-(PhNHN=)-	74–80	255 (decomp.)	Acetic acid	231
B	Unsubstituted	2(*p*-MeC_6H_4NHN=)-	74–80	261–263	Acetic acid	231
B	Unsubstituted	2-(*m*-MeC_6H_4NHN=)-	74–80	252	Acetic acid	231
B	Unsubstituted	2-(*p*-ClC_6H_4NHN=)-	74–80	284	Acetic acid	231
A	Unsubstituted	2-(*p*-$O_2NC_6H_4NHN$=)-	73	283–284	1-Propanol	131, 211
A	Unsubstituted	2-(*p*-$H_2NSO_2C_6H_4NHN$=)-	71	242–245	Ethanol	211
A	Unsubstituted	2-(*p*-$HO_2CC_6H_4NHN$=)-	72	307.5–308	Ethanol	211
A	Unsubstituted	2-(*p*-$HSO_3C_6H_4NHN$=)-	66	>330	Water	211
C	6,7-di-Me-	2-(*p*-$MeOC_6H_4NHN$=)-6,7-di-Me-	24	228–229	Dioxane	134
C	6,7-di-Me-	2-(*p*-BrC_6H_4NHN=)-6,7-di-Me-	56	262–263	Dimethylformamide	134

[a] *A* = $Ar\overset{+}{N}_2Cl^-$, NaOAc, AcOH, MeOH, H_2O/(10–15°) (reaction time not specified); *B* = $Ar\overset{+}{N}_2Cl^-$, NaOAc, MeOH, H_2O (temp. and reaction time not specified); *C* = $Ar\overset{+}{N}_2Cl^-$, NaOAc, Ac_2O, AcOH, MeOH/(18–20°)(18–48 hr).

TABLE 6.74. DIAZO COUPLING REACTIONS OF 9*H*-IMIDAZO[1,2-*a*]BENZIMIDAZOLE DERIVATIVES

Reaction	9*H*-Imidazo[1,2-*a*]benzimidazole		Yield	m.p.	Solvent of	
conditions[a]	Substrate	Product	(%)	(°C)	crystallization	Ref.
A	2-Ph-9-Me-	2-Ph-3-(p-$BrC_6H_4N{=}N$)-9-Me-	84	295	Benzene	227
B	2-Ph-9-Me-	2-Ph-3-(p-$HO_2CC_6H_4N{=}N$)-9-Me-	77	306	Dimethylformamide	227
C	2-PhNH-9-Me-[b]	2-PhNH-3-(p-$BrC_6H_4N{=}N$)-9-Me-	67	244	Ethanol	174
C	2-(p-$O_2NC_6H_4NH$)-9-Me-[b]	2-(p-$O_2NC_6H_4NH$)-3-(p-$BrC_6H_4N{=}N$)-9-Me-	92	302	Dimethylformamide	174
C	2-(p-ClC_6H_4NH)-9-Me-[b]	2-(p-ClC_6H_4NH)-3-(p-$BrC_6H_4N{=}N$)-9-Me-	96	193	Benzene	174
C	2-(p-$EtO_2CC_6H_4NH$)-9-Me-[b]	2-(p-$EtO_2CC_6H_4NH$)-3-(p-$BrC_6H_4N{=}N$)-9-Me-	87	245	Dimethylformamide	174
D	2-[PhN(Me)]-9-Me-[b]	2-[PhN(Me)]-3-(p-$BrC_6H_4N{=}N$)-9-Me-	54	195	Ethanol	174

[a] $A = p\text{-}BrC_6H_4\overset{+}{N}_2Cl^-$, AcOH/(100°)(5–10 min); $B = p\text{-}HO_2CC_6H_4\overset{+}{N}_2Cl^-$, ethanol, H_2O/(room temp.)(0.5 hr); $C = p\text{-}BrC_6H_4\overset{+}{N}_2Cl^-$, NaOAc, AcOH/(room temp.)(1 hr); $D = p\text{-}BrC_6H_4\overset{+}{N}_2Cl^-$, NaOAc, H_2O/(room temp.)(1 hr).
[b] Hydrochloride.

TABLE 6.75. DIAZO COUPLING REACTIONS OF 4*H*-IMIDAZO[1,5-*a*]BENZIMIDAZOLE DERIVATIVES

Reaction	4*H*-Imidazo[1,5-*a*]benzimidazole		Yield	m.p.	Solvent of	
contitions[a]	Substrate	Product	(%)	(°C)	crystallization	Ref.
A	3-Ph-4-Me-	1-(p-$BrC_6H_4N{=}N$)-3-Ph-4-Me-	50	200–201	Methanol–water–dimethylformamide	221
A	3-Ph-4-Me-	1-(p-$MeOC_6H_4N{=}N$)-3-Ph-4-Me-	43	186–187	Ethanol	221
B	1-Ph-4-$PhCH_2$-	1-Ph-3-(p-$BrC_6H_4N{=}N$)-4-$PhCH_2$-	46	221–223	Dimethylformamide	193
B	1-Ph-4-$PhCH_2$-	1-Ph-3-(p-$MeOC_6H_4N{=}N$)-4-$PhCH_2$-	30	183–184	—[b]	193

[a] $A = Ar\overset{+}{N}_2Cl^-$, AcOH, MeOH (20°) (12 hr); $B = Ar\overset{+}{N}_2Cl^-$, Ac_2O, AcOH, MeOH/(20°) [7 days (in the dark)].
[b] Solvent of crystallization not specified.

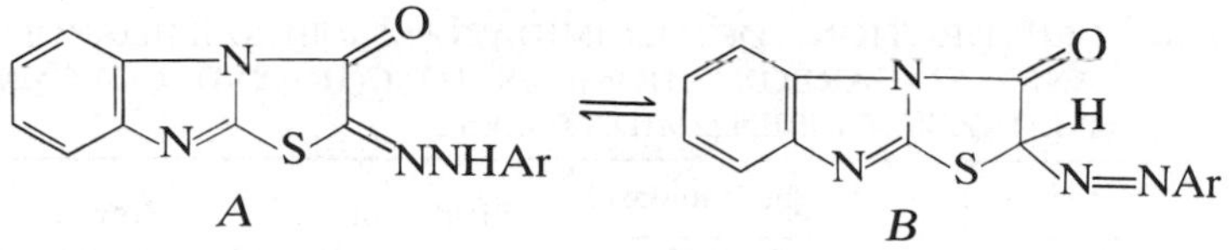

(6.404)

Scheme 6.88

have been principally exploited for the synthesis of azo dyestuffs containing the 4*H*-pyrazolo[2,3-*a*]benzimidazole nucleus.[198,199,205,230,240]

C(2)-Aryl[227] and *C*(2)-amino[174] 9*H*-imidazo[1,2-*a*]benzimidazoles couple with aryldiazonium salts at the *C*(3) position under weakly basic or neutral conditions affording the anticipated arylazo derivatives in good to excellent yield (Table 6.74). The lower tendency of the imidazole ring in 1*H*-imidazo[1,2-*a*]benzimidazoles to undergo electrophilic substitution (see before) accounts for the failure[157] of 1-methyl-2-phenyl-1*H*-imidazo[1,2-*a*]benzimidazole to couple at the *C*(3) position with aryldiazonium salts. The coupling reactions of 4*H*-imidazo[1,5-*a*]benzimidazoles with aryldiazonium salts in weakly acidic or basic solution are apparently less efficient than those of their 9*H*-imidazo[1,2-*a*]benzimidazole counterparts and result in only moderate yields (Table 6.75) of the expected *C*(1)[221] or *C*(3)[193] arylazo products.

Reactions with Nucleophiles

The generally electron-rich character of 6-5-5 fused benzimidazoles with one additional heteroatom is not conducive to nucleophilic substitution. The ring systems in such molecules tend therefore to be inert to nucleophilic reagents or, if subjected to forcing conditions undergo ring scission rather than simple substitution. Attached substituents generally exhibit orthodox behavior toward nucleophilic attack.

DEPROTONATION. The dehydration of readily accessible 4*H*-imidazo[1,5-*a*]benzimidazole *C*(1) and *C*(3) carboxaldoximes using acetic anhydride alone or in the presence of sodium acetate allows the synthesis in high yield (Table 6.76) of the respective nitriles.[193,216]

HYDROXYLATION AND RELATED REACTIONS. The stability of the 2,3-dihydrooxazolo[3,2-*a*]benzimidazole ring system to hydrolysis is implicit in its survival under aqueous alkaline conditions, which serve to convert an attached amide substituent into a carboxyl group.[91] The alkaline hydrolysis of thiazolo[3,2-*a*]benzimidazole-2-carboxylic esters to give the corresponding carboxylic acids in high yield[115] likewise testifies to the inertness of the thiazolo[3,2-*a*]benzimidazole ring system to hydrolytic attack. On the other

TABLE 6.76. DEHYDRATION OF 4*H*-IMIDAZO[1,5-*a*]BENZIMIDAZOLE *C*(1)- AND *C*(3)-CARBOXALDOXIMES TO *C*(1)- AND *C*(3)-CYANO-4*H*-IMIDAZO[1,5-*a*]BENZIMIDAZOLES

Reaction conditions[a]	4*H*-Imidazo[1,5-*a*]benzimidazole		Yield (%)	m.p. (°C)	Solvent of crystallization	Ref.
	Substrate	Product				
A	1-(HON=CH)-3,4-di-Me-	1-CN-3, 4-di-Me-	97	179–179.5	Ethanol	216
A	1-(HON=CH)-3-Ph-4-Me-	1-CN-3-Ph-4-Me-	88	192–194	Acetone	216
B	1-Ph-3-(HON=CH)-4-CH_2Ph-	1-Ph-3-CN-4-CH_2Ph-	91	204.5–206	Acetic acid	193

[a] *A* = Ac_2O, NaOAc/(reflux)/(2.5 hr); *B* = Ac_2O/(reflux)/(3 hr).

hand, the *C*(3) carbonyl substituent in 2,3-dihydrothiazolo[3,2-*a*]benzimidazol-3-ones confers hydrolytic instability, which, under alkaline conditions, leads by cleavage of the *C*(3)–*N*(4) bond to 2-benzimidazolylthioacetic acid derivatives.[129,138,148,241] Conversely, the acidic hydrolysis of 2,3-dihydrothiazolo[3,2-*a*]benzimidazol-3-ones occurs at the *N*(9)–*C*(9a) bond giving good yields of *N*-(2-aminophenyl)thiazolidine-2,4-diones.[146,149] The contrasting stability of 1*H*,3*H*-thiazolo[3,4-*a*]benzimidazol-2-ones to acidic hydrolysis is demonstrated by their formation from 1-imino-1*H*,3*H*-thiazolo[3,4-*a*]benzimidazoles in acid media.[152,153]

Ester[208] and acylamino[200,242] substituents can be hydrolytically removed from 4*H*-pyrazolo[2,3-*a*]benzimidazoles without disruption of the ring system. 9*H*-Imidazo[1,2-*a*]benzimidazoles are likewise moderately stable to acid- and base-catalyzed hydrolysis under conditions that effect the removal of an *N*(1)- quaternary methyl group[165] or promote the loss of ester,[173] acyl,[171,172] and hydroxyalkyl[164] side- chains from the *C*(2) or *C*(3) positions. However, the presence of a *C*(3) nitroso substituent[185,227] or an *N*(1) quaternary center[162a,227] renders the 9*H*-imidazo[1,2-*a*]benzimidazole ring system susceptible to hydrolytic cleavage at the *N*(1)–*C*(2) bond with formation of a 2-iminobenzimidazoline derivative or the 2-benzimidazolone formed by its further hydrolysis. The carbonyl substituent in 2,3-dihydro-9*H*-imidazo[1,2-*a*]benzimidazol-2-ones acts as a focal point for both acid and base-catalyzed hydrolysis, which again leads to 2-iminobenzimidazoline or 2-benzimidazolone formation, respectively.[180] The acid- or base-promoted cleavage of the imidazolinone ring in 2,3-dihydro-9*H*-imidazo-[1,2-*a*]benzimidazol-3-ones, on the other hand, results in the formation of (2-benzimidazolyl)aminoacetic acid derivatives.[6,182] 2,3-Dihydro-9*H*-imidazo[1,2-*a*]benzimidazole-2,3-diones are selectively hydrolyzed at the *C*(3)–*N*(4) bond under alkaline conditions giving the respective 2-oxalylaminobenzimidazoles.[185]

Though stable to alkaline hydrolysis, 4*H*-imidazo[1,5-*a*]benzimidazoles are cleaved at the *C*(1)–*N*(9) bond in acidic media with formation of 2-(2-

aminoalkyl)benzimidazole derivatives.[215] 1*H*,3*H*-Imidazo[3,4-*a*]benzimidazoles are effectively *N*,*N*-acetals of formaldehyde and as such are also hydrolyzed under acidic conditions to 2-(2-aminoalkyl)benzimidazoles.[188] *C*(1)-Cyano groups in 4*H*-imidazo[1,5-*a*]benzimidazoles undergo orthodox triethylamine catalyzed addition of hydrogen sulfide to afford the corresponding thioamides [Scheme 6.89; (**6.405**)] in essentially quantitative yield.[216]

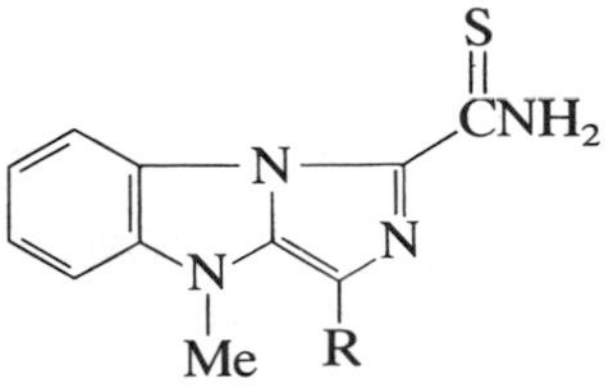

(6.405)

R	Yield (%)	m.p. (°C) (from AcOH)
Me	99	242–244
Ph	quant.	249.5–250.5

Scheme 6.89

Amination. 2,3-Dihydrothiazolo[3,2-*a*]benzimidazol-3-ones exhibit amide-like behavior toward carbonyl reagents such as hydrazine[231] and phenylhydrazine[130,231] undergoing ring-opening to 2-benzimidazolylthioacethydrazides rather than condensation at the carbonyl group. *C*(2)-Acetyl derivatives of thiazolo[3,2-*a*]benzimidazoles, on the other hand, form the usual condensates (oximes, etc.) with carbonyl reagents.[115] Acid chlorides of the thiazolo[3,2-*a*]benzimidazole series are also aminated in standard fashion to carboxamide derivatives.[115]

The 4*H*-pyrazolo[2,3-*a*]benzimidazole ring system is stable to forcing aminolysis under conditions[199] that convert an ester substituent in high yield (Table 6.77) into a primary carboxamide group. 3-Acylamino-4*H*-pyrazolo-[2,3-*a*]benzimidazoles are oxidatively aminated at the *C*(3) position to afford moderate yields (Table 6.77) of 3,3-diamino-3*H*-pyrazolo[2,3-*a*]-benzimidazoles (**6.407**), which are convertible by base-catalyzed elimination of the acylamino substituent into azomethine dyestuffs of the type (**6.408**).[200,242]

C(3)-Bromo substituents in 9*H*-imidazo[1,2-*a*]benzimidazoles are somewhat surprisingly, prone to nucleophilic displacement and react readily with secondary amines to give high yields (Table 6.78) of the corresponding 3-amino-9*H*-imidazo[1,2-*a*]benzimidazole derivatives.[162a] Despite the electron-donating (and hence deactivating) character of the 9*H*-imidazo-[1,2-*a*]benzimidazole ring system, formyl[165] and nitroso[227] substituents at the *C*(3) position participate in orthodox condensation reactions with arylamines affording the expected anils and azo compounds in moderate to high yield (Table 6.78).

TABLE 6.77. AMINATION REACTIONS OF 4*H*-PYRAZOLO[2,3-*a*]-BENZIMIDAZOLE DERIVATIVES

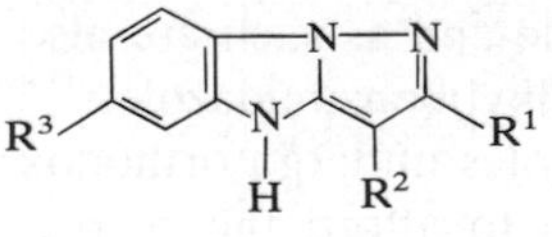

(**6.406**) 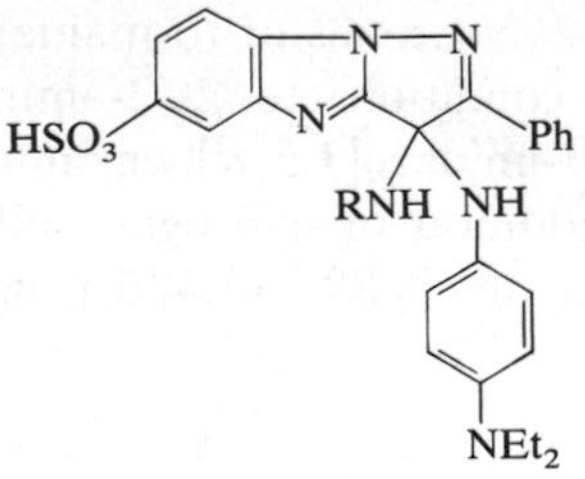 (**6.407**) 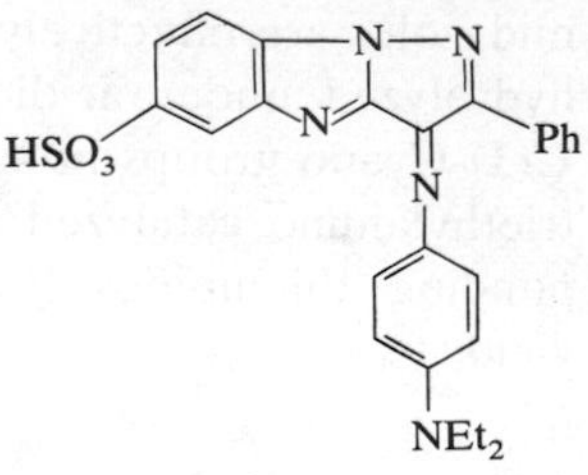 (**6.408**)

Starting material	Reaction conditions[a]	Product	Yield (%)	m.p. (°C)	Ref.
(**6.406**; $R^1 = CO_2Me$, $R^2 = R^3 = H$)	*A*	(**6.406**; $R^1 = CONH_2$, $R^2 = R^3 = H$)	90	272[b]	199
(**6.406**; $R^1 = Ph$, $R^2 = NHCOMe$, $R^3 = SO_3H$)	*B*	(**6.407**; R = COMe)	54	230 (decomp.)	200, 242
(**6.406**; $R^1 = Ph$, $R^2 = NHCOEt$, $R^3 = SO_3H$)	*B*	(**6.407**; R = COEt)	57	220 (decomp.)	200, 242
(**6.407**; R = COMe)	*C*	(**6.408**)	78	>315	200, 242
(**6.407**; R = COEt)	*C*	(**6.408**)	78	—	200, 242

[a] *A* = 25% NH_4OH/(80–85°, autoclave)(8 hr); *B* = *p*-$Et_2NC_6H_4NH_2$, 30% H_2O_2, Na_2CO_3, H_2O/(room temp.)(1 hr); *C* = NaOMe, MeOH/(100°)(few min).
[b] Crystallized from methanol.

TABLE 6.78. AMINATION REACTIONS OF 9*H*-IMIDAZO[1,2-*a*]BENZIMIDAZOLE DERIVATIVES

Reaction conditions[a]	9*H*-Imidazo[1,2-*a*]benzimidazole		Yield (%)	m.p. (°C)	Solvent of crystallization	Ref.
	Substrate	Product				
A	2-Ph-3-Br-9-Me-	2-Ph-3-(1-piperidyl)-9-Me-	100	134–135	Light petroleum	162a
B	2-Ph-3-Br-9-Me-	2-Ph-3-(1-morpholinyl)-9-Me-	90	212–213	Light petroleum	162a
C	2,9-di-Me-3-CHO-	2,9-di-Me-3-(*o*-$HOC_6H_4N{=}CH$)-	83	292 (decomp.)	Dimethyl-formamide	165
C	2-Ph-3-CHO-9-Me-	2-Ph-3-(*o*-HOC_6H_4-N=CH)-9-Me-	38	245	Benzene	165
C	2,9-di-Me-3-CHO-	2,9-di-Me-3-(*p*-$O_2NC_6H_4$-N=CH)-	98	142	Ethanol	165
D	2-Ph-3-CHO-9-Me-	2-Ph-3-(*p*-$O_2NC_6H_4$-N=CH)-9-Me-	76	230	Benzene	165
E	2-Ph-3-NO-9-Me-	2-Ph-3-(*p*-$HO_2CC_6H_4$-N=N)-9-Me-	57	305–306	Dimethyl-formamide	165

[a] *A* = piperidine, dimethylformamide/(reflux)(2 hr); *B* = morpholine, dimethylformamide/(reflux)(2 hr); *C* = $ArNH_2$, EtOH/(reflux)(2–10 hr); *D* = *p*-$O_2NC_6H_4NH_2$/[150° (melt)](5 hr); *E* = *p*-$HO_2CC_6H_4NH_2$, AcOH/(reflux)(2 hr).

TABLE 6.79. REACTIONS OF 2,3-DIHYDROTHIAZOLO[3,2-*a*]BENZIMIDAZOL-3-ONE DERIVATIVES WITH CARBANIONIC REAGENTS[a]

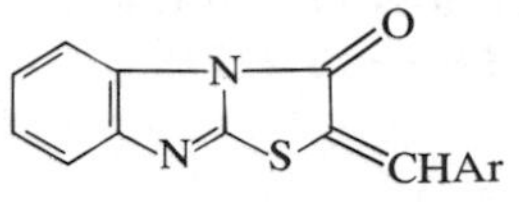

(**6.409**) 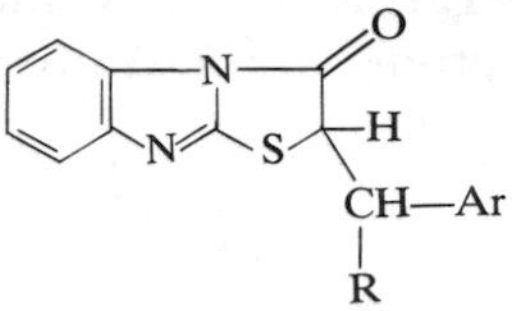 (**6.410**) 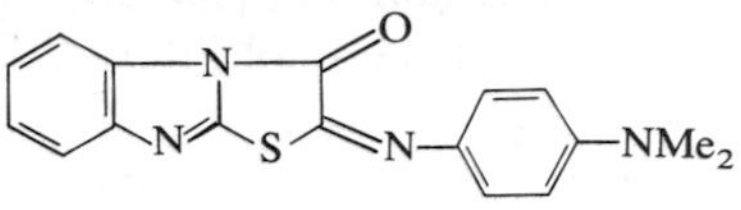 (**6.411**) 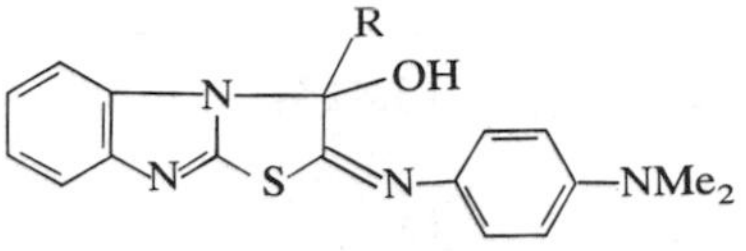 (**6.412**) 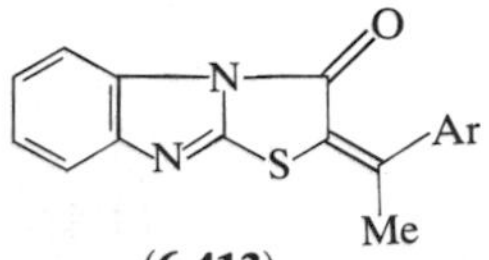 (**6.413**)

Starting material	Reaction conditions[b]	Product	Yield (%)	m.p. (°C)[c]
(**6.409**; Ar = Ph)	*A*	(**6.410**; R = Ar = Ph)	78	165–166
(**6.409**; Ar = Ph)	*A*	(**6.410**; R = *p*-$MeOC_6H_4$, Ar = Ph)	51	134
(**6.409**; Ar = *p*-$MeOC_6H_4$)	*A*	(**6.410**; R = Ph, Ar = *p*-$MeOC_6H_4$)	66	—
(**6.409**; Ar = Ph)	*A*	(**6.410**; R = Et, Ar = Ph)	76	120
(**6.409**; Ar = *p*-ClC_6H_4)	*A*	(**6.410**; R = Ph, Ar = *p*-ClC_6H_4)	79	150
(**6.409**; Ar = *o*-ClC_6H_4)	*A*	(**6.410**; R = Ph, Ar = *o*-ClC_6H_4)	70	160–161
(**6.411**)	*B*	(**6.412**; R = Et)	60	205–207
(**6.411**)	*B*	(**6.412**; R = Ph)	71	202–203
(**6.411**)	*B*	(**6.412**; R = *p*-MeC_6H_4)	65	190–191
(**6.409**; Ar = *o*-$MeOC_6H_4$)	*C*	(**6.413**; Ar = *o*-$MeOC_6H_4$)	80	138–140
(**6.409**; Ar = *o*-$EtOC_6H_4$)	*C*	(**6.413**; Ar = *o*-$EtOC_6H_4$)	70	140–142
(**6.409**; Ar = *o*-ClC_6H_4)	*C*	(**6.413**; Ar = *o*-ClC_6H_4)	74	141

[a] From Ref. 231.
[b] *A* = RMgX, benzene, ether/(reflux)(2 hr); *B* = RMgX, benzene/(reflux)(20 min); *C* = CH_2N_2, ether, chloroform/(room temp.)(12 hr).
[c] Crystallized from ethanol.

Reactions with Anionic Reagents. The mode of reaction (Table 6.79) of 2,3-dihydrothiazolo[3,2-*a*]benzimidazol-3-one derivatives with Grignard reagents is markedly dependent on the nature of the substitution at the *C*(2) position. *C*(2)-Arylidene derivatives (**6.409**) react by addition to the carbon–carbon double bond rather than the carbonyl group giving Michael adducts of the type (**6.410**).[231] In contrast, 2-imino-2,3-dihydrothiazolo[3,2-*a*]benzimidazol-3-ones such as (**6.411**), for which Michael addition is precluded, undergo preferential reaction at the carbonyl group affording good yields (Table 6.79) of the hydroxy products (**6.412**).[231] The reaction[231] of *C*(2)-arylidene 2,3-dihydrothiazolo[3,2-*a*]benzimidazol-3-ones (**6.409**) with diazomethane to give high yields (Table 6.79) of methylated products (**6.413**) can be rationalized by a course (Scheme 6.90) involving initial Michael addition [(**6.414**) → (**6.415**)] followed by nitrogen loss and ketonization of the enol intermediate produced [(**6.415**) → (**6.416**) → (**6.417**)].

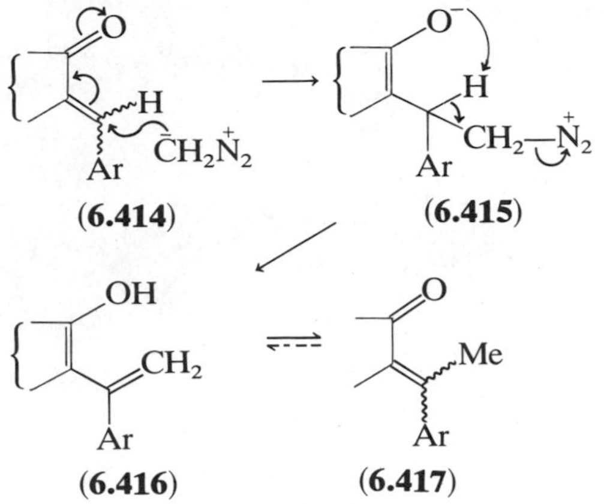

Scheme 6.90

The electrophilic reactivity of attached carbonyl and nitroso substituents appears to be little diminished by the electron-rich and hence potentially deactivating character of the 9*H*-imidazo[1,2-*a*]benzimidazole ring system. *C*(3) Formyl and acetyl substituents react readily[164] with Grignard reagents giving the anticipated carbinols in good to excellent yield (Table 6.80), and participate efficiently (Table 6.80) in aldol-type condensation reactions with active methylene compounds of various types.[165,214] The base-catalyzed condensation of 3-nitroso-9*H*-imidazo[1,2-*a*]benzimidazoles with substrates such as phenylacetonitrile likewise leads to azomethine formation in moderate yield (Table 6.80).[227] Both carbonyl substituents in 9-benzyl-2,3-dihydro-9*H*-imidazo[1,2-*a*]benzimidazole-2,3-dione react with ethylmagnesium bromide giving 9-benzyl-2,3-diethyl-2,3-dihydroxy-2,3-dihydro-9*H*-imidazo[1,2-*a*]benzimidazole in moderate yield (Table 6.80).[185]

Heating 3-bromo-9*H*-imidazo[1,2-*a*]benzimidazoles with sodium or potassium nitrite in dimethylformamide results in replacement of the halogen substituent to give the corresponding 3-nitro-9*H*-imidazo[1,2-*a*]-benzimidazoles in high yield (Table 6.81).[162a,163] 1-Bromo-4-methyl-3-phenyl-4*H*-imidazo[1,5-*a*]benzimidazole is likewise converted in high yield (92%) into 1-nitro-4-methyl-3-phenyl-4*H*-imidazo[1,5-*a*]benzimidazole (m.p. 233–234.5°) by heating with sodium nitrite in dimethyl sulfoxide.[221] The ease of these presumed nucleophilic displacement reactions is surprising when viewed in the light of the electron-rich character of the *C*(3) and *C*(1) positions in the 9*H*-imidazol[1,2-*a*]benzimidazole and 4*H*-imidazo[1,5-*a*]benzimidazole ring systems. The ready replacement of the halogen atom

TABLE 6.80. REACTIONS OF 9*H*-IMIDAZO[1,2-*a*]BENZIMIDAZOLES WITH CARBANIONIC REAGENTS

Reaction conditions[a]	9*H*-Imidazo[1,2-*a*]benzimidazole		Yield (%)	m.p. (°C)	Solvent of crystallization	Ref.
	Substrate	Product				
A	2,9-di-Me-3-CHO-	2,9-di-Me-3-PhCH(OH)-	90	199	Ethanol	164
A	2,9-di-Me-3-CHO-	2,9-di-Me-3-*p*-$Me_2NC_6H_4CH(OH)$-	75	193	Ethanol	164
A	2-Ph-3-CHO-9-Me-	2-Ph-3-*p*-$EtOC_6H_4CH(OH)$-9-Me-[b]	75	174–175	Ethanol	164
A	2-*p*-BrC_6H_4-3-CHO-9-Me-	2-*p*-BrC_6H_4-3-PhCH(OH-9-Me-[c]	70	183–184	Ethanol	164
A	2-α-naphthyl-3-CHO-9-Me-	2-α-naphthyl-3-PhCH(OH)-9-Me-[d]	90	197–199	Ethanol	164
A	2-Ph-3-CHO-9-CH_2Ph-	2-Ph-3-PhCH(OH)-9-CH_2Ph[e]	63	171–172	Ethanol	164
A	2,9-di-Me-3-CHO-	2,9-di-Me-3-PhC≡CCH(OH)-	75	197–198 (decomp.)	Benzene–light petroleum	164
B	2,9-di-Me-3-CHO-	2,9-di-Me-3-PhC≡CCH(OH)-	57	—	—	164
B	2-Ph-3-CHO-9-Me-	2-Ph-3-PhC≡CCH(OH)-9-Me-	71	192–193 (decomp.)	Ethanol	164
B	2-Ph-3-CHO-9-Et-[f]	2-Ph-3-PhC≡CCH(OH)-9-Et-[f]	68	208–209 (decomp.)	Ethanol	164
A	2,9-di-Me-3-COMe-	2,9-di-Me-3-PhC(OH)Me-	70	160–161	Ethanol	164
A	2-Ph-3-COMe-9-Me-	2-Ph-3-PhC(OH)Me-9-Me-[g]	85	156–157	Ethanol	164
A	2-*p*-BrC_6H_4-3-COMe-9-Me-	2-*p*-BrC_6H_4-3-PhC(OH)Me-9-Me-	63	137–138	Ethanol	164
B	2,9-di-Me-3-COMe-	2,9-di-Me-3-PhC≡CC(OH)Me-	55	122–123	Ethanol	164
A	2,3-dihydro-9-CH_2Ph-2,3-dione	2,3-dihydro-2,3-di-Et-2,3-di-OH-9-CH_2Ph-	52	112	Chloroform–light petroleum	185
C	2,9-di-Me-3-CHO-	2,9-di-Me-3-PhCOCH=CH-	80	204–205 (decomp.)	Ethanol	214
C	2,9-di-Me-3-CHO-	2,9-di-Me-3-*p*-$MeOC_6H_4COCH$=CH-	88	216–217 (decomp.)	Ethanol	214
C	2-Ph-3-CHO-9-Me-	2-Ph-3-PhCOCH=CH-9-Me-	85	168–169 (decomp.)	Dimethyl-formamide	214
C	2-Ph-3-CHO-9-Me-	2-Ph-3-*p*-$MeOC_6H_4COCH$=CH-9-Me-	89	178–179 (decomp.)	Ethanol–dimethyl-formamide	214
C	2-Ph-3-CHO-9-Me-	2-Ph-3-*m*-$O_2NC_6H_4COCH$=CH-9-Me-	34	272–273 (decomp.)	Dimethyl formamide	214

TABLE 6.80 (*Continued*)

Reaction conditions[a]	9*H*-Imidazo[1,2-*a*]benzimidazole		Yield (%)	m.p. (°C)	Solvent of crystallization	Ref.
	Substrate	Product				
C	2-Ph-3-CHO-9-Me-	2-Ph-3-(α-naphthoyl)CH═CH-9-Me-	84	164–165	Ethanol	214
C	2-Ph-3-CHO-9-Me-	2-Ph-3-(2-furoyl)CH═CH-9-Me-	98	221–222	Dimethyl formamide	214
D	2-Ph-3-CHO-9-Me-	2-Ph-3-O_2NCH═CH-9-Me-	47	189–189.5	Ethanol	165
E	2-Ph-3-NO-9-Me-	2-Ph-3-PhC(CN)═N-9-Me-[h]	50	183–184 (decomp.)	Dioxane–ether	227

[a] *A* = RMgX, tetrahydrofurane/(reflux)(3 hr); *B* = PhC≡CMgBr, ether/(reflux)(4 hr); *C* = ArCOMe, 40% NaOH aq., EtOH/(warm), then (room temp.)(few hr); *D* = $MeNO_2$, NH_4OAc/(reflux)(40 hr); *E* = $PhCH_2CN$, 5% NaOH aq., EtOH/(reflux)(20–30 min).

[b] Forms a hydrochloride, m.p. 286–287° (decomp.) (from ethanol).

[c] Forms a hydrochloride, m.p. 300° (decomp.) (from ethanol).

[d] Forms a hydrochloride, m.p. 290–293° (decomp.) (from ethanol).

[e] Forms a hydrochloride, m.p. 276–277° (decomp.) (from ethanol).

[f] 6,7-Dimethyl derivative.

[g] Forms a hydrochloride, m.p. 285° (decomp.) (from ethanol).

[h] Monohydrate.

TABLE 6.81. NITRODEBROMINATION REACTIONS OF 9*H*-IMIDAZO[1,2,*a*]-BENZIMIDAZOLE DERIVATIVES

Reaction conditions[a]	9*H*-Imidazo[1,2-*a*]benzimidazole Substrate	Product	Yield (%)	m.p. (°C)	Solvent of crystallization	Ref.
A	2-Ph-3-Br-9-Me-	2-Ph-3-NO_2-9-Me-	80	205	Acetone–ethanol	162a
B	2-*p*-BrC_6H_4-3-Br-9-Me-	2-*p*-BrC_6H_4-3-NO_2-9-Me-	65	271–272	Dimethyl-formamide	163
B	2-*p*-$O_2NC_6H_4$-3-Br-9-Me-	2-*p*-$O_2NC_6H_4$-3-NO_2-9-Me-	75	237	Ethanol–dimethyl-formamide	163
B	2-Ph-3-Br-9-Et-[b]	2-Ph-3-NO_2-9-Et-[b]	95	234–235 (decomp.)	Dimethyl formamide	163

[a] *A* = $NaNO_2$, dimethylformamide/(reflux)(1 hr); *B* = KNO_2, dimethylformamide/(reflux)(reaction time not specified).
[b] 6,7-Dimethyl derivative.

in the 3-bromo-4*H*-imidazo[1,5-*a*]benzimidazole by nitrite ion becomes even more intriguing when compared with its inertness to reagents such as ethanolic sodium hydroxide or ethanolic sodium ethoxide.[221] Further information on the mechanism of the nitrodebrominations undergone by 3-bromo-9*H*-imidazo[1,2-*a*]benzimidazoles and 1-bromo-4*H*-imidazo[1,5-*a*]benzimidazoles would therefore be welcome.

Oxidation

The stability of the 2,3-dihydrothiazolo[3,2-*a*]benzimidazole ring system to oxidation by chromium trioxide–pyridine is implicit in the use of this reagent for the oxidation of 3-hydroxy-2,3-dihydrothiazolo[3,2-*a*]benzimidazoles to 2,3-dihydrothiazolo[3,2-*a*]benzimidazol-3-ones.[94,103] The attempted peracid oxidation of 2,3-dihydrothiazolo[3,2-*a*]benzimidazol-3-ones to the corresponding sulfones results in scission to 2-benzimidazolone derivatives.[231]

Permanganate oxidation of 9*H*-imidazo[1,2-*a*]benzimidazoles and their 2,3-dihydro derivatives leads either to complete disruption[166] of the molecule or to the formation of azobenzimidazole derivatives.[165,180] The stability of 9*H*-imidazo[1,2-*a*]benzimidazoles to manganese dioxide oxidation, on the other hand, provides the means for their synthesis[157] from 2,3-dihydro-9*H*-imidazo[1,2-*a*]benzimidazoles and for the oxidation of propargyl alcohols of the 9*H*-imidazo[1,2-*a*]benzimidazole series to the respective ketones.[164] 2-Phenyl-2,3-dihydro-1*H*-imidazo[1,2-*a*]benzimidazole is reported[166] to be stable to dehydrogenation with reagents such as chloranil or *N*-bromosuccinimide. The attempted selenium dioxide oxidation of 2-methyl-9*H*-imidazo[1,2-*a*]benzimidazoles to the corresponding aldehydes leads to complex mixtures of products.[165] 4*H*-Imidazo[1,5-*a*]benzimidazoles are notorious for their ease of oxidation by atmospheric oxygen to unidentified blue substances.[192,224]

Reduction

The reduction[103,115] of 2-acylthiazolo[3,2-*a*]benzimidazoles to the corresponding carbinols in high yield (Table 6.82) using sodium borohydride, demonstrates the stability of the thiazolo[3,2-*a*]benzimidazole ring system to reducing agents of this type. The 2,3-dihydrothiazolo[3,2-*a*]benzimidazole ring system is likewise stable to reduction[128] by hydrazine in the presence of Raney nickel under conditions (Table 6.82) that generate an amino group from a *C*(7)-nitro substituent. The reductive scission[235] of 2-phenylhydrazono-2,3-dihydrothiazolo[3,2-*a*]benzimidazol-3-one to 2-amino-2,3-dihydrothiazolo[3,2-*a*]benzimidazol-3-one in moderate yield (Table 6.82) represents a potentially general method for the synthesis of such amino derivatives.

The catalytic reduction of readily accessible 3-nitroso-4*H*-pyrazolo[2,3-*a*]benzimidazoles provides a general method[199,200] for the synthesis in moderate to high yield (Table 6.83) of 3-amino-4*H*-pyrazolo[2,3-*a*]-benzimidazoles.

Metal-proton donor reagents such as zinc and acetic acid, and tin or stannous chloride in combination with hydrochloric acid reduce 3-nitroso- and 3-nitro-9*H*-imidazo[1,2-*a*]benzimidazoles to the corresponding amines.[175,243] These tend to be unstable relative to their open-chain nitrile tautomers but in some cases (Table 6.84) can be isolated in the free form or as hydrochlorides. 9-Benzyl-2-methyl-9*H*-imidazo[1,2-*a*]benzimidazole is debenzylated by sodium in liquid ammonia to afford 2-methyl-9*H*-imidazo-[1,2-*a*]benzimidazole in quantitative yield (Table 6.84).[166] In the analogous

TABLE 6.82. REDUCTION OF THIAZOLO[3,2-*a*]BENZIMIDAZOLE DERIVATIVES

Reaction conditions[a]	Thiazolo[3,2-*a*]benzimidazole		Yield (%)	m.p. (°C)	Solvent of crystallization	Ref.
	Substrate	Product				
A	2-COMe-	2-MeCH(OH)-	85	116–118	Benzene	103
A	2-COMe-3-Me-	2-MeCH(OH)-3-Me-	87	227–228	Dimethylformamide	115
A	$2\text{-}CO_2Et$-3-Me-	$2\text{-}CH_2OH$-3-Me-	86	194–195	Ethanol–water	103
B	$7\text{-}NO_2$-2,3-dihydro-	$7\text{-}NH_2$-2,3-dihydro-	57	229–230	Water	128
C	2-(PhNHN=)-2,3-dihydro-3-one	$2\text{-}NH_2$-2,3-dihydro-3-one	57	185	Ethanol	235

[a] *A* = $NaBH_4$, EtOH/(reflux)(2 hr); *B* = Raney-Ni, 80% $N_2H_4{\cdot}H_2O$, EtOH/(reflux)(30 min); *C* = $Na_2S_2O_4$, EtOH, H_2O/(reflux)(30 min).

TABLE 6.83. REDUCTION[a] OF 3-NITROSO-4*H*-PYRAZOLO[2,3-*a*]-BENZIMIDAZOLE DERIVATIVES

3-Nitroso-4*H*-pyrazolo[2,3-*a*]benzimidazole		Yield	m.p.	Solvent of	
Substrate	Product	(%)	(°C)	crystallization	Ref.
2-Me-	2-Me-3-NH_2-	54	>175 (decomp.)	Water	199
2-Ph-	2-Ph-3-NH_2-	67	—[b]	—[c]	199
2-CO_2H-	2-CO_2H-3-NH_2-	64	—[b]	—[c]	199
2-$CONH_2$-	2-$CONH_2$-3-NH_2-[d]	52	220 (decomp.)	Methanol	199
2-Ph-6-SO_3H-	2-Ph-3-NH_2-6-SO_3H-	89	260 (decomp.)	—[c]	200
2,6-di-CO_2H-	2,6-di-CO_2H-3-NH_2-	69	>260 (decomp.)	2-Propanol	199
2-CO_2H-6-SO_3H-	2-CO_2H-3-NH_2-6-SO_3H-	62	280 (decomp.)	—[c]	199

[a] H_2, Raney-Ni, 25% NH_4OH aq., MeOH or Pr^iOH/50°, 50 atm.
[b] Melting point not quoted.
[c] Solvent of crystallization not specified.
[d] Sulfate.

sodium and liquid ammonia reduction of 9-benzyl-2-phenyl-9*H*-imidazo-[1,2-*a*]benzimidazole[166] debenzylation is accompanied by reduction of the *C*(2)–*C*(3) double bond giving a mixture of 2-phenyl-9*H*-imidazo[1,2-*a*]-benzimidazole and its 2,3-dihydro derivative in good overall yield (Table 6.84). The reduction of both carbonyl substituents in 2,3-dihydro-9*H*-imidazo[1,2-*a*]benzimidazole-2,3-diones by lithium aluminum hydride provides a method for the synthesis of 2,3-diols of the 2,3-dihydro-9*H*-imidazo[1,2-*a*]benzimidazole series.[185] *C*(7)-Nitro substituted 2,3-dihydro-1*H*-imidazo[1,2-*a*]benzimidazoles are converted[176] into the corresponding amines (Table 6.84) by orthodox catalytic hydrogenation or by reduction with hydrazine in the presence of Raney nickel.

6.2.4. Practical Applications

Biological Properties

2,3-Dihydrooxazolo[3,2-*a*]benzimidazole derivatives are reported[92] to enhance the antibiotic activity of penicillins and cephalosporins and are also therapeutic agents[91] for certain disorders of the central nervous system.

Thiazolo[3,2-*a*]benzimidazole derivatives exhibit antibacterial activity[101] and their quaternary salts have been patented[244] as hypoglycaemic agents. 2,3-Dihydrothiazolo[3,2-*a*]benzimidazole derivatives are associated with a particularly wide range of biological properties including antitumor,[102,245]

TABLE 6.84. REDUCTION OF IMIDAZO[1,2-*a*]BENZIMIDAZOLE DERIVATIVES

Reaction conditions[a]	Imidazo[1,2-*a*]benzimidazole Substrate	Product	Yield (%)	m.p. (°C)	Solvent of crystallization	Ref.
A	2,9-di-Me-3-NO_2-9*H*-	2,9-di-Me-3-NH_2-9*H*-[b]	62	192.5	Benzene	175
B	2-Ph-3-NO-9-CH_2Ph-9*H*-	2-Ph-3-NH_2-9-CH_2Ph-9*H*-[c,d]	—[e]	217 (decomp.)	Ethanol–water	175
B	2-*p*-BrC_6H_4-3-NO-9-Me-9*H*-	2-*p*-BrC_6H_4-3-NH_2-9-Me-9*H*-[c,f]	—[e]	211 (decomp.)	Ethanol–water	175
C	2-Me-9-CH_2Ph-9*H*-	2-Me-9*H*-	quant.	194 (decomp.)	Benzene	166
C	2-Ph-9-CH_2Ph-9*H*-	2-Ph-9*H*- +	9	310	Dimethylformamide	166
		2 Ph-2,3-dihydro-1*H*-	90	221	Ethanol	
D	1-PhCH(Me)-7-NO_2-2,3-dihydro-1*H*-	1-PhCH(Me)-7-NH_2-2,3-dihydro-1*H*-	80	123–125	Chloroform–ligroin	176
E	1-$CHPh_2$-7-NO_2-2,3-dihydro-1*H*-	1-$CHPh_2$-7-NH_2-2,3-dihydro-1*H*-	34	228–230	Ligroin	176
F	9-Me-2,3-dihydro-9*H*-2,3-dione	2,3-di-OH-9-Me-2,3-dihydro-9*H*-	60	235–236	Ethanol	185
F	9-CH_2Ph-2,3-dihydro-9*H*-2,3-dione	2,3-di-OH-9-CH_2Ph-2,3-dihydro-9*H*-	70	184	Ethanol	185

[a] *A* = $SnCl_2$, HCl, EtOH/(reaction temp. and time not specified); *B* = Zn, AcOH/(room temp.)(2–2.5 hr); *C* = Na, NH_3 liq./10–15 min; *D* = H_2, Raney/Ni EtOH/55 psi; *E* = 100% N_2H_4, H_2O, Raney/Ni, EtOH, dimethylformamide/(100°)(1 hr); *F* = $LiAlH_4$, ether/(reflux)(4 hr).

[b] Forms a dihydrochloride, m.p. 297° (from ethanol–ether).

[c] Hydrochloride.

[d] Forms a picrate, m.p. 188° (decomp.) (from acetic acid), and a monoacetyl derivative, m.p. 212–213° (from ethanol–water).

[e] Yield not quoted.

[f] Forms a picrate, m.p. 280–281.5° (from acetic acid).

antiviral,[246] and antitubercular[109] activity, as well as having a depressant effect on the central nervous system.[109] 2,3-Dihydrothiazolo[3,2-*a*]-benzimidazol-3-one derivatives are also central nervous system depressants[141] and exhibit marked anticonvulsant,[136,139,235] antispasmodic,[139,149] and hypotensive[149] activity. In addition, 2,3-dihydrothiazolo[3,2-*a*]-benzimidazol-3-one derivatives are reported[133] to be effective antibacterial agents. The fungicidal,[99,101,133,247,248] herbicidal,[115b] insecticidal,[248] and plant-growth-regulating[144] properties of thiazolo[3,2-*a*]benzimidazoles and their 2,3-dihydro derivatives have also been reported. 1*H*,3*H*-Thiazolo[3,4-*a*]benzimidazole derivatives have been patented[152] as parasiticides and rodentocides.

4*H*-Pyrazolo[2,3-*a*]benzimidazole derivatives have attracted attention[236,237] as central nervous system stimulants. Depending on the nature of the substituents, 9*H*-imidazo[1,2-*a*]benzimidazoles can exert either a stimulating or a depressant effect on the central nervous system,[249] and consequently have useful sedative and analgetic properties.[155,160] Certain 1*H*- and 9*H*-imidazo[1,2-*a*]benzimidazole derivatives are associated with marked cardiovascular, and in particular hypotensive activity.[162b,164,177,214,250] 9*H*-Imidazo[1,2-*a*]benzimidazoles have also been patented as antiviral agents.[156] Derivatives of the 4*H*-imidazo[1,5-*a*]benzimidazole ring system are of interest because of their antitubercular activity.[216]

Dyestuffs

The thiazolo[3,2-*a*]benzimidazole ring system has been employed as a chromophoric unit in cyanine dyes,[135] 2,3-dihydrothiazolo[3,2-*a*]benzimidazol-3-ones[129,251] having found particular application in this respect. The utility of 2,3-dihydrothiazolo[3,2-*a*]benzimidazol-3-one derivatives as photographic sensitizing agents is emphasized in a number of publications.[228,232,252]

The 4*H*-pyrazolo[2,3-*a*]benzimidazole ring system has been used extensively as a chromophore in azomethine,[242,253] azo,[230,240] and metal complex[254] dyes. 4*H*-Pyrazolo[2,3-*a*]benzimidazole derivatives are also the subject of patents relating to photographic developers,[199,255] and color coupling[198,204,205,256] and photosensitizing[219] agents.

6.3. Tricyclic 6-5-5 Fused Benzimidazoles with Two Additional Heteroatoms

6-5-5 Fused benzimidazole structures incorporating two additional heteroatoms (Scheme 6.91) and Table 6.85) include the unique, silicon-containing, 2-sila-3*H*-thiazolo[3,2-*a*]benzimidazole (**6.418**) and 3-sila-2*H*-imidazo[1,2-*a*]benzimidazole (**6.419**) frameworks, and the sulfur-containing

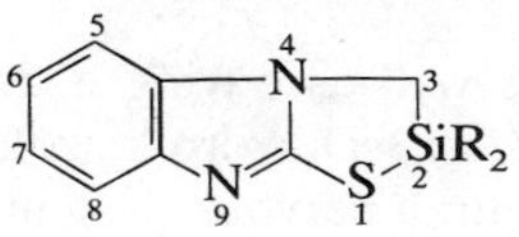

(**6.418**) 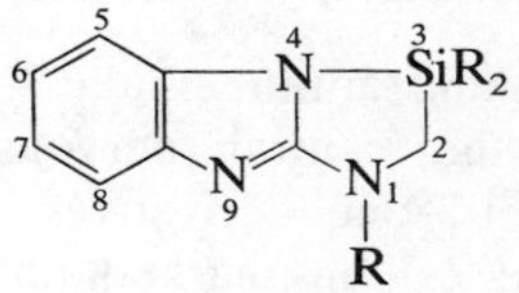 (**6.419**)

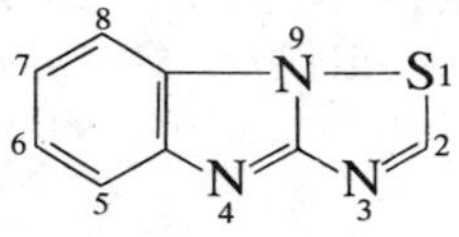

(**6.420**) 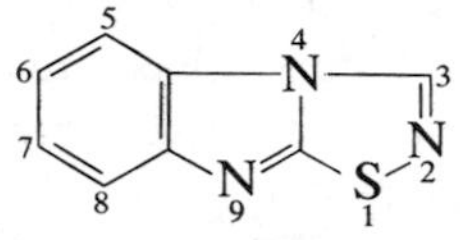 (**6.421**)

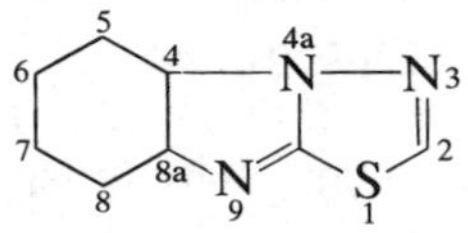

(**6.422**) 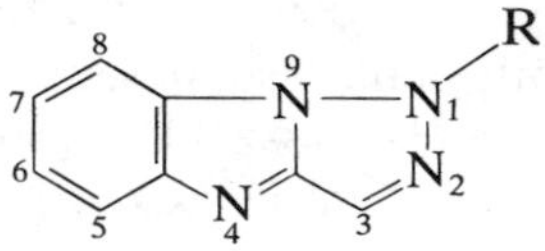 (**6.423**)

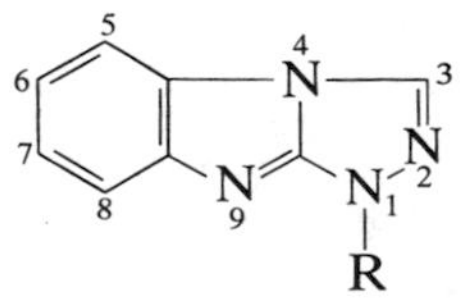

(**6.424**) 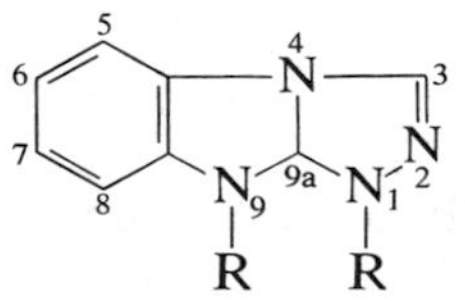 (**6.425**)

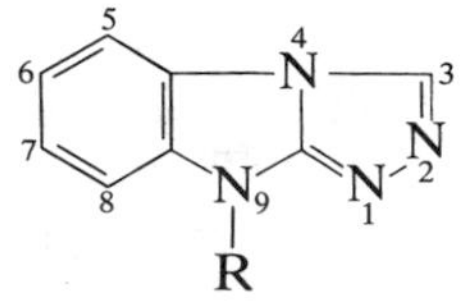

(**6.426**) 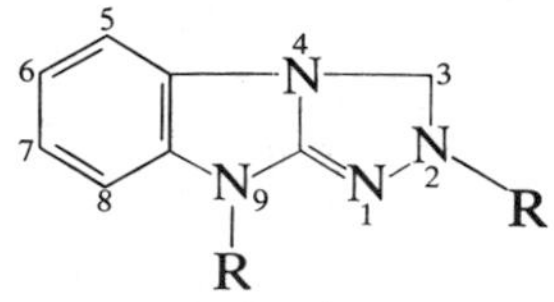 (**6.427**)

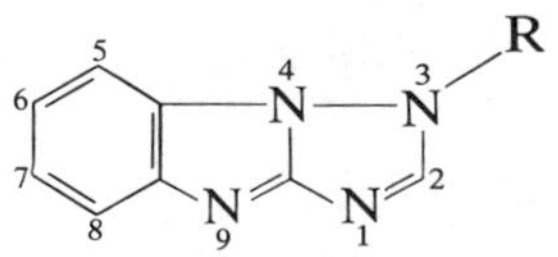

(**6.428**)

Scheme 6.91

TABLE 6.85. TRICYCLIC 6-5-5 FUSED BENZIMIDAZOLE RING SYSTEMS WITH TWO ADDITIONAL HETEROATOMS

Structure[a]	Name[b]
(**6.418**)	2-sila-3*H*-thiazolo[3,2-*a*]benzimidazole
(**6.419**)	3-sila-2*H*-imidazo[1,2-*a*]benzimidazole
(**6.420**)	1,2,4-thiadiazolo[2,3-*a*]benzimidazole
(**6.421**)	1,2,4-thiadiazolo[4,5-*a*]benzimidazole
(**6.422**)	4*a*,5,6,7,8,8*a*-hexahydro-1,3,4-thiadiazolo[3,2-*a*]benzimidazole
(**6.423**)	1*H*-1,2,3-triazolo[1,5-*a*]benzimidazole
(**6.424**)	1*H*-1,2,4-triazolo[4,3-*a*]benzimidazole
(**6.425**)	9,9*a*-dihydro-1*H*-1,2,4-triazolo[4,3-*a*]benzimidazole
(**6.426**)	9*H*-1,2,4-triazolo[4,3-*a*]benzimidazole
(**6.427**)	2,3-dihydro-9*H*-1,2,4-triazolo[4,3-*a*]benzimidazole
(**6.428**)	1*H*-1,2,4-triazolo[2,3-*a*]benzimidazole

[a] Cf. Scheme 6.91.
[b] Based on the Ring Index.

1,2,4-thiadiazolo[2,3-*a*]benzimidazole (**6.420**), 1,2,4-thiadiazolo[4,5-*a*]-benzimidazole (**6.421**), and 1,3,4-thiadiazolo[3,2-*a*]benzimidazole ring systems, the latter being known only in the 4*a*,5,6,7,8,8*a*-hexahydro form (**6.422**). Fully nitrogen-containing 6-5-5 fused benzimidazoles having two additional heteroatoms are represented by the 1*H*-1,2,3-triazolo[1,5-*a*]-benzimidazole (**6.423**), 1*H*- (**6.424**) and 9*H*- (**6.426**) 1,2,4-triazolo[4,3-*a*]-benzimidazole ring systems and their 9,9*a*- (**6.425**) and 2,3- (**6.427**) dihydro derivatives, and by the 1*H*-1,2,4-triazolo[2,3-*a*]benzimidazole ring system (**6.428**).

6.3.1. Synthesis

Ring-Closure Reactions of Benzimidazole Derivatives

Derivatives of the 2-sila-3*H*-thiazolo[3,2-*a*]benzimidazole ring system are accessible[257] (Scheme 6.92) by the base-catalyzed cyclization of condensates (**6.431**) derived by the reaction of 2-benzimidazolethiones (**6.429**) with bromomethyldimethylchlorosilane (**6.430**). Ring-formation of the type [(**6.431**) → (**6.432**)] is akin to that of 2,3-dihydrothiazolo[3,2-*a*]benzimidazoles from 2-(β-halogenoalkylthio)benzimidazoles (see page 101) and is effected in good yield (Table 6.86) using 1,8-bis(dimethylamino) naphthalene as the basic catalyst. As in the closely related 2-(β-halogenoalkylthio)benzimidazole cyclizations (see page 101) the factors governing the direction of ring-closure in substrates (**6.431**) unsymmetrically substituted in the benzene ring are as yet unknown.[257] The uncatalyzed condensation (Scheme 6.92) of bromomethyldimethylchlorosilane (**6.430**) with 2-aminobenzimidazoles (**6.433**) occurs at *N*(1) rather than at the amino group to

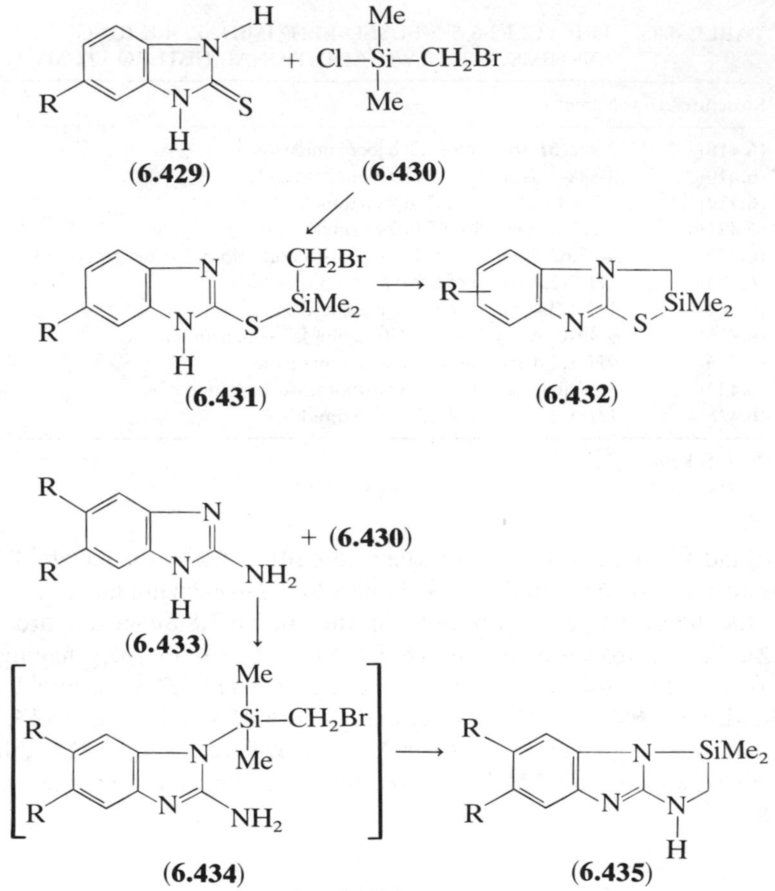

H Me

+ Cl—Si—CH_2Br

R N S Me

H

(**6.429**) (**6.430**)

N CH_2Br

R S $SiMe_2$ ⟶ R N $SiMe_2$

N N S

H

(**6.431**) (**6.432**)

R N

+ (**6.430**)

R N NH_2

H

(**6.433**)

Me

R Si—CH_2Br R

N Me N— $SiMe_2$ ⟶

R N NH_2 R N N

H

(**6.434**) (**6.435**)

Scheme 6.92

TABLE 6.86. SYNTHESIS OF 2-SILA-3*H*-THIAZOLO[3,2-*a*]BENZIMIDAZOLES AND 3-SILA-2*H*-IMIDAZO[1,2-*a*]-BENZIMIDAZOLES BY RING-CLOSURE REACTIONS OF BENZIMIDAZOLE DERIVATIVES

Starting materials	Reaction conditions[a]	Product	Yield (%)	m.p. (°C)	Ref.
(**6.431**; R = H)	*A*	(**6.432**; R = H)	>80	181–182	257
(**6.431**; R = NO_2)	*A*	(**6.432**; R = NO_2)[b]	65	165 (decomp.)	257
(**6.433**; R = H)	*B*	(**6.435**; R = H)	48	62–64	258
(**6.433**; R = Me)	*B*	(**6.435**; R = Me)	70	123–125	258
(**6.433**; R = Cl)	*B*	(**6.435**; R = Cl)	82	107–109	258
(**6.433**; R = NO_2)	*B*	(**6.435**; R = NO_2)	68	192–195	258

[a] *A* = 1,8-bisdimethylaminonaphthalene, tetrahydrofurane/(room temp.)(24 hr); *B* = tetrahydrofurane/(room temp.)(2–4 days).
[b] Position of the nitro substituent not established.

afford via presumed silylated intermediates (**6.434**), moderate to good yields (Table 6.86) of 3-sila-2*H*-imidazo[1,2-*a*]benzimidazole derivatives (**6.435**).[258] The structures of these compounds follow from their mode of ring-opening on reaction with acetic anhydride (see later). The uncatalyzed character of 3-sila-2*H*-imidazo[1,2-*a*]benzimidazole formation compared with the base-catalyzed nature of 2-sila-3*H*-thiazolo[3,2-*a*]benzimidazole synthesis reflects the difference in basicity of the nitrogen centers involved in these related ring-closure reactions.[258]

Readily accessible 2-thioacylaminobenzimidazoles are oxidatively cyclized, mainly in high yield (Table 6.87), to 1,2,4-thiadiazolo[2,3-*a*]benzimidazole derivatives [Scheme 6.93; (**6.436**) → (**6.437**)] by treatment[259] with bromine or *m*-chloroperbenzoic acid in chloroform solution. Reactions of this type have been principally exploited for the synthesis of 2-hetaryl-substituted 1,2,4-thiadiazolo[2,3-*a*]benzimidazoles (**6.437**) (Table 6.87).[259]

TABLE 6.87. SYNTHESIS OF 1,2,4-THIADIAZOLO[2,3-*a*]BENZIMIDAZOLE DERIVATIVES (**6.437**) BY OXIDATIVE RING-CLOSURE OF 2-THIOACYLAMINOBENZIMIDAZOLES (**6.436**)[a]

Starting material (**6.436**)	Reaction conditions[b]	Product (**6.437**)	Yield (%)	m.p. (°C)[c]
Ph	*A*	Ph	60	235–236
p-MeC_6H_4	*A*	p-MeC_6H_4	50	251.5–253
p-$Bu^tC_6H_4$	*A*	p-$Bu^tC_6H_4$	90	217.5–218
p-$F_3CC_6H_4$	*A*	p-$F_3CC_6H_4$	70	249–250
p-ClC_6H_4	*A*	p-ClC_6H_4	60	230–232
2-Naphthyl	*A*	2-Naphthyl	85	258–259
2-Furyl	*A*	2-Furyl	50	203–204.5
2-Thienyl	*A*	2-Thienyl	70	229–230
3-Thienyl	*A*	3-Thienyl	20	218–218.5
2-Thiazolyl	*A*	2-Thiazolyl	60	244–244.5
4-Thiazolyl	*A*	4-Thiazolyl	90	252.5–254.5
5-Thiazolyl	*A*	5-Thiazolyl	90	227 (decomp.)
2-Pyridyl	*B*	2-Pyridyl	50	258–260
3-Pyridyl	*A*	3-Pyridyl	85	258
2-Pyrazinyl	*B*	2-Pyrazinyl	—[d]	245–246.5
3-Isothiazolyl	*A*	3-Isothiazolyl	95	—[e]
4-Isothiazolyl	*A*	4-Isothiazolyl	80	225.5–226 (decomp.)
5-Isothiazolyl	*A*	5-Isothiazolyl	75	245–245.5
1,2,3-Thiadiazol-4-yl	*A*	1,2,3-Thiadiazol-4-yl	80	256–257 (decomp.)
1,2,5-Thiadiazol-3-yl	*A*	1,2,5-Thiadiazol-3-yl	60	226.5–229.5

[a] From Ref. 259.
[b] $A = Br_2$, $CHCl_3$/(room temp.)(3 hr); B = *m*-chloroperbenzoic acid, $CHCl_3$ (room temp.)(15 hr).
[c] Solvent of crystallization not specified.
[d] Yield not quoted.
[e] Melting point not quoted.

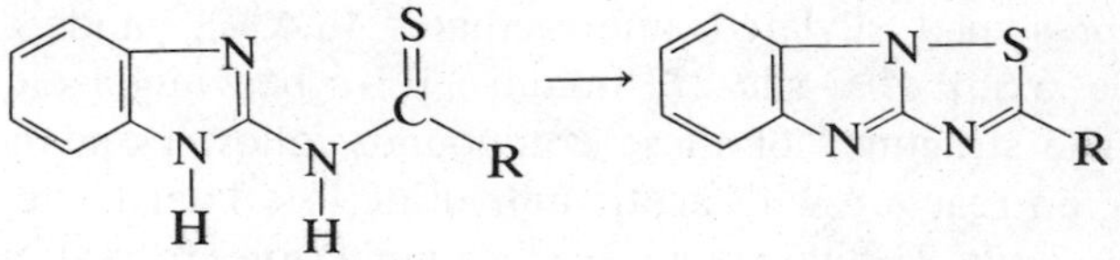

(6.436) (6.437)

Scheme 6.93

The orthodox thermal condensation (Scheme 6.94) of benzimidazole-2-sulfonamide (**6.438**) with ethyl orthoformate provides an efficient method for the synthesis[260] of the 1,2,4-thiadiazolo[4,5-*a*]benzimidazole sulfone (**6.439**). Unfortunately, attempts to generalize this straightforward synthetic procedure were unsuccessful. A more general if less orthodox synthetic approach (Scheme 6.94) to 1,2,4-thiadiazolo[4,5-*a*]benzimidazole derivatives (**6.442**) is provided by the uncatalyzed condensation of 2-thiocyanatobenzimidazoles (**6.440**) with imidazole derivatives (**6.441**).[261] Despite the unanticipated[261a] course followed in these reactions, the structures of the

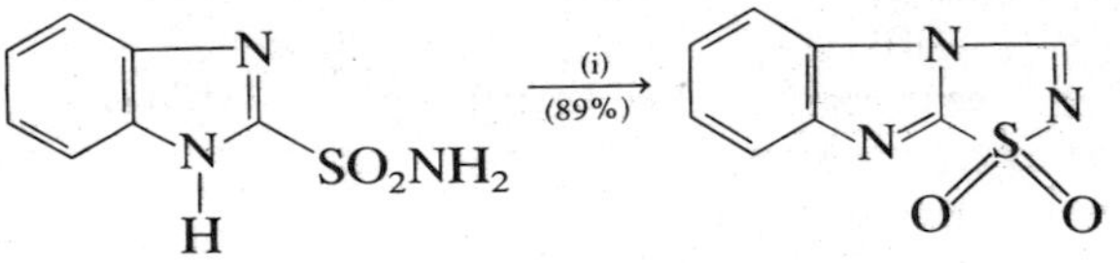

(6.438) (6.439) (m.p. 332–335°)

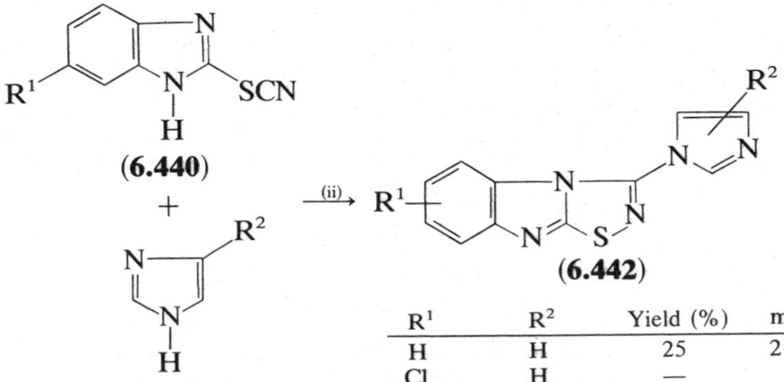

(6.440) + (6.441) → (6.442)

R^1	R^2	Yield (%)	m.p. (°C)
H	H	25	213–214
Cl	H	—	—
NO_2	H	—	—
COPh	H	—	—
H	Ph	—	—
H	$p\text{-}ClC_6H_4$	—	—
H	$p\text{-}O_2NC_6H_4$	—	—
Et	Et	—	—

(i) $(EtO)_3CH$/(reflux)(30 min)
(ii) acetone or acetonitrile/(room temp.)(14 hr)

Scheme 6.94

products (**6.442**) are established beyond doubt by X-ray analysis[261a] of the parent compound (**6.442**; $R^1 = R^2 = H$). Transformations of the type [(**6.440**) + (**6.441**) → (**6.442**)] suffer from the disadvantage[261] of giving rise to isomer mixtures when unsymmetrically substituted thiocyanatobenzimidazoles or imidazoles are used as reactants but merit further study in terms of their possible application to the general synthesis of 1,2,4-thiadiazolo-[4,5-*a*]benzimidazoles containing substituents other than imidazolyl. Further study of such reactions in a mechanistic sense is warranted by the lack of information on the nature of the formal S → N shift and oxidation required to account for 1,2,4-thiadiazolo[4,5-*a*]benzimidazole formation.

What would appear to be the only reported example of 1,2,3-triazolo[1,5-*a*]benzimidazole formation[262] is represented (Scheme 6.95) by the *N*-bromosuccinimide mediated ring-closure of the hydrazone (**6.443**) to the hydrobromide (**6.444**), which affords the parent base (**6.445**) on treatment with aqueous pyridine. Syntheses of the isosteric 1*H*- and 9*H*-1,2,4-triazolo-[4,3-*a*]benzimidazole ring systems in contrast are well documented and are invariably effected by ring-closure of 2-benzimidazolylhydrazine derivatives (Scheme 6.96 and Table 6.88).[263–268] 1*H*-1,2,4-Triazolo[4,3-*a*]benzimidazole formation is illustrated by the acid-catalyzed cyclization of the hydrazone (**6.447**) [readily accessible by the reaction of the 2-benzimidazolylhydrazine (**6.446**) with ethyl orthoacetate] in moderate yield (Table 6.88) to 1,3-dimethyl-1*H*-1,2,4-triazolo[4,3-*a*]benzimidazole

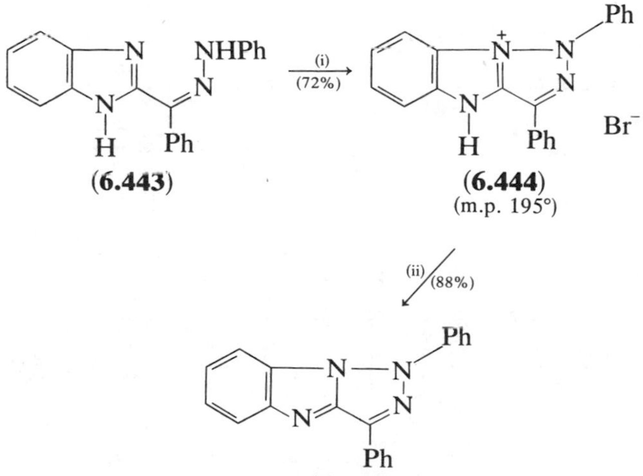

(**6.443**) —(i) (72%)→ (**6.444**) (m.p. 195°) —(ii) (88%)→ (**6.445**) (m.p. 148°)

(i) N-bromosuccinimide, EtOAc/room temp. (reaction time not specified)
(ii) pyridine-water

Scheme 6.95

TABLE 6.88. SYNTHESIS OF 1*H*- AND 9*H*-1,2,4-TRIAZOLO[4,3-*a*]BENZIMIDAZOLES BY RING-CLOSURE OF 2-BENZIMIDAZOLYLHYDRAZINE DERIVATIVES

Starting material	Reaction conditions[a]	Product	Yield (%)	m.p. (°C)	Solvent of crystallization	Ref.
(**6.447**)	*A*	(**6.448**)	40	73–74	—[b]	263
(**6.446**; $R = R^1 = H$)	*B*	(**6.450**; $R = H$, $R^1 = Me$)	93	231–232	Butanol	264
(**6.446**; $R = R^1 = H$)	*B*	(**6.450**; $R = H$, $R^1 = Me$)	84	—	—	265
(**6.446**; $R = R^1 = H$)	*B*	(**6.450**; $R = H$, $R^1 = Me$)	92	233	—[b]	263
(**6.446**; $R = R^1 = H$)	*C*	(**6.450**; $R = H$, $R^1 = Et$)	79	260	Butanol	264, 265
(**6.446**; $R = Me$, $R^1 = H$)	*B*	(**6.450**; $R = R^1 = Me$)	42	177	—[b]	263
(**6.446**; $R = R^1 = H$)	*D*	(**6.449**; $R = H$)	—[c]	284 (decomp.)	Ethanol–water	266
(**6.446**; $R = R^1 = H$)	*E*	(**6.449**; $R = H$)	59	275 (decomp.)	—[d]	265
(**6.446**; $R = R^1 = H$)	*F*	(**6.449**; $R = H$)	95	282–283 (decomp.)	Ethanol–water	267
(**6.446**; $R = Me$, $R^1 = H$)	*F*	(**6.449**; $R = Me$)	quant.	263–265	Ethanol–water	267
(**6.446**; $R = CH_2Ph$, $R^1 = H$)	*F*	(**6.449**; $R = CH_2Ph$)	quant.	216–218	Ethanol–water	267
(**6.446**; $R = Ph$, $R^1 = H$)	*G*	(**6.449**; $R = Ph$)	68	237–241	Dimethylformamide–water	268

[a] *A* = toluene-*p*-sulfonic acid, xylene/(reflux)(4 hr); *B* = $(EtO)_3CMe$, xylene/(reflux)(4–5 hr); *C* = $(EtO)_3CEt$, xylene/(reflux)(3–4 hr); *D* = CS_2, KOH, EtOH/heat until H_2S evolution ceases; *E* = PhN=C=S, trichlorobenzene/(reflux)(2.5 hr); *F* = CS_2, dioxane or pyridine/(reflux)(1–2 hr); *G* = CS_2, pyridine/(reflux)(4 hr).

[b] Purified by sublimation.

[c] Yield not quoted.

[d] Purified by precipitation from sodium hydroxide solution with acetic acid.

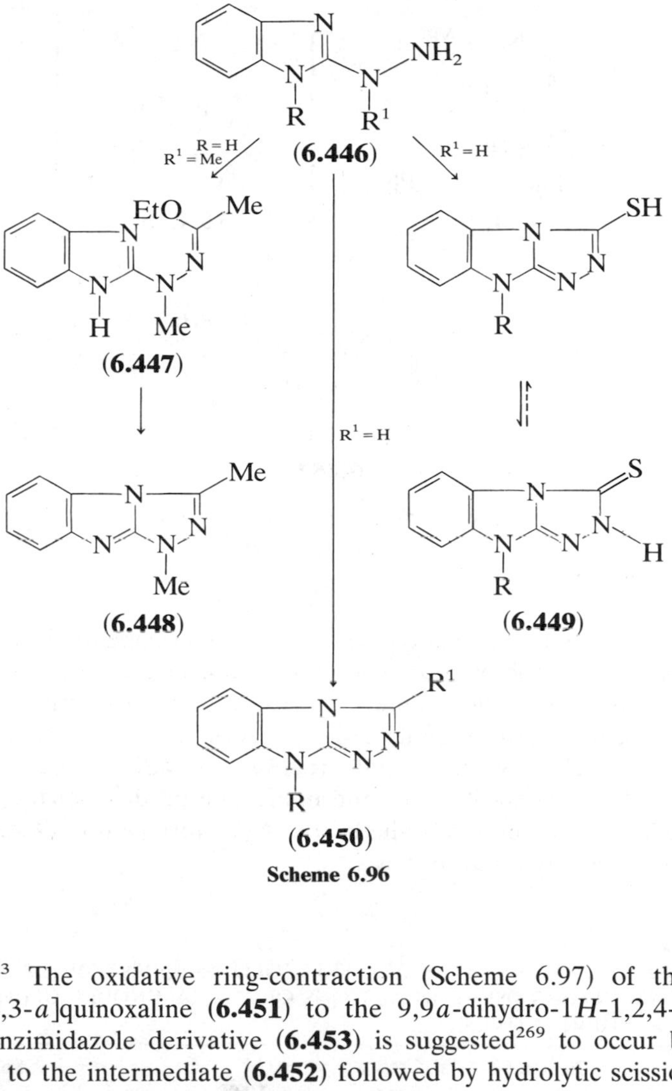

(**6.446**)

(**6.447**)

(**6.448**)

(**6.449**)

(**6.450**)

Scheme 6.96

(**6.448**).[263] The oxidative ring-contraction (Scheme 6.97) of the 1,2,4-triazolo[4,3-*a*]quinoxaline (**6.451**) to the 9,9*a*-dihydro-1*H*-1,2,4-triazolo-[4,3-*a*]benzimidazole derivative (**6.453**) is suggested[269] to occur by initial oxidation to the intermediate (**6.452**) followed by hydrolytic scission at the *C*(4)–*N*(5) bond and then recyclization at *C*(1). 9*H*-1,2,4-Triazolo[4,3-*a*]-benzimidazole derivatives are readily synthesized in moderate to high yield (Table 6.88) by the direct thermal ring-closure of 2-benzimidazolylhydrazine (**6.446**; $R = R^1 = H$) and its ring *N*-alkyl derivatives with ortho esters [Scheme 6.96; (**6.446**; R = H or alkyl, $R^1 = H$) → (**6.450**; R = H or alkyl, R^1 = alkyl)].[263–265] Correspondingly, reaction of 2-benzimidazolylhydrazine and its ring *N*-alkyl and aryl derivatives with carbon disulfide in the presence of potassium hydroxide[266] or pyridine,[267,268] or with phenyl isothiocyanate[265] results in the formation (Table 6.88) of tautomeric 2,3-dihydro-9*H*-1,2,4-triazolo[4,3-*a*]benzimidazole-3-thiones [Scheme 6.96; (**6.446**; R = H, alkyl, or aryl, $R^1 = H$) → (**6.449**; R = H, alkyl, or aryl)]. The

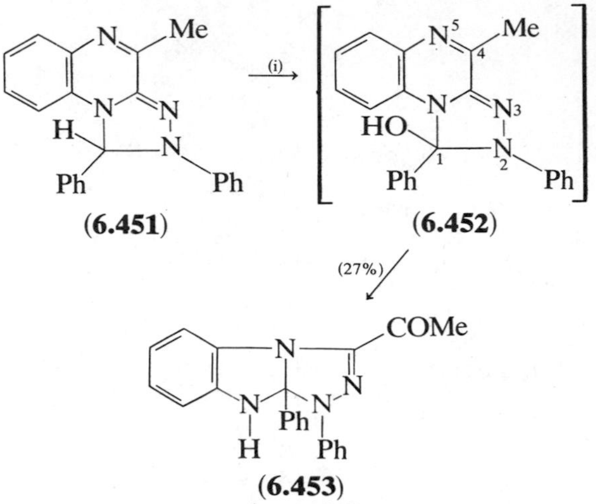

(**6.451**) (i) → [(**6.452**)] (27%) → (**6.453**)

(m.p. 210°)

(i) O_2, light petroleum/(reflux)(8 hr)

Scheme 6.97

acylative cyclization of readily synthesized *N*-acylamino 2-aminobenzimidazoles, using acetic or propionic anhydrides, or benzoyl chloride, provides a general, high yield (Table 6.89) route to *C*(2)-substituted *N*-acyl-1*H*-1,2,4-triazolo[2,3-*a*]benzimidazoles, deacylation of which affords the parent heterocycles [Scheme 6.98; (**6.454**) → (**6.455**) → (**6.456**)].[270] The position of the acyl substituent in the immediate products of these cyclization reactions was not established, the *N*(1) formulation (**6.455**) being assigned on a purely arbitrary basis.[270]

TABLE 6.89. SYNTHESIS OF 1*H*-1,2,4-TRIAZOLO[2,3-*a*]BENZIMIDAZOLES BY RING-CLOSURE REACTIONS OF 1,2-DIAMINOBENZIMIDAZOLE DERIVATIVES[a]

Starting material[b]	Reaction conditions[c]	Product	Yield (%)	m.p. (°C)	Solvent of crystallization
(**6.454**; R = Me)	*A*	(**6.455**; R = Me)	71	154–155	Acetonitrile
(**6.455**; R = Me)	*B*	(**6.456**; R = Me)	80	258–259	Acetonitrile
(**6.454**; R = Et)	*A*	(**6.455**; R = Et)	71	111–113	—[d]
(**6.455**; R = Et)	*B*	(**6.456**; R = Me)	66	198–200	Ethyl acetate
(**6.454**; R = Ph)	*C*	(**6.455**; R = Ph)	83	230–232	Acetonitrile
(**6.455**; R = Ph)	*D*	(**6.456**; R = Ph)	60	310–315	Ethanol

[a] From Ref. 270.

[b] Hydrobromides.

[c] *A* = $(RCO)_2O$/(reflux)(5 hr); *B* = conc. HCl/(reflux)(2 hr); *C* = PhCOCl/(reflux)(5 hr); *D* = 10% NaOH aq./(reflux)(2 hr).

[d] Solvent of crystallization not specified.

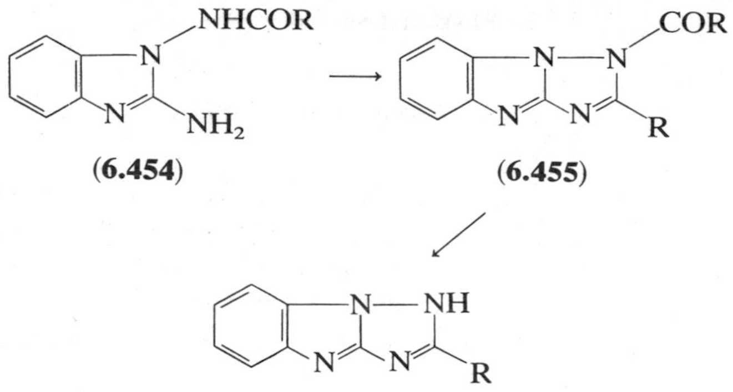

(6.454) (6.455)

(6.456)

Scheme 6.98

Ring-closure Reactions of Other Heterocycles

Ring-closure reactions of heterocyclic substrates other than benzimidazole derivatives to 6-5-5 fused benzimidazole frameworks having two additional heteroatoms are rare. Perhaps the best documented[271] process in this category is the brominative cyclization (Scheme 6.99) of 2-(2-cyclohexenylamino)-1,3,4-thiadiazole derivatives (**6.457**) to give the hydrobromides of 4*a*,5,6,7,8,8*a*-hexahydro-1,3,4-thiadiazolo[3,2-*a*]benzimidazoles (**6.458**) in good yield. Ring-closure of this type represents the only method for the construction of the 1,3,4-thiadiazolo[3,2-*a*]-benzimidazole ring system reported to date.

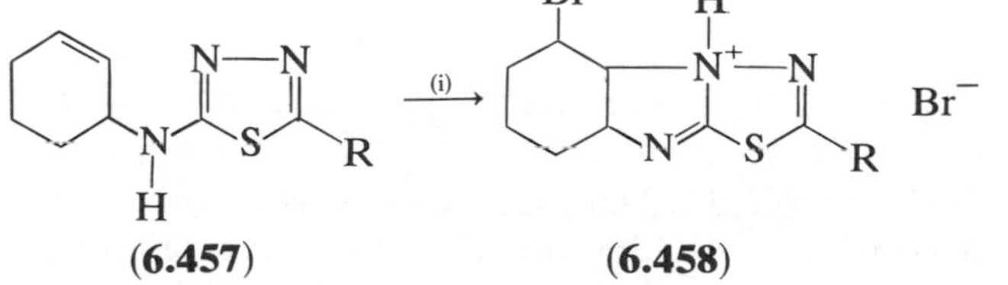

(6.457) (6.458)

R	Yield (%)	m.p. (°C)
Me	69	159–160
$CH_2CH_2CO_2H$	78	194–195
CH_2Ph	79	124–125
Ph	60	181–182

(i) Br_2, $CHCl_3$/(O°)(few min)

Scheme 6.99

6.3.2. Physicochemical Properties

Spectroscopic Studies

INFRARED SPECTRA. The infrared spectra (Table 6.90)[258] of 3-sila-2*H*-imidazo[1,2-*a*]benzimidazole derivatives (**6.459**) contain NH absorption at 3400–2800 cm^{-1} and characteristic bands in the ranges 1435–1430 and 1261–1254 cm^{-1} attributable to the asymmetric and symmetric Me–Si stretching modes, respectively. The *NH*-substituent in *N*(9)-unsubstituted

TABLE 6.90. INFRARED SPECTRA[a] OF 3-SILA-2*H*-IMIDAZO[1,2-*a*]-BENZIMIDAZOLE DERIVATIVES (**6.459**)[b]

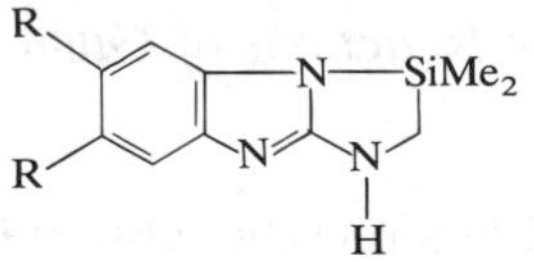

R
R
N—SiMe₂
N
N
H

(**6.459**)

(**6.459**) R	ν_{max} (cm^{-1}) NH	SiMe[c]	SiMe[d]
H	3400–2800	1430	1261
Me	3400–2800	1435	1258
Cl	3400–2800	1431	1254
NO_2	3400–2800	1432	1257

[a] Measured for solid dispersions in KBr.
[b] From Ref. 258.
[c] Asymmetric stretching vibration.
[d] Symmetric stretching vibration.

9*H*-1,2,4-triazolo[4,3-*a*]benzimidazoles gives rise to broad IR absorption (Table 6.91) in the range 3400–2600 cm^{-1}.[263,272] The acyl substituent in *N*-acyl-1*H*-1,2,4-triazolo[2,3-*a*]benzimidazoles is associated with uniform IR carbonyl absorption at 1700 cm^{-1}.[270] 9*H*-1,2,4-Triazolo[4,3-*a*]benzimidazoles give rise to IR C═N absorption (Table 6.91) at lower frequencies (1650–1620 cm^{-1}) than that (1660 cm^{-1}) (Table 6.91) of 1*H*-1,2,4-triazolo-[4,3-*a*]benzimidazoles, thus allowing the differentiation of the two structural types. The presence of NH absorption at ca. 3100 cm^{-1} and C═S absorption at 1517–1488 cm^{-1} in the IR spectra (Table 6.91) of 2,3-dihydro-9*H*-1,2,4-triazolo[4,3-*a*]benzimidazole-3-thiones is consistent with the preferential existence of these molecules in the thione tautomeric form (**6.462**) both in the solid state and in solution.

TABLE 6.91. INFRARED SPECTRA OF 1*H*- AND 9*H*-1,2,4-TRIAZOLO[4,3-*a*]BENZIMIDAZOLE DERIVATIVES (**6.460**), (**6.461**), AND (**6.462**)

(**6.460**) (**6.461**) (**6.462**)

Compound	R^1	R^2	R^3	R^4	Medium	$\nu_{max}(cm^{-1})$ NH	C=N	C=S	Others	Ref.
(**6.460**)	—	—	—	—	KBr	—	1660	—	—	263
(**6.461**)	Me	H	H	H	KBr	3060–2650	1630	—	1620	263
(**6.461**)	Me	Me	H	H	KBr	—	1630	—	—	263
(**6.461**)	Me	H	H	NO_2	KBr	3100	1620	—	1520, 1340	11
(**6.461**)	Me	H	NO_2	NO_2	KBr	3100	1650	—	1540, 1340	11
(**6.461**)	Me	H	H	Br	KBr	3160–2600	1625	—	—	11
(**6.461**)	Me	H	Br	Br	KBr	3089	1640	—	—	272
(**6.461**)	Me	H	H	NH_2	KBr	3400, 3210	1620[a]	—	—	272
(**6.461**)	Me	H	H	D	KBr	3060–2630	—	—	—	272
(**6.461**)	MeS	Me	H	H	Nujol	—	1651	—	—	267
(**6.462**)	H	H	—	—	CS_2, Nujol	3283[b]	1650[c]	1494[c]	—	267
(**6.462**)	Me	H	—	—	CS_2, Nujol	3078[b]	1652[c]	1517, 1494[c]		267
(**6.462**)	Me	CH_2N	—	—	Nujol	—	1652	1512	—	267
(**6.462**)	CH_2N	CH_2N	—	—	Nujol	—	1649	1488	—	267

[a] May be reassigned to NH def.
[b] Measured for a solution in CS_2.
[c] Measured for a suspension in Nujol.

TABLE 6.92. ULTRAVIOLET SPECTRA OF 1*H*- AND 9*H*-1,2,4-TRIAZOLO[4,3-*a*]-BENZIMIDAZOLE DERIVATIVES (**6.463**), (**6.464**), (**6.465**), AND (**6.466**).

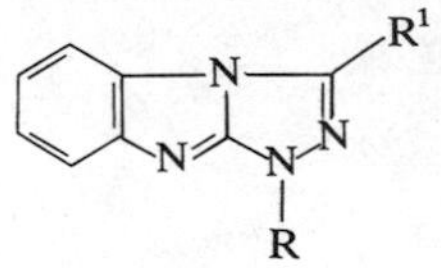

(**6.463**) 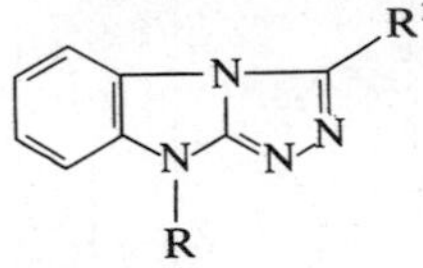 (**6.464**)

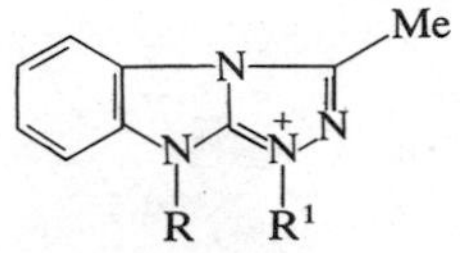

(**6.465**) 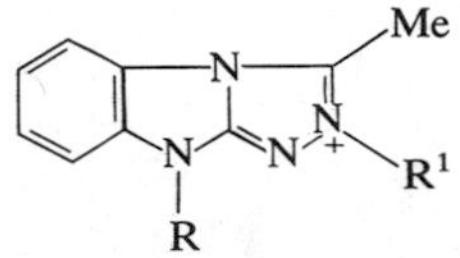 (**6.466**)

Compound	R	R^1	Solvent	λ_{max}(nm)(log ε)	Ref.
(**6.463**)	H	Me	Dioxane	222(4.49), 232 sh (4.30), 237 sh (4.24), 257(3.91), 263 sh (3.87), 293(3.67), 301 sh (3.62)	226
(**6.464**)	H	Me	EtOH	215(4.55), 230 sh (4.10), 240 sh (3.72), 289(3.70), 294(3.70)	263
(**6.464**)	H	Me	MeOH	214(4.78), 232(4.29), 288(3.69), 293(3.70)	265
(**6.464**)	H	Me	H_2O	212(4.56), 286(3.52), 290(3.52)	263
(**6.464**)	H	Me	$CHCl_3$	291(3.49), 295(3.49)	263
(**6.462**)⇌ (**6,464**)	H	Me	EtOH-HCl	275(3.60), 281.5(3.68), 292.5(3.46), 297.5(3.45)	273
(**6.463**)⇌ (**6.464**)	H	Me	CF_3CO_2H	269(3.62), 276(3.67), 287(3.46), 295 sh (3.40)	273
(**6.464**)	H	Et	MeOH	214(4.79), 232(4.30), 288(3.74), 294(3.75)	265
(**6.463**)	Me	Me	Dioxane	225(4.38), 236(4.20), 241(4.18), 261(3.95), 266(3.97), 287 sh (3.41), 296(3.54), 306(3.49)	226
(**6.463**)	Me	Me	EtOH	216 sh (4.45), 220(4.45), 233 sh (3.93), 237 sh (3.91), 253(3.72), 260(3.71), 291(3.30), 300(3.28)	263
(**6.463**)	Me	Me	H_2O	214 sh (4.43), 219(4.44), 234 sh (3.90), 250(3.59), 256 sh (3.58), 289(3.20), 296(3.20)	263
(**6.463**)	Me	Me	$CHCl_3$	295(3.30), 303(3.26)	263
(**6.463**)	Me	Me	EtOH-HCl	267(3.33), 275(3.54), 282.5(3.57), 292 sh (2.63)	273
(**6.463**)	Me	Me	CF_3CO_2H	264 sh (3.33), 272(3.50), 278(3.54), 287 sh (2.67)	273
(**6.464**)	Me	Me	Dioxane	224(4.31), 234(4.10), 251 sh (3.80), 300(3.84)	226
(**6.464**)	Me	Me	EtOH	216(4.48), 231 sh (3.80), 245 sh (3.52), 291(3.49), 297(3.50)	263
(**6.464**)	Me	Me	H_2O	215(4.49), 288(3.32), 296(3.32)	263
(**6.464**)	Me	Me	EtOH-HCl	276(3.43), 283(3.55), 297(3.48)	273

TABLE 6.92 (*Continued*)

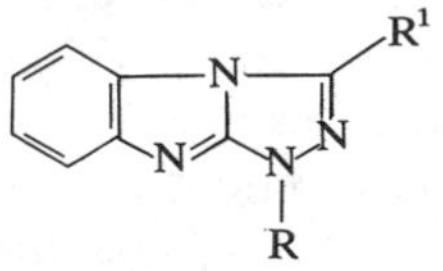

(**6.463**) 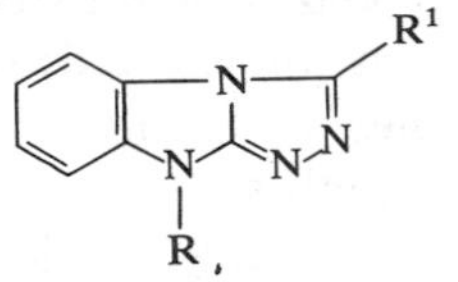 (**6.464**)

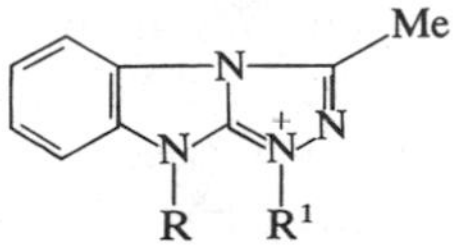

(**6.465**) 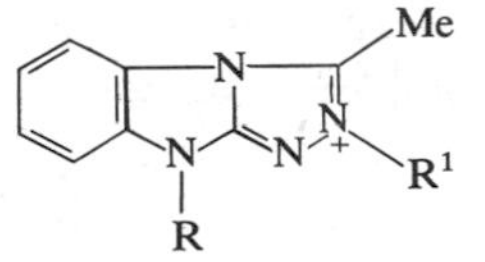 (**6.466**)

Compound	R	R^1	Solvent	λ_{max}(nm)(log ε)	Ref.
(**6.464**)	Me	Me	CF_3CO_2H	272.5(3.45), 279(3.55), 295(3.34)	273
(**6.465**)	Me	Me	EtOH	263 sh (3.42), 277(3.65), 284(3.67), 297 sh (2.72)	273
(**6.465**)	Me	Me	CF_3CO_2H	265 sh (3.53), 274(3.68), 281(3.69), 303 sh (2.74)	273
(**6.465**)	Me	Et	EtOH	269 sh (3.43), 277(3.62), 284(3.66), 296 sh (2.51)	273
(**6.465**)	Me	Et	CF_3CO_2H	267 sh (3.49), 274(3.63), 282(3.68), 300 sh (2.18)	273
(**6.465**)	Et	Me	EtOH	265 sh (3.35), 277(3.64), 284(3.68), 300 sh (2.61)	273
(**6.465**)	Et	Me	CF_3CO_2H	267 sh (3.49), 274(3.65), 282(3.69)	273
(**6.466**)	Me	Me	EtOH	293(3.63), 300(3.62)	273
(**6.466**)	Me	Me	CF_3CO_2H	294(3.53), 302(3.50)	273
(**6.466**)	Me	Et	EtOH	297(3.65), 301(3.64)	273
(**6.466**)	Me	Et	CF_3CO_2H	295(3.59), 303(3.63)	273
(**6.466**)	Et	Me	EtOH	295(3.64), 302(3.65)	273
(**6.466**)	Et	Me	CF_3CO_2H	295(3.55), 304(3.52)	273

ULTRAVIOLET SPECTRA. The UV spectra (Table 6.92) of 1(9)*H*-1,2,4-triazolo[4,3-*a*]benzimidazoles and their *N*(1) and *N*(9) alkyl derivatives are typified by a series of intense absorption bands indicative of the delocalized aromatic character of these molecules. In accord with their more extensively conjugated structures, *N*(9)-alkyl 9*H*-1,2,4-triazolo[4,3-*a*]benzimidazoles tend to absorb more intensely at longer wavelength than their *N*(1)-alkyl 1*H*-1,2,4-triazolo[4,3-*a*]benzimidazole counterparts. This distinguishing feature has been utilized[226,263] for the study of the tautomeric equilibrium in 3-methyl-1(9)*H*-1,2,4-triazolo[4,3-*a*]benzimidazole, the position of which is found[226,263] to be markedly sensitive to solvent effects, nonpolar media (e.g., dioxane) favoring the *N*(1)*H* tautomer, and polar media (e.g., ethanol, water) the *N*(9)*H* isomer. Protonation[273] of 3-methyl-1(9)*H*-1,2,4-triazolo-[4,3-*a*]benzimidazole and its *N*(1) and *N*(9) methyl derivatives has a dramatic, though not unexpected effect (Table 6.92) on the UV absorption of these molecules, the bands at shorter wavelength disappearing and those at longer wavelength suffering a marked reduction in intensity. Alkylation of the *N*(1) position in *N*(9)-alkyl-9*H*-1,2,4-triazolo[4,3-*a*]benzimidazoles has a similar effect (Table 6.92) which is heightened by quaternization of these molecules at the *N*(2) position. The site of *N*-alkylation in 9-alkyl-9*H*-1,2,4-triazolo[4,3-*a*]benzimidazoles can therefore be established unambiguously using UV spectroscopy.[273] Comparison of the UV spectra (Table 6.92) of model *N*(1) and *N*(2) alkyl 3,9-dimethyl-9*H*-1,2,4-triazolo[4,3-*a*]benzimidazolium cations with the UV spectra of 3-methyl-1(9)*H*-1,2,4-triazolo-[4,3-*a*]benzimidazole and its *N*(1) and *N*(9) methyl derivatives in acidic media (Table 6.92) likewise allows the site of protonation in these molecules to be determined (see later).[273]

Nuclear Magnetic Resonance Spectra. The protons of the dimethylsilyl substituent in 3-sila-2*H*-imidazo[1,2-*a*]benzimidazoles resonate as a well-defined six proton singlet at δ 0.15–0.19 (Table 6.93), while those of the *C*(2) methylene group give rise to uniform singlet absorption in the range δ 2.44–2.49.[258] The enhanced deshielding[260] of the *C*(3) proton in 1,2,4-thiadiazolo[4,5-*a*]benzimidazole sulfone [cf. page 224, Scheme 6.94; (**6.439**)], which absorbs as a singlet at δ 9.05, is consistent with the combined effects of delocalization in the 1,2,4-thiadiazolo[4,5-*a*]benzimidazole ring system and electron-withdrawal by the sulfone substituent.

The successful analysis[274] of the ^{1}H NMR spectra (Table 6.94) of 3-methyl-1(9)*H*-1,2,4-triazolo[4,3-*a*]benzimidazole and its *N*(1) and *N*(9)

TABLE 6.93. ^{1}H NMR SPECTRA[a,b] OF 3-SILA-2*H*-IMIDAZO[1,2-*a*]BENZIMIDAZOLE DERIVATIVES (**6.465**)[c]

R, R, N, SiMe$_2$, N, N, H

(**6.459**)

(**6.459**)	R	ArH	NH	CH$_2$	Si(Me)	Others
	H	7.00–7.50m	3.87	2.48	0.15	—
	Me	7.12	3.83br	2.44	0.17	2.22
	Cl	7.33	3.70br	2.49	0.19	—
	NO$_2$	7.61	3.81	2.49	0.17	—

[a] δ in ppm measured for solutions in $(CD_3)_2SO$ from TMS.

[b] Signals are sharp singlets unless denoted as br = broad; m = multiplet.

[c] From Ref. 258.

TABLE 6.94. ^{1}H NMR SPECTRA[a,b] *of* 1*H*- AND 9*H*-1,2,4-TRIAZOLO[4,3-*a*]BENZIMIDAZOLE DERIVATIVES (**6.467A**) AND (**6.467B**)

(**6.467A**) (**6.467B**)

Compound	R^1	R^2	R^3	Solvent[c]	H(5)	H(6)	H(7)	H(8)	Me(3)	NMe	Others	Ref.
(**6.467A**) ⇌ (**6.467B**)	H	H	H	*A*	7.78[d,e,f]	7.15[d,g,h]	7.32[e,g,i]	7.44[f,h,i]	2.73	—	—	274
(**6.467A**) ⇌ (**6.467B**)	H	H	H	*A*	7.79[e,f]	—[j]	7.32[e,i]	7.43[f,i]	2.73	—	—	272
(**6.467A**) ⇌ (**6.467B**)	H	H	H	*A*	← 7.53–8.00m →				2.74	—	—	263
(**6.467A**) ⇌ (**6.467B**)	H	H	D	*A*	7.78[e,f]	—	7.31[e,i]	7.43[f,i]	2.72	—	—	272
(**6.467A**) ⇌ (**6.467B**)	H	H	Br	*B*	8.19[e,f]	—	7.88[e,i]	7.70[f,i]	3.07	—	—	272
(**6.467A**) ⇌ (**6.467B**)	H	Br	Br	*B*	8.20[e]	—	7.90[e]	—	3.04	—	—	272
(**6.467A**) ⇌ (**6.467B**)	H	H	NO_2	*B*	8.99[e,f]	—	8.70[e,i]	8.00[f,i]	3.22	—	—	272
(**6.467A**) ⇌ (**6.467B**)	H	NO_2	NO_2	*B*	9.55[e]	—	9.40[e]	—	3.30	—	—	272
(**6.467A**) ⇌ (**6.467B**)	H	H	NH_2	*A*	7.04[e,f]	—	6.73[e,i]	6.60[f,i]	2.65	—	—	272
(**6.467A**) ⇌ (**6.467B**)	H	H	NH_2	*B*	8.25[e,f]	—	7.93[e,i]	7.68[f,i]	2.93	—	—	272
(**6.467A**)	Me	H	H	*A*	7.71[d,e,f]	7.07[d,g,h]	7.28[d,g,i]	7.49[f,h,i]	2.66	3.73	—	274
(**6.467A**)	Me	H	H	*A*	← 7.00–8.00m →				2.71	3.74	—	263
(**6.467A**)	Me	H	H	*C*	← —[j] →				2.68	3.84	—	263
(**4.467B**)	Me	H	H	*A*	7.69[d,e,f]	7.18[d,g,h]	7.38[e,g,i]	7.47[f,h,i]	2.65	3.60	—	274
(**6.467B**)	Me	H	H	*A*	← 7.00–8.00m →				2.68	3.64	—	263
(**6.467B**)	Me	H	H	*C*	← —[j] →				2.76	3.71	—	263
(**6.467B**)	COMe	H	H	*A*	← 7.00–7.80m →			8.30m	2.74[k]	—	2.82[k,l]	263
(**6.467B**)	COMe	H	H	*C*	← 7.30–7.60m →			8.45m	2.82[k]	—	2.96[k,l]	263

[a] δ in ppm measured from TMS.
[b] Signals are sharp singlets unless denoted as m = multiplet
[c] *A* = $(CD_3)_2SO$; *B* = CF_3CO_2H; *C* = $CDCl_3$.
[d] $J_{5,6} = 8.00–8.03$ Hz.
[e] $J_{5,7} = 1.17–2.00$ Hz.
[f] $J_{5,8} = 0.50–0.67$ Hz.
[g] $J_{6,7} = 7.41–7.67$ Hz.
[h] $J_{6,8} = 0.95–1.13$ Hz.
[i] $J_{7,8} = 8.16–9.30$ Hz.
[j] δ values not quoted.
[k] These signal assignments may be interchanged.
[l] CO*Me*.

TABLE 6.95. ^{13}C NMR SPECTRA[a] OF 1*H*- AND 9*H*-1,2,4-TRIAZOLO[4,3-*a*]BENZIMIDAZOLE DERIVATIVES (**6.468**), (**6.469**), (**6.470**), AND (**6.471**)[b]

(**6.468**) (**6.469**) (**6.470**) (**6.471**)

Compound	R^1	R^2	Solvent[c]	C(3)	C(4a)	C(5)	C(6)	C(7)	C(8)	C(8a)	C(9)	Me	NAlk
(**6.468**) ⇌ (**6.469**)	H	H	*A*	139.1	123.5	111.7	119.4	124.5	114.1	142.7	156.0	11.3	—
(**6.468**) ⇌ (**6.469**)	H	H	*B*	—[d]	123.6	114.4	126.5	130.7	116.0	137.7	—[d]	11.4	—
(**6.468**) ⇌ (**6.469**)	H	D	*A*	139.0	123.4	111.6	—	124.4	114.0	142.5	155.8	11.4	—
(**6.468**) ⇌ (**6.469**)	H	NO_2	*A*	138.9	123.9	108.8	139.8	120.6	116.5	152.8	158.5	11.6	—
(**6.468**) ⇌ (**6.469**)	H	NO_2	*B*	145.5	123.4	111.7	142.5	126.0	116.7	142.7	152.2	11.6	—
(**6.468**)	Me	H	*A*	137.2	125.6	110.9	118.1[e]	123.9	117.5[e]	150.1	154.7	11.4	34.0[b]
(**6.468**)	Me	H	*B*	144.0	124.2	114.1	126.7	130.0	115.9	137.5	148.0	11.5	36.9[f]
(**6.468**)	Me	H	*A*	140.3	121.7	111.9	120.3	124.8	110.1	138.4	156.0	11.3	29.0[g]
(**6.469**)	Me	H	*B*	144.8	123.2	114.7	126.2	130.4	113.3	139.4	151.4	11.2	31.3[g]
(**6.470**)	Me	Me	*A*	140.7	121.4	112.4	123.7	127.0	113.6	137.7	146.2	11.2	36.6[f] 30.8[g]
(**6.470**)	Me	Me	*B*	144.0	123.4	114.5	126.8	129.9	113.7	139.6	147.7	11.9	37.9[f] 31.9[g]

(**6.470**)	Me	Et	*A*	141.0	121.5	112.4	123.7	127.0	113.5	137.8	145.8	11.2	30.7[g] 14.4[h] 44.6[i]
(**6.470**)	Et	Me	*A*	140.9	121.7	112.5	123.8	127.1	113.7	136.7	145.7	11.1	36.5[f] 14.8[h] 39.2[i]
(**6.471**)	Me	Me	*A*	142.5	121.4	111.8	122.3	128.0	114.3	137.7	151.0	10.7	29.9[g] 38.2[j]
(**6.471**)	Me	Me	*B*	143.5	123.3	115.2	125.2	130.7	113.0	139.6	152.8	12.0	30.8[g] 39.5[j]
(**6.471**)	Me	Et	*A*	141.8	121.4	111.7	122.4	128.0	114.3	137.8	151.4	10.5	29.8[g] 14.1[h] 46.1[i]
(**6.471**)	Et	Me	*A*	142.5	126.6	112.0	122.4	128.1	114.5	136.9	150.4	10.7	38.2[j] 13.0[h] 39.6[i]

[a] δ in ppm measured from TMS.
[b] From Ref. 273.
[c] $A = (CD_3)_2SO$; $B = CF_3CO_2H$.
[d] Not observable.
[e] These signal assignments may be interchanged.
[f] Me(1).
[g] Me (9).
[h] Me of Et group.
[i] CH_2 of Et group.
[j] Me (2).

methyl derivatives using the LAOCN program in conjunction with nuclear Overhauser effects (NOE) allows the assignment of chemical shifts and coupling constants to the benzenoid protons in these molecules. The attempted investigation[274] of the tautomeric equilibrium in 3-methyl-1(9)*H*-1,2,4-triazolo[4,3-*a*]benzimidazole by ^{1}H NMR spectroscopy was unsuccessful. However, this equilibrium is readily studied[273] by means of ^{13}C NMR spectroscopy (Table 6.95) which demonstrates the predominance of the 9*H*-tautomer (**6.469**; $R^1 = R^2 = H$) to the extent of 60–70% in dimethyl sulfoxide. The value of the ^{13}C NMR method for the investigation of such

TABLE 6.96. ^{1}H NMR SPECTRA[a,b] OF 1*H*-1,2,4-TRIAZOLO[2,3-*a*]-BENZIMIDAZOLE DERIVATIVES (**6.472**)[c]

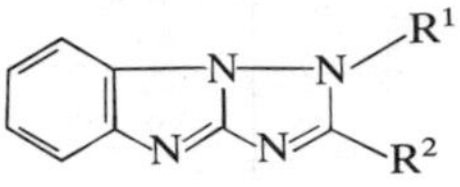

(**6.472**)

(**6.472**)	R^1	R^2	Solvent[d]	H(5)	H(6)	H(7)	H(8)	Me	CH_2	Others
	H	Me	*A*	←—— 7.50 m ——→				2.50[e]	—	—
	COMe	Me	*B*	←— 7.40 m —→			8.50 m[f]	2.83[e]	—	—
								2.46[g]		
	H	Et	*A*	←—— 7.55 m ——→				1.40 t[h,i]	2.85q[j,i]	—
	COEt	Et	*B*	←— 7.55 m —→			8.55[f]	1.37t[h,i]	2.90q[j,i]	
								1.42 t[k,i]	3.40q[k,i]	
	H	Ph	*B*	←—— 7.54 m ——→				—	—	8.00 m[m]
	COPh	Ph	*B*	←— 7.37 m —→			8.43 m[f]	—	—	7.54 m[m]
										8.04 m[n]
	CH_2Ph	Ph	*C*	←—— 7.20 m ——→				—	5.38[o]	7.30 m[m]
										7.82[p]

[a] δ in ppm measured from TMS.
[b] Signals are sharp singlets unless denoted as t = triplet; q = quartet; m = multiplet.
[c] From Ref. 270.
[d] $A = CD_3CO_2D$; $B = CDCl_3$; $C = CCl_4$.
[e] Me (2).
[f] May be reassigned to H(5).
[g] COMe.
[h] Me of Et group.
[i] Coupling constant not quoted.
[j] CH_2 of Et group.
[k] Me of COEt group.
[l] CH_2 of COEt group.
[m] Ph (2).
[n] CO*Ph*.
[o] *CH*$_2$Ph.
[p] CH_2*Ph*.

tautomeric equilibria resides in its applicability in solvent systems inappropriate for UV spectroscopic studies (see before) or dipole moment measurements (see later). The site of protonation and quaternization in 1*H*- and 9*H*-1,2,4-triazolo[4,3-*a*]benzimidazoles can be determined on the basis of the consequent shielding effect of ca. 10 ppm on the ^{13}C chemical shifts (Table 6.95) of carbon atoms adjacent to the nitrogen centers involved.[273] The enhanced deshielding of one of the benzenoid protons [*H*(5) or *H*(8)] in 1-acyl-1*H*-1,2,4-triazolo[2,3-*a*]benzimidazoles (Table 6.96) is attributed[270] to a long-range anisotropic effect of the *N*-acyl substituent.

General Studies

Dipole Moments. Dipole measurements provide a useful means for the study of tautomeric equilibria in 1(9)*H*-1,2,4-triazolo[4,3-*a*]benzimidazoles. For example, comparison of the dipole moments (Table 6.97) of 1,3-dimethyl-1*H*-1,2,4-triazolo[4,3-*a*]benzimidazole (**6.473**; R = Me) and 3,9-dimethyl-9*H*-1,2,4-triazolo[4,3-*a*]benzimidazole (**6.474**; R = Me) measured in dioxane with that of 3-methyl-1(9)*H*-1,2,4-triazolo[4,3-*a*]benzimidazole taken in the same solvent, indicates the tautomeric equilibrium of the latter compound in dioxane to favor the 1*H*-tautomer (**6.473**; R = H). The contrasting predominance of the 9*H*-tautomer (**6.474**; R = H) in ethanol revealed by UV studies (see before), is a reflection of the solvent-dependent nature of the equilibrium process [(**6.473**; R = H) ⇌ (**6.474**; R = H)].

TABLE 6.97. DIPOLE MOMENTS[a] OF 1*H*- AND 9*H*-1,2,4-TRIAZOLO[4,3-*a*]-BENZIMIDAZOLE DERIVATIVES (**6.473**) AND (**6.474**)[b]

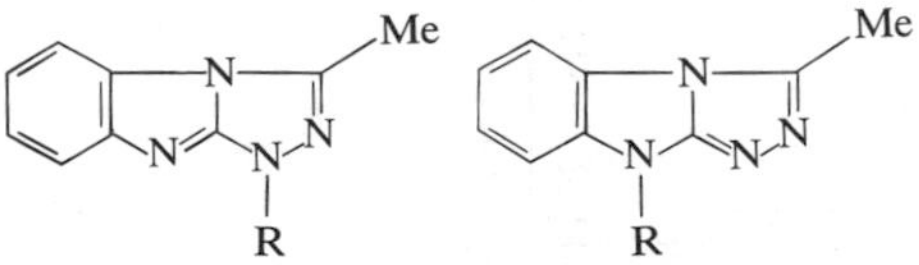

(**6.473**) (**6.474**)

Compound	R	μ(D)
(**6.473**) ⇌ (**6.474**)	H	3.20
(**6.473**)	Me	3.55
(**6.474**)	Me	5.12

[a] Measured in dioxane at 25°.
[b] From Ref. 226.

TABLE 6.98. ALKYLATION AND ACYLATION REACTIONS OF 1(9)*H*-1,2,4-TRIAZOLO[4,3-*a*]BENZIMIDAZOLES

Reaction conditions[a]	1(9)*H*-1,2,4-Triazolo[4,3-*a*]benzimidazole: Substrate	Product	Yield (%)	m.p. (°C)	Solvent of crystallization	Ref.
A	3-Me-1(9)*H*-	1,3-di-Me-1*H*-	55	73–74	—[b]	263
		+				
		3,9-di-Me-9*H*-	45	177	—[b]	
B	3-Me-1(9)*H*-	1,3-di-Me-1*H*-	70	—	—	263
		+				
		3,9-di-Me-9*H*-	30	—	—	
C	2,3-Dihydro-1(9)*H*-3-thione	3-MeS-9-Me-9*H*-	90	104–105	Ethanol–water	267
C	9-Me-2,3-dihydro-9*H*-3-thione	3-MeS-9-Me-9*H*-	90	—	—	267
D	3-Me-1(9)*H*-	3-Me-9-COMe-9*H*-	—[c]	214–216	—[d]	263
E	2,3-Dihydro-1(9)*H*-3-thione	2,9-di-CH_2N(pyrrolidino)-2,3-dihydro-9*H*-3-thione	90	144–146	Ethanol	272
E	2,3-Dihydro-1(9)*H*-3-thione	2,9-di-CH_2N(piperidino)-2,3-dihydro-9*H*-3-thione	quant.	134–136	Ethanol	272

E	2,3-Dihydro-1(9)*H*-3-thione	2,9-di-CH_2N 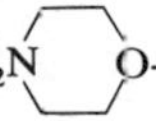 O-2,3-dihydro-9*H*-3-thione	90	152–154	Ethanol	272
E	9-Me-2,3-dihydro-1(9)*H*-3-thione	9-Me-2-CH_2N  -2,3-dihydro-9*H*-3-thione	90	146–148	Ethanol	272
E	9-Me-2,3-dihydro-1(9)*H*-3-thione	9-Me-2-CH_2N 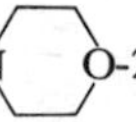 O-2,3-dihydro-9*H*-3-thione	80	176–178	Ethanol	272
E	9-CH_2Ph-2,3-dihydro-1(9)*H*-3-thione	9-CH_2Ph-2-CH_2N  -2,3-dihydro-9*H*-3-thione	80	171–173	Ethanol	272
E	9-CH_2Ph-2,3-dihydro-1(9)*H*-3-thione	9-CH_2Ph-2-CH_2N 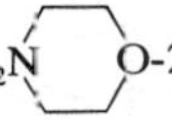 O-2,3-dihydro-9*H*-3-thione	80	167–168	Ethanol	272
E	9-CH_2NMe_2-2,3-dihydro-1(9)*H*-3-thione	9-CH_2NMe_2-2-CH_2NMe_2-2,3-dihydro-9*H*-3-thione	quant.	153–154	Ethanol	272

[a] *A* = MeI, NaOEt, EtOH/(reflux)(7 hr); *B* = Me_2SO_4, NaOH, H_2O/(reflux)(6 hr); *C* = MeI, EtOH/(reflux)(10–15 min); *D* = Ac_2O, pyridine (reaction temp. and time not specified). *E* = R_2N, 40% HCHO aq. (reaction temp. and time not specified).

[b] Purified by sublimation.

[c] Yield not quoted.

[d] Solvent of crystallization not specified.

6.3.3. Reactions

The brevity of the following account is a measure of the general lack of information on the chemical behavior of the several 6-5-5 fused benzimidazole ring systems having two additional heteroatoms.

Reactions with Electrophiles

Changes in the UV and ^{13}C NMR absorption[273] of 1*H*- and 9*H*-1,2,4-triazolo[4,3-*a*]benzimidazoles with decrease in pH are consistent with the ready protonation and hence basic character of these molecules. In the case of 1,3-dimethyl-1*H*-1,2,4-triazolo[4,3-*a*]benzimidazole these changes are interpretable[273] in terms of exclusive protonation at *N*(9). The UV and ^{13}C NMR spectra of 3-methyl-1(9)*H*-1,2,4-triazolo[4,3-*a*]benzimidazole and its *N*(9)-methyl derivative in acidic media, on the other hand, indicate the presence of a mixture of the two monocations derived by competing protonation at *N*(1) and *N*(2).[273]

The product (m.p. 130–132°), obtained in moderate yield (50%) by the sodium hydride-mediated benzylation[270] of 2-phenyl-1*H*-1,2,4-triazolo[2,3-*a*]benzimidazole has been arbitrarily assigned an *N*(1)-benzyl structure, though the *N*(3) or *N*(4) position for the benzyl group appears equally likely. In accordance with its tautomeric character, 3-methyl-1(9)*H*-1,2,4-triazolo[4,3-*a*]benzimidazole undergoes competing base-catalyzed methylation[263] at the *N*(1) and *N*(9) positions in good overall yield (Table 6.98). The uncatalyzed reaction[267] of 2,3-dihydro-1(9)*H*-1,2,4-triazolo[4,3-*a*]-benzimidazole-3-thione with methyl iodide results in methylation at both sulfur and nitrogen giving 9-methyl-3-thiomethyl-9*H*-1,2,4-triazolo[4,3-*a*]-benzimidazole in high yield (Table 6.98). ^{13}C NMR spectroscopy provides a highly sensitive method for determining the site of quaternization in 1*H*- and 9*H*-1,2,4-triazolo[4,3-*a*]benzimidazole derivatives.[273]

3-Sila-2*H*-imidazo[1,2-*a*]benzimidazoles are unstable to acetylation which promotes cleavage of the silicon–nitrogen bond in these molecules giving unstable silyl carboxylates convertible by hydrolysis into isolable siloxanes [e.g., Scheme 6.100; (**6.475**) → (**6.476**) → (**6.477**)].[258] Acylative ring-opening of this type serves to establish the gross structure of the 3-sila-2*H*-imidazo[1,2-*a*]benzimidazole ring system. The 1(9)*H*-1,2,4-triazolo-[4,3-*a*]benzimidazole ring system is stable to ring-opening by reagents (e.g., acetic anhydride–pyridine), which effect acetylation at the *N*(9) position (Table 6.98).[263] 2,3-Dihydro-9*H*-1,2,4-triazolo[4,3-*a*]benzimidazole-3-thiones are aminoalkylated in high yield (Table 6.98) at the *N*(2) and *N*(9) positions under the conditions of the Mannich reaction.[272]

The nitration of 3-(1-imidazolyl)-1,2,4-thiadiazolo[4,5-*a*]benzimidazole occurs at both the benzene nucleus and the imidazole substituent giving

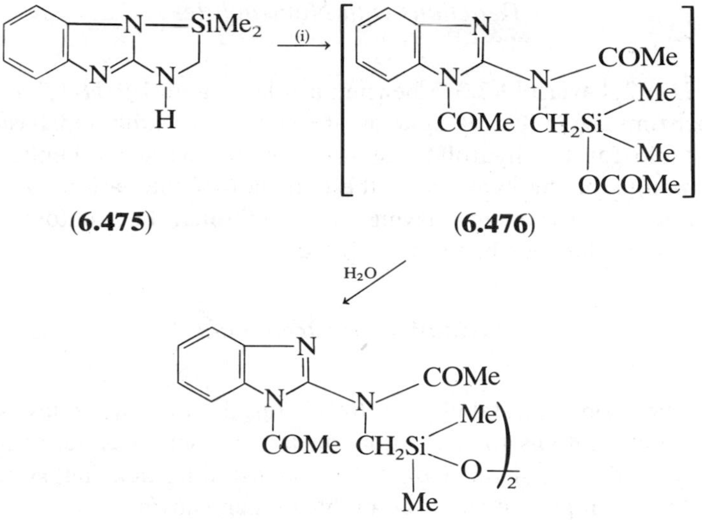

(6.475) **(6.476)**

(6.477)

(i) $Ac_2O/(50–55°)(1 hr)$

Scheme 6.100

isomer mixtures of unestablished constitution.[261] 3-Methyl-1(9)*H*-1,2,4-triazolo[4,3-*a*]benzimidazole undergoes bromination and nitration sequentially at the *C*(6) and *C*(8) positions in the benzene ring in good yield (Table 6.99).[272] The amino substituent in 6-amino-3-methyl-1(9)*H*-1,2,4-triazolo-[4,3-*a*]benzimidazole can be diazotized[272] under standard conditions (Table 6.99) to afford a diazonium salt which exhibits orthodox chemical reactivity (e.g., toward hypophosphorus acid reduction—Table 6.99).[272]

TABLE 6.99. BROMINATION, NITRATION, AND DIAZOTIZATION REACTIONS OF 1(9)*H*-1,2,4-TRIAZOLO[4,3-*a*]BENZIMIDAZOLE DERIVATIVES

Reaction conditions[a]	1(9)*H*-1,2,4-Triazolo[4,3-*a*]benzimidazole		Yield (%)	m.p. (°C)	Solvent of crystallization	Ref.
	Substrate	Product				
A	3-Me-1(9)*H*-	3-Me-6-Br-1(9)*H*- + 3-Me-6,8-di-Br-1(9)*H*-	51 49	288 >300	—[b] Methanol	272
B	3-Me-1(9)*H*-	3-Me-6-NO_2-1(9)*H*-	86	>350	—[c]	272
C	3-Me-1(9)*H*-	3-Me-6,8-di-NO_2-1(9)*H*-	86	>350	—[c]	272
D	3-Me-6-NH_2-1(9)*H*-	3-Me-6-D-1(9)*H*-	80	230	—[c]	272

[a] *A* = Br_2, NaOAc, AcOH/(0°) then (room temp.)(5 hr); *B* = conc. HNO_3, conc. H_2SO_4/(−10 to −5°)(0.5 hr); *C* = conc. HNO_3, conc. H_2SO_4/(−10 to −5°)(3 hr); *D* = $NaNO_2$, DCl/(−10°)(1 hr), then H_3PO_2, D_2O/(room temp.)(4 hr).

[b] Purified by sublimation.

[c] Solvent of crystallization not specified.

Reactions with Nucleophiles

The 1*H*-1,2,4-triazolo[2,3-*a*]benzimidazole[270] and 1(9)*H*-1,2,4-triazolo-[4,3-*a*]benzimidazole[267] ring systems are stable to acidic and basic conditions suitable for the hydrolytic removal of *N*-acyl substituents. Routine transformations of this type apart, the behavior of the various 6-5-5 fused benzimidazole ring systems having two additional heteroatoms to nucleophilic attack, has not been investigated.

Oxidation and Reduction

Information on the stability of 6-5-5 fused benzimidazoles with two additional heteroatoms toward oxidation and reduction is very limited. In one study[259] the 1,2,4-thiadiazolo[2,3-*a*]benzimidazole ring system, and specifically its component sulfur atom, have been shown to be unaffected by peracid oxidation under conditions that convert a thiomethyl substituent into the *S*-oxide. The conversion[261] of 3-(1-imidazolyl)-1,2,4-thiadiazolo-[4,5-*a*]benzimidazole into 2-mercaptobenzimidazole on treatment with lithium aluminum hydride, on the other hand, illustrates the susceptibility of 1,2,4-thiadiazolo[4,5-*a*]benzimidazoles to reductive ring scission. In contrast, 3-methyl-6-nitro-1(9)*H*-1,2,4-triazolo[4,3-*a*]benzimidazole can be catalytically reduced in high yield (90%) to the amine (m.p. 172°) without disruption of the ring system.[272]

6.3.4. Practical Applications

Derivatives of the 1,2,4-thiadiazolo[2,3-*a*]benzimidazole,[259] 1,2,4-thiadiazolo[4,3-*a*]benzimidazole,[275] 1,2,4-thiadiazolo[4,5-*a*]benzimidazole,[261] and 1,2,4-triazolo[4,3-*a*]benzimidazole,[276] ring systems have found application as antifungal agents. 1(9)*H*-1,2,4-Triazolo[4,3-*a*]-benzimidazole derivatives have been patented as antifogging agents for photographic emulsions.[264,277]

6.4. Tricyclic 6-5-5 Fused Benzimidazoles with Three Additional Heteroatoms

The ephemeral tetrazolo[1,5-*a*]benzimidazole ring system [Scheme 6.101; (**6.478**)] is the sole representative of the class of 6-5-5 fused benzimidazoles having three additional heteroatoms. To date, tetrazolo[1,5-*a*]-benzimidazoles have been encountered[278] only as the unstable ring tautomers in azidoazomethine–tetrazole equilibria[279] involving 2-azido-

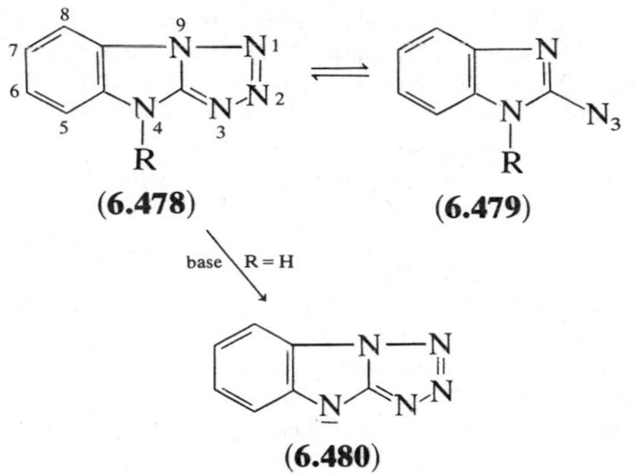

(6.478) ⇌ (6.479)

base, R = H

(6.480)

Scheme 6.101

benzimidazoles [e.g., Scheme 6.101; (**6.478**; R = COMe) ⇌ (**6.479**; R = COMe)] or as stable anions[278] derived by deprotonation and spontaneous ring-closure of *N*-unsubstituted 2-azidobenzimidazoles [Scheme 6.101; (**6.479**; R = H) ⇌ (**6.478**; R = H) → (**6.480**)].

REFERENCES

1. A. Bistrzycki and K. Fassler, *Helv. Chim. Acta*, **6,** 519 (1923).
2. W. Ried and G. Isenbruck, *Chem. Ber.*, **105,** 337 (1972); *ibid.*, *Chem. Ber.*, **105,** 353 (1972).
3. R. L. Williams and R. Brown, *NASA Contract Rep.* 1976, NASA-CR-148483; *Sci. Tech. Aerosp. Rep.*, **14,** abs. no. N76-28418 (1976); *Chem. Abstr.*, **86,** 72518 (1977).
4. W. Ried and H. Knorr, *Ann. Chem.*, **1976,** 284.
5. W. Ried, A. H. Schmidt, and W. Kuhn, *Chem. Ber.*, **104,** 2622 (1971).
6. Y. Shiokawa and S. Ohki, *Chem. Pharm. Bull.* (*Tokyo*), **21,** 981 (1973); *Chem. Abstr.*, **79,** 66246 (1973).
7. Z. V. Esayan, L. A. Manucharova, and G. T. Tatevosyan, *Arm. Khim. Zh.*, **25,** 345 (1972); through *Chem. Abstr.*, **77,** 139889 (1972).
8. G. T. Tatevosyan and Z. V. Esayan, *USSR Patent* 449,054; *Chem. Abstr.*, **82,** 57699 (1975).
9. N. N. Bortnick and M. F. Fegley, *U.S. Patent* 2,993,046; *Chem. Abstr.*, **56,** 4779b (1962).
10. W. Ried and R. Lantzsch, *Ann. Chem.* **750,** 97 (1971).
11. R. M. Palei and P. M. Kochergin, *Chem. Heterocycl. Compd.*, **1967,** 431; *Khim. Geterotsikl. Soedin.*, **1967,** 536; *Chem. Abstr.*, **68,** 49511 (1968).
12. F. S. Babichev and A. F. Babicheva, *Chem. Heterocycl. Compd.*, **1967,** 723; *Khim. Geterotsikl. Soedin.*, **1967,** 917; *Chem. Abstr.*, **69,** 60031 (1968); *Khim. Geterotsikl Soedin.*, **1967,** 187; through *Chem. Abstr.*, **67,** 82157 (1967).
13. F. S. Babichev, G. F. Kutrov, and M. Y. Kornilov, *Ukr. J. Chem.*, **34,** 35 (1968); *Ukr. Khim. Zh.*, **34,** 1020 (1968); *Chem. Abstr.*, **70,** 68249 (1969).
14. M. Y. Kornilov, G. G. Dyadyusha, and F. S. Babichev, *Chem. Heterocycl. Compd.*, **1968,** 654; *Khim. Geterotsikl. Soedin*, **1968,** 905; *Chem. Abstr.*, **70,** 115067 (1969).
15. R. M. Palei and P. M. Kochergin, *Chem. Heterocycl. Compd.*, **1969,** 641; *Khim. Geterotsikl. Soedin.*, **1969,** 865; *Chem. Abstr.*, **72,** 111372 (1970).

16. R. M. Palei and P. M. Kochergin, *Chem. Heterocycl. Compd.*, **1969,** 812; *Khim. Geterotsikl. Soedin,* **1969,** 1075; *Chem. Abstr.*, **72,** 132611 (1970).
17. P. M. Kochergin, Y. N. Sheinker, A. A. Druzhinina, R. M. Palei, and L. M. Alekseeva, *Chem. Heterocycl. Compd.*, **7,** 771 (1971); *Khim. Geterotsikl. Soedin,* **7,** 826 (1971); *Chem. Abstr.*, **76,** 25170 (1972).
18. R. M. Palei and P. M. Kochergin, *Chem. Heterocycl. Compd.*, **1970,** 531; *Khim. Geterotsikl. Soedin.*, **1970,** 572; *Chem. Abstr.*, **73,** 87846 (1970).
19. R. M. Palei and P. M. Kochergin, *Chem. Heterocycl. Compd.*, 1972, 368; *Khim. Geterotsikl. Soedin.*, **1972,** 403; *Chem. Abstr.*, **77,** 88388 (1972).
20. P. M. Kochergin, A. A. Druzhinina, and R. M. Palei, *Chem. Heterocycl. Compd.*, **1976,** 1274; *Khim. Geterotsikl. Soedin.*, **1976,** 1549; *Chem. Abstr.*, **86,** 106474 (1977).
21. V. A. Kovtunenko and F. S. Babichev, *Ukr. J. Chem.*, **38,** 46 (1972); *Ukr. Khim. Zh.*, **38,** 1244 (1972); *Chem. Abstr.*, **78,** 72007 (1973).
22. P. V. Tkachenko, I. I. Popov, A. M. Simonov, and Y. V. Medvedev, *Chem. Heterocycl. Compd.*, **1976,** 805; *Khim. Geterotsikl. Soedin.*, **1976,** 972; *Chem. Abstr.*, **85,** 159980 (1976).
23. H. Ogura and K. Kikuchi, *J. Org. Chem.*, **37,** 2679 (1972).
24. I. Zugravescu, J. Herdan, and I. Druta, *Rev. Roum. Chim.*, **19,** 649 (1974); *Chem. Abstr.*, **81,** 25603 (1974).
25. R. M. Acheson and M. S. Verlander, *J. Chem. Soc. Perkin Trans. I*, **1973,** 2348.
26. R. M. Acheson, M. W. Foxton, P. J. Abbot, and K. R. Mills, *J. Chem. Soc. C*, **1967,** 882.
27. R. M. Acheson and W. R. Tully, *J. Chem. Soc. C*, **1968,** 1623.
28. R. M. Acheson and M. S. Verlander, *J. Chem. Soc. Perkin Trans. I*, **1974,** 430.
29. R. M. Acheson and M. S. Verlander, *J. Chem. Soc. Perkin Trans. I*, **1972,** 1577.
30. K. A. Suerbaev and C. S. Kadyrov, *Chem. Heterocycl. Compd.*, **1974,** 989; *Khim. Geterotsikl. Soedin.*, **1974,** 1137; *Chem. Abstr.*, **81,** 152107 (1974).
31. R. C. De Selms, *J. Org. Chem.* **27,** 2165 (1962).
32. A. R. Freedman, D. S. Payne, and A. R. Day, *J. Heterocycl. Chem.*, **3,** 257 (1966).
33. W. C. Aten and K. H. Buechel, *Z. Naturforsch. B*, **25,** 928 (1970); *Chem. Abstr.*, **74,** 3555 (1971).
34. A. Bistrzycki and W. Schmutz, *Ann. Chem.*, **415,** 1 (1918).
35. W. Reppe et al., *Ann. Chem.*, **596,** 209 (1955).
36. A. Botta and C. Rasp, *Ger. Patent*, 2,435,406; *Chem. Abstr.*, **84,** 164782 (1976).
37. R. Meyer and H. Luders, *Ann. Chem.*, **415,** 29 (1918).
38. J. L. Aubagnac, J. Elguero, and R. Robert, *Bull. Soc. Chim. Fr.*, **1972,** 2868.
39. M. V. Betrabet and G. C. Chakravarti, *J. Indian Chem. Soc.*, **7,** 191 (1930).
40. J. Stanek and V. Wollrab, *Monatsh. Chem.*, **91,** 1064 (1960); *Chem. Abstr.*, **55,** 16520c (1961).
41. I. I. Chizhevskaya, N. N. Khovratovich, and Z. M. Grabovskaya, *Chem. Heterocycl. Compd.*, **1968,** 329; *Khim. Geterotsikl. Soedin.*, **1968,** 443; *Chem. Abstr.*, **69,** 86906 (1968).
42. H. A. Naik, V. Purnaprajna, and S. Seshadri, *Indian J. Chem. Sect. B*, **15,** 338 (1977); *Chem. Abstr.*, **87,** 201406 (1977).
43. W. W. Paudler and A. G. Zeiler, *J. Org. Chem.* **34,** 2138 (1969).
44. H. H. Zoorob, H. A. Hammouda, and E. Ismail, *Z. Naturforsch. B*, **32B,** 443 (1977); *Chem. Abstr.*, **87,** 53158 (1977).
45. I. Zugravescu, J. Herdan, and I. Druta, *Rev. Roum. Chim.*, **19,** 659 (1974); *Chem. Abstr.*, **81,** 25603 (1974).
46. J. W. Lown and B. E. Landberg, *Can. J. Chem.*, **53,** 3782 (1975).
47. O. Meth-Cohn and H. Suschitzky. *Adv. Heterocycl. Chem.*, **14,** 211 (1972).
48. M. D. Nair and R. Adams, *J. Am. Chem. Soc.*, **83,** 3518 (1961).
49. O. Meth-Cohn and H. Suschitzky, *J. Chem. Soc.*, **1963,** 4666.
50. D. P. Ainsworth and H. Suschitzky, *J. Chem. Soc.* (*C*), **1966,** 111.
51. R. Garner and H. Suschitzky, *J. Chem. Soc.* (*C*), **1967,** 74.

52. O. Meth-Cohn, *J. Chem. Soc.* (*C*), **1971,** 1356.
53. H. Takahashi, S. Sato, and H. Otomasu, *Chem. Pharm. Bull.* (*Tokyo*), **22,** 1921 (1974); *Chem. Abstr.*, **82,** 43248 (1975).
54. H. Mohrle and J. Gerloff, *Arch. Pharm.*, **311,** 381 (1978); *Chem. Abstr.*, **89,** 109231 (1978).
55. H. Mohrle and H. J. Hemmerling, *Arch. Pharm.*, **311,** 586 (1978); *Chem. Abstr.*, **89,** 163531 (1978).
56. O. Meth-Cohn, *J. Chem. Soc.*, **1964,** 5245.
57. O. Meth-Cohn, R. K. Smalley, and H. Suschitzky, *J. Chem. Soc.*, **1963,** 1666.
58. M. J. Libeer, H. Depoorter, and G. G. Van Mierlo, *U.S. Patent,* 3,931,156; *Chem. Abstr.*, **85,** 64804 (1976); *Belgian Patent,* 618,235; *Chem. Abstr.*, **58,** 14164f (1963).
59. R. K. Grantham and O. Meth-Cohn, *J. Chem. Soc.* (*C*), **1969,** 1444.
60. J. Martin, O. Meth-Cohn, and H. Suschitzky, *Chem. Commun.*, **1971,** 1319.
61. H. Suschitzky and M. E. Sutton, *Tetrahedron,* **24,** 4581 (1968).
62. R. Fielden, O. Meth-Cohn, and H. Suschitzky, *J. Chem. Soc. Perkin Trans. I,* **1973,** 696.
63. D. J. Neadle and R. J. Pollitt, *J. Chem. Soc.* (*C*), **1969,** 2127.
64. J. D. Loudon and G. Tennant, *Quart. Rev.* (*London*), **18,** 389 (1964); P. N. Preston and G. Tennant, *Chem. Rev.*, **72,** 627 (1972).
65. R. K. Grantham and O. Meth-Cohn, *J. Chem. Soc.* (*C*), **1969,** 70.
66. D. P. Ainsworth, O. Meth-Cohn, and H. Suschitzky, *J. Chem. Soc.* (*C*), **1968,** 923.
67. H. Kawamoto, T. Matsuo, S. Morosawa, and A. Yokoo, *Bull. Chem. Soc.* (*Tokyo*), **46,** 3898 (1973); *Chem. Abstr.*, **80,** 82924 (1974).
68. R. K. Grantham, O. Meth-Cohn, and M. A. Naqui, *J. Chem. Soc.*, (*C*), **1969,** 1438.
69. J. W. Clark-Lewis, K. Moody, and M. J. Thompson, *Austr. J. Chem.*, **23,** 1249 (1970).
70. F. S. Babichev, G. P. Kutrov, and M. Y. Kornilov, *Ukr. J. Chem.*, **36,** 37 (1970); *Ukr. Khim. Zh.*, **36,** 909 (1970); *Chem. Abstr.*, **74,** 76367 (1971).
71. R. C. Perera, R. K. Smalley, and L. G. Rogerson, *J. Chem. Soc.* (*C*), **1971,** 1348.
72. L. M. Alekseeva, G. G. Dvoryantseva, I. V. Persianova, Y. N. Sheinker, R. M. Palei, and P. M. Kochergin, *Chem. Heterocycl. Compd.*, **1972,** 1023; *Khim. Geterotsikl. Soedin.*, **1972,** 1132; *Chem. Abstr.*, **77,** 152063 (1972).
73. R. Fielden, O. Meth-Cohn, and H. Suschitzky, *J. Chem. Soc. Perkin Trans. I,* **1973,** 705.
74. F. S. Babichev, N. T. N. Fyong, and M. Y. Kornilov, *Ukr. J. Chem.*, **36,** 54 (1970); *Ukr. Khim. Zh.*, **36,** 819 (1970); *Chem. Abstr.*, **74,** 76366 (1971).
75. A. A. Druzhinina, P. M. Kochergin, R. M. Palei, and O. N. Minailova, *Chem. Heterocycl. Compd.*, **1977,** 180; *Khim. Geterotsikl. Soedin.*, **1977,** 225; *Chem. Abstr.*, **87,** 5863 (1977).
76. O. Meth-Cohn and R. K. Grantham, *J. Chem. Soc.* (*C*), 1971, 1354.
77. L. M. Alekseeva, G. G. Dvoryantseva, Y. N. Sheinker, A. A. Druzhinina, R. M. Palei, and P. M. Kochergin, *Chem. Heterocycl. Compd.*, **1976,** 64; *Khim. Geterotsikl. Soedin.*, 1976, 70; *Chem. Abstr.*, **84,** 163946 (1976).
78. O. S. Anisimova, Y. N. Sheinker, R. M. Palei, P. M. Kochergin, and V. S. Ponomar, *Chem. Heterocycl. Compd.*, **1975,** 982; *Khim. Geterotsikl. Soedin.*, **1975,** 1124; *Chem. Abstr.*, **83,** 206163 (1975).
79. M. Y. Kornilov, G. P. Kutrov, and F. S. Babichev, *Ukr. Khim. Zh.*, **42,** 1218 (1976); through *Chem. Abstr.*, **86,** 89696 (1977).
80. L. I. Savranskii, V. A. Kovtunenko, and F. S. Babichev, *Chem. Heterocycl. Compd.*, **1974,** 230; *Khim. Geterotsikl. Soedin.*, **1974,** 261; *Chem. Abstr.*, **80,** 145161 (1974).
81. F. S. Babichev and N. T. N. Fyong, *Ukr. J. Chem.*, **35,** 24 (1969); *Ukr. Khim. Zh.*, **35,** 932 (1969); *Chem. Abstr.*, **72,** 3432 (1970).
82. L. L. Linclon and L. G. S. Brooker, *Brit. Patent,* 1,054,107; *Chem. Abstr.*, **66,** 96228 (1967).
83. E. B. Mullock, R. Searby, and H. Suschitzky, *J. Chem. Soc.* (*C*), **1970,** 829.
84. J. Martin, O. Meth-Cohn, and H. Suschitzky, *Tetrahedron Lett.*, **1973,** 4495.

85. N. M. Bortnick and M. F. Fegley, *U.S. Patent*, 2,957,885; *Chem. Abstr.*, **55,** 4535h (1961).
86. A. R. Katritzky and J. M. Lagowski in *Chemistry of the Heterocyclic N-Oxides*, Academic, New York, 1971, Chap. 4, p. 366.
87. N. M. Bortnick and M. F. Fegley, *U.S. Patent*, 2,993,046; *Chem. Abstr.*, **56,** 4779b (1962).
88. G. L. Hiller, *U.S. Patent*, 882,018; *Chem. Abstr.*, **74,** 93451 (1971); *Belgian Patent*, 659,415; *Chem. Abstr.*, **64,** 6802h (1966).
89. R. G. Willis and W. R. Schleigh, *U.S. Patent*, 3,850,638; *Chem. Abstr.*, **82,** 105176 (1975).
90. P. L. De Benneville and W. D. Neiderhauwer, *U.S. Patent*, 3,267,082; *Chem. Abstr.*, **65,** 18720g (1966).
91. F. Benigni, L. Trevisan, and S. Freddi, *Farm. Ed. Sci.*, **31,** 901 (1976); *Chem. Abstr.*, **86,** 106473 (1977); *Ger. Patent*, 2,746,042; *Chem. Abstr.*, **89,** 24313 (1978).
92. F. Benigni and L. Trevisan, *Ger. Patent*, 2,517,270; *Chem. Abstr.*, **86,** 55454 (1977); *Belg. Patent*, 835,570; *Chem. Abstr.*, **86,** 55443 (1977).
93. R. J. Hayward, M. Htay, and O. Meth-Cohn, *Chem. Ind.*, **1977,** 373.
94. H. Ogura, T. Itoh, and Y. Shimada, *Chem. Pharm. Bull.* (*Tokyo*), **16,** 2167 (1968); *Chem. Abstr.*, **70,** 68261 (1969).
95. S. Singh, H. Singh, M. Singh, and K. S. Narang, *Indian J. Chem.*, **8,** 230 (1970); *Chem. Abstr.*, **72,** 132624 (1970).
96. V. K. Chadha, K. S. Sharma, and H. K. Pujari, *Indian J. Chem.*, **9,** 913 (1971); *Chem. Abstr.*, **75,** 151730 (1971).
97. J. Mohan and H. K. Pujari, *Indian J. Chem.*, **10,** 274 (1972); *Chem. Abstr.*, **77,** 88400 (1972).
98. J. Mohan, V. K. Chadha, and H. K. Pujari, *Indian J. Chem.*, **11,** 1119 (1973); *Chem. Abstr.*, **80,** 108448 (1974).
99. K. C. Joshi, V. N. Pathak, and P. Arya, *Agric. Biol. Chem.*, **41,** 543 (1977); *Chem. Abstr.*, **87,** 53140 (1977).
100. A. N. Krasovskii and P. M. Kochergin, *Chem. Heterocycl. Compds.*, **1969,** 243; *Khim. Geterotsikl. Soedin.*, **1969,** 321; *Chem. Abstr.*, **71,** 22067 (1969).
101. V. K. Chadha, H. S. Chaudhary, and H. K. Pujari, *Indian J. Chem.*, **7,** 769 (1969); *Chem. Abstr.*, **71,** 101780 (1969).
102. S. C. Bell and P. H. L. Wei, *J. Med. Chem.*, **19,** 524 (1976); *Chem. Abstr.*, **84,** 130238 (1976).
103. A. E. Alper and A. Taurins, *Can. J. Chem.*, **45,** 2903 (1967).
104. H. Andersag and K. Westphal, *Chem. Ber.*, **70B,** 2035 (1937).
105. P. M. Kochergin and A. N. Krasovskii, *Chem. Heterocycl. Compds.*, **1966,** 726; *Khim. Geterotsikl. Soedin.*, **1966,** 945; *Chem. Abstr.*, **66,** 115642 (1967).
106. A. N. Krasovskii and P. M. Kochergin, *Chem. Heterocycl. Compds.*, **1967,** 709; *Khim. Geterotsikl. Soedin.*, **1967,** 899; *Chem. Abstr.*, **68,** 105105 (1968).
107. H. Ogura, T. Itoh, and K. Kikuchi, *J. Heterocycl. Chem.*, **6,** 797 (1969).
108. H. Alper, E. C. H. Keung, and R. A. Partis, *J. Org. Chem.*, **36,** 1352 (1971); H. Alper, A. E. Alper, and A. Taurins, *J. Chem. Ed.*, **47,** 222 (1970).
109. P. H. L. Wei and S. C. Bell, *U.S. Patent*, 3,704,239; *Chem. Abstr.*, **78,** 43482 (1973); *U.S. Patent*, 3,775,426; *Chem. Abstr.*, **80,** 70807 (1974).
110. A. R. Todd, F. Bergel, and Karimullah, *Chem. Ber.*, **69B,** 217 (1936).
111. G. de Stevens and A. Halamandaris, *J. Am. Chem. Soc.*, **79,** 5710 (1957).
112. S. N. Kukota, M. O. Lozinskii, and P. S. Pelkis, *Ukr. Khim. Zh.*, **42,** 1162 (1976); through *Chem. Abstr.*, **86,** 89695 (1977).
113. M. Baboulene and G. Sturtz, *Compt. Rend.*, **284,** 799 (1977); *Chem. Abstr.*, **87,** 117807 (1977).
114. R. E. Moser, L. J. Powers, and Z. S. Ariyan, *Ger. Patent*, 2,701,853; *Chem. Abstr.*, **87,** 135310 (1977).

115. (a) J. J. D'Amico, R. H. Campbell, and E. C. Guinn, *J. Org. Chem.*, **29,** 865 (1964); (b) J. J. D'Amico, *U.S. Patent*, 3,225,059; *Chem. Abstr.*, **64,** 8193d (1966).
116. Y. Akasaki and A. Ohno, *J. Am. Chem. Soc.*, **96,** 1957 (1974).
117. A. N. Krasovskii, P. M. Kochergin, and T. E. Kozlovskaya, *Chem. Heterocycl. Compds.*, **7,** 363 (1971); *Khim. Geterotsikl. Soedin.*, **7,** 393 (1971); *Chem. Abstr.*, **76,** 14433 (1972); P. M. Kochergin and A. N. Krasovskii, *USSR Patent*, 230,823; through *Chem. Abstr.*, **70,** 68375 (1969).
118. (a) P. M. Kochergin, E. A. Priimenko, V. S. Ponomar, M. V. Povstyanoi, A. A. Tkachenko, I. A. Mazur, A. N. Krasovskii, E. G. Knysh, and M. I. Yurchenko, *Chem. Heterocycl. Compds.*, **1969,** 135; *Khim. Geterotsikl. Soedin.*, **1969,** 177; *Chem. Abstr.*, **71,** 13065 (1969); (b) A. N. Krasovskii, P. M. Kochergin, and L. V. Samoilenko, *Chem. Heterocycl. Compds.*, **1970,** 766; *Khim. Geterotsikl. Soedin.*, **1970,** 827; *Chem. Abstr.*, **73,** 109740 (1970).
119. I. Iwai and T. Hiraoka, *Chem. Pharm. Bull.* (*Tokyo*), **12,** 813 (1964); *Chem. Abstr.*, **61,** 9487g (1964); *Jap. Patent*, 5099 ('66); through *Chem. Abstr.*, **65,** 2272d (1966).
120. (a) K. K. Balasubramanian and R. Nagarajan, *Synthesis*, **1976,** 189; (b) K. K. Balasubramanian and B. Venugopalan, *Tetrahedron Lett.*, **1974,** 2643.
121. A. Davidson, I. E. P. Murray, P. N. Preston, and T. J. King, *J. Chem. Soc. Perkin Trans. I*, **1979,** 1239.
122. W. Ried, W. Merkel, S. W. Park, and M. Drager, *Ann. Chem.*, 1975, 79; S. W. Park, W. Ried, and W. Schuckmann, *ibid.*, 1977, 106.
123. A. N. Krasovskii and P. M. Kochergin, *Chem. Pharm. J.*, **2,** 545 (1968); *Khim. Farm. Zh.*, **2,** 18 (1968); *Chem. Abstr.*, **70,** 47349 (1969).
124. S. L. Mukherjee, G. Bagavant, V. S. Dighe, and S. Somasekhara, *Current Sci.* (*India*), **32,** 454 (1963); *Chem. Abstr.*, **59,** 15275h (1963).
125. K. Hideg, O. Hankovszky, E. Palosi, G. Hajos, L. Szporny, *Ger. Patent*, 2,429,290; *Chem. Abstr.*, **82,** 156307 (1975).
126. B. Stanovnik and M. Tisler, *Angew. Chem. Int. Edn.*, **5,** 605 (1966).
127. P. M. Kochergin, A. N. Krasovskii, and A. B. Roman, *USSR Patent*, 355,174; through *Chem. Abstr.*, **78,** 72142 (1973).
128. E. A. Kuznetsova, S. V. Zhuravlev, T. N. Stepanova, and V. S. Troitskaya, *Chem. Heterocycl. Compds.*, **1972,** 156; *Khim. Geterotsikl. Soedin.*, **1972,** 177; *Chem. Abstr.*, **76,** 153674 (1972).
129. G. F. Duffin and J. D. Kendall, *J. Chem. Soc.*, **1956,** 361; **1951,** 734; *British Patent*, 634,951; *Chem. Abstr.*, **44,** 9287c (1950).
130. A. L. Misra, *J. Org. Chem.*, **23,** 897 (1958).
131. I. I. Chizhevskaya, L. I. Gapanovich, and L. V. Poznyak, *J. Gen. Chem. USSR*, **33,** 945 (1963); *Zh. Obshch. Khim.*, **33,** 945 (1963); *Chem. Abstr.*, **59,** 8725a (1963).
132. I. I. Chizhevskaya, L. I. Gapanovich, and L. V. Poznyak, *J. Gen. Chem. USSR*, **35,** 1282 (1965); *Zh. Obshch. Khim.*, **35,** 1276 (1965); *Chem. Abstr.*, **63,** 11539*f* (1965).
133. H. S. Chaudhary, C. S. Panda, and H. K. Pujari, *Indian J. Chem.*, **8,** 10 (1970); *Chem. Abstr.*, **72,** 121432 (1970).
134. A. N. Krasovskii, P. M. Kochergin, and A. B. Roman, *Chem. Heterocycl. Compds.*, **7,** 767 (1971); *Khim. Geterotsikl. Soedin.*, **7,** 822 (1971); *Chem. Abstr.*, **76,** 25169 (1972).
135. P. N. Dhal and A. Nayak, *J. Indian Chem. Soc.*, **52,** 1193 (1975); *Chem. Abstr.*, **85,** 22737 (1976).
136. S. P. Singh, S. S. Parmar, and B. R. Pandey, *J. Heterocycl. Chem.*, **14,** 1093 (1977).
137. I. I. Chizhevskaya, N. N. Khovratovich, and Z. M. Grabovskaya, *Chem. Heterocycl. Compds.*, **1968,** 329; *Khim. Geterotsikl. Soedin.*, **1968,** 443; *Chem. Abstr.*, **69,** 86906 (1968).
138. J. A. vanAllan, *J. Org. Chem.*, **21,** 24 (1956).
139. J. M. Singh, *J. Med. Chem.*, **12,** 962 (1969); *Chem. Abstr.*, **71,** 91378 (1969).
140. H. W. Stephen and F. J. Wilson, *J. Chem. Soc.*, **1926,** 2531.
141. P. H. L. Wei and S. C. Bell, *U.S. Patent*, 3,475,424; *Chem. Abstr.*, **72,** 31850 (1970).

142. Y. Akasaki, M. Hatano, and M. Fukuyama, *Tetrahedron Lett.*, **1977,** 275; A. Ohno, T. Morishita, and S. Oka, *Bioorg. Chem.*, **5,** 383 (1976); through *Chem. Abstr.*, **86,** 116700 (1977).
143. W. D. Ollis and C. A. Ramsden, *Adv. Heterocycl. Chem.*, **19,** 1 (1976).
144. G. Hasegawa and A. Kotani, *Jap. Patent*, 74 95,997; through *Chem. Abstr.*, **82,** 156299 (1975).
145. K. C. Liu, J. Y. Tuan, and B. J. Shih, *J. Chin. Chem. Soc. (Taipei)*, **24,** 65 (1977); *Chem. Abstr.*, **87,** 135200 (1977).
146. N. M. Turkevich and O. F. Lymar, *Ukr. Khim. Zh.*, **27,** 503 (1961); through *Chem. Abstr.*, **56,** 7296c (1962).
147. G. C. A. Bellinger, A. Davidson, T. J. King, A. McKillop, and P. N. Preston, *Tetrahedron Lett.*, **1978,** 2621.
148. M. Lacova and F. Volna, *Acta Fac. Rerum Natur. Univ. Comenianae, Chim.*, **1972,** 1; *Chem. Abstr.*, **79,** 18638 (1973).
149. N. M. Turkevich and O. F. Lymar, *J. Gen. Chem. USSR*, **31,** 1523 (1961); *Zh. Obshch. Khim.*, **31,** 1635 (1961); *Chem. Abstr.*, **55,** 23503h (1961).
150. E. I. Grinblat and I. Y. Postovskii, *J. Gen. Chem. USSR*, **31,** 357 (1961); *Zh. Obshch. Khim.*, **31,** 394 (1961); *Chem. Abstr.*, **55,** 22298i (1961).
151. M. O. Lozinskii, A. F. Shivanyuk, and P. S. Pelkis, *Chem. Heterocycl. Compds.*, **7,** 869 (1971); *Khim. Geterotsikl. Soedin.*, **7,** 930 (1971); *Chem. Abstr.*, **76,** 34200 (1972); *Chem. Heterocycl. Compds.*, **7,** 439 (1971); *Khim. Geterotsikl. Soedin*, **7,** 471 (1971); *Chem. Abstr.*, **76,** 25184 (1972); *Dopov. Akad. Nauk Ukr. RSR Ser. B*, **31,** 1096 (1969); through *Chem. Abstr.*, **73,** 14767 (1970); *USSR Patent* 256,774; through *Chem. Abstr.*, **72,** 132733 (1970).
152. R. D. Haugwitz, B. V. Maurer, and V. L. Narayanan, *J. Org. Chem.*, **39,** 1359 (1974); R. D. Haugwitz and V. L. Narayanan, *U.S. Patent*, 3,665,007; *Chem. Abstr.*, **77,** 62001 (1972); *U.S. Patent*, 3,819,618; *Chem. Abstr.*, **81,** 105520 (1974); *Ger. Patent*, 2,121,395; *Chem. Abstr.*, **76,** 85817 (1972); *Ger. Patent*, 2,202,704; *Chem. Abstr.*, **77,** 140081 (1972).
153. R. D. Haugwitz, B. V. Maurer, and V. L. Narayanan, *Chem. Commun.*, **1971,** 1100.
154. Y. Tamura, H. Hayashi, and M. Ikeda, *J. Heterocycl. Chem.*, **12,** 819 (1975).
155. H. Ogura, H. Takayanagi, Y. Yamazaki, S. Yonezawa, H. Takagi, S. Kobayashi, T. Kamioka, and K. Kamoshita, *J. Med. Chem.*, **15,** 923 (1972); *Chem. Abstr.*, **77,** 160002 (1972).
156. H. Ogura, *Ger. Patent* 2,003,825; *Chem. Abstr.*, **74,** 53787 (1971).
157. V. A. Anisimova, A. M. Simonov, and A. F. Pozharskii, *Chem. Heterocycl. Compds.*, **1973,** 731; *Khim. Geterotsikl. Soedin.*, **1973,** 797; *Chem. Abstr.*, **79,** 92108 (1973).
158. V. S. Ponomar and P. M. Kochergin, *Chem. Heterocycl. Compds.*, **1972,** 229; *Khim. Geterotsikl. Soedin.*, **1972,** 253; *Chem. Abstr.*, **76,** 140650 (1972); *USSR Patent*, 230,827; through *Chem. Abstr.*, **70,** 87816 (1969); V. S. Ponomar and N. G. Kasyanenko, *Khim. Issled. Farm.*, **1970,** 52; through *Chem. Abstr.*, **76,** 3757 (1972).
159. V. P. Arya, F. Fernandes, and V. Sudarsanam, *Indian J. Chem.*, **10,** 598 (1972); *Chem. Abstr.*, **78,** 4212 (1973).
160. H. Ogura, T. Itoh, H. Takayanagi, Y. Yamazaki, and H. Takagi, *Ger. Patent*, 2,131,330; *Chem. Abstr.*, **76,** 85816 (1972).
161. A. M. Simonov and P. M. Kochergin, *Chem. Heterocycl. Compd.*, **1965,** 210; *Khim. Geterotsikl. Soedin. Akad. Nauk Latv. SSR*, **1965,** 316; *Chem. Abstr.*, **63,** 6994d (1965); *Khim. Geterotsikl. Soedin. Sb.* 1: *Azotsoderzhashchie Geterotsikl.*, **1967,** 133; through *Chem. Abstr.*, **70,** 96712 (1969).
162. (a) A. M. Simonov and V. A. Anisimova, *Chem. Heterocycl. Compd.*, **1968,** 801; *Khim. Geterotsikl. Soedin.*, **1968,** 1102; *Chem. Abstr.*, **70,** 77868 (1969); (b) A. M. Simonov, A. A. Belous, V. A. Anisimova, and S. V. Ivanovskaya, *Chem. Pharm. J.*, **3,** 4 (1969); *Khim. Farm. Zh.*, **3,** 7 (1969); *Chem. Abstr.*, **71,** 81267 (1969).
163. V. A. Anisimova and A. M. Simonov, *Chem. Heterocycl. Compd.*, **1975,** 222; *Khim. Geterotsikl. Soedin.*, **1975,** 258; *Chem. Abstr.*, **82,** 170796 (1975).

164. V. A. Anisimova, N. I. Avdyunina, A. M. Simonov, G. V. Kovalev, and S. M. Gofman, *Chem. Heterocycl. Compd.*, **1976,** 114; *Khim. Geterotsikl. Soedin.*, **1976,** 126; *Chem. Abstr.*, **85,** 21209 (1976).
165. A. M. Simonov, V. A. Anisimova, and L. E. Grushina, *Chem. Heterocycl. Compd.*, **1970,** 778; *Khim. Geterotsikl. Soedin.*, **1970,** 838; *Chem. Abstr.*, **73,** 109739 (1970).
166. V. A. Anisimova, A. M. Simonov, and T. A. Borisova, *Chem. Heterocycl. Compd.*, **1973,** 726; *Khim. Geterotsikl. Soedin.*, **1973,** 791; *Chem. Abstr.*, **79,** 105140 (1973).
167. L. M. Werbel and M. L. Zamora, *J. Heterocycl. Chem.*, **2,** 287 (1965).
168. F. Kröhnke, *Angew. Chem. Int. Edn.*, **2,** 232 (1963).
169. I. I. Popov, P. V. Tkachenko, and A. M. Simonov, *Chem. Heterocycl. Compd.*, **1975,** 347; *Khim. Geterotsikl. Soedin.*, **1975,** 396; *Chem. Abstr.*, **83,** 28154 (1975); *USSR Patent*, 414,260; through *Chem. Abstr.*, **80,** 133437 (1974).
170. I. I. Popov, P. V. Tkachenko, and A. M. Simonov, *Chem. Heterocycl. Compd.*, **1975,** 461; *Khim. Geterotsikl. Soedin.*, **1975,** 523; *Chem. Abstr.*, **83,** 79148 (1975).
171. Y. V. Koshchienko, G. M. Suvorova, and A. M. Simonov. *Chem. Heterocycl. Compd.*, **1975,** 124; *Khim. Geterotsikl. Soedin.*, **1975,** 140; *Chem Abstr.*, **82,** 140013 (1975); *Chem. Heterocycl. Compd.*, 1977, 94; *Khim. Geterotsikl. Soedin.*, **1977,** 111; *Chem. Abstr.*, **86,** 189804 (1977).
172. V. A. Anisimova and A. M. Simonov, *Chem. Heterocycl. Compd.*, **1976,** 110; *Khim. Geterotsikl. Soedin*, **1976,** 121; *Chem. Abstr.*, **85,** 21208 (1976).
173. A. M. Simonov, V. A. Anisimova, and T. A. Borisova, *Chem. Heterocycl. Compd.*, **1973,** 99; *Khim. Geterotsikl. Soedin.*, **1973,** 111; *Chem. Abstr.*, **78,** 97556 (1973).
174. A. M. Simonov, T. A. Kuzmenko, and L. G. Nachinennaya, *Chem. Heterocycl. Compd.*, **1975,** 1188; *Khim. Geterotsikl. Soedin.*, **1975,** 1394; *Chem. Abstr.*, **84,** 43937 (1976).
175. A. M. Simonov and V. A. Anisimova, *Chem. Heterocycl. Compd.*, **7,** 632 (1971); *Khim. Geterotsikl. Soedin.*, **7,** 673 (1971); *Chem. Abstr.*, **76,** 126866 (1972).
176. R. J. North and A. R. Day, *J. Heterocycl. Chem.*, **6,** 655 (1969).
177. A. C. White and R. M. Black, *U.S. Patent*, 3,989,709; *Chem. Abstr.*, **86,** 72694 (1977).
178. A. Hunger, J. Kebrle, A. Rossi, and K. Hoffmann, *Helv. Chim. Acta*, **44,** 1273 (1961).
179. T. Aka, Y. Watanabe, and M. Aida, *Yakugaku Zasshi*, **86,** 665 (1966); *Chem. Abstr.*, **65,** 16959h (1966).
180. T. A. Borisova, A. M. Simonov, and V. A. Anisimova, *Chem. Heterocycl. Compd.*, **1973,** 736; *Khim. Geterotsikl. Soedin.*, **1973,** 803; *Chem. Abstr.*, **79,** 92104 (1973).
181. F. Troxler and H. P. Weber, *Helv. Chim. Acta*, **57,** 2356 (1974).
182. C. W. Bird, *J. Chem. Soc.*, **1964,** 5284.
183. C. W. Bird and J. D. Twibell, *Tetrahedron*, **28,** 2813 (1972).
184. C. W. Bird, M. W. Kaczmar, and C. K. Wong, *Tetrahedron*, **30,** 2549 (1974).
185. A. M. Simonov, Y. V. Koshchienko, G. M. Suvorova, B. A. Tertov, and E. N. Malysheva, *Chem. Heterocycl. Compd.*, **1976,** 1151; *Khim. Geterotsikl. Soedin.*, **1976,** 1391; *Chem. Abstr.*, **86,** 72521 (1977); A. M. Simonov, G. M. Suvorova, and Y. V. Koshchienko, *USSR Patent*, 485,116; through *Chem. Abstr.*, **84,** 31068 (1976).
186. M. O. Lozinskii, A. F. Shivanyuk, and P. S. Pelkis, *USSR Patent*, 261,385; through *Chem. Abstr.*, **73,** 14848 (1970).
187. H. J. Scholl and E. Klause, *Ger. Patent*, 2,210,884; *Chem. Abstr.*, **79,** 137149 (1973); H. J. Scholl, E. Klause, F. Grewe, and I. Hammann, *U.S. Patent*, 3,934,019; *Chem. Abstr.*, **85,** 46674 (1976); *Ger. Patent*, 2,065,977; *Chem. Abstr.*, **87,** 201537 (1977).
188. H. Schubert, H. Lettau, and J. Fischer, *Tetrahedron*, **30,** 1231 (1974).
189. V. M. Aryuzina and M. N. Shchukina, *Chem. Heterocycl. Compd.*, **1966,** 460; *Khim. Geterotsikl. Soedin.*, **1966,** 605; *Chem. Abstr.*, **66,** 94952 (1967).
190. V. M. Aryuzina and M. N. Shchukina, *Chem. Heterocycl. Compd.*, **1968,** 375; *Khim. Geterotsikl. Soedin.*, **1968,** 506; *Chem. Abstr.*, **69,** 96572 (1968).
191. V. M. Aryuzina and M. N. Shchukina, *Chem. Heterocycl. Compd.*, **1968,** 377; *Khim. Geterotsikl. Soedin.*, **1968,** 509; *Chem. Abstr.*, **69,** 96573 (1968).
192. V. M. Aryuzina, and M. N. Shchukina, *Chem. Heterocycl. Compd.*, **1970,** 486; *Khim. Geterotsikl. Soedin.*, **1970,** 525; *Chem. Abstr.*, **73,** 87845 (1970).

193. V. M. Aryuzina and M. N. Shchukina, *Chem. Heterocycl. Compd.*, **1973,** 366; *Khim. Geterotsikl. Soedin.*, **1973,** 395; *Chem. Abstr.*, **78,** 147870 (1973).
194. H. Schubert and J. Fischer, *Z. Chem.*, **11,** 9 (1971); *Chem. Abstr.*, **74,** 111961 (1971).
195. A. J. Hubert, *J. Chem. Soc.* (*C*), **1969,** 1334.
196. B. Rudner, *U.S. Patent*, 2,790,172; *Chem. Abstr.*, **51,** 13934b (1957).
197. G. S. Chekrii and I. V. Smolanka, *Ukr. J. of Chem.*, **40,** 72 (1974); *Ukr. Khim. Zh.*, **40,** 635 (1974); *Chem. Abstr.*, **81,** 105396 (1974); cf. also T. A. Krasnitskaya, I. V. Smolanka, and A. L. Vais, *Chem. Heterocycl. Compd.*, **1973,** 395; *Khim. Geterotsikl. Soedin.*, **1973,** 424; *Chem. Abstr.*, **78,** 159509 (1973).
198. K. H. Menzel, O. Wahl, and W. Pelz, *Ger. Patent*, 1,070,030; *Chem. Abstr.*, **55**, 23138i (1961).
199. O. Wahl and K. H. Menzel, *Ger. Patent*, 1,099,349; *Chem. Abstr.*, **56,** 10333a (1961).
200. K. H. Menzel and W. Pueschel, *Mitt. Forschungslab. Agfa-Gevaert AG Leverkusen-Muenchen*, **4,** 376 (1964); *Chem. Abstr.*, **64,** 3734c (1966).
201. M. Yoshida, A. Okumura, and Y. Aotani, *Ger. Patent*, 2,156,111; *Chem. Abstr.*, **77,** 101610 (1972).
202. K. Loeffler and K. H. Menzel, *Belg. Patent*, 621,241; *Chem. Abstr.*, **59,** 7534h (1963).
203. K. H. Menzel and R. Puetter, *Belg. Patent*, 643,802; *Chem. Abstr.*, **63,** 4440e (1965).
204. H. Schellenberger, W. Pueschel, K. Loeffler, W. Pelz, and K. H. Menzel, *Ger. Patent*, 1,116,534; *Chem. Abstr.*, **56,** 10329h (1962).
205. M. Fujiwara and Y. Kojima, *Jap. Patent*, 71 10,068; through *Chem. Abstr.*, **75,** 22490 (1971).
206. *Brit. Patent*, 1,241,069; *Chem. Abstr.*, **76,** 87183 (1972).
207. S. Mignonac-Mondon, J. Elguero, and R. Lazaro, *Compt. Rend.* (*C*), **276,** 1533 (1973).
208. M. A. Khan and V. L. T. Ribeiro, *Heterocycles*, **6,** 979 (1977); *Chem. Abstr.*, **87,** 135188 (1977).
209. R. Gompper and F. Effenberger, *Chem. Ber.*, **92,** 1928 (1959).
210. V. M. Dziomko and A. V. Ivashchenko, *J. Gen. Chem. USSR*, **43,** 1322 (1973); *Zh. Obshch. Khim.*, **43,** 1330 (1973); *Chem. Abstr.*, **79,** 60988 (1973).
211. I. I. Chizhevskaya, M. I. Zavadskaya, and N. N. Khovratovich, *Chem. Heterocycl. Compd.*, **1968,** 730; *Khim. Geterotsikl. Soedin.*, **1968,** 1008; *Chem. Abstr.*, **70,** 68257 (1969).
212. L. M. Alekseeva, E. M. Peresleni, Y. N. Sheinker, P. M. Kochergin, A. N. Krasovskii, and B. V. Kurmaz, *Chem. Heterocycl. Compd.*, **1972,** 1017; *Khim. Geterotsikl. Soedin.*, **1972,** 1125; *Chem. Abstr.*, **77,** 139322 (1972).
213. N. I. Avdyunina, V. A. Anisimova, and A. M. Simonov, *Chem. Heterocycl. Compd.*, **1974,** 1389; *Khim. Geterotsikl. Soedin.*, **1974,** 1577; *Chem. Abstr.*, **83,** 114295 (1975).
214. V. A. Anisimova, N. I. Avdyunina, A. M. Simonov, G. V. Kovalev, and S. M. Gofman, *Chem. Heterocycl. Compd.*, **1976,** 1365; *Khim. Geterotsikl. Soedin.*, **1976,** 1660; *Chem. Abstr.*, **86,** 171326 (1977).
215. V. M. Aryuzina and M. N. Shchukina, *Chem. Heterocycl. Compd.*, **1968,** 806; *Khim. Geterotsikl. Soedin.*, **1968,** 1108; *Chem. Abstr.*, **70,** 77869 (1969).
216. V. M. Aryuzina and M. N. Shchukina, *Chem. Pharm. J.*, **6,** 218 (1972); *Khim. Farm. Zh.*, **6,** 22 (1972); *Chem. Abstr.*, **77,** 48403 (1972).
217. V. A. Grin and N. G. Krasyanenko, *Khim. Issled. Farm.*, **1970,** 18; through *Chem. Abstr.*, **76,** 52165 (1972).
218. T. V. Bobkova and I. A. Soloveva, *Tr. Vses. Nauch. Issled Proekt. Inst. Khim. Fotogr. Prom.*, **1968,** 60; *Chem. Abstr.*, **74,** 127523 (1971).
219. K. Shiba, A. Sato, and M. Hinata, *Ger. Patent*, 2,318,761; *Chem. Abstr.*, **80,** 9061 (1974).
220. V. M. Dziomko and A. V. Ivashchenko, *J. Gen. Chem. USSR*, **45,** 405 (1975); *Zh. Obshch. Khim.*, **45,** 418 (1975); *Chem. Abstr.*, **82,** 105798 (1975).
221. V. M. Aryuzina and M. N. Shchukina, *Chem. Heterocycl. Compd.*, **1972,** 361; *Khim. Geterotsikl. Soedin.*, **1972,** 396; *Chem. Abstr.*, **77,** 88396 (1972).

222. G. G. Dvoryantseva, L. M. Alekseeva, T. N. Ulyanova, Y. N. Sheinker, P. M. Kochergin, and A. N. Krasovskii, *Chem. Heterocycl. Compd.*, **7,** 875 (1971); *Khim. Geterotsikl. Soedin.*, **7,** 937 (1971); *Chem. Abstr.*, **75,** 156761 (1971).
223. J. Elguero, A. Fruchier, L. Knutsson, R. Lazaro, and J. Sandstrom, *Can. J. Chem.* **52,** 2744 (1974).
224. G. G. Dvoryantseva, T. N. Ulyanova, G. P. Syrova, Y. N. Sheinker, V. M. Aryuzina, T. P. Sycheva, and M. N. Shchukina, *Teor. Eksp. Khim.*, **6,** 23 (1970); through *Chem. Abstr.*, **73,** 29063 (1970).
225. O. S. Anisimova, Y. N. Sheinker, P. M. Kochergin, and A. N. Krasovskii, *Chem. Heterocycl. Compd.*, **1974,** 674; *Khim. Geterotsikl. Soedin.*, **1974,** 778; *Chem. Abstr.*, **81,** 151141 (1974).
226. J. P. Fayet, M. C. Vertut, P. Mauret, J. De Mendoza, and J. Elguero, *J. Heterocycl. Chem.*, **12,** 197 (1975).
227. A. M. Simonov, V. A. Anisimova, and N. K. Shub, *Chem. Heterocycl. Compd.*, **1970,** 909; *Khim. Geterotsikl. Soedin.*, **1970,** 977; *Chem. Abstr.*, **74,** 76372 (1971).
228. D. J. Fry, G. E. Ficken, C. J. Palles, and A. W. Yates, *Brit. Patent,* 1, 392,499; *Chem. Abstr.*, **83,** 149115 (1975).
229. R. M. Acheson, J. K. Stubbs, C. A. R. Baxter, and D. E. Kuhla, *Ger. Patent,* 2,444,890; *Chem. Abstr.*, **83,** 79230 (1975).
230. G. Wolfrum, R. Puetter, and K. H. Menzel, *Ger. Patent,* 1,234,891; *Chem. Abstr.*, **66,** 105886 (1967).
231. A. Mustafa, M. I. Ali, and A. Abou-State, *Ann. Chem.*, **740,** 132 (1970).
232. I. I. Chizhevskaya, L. I. Gapanovich, and R. S. Kharchenko, *Puti Sin. Izyskaniya Protivoopukholevykh Prep. Tr. Simp. Khim. Protivoopukholevykh Veshchestv.*, **1967,** 62; *Chem. Abstr.*, **70,** 106421 (1969); I. I. Chizhevskaya and M. I. Zavadskaya, *Khim. Geterotsikl, Soedin.*, **1971,** 93; through *Chem. Abstr.*, **77,** 164592 (1972).
233. E. J. Poppe, *Brit. Patent,* 1,072,384; *Chem. Abstr.*, **67,** 69446 (1967).
234. B. A. Porai-Koshits, I. Y. Kvitko, and E. A. Shutkova, *Latvijas PSR Zinatnu Akad. Vestis Kim. Ser.*, **1965,** 587; through *Chem. Abstr.*, **64,** 8168h (1966).
235. J. M. Singh, *J. Med. Chem.*, **13,** 1018 (1970); *Chem. Abstr.*, **73,** 109742 (1970).
236. K. H. Menzel, W. Wirth, and H. Kreiskott, *Ger. Patent,* 1,178,075; *Chem. Abstr.*, **61,** 14681h (1964).
237. *Fr. Patent,* M 3709; *Chem. Abstr.*, **66,** 79557 (1967).
238. T. A. Kuzmenko, V. A. Anisimova, N. I. Avdyunina, and A. M. Simonov, *Khim. Geterotsikl. Soedin.*, **1978,** 522; through *Chem. Abstr.*, **89,** 43247 (1978).
239. S. Kano, *Yakugaku Zasshi,* **92,** 927 (1972); through *Chem. Abstr.*, **77,** 114304 (1972); S. Kano and T. Noguchi, *Jap. Patent,* 73 38,720; through *Chem. Abstr.*, **81,** 3944 (1974).
240. W. Scholl and G. Dittmar, *Belgian Patent,* 626,394; through *Chem. Abstr.*, **60,** 13360c (1964); H. Wunderlich and K. H. Menzel, *Belgian Patent,* 642,347; through *Chem. Abstr.*, **63,** 4428b (1965); *Brit. Patent,* 951,113; *Chem. Abstr.*, **62,** 6599c (1965); N. Yamada and T. Kikuchi, *Jap. Patent,* 69 08,118; through *Chem. Abstr.*, **71,** 62278 (1969); *Jap. Patent,* 69 08,119; through *Chem. Abstr.*, **71,** 62277 (1969).
241. G. Hasegawa and A. Kotani, *Jap. Patent,* 75 52,065; through *Chem. Abstr.*, **83,** 206268 (1975).
242. K. H. Menzel, *Belgian Patent,* 658,107; *Chem. Abstr.*, **64,** 8362d (1966).
243. A. M. Simonov, V. A. Anisimova, and Y. V. Koshchienko, *Chem. Heterocycl. Compd.*, **1969,** 140; *Khim. Geterotsikl. Soedin.*, **1969,** 184; *Chem. Abstr.*, **71,** 3324 (1969).
244. D. E. Kuhla, *U.S. Patent,* 3,860,718; *Chem. Abstr.*, **82,** 140133 (1975).
245. R. L. Fenichel, F. J. Gregory, and H. E. Alburn, *Brit. J. Cancer,* **33,** 329 (1976); *Chem. Abstr.*, **85,** 298 (1976).
246. P. Schauer, M. Likar, B. Stanovnik, and M. Tisler, *Biol. Vestn.*, **20,** 65 (1972); *Chem. Abstr.*, **79,** 45759 (1973).
247. J. Mohan, V. K. Chadha, H. S. Chaudhary, B. D. Sharma, H. K. Pujari, and L. N. Mohapatra, *Indian J. Exp. Biol.*, **10,** 37 (1972); through *Chem. Abstr.*, **77,** 43730 (1972).

248. H. J. Scholl, E. Klauke, F. Grewe, and I. Hammann, *Ger. Patent*, 2,062,348; *Chem. Abstr.*, **77,** 114391 (1972).
249. S. V. Ivanovskaya, *Sb. Nauch. Rab. Volgograd. Gos. Med. Inst.*, **22,** 139 (1969); through *Chem. Abstr.*, **75,** 47311 (1971); G. V. Kovalev, I. S. Morozov, and I. N. Tyurenkov, *Farmakol. Toksikol. (Moscow)*, **37,** 558 (1974); *Chem. Abstr.*, **82,** 361 (1975).
250. S. V. Ivanovskaya, *Sb. Nauch. Rab., Volgograd. Med. Inst.*, **21,** 175 (1968) through *Chem. Abstr.*, **73,** 75494 (1970); *Sb. Nauch. Rab. Volgograd. Gos. Med. Inst.*, **22,** 142 (1969); through *Chem. Abstr.*, **75,** 33574 (1971); *Sb. Nauch. Rab., Volgograd. Gos. Med. Inst.*, **23,** 228 (1970); through *Chem. Abstr.*, **75,** 47283 (1971); G. V. Kovalev, S. M. Gofman, S. V. Ivanovskaya, M. V. Panshina, V. I. Petrov, A. M. Simonov, and I. N. Tyurenkov, *Pharmacol. and Toxicol. (Moscow)*, **36,** 88 (1973); *Farmakol, and Toksikol. (Moscow)*, **36,** 232 (1973); *Chem. Abstr.*, **78,** 154693 (1973); M. V. Panshina, *Mater., Povolzh. Konf. Fiziol. Uchastiem Biokhim., Farmakol. Morfol. 6th* **2,** 49 (1973); through *Chem. Abstr.*, **82,** 80662 (1975); I. N. Tyurenkov, *Mater., Povolzh. Konf. Fiziol. Uchastiem Biokhim., Farmakol. Morfol. 6th* **2,** 63 (1973); through *Chem. Abstr.*, **82,** 118900 (1975).
251. J. D. Kendall and G. F. Duffin, *Brit. Patent*, 730,489; *Chem. Abstr.*, **49,** 15580b (1955); *Brit. Patent*, 734,792; *Chem. Abstr.*, **50,** 1502h (1956); J. D. Kendall, G. F. Duffin, and H. R. J. Waddington, *Brit. Patent*, 743,133; *Chem. Abstr.*, **51,** 899b (1957); J. D. Kendall and G. F. Duffin, *Brit. Patent*, 749,189; *Chem. Abstr.*, **51,** 904g (1957); *Brit. Patent*, 749,190; *Chem. Abstr.*, **51,** 902e (1957); *Brit. Patent*, 749,193; *Chem. Abstr.*, **50,** 16492f (1956); H. R. J. Waddington, G. F. Duffin, and J. D. Kendall, *Brit. Patent*, 785,334; *Chem. Abstr.*, **52,** 6030g (1958); G. F. Duffin, D. J. Fry, and J. D. Kendall, *Brit. Patent*, 785,939; *Chem. Abstr.*, **52,** 10777e (1958); E. J. Poppe, *Veroeffentl. Wiss. Photo-Lab. Wolfen*, **10,** 115 (1965); *Chem. Abstr.*, **65,** 9995b (1966).
252. *Belgian Patent*, 668,594; *Chem. Abstr.*, **65,** 3213b (1966); E. J. Poppe, *East Ger. Patent*, 49,396; *Chem. Abstr.*, **66,** 50702 (1967); *Ger. Patent*, 1,235,738; *Chem. Abstr.*, **66,** 110064 (1967).
253. T. V. Bobkova and I. A. Soloveva, *Tr. Vses. Nauch.-Issled. Proekt. Inst. Khim. Fotogr. Prom.*, **1968,** 60; through *Chem. Abstr.*, **74,** 127523 (1971).
254. *Brit. Patent*, 927,614; *Chem. Abstr.*, **62,** 2853d (1965); R. Mersch and F. Muenz, *Ger. Patent*, 1,225,320; through *Chem. Abstr.*, **66,** 3810 (1967); B. Sohngen and A. Brack, *Ger. Patent*, 2,415,055; *Chem. Abstr.*, **84,** 32599 (1976).
255. G. Schaum and K. H. Menzel, *Ger. Patent*, 1,158,836; *Chem. Abstr.*, **60,** 6388b (1964); H. Vetter, W. Pueschel, A. Melzer, and M. Peters, *Ger. Patent*, 2,505,248; through *Chem. Abstr.*, **85,** 200531 (1976).
256. K. H. Menzel and H. Ulrich, *Ger. Patent*, 1,127,220; *Chem. Abstr.*, **57,** 16811d (1962); M. Iwama, I. Inoue, and T. Hanzawa, *Ger. Patent*, 1,804,167; *Chem. Abstr.*, **73,** 16321 (1970); K. Shiba, M. Hinata, S. Kubodera, Y. Hayakawa, and S. Moriuchi, *Ger. Patent*, 2,323,462; *Chem. Abstr.*, **80,** 114771 (1974); M. Fujihara, Y. Takai, T. Endo, and T. Masukawa, *Jap. Patent*, 74 53,435; through *Chem. Abstr.*, **81,** 162088 (1974); K. Shiba, M. Hinata, R. Oki, and T. Shishido, *Ger. Patent*, 2,414,869; *Chem. Abstr.*, **82,** 9976 (1975); A. Arai, K. Shiba, M. Yamada, N. Furutachi, and K. Nakamura, *Ger. Patent*, 2,528,845; *Chem. Abstr.*, **85,** 134259 (1976); A. Arai, K. Nakamura, M. Yamada, and N. Furutachi, *Ger. Patent*, 2,532,225; *Chem. Abstr.*, **85,** 134266 (1976); T. Endo, S. Sato, S. Kikuchi, K. Takabe, H. Imamura, T. Kozima, and T. Usui, *Ger. Patent*, 2,607,040; *Chem. Abstr.*, **85,** 200534 (1976); T. Endo, S. Sato, S. Kikuchi, K. Takabe, H. Imamura, T. Kozima, and T. Usui, *Ger. Patent*, 2,607,648; *Chem. Abstr.*, **85,** 184815 (1976); T. Kojima, S. Sato, K. Takabe, T. Endo, H. Sugita, and H. Imamura, *Jap. Patent*, 76 112,341; through *Chem. Abstr.*, **86,** 56766 (1977); K. Takabe, S. Sato, T. Kojima, T. Endo, and T. Usui, *Jap. Patent*, 76 112,344; through *Chem. Abstr.*, **86,** 56765 (1977); E. Boeckly, W. Himmelmann, E. Meier, W. Sauertag, I. Boie, and P. Bergthaller, *Ger. Patent*, 2,517,408; *Chem. Abstr.*, **87,** 76356 (1977).
257. H. Alper and M. S. Wolin, *J. Org. Chem.*, **40,** 437 (1975).

258. H. Alper and M. S. Wolin, *J. Organometal. Chem.*, **99,** 385 (1975).
259. C. C. Beard, *Ger. Patent,* 2,446,119; *Chem. Abstr.*, **83,** 28234 (1975); *U.S. Patent,* 3,976,654; *Chem. Abstr.*, **86,** 5464 (1977); *U.S. Patent,* 4,009,164; *Chem. Abstr.*, **86,** 189943 (1977).
260. B. Stanovnik and M. Tisler, *Arch. Pharm.*, **300,** 322 (1967); *Chem. Abstr.*, **67,** 82161 (1967).
261. (a) R. D. Haugwitz, B. Toeplitz, and J. Z. Gougoutas, *Chem. Commun.*, **1977,** 736; (b) R. D. Haugwitz and V. L. Narayanan, *U.S. Patent,* 3,864,353; *Chem. Abstr.*, **82,** 156323 (1975).
262. A. Messmer and A. Gelleri, *Angew. Chem. Int. Edn.*, **6,** 261 (1967).
263. J. De. Mendoza and J. Elguero, *Bull. Soc. Chim. France,* **1974,** 1675.
264. J. A. Van Allan, *U.S. Patent,* 2,891,862; *Chem. Abstr.*, **54,** 4224a (1960).
265. G. A. Reynolds and J. A. van Allan, *J. Org. Chem.*, **24,** 1478 (1959).
266. J. D. Bower and F. P. Doyle, *J. Chem. Soc.*, **1957,** 727.
267. N. P. Bednyagina and I. N. Getsova, *J. Org. Chem. USSR,* **1,** 135 (1965); *Zh. Organ. Khim.*, **1,** 139 (1965); *Chem. Abstr.*, **62,** 16234d (1965).
268. G. N. Tyurenkova and N. P. Bednyagina, *J. Org. Chem. USSR,* **1,** 132 (1965); *Zh. Organ. Khim.*, **1,** 136 (1965); *Chem. Abstr.*, **62,** 16234e (1965).
269. H. Daniel, *Chem. Ber.*, **102,** 1028 (1969).
270. R. I. Fu Ho and A. R. Day, *J. Org. Chem.*, **38,** 3084 (1973).
271. G. S. Chekrii and I. V. Smolanka, *Ukr. J. Chem.*, **40,** 39 (1974); *Ukr. Khim. Zh.*, **40,** 262 (1974); *Chem. Abstr.*, **80,** 133355 (1974).
272. J. De Mendoza, P. Rull, and M. L. Castellanos, *Afinidad,* **35,** 197 (1978); *Chem. Abstr.*, **89,** 129455 (1978).
273. R. Faure, E. J. Vincent, J. Elguero, J. De Mendoza, and P. Rull, *Bull. Soc. Chim. France,* **1978,** II, 273.
274. J. De Mendoza and M. C. Pardo, *An. Quim.*, **71,** 434 (1975); *Chem. Abstr.*, **83,** 205356 (1975).
275. Y. Yasuda, Y. Soeda, A. Ueda, S. Kano, and Y. Kato, *Jap. Patent,* 74 08,852; through *Chem. Abstr.*, **81,** 164732 (1974).
276. Y. Yasuda, Y. Soeda, A. Ueda, S. Kano, and K. Kato, *Jap. Patent,* 74 11,063; through *Chem. Abstr.*, **83,** 73454 (1975).
277. *Belgian Patent,* 559,022; through *Chem. Abstr.*, **54,** 132a (1960).
278. E. Alcalde and R. M. Claramunt, *Tetrahedron Lett.*, **1975,** 1523; cf. also J. D. Bower and F. P. Doyle, *J. Chem. Soc.*, **1957,** 727.
279. R. N. Butler, *Chem. Ind.*, **1973,** 371.

CHAPTER 7

Condensed Benzimidazoles of Type 6-5-6

G. TENNANT

7.1 Tricyclic 6-5-6 Fused Benzimidazoles with No Additional Heteroatom

Benzimidazole-derived 6-5-6 fused heterocycles with no additional heteroatoms conform to a single skeletal type (Scheme 7.1 and Table 7.1) represented by the fully unsaturated pyrido[1,2-*a*]benzimidazole ring system (**7.1**) and its various dihydro (**7.2**)–(**7.7**), tetrahydro (**7.8**)–(**7.10**), and hexahydro (**7.11**) derivatives. The chemistry of pyrido[1,2-*a*]benzimidazole derivatives was briefly reviewed by Mosby in 1961.[1a]

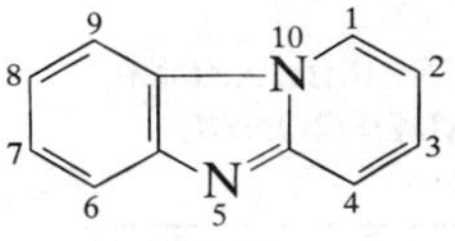

(**7.1**) 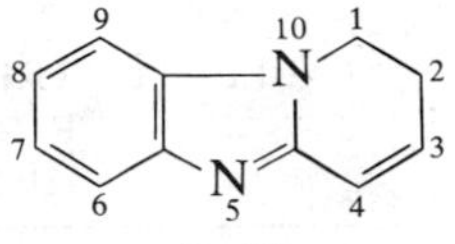 (**7.2**)

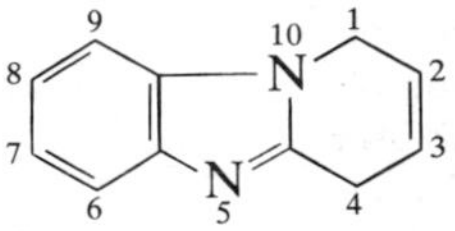

(**7.3**) 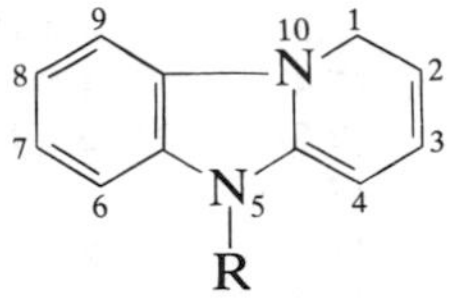 (**7.4**)

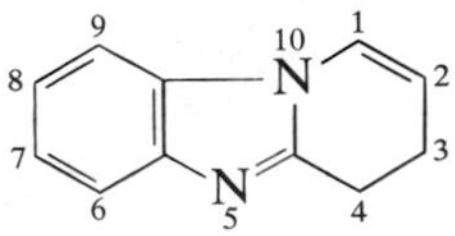

(**7.5**) 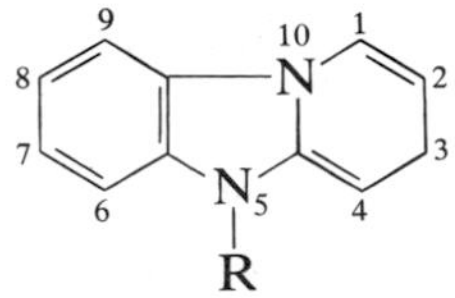 (**7.6**)

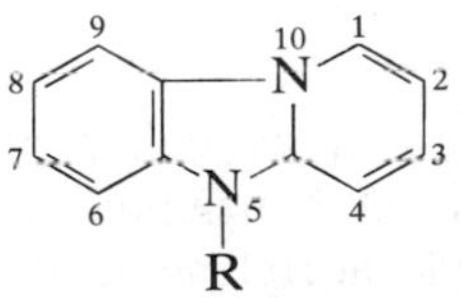

(**7.7**) 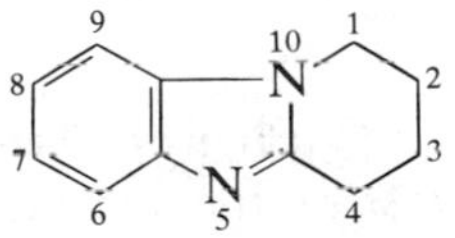 (**7.8**)

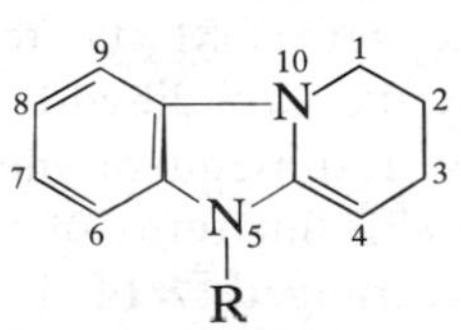

(**7.9**) 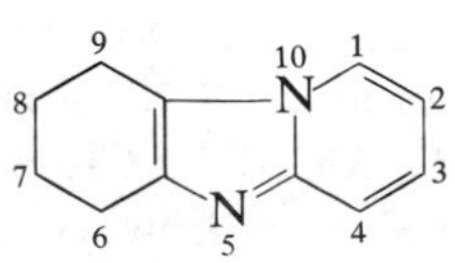 (**7.10**)

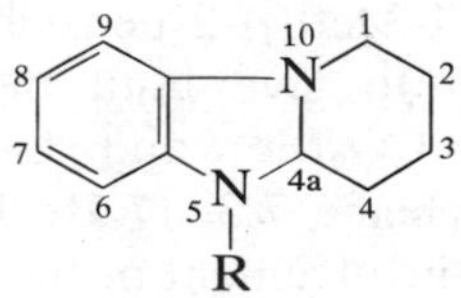

(**7.11**)

Scheme 7.1

TABLE 7.1. TRICYCLIC 6-5-6 FUSED BENZIMIDAZOLE RING SYSTEMS WITH NO ADDITIONAL HETEROATOMS

Structure[a]	Name[b]
(**7.1**)	Pyrido[1,2-*a*]benzimidazole
(**7.2**)	1,2-Dihydropyrido[1,2-*a*]benzimidazole
(**7.3**)	1,4-Dihydropyrido[1,2-*a*]benzimidazole
(**7.4**)	1,5-Dihydropyrido[1,2-*a*]benzimidazole
(**7.5**)	3,4-Dihydropyrido[1,2-*a*]benzimidazole
(**7.6**)	3,5-Dihydropyrido[1,2-*a*]benzimidazole
(**7.7**)	4*a*,5-Dihydropyrido[1,2-*a*]benzimidazole
(**7.8**)	1,2,3,4-Tetrahydropyrido[1,2-*a*]benzimidazole
(**7.9**)	1,2,3,5-Tetrahydropyrido[1,2-*a*]benzimidazole
(**7.10**)	6,7,8,9-Tetrahydropyrido[1,2-*a*]benzimidazole
(**7.11**)	1,2,3,4,4*a*,5-Hexahydropyrido[1,2-*a*]benzimidazole

[a] Cf. Scheme 7.1.
[b] Based on the Ring Index.

7.1.1. Synthesis

Ring-closure Reactions of Benzimidazole Derivatives

Pyrido[1,2-*a*]benzimidazole derivatives are formed as minor products (Table 7.2) of the uncatalyzed thermal reactions of benzimidazoles with acetylenic esters [methyl propiolate, dimethyl acetylenedicarboxylate (DMAD)].[1b] Benzimidazole in particular reacts with three molecules of DMAD to give a 1,5-dihydropyrido[1,2-*a*]benzimidazole derivative [Scheme 7.2; (**7.12**; R = R^1 = H) → (**7.13**; R = $MeO_2C\overset{|}{C}{=}CHCO_2Me$)].[2] In the analogous reaction of 1-methylbenzimidazole with DMAD, the 1,5-dihydro product (**7.13**; R = Me) is accompanied by the 4a,5-dihydro isomer (**7.14**; R = Me, R^1 = CO_2Me, R^2 = H) formed by a subsequent methoxy-carbonyl shift.[2,3] Low yields (Table 7.2) of separable mixtures of unrearranged (**7.14**; R = R^1 = alkyl, R^2 = CO_2Me) and rearranged (**7.14**; R = R^2 = alkyl, R^1 = CO_2Me) 4a,5-dihydropyrido[1,2-*a*]benzimidazoles are likewise produced in the thermal cycloaddition reactions of DMAD with simple 1,2-dialkylbenzimidazoles.[1b,2–4] 1-Methyl-2-benzylbenzimidazole, on the other hand, reacts[5] with DMAD to give both the direct cycloadduct (**7.14**; R = Me, R^1 = CH_2Ph, R^2 = CO_2Me) and the 1,5-dihydropyrido[1,2-*a*]-benzimidazole derivative [Scheme 7.3; (**7.16**; R^1 = Me, R^2 = Ph, R^4 = R^5 = CO_2Me)] resulting from the involvement of the benzyl substituent. Addition involving the side chain also intervenes in the thermal reactions (Scheme 7.3) of acetylenic esters (methyl propiolate, DMAD) with benzimidazoles (**7.15**; R = H or Me, R^2 = CO_2Et or CN) having an active methylene

TABLE 7.2. SYNTHESIS OF PYRIDO[1,2-*c*]BENZIMIDAZOLE DERIVATIVES BY RING-CLOSURE REACTIONS OF BENZIMIDAZOLES WITH ACETYLENIC ESTERS AND METHYL VINYL KETONE

Starting material ($R \rightarrow R^5$ unspecified = H)	Reaction conditions[a]	Product ($R \rightarrow R^4$ unspecified = H)	Yield (%)	m.p. (°C)	Solvent of crystallization	Ref.
(**7.12**)	*A*	(**7.13**; R = $MeO_2CC(=CHCO_2Me)$)	8	191–192	Methanol	2
(**7.12**; R = Me)	*B*	(**7.13**; R = Me)	23	225–226	Methanol	2, 3
		(**7.14**; R = Me, $R^1 = CO_2Me$)	3	173–174	Methanol	
(**7.12**; R = Me)	*C*	(**7.14**; R = Me, $R^1 = CO_2Me$)	11	—	—	3
(**7.12**; R = R^1 = Me)	*D*	(**7.14**; R = R^1 = Me, $R^2 = CO_2Me$) +	1.5	138–139	Methanol	2, 3
		(**7.14**; R = R^2 = Me, $R^1 = CO_2Me$)	0.6	161–162	Methanol	
(**7.12**; R = Me, R^1 = Et)	*E*	(**7.14**; R = Me, R^1 = Et, $R^2 = CO_2Me$)	9	144–146	Methanol	2, 3
(**7.12**; R = Et, R^1 = Me)	*D*	(**7.14**; R = Et, R^1 = Me, $R^2 = CO_2Me$) +	—[b]	146–147	Methanol	2, 3
		(**7.14**; R = Et, $R^1 = CO_2Me$, R^2 = Me)	—[c]			
(**7.12**; R = Me, $R^1 = Pr^i$)	*F*	(**7.14**; R = Me, $R^1 = Pr^i$, $R^2 = CO_2Me$)	65	115–117[d]	Methanol	4
(**7.12**; R = Me, $R^1 = CH_2Ph$)	*G*	(**7.14**; R = Me, $R^1 = CH_2Ph$, $R^2 = CO_2Me$)	13.6	133	Methanol	5
		(**7.16**; R^1 = Me, R^2 = Ph, $R^3 = R^4 = R^5 = CO_2Me$)	—[c]	175	Ethyl acetate	
(**7.12**; R = $CH\overset{t}{=}CHCO_2Me$, $R^1 = Bu^t$)	*F*	(**7.14**; R = $CH\overset{t}{=}CHCO_2Me$, $R^1 = Bu^t$, $R^2 = CO_2Me$)	29	200–201	Methanol	1, 4
(**7.12**; R = $CH\overset{t}{=}CHCO_2Me$, R^1 = Ph)	*F*	(**7.14**; R = $CH\overset{t}{=}CHCO_2Me$, $R^1 = CO_2Me$, R^2 = Ph)	14	190–191	Methanol	1, 4
(**7.15**; R^2 = CN)	*H*	(**7.16**; $R^1 = CH\overset{t}{=}CHCO_2Me$, R^2 = CN, $R^4 = CO_2Me$) +	2.6	159–160	Methanol	6
		(**7.17**; $R^1 = CH\overset{t}{=}CHCO_2Me$, R^2 = CN) +	19	269–271	Methanol	
		(**7.18**; $R^1 = CH\overset{t}{=}CHCO_2Me$, R^2 = CN)	0.16	263–266	Methanol	
(**7.15**; R^2 = CN)	*F*	(**7.17**; R^2 = CN, $R^4 = CO_2Me$)	5	255–256	Methanol	6
(**7.15**; R^2 = CN)	*I*	(**7.18**; R^2 = CN, $R^3 = CO_2Me$)	32	290–292	Dimethylformamide	7
(**7.15**; R^1 = Me, R^2 = CN)	*H*	(**7.16**; R^1 = Me, R^2 = CN, $R^4 = CO_2Me$)	19	171.5–172.5	Methanol	6
(**7.15**; R^1 = Me, R^2 = CN)	*F*	(**7.16**; R^1 = Me, R^2 = CN, $R^3 = R^4 = R^5 = CO_2Me$)	11	213.5–215.5	Methanol	6

TABLE 7.2 (*Continued*)

Starting material ($R \rightarrow R^5$ unspecified = H)	Reaction conditions[a]	Product ($R \rightarrow R^4$ unspecified = H)	Yield (%)	m.p. (°C)	Solvent of crystallization	Ref.
(**7.15**; $R^1 = Me$, $R^2 = CN$)	*I*	(**7.18**; $R^1 = Me$, $R^2 = CN$, $R^3 = CO_2Me$)	30	239	Dimethylformamide	7
(**7.15**; $R^2 = CO_2Et$)	*H*	(**7.17**; $R^1 = CH \overset{t}{=} CHCO_2Me$, $R^3 = CO_2Me$)	6	260 (decomp.)	Methanol	6
(**7.15**; $R^2 = CO_2Et$)	*F*	(**7.16**; $R^2 = CO_2Et$, $R^3 = R^4 = R^5 = CO_2Me$)	7.8	212–213 (decomp.)	Methanol	6
(**7.15**; $R^2 = CO_2Et$)	*I*	(**7.18**; $R^2 = CO_2Et$, $R^3 = CO_2Me$)	16	223–224	Dimethylformamide	7
(**7.15**; $R^1 = Me$, $R^2 = CO_2Et$)	*H*	(**7.16**; $R^1 = Me$, $R^2 = CO_2Et$, $R^4 = CO_2Me$) +	25	172–174	Methanol	6
		(**7.17**; $R^1 = Me$, $R^3 = CO_2Me$)	13	292–293	Methanol	
(**7.15**; $R^1 = Me$, $R^2 = CO_2Et$)	*F*	(**7.16**; $R^1 = Me$, $R^2 = CO_2Et$, $R^3 = R^4 = R^5 = CO_2Me$)	15	146–147	Methanol	6
(**7.29**; $R^1 = Me$)	*J*	(**7.30**; $R^1 = Me$) +	18	202 (decomp.)	Ethanol	11
		(**7.31**; R = H)[e] or	21	119–121	Acetic acid	
		(**7.31**; R = Et)[f]	16	166–168	Ethanol	
(**7.29**; $R^1 = Et$, $R^2 = Cl$)	*J*	(**7.30**; $R^1 = Et$, $R^2 = Cl$)	26	232–233 (decomp.)	Ethanol acetonitrile	11
(**7.29**; $R^1 = Ph$)	*K*	(**7.30**; $R^1 = Ph$)	37	213–214	Ethanol	11

[a] $A = MeO_2CC{\equiv}CCO_2Me$, benzene/(reflux)(16 hr); $B = MeO_2CC{\equiv}CCO_2Me$, acetonitrile/(reflux)(14 hr); $C = MeO_2CC{\equiv}CCO_2Me$, toluene/(0°)(4 hr), then (room temp.)(5 days); $D = MeO_2CC{\equiv}CCO_2Me$, tetrahydrofuran/(room temp.)(2–4 days); $E = MeO_2CC{\equiv}CCO_2Me$, acetonitrile/(room temp.)(4 days); $F = MeO_2CC{\equiv}CCO_2Me$, acetonitrile/(reflux)(1–5 days); $G = MeO_2CC{\equiv}CCO_2Me$, acetonitrile/(reflux)(5 hr); $H = HC{\equiv}CCO_2Me$/(reflux)-(8–12 days); $I = MeO_2CC{\equiv}CCO_2Me$, dimethylformamide/(100°)(0.5–1 hr); $J = MeCOCH{=}CH_2$, acetonitrile/(room temp.)(2–4 weeks), then pyridine/boil 1 min, and treat with $HClO_4$; $K = MeCOCH{=}CH_2$, acetonitrile/(room temp.)(10 days), then 2,6-lutidine/(boil 1 min, and treat with $HClO_4$).

[b] Mixture not separated.

[c] Yield not quoted.

[d] With remelting at 127–128°.

[e] By-product using acetic acid in the workup.

[f] By-product using ethanol in the workup.

Scheme 7.2

Scheme 7.3

substituent at the $C(2)$ position.[6,7] Depending on the nature of the benzimidazole substrate, the acetylenic ester used as reagent, and the reaction conditions, three types of pyrido[1,2-*a*]benzimidazole product result in low yield (Table 7.2) from such reactions. Broadly, reaction with DMAD in acetonitrile affords 1,5-dihydropyrido[1,2-*a*]benzimidazoles of the type (**7.16**; $R^2 = CO_2Et$ or CN, $R^3 \rightarrow R^5 = CO_2Me$),[6] whereas the use of dimethylformamide as the solvent leads to products formulated on the basis of their ^1H NMR absorption as pyrido[1,2-*a*]benzimidazol-1(5*H*)-ones (**7.18**; $R^2 = CO_2Et$ or CN, $R^3 = CO_2Me$).[7] In further contrast, reaction with methyl propiolate in acetonitrile,[6] tends to convert 2-ethoxycarbonylmethyl- and 2-cyanomethylbenzimidazoles into pyrido[1,2-*a*]benzimidazol-3(5*H*)-ones (**7.17**), though an exception is provided by 2-cyanomethyl-1-methylbenzimidazole, which under these conditions yields the 1,5-dihydropyrido-[1,2-*a*]benzimidazole derivative (**7.16**; $R^1 = Me$, $R^2 = CN$, $R^3 = R^5 = H$, $R^4 = CO_2Me$).[6] The reaction[6] of 2-cyanomethylbenzimidazole with methyl propiolate in acetonitrile is also apparently exceptional in giving the 1,5-dihydropyrido[1,2-*a*]benzimidazole (**7.16**; $R^1 = CH{=}CHCO_2Me$, $R^2 = CN$, $R^3 \rightarrow R^5 = H$) and the pyrido[1,2-*a*]benzimidazol-1(5*H*)-one (**7.18**; $R^1 = CH{=}CHCO_2Me$, $R^2 = CN$, $R^3 = H$) as well as the anticipated pyrido[1,2-*a*]-benzimidazol-3(5*H*)-one (**7.17**; $R^1 = CH{=}CHCO_2Me$, $R^2 = CN$, $R^3 = R^4 = H$). The reaction of the benzimidazole derivatives (**7.15**; $R^1 = H$ or Me, $R^2 = CO_2Et$ or CN) with acetylenic esters to give 1,5-dihydropyrido[1,2-*a*]benzimidazoles (**7.16**) and pyrido[1,2-*a*]benzimidazol-3(5*H*)-ones (**7.17**) is explicable (Scheme 7.4) in terms of the formation of a common zwitterionic intermediate (**7.20**), which can cyclize directly [Scheme 7.4; (**7.20**) → (**7.22**) → (**7.24**) → (**7.17**)] or after reaction with a second molecule of acetylenic ester [Scheme 7.4; (**7.20**) → (**7.21**) → (**7.23**) → (**7.16**)] giving the observed products. Pyrido[1,2-*a*]benzimidazol-1(5*H*)-one formation, on the other hand, is the obvious outcome of initial Michael addition of the methylene center in the benzimidazole at the acetylenic triple bond, followed by cyclization of the adduct produced [Scheme 7.4; (**7.15**) → (**7.19**) → (**7.18**)]. The reason for preferential pyrido[1,2-*a*]benzimidazol-1(5*H*)-one formation (and hence Michael addition) in dimethylformamide[7] compared with 1,5-dihydropyrido[1,2-*a*]benzimidazole or pyrido[1,2-*a*]benzimidazol-3(5*H*)-one (and hence zwitterion) formation in acetonitrile[6] is not clear but may be a consequence of the greater basicity of the former solvent compared with the latter. The reactions (Scheme 7.5) of benzimidazolium ylides of the type (**7.25**) with methyl propiolate or dimethyl acetylenedicarboxylate are reported to give low yields of products variously formulated as pyrido-[1,2-*a*]benzimidazol-1(5*H*)-ones (**7.26**)[8,9] or benzimidazolium betaines (**7.27**).[9] However, more recent evidence[10] that the product of the reaction of the ylide (**7.25**; $R = Et$, $R^1 = Me$) with DMAD is in fact the pyrroloquinoxalinone (**7.28**) implies similar structures for the remaining products (Scheme 7.5) of such reactions. The prolonged reaction (Scheme 7.6) of 1-alkyl-2-ethoxycarbonylmethylbenzimidazolium perchlorates (**7.29**) with

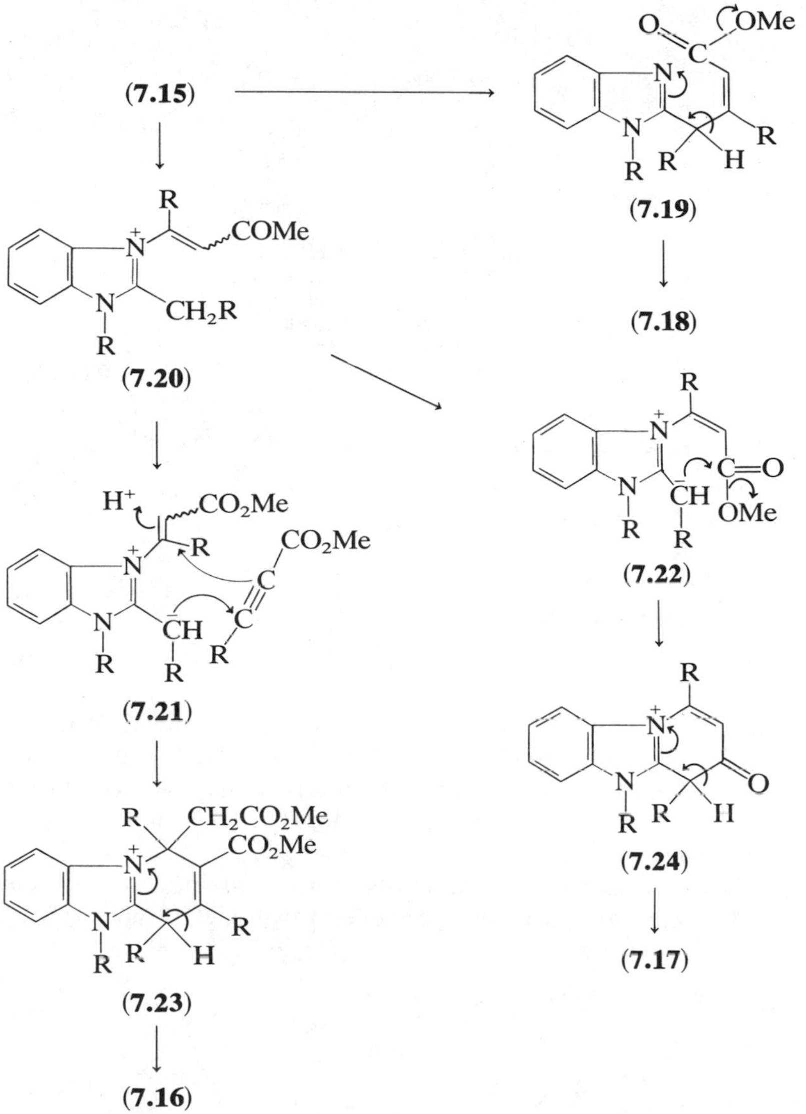

Scheme 7.4

methyl vinyl ketone (MVK) at ambient temperature in acetonitrile affords low yields (Table 7.2) of condensates formulated on the basis of their ^1H NMR absorption as 5-alkyl-1,2-dihydropyrido[1,2-*a*]benzimidazolium perchlorates (**7.30**).[11] The orientation of these products suggests that the initial step in their formation involves either aldol-type condensation between the carbonyl group in MVK and the *C*(2) methylene center in the benzimidazolium salt or more probably addition of *N*(3) in the latter at the double

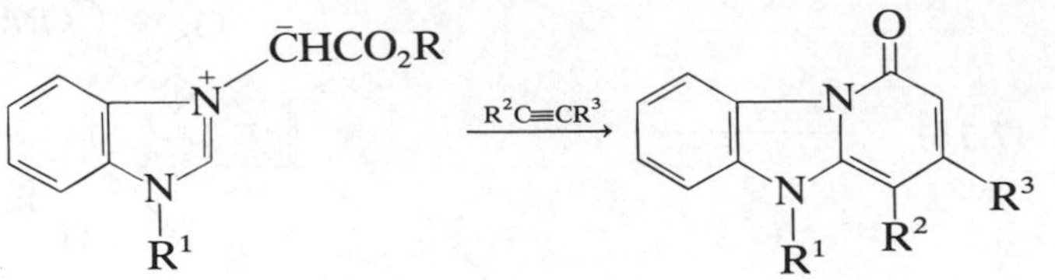

$R^2C{\equiv}CR^3$

(7.25) **(7.26)**

R^1	R^2	R^3
Me	CO_2Et	H
CH_2Ph	CO_2Et	H
Me	CO_2Me	CO_2Me
Et	CO_2Me	CO_2Me

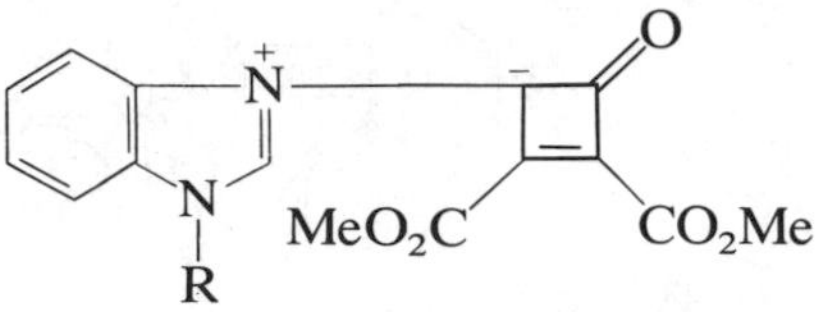

(7.27)
(R = Me or CH_2Ph)

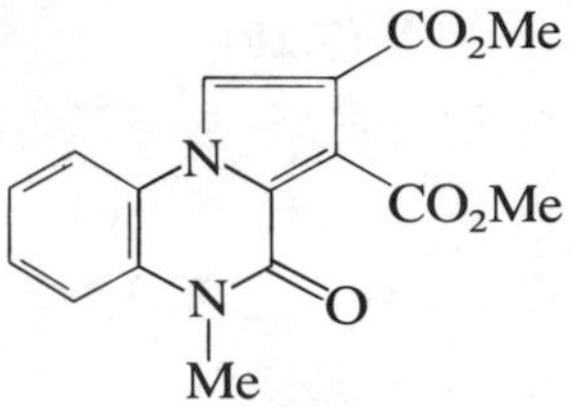

(7.28)

Scheme 7.5

bond in MVK. Competing Michael addition between the *C*(2) methylene center in the benzimidazolium salt and MVK is indicated by the isolation of the adducts (**7.31**; R = H or Et) when the reaction mixture derived from the benzimidazolium perchlorate (**7.29**; R^1 = Me, R^2 = H) is subjected to aqueous or ethanolic workup.[11]

Pyrido[1,2-*a*]benzimidazoles variously substituted in both the benzene and pyridine rings are generally accessible in high yield (Table 7.3) by the

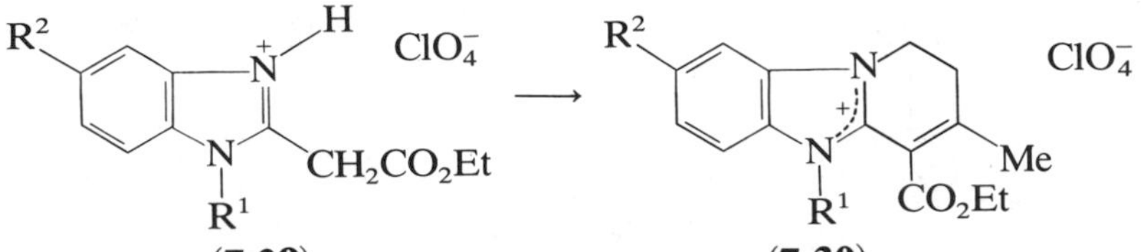

(7.29) **(7.30)**

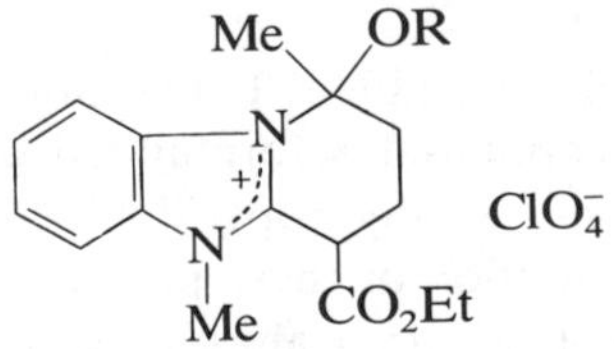

(7.31)

Scheme 7.6

TABLE 7.3. SYNTHESIS OF PYRIDO[1,2-*a*]BENZIMIDAZOLE DERIVATIVES BY RING-CLOSURE REACTIONS OF BENZIMIDAZOLES WITH β-DICARBONYL COMPOUNDS AND RELATED PROCESSES

Starting materials ($R \rightarrow R^5$ unspecified = H)	Reaction conditions[a]	Product ($R \rightarrow R^5$ unspecified = H)	Yield (%)	m.p. (°C)	Solvent of crystallization	Ref.
(**7.32**; $R^3 = Me$) + (**7.33**; $R^4 = CN$)	*A*	(**7.34**; $R^3 = Me$, $R^4 = CN$)	59	218–219	Dimethylformamide–water	12
(**7.32**; $R^1 = R^3 = Me$) + (**7.33**; $R^4 = CN$)	*A*	(**7.34**; $R^1 = R^3 = Me$, $R^4 = CN$)	92	235–237	Dimethylformamide–water	12
(**7.32**; $R^1 = R^3 = Me$) + (**7.33**; $R^4 = CN$, $R^5 = Me$)	*A*	(**7.34**; $R^1 = R^3 = R^5 = Me$, $R^4 = CN$)	39	204	Dimethylformamide–water	12
(**7.32**; $R^1 = R^3 = Me$) + (**7.33**; $R^4 = CN$, $R^5 = Cl$)	*A*	(**7.34**; $R^1 = R^3 = Me$, $R^4 = CN$, $R^5 = Cl$)	40	212–218	Dimethylformamide–water	12
(**7.32**; $R^1 = R^3 = Me$) + (**7.33**; $R^4 = CN$, $R^5 = OMe$)	*A*	(**7.34**; $R^1 = R^3 = Me$, $R^4 = CN$, $R^5 = OMe$)	65	214–215	Dimethylformamide–water	12
(**7.32**; $R^1 = R^3 = Me$) + (**7.33**; $R^4 = CN$, $R^5 = CO_2Me$)	*A*	(**7.34**; $R^1 = R^3 = Me$, $R^4 = CN$, $R^5 = CO_2Me$)	90	295–305	Dimethylformamide–water	12
(**7.32**; $R^1 = R^3 = Me$) + (**7.33**; $R^4 = CN$)[b]	*A*	(**7.34**; $R^1 = R^3 = Me$, $R^4 = CN$)[b]	44	295–297	Dimethylformamide–water	12
(**7.32**; $R^1 = R^3 = Me$) + (**7.33**; $R^4 = CN$)[c]	*A*	(**7.34**; $R^1 = R^3 = Me$, $R^4 = CN$)[c]	57	304	Dimethylformamide–water	12
(**7.32**; $R^1 = R^2 = R^3 = Me$) + (**7.33**; $R^4 = CN$)	*A*	(**7.34**; $R^1 = R^2 = R^3 = Me$, $R^4 = CN$)	49	285–286	Dimethylformamide–water	12
(**7.32**; $R^1 = R^3 = Me$, $R^2 = CH_2CO_2Et$) + (**7.33**; $R^4 = CN$)	*A*	(**7.34**; $R^1 = R^3 = Me$, $R^2 = CH_2CO_2Et$, $R^4 = CN$)	62	206–207	Dimethylformamide–water	12

TABLE 7.3 (*Continued*)

Starting materials ($R \rightarrow R^5$ unspecified = H)	Reaction conditions[a]	Product ($R \rightarrow R^5$ unspecified = H)	Yield (%)	m.p. (°C)	Solvent of crystallization	Ref.
(**7.32**; $R^2 = NO_2$) + (**7.33**; $R^4 = CN$)	*A*	(**7.34**; $R^2 = NO_2$, $R^4 = CN$)	75	—[d]	Dimethylformamide–water	12
(**7.32**; $R^1 = R^3 = Me$) + (**7.33**; $R^4 = NO_2$)	*B*	(**7.34**; $R^1 = R^3 = Me$, $R^4 = NO_2$)	48	248–250	Acetonitrile	13
(**7.32**; $R^1 = Me$, $R^3 = OEt$) + (**7.33**; $R^4 = CN$)	*C*	(**7.34**; $R^1 = OH$, $R^3 = Me$, $R^4 = CN$)	89	300 (decomp.)	—[e]	14
(**7.32**; $R^1 = OEt$, $R^3 = CO_2Et$)[c] + (**7.33**; $R^4 = CN$)	*D*	(**7.34**; $R^1 = OH$, $R^3 = CO_2Et$, $R^4 = CN$)	85	293	Dimethylformamide	15
(**7.32**; $R^1 = OMe$, $R^3 = CH_2CO_2Me$) + (**7.33**; $R^4 = CN$)	*E*	(**7.34**; $R^1 = OH$, $R^3 = CH_2CO_2Me$, $R^4 = CN$)	89	275	—[e]	15
(**7.32**; $R^1 = R^3 = Me$) + (**7.33**; $R = Me$, $R^4 = CN$)[g]	*F*	(**7.35**; $R^1 = R^3 = Me$, $R^4 = CN$)	88	249–250	Nitromethane–ethanol	16
(**7.32**; $R^1 = R^2 = R^3 = Me$) + (**7.33**; $R = Me$, $R^4 = CN$)[g]	*F*	(**7.35**; $R^1 = R^2 = R^3 = Me$, $R^4 = CN$)	91	246	Nitromethane–ethanol	16
(**7.32**; $R^1 = Me$, $R^3 = CO_2Me$) + (**7.33**; $R^4 = CN$)[g]	*F*	(**7.35**; $R^1 = Me$, $R^3 = CO_2Me$, $R^4 = CN$)	86	265–267	Nitromethane–ethanol	16
(**7.32**; $R^1 = Me$, $R^3 = Ph$) + (**7.33**; $R^4 = CN$)[g]	*F*	(**7.35**; $R^1 = Me$, $R^3 = Ph$, $R^4 = CN$) + (**7.35**; $R^1 = Ph$, $R^3 = Me$, $R^4 = CN$)	28 / 57	232–238[h]	Nitromethane–ethanol	16
(**7.36**) + (**7.37**)	*G*	(**7.39**)	62	161–163	Ethanol	18

(**7.40**; $R^1 = Me$)	*H*	(**7.43**; $R^1 = Me$)	78	88–90	Diisopropyl ether	19
(**7.40**; $R^1 = Pr^i$)	*H*	(**7.43**; $R^1 = Pr^i$)	84	77–79	Diisopropyl ether	19
(**7.40**; $R^1 = R^2 = Me$)	*H*	(**7.43**; $R^1 = R^2 = Me$)	73	77–79	Diisopropyl ether	19
(**7.42**; $R^1 = R^2 = Me$)	*I*	(**7.43**; $R^1 = R^2 = Me$)	84	77–79	Acetone	19
(**7.40**; $R = Me$)	*H*	(**7.43**; $R = Me$)	34	100–102	Acetone	19
(**7.40**; $R = R^1 = Me$)	*H*	(**7.43**; $R = R^1 = Me$)	72	109–110	Acetone	19
(**7.40**; $R = Me, R^1 = Ph$)	*H*	(**7.43**; $R = Me, R^1 = Ph$)	32	171–172	Acetone	19
(**7.40**; $R = R^1 = R^2 = Me$)	*H*	(**7.43**; $R = R^1 = R^2 = Me$)	65	157–159	Acetone	19
(**7.44**)	*J*	(**7.47**)	89	148–150	Benzene–light petroleum	20
(**7.44**; $R^1 = Me$)	*J*	(**7.47**; $R^1 = Me$)	92	120–124	Benzene–light petroleum	20
(**7.44**; $R^1 = Cl$)	*J*	(**7.47**; $R^2 = Cl$)	90	163	Benzene–light petroleum	20
(**7.44**; $R^1 = OMe$)	*J*	(**7.47**; $R^2 = OMe$)	92	159–161	Benzene–light petroleum	20
(**7.44**; $R^1 = NO_2$)	*J*	{ (**7.47**; $R^1 = NO_2$) +	54	212–214	Benzene–light petroleum	20
		(**7.47**; $R^2 = NO_2$) }	30	144–148	Benzene–light petroleum	
(**7.44**; $R^1 = CO_2Me$)	*J*	(**7.47**; $R^2 = CO_2Me$)	61	163	Benzene–light petroleum	20
(**7.44**)	*K*	(**7.47**; $R^3 = Me$)	76	94	Benzene–light petroleum	20
(**7.44**)	*L*	(**7.47**; $R^3 = Et$)	73	134	Benzene–light petroleum	20
(**7.48**; $R = CO_2Et$) + (**7.49**) }	*M*	(**7.54**; $R = CO_2Et$)	97	225–227	Benzene (decomp.)	21

TABLE 7.3 (*Continued*)

Starting materials ($R \rightarrow R^5$ unspecified = H)	Reaction conditions[a]	Product ($R \rightarrow R^5$ unspecified = H)	Yield (%)	m.p. (°C)	Solvent of crystallization	Ref.
(**7.48**; R = $CONH_2$) + (**7.49**)	*M*	(**7.54**; R = $CONH_2$)	91	273–275 (decomp.)	Acetic acid–water	21
(**7.48**; R = CN) + (**7.49**)	*N*	(**7.54**; R = CN)	50	315	Ethanol	21
(**7.55**; R = H) + (**7.49**)	*M*	(**7.57**)	10	217–219	Acetone	21
(**7.55**; R = H) + (**7.49**)	*O*	(**7.57**)	40	—	—	21
(**7.55**; R = COMe) + (**7.49**)	*M*	(**7.57**)	20	—	—	21
(**7.58**)	*P*	(**7.59**)	72	—[d]	—[e]	22
(**7.60**)	*Q*	(**7.62**)	79	165	Ethyl acetate–light petroleum	23

[a] A = NaOEt, EtOH/(100°)(2 hr); *B* = 70% $HClO_4$ aq./(reflux)(15 min); *C* = piperidine, methanol/(30–40°)(1 hr), then (reflux)(14 hr); *D* = piperidine, ethanol, dimethylformamide/(reflux)(10 hr); *E* = Et_3N, methanol, dimethylformamide/(reflux)(10 hr); *F* = 140–180°/15–90 min; *G* = ethanol/(reflux)-(4 hr); *H* = toluene-*p*-sulfonic acid/(250–270°)(20–30 min); *I* = toluene-*p*-sulfonic acid/(260°)(15 min); *J* = Ac_2O, Et_3N/(reflux)(16 hr); *K* = $(EtCO)_2O$, Et_3N/(reflux)(16 hr); *L* = $(Pr^nCO)_2O$, Et_3N/(reflux)(16 hr); *M* = AcOH/(room temp.)(3–5 days); *N* = AcOH/(room temp.)(14 hr); *O* = Ac_2O/(room temp.)(4 days); *P* = PCl_5 (reaction conditions not specified). *Q* = AcOH/(75°)(2 hr).

[b] 9-Amino-7-chloro derivative.

[c] 7,8-Dichloro derivative.

[d] Melting point not quoted.

[e] Solvent of crystallization not specified.

[f] Sodium salt.

[g] Perchlorate.

[h] Isomer mixture.

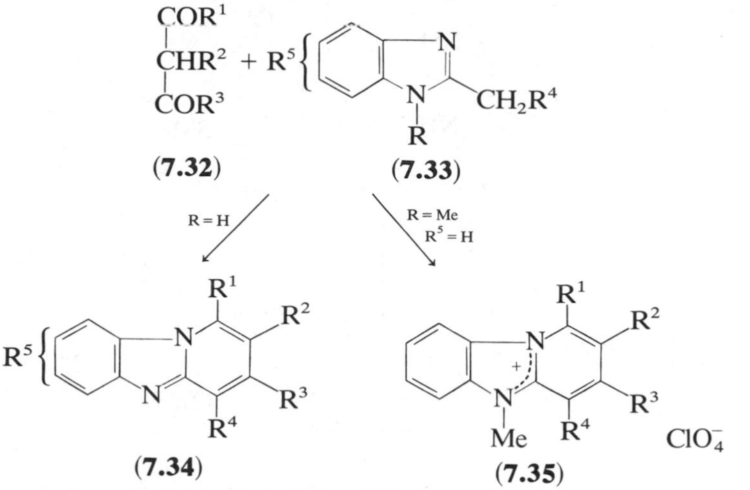

Scheme 7.7

base- or acid-catalyzed condensation of β-dicarbonyl compounds with N-unsubstituted benzimidazoles having an active methylene substituent at the $C(2)$ position [Scheme 7.7; (**7.32**) + (**7.33**; $R^4 = CN$ or NO_2) → (**7.34**; $R^4 = CN$ or NO_2)].[12–15] Effective catalysts for these reactions include sodium ethoxide,[12] piperidine,[14,15] triethylamine,[15] and perchloric acid.[13] The use of unsymmetrically substituted benzimidazoles or β-dicarbonyl compounds in condensation reactions of the type [Scheme 7.7; (**7.32**) + (**7.33**) → (**7.34**)] is complicated by the possibility of isomer formation as a result of the availability of several modes of ring-closure. In the few examples of such reactions studied[14,15] only a single pyrido[1,2-*a*]benzimidazole product was isolated and its orientation assigned without unequivocal proof of structure. Obviously, more extensive studies of the scope and orientational preference of these synthetically useful reactions are warranted. N-Substituted benzimidazoles having an activated $C(2)$ methylene substituent also condense readily with β-dicarbonyl compounds under acidic conditions, the products of these reactions being pyrido[1,2-*a*]benzimidazolium salts. For example, 2-cyanomethyl-1-methylbenzimidazolium perchlorate reacts smoothly on heating with a variety of β-dicarbonyl compounds giving high yields (Table 7.3) of the corresponding 4-cyano-5-methylpyrido[1,2-*a*]benzimidazolium perchlorates [Scheme 7.7; (**7.32**) + (**7.33**; $R = Me$, $R^4 = CN$, $R^5 = H$) → (**7.35**; $R^4 = CN$)].[16] The orientation of the pyrido[1,2-*a*]benzimidazolium salts obtained when unsymmetrically substituted β-dicarbonyl compounds are employed as substrates indicate that the structure of the final product in these reactions is dictated by preferential initial condensation between the $C(2)$ methylene substituent in the benzimidazole and the more electrophilic carbonyl center in the β-dicarbonyl compound. A general synthesis of pyrido[1,2-*a*]benzimidazole derivatives involving the condensation of $C(2)$

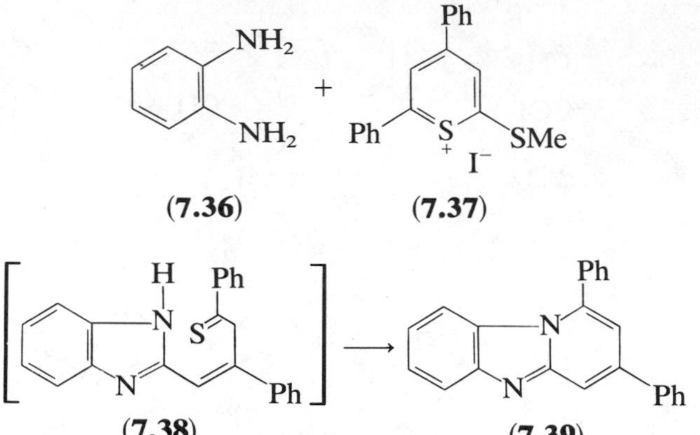

(7.36) (7.37)

(7.38) (7.39)

Scheme 7.8

unsubstituted benzimidazoles with ethoxymethylenemalononitrile and related reagents has recently been described.[17] 1,3-Diphenylpyrido[1,2-*a*]-benzimidazole is the end-product of the reaction (Scheme 7.8) of *ortho*-phenylenediamine with the thiapyrilium salt (**7.37**).[18] This transformation proceeds in good yield (Table 7.3) and may be rationalized in terms of the intermediate formation and cyclization of the benzimidazole derivative (**7.38**).

Cyclic β-enaminoketones derived from *ortho*-phenylenediamine cyclize in the presence of a catalytic amount of toluene-*p*-sulfonic acid to provide a moderately efficient (Table 7.3) general route to otherwise rarely encountered 3,4-dihydropyrido[1,2-*a*]benzimidazole derivatives [Scheme 7.9; (**7.40**) → → → (**7.43**)].[19] Omission of the acid catalyst in the cyclization of

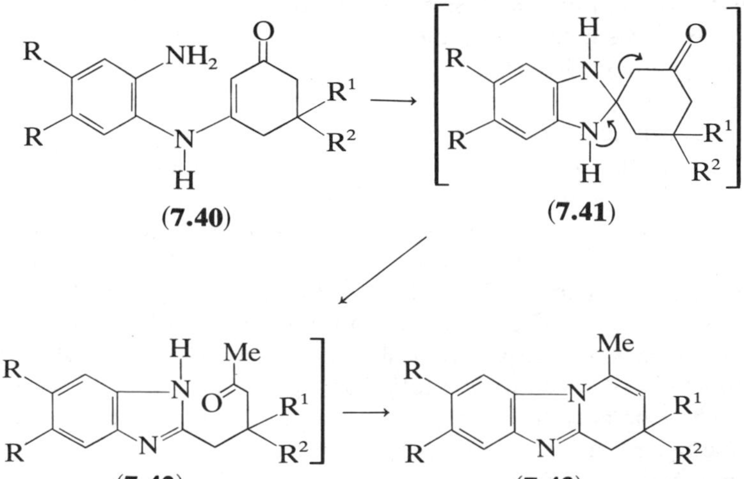

(7.40) (7.41)

(7.42) (7.43)

Scheme 7.9

the compound (**7.40**; R = H, $R^1 = R^2$ = Me) results in the formation of the 2-(δ-oxoalkyl)benzimidazole (**7.42**; R = H, $R^1 = R^2$ = Me), which is smoothly cyclized by treatment with toluene-*p*-sulfonic acid to give the 3,4-dihydropyrido[1,2-*a*]benzimidazole (**7.43**; R = H, $R^1 = R^2$ = Me). These observations prompt[19] a mechanism (Scheme 7.9) for 3,4-dihydropyrido[1,2-*a*]benzimidazole formation involving initial ring-closure to and subsequent ring-opening of a spiro-benzimidazole intermediate (**7.41**) followed by cyclization of the 2-(δ-oxoalkyl)benzimidazole (**7.42**) produced. Pyrido[1,2-*a*]benzimidazol-1(4*H*)-ones are generally available, mostly in high yield (Table 7.3), by the acylative ring-closure of readily accessible 2-(β-oxopentyl)benzimidazole derivatives with acid anhydrides in the presence of triethylamine or sodium acetate as catalyst [Scheme 7.10; (**7.44**) → → (**7.47**)].[20] Ring-formation in these reactions may be envisaged as occurring either by cyclization of a mixed anhydride intermediate (**7.45**) or by aldol condensation in an *N*-acyl derivative (**7.46**). In either case the use of substrates (**7.44**) unsymmetrically substituted in the benzene ring can lead to two possible products, depending on which of the benzimidazole nitrogen atoms is involved in the final ring-closure of (**7.45**) or the initial *N*-acylation to give (**7.46**). In practice[20] ring-formation occurs almost exclusively through

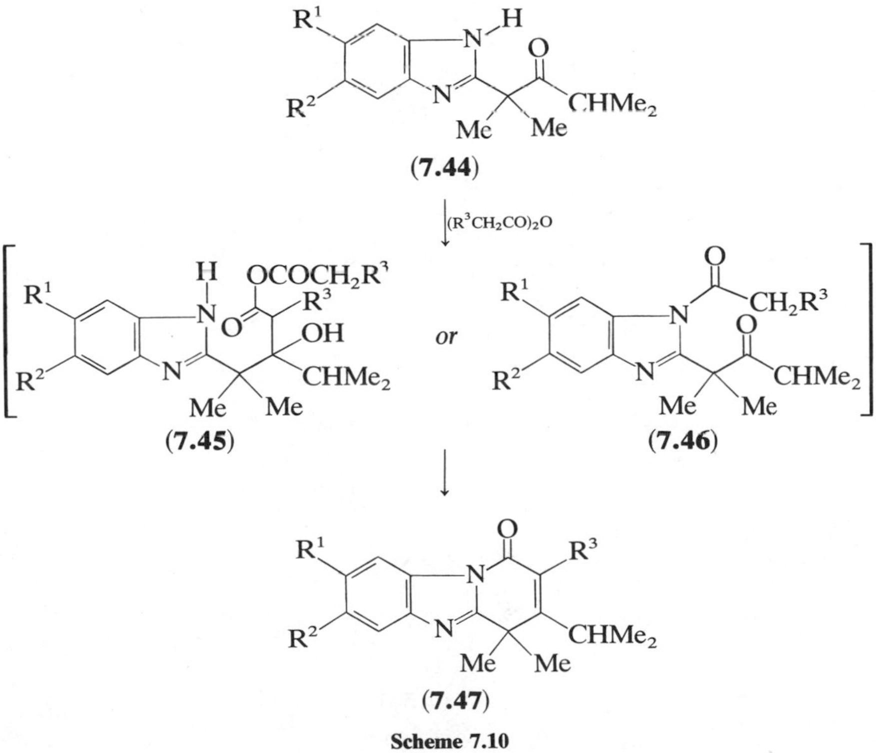

Scheme 7.10

condensation at the benzimidazole nitrogen atom *para* to the benzene substituent irrespective of whether this is electron-donating or moderately electron-withdrawing. The latter result is surprising, implying as it does that acylative attack occurs preferentially at the apparently less basic of the two available nitrogen centers. Only in the case of the powerfully electron-withdrawing nitro substituent does competing ring-closure at the nitrogen atom *meta* to the substituent predominate, and even here condensation

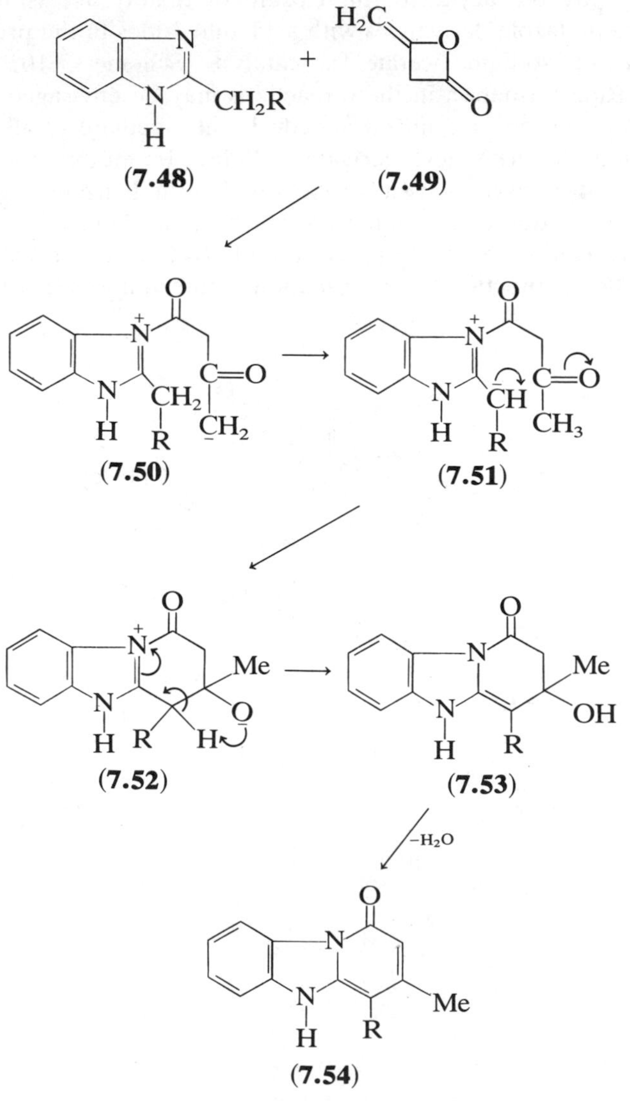

(7.48) **(7.49)**

(7.50) **(7.51)**

(7.52) **(7.53)**

$-H_2O$

(7.54)

Scheme 7.11

through the *para* nitrogen atom still occurs to a marked extent.[20] The reaction of benzimidazoles containing an active methylene center at the *C*(2) position with diketene in acetic acid solution provides a convenient method for the synthesis in moderate to high yield (Table 7.3) of pyrido[1,2-*a*]benzimidazol-1(5*H*)-ones [Scheme 7.11; (**7.48**; R = CO_2Et, $CONH_2$, or CN) + (**7.49**) → → → (**7.54**; R = CO_2Et, $CONH_2$, or CN)].[21] Pyrido[1,2-*a*]-benzimidazole synthesis by this means bears an obvious relationship to that through *C*(2)-methylene-substituted benzimidazoles and β-dicarbonyl compounds (see before) and is suggested[21] to occur via aldol-type ring-closure in an initially formed betaine intermediate [Scheme 7.11; (**7.48**) + (**7.49**) → (**7.50**) → (**7.51**) → → → (**7.54**)]. A similar benzimidazole betaine intermediate is postulated[21] to account for the annelation of *C*(2)-unsubstituted benzimidazoles by diketene in acetic acid or acetic anhydride which provides synthetic access, albeit in only low yield (Table 7.3), to 4a,5-dihydropyrido-[1,2-*a*]benzimidazol-1,3(2*H*, 4*H*)-diones [Scheme 7.12; (**7.55**) + (**7.49**) → (**7.56**) → (**7.57**)].

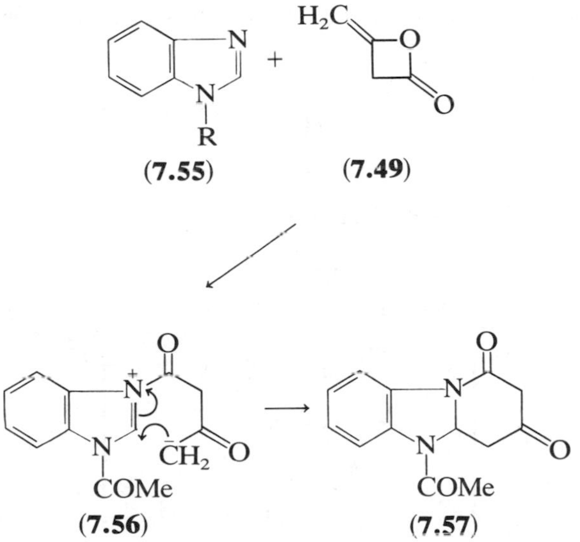

H_2C O O N N R + (**7.55**) (**7.49**)

N N COMe CH_2 O O (**7.56**) N N COMe O O (**7.57**)

Scheme 7.12

1,2,3,4-Tetrahydropyrido[1,2-*a*]benzimidazoles are simply and often efficiently obtained by acylative or alkylative ring-closure in benzimidazoles suitably functionalized at the *C*(2) position. Acylative processes of this type are illustrated (Scheme 7.13) by the phosphorus pentachloride-mediated cyclization of the 2-benzimidazolylbutyric acid derivative (**7.58**) to give the 3,4-dihydropyrido[1,2-*a*]benzimidazol-1(2*H*)-one derivative (**7.59**) in good yield (Table 7.3).[22] The transformation (Scheme 7.14)[23] of the diamide (**7.60**)

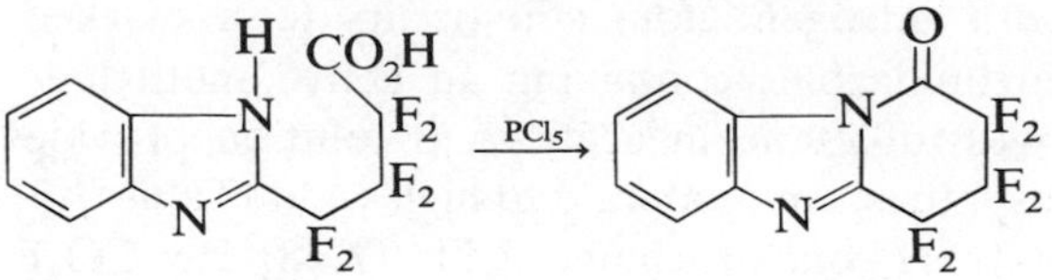

$\xrightarrow{PCl_5}$

(7.58) **(7.59)**

Scheme 7.13

in good yield (Table 7.3) into the 3,4-dihydropyrido[1,2-*a*]benzimidazol-1(2*H*)-one (**7.62**) simply on warming in acetic acid is explicable in terms of the intermediate formation and cyclization of the 2-benzimidazolyl-butyramide (**7.61**) and demonstrates the facility of such acylative ring-closure. Alkylative ring-closure (Scheme 7.15) in 4-(2-benzimidazolyl)-1-butanol (**7.65**) provides a simple rationale for the reaction of *ortho*-phenylenediamine (**7.63**) with δ-valeronitrile (**7.64**) to give 1,2,3,4-tetrahydropyrido[1,2-*a*]benzimidazole (**7.67**).[24] Despite the low yield obtained (Scheme 7.15), this transformation represents probably the most convenient method for the synthesis of the tetrahydropyrido[1,2-*a*]benzimidazole

(7.60)

↓

(7.61)

↓

(7.62)

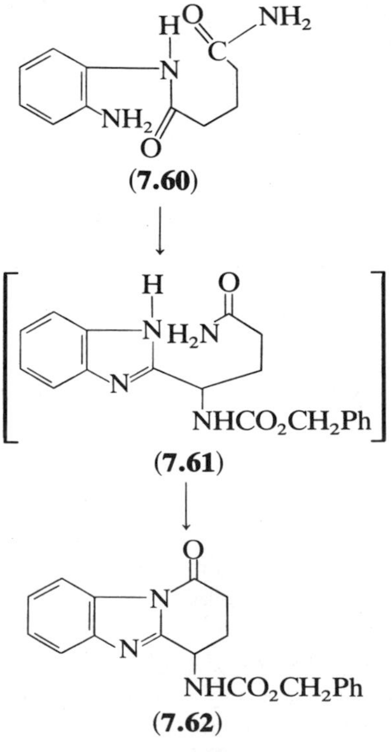

Scheme 7.14

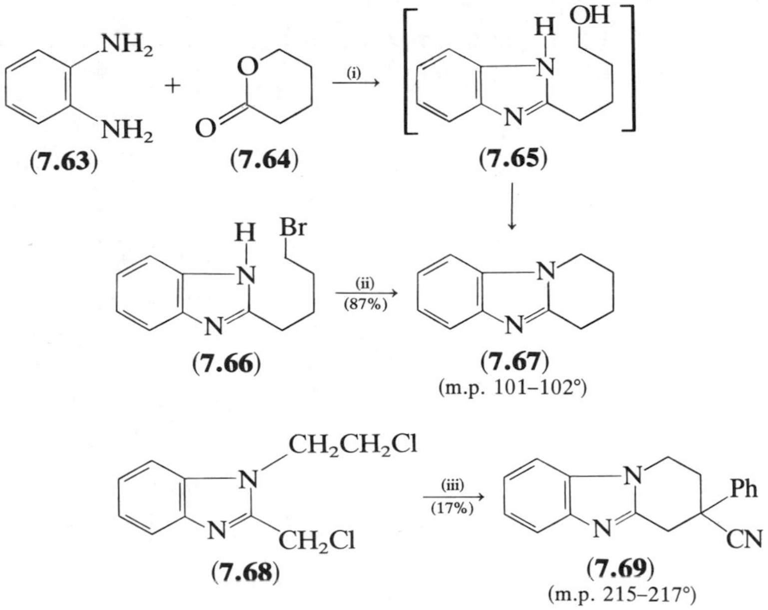

(7.63) (7.64) (7.65) (7.66) (7.67) (m.p. 101–102°) (7.68) (7.69) (m.p. 215–217°)

(i) 230°/16 h
(ii) KOH, EtOH/(room temp.)(20 min)
(iii) $PhCH_2CN$, $NaNH_2$, toluene/(room temp.) 30 min., then (reflux)(6 hr)

Scheme 7.15

(**7.67**), which is also accessible in high yield (87%) by the potassium hydroxide catalyzed cyclization (Scheme 7.15) of 2-(4-bromobutyl)benzimidazole (**7.66**).[25] The latter synthetic approach, unlike that based on 1-(2-substituted aryl)piperidine cyclizations (see later), is applicable to the unequivocal synthesis of 1,2,3,4-tetrahydropyrido[1,2-*a*]benzimidazoles substituted in the piperidine ring. The sodamide-induced dehalogenative ring-closure (Scheme 7.15) of 1-(2-chloroethyl)-2-chloromethylbenzimidazole (**7.68**), though inefficient, is of interest in affording a 1,2,3,4-tetrahydropyrido[1,2-*a*]benzimidazole (**7.69**) useful as the precursor of a compound related to the drug Demerol.[26]

Ring-closure Reactions of Other Heterocycles

Fully unsaturated pyrido[1,2-*a*]benzimidazoles nitrated in the benzene ring (**7.74**; R^1 and/or $R^2 = NO_2$) are obtained in moderate to high yield (Table 7.4) by the thermal cyclization (Scheme 7.16) of 2-(2-nitroarylamino)pyridines (**7.72**; R^1 and/or $R^2 = NO_2$) which can be preformed or prepared *in situ* by the condensation of 2-nitrochlorobenzene derivatives (**7.70**; R^1 and/or $R^2 = NO_2$) with 2-aminopyridines (**7.71**).[27–29] Cyclization of

TABLE 7.4. SYNTHESIS OF PYRIDO[1,2-*a*]BENZIMIDAZOLES BY RING-CLOSURE REACTIONS OF PYRIDINE DERIVATIVES

Starting materials ($R \rightarrow R^4$ unspecified = H)	Reaction conditions[a]	Product ($R \rightarrow R^4$ unspecified = H)	Yield (%)	m.p. (°C)	Solvent of crystallization	Ref.
(**7.72**; $R^1 = NO_2$)	*A*	(**7.74**; $R^1 = NO_2$	76	262–263	Nitrobenzene	28
(**7.70**; $R^2 = NO_2$) + (**7.71**)	*B*	(**7.74**; $R^2 = NO_2$)	12	272	Ethylene glycol monoethylether	29
(**7.72**; $R^2 = NO_2$)	*C*	(**7.74**; $R^2 = NO_2$)	32	—	—	29
(**7.72**; $R^1 = R^2 = NO_2$)	*D*	(**7.74**; $R^1 = R^2 = NO_2$)	83	>300	Nitrobenzene	27
(**7.72**; $R^1 = R^2 = NO_2$, $R^3 = Me$)	*E*	(**7.74**; $R^1 = R^2 = NO_2$, $R^3 = Me$)	—[b]	256–260	Nitrobenzene	27
(**7.75**)	*F*	(**7.80**)	5–10	179	Ethanol	30
(**7.75**)	*G*	(**7.80**)	70	—	—	30
(**7.75**; $R^4 = Me$)	*G*	(**7.80**; $R^4 = Me$)	90	118–120	Ethanol or heptane	30
(**7.75**; $R^2 = NO_2$)	*H*	(**7.80**; $R^2 = NO_2$)	12	237–239	Ethanol	32
(**7.75**; $R^2 = NO_2$)	*I*	(**7.80**; $R^2 = NO_2$)	10	—	—	31, 34
(**7.84**)	*J*	(**7.85**)	50	—	—	32
(**7.75**; $R^4 = NO_2$)	*K*	(**7.80**; $R^4 = NO_2$)	56	223–225	Ethanol/water	33, 35
(**7.75**; $R^3 = NO_2$)	*L*	(**7.80**; $R^3 = NO_2$)	8	228–230	—[c]	35
(**7.75**; $R^1 = Me$, $R^2 = NO_2$)	*L*	(**7.80**; $R^1 = Me$, $R^2 = NO_2$)	20	215–217	Methanol	31
(**7.75**; $R^2 = NO_2$, $R^3 = Me$)	*L*	(**7.80**; $R^2 = NO_2$, $R^3 = Me$	28	240–242	Ethanol	31
(**7.75**; $R^2 = CO_2Et$)	*L*	(**7.80**; $R^2 = CO_2Et$)[d]	<1	233–235	Acetone	33
(**7.75**; $R^1 = R^3 = Me$, $R^2 = CO_2Et$)	*L*	(**7.80**; $R^1 = R^3 = Me$, $R^2 = CO_2Et$)[d]	<1	271–271.5	Acetone	33, 34, 35
(**7.81**)	*M*	(**7.83**)[e]	82	>300	Ethanol	36
(**7.86**) + (**7.87**)	*N*	(**7.90**)[f]	65	95–96	Ether–light petroleum	37
(**7.88**)	*O*	(**7.90**)	80	—	—	37
(**7.86**) + (**7.87**; $R^2 = Me$)	*N*	(**7.90**; $R^2 = Me$)[g,h]	48	235–237	—[c]	37
(**7.86**) + (**7.87**; $R^2 = Me$)	*P*	(**7.90**; $R^2 = Me$)	46	—	—	37
(**7.86**) + (**7.87**; $R^1 = NO_2$)	*Q*	(**7.90**; $R^1 = NO_2$)[i]	30	219–220	Ethanol	32, 37
(**7.86**) + (**7.87**; $R^1 = Br$)	*R*	(**7.90**; $R^1 = Br$)[j]	73	148–149	Light petroleum (b.p. 100–120°)	37
(**7.86**) + (**7.87**; $R^1 = Br$)	*P*	(**7.90**; $R^1 = Br$)	17	—	—	37
(**7.86**) + (**7.87**; $R^1 = R^2 = Br$)	*R*	(**7.90**; $R^1 = R^2 = Br$)[k]	37	159–160	Light petroleum (b.p. 100–120°)	37
(**7.86**) + (**7.87**; $R^1 = R^2 = Br$)	*P*	(**7.90**; $R^1 = R^2 = Br$)	30	—	—	37
(**7.91**)	*S*	(**7.92**)[l]	74	181–182	—[c]	38

(**7.93**) + (**7.94**)	*T*	(**7.95**)[m]	70	260 (decomp.)	—[n]	39
(**7.93** R = Me) + (**7.94**)	*T*	(**7.95**; R^1 = Me) +	—[b]	273[o]	Methanol	39
		(**7.95**; R^2 = Me)	—[b]	222[o]	Methanol	
(**7.97**) + (**7.98**)	*U*	(**7.101**)	61	152–153	Light petroleum (b.p. 40–60°)	40
(**7.96**) + (**7.98**; R^2 = Me)	*U*	(**7.101**; R^2 = Me)	69	167–168	Light petroleum (b.p. 40–60°)	41
(**7.96**) + (**7.98**; R^1 = Me)	*U*	(**7.101**; R^1 = Me) +	10	186–187	Light petroleum (b.p. 40–60°)	41
		(**7.101**; R^3 = Me)	45	106–107	Light petroleum (b.p. 40–60°)	
(**7.96**) + (**7.98**; R^2 = Et)	*U*	(**7.101**; R^2 = Et)	58	127–128	Light petroleum (b.p. 40–60°)	41
(**7.96**) + (**7.98**; R^1 = R^3 = Me)	*U*	(**7.101**; R^1 = R^3 = Me)	62	124–125	Light petroleum (b.p. 40–60°)	41

[a] *A* = naphthalene or diphenyl/300–310° (reaction time not specified); *B* = $CaCO_3$, ethylene glycol monoethyl ether/(reflux)(16 hr); *C* = ethylene glycol monoethyl ether/(reflux)(16 hr); *D* = *N,N*-dimethylaniline/(reflux)(2–5 hr); *E* = phenol, xylene/heat (reaction conditions not specified); *F* = polyphosphoric acid/(150°)(reaction time not specified); *G* = $h\nu$, Pyrex filter, EtOH/24 hr; *H* = 98% orthophosphoric acid/(160°)(30 min); *I* = pyrophosphoric acid/heat (reaction conditions not specified); *J* = 0.67 *M* HCl, EtOH/(reflux)(18 hr); *K* = pyrophosphoric acid/(110–160°)(15 min); *L* = pyrophosphoric acid/(140–180°)(15 min); *M* = $NaNO_2$, H_2SO_4 aq./(5–20°)(16 hr, then reflux 45 min); *N* = no solvent/(reflux)(15 min); *O* = hydrogen bromide–acetic acid/(room temp.)(1 hr); *P* = Na_2CO_3, EtOH/(reflux)(3–18 hr); *Q* = no solvent/(120–130°)(2 hr); *R* = EtOH/(reflux)(26 hr); *S* = P_2O_5 (reaction conditions not specified); *T* = AcOH, H_2O/(reflux)(5 min); *U* = $LiClO_4$, acetonitrile/(0.55 v)(4 hr).

[b] Yield not quoted.

[c] Solvent of crystallization not specified.

[d] Methiodide.

[e] Forms a picrate, m.p. 196–198° (decomp.).

[f] Forms a hydrate (in air) m.p. 56–58°, and a picrate, m.p. 256–258° (from ethanol–acetic acid).

[g] Hydrochloride; free base has m.p. 85–87°.

[h] Forms a dihydrate, m.p. 60–61° (from water) and a picrate, m.p. 155–158 (from ethanol).

[i] Forms a picrate, m.p. 216–219° (decomp.) (from ethanol–acetic acid).

[j] Forms a hydrochloride, m.p. 260–262°, and a picrate, m.p. 264–265° (decomp.) (from acetic acid).

[k] Forms a picrate, m.p. 167–169° (from ethanol–acetic acid).

[l] Forms a methiodide, m.p. 246–247°.

[m] Forms a hydrochloride, m.p. 243° (decomp.).

[n] Purified by sublimation.

[o] These m.p. assignments may be reversed.

Scheme 7.16

the isolated 2-(2-nitroarylamino)pyridine (**7.72**) is achieved by heating in the melt,[27] in a high-boiling medium such as nitrobenzene or *N*,*N*-dimethylaniline,[27] naphthalene or diphenyl,[28] or ethylene glycol monoethyl ether,[29] or with phenol in xylene.[27] Conversely, *in situ* formation of the nitropyrido[1,2-*a*]benzimidazole (**7.74**; R^1 and/or $R^2 = NO_2$) is accomplished by heating the nitrochlorobenzene and aminopyridine components in ethylene glycol monoethyl ether in the presence of calcium carbonate.[29] Ring-closure of the 2-(2-nitroarylamino)pyridine is logically explained in terms of the intramolecular nucleophilic displacement of the *ortho*-nitro substituent by the NH center in an iminopyridine tautomeric form [Scheme 7.16; (**7.72**)⇌(**7.73**)→(**7.74**)]. Pyrido[1,2-*a*]benzimidazole derivatives are also formed in low yield (Table 7.4) when 1-(2-pyridyl)benzo-1,2,3-triazoles are heated in polyphosphoric acid under the conditions of the Graebe–Ullmann reaction.[30–35] These transformations may be viewed[31,35] as ionic processes involving ring-opening to diazonium intermediates convertible by intramolecular displacement of the diazonium substituent into the observed products [Scheme 7.17; (**7.75**)→(**7.76**)⇌(**7.78**)→(**7.80**)]. By way of contrast, the photolytic conversion of 1-(2-pyridyl)benzo-1,2,3-triazoles into pyrido[1,2-*a*]benzimidazoles proceeds in high yield (Table 7.4) and is formulated[30] as a radical-mediated process [Scheme 7.17; (**7.75**)→(**7.77**)↔(**7.79**)→(**7.80**)]. The formation and cyclization of a diazonium cation intermediate accounts for the diazotative conversion (Scheme 7.18) of the 2-arylaminopyridine

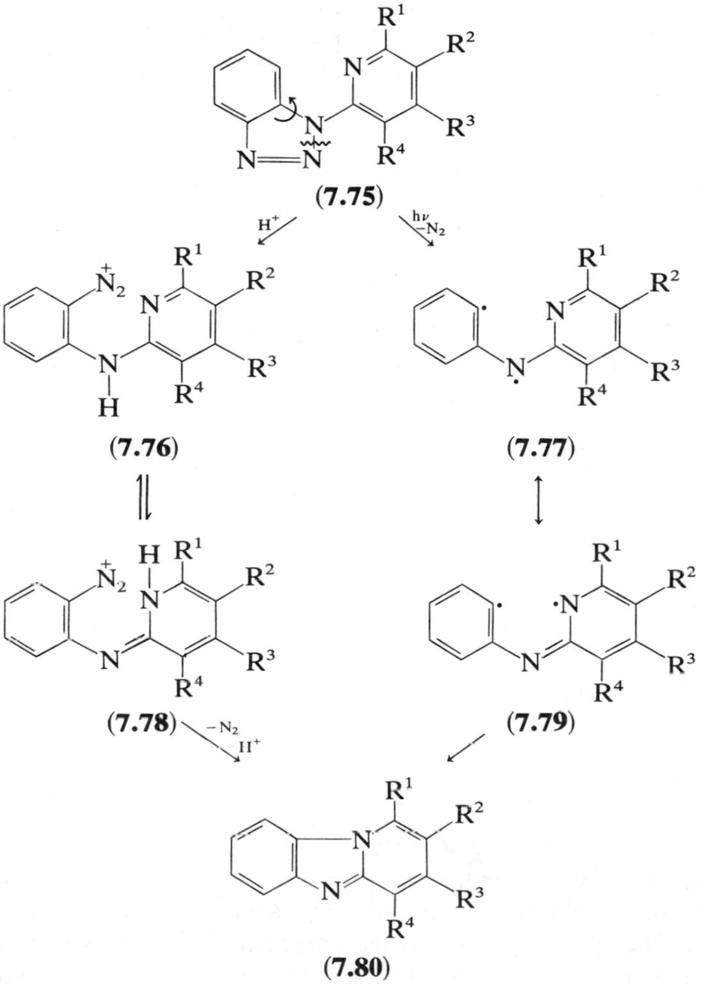

(7.75)

(7.76) (7.77)

(7.78) (7.79)

(7.80)

Scheme 7.17

derivative (**7.81**) in high yield (Table 7.4) into the pyrido[1,2-*a*]benzimidazolium salt (**7.83**).[36] The direct cyclization (Scheme 7.19) of the 2-arylaminopyridine derivative (**7.84**) to the pyrido[1,2-*a*]benzimidazole (**7.85**) under acidic conditions,[32] on the other hand, is unusual in that the displacement of an amino substituent from a deactivated benzene nucleus appears to be involved.

Pyrido[1,2-*a*]benzimidazoles are also the end-products of amine–carbonyl condensations in a variety of pyridine derivatives. 6,7,8,9-Tetrahydropyrido-[1,2-*a*]benzimidazoles in particular, are generally accessible in moderate yield (Table 7.4) by the reaction of 2-aminopyridine derivatives with 2-chlorocyclohexanone in boiling ethanol, alone, or in the presence of sodium

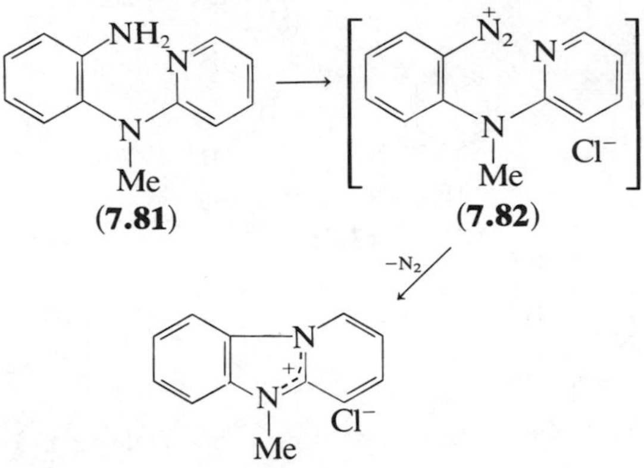

(7.81) (7.82) (7.83)

Scheme 7.18

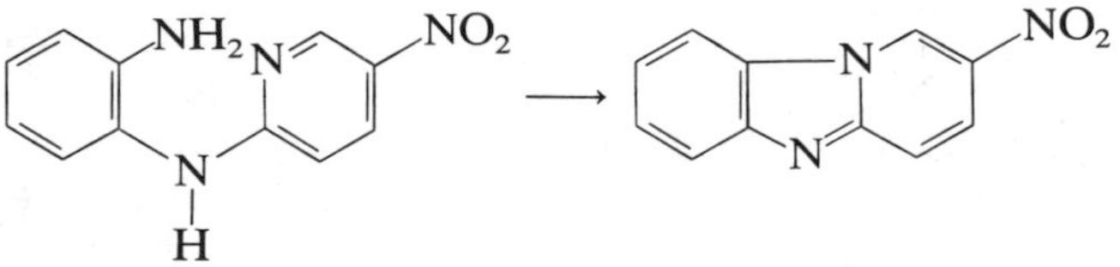

(7.84) (7.85)

Scheme 7.19

carbonate [Scheme 7.20; (**7.86**)+(**7.87**)→→(**7.90**)].[32,37] The intermediacy of tautomeric 2-pyridylaminocyclohexanone derivatives [(**7.88**)⇌(**7.89**)] in these reactions is indicated by the successful acid-catalyzed cyclization of the compound (**7.88**; $R^1 = R^2 = H$) to the parent 6,7,8,9-tetrahydropyrido[1,2-*a*]benzimidazole (**7.90**; $R^1 = R^2 = H$) in high yield (Table 7.4).[37] Pyrido[1,2-*a*]benzimidazole itself is efficiently formed (Table 7.4) by the cyclodehydration of 1-(2-aminophenyl)pyridin-2(1*H*)-one using phosphorus pentoxide [Scheme 7.21; (**7.91**)→(**7.92**)].[38] 2-Aminopyridine is reported[39] to condense with 1,4-benzoquinones in hot aqueous acetic acid to give good yields (Table 7.4) of 8-hydroxypyrido[1,2-*a*]benzimidazoles [Scheme 7.22; (**7.93**)+(**7.94**)→(**7.95**)]. The position of the hydroxyl group in these products, if correct, is indicative of preferential initial condensation between the amino substituent of 2-aminopyridine with one of the carbonyl groups of the quinone. Alkyl-substituted pyrido[1,2-*a*]benzimidazole derivatives are formed in good yield (Table 7.4) when 2,4,6-tri-tert.-butylaniline is electrolytically oxidized in the presence of pyridine and its *C*(3) and *C*(4) alkyl derivatives.[40,41] These reactions are suggested[41] to involve the *in situ* formation of 2,4,6-tri-tert.-butylphenylnitrenium ion [Scheme 7.23; (**7.96**)→(**7.97**)] and its nucleophilic substitution by the alkylpyridine to afford a

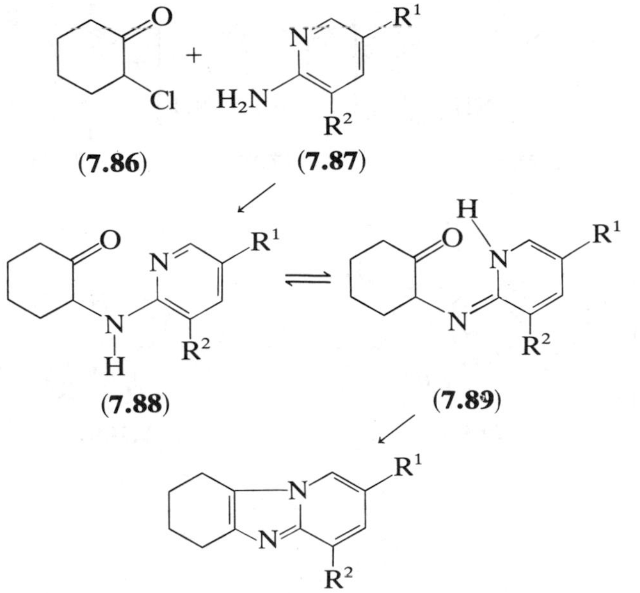

(**7.86**) (**7.87**)

(**7.88**) (**7.89**)

(**7.90**)

Scheme 7.20

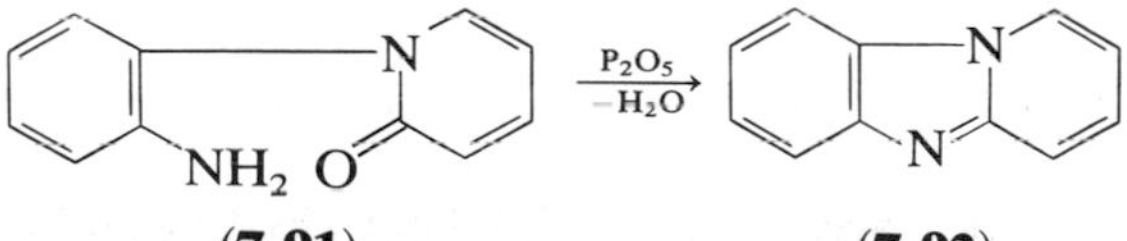

P_2O_5
$-H_2O$

(**7.91**) (**7.92**)

Scheme 7.21

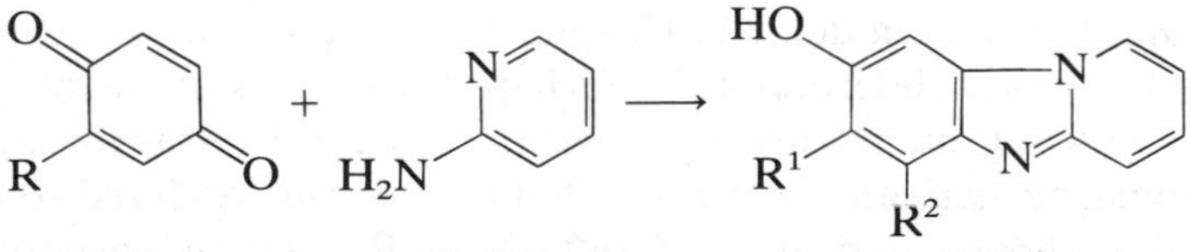

(**7.93**) (**7.94**) (**7.95**)

Scheme 7.22

cationic intermediate (**7.99**) convertible by cyclization into an unstable dihydro product (**7.100**) and thence by further oxidation into the observed alkylpyrido[1,2-*a*]benzimidazole (**7.101**). Consistent with this mechanism is the finding[41] that pyrido[1,2-*a*]benzimidazole formation fails for pyridines having a single *C*(2) alkyl group, thus demonstrating steric inhibition of electrophilic attack by the nitrenium cation on the piperidine derivative.

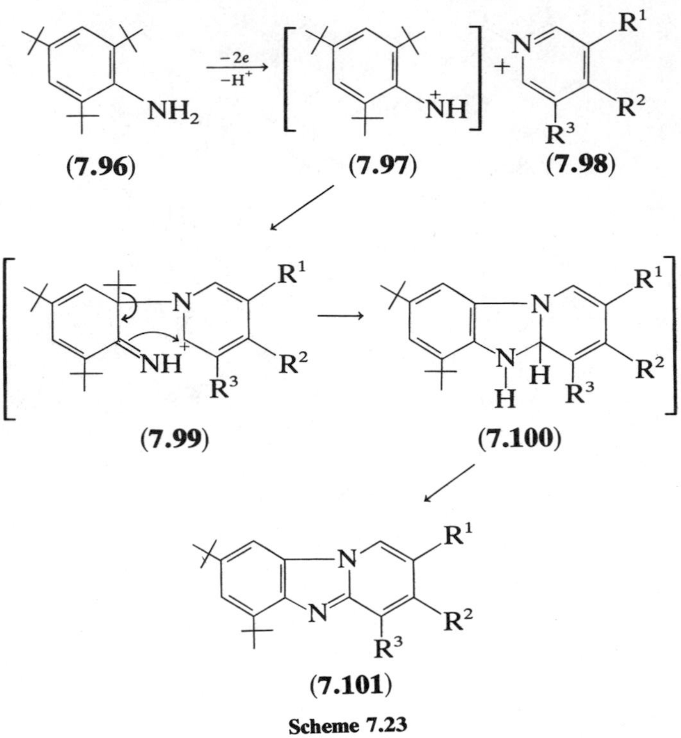

(7.96) **(7.97)** **(7.98)**

(7.99) **(7.100)**

(7.101)

Scheme 7.23

General synthetic access to 1,2,3,4-tetrahydropyrido[1,2-*a*]benzimidazoles is provided by the cyclization of variously *ortho*-substituted 1-arylpiperidine derivatives (Scheme 7.24) using a range of reagents and reaction conditions (Table 7.5).[42] Analogous ring-closure reactions of *ortho*-substituted 1-arylpyrrolidines to 2,3-dihydro-1*H*-pyrrolo[1,2-*a*]benzimidazoles are discussed in detail in Chapter 6 (cf. section 6.1.1, "Ring-closure Reactions of Other Heterocycles"). Perhaps the simplest mode of cyclization of an *ortho*-substituted 1-arylpiperidine to a 1,2,3,4-tetrahydropyrido[1,2-*a*]benzimidazole derivative is represented by the acid-catalyzed conversion[43] of the piperidinone derivative (**7.102**; $R^1 \rightarrow R^3 = H$) in unspecified yield (Table 7.5) into 1,2,3,4-tetrahydropyrido[1,2-*a*]benzimidazole (**7.105**; $R^1 \rightarrow R^3 = H$). However, a more convenient general procedure giving good yields (Table 7.5) of 1,2,3,4-tetrahydropyrido[1,2-*a*]benzimidazoles involves the oxidative cyclization of 1-(2-aminophenyl)piperidines or their *N*-acyl derivatives [Scheme 7.24; (**7.103**; R = H, alkyl, or acyl) → (**7.105**)].[42,44–49] The reportedly successful use of persulfuric acid[44a] to effect such oxidative cyclizations could not be substantiated.[44b] However, oxidative ring-closure is readily accomplished using pertrifluoroacetic acid,[45] or performic acid,[46–49] in each case generated *in situ* by the reaction of trifluoroacetic acid or

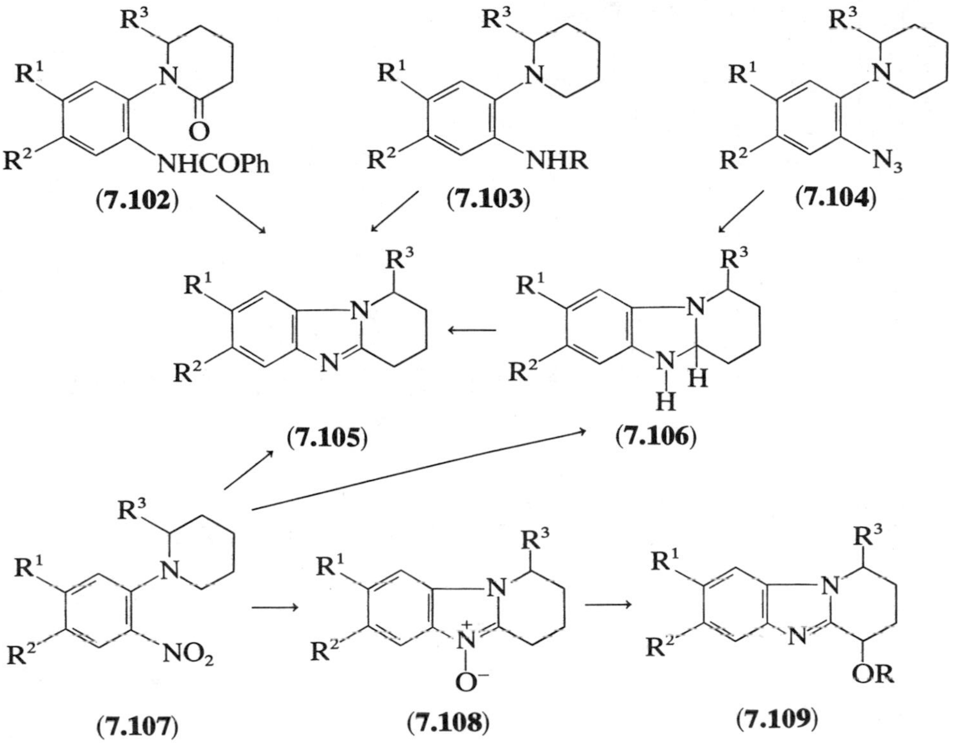

(7.102) (7.103) (7.104)

(7.105) (7.106)

(7.107) (7.108) (7.109)

Scheme 7.24

formic acid with hydrogen peroxide. The oxidative cyclization of 1-(2-aminophenyl)piperidines by peracids may be viewed[45] as involving the initial formation of nitroso intermediates convertible by subsequent cyclodehydration into the observed 1,2,3,4-tetrahydropyrido[1,2-*a*]-benzimidazoles. Tentative support for this mechanism is provided by the reaction (Scheme 7.25) of nitrobenzene (**7.110**) with *N*-lithiopiperidide (**7.111**) to give 1,2,3,4-tetrahydropyrido[1,2-*a*]benzimidazole (**7.114**) via the plausible intermediacy of 1-(2-nitrosophenyl)piperidine (**7.112**).[50] As in the case of *ortho*-aminated 1-arylpyrrolidines (cf. Chapter 6, section 6.1.1, "Ring-closure Reactions of Other Heterocycles"), the oxidative transformation of *ortho*-aminated 1-arylpiperidines by peracids into 1,2,3,4-tetrahydropyrido[1,2-*a*]benzimidazoles may also be rationalized[42,46] in terms of the initial formation and subsequent Polonovski rearrangement of a piperidine *N*-oxide intermediate. Support for the involvement of the latter has been adduced from the successful reductive cyclization of 1-(2-nitrophenyl)piperidine 1-*N*-oxide (**7.113**) to 1,2,3,4-tetrahydropyrido[1,2-*a*]-benzimidazole (**7.114**) in good yield (Table 7.5).[51] However, this result is rendered irrelevant by the probable reductive removal of the *N*-oxide

TABLE 7.5. SYNTHESIS OF 1,2,3,4-TETRAHYDROPYRIDO[1,2-*a*]BENZIMIDAZOLES AND 1,2,3,4,4a, 5-HEXAHYDROPYRIDO[1,2-*a*]-BENZIMIDAZOLES BY RING-CLOSURE REACTIONS OF *N*-(2-SUBSTITUTED PHENYL)PIPERIDINE DERIVATIVES

Starting materials ($R \rightarrow R^3$ unspecified = H)	Reaction conditions[a]	Product ($R \rightarrow R^4$ unspecified = H)	Yield (%)	m.p. (°C)	Solvent of crystallization	Ref.
(**7.102**)	*A*	(**7.105**)	—[b]	—[c]	—[d]	43
(**7.103**)	*B*	(**7.105**)	58	99–100	Cyclohexane	45
(**7.103**; R = COPh)	*B*	(**7.105**)	35	—	—	45
(**7.103**)	*C*	(**7.105**)	85–95	—[c]	—[d]	46
(**7.103**; R = COPh)	*C*	(**7.105**)	85–95	—[c]	—[d]	46
(**7.110**) + (**7.111**)	*D*	(**7.114**)	16	100–101	Cyclohexane	50
(**7.113**)	*E*	(**7.114**)	82	—	—	51
(**7.113**)	*F*	(**7.114**)	34	—	—	51
(**7.103**)	*G*	(**7.105**)	quant.	101	Cyclohexane	52
(**7.103**; R = COPh)	*G*	(**7.105**)	quant.	—	—	52
(**7.103**; R = CHO)	*G*	(**7.105**)	89	—	—	53
(**7.103**; R = COMe)	*G*	(**7.105**)	33	—	—	53
(**7.103**; R = $CONH_2$)	*G*	(**7.105**)	76	—	—	53
(**7.107**)	*H*	(**7.105**)	62	100	—[d]	54
(**7.107**)	*I*	(**7.105**)	83	—[c]	—[d]	55
(**7.104**)	*J*	(**7.105**)[e]	77	101–102	Ethyl acetate	29
(**7.123**)	*K*	(**7.125**)	92	—[c]	—[d]	60
(**7.103**; R^2 = Me)	*B*	(**7.105**; R^2 = Me)	60	126	Cyclohexane	45
(**7.107**; R^2 = Me)	*H*	(**7.105**; R^2 = Me)	55	126	—[d]	54
(**7.107**; R^2 = Me)	*L*	(**7.105**; R^2 = Me)	quant.	126	—[d]	44b
(**7.107**; R^2 = CF_3)	*H*	(**7.105**; R^2 = CF_3)	65	140	—[d]	54
(**7.107**; R^2 = Me)	*H*	(**7.105**; R^2 = Me)	55	126	—[d]	54
(**7.107**; R^2 = Me)	*L*	(**7.105**; R^2 = Me)	quant.	126	—[d]	44b
(**7.107**; R^2 = CF_3)	*H*	(**7.105**; R^2 = CF_3)	65	140	—[d]	54
(**7.107**; R^2 = CF_3)	*M*	(**7.105**; R^2 = CF_3)	65	134–135	—[d]	58
(**7.104**; R^2 = CF_3)	*N*	(**7.105**; R^2 = CF_3)	—[b]	—	—	58
(**7.104**; R^2 = CF_3)	*J*	(**7.105**; R^2 = CF_3)	—[b]	140–141	—[d]	60
(**7.103**; R^2 = Cl)	*B*	(**7.105**; R^2 = Cl)	66	152	Cyclohexane	45

(**7.103**; $R^2 = Cl$)	*C*	(**7.105**; $R^2 = Cl$)	—[b]	150	Light petroleum (b.p. 100–120°)	49
(**7.107**; $R^2 = Cl$)	*H*	(**7.105**; $R^2 = Cl$)	70	152	—[d]	54
(**7.104**; $R^2 = Cl$)	*J*	(**7.105**; $R^2 = Cl$)[e]	82	153–154	Ethylene dichloride	29
(**7.104**; $R^2 = Cl$)	*J*	(**7.105**; $R^2 = Cl$)	—[b]	153	—[d]	60
(**7.103**; $R^1 = Cl$)	*C*	(**7.105**; $R^1 = Cl$)	—[b]	148	Light petroleum (b.p. 100–120°)	49
(**7.104**; $R^1 = Cl$)	*J*	(**7.105**; $R^1 = Cl$)	—[b]	146	—[d]	60
(**7.107**; $R^1 = Cl$)	*O*	(**7.105**; $R^1 = Cl$)	—[b]	146	—[d]	57
(**7.107**; $R^2 = Br$)	*J*	(**7.105**; $R^2 = Br$)	—[b]	163		60
(**7.104**; $R^2 = F$)	*J*	(**7.105**; $R^2 = F$)	—[b]	110	—[d]	60
(**7.104**; $R^1 = R^2 = Cl$)	*J*	(**7.105**; $R^1 = R^2 = Cl$)	—[b]	184	—[d]	60
(**7.127**)	*P*	(**7.125**; $R = Cl$)	60	188	—[d]	67
(**7.124**; $R = F$)	*Q*	(**7.125**; $R = F$)	5	164	—[d]	64
(**7.103**; $R^2 = NO_2$)	*B*	(**7.105**; $R^2 = NO_2$)	95	219–220	Methyl ethyl ketone	45
(**7.103**; $R = COMe$, $R^2 = NO_2$)	*C*	(**7.105**; $R^2 = NO_2$)	80–90	219	—[d]	47
(**7.103**; $R^2 = NO_2$)	*Q*	(**7.105**; $R^2 = NO_2$)	50	219–220	Benzene or ethanol	44a
(**7.107**; $R^2 = NO_2$)	*H*	(**7.105**; $R^2 = NO_2$)	82	209	—[d]	54
(**7.107**; $R^2 = NO_2$)	*L*	(**7.105**; $R^2 = NO_2$)	55	220–222	—[d]	44b
(**7.107**; $R^2 = NO_2$)	*R*	(**7.105**; $R^2 = NO_2$)	—[b]	219–220	Benzene or ethanol	43
(**7.104**; $R^2 = NO_2$)	*J*	(**7.105**; $R^2 = NO_2$)	80	218.5–219.5	Chlorobenzene	29
(**7.103**; $R = COMe$, $R^1 = NO_2$)	*C*	(**7.105**; $R^1 = NO_2$)	80–90	218	—[d]	47
(**7.104**; $R^1 = NO_2$)	*N*	(**7.105**; $R^1 = NO_2$)	—[b]	—[c]	—[d]	58
(**7.104**)[g]	*J*	(**7.105**)[g,h]	—[b]	107–108	Light petroleum b.p. 80–100°	29
(**7.104**; $R^1 = N(CH_2)_5$, $R^2 = NO_2$)	*S*	(**7.105**; $R^1 = N(CH_2)_5$, $R^2 = NO_2$)	24	137–140	Methyl ethyl ketone	29
(**7.107**; $R^2 = CN$)	*M*	(**7.105**; $R^2 = CN$)	65	175	—[d]	58
(**7.104**; $R^2 = CN$)	*N*	(**7.105**; $R^2 = CN$)	—[b]	—	—[d]	58
(**7.104**; $R^2 = CN$)	*J*	(**7.105**; $R^2 = CN$)	—[b]	176	—[d]	60
(**7.107**; $R^2 = CO_2H$)	*H*	(**7.105**; $R^2 = CO_2H$)	88	298	—[d]	54
(**7.107**; $R^2 = CO_2H$)	*M*	(**7.105**; $R^2 = CO_2Et$)	—[b]	125	—[d]	58
(**7.104**; $R^2 = CO_2H$)	*N*	(**7.105**; $R^2 = CO_2H$)	—[b]	298	Diglyme	58

TABLE 7.5 (*Continued*)

Starting materials ($R \rightarrow R^3$ unspecified = H)	Reaction conditions[a]	Product ($R \rightarrow R^4$ unspecified = H)	Yield (%)	m.p. (°C)	Solvent of crystallization	Ref.
(**7.104**; $R^2 = SO_2N$(piperidine ring))	*J*	(**7.105**; $R^2 = SO_2N$(piperidine ring))	—[b]	229	—[d]	60
(**7.103**; R = COMe, R^2 = NHCOMe)	*C*	(**7.105**; R^2 = NHCOMe)	60	222	—[d]	48
(**7.107**; R^2 = NHCOMe)	*H*	(**7.105**; R^2 = NHCOMe)	40	222	—[d]	54
(**7.104**; R^2 = NHCOMe)	*J*	(**7.105**; R^2 = NHCOMe)	—[b]	218–220	Chlorobenzene	29
(**7.103**; R = COMe, R^1 = NHCOMe)	*C*	(**7.105**; R^1 = NHCOMe)	84	238	—[d]	48
(**7.107**; R^1 = NHCOMe)	*H*	(**7.105**; R^1 = NHCOMe)	60	238	—[d]	54
(**7.131**; R = Me)	*T*	(**7.132**; R = Me)	14	263–264	Methanol	66
(**7.131**; R = N(piperidine ring))	*T*	(**7.132**; R = N(piperidine ring))[i]	10	231–232	Chloroform–light petroleum	66
(**7.104**; R^3 = Me)	*S*	(**7.105**; R^3 = Me)	50	98	—[d]	59
(**7.107**)	*U*	(**7.109**; R = H)	73	170	Ethyl acetate	56
(**7.107**; $R^2 = NO_2$)	*U*	(**7.109**; R = COMe)	85	160	Ethyl acetate – light petroleum b.p. 60–80°	56
(**7.107**)	*V*	(**7.108**)[j]	61	202–204	Ethyl acetate	55
(**7.107**; R^2 = Cl)	*V*	(**7.108**; R^2 = Cl)	27	131	Ethyl acetate	55
(**7.107**; $R^2 = NO_2$)	*V*	(**7.108**; $R^2 = NO_2$)[k]	20	228	Ethyl acetate	55
(**7.118**; Ar = Ph)	*W*	(**7.122**; Ar = Ph)	84	288 (decomp.)	—[l]	63
(**7.118**; Ar = 2-ClC_6H_4)	*W*	(**7.122**; Ar = 2-ClC_6H_4)[m]	86	244–246	—[l]	63
(**7.118**; Ar = 4-ClC_6H_4)	*W*	(**7.122**; Ar = 4-ClC_6H_4)	90	252	—[l]	63
(**7.118**; Ar = 2-Cl-5-$O_2NC_6H_4$)	*W*	(**7.122**; Ar = 2-Cl-5-$O_2NC_6H_4$)	80	243 (decomp.)	—[l]	63
(**7.117**)	*X*	(**7.121**)	—[b]	> 300 (decomp.)	Acetic acid–water	61

(**7.117**)	*X*	(**7.121**)[n]	34	360 (decomp.)	Water	62
(**7.129**)	*Y*	(**7.130**)	38	304–306 (decomp.)	—[d]	65
(**7.104**)	*N*	(**7.106**)	quant.	119	Ethyl acetate – light petroleum	58
(**7.104**; $R^2 = NO_2$)	*Z*	(**7.106**; $R^2 = NO_2$)	quant.	129	Ethyl acetate – light petroleum	58
(**7.104**; $R^2 = NO_2$)	*N*	(**7.106**; $R^2 = NO_2$)	75	—	—	58
(**7.107**; $R^2 = NO_2$)	*M*	(**7.106**; $R^2 = NO_2$)	50–60	—	—	58
(**7.115**)	*G*	(**7.116**)	69	137–138	Ethanol – water	52

[a] $A = 2M\ H_2SO_4$/(reflux)(2 hr); $B = 30\%\ H_2O_2$, CF_3CO_2H, CH_2Cl_2/(reflux)(15–30 min); $C = 30\%\ H_2O_2$, 98% HCO_2H/(100°)(10–15 min); $D =$ ether/(−50°)(70 min); $E =$ Sn, HCO_2H/(reflux)(3 hr); $F =$ Zn, NH_4Cl, H_2O/(room temp.)(6 hr); $G =$ HgO, EDTA, EtOH—H_2O (1 : 1)/(room temp.)(5–30 min); $H =$ sand/(220–240°)(0.5–4 hr); $I = h\nu$, HCl aq., MeOH/(room temp.)(66 hr); $J = PhNO_2$/165–175° (reaction time not specified); $K =$ conc. HCl, EtOH/(reflux)(4 hr); $L = TiCl_2$, conc. HCl/(80°)(1 hr); $M = (EtO)_3P$/(reflux)(8–10 hr); $N =$ diethylene glycol dimethyl ether/(reflux)(10 min); $O =$ electrolysis in aqueous ethanol; $P = SO_2Cl_2$/room temp. (reaction time not specified); $Q = H_2O_2$, conc. H_2SO_4/(room temp.)(20 hr); $R = SnCl_2$, conc. HCl/(room temp.)(24 hr); $S = PhNO_2$/(170–180°)(0.5 hr); $T = h\nu$, $CHCl_3$; $U = Ac_2O$, $ZnCl_2$/(reflux)(4 hr); $V =$ conc. HCl/(110–160°)(7–72 hr); $W =$ conc. HCl, EtOH/(room temp.)(14 hr); $X =$ alloxan, conc. HCl, EtOH/(room temp.)(14–72 hr); $Y = SO_2$, H_2O/(70°)(0.5 hr), or conc. HCl, EtOH/warm; $Z = h\nu$, benzene/(room temp.)(18 hr).

[b] Yield not specified.

[c] Melting point not quoted.

[d] Solvent of crystallization not specified.

[e] Forms a picrate, m.p. 229–230° (from ethylene glycol monoethyl ether).

[f] Forms a hydrochloride, m.p. 295–296 (from water).

[g] 6-Nitro derivative.

[h] Forms a hydrochloride, m.p. 258–260°.

[i] Benzenesulfonate salt; free base has m.p. 191°.

[j] Hydrochloride hydrate.

[k] Hydrochloride.

[l] Precipitation from ethanol by acetone.

[m] Hemihydrate.

[n] Hydrate.

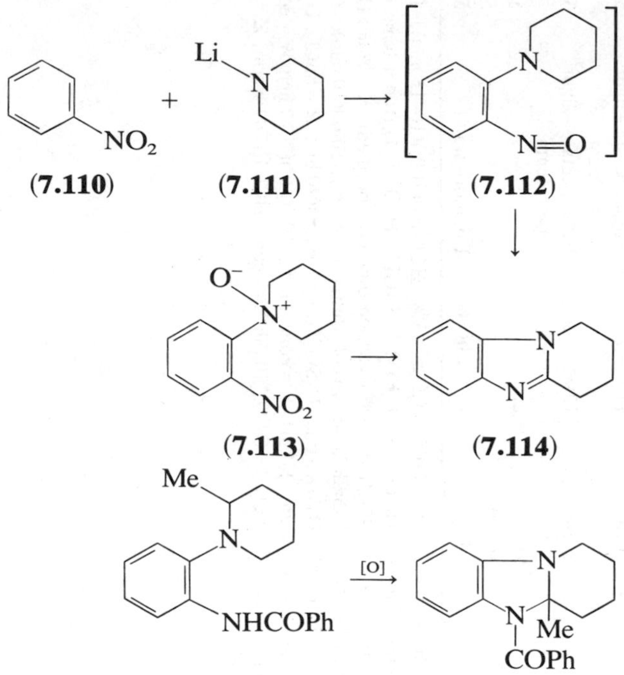

(**7.110**) (**7.111**) (**7.112**)

(**7.113**) (**7.114**)

(**7.115**) (**7.116**)

Scheme 7.25

substituent prior to ring-closure, and the known[43] propensity of 1-(2-nitrophenyl)piperidines to undergo reductive cyclization to 1,2,3,4-tetrahydropyrido[1,2-*a*]benzimidazoles under standard conditions (see later). 1-(2-Aminophenyl)piperidine and its *N*-acyl derivatives are cyclized to 1,2,3,4-tetrahydropyrido[1,2-*a*]benzimidazole in essentially quantitative yield (Table 7.5) by oxidation with mercuric oxide in combination with ethylenediaminetetraacetic acid (EDTA).[52,53] The similar oxidative ring-closure of the methyl-substituted piperidine derivative (**7.115**), on the other hand, affords the 1,2,3,4,4*a*,5-hexahydropyrido[1,2-*a*]benzimidazole (**7.116**) in good yield (Table 7.5).[52]

A number of useful methods for the synthesis of 1,2,3,4-tetrahydropyrido[1,2-*a*]benzimidazoles are based on the ring-closure of 1-(2-nitrophenyl)piperidine derivatives under a variety of conditions. For example, 1,2,3,4-tetrahydropyrido[1,2-*a*]benzimidazoles (**7.105**) are formed in moderate to good yield (Table 7.5) simply by heating the corresponding 1-(2-nitrophenyl)piperidine derivatives (**7.107**) in sand at 220–240°.[54] Cyclization of this type is promoted by the presence of electron-withdrawing groups in the benzene nucleus of the 1-(2-nitrophenyl)piperidine substrate. Conversely, the presence of electron-donating groups results in longer reaction times and lower yields. The thermal transformation of 1-(2-nitrophenyl)-piperidine derivatives into 1,2,3,4-tetrahydropyrido[1,2-*a*]benzimidazoles can be shown not to involve nitrene intermediates and is suggested[54] to occur by cyclodehydration to the corresponding 1,2,3,4-tetrahydropyrido-[1,2-*a*]benzimidazole 5-*N*-oxides [Scheme 7.24; (**7.107**) → (**7.108**)] followed by thermal deoxygenation. 1,2,3,4-Tetrahydropyrido[1,2-*a*]benzimidazole 5-*N*-oxides are in fact formed in moderate yield (Table 7.5) by the cyclodehydration of 1-(2-nitrophenyl)piperidine derivatives in hot concentrated hydrochloric acid.[55] In contrast, the photolysis of 1-(2-nitrophenyl)-piperidine in methanolic hydrochloric acid results in complete deoxygenation giving 1,2,3,4-tetrahydropyrido[1,2-*a*]benzimidazole in high yield (Table 7.5).[55] 4-Acetoxy- or 4-hydroxy-1,2,3,4-tetrahydropyrido[1,2-*a*]-benzimidazoles are formed in good yield (Table 7.5) when 1-(2-nitrophenyl)-piperidine derivatives are heated under reflux with zinc chloride in acetic anhydride.[56] These cyclization reactions are explicable[56] in terms of the intermediacy of 1,2,3,4-tetrahydropyrido[1,2-*a*]benzimidazole 5-*N*-oxides convertible into the observed products by further reaction with acetic anhydride with or without accompanying hydrolysis. 1,2,3,4-Tetrahydro-pyrido[1,2-*a*]benzimidazole 5-*N*-oxides are also postulated[44b] as intermediates in the reductive cyclization of 1-(2-nitrophenyl)piperidine derivatives to tetrahydropyrido[1,2-*a*]benzimidazoles using titanous chloride. These reactions proceed cleanly and in high yield (Table 7.5), and represent probably the most convenient method for the synthesis of 1,2,3,4-tetra-hydropyrido[1,2-*a*]benzimidazoles. The reductive formation of the latter from 1-(2-nitrophenyl)piperidines can also be effected electrolytically[57] or using reducing agents such as stannous chloride[44a] and under these conditions is believed to involve the formation and cyclodehydration of transient 1-(2-nitrosophenyl)piperidine intermediates. 1-(2-Nitrophenyl)piperidine derivatives are also reductively cyclized in high yield (Table 7.5), via isolable 1,2,3,4,4a,5-hexahydropyrido[1,2-*a*]benzimidazole intermediates, to 1,2,3,4-tetrahydropyrido[1,2-*a*]benzimidazoles using trimethyl phosphite as the reducing agent [Scheme 7.24; (**7.107**) → (**7.106**) → (**7.105**)].[58] In related processes (Scheme 7.24), 1,2,3,4-tetrahydropyrido[1,2-*a*]benzimidazoles (**7.105**) are formed in high yield (Table 7.5) by the thermolysis of 1-(2-azidophenyl)piperidine derivatives (**7.104**) in high-boiling solvents such as nitrobenzene[29,59,60] and diethylene glycol dimethyl ether.[58] These reactions may be viewed as involving cyclizative insertion in initial nitrene intermediates to give 1,2,3,4,4a,5-hexahydropyrido[1,2-*a*]benzimidazole derivatives (**7.106**), which in some instances[58] can be isolated, but in most cases are spontaneously oxidized to product (**7.105**).

1,2,3,4-Tetrahydropyrido[1,2-*a*]benzimidazoles are also the end-products of a miscellany of reactions which share the common feature of ring-closure between an unsaturated *ortho*-substituent and the *C*(2) methylene center in the hetero ring of a 1-arylpiperidine derivative. Cyclization reactions of this type are exemplified (Scheme 7.26) by the acid-catalyzed formation of the

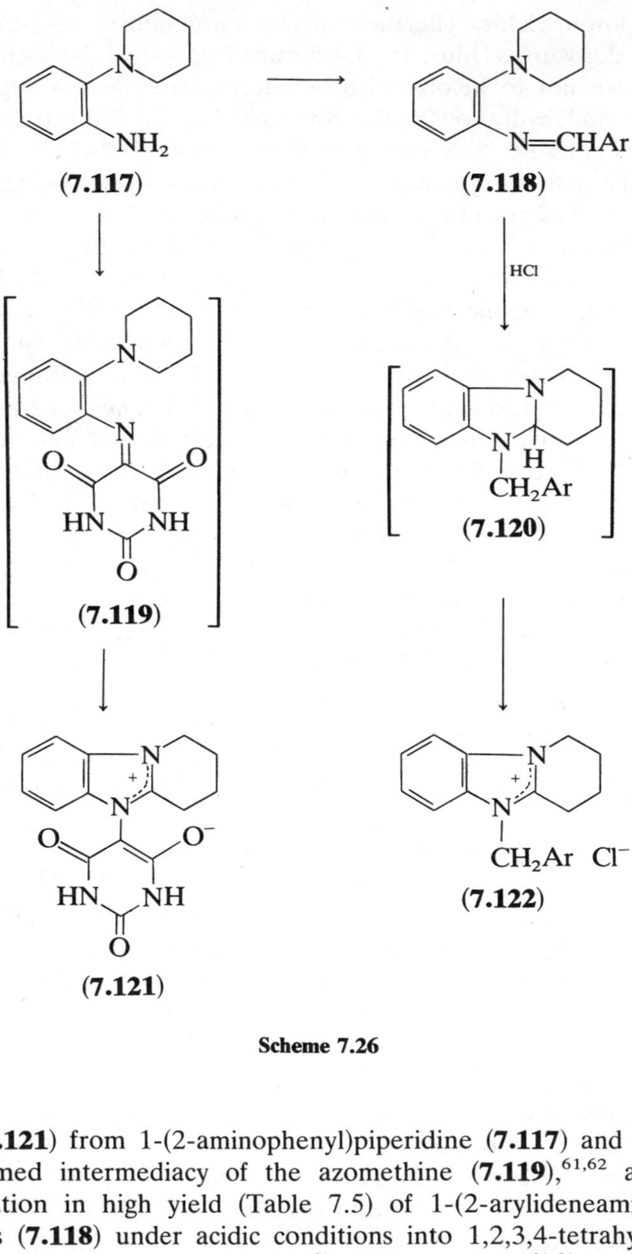

(7.117) (7.118) (7.119) (7.120) (7.121) (7.122)

Scheme 7.26

betaine (**7.121**) from 1-(2-aminophenyl)piperidine (**7.117**) and alloxan via the presumed intermediacy of the azomethine (**7.119**),[61,62] and by the transformation in high yield (Table 7.5) of 1-(2-arylideneaminophenyl)-piperidines (**7.118**) under acidic conditions into 1,2,3,4-tetrahydropyrido-[1,2-*a*]benzimidazolium salts (**7.122**).[63] Labeling studies[61,63] suggest that the acid-catalyzed ring-closure of the azomethines (**7.118**) and (**7.119**) to the tetrahydropyrido[1,2-*a*]benzimidazoles (**7.122**) and (**7.121**) involves the formation and disproportionation of 1,2,3,4,4a,5-hexahydropyrido[1,2-*a*]-benzimidazole intermediates of the type (**7.120**). Interaction between an *ortho*-diazo substituent and the *C*(2) methylene center in the hetero ring of

(**7.123**) $\xrightarrow{H^+}$ (**7.125**)

(**7.124**) $\xrightarrow{N_2H_4}$ [(**7.126**)]

(**7.127**) $\xrightarrow{SO_2Cl_2}$ [(**7.128**)]

(**7.129**) $\xrightarrow{H^+}$ (**7.130**)

(**7.131**) $\xrightarrow{h\nu}$ (**7.132**)

Scheme 7.27

1-arylpiperidines provides a further mode of ring-closure leading to 1,2,3,4-tetrahydropyrido[1,2-*a*]benzimidazoles. Representative of such reactions (Scheme 7.27) are the acid-catalyzed conversion of the azobenzene derivative (**7.123**; R = H) in high yield (Table 7.5) into 1,2,3,4-tetrahydropyrido-[1,2-*a*]benzimidazole (**7.125**; R = H),[61] and the unusual formation[64] (Table 7.5) of its tetrafluoro derivative (**7.125**; R = F) by the reaction of the piperidine *N*-oxide (**7.124**; R = F) with hydrazine, a transformation suggested[64] to involve the diimide intermediate (**7.126**). Cyclization of the latter type is further illustrated by the acid-catalyzed conversion of the diazo sulfonate (**7.129**) in moderate yield (Table 7.5) into the 1,2,3,4-tetrahydro-pyrido[1,2-*a*]benzimidazole 5-*N*-imine (**7.130**).[65] Other processes leading to 1,2,3,4-tetrahydropyrido[1,2-*a*]benzimidazoles include the photocyclization of quinonediimines [Scheme 7.27; (**7.31**) → (**7.132**)][66] and the ring-closure of 1-(2-aminophenyl)piperidine using thionyl chloride [Scheme 7.27; (**7.127**; R = H) → → (**7.125**; R = Cl)],[67] the latter reaction involving the possible intermediacy of the sulfonylamine derivative (**7.128**).

7.1.2. Physicochemical Properties

Spectroscopic Studies

INFARED SPECTRA (Tables 7.6 and 7.7). The NH tautomeric form of 1,5- and 3,5-dihydropyrido[1,2-*a*]benzimidazoles (**7.136**; R^1 = H), (**7.138**; R^1 = H), and (**7.139**; R^1 = H) is substantiated by the presence in their IR spectra (Table 7.7) of well-defined NH absorption in the range 3480–3255 cm^{-1}. The exceptionally low IR stretching frequencies (ca. 2190 cm^{-1}) (Table 7.7) of *C*(4) cyano groups in 1,5-dihydropyrido[1,2-*a*]benzimidazoles are attributed[6] to resonance interaction with the imidazole ring [Scheme 7.28; (**7.141**) ↔ (**7.142**)]. In contrast, *C*(4) cyano substituents in pyrido[1,2-*a*]-benzimidazol-1(5*H*)-ones[6,7,21] and pyrido[1,2-*a*]benzimidazol-3(5*H*)-ones[6] exhibit normal IR absorption (Table 7.7) as a result of competing resonance interaction between the imidazole ring and the *C*(1) or *C*(3) carbonyl substituent, e.g. [Scheme 7.28; (**7.143**; R^2 = CN) ↔ (**7.144**; R^2 = CN)]. It follows that the *C*(1) and *C*(3) carbonyl groups of pyrido[1,2-*a*]benzimidazol-1(5*H*)-ones[6,7,21] and pyrido[1,2-*a*]benzimidazol-3(5*H*)-ones[6] are associated with very low IR carbonyl absorption (ca. 1680–1640 cm^{-1}) (Table 7.7). Pyrido[1,2-*a*]benzimidazol-1(4*H*)-ones,[20] on the other hand, are distinguished by IR carbonyl absorption at significantly higher frequencies (ca 1700 cm^{-1}) (Table 7.7). Resonance interaction of the type [Scheme 7.28; (**7.141**) ↔ (**7.142**)] also accounts for the low IR carbonyl stretching frequencies of *C*(4) ester substituents in 1,5-dihydropyrido[1,2-*a*]benzimidazoles (Table 7.7).[6] Bands at ca. 1650–1640 cm^{-1} attributable to C=N absorption are characteristic of the IR spectra (Table 7.6) of fully unsaturated pyrido-[1,2-*a*]benzimidazole derivatives.

TABLE 7.6. INFRARED SPECTRA OF PYRIDO[1,2-*a*]BENZIMIDAZOLE DERIVATIVES (**7.133**) AND (**7.134**)

(**7.133**)

(**7.134**)

Compound	R^1	R^2	R^3	R^4	R^5	X	Medium	$\nu_{max}(cm^{-1})$			Ref.
								NH,OH	C=O	C=N	
(**7.133**)	H	H	H	H	—	—	KBr	—	—	1645	30
(**7.133**)	Me	H	H	H	—	—	KBr	—	—	1640	30
(**7.133**)	H	Ph	H	Ph	—	—	Nujol	—	—	1645	18
(**7.133**)	H	H	NH_2	H	—	—	Nujol	3430,3180	—	—	32
(**7.133**)	H	Me	CO_2H	Me	—	—	KBr	3500–2800br	1680	—	35
(**7.133**)	H	H	NO_2	H	—	—	Nujol	—	—	1640	31
(**7.133**)	H	Me	NO_2	H	—	—	Nujol	—	—	1640	31
(**7.133**)	H	H	NO_2	Me	—	—	Nujol	—	—	1640	31
(**7.134**)	Me	H	Me	H	H	ClO_4	KBr	—	—	1657	11a
(**7.134**)	Me	H	Ph	H	Ph	I	Nujol	—	—	1645	18
(**7.134**)	Me	CO_2Et	Me	H	H	ClO_4	KBr	—	1740	—	11a
(**7.134**)	Ph	CO_2Et	Me	H	H	ClO_4	KBr	—	1740	—	11a
(**7.134**)	Me	CO_2Me	CO_2Me	CO_2Me	CO_2Me	ClO_4	Nujol	—	1740	1650	2

TABLE 7.7. INFRARED SPECTRA OF DIHYDROPYRIDO[1,2-*a*]BENZIMIDAZOLE

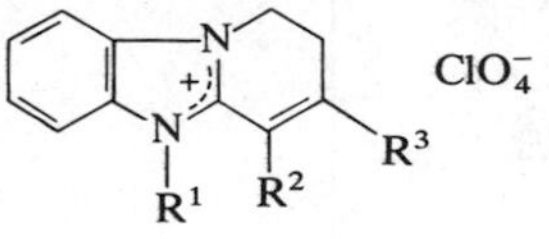

(**7.135**) 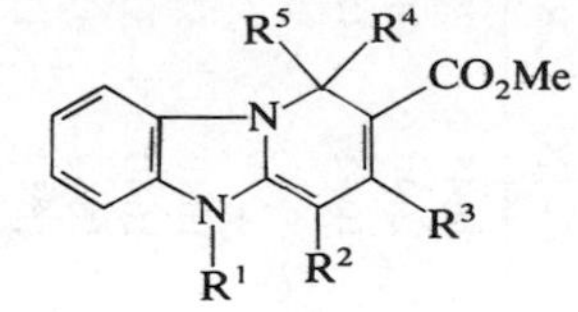 (**7.136**) 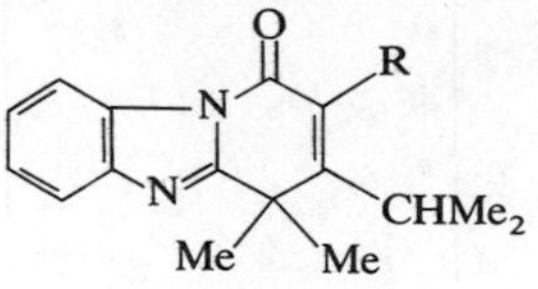 (**7.137**)

Compound	R	R^1	R^2	R^3	R^4
(**7.135**)	—	Me	H	Me	—
(**7.135**)	—	Me	CO_2Et	Me	—
(**7.135**)	—	Et	CO_2Et	Me	—
(**7.135**)	—	Ph	CO_2Et	Me	—
(**7.136**)	—	Me	CO_2Me	CO_2Me	CO_2Me
(**7.136**)	—	$MeO_2C\overset{\vert}{C}{=}CHCO_2Me$	CO_2Me	CO_2Me	CO_2Me
(**7.136**)	—	$MeO_2C\underset{\vert}{C}HCH_2CO_2Me$	CO_2Me	CO_2Me	CO_2Me
(**7.136**)	—	Me	CN	H	CH_2CO_2Me
(**7.136**)	—	$CH\overset{t}{=}CHCO_2Me$	CN	H	H
(**7.136**)	—	Me	CO_2Et	H	H
(**7.136**)	—	H	CO_2Et	CO_2Me	CO_2Me
(**7.136**)	—	Me	CO_2Et	CO_2Me	CO_2Me
(**7.136**)	—	Me	CN	CO_2Me	CO_2Me
(**7.137**)	Me	—	—	—	—
(**7.137**)	Et	—	—	—	—
(**7.138**)	—	$CH\overset{t}{=}CHCO_2Me$	CN	H	—
(**7.138**)	—	H	CN	Me	—
(**7.138**)	—	H	CN	CO_2Me	—
(**7.138**)	—	Me	CN	CO_2Me	—
(**7.138**)	—	COMe	CN	CO_2Me	—
(**7.138**)	—	H	CO_2Et	Me	—
(**7.138**)	—	H	CO_2Et	CO_2Me	—
(**7.138**)	—	COMe	CO_2Et	CO_2Me	—
(**7.138**)	—	H	$CONH_2$	Me	—
(**7.139**)	—	$CH\overset{t}{=}CHCO_2Me$	CN	H	H
(**7.139**)	—	H	CN	H	CO_2Me
(**7.139**)	—	Me	H	CO_2Me	H
(**7.139**)	—	$CH\overset{t}{=}CHCO_2Me$	H	CO_2Me	H
(**7.140**)	—	Me	Me	CO_2Me	—
(**7.140**)	—	Me	Et	CO_2Me	—
(**7.140**)	—	Me	CO_2Me	H	—

[a] 8-Chloro derivative.

[b] C=N or C=C.

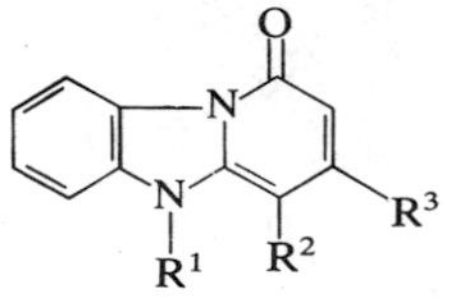

DERIVATIVES (**7.135**)–(**7.140**)

(**7.138**) 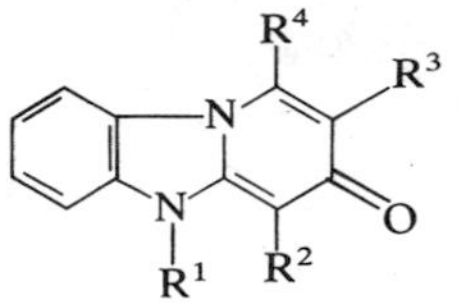 (**7.139**) 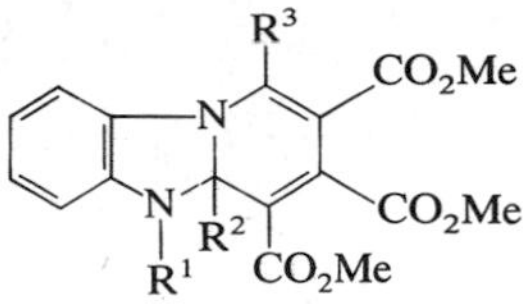 (**7.140**)

R^5	Medium	$\nu_{max}(cm^{-1})$ NH	C≡N	C=O	C=N	C=C	Ref.
—	KBr	—	—	—	1640	—	11a
—	KBr	—	—	1735	—	—	11a
—[a]	KBr	—	—	1725	—	—	11a
—	KBr	—	—	1740	—	—	11a
H	Nujol	—	—	1740, 1730	1670	—	2
H	Nujol	—	—	1740, 1700	1680	1640	2
H	Nujol	—	—	1740, 1730	1680	—	2
—	Nujol	—	2190	1748, 1669	←1630[b]→		6
CH_2CO_2Me	Nujol	—	2200	1733, 1723, 1671	1664[b]	1625[b]	6
CH_2CO_2Me	Nujol	—	—	1747, 1683	←1662[b]→		6
CH_2CO_2Me	Nujol	3255	—	1753, 1742, 1709	1660[b]	1624[b]	6
CH_2CO_2Me	Nujol	—	—	1770, 1734, 1720	1670[b]	1643[b]	6
CH_2CO_2Me	Nujol	—	2192	1742, 1717	1664[b]	1626[b]	6
—	KBr	—	—	1695	—	—	20
—	KBr	—	—	1700	—	—	20
—	Nujol	—	2222	1733, 1686	←1656[b]→		6
—	KBr	3480, 3120–2729 br	2240	1670	—	—	21
—	Nujol	—	2214	1724, 1660	—	—	7
—	Nujol	—	2220	1734, 1668	—	—	7
—	Nujol	—	2224	1730, 1678	—	—	7
—	KBr	3390	—	1683, 1640	—	—	21
—	Nujol	3300	—	1744, 1732, 1696	←1652[b]→		7
—	Nujol	—	—	1746, 1725, 1692, 1662	—	—	7
—	KBr	3380, 3220	—	1660 sh, 1645	—	—	21
—	Nujol	—	2214	1716, 1647	←1625[b]→		6
—	Nujol	3250–2500 br	2220	1736, 1668	—	—	6
—	Nujol	—	—	1703, 1655	←1620[b]→		6
—	Nujol	—	—	1740, 1707, 1649	—	—	6
—	Nujol	—	—	1730, 1710, 1690	—	—	2
—	Nujol	—	—	1730, 1690	—	—	2
—	Nujol	—	—	1740, 1705	—	—	2,3

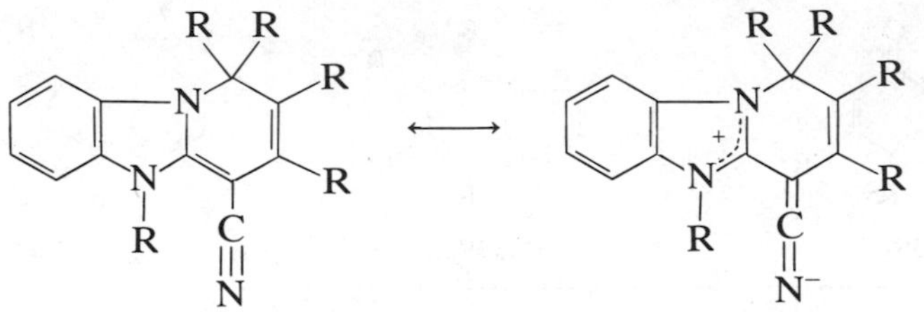

(**7.141**) ⟷ (**7.142**)

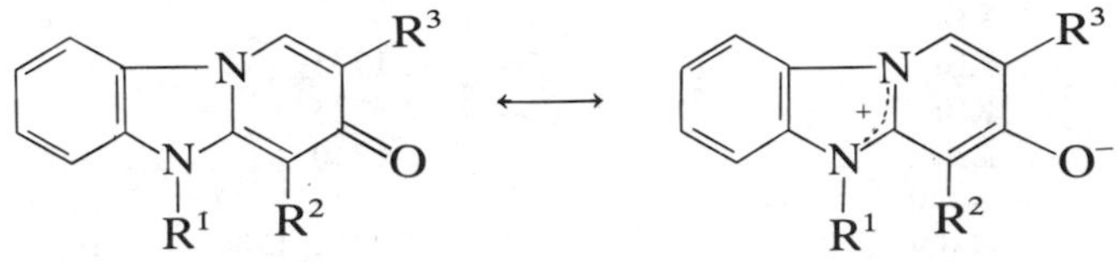

(**7.143**) ⟷ (**7.144**)

Scheme 7.28

Ultraviolet Spectra. The highly delocalized nature of the ring system in fully unsaturated pyrido[1,2-*a*]benzimidazoles is illustrated by their UV spectra (Table 7.8), which are typified by a series of intense absorption maxima in the range 240–390 nm. Not unexpectedly, the presence of unsaturated substituents at the *C*(1) and *C*(3) positions of the pyrido[1,2-*a*]-benzimidazole ring system results in a uniform shift of the UV maxima to longer wavelength (Table 7.8). On the other hand, quaternization of the *N*(5) position in pyrido[1,2-*a*]benzimidazoles appears to have little effect on the UV absorption (Table 7.8).

The UV spectra (Table 7.9) of 1,5-dihydropyrido[1,2-*a*]benzimidazoles feature three intense UV bands in the ranges 240–250, 300–320, and 390–430 nm.[2,5,6] Preferential protonation of 1,5-dihydropyrido[1,2-*a*]benzimidazoles at the *C*(4) position [Scheme 7.29; (**7.156**) → (**7.157**)] is indicated by the similarity of their UV absorption in acidic media (Table 7.9) to that of the benzimidazolium cation.[2,5,6] The largely similar UV absorption of pyrido[1,2-*a*]benzimidazol-1(5*H*)-ones (**7.149**)[6,7] and pyrido[1,2-*a*]benzimidazol-3(5*H*)-ones (**7.150**)[6] in both neutral and acidic media (Table 7.9), on the other hand, is consistent with the presence in these molecules of resonance interaction of the type [Scheme 7.28; (**7.143**) ↔ (**7.144**)]. The UV absorption (Table 7.9) of pyrido[1,2-*a*]benzimidazol-3(5*H*)-ones (**7.150**)[6] is distinguished from that of pyrido[1,2-*a*]benzimidazol-1(5*H*)-ones (**7.149**)[6,7] in being generally less complex. The ready protonation of 4a,5-dihydropyrido[1,2-*a*]benzimidazoles (**7.151**) accounts for the reduction in the complexity, and the hypsochromic shift, of the UV absorption of these molecules (Table 7.9) on changing from neutral to acidic solution.[2,5] The UV absorption of 1,2,3,4-tetrahydropyrido[1,2-*a*]benzimidazoles (Table 7.10), as might be expected, resembles that of simple benzimidazole derivatives.

TABLE 7.8. ULTRAVIOLET SPECTRA OF PYRIDO[1,2-*a*]BENZIMIDAZOLE DERIVATIVES (**7.145**)–(**7.147**)

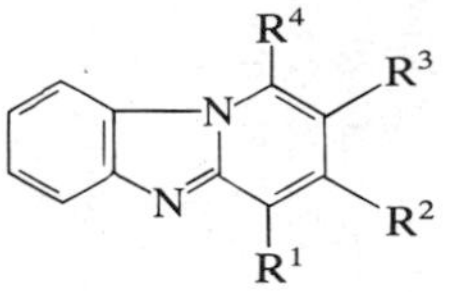

(**7.145**) 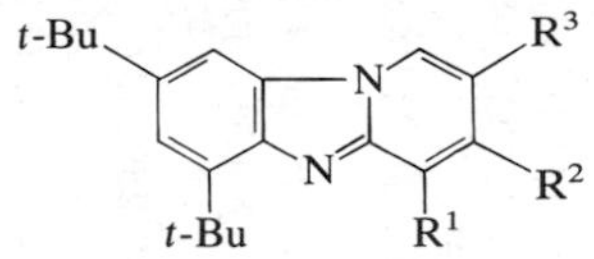 (**7.146**) 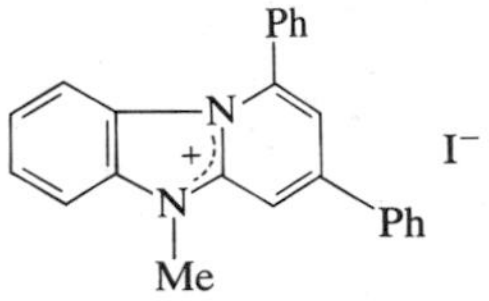 (**7.147**)

Compound	R^1	R^2	R^3	R^4	Solvent[a]	λ_{max}(nm)(log ε)	Ref.
(**7.145**)	H	H	H	H	*A*	240(4.45), 245(4.46), 258(3.99), 266(3.29)	30
(**7.145**)	Me	H	H	H	*A*	240(4.54), 245(4.53), 254(4.11), 266(4.00)	30
(**7.145**)	H	Ph	H	Ph	*A*	217(3.75), 270(4.06), 315(3.60), 355(3.54)	18
(**7.147**)	—	—	—	—	*A*	225(4.04), 272(3.95), 317(3.84), 355(3.70)	18
(**7.146**)	H	H	H	—	*B*	248(4.64), 255(4.78), 265(4.34), 274(4.26), 296(3.56), 308(3.70), 320(3.74), 344(3.60), 358(3.68), 375(3.64), 396(3.37)	40
(**7.146**)	Me	H	H	—	*B*	254(4.77), 264(4.38), 272(4.23), 294(3.57), 305(3.72), 317(3.69), 349(3.74), 364(3.72), 283(3.43)	41
(**7.146**)	H	Me	H	—	*B*	249(4.60), 256(4.75), 266(4.36), 275(4.31), 295(3.58), 307(3.70), 319(3.59), 355(3.62), 370(3.57)	41
(**7.146**)	H	H	Me	—	*B*	256(4.74), 266(4.31), 275(4.17), 299(3.51), 311(3.68), 324(3.77), 358(3.69), 374(3.66), 394(3.36)	41
(**7.146**)	H	Et	H	—	*B*	249(4.61), 256(4.75), 266(4.38), 275(4.33), 295(3.60), 307(3.71), 319(3.61), 354(3.66), 368(3.61)	41
(**7.146**)	Me	H	Me	—	*B*	256(4.73), 273(4.13), 308(3.69), 321(3.74), 350(3.76), 363(3.73)	41
(**7.145**)	H	H	NH_2	H	*A*	249(4.55), 315(3.81), 328(3.85)	32
(**7.145**)	H	H	NO_2	H	—[b]	228(4.52), 298(4.47), 333(3.98), 351(3.85), 389(3.41)	31,32
(**7.145**)	CN	Me	H	Me	—[b]	212(4.03), 257(4.64), 265(4.41), 275(4.42), 304(3.66), 317(3.56)	12

[a] *A* = ethanol; *B* = cyclohexane.
[b] Solvent not specified.

NUCLEAR MAGNETIC RESONANCE SPECTRA. Conjugative interaction with *N*(5) would be expected to confer enhanced electron deficiency on the *C*(1) and *C*(3) positions in fully unsaturated pyrido[1,2-*a*]benzimidazoles. In accord with this expectation *H*(1) in pyrido[1,2-*a*]benzimidazole derivatives resonates uniformly at lower field (Table 7.11) than the other pyridine protons, followed closely by *H*(3). Predictably *H*(6) is the most deshielded

TABLE 7.9. ULTRAVIOLET SPECTRA OF DIHYDROPYRIDO[1,2-*a*]BENZIMIDAZOLE DERIVATIVES (**7.148**)–(**7.151**)

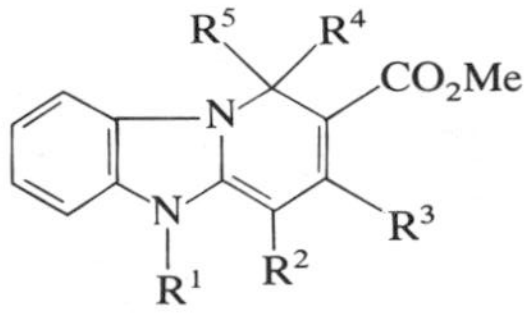

(**7.148**) 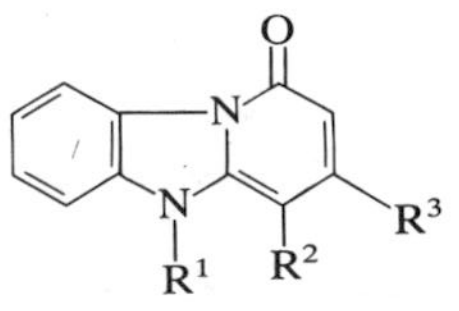 (**7.149**) 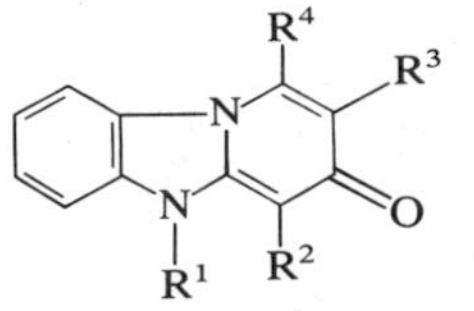 (**7.150**) 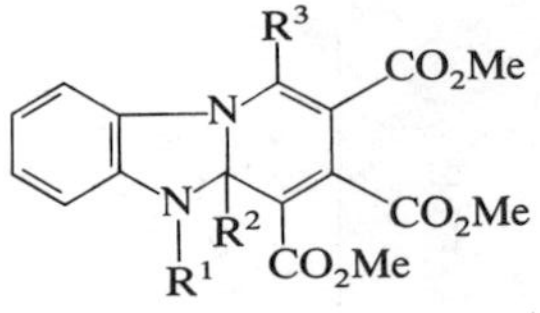 (**7.151**)

Compound	R^1	R^2	R^3	R^4	R^5	Solvent[a]	λ_{max} (nm) (log ϵ)	Ref.
(**7.148**)	Me	CO_2Me	CO_2Me	CO_2Me	H	*A*	254(4.03), 316.5(4.47), 400(4.29)	2
(**7.148**)	$MeO_2C\overset{\vert}{C}{=}CHCO_2Me$	CO_2Me	CO_2Me	CO_2Me	H	*A*	242 inf (3.85), 265 inf (3.70), 309(4.16), 403(4.03)	2
(**7.148**)	$MeO_2C\overset{\vert}{C}HCH_2CO_2Me$	CO_2Me	CO_2Me	CO_2Me	H	*A*	250.5 inf (3.93), 316.5(4.41), 397(4.21)	2
(**7.148**)	Me	CO_2Et	H	H	CH_2CO_2Me	*A*	213(4.38), 250(4.11), 273 inf (3.84), 315(4.58), 398(4.35)	6
(**7.148**)	Me	CO_2Et	H	H	CH_2CO_2Me	*B*	211(4.41), 267 inf (3.92), 315(4.58), 398(4.35)	6
(**7.148**)	Me	CN	H	H	CH_2CO_2Me	*A*	211 inf (4.26), 220(4.37), 248(4.24), 262 inf (4.06), 298.5(4.31), 409(4.34)	6
(**7.148**)	Me	CN	H	H	CH_2CO_2Me	*B*	216 inf (4.28), 276(4.27)	6
(**7.148**)	$CH\overset{t}{=}CHCO_2Me$	CN	H	H	CH_2CO_2Me	*A*	213(4.51), 242(4.38), 259(4.37), 280(4.42), 293 inf (4.23), 428(4.53)	6
(**7.148**)	$CH\overset{t}{=}CHCO_2Me$	CN	H	H	CH_2CO_2Me	*B*	206(4.49), 268 inf (4.27), 282(4.29), 365(3.69)	6
(**7.148**)	H	CO_2Et	CO_2Me	CO_2Me	CH_2CO_2Me	*A*	211(4.26), 217 inf (4.24), 225 inf (4.19), 248(4.18), 308(4.58), 412(4.29)	6
(**7.148**)	H	CO_2Et	CO_2Me	CO_2Me	CH_2CO_2Me	*B*	215 inf (4.25), 247(3.92), 265(3.82), 272(3.84), 281(3.82), 307(4.07)	6
(**7.148**)	Me	CO_2Et	CO_2Me	CO_2Me	CH_2CO_2Me	*A*	208(4.25), 228(4.19), 286(4.43), 330(4.30), 398(3.63)	6
(**7.148**)	Me	CO_2Et	CO_2Me	CO_2Me	CH_2CO_2Me	*B*	210(4.53), 236(4.17), 269(4.15), 283 inf (4.00)	6

(7.148)	Me	CN	CO_2Me	CO_2Me	CH_2CO_2Me	*A*	211 inf (4.09), 228(4.30), 283(4.56), 314 inf (4.10), 323(4.22), 400(3.50)	6
(7.148)	Me	CN	CO_2Me	CO_2Me	CH_2CO_2Me	*B*	233(4.16), 268(4.07), 287(4.11), 305 inf (3.92), 321 inf (3.85)	6
(7.148)	Me	Ph	CO_2Me	CO_2Me	CH_2CO_2Me	*A*	220 inf (4.51), 275(4.20), 281(4.19), 315(4.18), 480(4.25)	5, 6
(7.148)	Me	Ph	CO_2Me	CO_2Me	CH_2CO_2Me	*B*	275(4.22), 282(4.19)	5, 6
(7.149)	$CH\overset{t}{=}CHCO_2Me$	CN	H	—	—	*A*	211(4.41), 224(4.46), 258(4.37), 281(4.32), 358 inf (4.27), 371(4.39), 400(3.46)	6
(7.149)	$CH\overset{t}{=}CHCO_2Me$	CN	H	—	—	*B*	242(4.44), 259(4.39), 269 inf (4.33), 304(3.70), 354 inf (4.23), 365(4.32), 402 inf (3.36)	6
(7.149)	H	CO_2Et	CO_2Me	—	—	*A*	228(4.26), 241(4.30), 273(4.00), 298(4.45), 357(4.21)	7
(7.149)	COMe	CO_2Et	CO_2Me	—	—	*A*	236(4.31), 299(4.20), 359(4.18)	7
(7.149)	H	CN	CO_2Me	—	—	*A*	228(4.28), 255(4.37), 284(4.33), 387(4.00)	7
(7.149)	Me	CN	CO_2Me	—	—	*A*	229(4.35), 253(4.40), 295(4.33), 385(4.04)	7
(7.149)	COMe	CN	CO_2Me	—	—	*A*	223(4.28), 247(4.24), 264(4.25), 377(4.03)	7
(7.150)	Me	H	CO_2Me	H	—	*A*	212(4.42), 230(4.34), 261(4.67), 306 inf (3.97), 315(4.09), 336(3.97)	6
(7.150)	Me	H	CO_2Me	H	—	*B*	210 inf (4.08), 217(4.11), 258 inf (4.29), 267(4.31), 307(3.77), 316 inf (3.76)	6
(7.150)	$CH\overset{t}{=}CHCO_2Me$	H	CO_2Me	H	—	*A*	211(4.39), 219 inf (4.35), 259(4.63), 280 inf (4.29), 290 inf (4.22), 327(4.38)	6
(7.150)	$CH\overset{t}{=}CHCO_2Me$	H	CO_2Me	H	—	*B*	220 inf (4.31), 251(4.53), 262(4.50), 275(4.44), 329(4.20)	6
(7.150)	$CH\overset{t}{=}CHCO_2Me$	CN	H	H	—	*A*	211(4.39), 249(4.62), 256 inf (4.51), 274(4.29), 283(4.27), 318(4.34)	6
(7.150)	$CH\overset{t}{=}CHCO_2Me$	CN	H	H	—	*B*	212(4.49), 249(4.48), 261(4.44), 282(4.32), 323(4.09)	6
(7.150)	H	CN	H	CO_2Me	—	*A*	211(4.29), 230 inf (4.13), 270(4.47), 283 inf (4.41), 308 inf (4.00), 416(3.92)	6

TABLE 7.9 (*Continued*)

(**7.148**) (**7.149**) (**7.150**) (**7.151**)

Compound	R^1	R^2	R^3	R^4	R^5	Solvent[a]	λ_{max} (nm) (log ϵ)	Ref.
(**7.150**)	H	CN	H	CO_2Me	—	*B*	215 inf (4.32), 233(4.28), 256(4.51), 271(4.55), 291 inf (4.09), 390(4.02)	6
(**7.151**)	Me	Me	CO_2Me	—	—	*A*	220(4.31), 256(4.18), 316(3.88), 441(3.96)	2
(**7.151**)	Me	Me	CO_2Me	—	—	*B*	243(4.27), 295(4.02)	2
(**7.151**)	Me	Et	CO_2Me	—	—	*A*	220(4.29), 254(4.17), 332.5(3.83), 457(3.90)	2
(**7.151**)	Me	Et	CO_2Me	—	—	*B*	253(4.27), 296(4.03)	2
(**7.151**)	Me	*i*-Pr	CO_2Me	—	—	*A*	—[b]	4
(**7.151**)	Me	*i*-Pr	CO_2Me	—	—	*B*	—[b]	4
(**7.151**)	Me	CH_2Ph	CO_2Me	—	—	*A*	224 inf (4.38), 254(4.22), 317(3.88), 450(3.88)	5
(**7.151**)	Me	CH_2Ph	CO_2Me	—	—	*B*	244(4.27), 296(4.00)	5
(**7.151**)	$CH\overset{t}{=}CHCO_2Me$	Bu^t	CO_2Me	—	—	*B*	—[b]	4
(**7.151**)	$CH\overset{t}{=}CHCO_2Me$	Bu^t	CO_2Me	—	—	*B*	—[b]	4
(**7.151**)	Me	CO_2Me	H	—	—	*A*	236(4.24), 271.5(4.19), 301(4.01), 437(3.95)	2
(**7.151**)	Me	CO_2Me	H	—	—	*B*	234.5(4.44), 236.5(4.43), 269(4.06), 299.5(3.95), 345(3.90), 436(3.97)	2
(**7.151**)	Me	CO_2Me	Ph	—	—	*A*	—[b]	4
(**7.151**)	Me	CO_2Me	Ph	—	—	*B*	—[b]	4

[a] *A* = MeOH; *B* = MeOH–70% $HClO_4$ aq.
[b] UV data available in a supplementary publication.

TABLE 7.10. ULTRAVIOLET SPECTRA OF TETRAHYDROPYRIDO[1,2-*a*]BENZIMIDAZOLE DERIVATIVES (**7.152**)–(**7.155**)

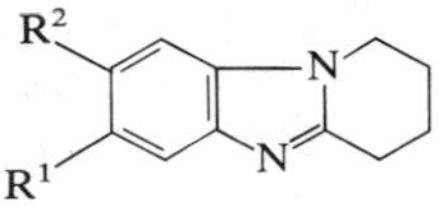

(**7.152**) 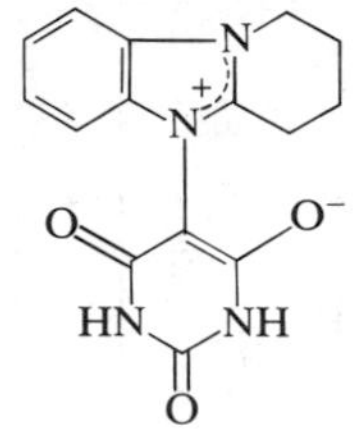 (**7.153**) 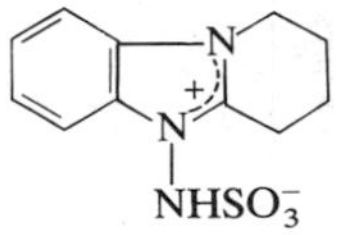 (**7.154**) 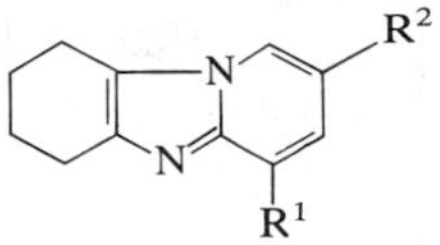 (**7.155**)

Compound	R^1	R^2	Solvent[a]	$\lambda_{max.}$(nm)(log ε)	Ref.
(**7.152**)	H	H	*A*	248.5–251(3.76), 254(3.78), 276.5(3.75), 283(3.81)	24, 50
(**7.152**)	H	H	*A*	249(3.76), 275(3.75), 282(3.76)	45
(**7.152**)	H	H	*A*	213.1(4.42), 254.6(3.71), 276.2(3.71), 282.4(3.73)	48
(**7.152**)	Me	H	*A*	249(3.76), 275(3.75), 282(3.76)	45
(**7.152**)	Cl	H	*A*	249(3.76), 275(3.75), 282(3.76)	45
(**7.152**)	NO_2	H	*A*	241(4.29)	45
(**7.152**)	Me	$NHSO_2Ph$	*A*	262(3.95), 285(3.90), 288(3.90), 295(3.91)	66
(**7.152**)	Me	$NHSO_2Ph$	*B*	288(4.24), 304(4.01)	66
(**7.152**)	 N-piperidino	$NHSO_2Ph$	*A*	213(4.61), 258(3.88), 299(4.01)	66
(**7.152**)	 N-piperidino	$NHSO_2Ph$	*B*	307(4.03)	66
(**7.153**)	—	—	*C*	249.5(4.34), 270.5(3.95), 277.5(3.93)	63
(**7.154**)	—	—	—[b]	206(4.01), 245.4(3.50), 269.6(3.52), 276.4(3.51)	65
(**7.155**)	H	H	—[b]	235(4.45), 281(3.62), 316(3.66)	37
(**7.155**)	H	Br	—[b]	245(4.45), 302(3.68), 332(3.70)	37
(**7.155**)	H	NO_2	*A*	281–283(4.36)	32
(**7.155**)	Br	Br	—[b]	245(4.45), 290(3.68), 325(3.70)	37

[a] *A* = EtOH; *B* = EtOH–NaOH aq.; *C* = H_2O.
[b] Solvent not specified.

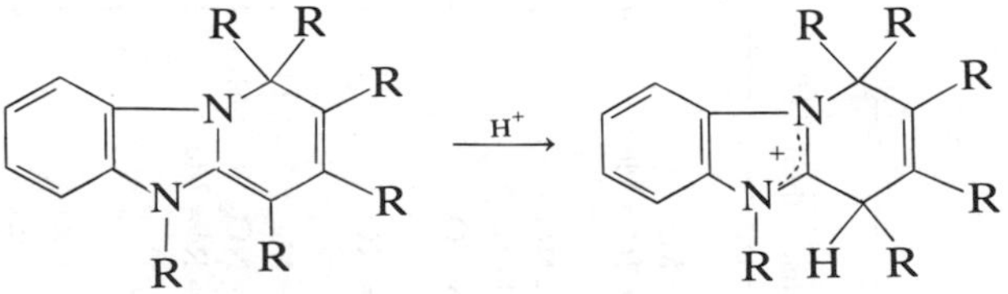

(**7.156**) $\xrightarrow{H^+}$ (**7.157**)

Scheme 7.29

TABLE 7.11. ^{1}H NMR SPECTRA[a,b] OF PYRIDO[1,2-*a*]BENZIMIDAZOLE DERIVATIVES (**7.158**)–(**7.160**)

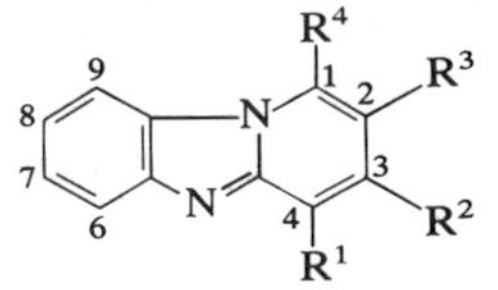

(**7.158**) 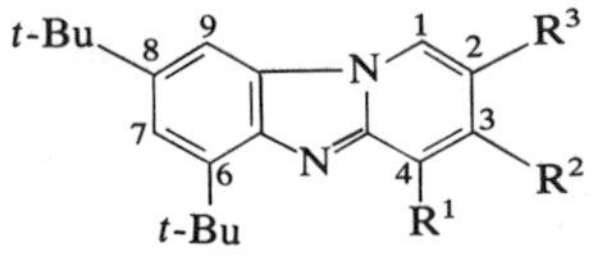 (**7.159**) 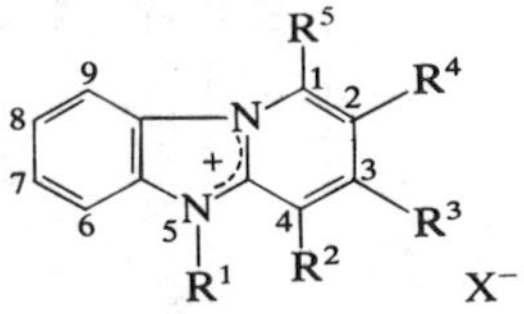 (**7.160**)

Compound ($R^1 \rightarrow R^5$ unspecified = H)	Solvent[c]	H(1)	H(2)	H(3)	H(4)	H(5)	H(6)	H(7)	H(8)	H(9)	Others	Ref.
(**7.158**)	*A*	9.04		← 7.50m →		—		← 7.50m →		8.29	—	30
(**7.158**; R^1 = Me)	*A*	8.95	6.96	7.40	2.60[d]	—	8.30	← 7.40m →		7.88	—	30
(**7.158**; R^2 = R^4 = Me)	*B*	2.88[d]	6.32br	2.36[d]	7.01–8.11m	—		← 7.01–8.11m →			—	19
(**7.158**; R^1 = NO_2)	*D*	9.62d[e]	7.90–8.23m	9.31d[f]	—	—	8.40–8.60m[g]	← 7.90–8.23m →		7.90–8.23m[g]	—	35
(**7.158**; R^2 = NO_2)	*A*	9.26d[e]	8.41q[h]	—	—[i]	—	7.85–8.03m	← 7.20–7.40m →			—	35
(**7.158**; R^3 = NO_2)	*D*	10.10d[j]	—	8.9d[j]	8.2d[k]	—	8.50m[g]	← 7.70–8.00m →		7.70–8.00m[g]	—	31, 32
(**7.158**; R^2 = Me, R^3 = NO_2)	*D*	—[i]	—	2.9[d]	—[i]	—		← —[i] →			—	31
(**7.158**; R^3 = NO_2, R^4 = Me)	*D*	3.5[d]		← —[i] →		—		← —[i] →			—	31
(**7.158**; R^2 = R^4 = Me, R^3 = CO_2H)	*D*	2.80[d]	—	2.63[d]	6.78	—		← 7.02–7.48m →			8.40[t]	35
(**7.159**)	*B*	8.30[e,m,n]	6.66[e,f,o]	7.25[f,m,p]	7.68[n,o,p]	—	1.67[q]	7.49d[j]	1.42[q]	7.67d[j]	—	40
(**7.159**; R^1 = Me)	*B*	8.19[e,m]	6.54[e,f]	6.99[f,m,r]	2.60[d]	—	1.67[q]	7.43d[j]	1.41[q]	7.61d[j]	—	41
(**7.159**; R^2 = Me)	*B*	8.35[e]	6.54[e,o]	2.34[d]	7.38[o,s]	—	1.62[q]	7.43d[j]	1.40[q]	7.68d[j]	—	41
(**7.159**; R^3 = Me)	*B*	8.14[m,t]	2.28[d]	7.08[p]	7.57[p]	—	1.66[q]	7.45d[j]	1.42[q]	7.63d[j]	—	41
(**7.159**; R^2 = Et)	*B*	8.25[e,n]	6.53[e,o]	1.27[u] –2.66[v]	7.44[n,o]	—	1.66[q]	7.44d[j]	1.42[q]	7.61d[j]	—	41
(**7.159**; R^1 = R^3 = Me)	*B*	7.97[m,t]	2.22[d]	6.85[m,r]	2.56[d]	—	1.65[q]	7.38d[j]	1.40[q]	7.56d[j]	—	41
(**7.160**; R^1 = R^3 = Me, X = ClO_4)	*C*	9.42d[w]	—[i]	2.6[d]	—[i]	4.04[x]		← —[i] →		8.54d[w]	—	11a

(**7.160**; R^1 = Me, R^3 = R^5 = Ph, X = I)	*D*	← 6.7–8.3m →				4.35[x]	← 6.7–8.3m →	—	18
(**7.160**; R^1 = R^3 = Me, R^2 = CO_2Et, X = ClO_4)	*C*	9.70d[w]	7.82d[w]	2.67[d]	—[i]	4.03[x]	← —[i] →	—	11a
(**7.160**; R^1 = Ph, X = ClO_4 R^2 = CO_2Et, R^3 = Me, X = ClO_4)	*C*	9.90d[w]	—[i]	2.63[d]	← —[i] →			—	11a
(**7.160**; R^1 = R^3 = R^5 = Me, R^2 = CN, X = ClO_4)	*D*	2.88[d]	—[i]	2.46[d]	—	← —[i] →		—	16
(**7.160**; R^1 = R^3 = R^4 = R^5 = Me, R^2 = CN, X = ClO_4)	*D*	2.98[d]	—[i]	2.57[d]	—	← —[i] →		—	16
(**7.160**; R^1 = R^5 = Me, R^2 = CN, R^3 = Ph, X = ClO_4)	*D*	2.99[d]	← —[i] →		—	← —[i] →		—	16
(**7.160**; R^1 = R^3 = Me, R^2 = CN, R^5 = Ph, X = ClO_4)	*D*	← —[c] →		2.52[d]	—	← —[i] →		—	16
(**7.160**; R^1 = R^5 = Me, R^2 = CN, R^3 = CO_2Me, X = ClO_4)	*D*	3.00[d]	← —[i] →		—	← —[i] →		—	16

[a] *A* = Me_2SO; *B* = $CDCl_3$; *C* = $(CD_3)_2SO$; *D* = CF_3CO_2H.
[b] δ values in ppm measured from TMS.
[c] Signals are sharp singlets unless denoted as: d = doublet; q = quartet; m = multiplet.
[d] *C*(Me).
[e] $J_{H(1)-H(2)} = 6.5$–7.5 Hz.
[f] $J_{H(2)-H(3)} = 6.5$–8.0 Hz.
[g] These signal assignments may be reversed.
[h] $J_{H(1)-H(2)} = 7.5$ Hz; $J_{H(2)-H(4)} \simeq 2$ Hz.
[i] δ values not specified.
[k] $J = 2$Hz.
[l] $J = 10$ Hz.
[m] $J_{H(1)-H(3)} = 1.3$–1.7 Hz.
[n] $J_{H(1)-H(4)} = 1.0$ Hz.
[o] $J_{H(2)-H(4)} = 1.5$ Hz.
[p] $J_{H(3)-H(4)} = 9.5$ Hz.
[q] Bu^t.
[r] $J_{Me(4)-H(3)} = 1.0$–1.2 Hz.
[s] $J_{Me(3)-H(4)} = 1.0$ Hz.
[t] $J_{Me(2)-H(1)} = 1.0$ Hz.
[u] Me of Et group.
[v] CH_2 of Et group.
[w] $J = 7$ Hz.
[x] *N*(Me).

TABLE 7.12. ^{1}H NMR SPECTRA[a,b] OF DIHYDROPYRIDO[1,2-*a*]BENZIMIDAZOLE

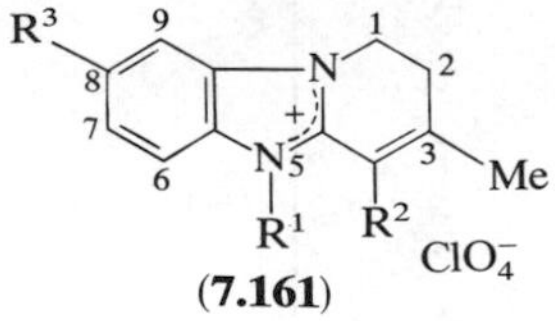

(**7.161**) ClO_4^-

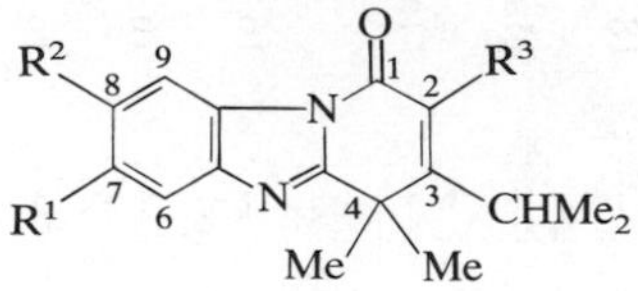

(**7.162**)

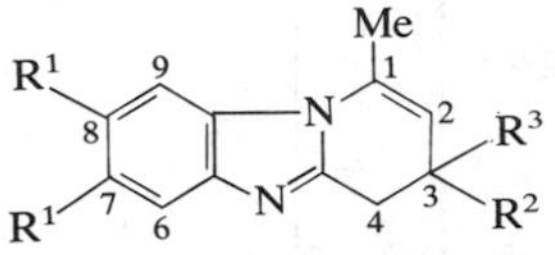

(**7.165**)

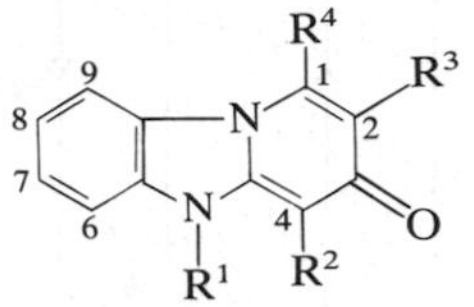

(**7.166**)

Compound ($R^1 \rightarrow R^5$ unspecified = H)	Solvent[c]	H(1)	H(2)	H(3)	H(4)
(**7.161**; R^1 = Me)	*A*	4.40t[d]	2.85t[d]	2.20[e]	6.7 m
(**7.161**; R^1 = Me, R^2 = CO_2Et)	*B*	4.60t[d]	3.00t[d]	2.44[e]	—
(**7.161**; R^1 = Et, R^2 = CO_2Et, R^3 = Cl)	*B*	4.26t[d]	3.02t[d]	2.36[e]	—
(**7.161**; R^1 = Ph, R^2 = CO_2Et)	*B*	4.70t[d]	3.15t[d]	2.37[e]	—
(**7.162**)	*C*	—	←— [g] —→		
(**7.162**; R^2 = Me)	*C*	—	←— [g] —→		
(**7.162**; R^1 = OMe)	*C*	—	←— [g] —→		
(**7.162**; R^1 = Cl)	*C*	—	←— [g] —→		
(**7.162**; R^1 = CO_2Me)	*C*	—	←— [g] —→		
(**7.162**; R^1 = NO_2)	*C*	—	←— [g] —→		
(**7.162**; R^2 = NO_2)	*C*	—	←— [g] —→		
(**7.162**; R^3 = Me)	*C*	—	2.22[e]	2.95m[h,j] 1.35d[h,j]	1.70[e]
(**7.162**; R^3 = Et)	*C*	—	1.22t[h,l] 2.75q[h,m]	2.75m[h,j] 1.35d[h,k]	1.70[e]
(**7.163**; R^1 = Me, R^2 = R^3 = R^4 = CO_2Me)	*C*	6.26	—	—	—
(**7.163**; R^1 = $MeO_2C\overset{\vert}{C}$ = $CHCO_2Me$, R^2 = R^3 = R^4 = CO_2Me)	*C*	6.31	—	—	—
(**7.163**; R^1 = $MeO_2C\overset{\vert}{C}HCH_2CO_2Me$, R^2 = R^3 = R^4 = CO_2Me)	*C*	6.24	—	—	—

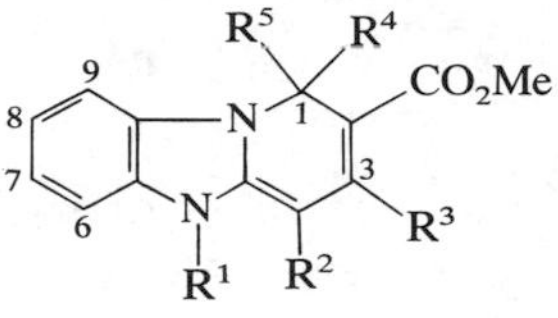

(**7.163**)

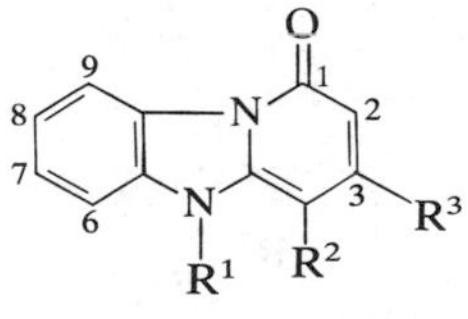

(**7.164**)

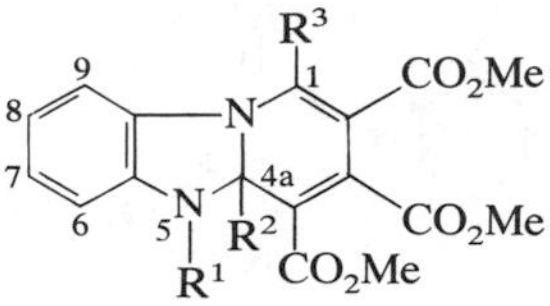

(**7.167**)

H(4a)	H(5)	H(6)	H(7)	H(8)	H(9)	OMe	Ref.
—	3.86[f]	←——— —[g] ———→				—	11a
—	3.90[f]	←——— —[g] ———→				—	11a
—	—[g]	8.10d[h]	7.70dd[h,i]	—	8.20[i]	—	11a
—	—	←——— —[g] ———→				—	11a
—	—	7.75m	7.30m	7.30m	8.30m	—	20
—	—	7.55d[h]	7.20dd[h,i]	—	8.20d[i]	—	20
—	—	7.30d[i]	—	7.00dd[h,i]	8.25d[h]	—	20
—	—	7.80d[i]	—	7.35dd[h,i]	8.30d[h]	—	20
—	—	8.55d[i]	—	8.15dd[h,i]	8.40d[h]	—	20
—	—	8.30d[i]	—	8.30dd[h,i]	8.55d[h]	—	20
—	—	7.80d[h]	8.33dd[h,i]	—	9.20d[i]	—	20
—	—	←——— 7.35–8.40m ———→				—	20
—	—	←——— 7.38–8.40m ———→				—	20
—	3.71[m]	←——— 7.37m ———→				3.02 3.71 3.77 3.97	2
—	6.63[n]	←——— 7.39m ———→				3.67 3.71 3.78 3.80 3.90 3.92	2
—	5.76dd[o,p]	←——— 7.26m ———→				3.50 3.70 3.70 3.77 3.80 3.94	2

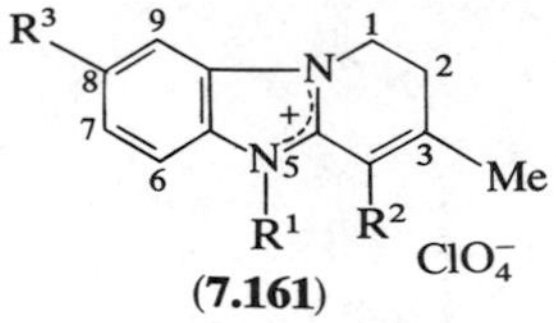

(**7.161**) 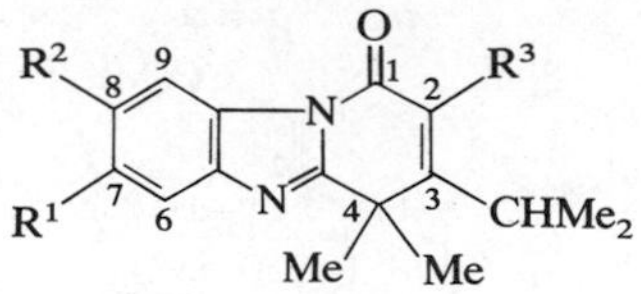 (**7.162**)

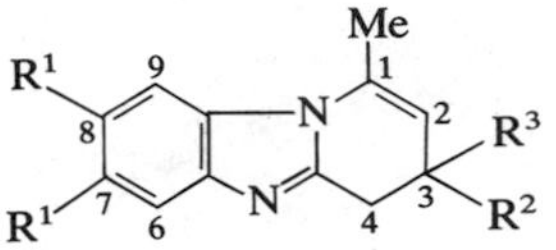

(**7.165**) 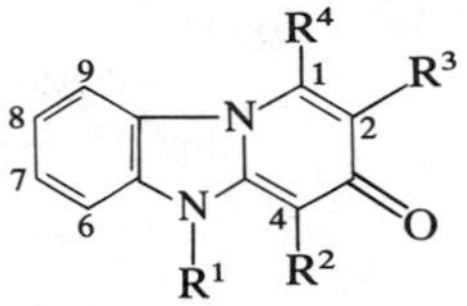 (**7.166**)

Compound ($R^1 \rightarrow R^5$ unspecified = H)	Solvent[c]	H(1)	H(2)	H(3)	H(4)
(**7.163**; $R^1 = Me$, $R^2 = CN$, $R^4 = CH_2CO_2Me$)	*C*	6.05t[q] 2.73d[q,r]	—	7.53	—
(**7.163**; $R^1 = CH\overset{t}{=}CHCO_2Me$, $R^2 = CN$, $R^4 = CH_2CO_2Me$)	*C*	6.08t[q] 2.76d[q,r]	—	7.52	—
(**7.163**; $R^1 = Me$, $R^2 = CO_2Et$, $R^4 = CH_2CO_2Me$)	*C*	6.09t[q] 2.50–2.95m[q,r,u]	—	8.17	1.20–1.50m[l] 4.27q[m,h]
(**7.163**; $R^2 = CO_2Et$, $R^3 = R^4 = CO_2Me$, $R^5 = CH_2CO_2Me$)	*B*	3.18d[r,u] 3.55d[r,u]	—	—	1.21t[h,l] 4.19q[h,m]
(**7.163**; $R^1 = Me$, $R^2 = CO_2Et$, $R^3 = R^4 = CO_2Me$, $R^5 = CH_2CO_2Me$)	*C*	3.15	—	—	1.30t[h,l] 4.13q[h,m]
(**7.163**; $R^1 = Me$, $R^2 = Ph$, $R^3 = R^4 = CO_2Me$, $R^5 = CH_2CO_2Me$)	*C*	3.38d[r,s]	—	—	7.20–7.35m
(**7.163**; $R^1 = Me$, $R^2 = CN$, $R^3 = R^4 = CO_2Me$, $R^5 = CH_2CO_2Me$)	*C*	3.21	—	—	—
(**7.164**; $R^1 = CH\overset{t}{=}CHCO_2Me$, $R^2 = CN$)	*C*	—	6.30d[h]	7.95d[h]	—
(**7.164**; $R^2 = CO_2Et$, $R^3 = Me$)	*C*	—	6.07	2.60[e]	1.44t[g,k] 4.42q[g,l]
(**7.164**; $R^2 = CN$, $R^3 = Me$)	*D*	—	6.95	2.80[e]	—
(**7.164**; $R^2 = CONH_2$, $R^3 = Me$)	*D*[w]	—	6.92	2.90[e]	—

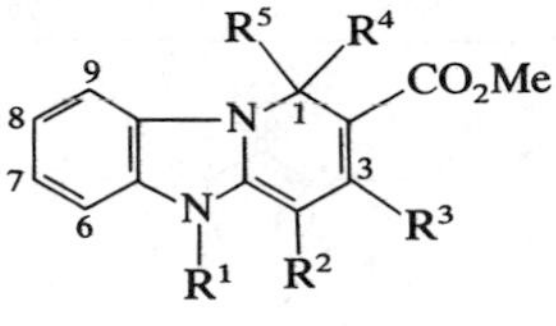

(7.163)

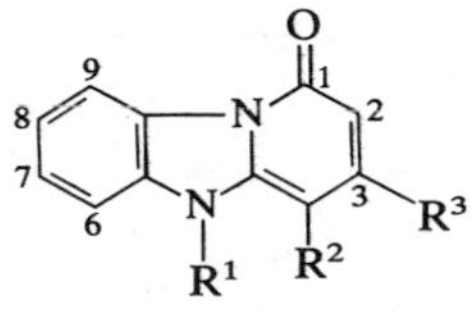

(7.164)

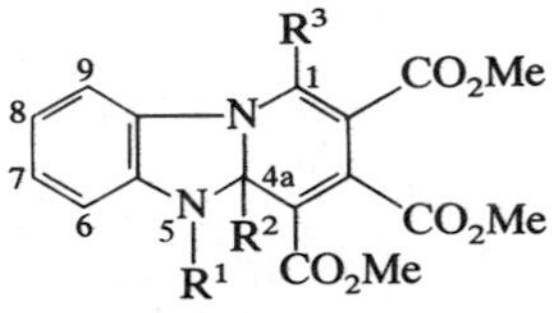

(7.167)

H(4a)	H(5)	H(6)	H(7)	H(8)	H(9)	OMe	Ref.
—	3.96[f]	←	7.15-7.60m	→		3.53 3.78	6
—	6.45d[n,s] 8.67d[t,s]	←	7.30-7.80m	→		3.55 3.82 3.90	6
—	3.83[f]	←	7.33–7.70m	→		3.51 3.82	6
—	—	←	7.30–7.80m	→		3.42 3.62 3.76 3.76	6
—	4.04[f]	←	6.80–7.30m	→		3.38 3.40 3.74 3.78	6
—	3.77[f]	←	7.20–7.35m	→	7.07m	2.84 3.43 3.66 3.66	5
—	4.80[f]	←	6.80–7.40m	→		3.60 3.75 3.80 3.86	6
—	8.60d[t,s] 6.76d[n,s]	←	7.40–8.10m	→	8.81m	3.85	6
—	12.06–12.40br[v]	←	7.40–7.55m	→	8.72–8.97m	—	21
—	—	←	7.50–7.92m	→	8.44–8.73m	—	21
—	—	←	7.50–7.95m	→	8.50–8.75m	—	21

(7.161) ClO_4^-

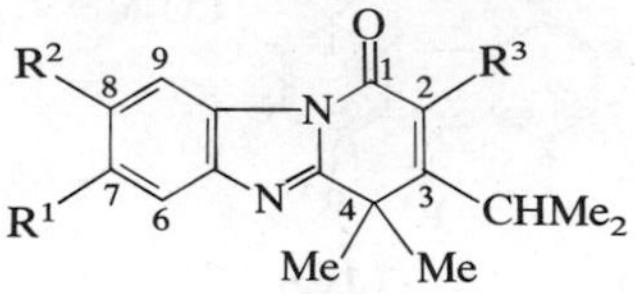

(7.162)

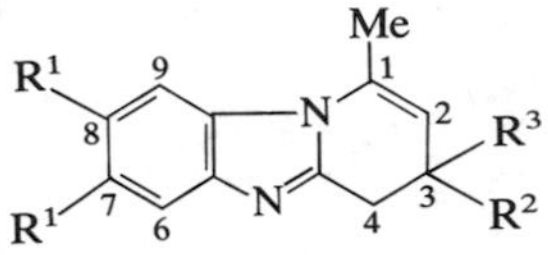

(7.165)

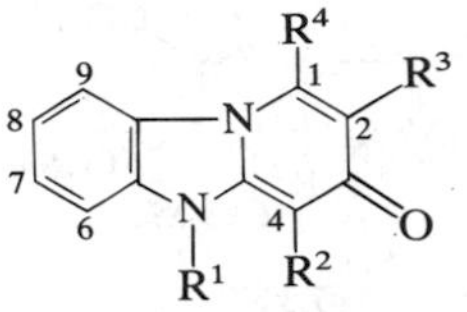

(7.166)

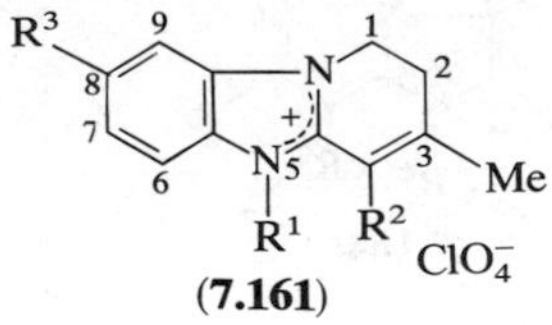

Compound ($R^1 \rightarrow R^5$ unspecified = H)	Solvent[c]	H(1)	H(2)	H(3)	H(4)
(**7.164**; $R^2 = CN$, $R^3 = CO_2Me$)	*B*	—	6.40	3.90[x]	—
(**7.164**; $R^1 = COMe$, $R^2 = CO_2Et$, $R^3 = CO_2Me$)	*B*	—	6.37	3.86[x]	4.22q
(**7.165**; $R^2 = Me$)	*C*	2.48br[e]	5.12m	← —[y] →	
(**7.165**; $R^1 = R^2 = Me$)	*C*	2.45br[e]	5.09m	← —[y] →	
(**7.165**; $R^2 = Pr^i$)	*C*	2.50br[e]	5.22m	← —[y] →	
(**7.165**; $R^1 = Me$)	*C*	2.48br[e]	5.21m	← —[y] →	
(**7.165**; $R^1 = Me$, $R^2 = Ph$)	*C*	2.54br[e]	5.35m	← —[y] →	
(**7.165**; $R^2 = R^3 = Me$)	*C*	2.48d[e,z]	5.08m	← —[y] →	
(**7.165**; $R^1 = R^2 = R^3 = Me$)	*C*	2.43br[e]	5.00m	← —[y] →	
(**7.166**; $R^1 = Me$, $R^3 = CO_2Me$)	*C*	8.80	3.98[x]	—	6.19
(**7.166**; $R^1 = CH\overset{t}{=}CHCO_2Me$, $R^3 = CO_2Me$)	*C*	8.89	—	—	6.63
(**7.166**; $R^2 = CN$, $R^4 = CO_2Me$)	*C*	3.95[x]	6.53	—	—
(**7.166**; $R^1 = CH\overset{t}{=}CHCO_2Me$, $R^2 = CN$)	*B*	8.83d[d]	6.54d[d]	—	—
(**7.167**; $R^1 = R^2 = Me$, $R^3 = CO_2Me$)	*C*	—	—	—	—
(**7.167**; $R^1 = R^3 = Me$, $R^2 = CO_2Me$)	*C*	2.96[e]	—	—	—
(**7.167**; $R^1 = Me$, $R^2 = Et$, $R^3 = CO_2Me$)	*C*	—	—	—	—
(**7.167**; $R^1 = Me$, $R^2 = Pr^i$, $R^3 = CO_2Me$)	*C*	—	—	—	—
(**7.167**; $R^1 = CH\overset{t}{=}CHCO_2Me$, $R^2 = Bu^t$, $R^3 = CO_2Me$)	*C*	—	—	—	—

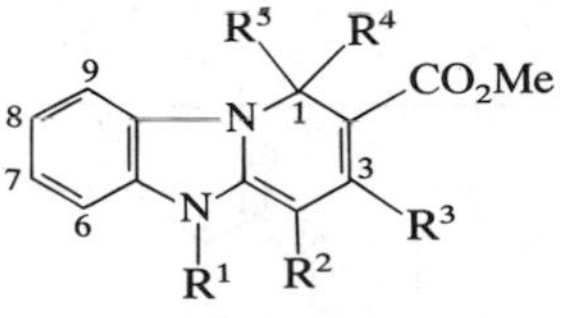

(7.163) 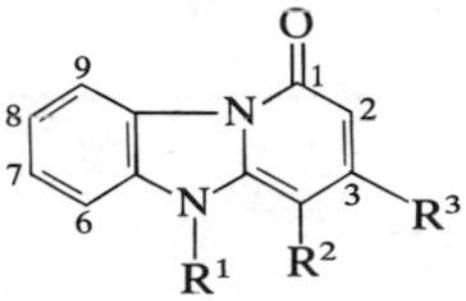 (7.164) 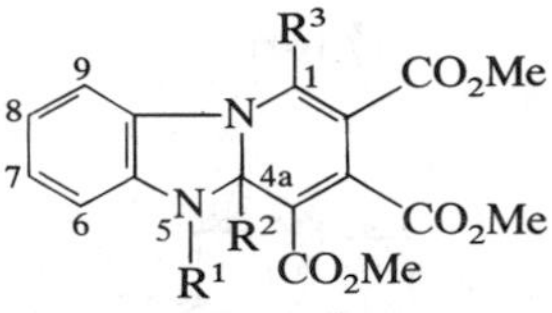 (7.167)

H(4a)	H(5)	H(6)	H(7)	H(8)	H(9)	OMe	Ref.
—	← —[g] →				8.50m	—	7
—	2.56 ←	—[g]			→	—	7
—	—	←	—[y]		→	—	19
—	—	←	—[y]		→	—	19
—	—	←	—[y]		→	—	19
—	—	←	—[y]		→	—	19
—	—	←	—[y]		→	—	19
—	—	←	—[y]		→	—	19
—	—	←	—[y]		→	—	19
—	3.58[f]	←	7.20–7.80m		→	—	6
—	6.35[n,s]	←	7.35–7.90m		→	3.90	6
	8.06[t,s]					4.00	
—	—	←	7.30–7.80m		→	8.63m	6
—	6.68d[n,s]	←	7.85–8.30m		→	3.82	6
	8.50d[t,s]		7.40–7.70m				
1.44[e]	3.09[f]	←	6.55–7.10m		→	3.74	2
						3.77	
						3.81	
						4.04	
—	3.08[f]	6.65d	6.77t[d]	7.03t[d]	7.20d	3.72	3
						–3.80	
0.76t[d,l]	3.17	←	6.50–7.20m		→	3.78	2
2.04q[d,m]						3.78	
						3.83	
						4.08	
0.90t[k]	3.14	←	6.54[d],	6.60–6.80,	7.10m →	3.77	4
						3.80	
						3.83	
						4.09	
1.05[aa]	5.81[n,s]	←	6.75–7.45m		→	3.79	1b, 4
	7.95d[t,s]					3.79	
						3.82	
						3.85	
						4.10	

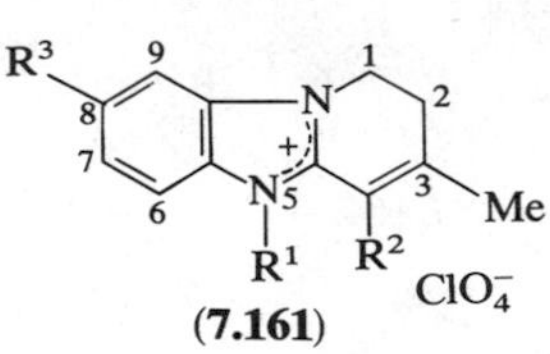

(7.161) 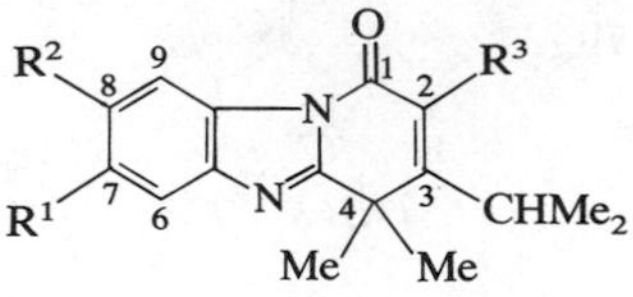 **(7.162)**

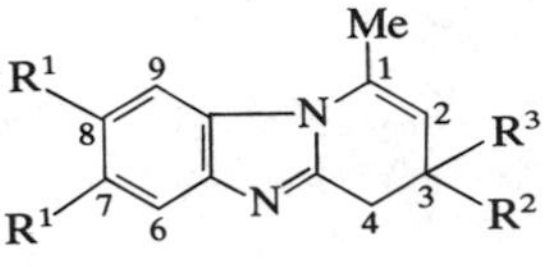

(7.165) 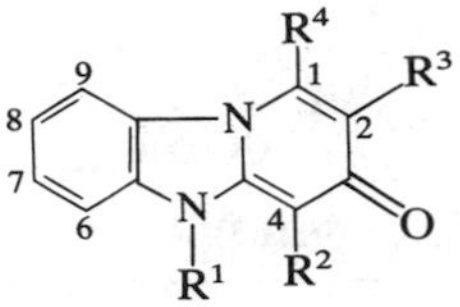 **(7.166)**

Compound ($R^1 \rightarrow R^5$ unspecified = H)	Solvent[c]	H(1)	H(2)	H(3)	H(4)
(**7.167**; $R^1 = Me$, $R^2 = CH_2Ph$, $R^3 = CO_2Me$)	*C*	—	—	—	—
(**7.167**; $R^1 = Me$, $R^2 = CO_2Me$)	*C*	8.26	—	—	—
(**7.167**; $R^1 = CH\overset{t}{=}CHCO_2Me$, $R^2 = CO_2Me$, $R^3 = Ph$)	*C*	7.60	—	—	—

[a] δ values in ppm measured from TMS.
[b] Signals are sharp singlets unless denoted as br = broad; d = doublet; dd = double doublet; t = triplet; q = quartet; m = multiplet.
[c] *A* = CD_3CN; *B* = $(CD_3)_2SO$: *C* = $CDCl_3$; *D* = CF_3CO_2H.
[d] $J = 7.5–8.1$ Hz.
[e] C(Me).
[f] N(Me).
[g] δ values not specified.
[h] $J = 7.10$ Hz.
[i] $J = 1–3$ Hz.
[j] C*H*Me$_2$.
[k] CH*Me*$_2$.
[l] Me of Et group.
[m] CH_2 of Et group.
[n] C=C*H*CO$_2$Me.
[o] C*H*CO$_2$Me.
[p] $J = 4$ and 10 Hz.
[q] $J_{CH_2\text{-}H(1)} = 5.1–5.2$ Hz.
[r] C*H*$_2$CO$_2$Me.

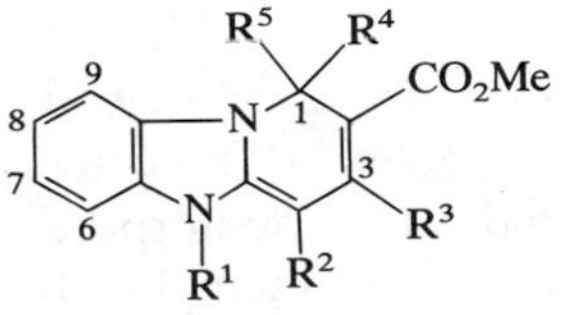

(7.163)

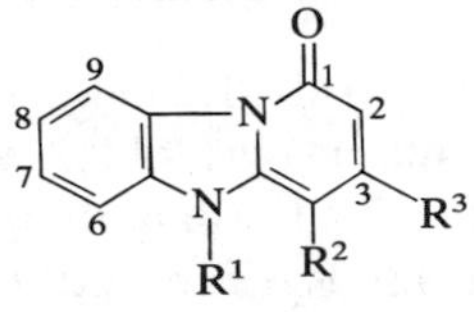

(7.164)

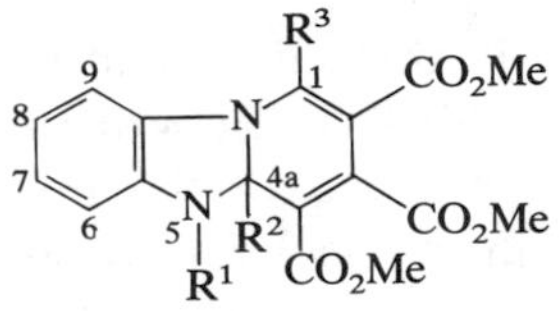

(7.167)

H(4a)	H(5)	H(6)	H(7)	H(8)	H(9)	OMe	Ref.
2.90d	3.10	←		6.80–7.40m	→	3.53	5
3.32d				6.30–6.70m		3.72	
						3.78	
						3.98	
—	3.11[f]	←		6.67–7.19	→	3.72	2
						3.80	
						3.80	
						3.91	
—	5.48d[n,s]	5.54dd[d,i]	6.60m	7.00m	7.16dd[d,i]	3.40	1b, 4
	7.97[t,s]					3.75	
						3.75	
						3.88	
						3.93	

[s] $J = 12.0–14.0$ Hz.
[t] $CH{=}C$.
[u] $J_{gem} = 13.9–14.2$ Hz.
[v] NH.
[w] δ Values in ppm measured from sodium 3-trimethylsilyl-1-propane sulfonate.
[x] OMe.
[y] Other signals in the ranges δ 0.93–3.86 and δ 7.10–7.88 not assigned.
[z] $J_{Me(1)\text{-}H(2)}$
[aa] Bu^t.

of the benzenoid protons in pyrido[1,2-*a*]benzimidazoles (Table 7.11). *C*(1) and *C*(3) methyl substituents in pyrido[1,2-*a*]benzimidazoles, as expected, are more deshielded (Table 7.11) than their *C*(2) and *C*(4) counterparts. Allylic coupling (H = 1.0 Hz) is observed (Table 7.11) for methyl substituents at the *C*(2), *C*(3), and *C*(4) positions in pyrido[1,2-*a*]benzimidazoles.

The enhanced deshielding (Table 7.12) of the *N*(5) methyl substituent in the 1,5-dihydropyrido[1,2-*a*]benzimidazole [Scheme 7.30; (**7.177**)] is attributed[2] to resonance interaction involving the nitrogen atoms of the imidazole ring [Scheme 7.30; (**7.177**) ↔ (**7.178**)]. *H*(2) in pyrido[1,2-*a*]benz-

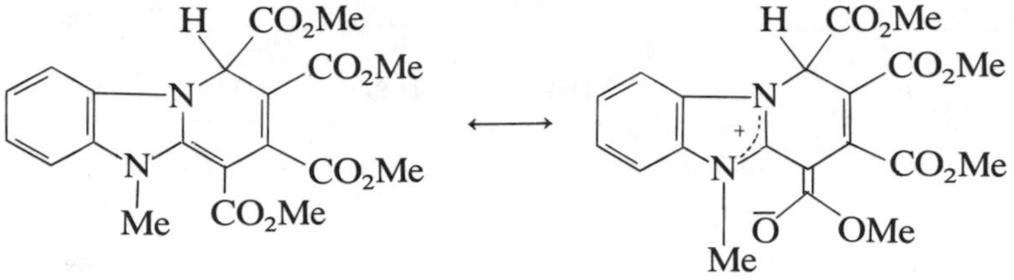

(**7.177**) (**7.178**)

Scheme 7.30

imidazol-1(5*H*)-ones (**7.164**) and pyrido[1,2-*a*]benzimidazol-3(5*H*)-ones (**7.166**) absorbs at higher field (Table 7.12) than *H*(3) or *H*(1), respectively, the particularly low field resonance of the latter being explained[6] in terms of deshielding associated with resonance interaction of the type [Scheme 7.28; (**7.143**) ↔ (**7.144**)]. Pyrido[1,2-*a*]benzimidazol-1(4*H*)-ones (**7.162**) and pyrido[1,2-*a*]benzimidazol-1(5*H*)-ones (**7.164**) are readily differentiated from their *C*(3)-keto counterparts on the basis of the enhanced deshielding of *H*(9) (Table 7.12) as a result of the anisotropic effect of the *C*(1) carbonyl group.

H(1) in 1,2,3,4-tetrahydropyrido[1,2-*a*]benzimidazoles (**7.168**) resonates (Table 7.13) at lower field (δ 3.8–4.4) than *H*(4) (δ 2.9–3.2), and both *H*(1) and *H*(4) are more deshielded than *H*(2) or *H*(3) (δ 1.8–2.6). The anisotropic effect of the *N*-oxide substituent in 1,2,3,4-tetrahydropyrido[1,2-*a*]-benzimidazole 5-*N*-oxides results in a deshielding of *H*(4) (Table 7.13), while both *H*(1) and *H*(4) in 1,2,3,4,-tetrahydropyrido[1,2-*a*]benzimidazolium salts (**7.170**) resonate at somewhat lower field (Table 7.13) compared with *H*(1) and *H*(4) in the parent 1,2,3,4-tetrahydropyrido[1,2-*a*]-benzimidazoles. Ring-chain tautomerism [Scheme 7.31; (**7.179**) ⇌ (**7.180**)] accounts for the presence in the ^{1}H NMR spectrum of the 1-hydroxy-1,2,3,4-tetrahydropyrido[1,2-*a*]benzimidazole derivative (**7.179**), of signals attributable to the open-chain keto isomer (**7.180**).[19] ^{13}C NMR (Table 7.14) allows the unambiguous assignment of structure to 1,5-dihydro- and 4a,5-dihydropyrido[1,2-*a*]benzimidazoles derived by the cycloaddition of dimethyl acetylenedicarboxylate to benzimidazole derivatives.[3]

TABLE 7.13. ^{1}H NMR SPECTRA[a,b] OF TETRAHYDRO- AND HEXAHYDROPYRIDO[1,2-*a*]BENZIMIDAZOLE DERIVATIVES (**7.168**)–(**7.174**)

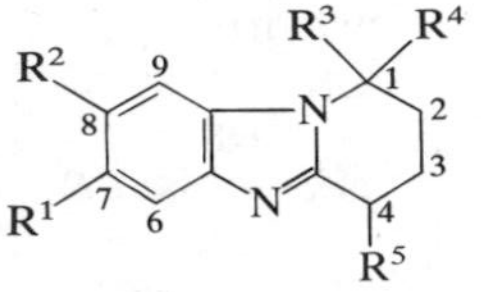

(**7.168**) 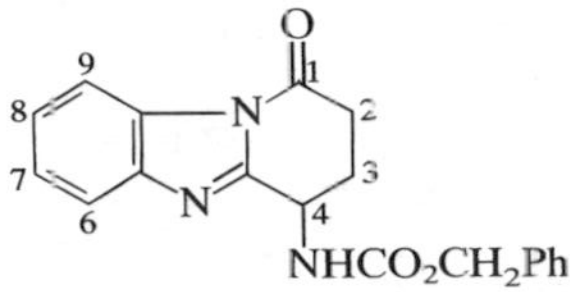 (**7.169**) 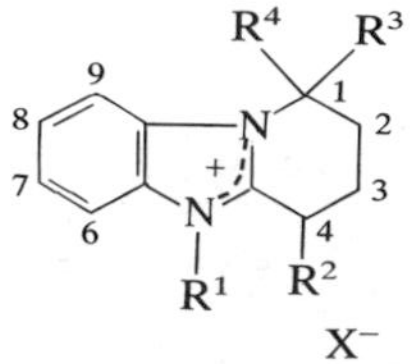 (**7.170**) 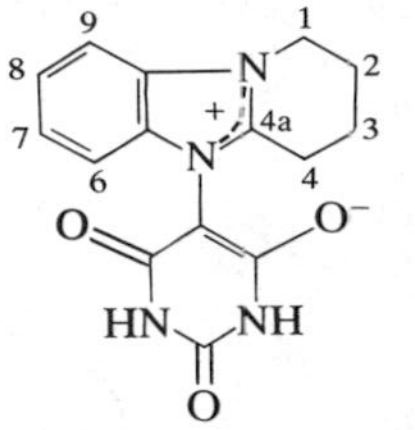 (**7.171**) 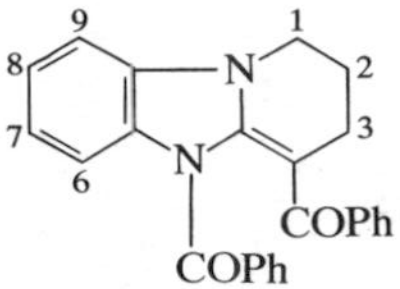 (**7.172**) 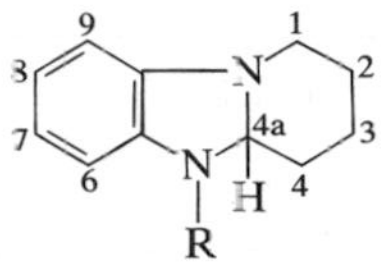 (**7.173**) 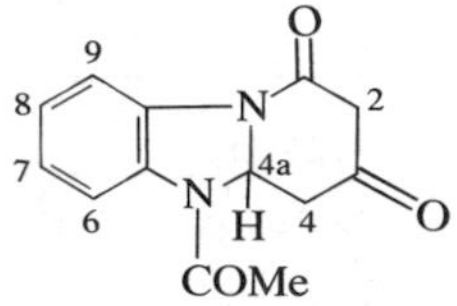 (**7.174**)

Compound ($R^1 \rightarrow R^5$ unspecified = H)	Solvent[a]	H(1)	H(2)	H(3)	H(4)	H(4a)	H(6)	H(7)	H(8)	H(9)	Others	Ref.
(**7.168**; *N*(5)-oxide)	*A*	4.32brt[e]	← 2.00–2.30m →		3.27brt[e]	—	← 7.63–7.76m →				—	55
(**7.168**; $R^1 = Cl$)	*B*	3.82–4.12t[f]	← 1.85–2.20m →		2.90–3.17[f]	—	← 7.10–7.75m →				—	49
(**7.168**; $R^1 = Cl$, *N*(5)-oxide)	*B*	4.15brt[e]	← 1.80–2.26m →		3.20brt[e]	—	7.90d[g]	← 7.27–7.39m →			—	55
(**7.168**; $R^2 = Cl$)	*B*	3.82–4.12t[f]	← 1.85–2.20m →		2.90–3.17t[f]	—	← 7.10–7.75m →				—	49
(**7.168**; $R^1 = NO_2$; *N*(5)-oxide)	*A*	4.44brt[e]	← 1.96–2.37m →		3.35brt[e]	—	8.71d[g]	—	8.48dd[g,h]	7.98d[h]	—	55
(**7.168**; $R^1 = NO_2$; $R^2 = Cl$)	*B*	3.95–4.26t[f]	← 1.85–2.30m →		2.97–3.27t[f]	—	8.22	—	—	7.40	—	49

TABLE 7.13 (*Continued*)

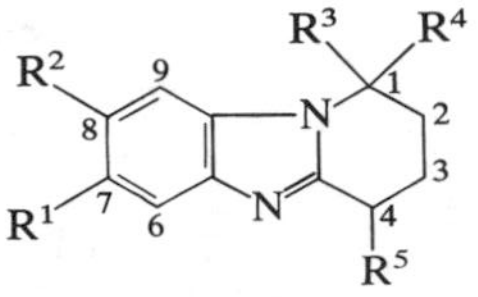

(**7.168**) 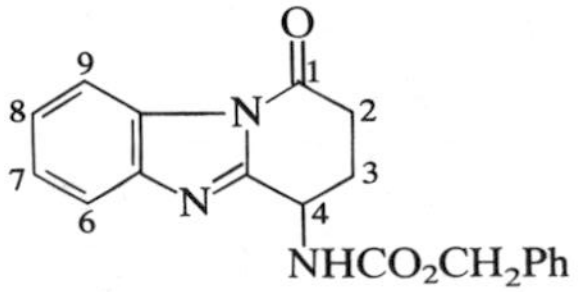 (**7.169**) 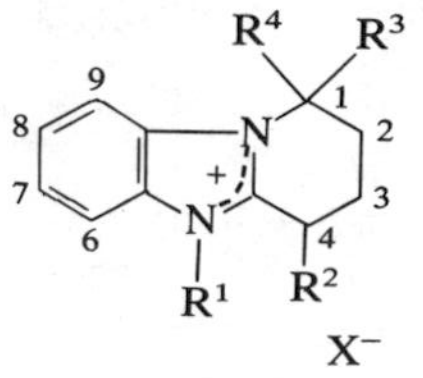 (**7.170**) 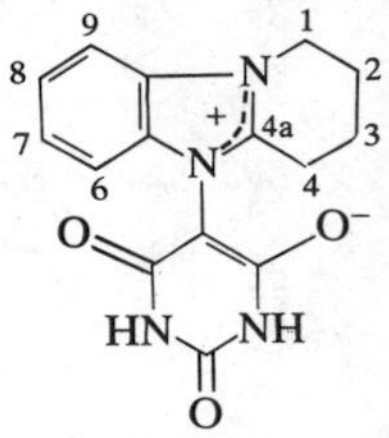 (**7.171**)

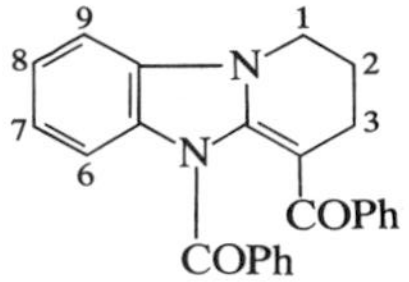

(**7.172**) 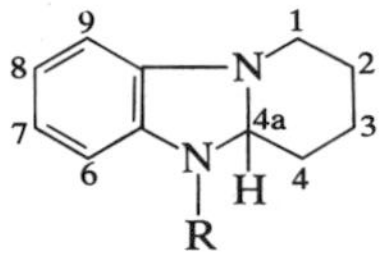 (**7.173**) 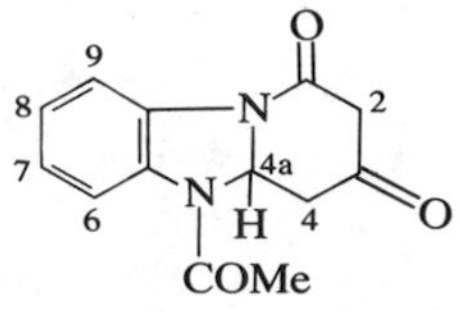 (**7.174**)

Compound ($R^1 \rightarrow R^5$ unspecified = H)	Solvent[a]	H(1)	H(2)	H(3)	H(4)	H(4a)	H(6)	H(7)	H(8)	H(9)	Others	Ref.
(**7.168**; $R^1 = Cl$, $R^2 = NO_2$)	*B*	4.06–4.32t[e]	← 2.00–2.35m →		3.05–3.35t[f]	—	8.02	—	—	7.80	—	49
(**7.168**; $R^1 = NO_2$, $R^2 = NH_2$)	*B*	3.92–4.25m	← 1.85–2.10m →		2.98–3.18m	—	8.55	—	—	7.75	—	49
(**7.168**; $R^1 = NO_2$, $R^2 = NHCOMe$)	*B*	4.35–4.60m	← 2.15–2.42m →		3.30–3.55m	—	9.08	—	—	8.80	2.50[i]	49
(**7.168**; $R^1 = NO_2$, $R^2 = N_3$)	*B*	4.40–4.70m	← 2.15–2.60m →		3.35–3.60m	—	8.59	—	—	7.70	—	49
(**7.168**; $R^1 = Me$, $R^2 = NHSO_2Ph$)	*B*	4.04t[f]	← 2.05br →		3.08t[f]	—	← 7.20–7.70m[j] →				1.88[k] 6.60br[l]	66
(**7.168**; R^1 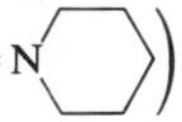 = N-piperidino)	*B*	4.04t[f]	← 2.31m →		3.01t[f]	—	← 7.20–7.80m[i] →				8.32br[l]	66

(**7.168**; $R^5 = OH$)	*B*	4.01brt[f]	2.21br	2.21br	5.15brt[f]	—	7.80m	←7.10–7.40m→			7.30br[m]	56
(**7.168**; $R^1 = NO_2$, $R^2 = OCOMe$)	*B*	4.28br	2.35br	2.35br	6.26br	—	6.56d[g]	—	8.16dd[g,h]	7.43d[g]	2.19[i]	56
(**7.168**; $R^5 = COPh$)	*B*	3.95t[f]	1.94q[e]	2.23t[m] 2.68t[o,e]	5.10t[f]	—	8.03[f]	←7.00–7.75m[j]→			—	52
(**7.169**)	*B*	—	3.00m	2.13m 2.80m	5.10m	—	←7.30–8.60m→				5.32[q] 6.10[l]	23
(**7.170**; $R^1 = CH_2Ph$, $X = Cl$)	*A*	4.50m	←1.92m→		3.40m	—	←7.20–7.90m→				5.69[q] 7.32[r]	62
(**7.170**; $R^1 = 2\text{-}ClC_6H_4CH_2$, $X = Cl$)	*A*	4.58m	←1.94m→		3.38m	—	←6.80–7.90m[i]→				5.75[q]	62
(**7.170**; $R^1 = 4\text{-}ClC_6H_4CH_2$, $X = Cl$)	*A*	4.52m	←1.80m→		3.38m	—	←6.70–7.90m[i]→				5.63[q]	62
(**7.170**; $R^1 = 2\text{-}Cl\text{-}5\text{-}O_2NC_6H_3CH_2$, $X = Cl$)	*A*	4.54m	←1.90m→		3.30m	—	←6.70–7.90m[i]→				5.73[q]	62
(**7.170**; $R^1 = R^3 = Me$, $R^2 = CO_2Et$, $R^4 = OH$, $X = ClO_4$)	*C*	1.80[k]	←2.20–2.80m→		4.50m	—	←7.50–8.30m→				1.23t[t,u] 3.95[v] 4.20q[w,v] 5.40br[m]	11a
(**7.170**; $R^1 = R^3 = Me$, $R^2 = CO_2Et$, $R^4 = OEt$, $X = ClO_4$)	*D*	1.87	←2.00–2.70m→		4.85br	—	←7.50–8.20m→				1.07[t,u] 1.19[t,u] 2.89q[w,x] 3.68q[w,x] 4.03[v] 4.25q[u,w]	11a
(**7.171**)	*E*	4.34br	←2.27br→		3.13br	—	←7.60m→				—	60
(**7.171**)	*E*	4.42m	←2.33m→		3.23m	—	←7.50–7.83m→				—	63
(**7.170**; $R^1 = NHSO_3^-$)	*F*	3.78brt[f]	←1.70quin→		3.00brt[f]	—	←7.20–8.30m→				7.80[l]	65
(**7.172**)	*B*	4.10[f]	2.15q[f]	2.88[f]	—	—	←6.97–7.73m, 8.20m[j]→				—	52
(**7.173**; $R = H$)	*B*	—	—	—	2.70 3.60brd[z]	4.50br[y]	6.33m	←6.60m→			3.70br[l]	58
(**7.173**; $R = 2\text{-}Cl\text{-}5\text{-}O_2NC_6H_3CH_2$)	*B*	2.35m 3.32m	←1.35br→			3.94t	←6.10–7.70m→				4.15q[q]	62
(**7.174**)	*E*	—	2.56q[aa]	—	3.02–3.95m	5.88–6.62m	←7.10–7.50m→		7.90–8.40m		2.61[i]	21

(*Footnotes overleaf*)

(*Footnotes to Table 7.13*)

[a] $A = D_2O$: $B = CDCl_3$; $C = CD_3CN$; $D = (CD_3)_2SO$: $E = CF_3CO_2H$; $F = C_5D_5N$ (deuteriopyridine).

[b] δ values in ppm measured from TMS.

[c] Signals are sharp singlets unless denoted as br = broad; d = doublet; dd = double doublet; t = triplet; q = quartet; m = multiplet.

[d] Hydrochloride hydrate.

[e] $J = 5.5$–6.0 Hz.

[f] J value not quoted.

[g] $J = 1.5$–2.3 Hz.

[h] $J = 9.1$–9.5 Hz.

[i] COMe.

[j] Includes Ph.

[k] *C*(Me).

[l] NH.

[m] OH.

[n] *H*(3) of keto form.

[o] *H*(3) of enol form.

[p] *H*(6) of keto form.

[q] CH_2.

[r] Ph*H*.

[s] Hemihydrate.

[t] Me of Et group.

[u] $J = 7.0$ Hz.

[v] NMe.

[w] CH_2 of Et group.

[x] $J = 7.0$ and 9.0 Hz.

[y] Resolved at 100 MHz into a doublet of doublets, $J = 9.0$ and 3.0 Hz.

[z] $J = 12.0$ Hz.

[aa] 19.0 Hz.

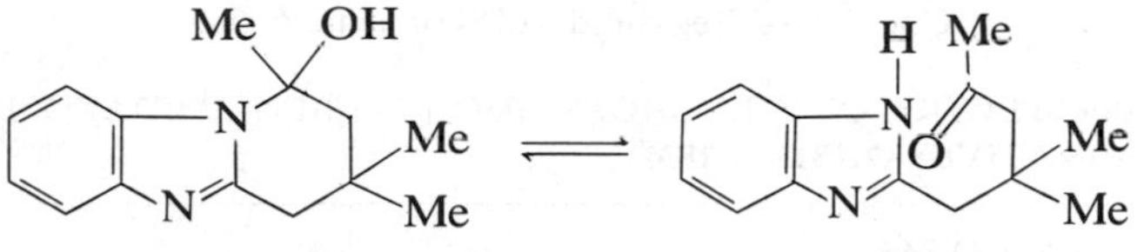

(**7.179**) ⇌ (**7.180**)

Scheme 7.31

TABLE 7.14. ^{13}C NMR SPECTRA[a] OF DIHYDROPYRIDO[1,2-*a*]BENZIMIDAZOLE DERIVATIVES (**7.175**) AND (**1.176**)[b]

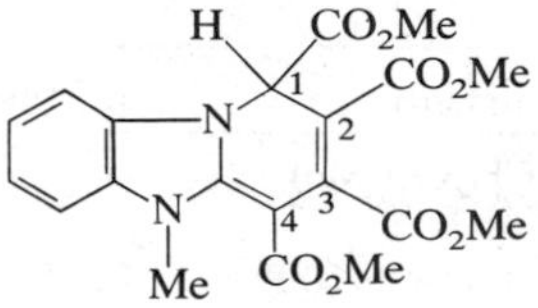

(**7.175**) 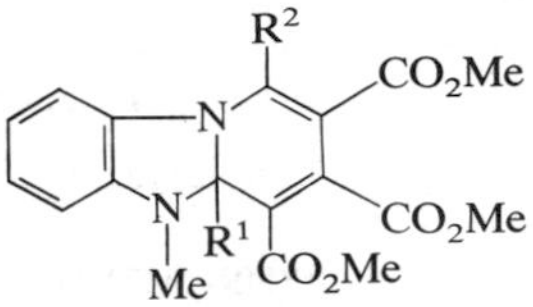 (**7.176**)

Compound	Solvent[c]	C(Ar)	$C(CO_2Me)$[d]	C=O	OMe	NMe	Others
(**7.175**)	*A*	109.4	97.1	162.4	50.3	34.9	53.5[e]
		110.0	123.6	163.2	51.3		
		123.4	129.0	167.2	51.8		
		123.6	133.9	167.9	52.6		
			149.9				
			149.0				
(**7.176**; $R^1 = H$,	*B*	111.0	103.9	165.0	52.6	33.2	17.2[f]
$R^2 = CO_2Me$)		112.1	119.0	165.2	53.1		85.7[g]
		120.7	130.4	165.9	53.3		
		127.3	139.8	168.3	54.4		
(**7.176**; $R^1 = CO_2Me$,	*B*	108.2	104.0	165.0	50.8	34.9	86.5[g]
$R^2 = H$)		110.7	110.6	165.5	52.8		136.4[e]
		121.3	136.4	168.5	53.2		
		127.3			53.5		
		131.7					
		144.1					
(**7.176**; $R^1 = Me$,	*C*	109.0	104.8	165.8	51.5	33.8	17.7[h]
$R^2 = CO_2Me$)		113.6	150.5	167.6	52.2		84.9[g]
		119.8		168.1	52.6		143.9[e]
		125.0			52.9		
		131.4					
		143.9					

[a] δ values in ppm measured from TMS.
[b] From Ref. 3.
[c] $A = (CD_3)_2SO\text{-}CDCl_3$ (1:1); $B = CD_3NO_2$; $C = CDCl_3$.
[d] Includes unidentified signals.
[e] *C*(1).
[f] Me(4a).
[g] *C*(4a).
[h] Me(1).

TABLE 7.15. MASS SPECTRA OF DIHYDROPYRIDO[1,2-*a*]BENZIMIDAZOLE DERIVATIVES (**7.181**)–(**7.183**)[a]

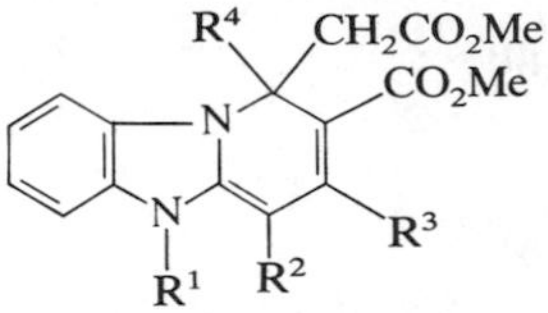

(**7.181**) 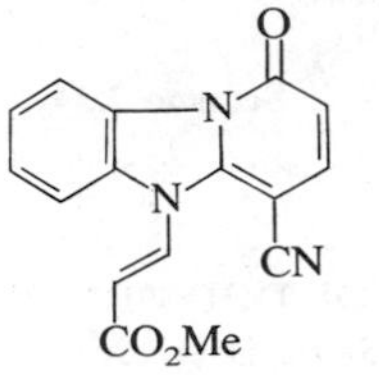 (**7.182**) 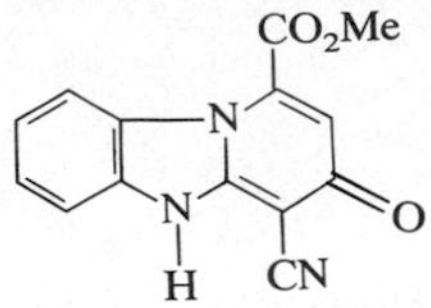 (**7.183**)

Compound	*m/e* (rel. abundance %) (assignment)	m* (metastable transition)
(**7.181**; $R^2 = CN$, $R^3 = R^4 = H$)	339(6.2) (M^+), 266(100) ($M - CH_2CO_2Me$)	—
(**7.181**; $R^1 = H$, $R^2 = CO_2Et$, $R^3 = R^4 = CO_2Me$)	485(9) (M^+), 454(3.3) ($M - OMe$), 426(100) ($M - CO_2Me$), 412(12) ($M = CH_2CO_2Me$)	425.5, 375, and 343 (485 → 454, 485 → 426, and 485 → 412)
(**7.181**; $R^1 = Me$, $R^2 = CO_2Et$, $R^3 = R^4 = CO_2Me$)	502(4) (M^+), 443(100) ($M - CO_2Me$), 429(30) ($M - CH_2CO_2Me$), and 398(5.6) ($M - CH_2CO_2Me - OMe$)	391 and 367 (502 → 443 and 502 → 429)
(**7.181**; $R^1 = Me$, $R^2 = CN$, $R^3 = R^4 = CO_2Me$)	455(5) (M^+), 396(100) ($M - CO_2Me$) 382(18) ($M - CH_2CO_2Me$), and 364(7)	345 and 335 (445 → 396 and 396 → 364)
(**7.182**)	293(100) (M^+), 265(2) ($M - CO$), 262(9.5) ($M - OMe$), and 234(17) ($M - CO_2Me$)	240 and 187 (293 → 265 and 293 → 234)
(**7.183**)	267(100) (M^+), 239(13) ($M - CO$), 236(10) ($M - OMe$), and 208(20) ($M - CO_2Me$)	214 and 208 (267 → 239 and 267 → 236)

[a] From Ref. 6.

MASS SPECTRA. The Mass spectra (Table 7.15) of 1,5-dihydropyrido[1,2-*a*]benzimidazoles (**7.181**) contain weak molecular ion peaks and base peaks corresponding to the loss of a *C*(1) substituent.[5,6] Loss of the *C*(1) ester substituent from substrates of the type (**7.181**; $R^4 = CO_2Me$) on electron impact results in the formation of stable aromatic cations.[5,6] Primary loss of CO is a characteristic feature of the mass spectra (Table 7.15) of pyrido[1,2-*a*]benzimidazol-1(5*H*)-ones [e.g. (**7.182**)] and pyrido[1,2-*a*]benzimidazol-3(5*H*)-ones [e.g. (**7.183**)]. The mass spectra of fully unsaturated pyrido[1,2-*a*]benzimidazoles,[40,41] 1,5-dihydropyrido[1,2-*a*]benzimidazoles,[5] and 1,2,3,4-tetrahydropyrido[1,2-*a*]benzimidazoles[63] have also been recorded.

7.1.3. Reactions

Reactions with Electrophiles

PROTONATION. The basicity of pyrido[1,2-*a*]benzimidazoles[27] and their 1,2,3,4-tetrahydro derivatives[45] is demonstrated by the ready solubility of these molecules in dilute mineral acids and their recovery unchanged on basification. The similarity of the UV spectra of 1,5-dihydropyrido[1,2-*a*]-benzimidazoles in acidic solution to that of the benzimidazolium cation is indicative of preferential protonation of the former molecules at the *C*(4) position.[2]

ALKYLATION. The presence of the acidic NH-center in *N*(5)-unsubstituted pyrido[1,2-*a*]benzimidazol-1(5*H*)-ones renders these molecules susceptible to orthodox methylation under basic conditions giving good yields (Table 7.16) of the corresponding *N*(5) methyl derivatives [e.g. Scheme 7.32, (**7.184**)].[7] The uncatalyzed alkylation of pyrido[1,2-*a*]benzimidazoles[16,18,19,33] and their 1,2,3,4-tetrahydro derivatives[29,60,62] also occurs at the *N*(5) position and affords good yields (Table 7.16) of the corresponding quaternary salts (**7.185**) and (**7.186**). Alkylation of this type can be achieved by heating with alkyl halides[18,29,33,60,62] and sulfates[16] in the absence of solvent[18,29,33,60,62] or in solvents such as acetone[60] or nitromethane.[16,60]

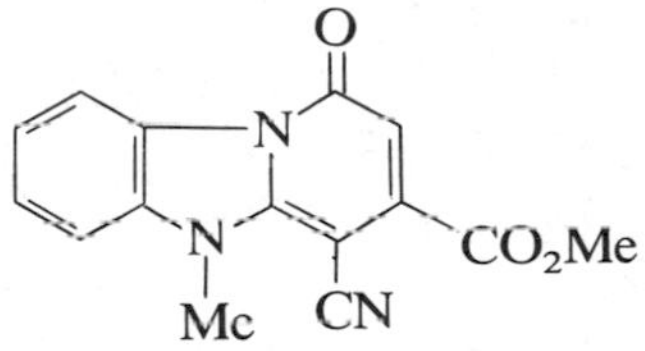

(**7.184**) 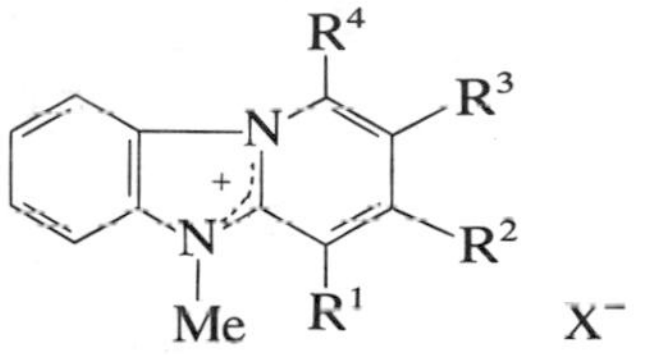 (**7.185**)

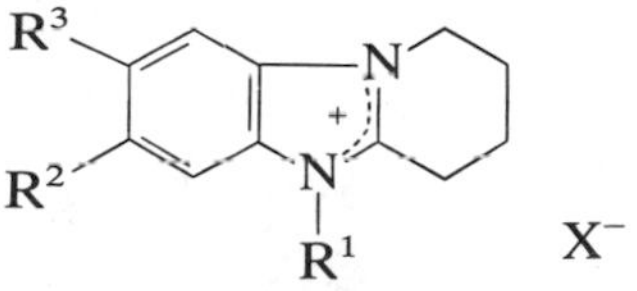

(**7.186**)

Scheme 7.32

ACYLATION. Hydroxyl[39] and amino[29,32] substituents in fully unsaturated pyrido[1,2-*a*]benzimidazoles behave in orthodox fashion toward acetylation, giving the corresponding acetoxy and acetamido derivatives, respectively (Table 7.17). The activation of *C*(3) methyl substituents in *N*(5)-substituted pyrido[1,2-*a*]benzimidazolium salts toward acylative condensation has been exploited for the synthesis of cyanine dyestuffs. Condensation reactions of this type are illustrated (Scheme 7.33) by the reaction of the pyrido[1,2-*a*]-benzimidazolium salt (**7.187**) with the benzothiazole derivative (**7.188**) to afford the cyanine (**7.189**).[11b]

N(5)-unsubstituted pyrido[1,2-*a*]benzimidazol-1(5*H*)-ones undergo orthodox acetylation at nitrogen with reagents such as acetyl chloride giving good yields (Table 7.17) of the *N*(5) acetyl derivatives.[7] In contrast *N*(5)-unsubstituted pyrido[1,2-*a*]benzimidazol-1(5*H*)-ones are formylated under

TABLE 7.16. ALKYLATION REACTIONS OF PYRIDO[1,2-*a*]BENZIMIDAZOLE DERIVATIVES

Alkylating agent[a]	Reaction time (hr)	Reaction temp. (°C)	Solvent	Product ($R^1 \rightarrow R^4$ unspecified = H)	Yield (%)	m.p. (°C)	Ref.
A	1.5	100	Dimethyl formamide	(**7.184**)	56	239[b]	7
B	16	100	—[c]	(**7.185**; X = I)	—[c]	246–247[e]	29
B	—[e]	—[g]	—[h]	(**7.185**; $R^3 = CO_2Et$, X = I)	—[c]	233–235[i]	33
B	10	—[j]	—[c]	(**7.185**; $R^2 = R^4$ = Ph, X = I)	70	280–282	18
C	3.5	—[k]	Trichloroethylene-nitromethane	(**7.185**; R^1 = CN, $R^2 = R^4$ = Me, X = $MeSO_4$)	70	207[l]	16
B	—[e]	—[g]	—[h]	(**7.185**; $R^2 = R^4$ = Me, $R^3 = CO_2Et$, X = I)	—[d]	271–271.5	33
B	—[e]	100	—[c]	(**7.186**; R^1 = Me, X = I)	—[d]	220–221[m]	29
B	30	—[k]	Acetone	(**7.186**; R^1 = Me, X = I)	—[d]	210	60
D	15	110	—[n]	(**7.186**; R^1 = Et, X = I)	—[d]	246	60
E	1	100	—[c]	(**7.186**; $R^1 = CH_2Ph$, X = Cl)	—[d]	288[o] (decomp.)	62
F	3	—[k]	Acetone	(**7.186**; $R^1 = (CH_2)_3OSO_3^-$)	—[d]	260	60
G	3	—[k]	Acetone	(**7.186**; $R^1 = (CH_2)_3SO_2NHCOMe$, X = Br)	—[d]	>260	60
H	4	—[k]	Acetone	(**7.186**; $R^1 = (CH_2)_4SO_2NHCOMe$, X = Br)	—[d]	206–208	60
I	5	—[k]	Acetone	(**7.186**; $R^1 = (CH_2)_2CONHSO_2Me$, X = Br)	—[d]	238	60
D	15	110	—[n]	(**7.186**; R^1 = Et, R^2 = Cl, X = I)	—[d]	>250	60
J	16	120	—[n]	(**7.186**; $R^1 = (CH_2)_2CO_2H$, R^2 = Cl, X = Br)	—[d]	228	60
F	2	120	—[n]	(**7.186**; $R^1 = (CH_2)_3OSO_3^-$, R^2 = Cl)	—[d]	>260	60
D	15	110	—[n]	(**7.186**; R^1 = Et, R^3 = Cl, X = I)	—[d]	250	60
D	15	110	—[n]	(**7.186**; R^1 = Et, R^2 = F, X = I)	—[c]	>250	60
D	15	110	—[n]	(**7.186**; R^1 = Et, R^2 = Br, X = I)	—[d]	>250	60
D	15	100	—[n]	(**7.186**; R^1 = Et, R^2 = CN, X = I)	—[d]	306	60

D	15	110	—[n]	($\mathbf{7.186}$; $R^1 = Et$, $R^3 = CN$, $X = I$)	—[d]	>260	60
B	1.5	—[k]	Acetone	($\mathbf{7.186}$; $R^1 = Me$, $R^2 = CF_3$, $X = I$)	—[d]	270	60
D	3	100	Nitromethane	($\mathbf{7.186}$; $R^1 = Et$, $R^2 = CF_3$, $X = I$)	—[d]	260	60
D	15	110	—[n]	($\mathbf{7.186}$; $R^1 = Et$, $R^2 = SO_2N(CH_2)_5$, $X = I$)	—[d]	>250	60
D	16	110	—[n]	($\mathbf{7.186}$; $R^1 = Et$, $R^2 = R^3 = Cl$, $X = I$)	—[d]	>250	60
K	4	110	—[n]	($\mathbf{7.186}$; $R^1 = (CH_2)_2OH$, $R^2 = R^3 = Cl$, $X = Br$)	—[d]	>250	60
I	3	—[k]	Acetone	($\mathbf{7.186}$; $R^1 = (CH_2)_2CONHSO_2Me$, $X = Br$)	—[d]	>260	60
D	15	110	—[n]	($\mathbf{7.186}$; $R^1 = Et$, $R^2 = Cl$, $R^3 = CN$, $X = I$)	—[d]	>250	60
D	15	110	—[n]	($\mathbf{7.186}$; $R^1 = Et$, $R^2 = F$, $R^3 = CN$, $X = I$)	—[d]	>250	60

[a] *A* = MeI, NaH; *B* = MeI; *C* = Me_2SO_4; *D* = EtI; *E* = $PhCH_2Cl$; *F* = $(CH_2)_3$–O–SO_2–O (cyclic); *G* = $MeCONHSO_2(CH_2)_3Br$; *H* = $MeCONHSO_2(CH_2)_4Br$; *I* = $MeSO_2NHCO(CH_2)_2Br$; *J* = $HO_2C(CH_2)_2Br$; *K* = $HO(CH_2)_2Br$.

[b] Crystallized from dimethylformamide.

[c] No cosolvent.

[d] Yield not quoted.

[e] Crystallized from water.

[f] Reaction time not specified.

[g] Reaction temperature not specified.

[h] Reaction solvent not specified.

[i] Crystallized from acetone.

[j] Room temperature.

[k] Reflux.

[l] Crystallized from ethanol.

[m] Crystallized from acetone–methanol.

[n] Sealed tube conditions.

[o] Crystallized from ether–ethanol.

TABLE 7.17. ACYLATION REACTIONS OF PYRIDO[1,2-*a*]BENZIMIDAZOLE DERIVATIVES

Reaction	Pyrido[1,2-*a*]benzimidazole				Solvent of	
conditions[a]	Substrate	Product	Yield (%)	m.p. (°C)	crystallization	Ref.
A	2-NH_2-	2-MeCONH-	—[b]	230–231	—[c]	32
B	6-NH_2-	6-MeCONH-[d]	—[b]	106–108	—[c]	29
C	8-HO-	8-$MeCO_2$-	quant.	209	Benzene	39
C	{6-Me-8-HO- + 7-Me-8-HO-}[e]	6-Me-8-$MeCO_2$-	30	139–140[f]	Benzene–light petroleum (b.p. 60–80°)	39
		7-Me-8-$MeCO_2$-	15	203–204[f]	Benzene–light petroleum (b.p. 60–80°)	
D	3-MeO_2C-4-CN-1(5*H*)-one	3-MeO_2C-4-CN-5-MeCO-1(5*H*)-one	73	296–298	Dichloromethane–ethanol	7
D	3-MeO_2C-4-EtO_2C-1(5*H*)-one	3-MeO_2C-4-EtO_2C-5-MeCO-1(5*H*)-one	60	144–146	Ethanol	7
E	3-Me-4-CN-1(5*H*)-one	2-CHO-3-Me-4-CN-1(5*H*)-one	96	320 (decomp.)	—[c]	14
F	3,7-di-Me-4-CN-1(5*H*)-one	2-CHO-3,7-di-Me-4-CN-1(5*H*)-one	98	300	—[c]	14
F	3-Me-4-H_2NCO-1(5*H*)-one	2-CHO-3-Me-4-H_2NCO-1(5*H*)-one	80	>350	—[c]	14
F	3-EtO_2C-4-CN-1(5*H*)-one	2-CHO-3-EtO_2C-4-CN-1(5*H*)-one	54	>200	—[c]	14
F	3-Bu^nO_2C-4-CN-1(5*H*)-one	2-CHO-3-Bu^nO_2C-4-CN-1(5*H*)-one	73	>300	—[c]	14

[a] *A* = Ac_2O/(40°)(reaction time not specified); *B* = Ac_2O,AcOH/(reflux)(reaction time not specified); *C* = Ac_2O/(100°)(2 hr); *D* = AcCl, diisopropylamine/(room temp.)(1 hr); *E* = $POCl_3$, dimethylformamide, $CHCl_3$/(30–80°)(1 hr); *F* = $POCl_3$, dimethylformamide, *N*-methylpyrrolidone/(10°)(30 min), then (30–60°)(1 hr), and finally (70°)(45 min).

[b] Yield not specified.

[c] Solvent of crystallization not specified.

[d] Hydrate; the anhydrous base has m.p. 142–144°.

[e] Mixture.

[f] These m.p. assignments may be reversed.

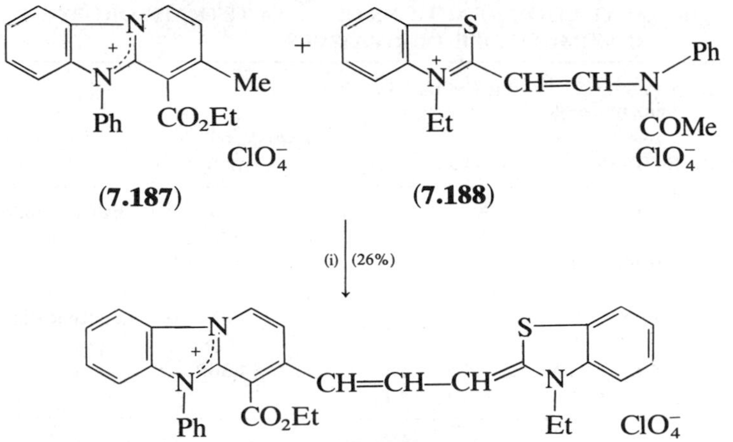

(7.187) (7.188)

(i) (26%)

(7.189)

(m.p. 288°)

(i) 1,5-diazabicyclo[3,2,2]nonane, MeCN/(room temp.)(10 min)

Scheme 7.33

Vilsmeier–Haack conditions at the *C*(2) position thus providing high yield (Table 7.17) synthetic access to pyrido[1,2-*a*]benzimidazol-1(5*H*)-one 2-carboxaldehydes.[14] The application of acylative condensation of this type to the synthesis of dyestuffs is illustrated by the reaction (Scheme 7.34) of the pyrido[1,2-*a*]benzimidazol-1(5*H*)-one (**7.190**) with the cyclic imide chloride (**7.191**) to give the condensate (**7.192**).[68]

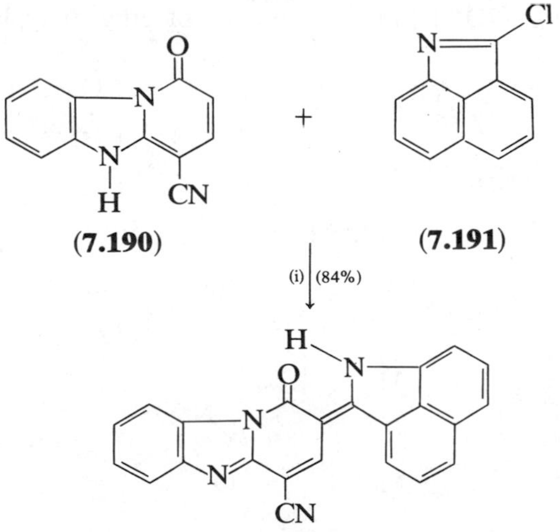

(7.190) (7.191)

(i) (84%)

(7.192)

(i) $PhNO_2$/80°

Scheme 7.34

TABLE 7.18. ACYLATION REACTIONS OF 1,2,3,4-TETRAHYDROPYRIDO[1,2-*a*] BENZIMIDAZOLE DERIVATIVES

Reaction conditions[a]	1,2,3,4-Tetrahydropyrido[1,2-*a*]-benzimidazole: Substrate	Product	Yield (%)	m.p. (°C)	Solvent of crystallization	Ref.
A	7-NH_2-	7-MeCONH-	—[a]	219.5–220	Ethyl acetate	29
B	9-NH_2-	9-MeCONH-	—[b]	236–237	Ethanol	29
C	Unsubstituted	—[c]	72	185–187	—[d]	42, 52
D	Unsubstituted	4-PhCO-	—[b]	168–169	—[d]	50, 52
E	8-O_2N-	—[e]	77	188	Benzene–light petroleum	43
F	4-CN-1,3-dione	2-(PhNHCH=)-4-CN-1,3-dione	94	>300	—[d]	70

[a] *A* = Ac_2O,AcOH/reflux (reaction time not specified); *B* = Ac_2O (reaction conditions not specified); *C* = PhCOCl, 10% NaOH aq./room temp. (reaction time not specified); *D* = PhCOCl, 10% NaOH aq. (reaction conditions not specified); *E* = PhCOCl, 15% NaOH aq. (reaction conditions not specified); *F* = $PhNH_2$, $(EtO)_3CH$, ethylene glycol/(150–180°)(20 min).
[b] Yield not quoted.
[c] 4,5-Dibenzoyl-1,2,3,5-tetrahydropyrido[1,2-*a*]benzimidazole.
[d] Solvent of crystallization not specified.
[e] 4,5-Dibenzoyl-8-nitro-1,2,3,5-tetrahydropyrido[1,2-*a*]benzimidazole.

The typical behavior of nuclear amino substituents in 1,2,3,4-tetrahydropyrido[1,2-*a*]benzimidazoles toward acylation (Table 7.18)[29] has been utilized in the synthesis of fused ring systems based on the 1,2,3,4-tetrahydropyrido-[1,2-*a*]benzimidazole nucleus. In particular (Scheme 7.35) 7-amino-1,2,3,4-tetrahydropyrido[1,2-*a*]benzimidazole (**7.193**) condenses smoothly with β-keto esters [e.g. (**7.194**)] in the presence of ethyl polyphosphate giving

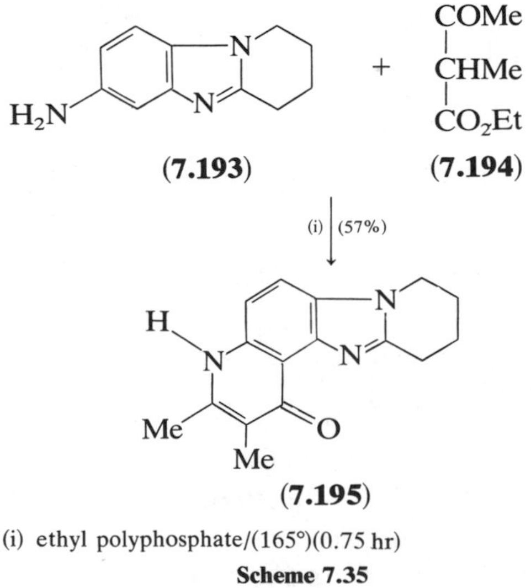

(7.193) + **(7.194)** —(i) (57%)→ **(7.195)**

(i) ethyl polyphosphate/(165°)(0.75 hr)

Scheme 7.35

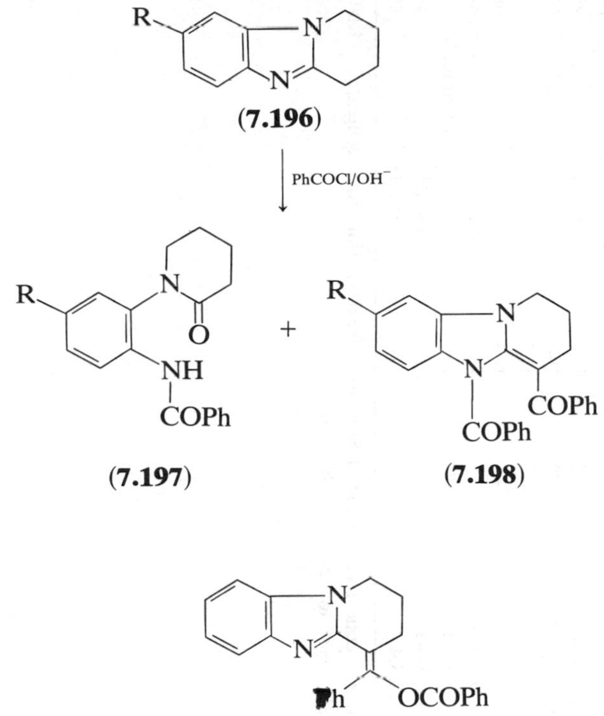

(**7.196**)

$PhCOCl/OH^-$

(**7.197**) (**7.198**)

(**7.199**)

Scheme 7.36

the corresponding fused pyridin-4(1*H*)-ones [e.g. (**7.195**)] in good yield.[69] The reaction (Scheme 7.36) of 1,2,3,4-tetrahydropyrido[1,2-*a*]benzimidazoles (**7.196**) with benzoyl chloride under alkaline conditions,[43,44a,50,52] as well as resulting in ring-opening to 1-(2-benzamidophenyl)piperidin-2(1*H*)-ones (**7.197**), affords products (Table 7.18) variously formulated as dibenzoyl derivatives (**7.198**)[52] and enol benzoates (**7.199**).[43]

Electrophilic Substitution Reactions. Information on the reactivity of pyrido[1,2-*a*]benzimidazoles toward electrophilic substitution is relatively sparse. Electrophilic halogenation, for example, appears to have been studied only in the single instance of the bromination of a 4a,5-dihydropyrido[1,2-*a*]benzimidazole to give a monobromo derivative of unestablished constitution.[2] Nitration (Table 7.19), on the other hand, has been more extensively investigated and in the case of pyrido[1,2-*a*]benzimidazole[29] affords the 8-nitro derivative in good yield. The nitration of 1,2,3,4-tetrahydropyrido[1,2-*a*]benzimidazoles[29,49,60] is also readily accomplished in good yield (Table 7.19) using nitric acid or potassium nitrate in conjunction with concentrated sulfuric acid and, depending on the substitution pattern of the benzene ring in the substrate, occurs at the *C*(7) or *C*(8)

Table 7.19. NITRATION REACTIONS OF PYRIDO[1,2-*a*]BENZIMIDAZOLE DERIVATIVES

Reaction conditions[a]	Pyrido[1,2-*a*]benzimidazole		Yield (%)	m.p. (°C)	Solvent of crystallization	Ref.
	Substrate	Product				
A	Unsubstituted	8-O_2N-	59	271–272	Methanol	29
B	1,2,3,4-tetrahydro-	8-O_2N-1,2,3,4-tetrahydro-	—[b]	—[c]	—[d]	60
C	7-Cl-1,2,3,4-tetrahydro-	7-Cl-8-O_2N-1,2,3,4-tetrahydro-	—[b]	171	Ethanol	49
B	7-Cl-1,2,3,4-tetrahydro-	7-Cl-8-O_2N-1,2,3,4-tetrahydro-	—[b]	194	—[d]	60
C	8-Cl-1,2,3,4-tetrahydro-	7-O_2N-8-Cl-1,2,3,4-tetrahydro-	—[b]	180	—[d]	49
B	7-Br-1,2,3,4-tetrahydro-	7-Br-8-O_2N-1,2,3,4-tetrahydro-	—[b]	184	—[d]	60
B	7-F-1,2,3,4-tetrahydro-	7-F-8-O_2N-1,2,3,4-tetrahydro-	—[b]	264	—[d]	60
D	7-MeCONH-1,2,3,4-tetrahydro-	7-MeCONH-8-O_2N-1,2,3,4-tetrahydro-	84	199–200	Methyl ethyl ketone	29
C	7-MeCONH-1,2,3,4-tetrahydro-	7-MeCONH-8-O_2N-1,2,3,4-tetrahydro-	88	204	Ethanol	49
C	8-MeCONH-1,2,3,4-tetrahydro-	8-MeCONH-7-O_2N-1,2,3,4-tetrahydro-	85	220	Ethanol	49

[a] *A* = conc. HNO_3, conc. H_2SO_4/(0–15°)(1 hr 20 min), then (40–50°)(15 min); *B* = conc. HNO_3, conc. H_2SO_4/(0–5°) (reaction time not specified); *C* = KNO_3, conc. H_2SO_4/(0–5°), then (room temp.)(1 hr); *D* = conc. HNO_3, conc. H_2SO_4/(−10 to −5°), then (0°)(1 hr).

[b] Yield not quoted.

[c] Melting point not quoted.

[d] Solvent of crystallization not specified.

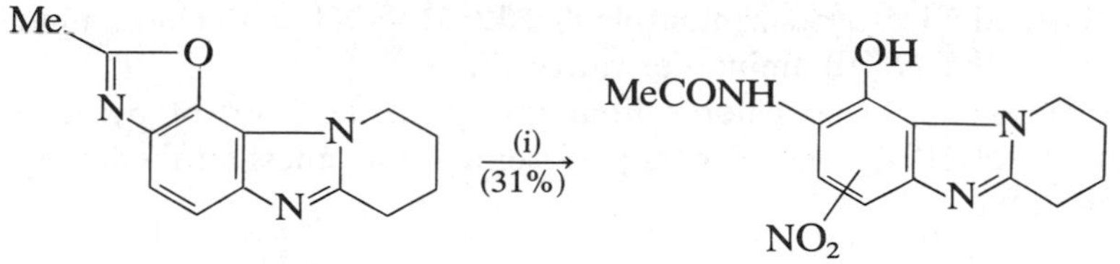

(i) (31%)

(7.200)

(7.201)

(i) KNO_3, conc. H_2SO_4/(20–25°)(16 hr)

[m.p. 235° (decomp.)]

Scheme 7.37

position. Nitration (Scheme 7.37) of the oxazolo-1,2,3,4-tetrahydropyrido-[1,2-*a*]benzimidazole (**7.200**) is accompanied by ring scission, the product being a nitro derivative (of unestablished orientation) of 8-acetamido-9-hydroxy-1,2,3,4-tetrahydropyrido[1,2-*a*]benzimidazole (**7.201**).[48]

Amino substituents in the benzene ring of pyrido[1,2-*a*]-benzimidazoles[27,28] and their 1,2,3,4-tetrahydro derivatives[29,48,49,60] can be diazotized under standard conditions to afford diazonium salts, which in some instances can be isolated, and in general undergo reactions (e.g., deamination) typical of such intermediates. The reactivity of the *C*(2) position in pyrido[1,2-*a*]benzimidazol-1(5*H*)-ones toward diazo coupling has been exploited for the synthesis of a wide range of azo dyestuffs.[71] The amination (Scheme 7.38) of pyrido[1,2-*a*]benzimidazole (**7.202**; R = H) by *O*-toluene-*p*-sulfonyl hydroxylamine is reported[72] to occur at the *N*(5) position giving 5-aminopyrido[1,2-*a*]benzimidazolium tosylate (**7.203**) in good yield. In contrast, the product derived by reaction (Scheme 7.38) of a

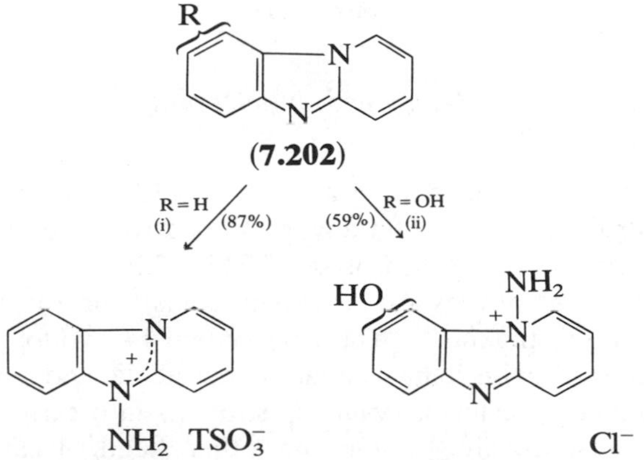

(7.202)

R = H (i) (87%)

R = OH (ii) (59%)

(7.203)

(7.204)

(m.p. 195–196°)

[m.p. 163° (decomp.)]

(i) NH_2OSO_2T, CH_2Cl_2/(room temp.)(0.25 hr)
(ii) NH_2Cl, $CHCl_3$/(10–20°)(48 hr)
[T = *p*-tolyl]

Scheme 7.38

hydroxypyrido[1,2-*a*]benzimidazole (**7.202**; R = OH) with chloramine is formulated as the *N*(10) amino derivative (**7.204**).[73]

The sulfonation and chlorosulfonation (Scheme 7.39) of pyrido[1,2-*a*]-benzimidazol-1(5*H*)-ones occurs preferentially at unestablished sites in the benzene ring.

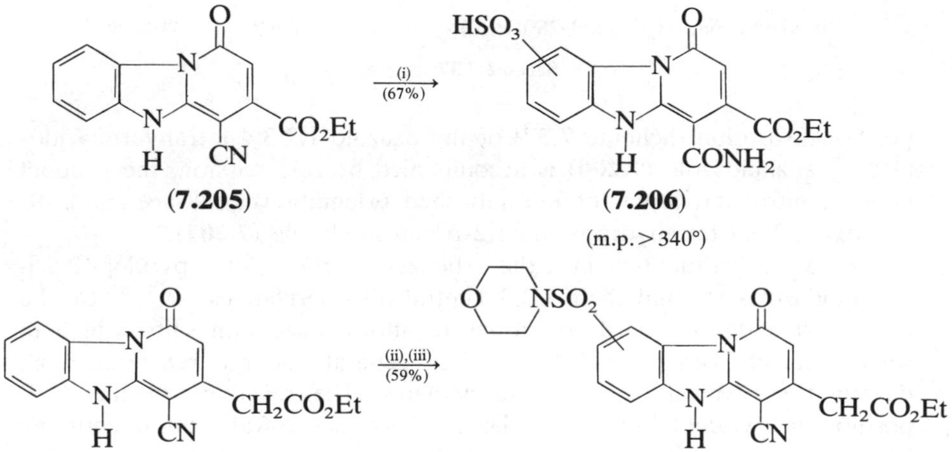

(**7.205**) —(i) (67%)→ (**7.206**) (m.p. > 340°)

(**7.207**) —(ii),(iii) (59%)→ (**7.208**) (m.p. > 300°)

(i) 23% oleum/(40°)(2 hr).
(ii) $ClSO_3H$/(80°)(4 hr).
(iii) morpholine, *N*-methylpyrrolidone/(20–30°)(1 hr)

Scheme 7.39

Reactions with Nucleophiles

The ring systems in fully unsaturated pyrido[1,2-*a*]benzimidazoles and their 1,2-dihydro- and 1,2,3,4-tetrahydro derivatives are stable to acidic conditions suitable for the hydrolysis (Table 7.20) of acetamido substituents,[29,48,49] or of nitriles to carboxylic acids[12] or for the hydrolytic removal of ester substituents.[11a] The ring system in pyrido[1,2-*a*]benzimidazole1(5*H*)-ones likewise remains intact under acidic and basic conditions, which serve to convert ester or cyano substituents into carboxyl or carboxamide groups, respectively.[14,15] On the other hand, 1,2,3,4-tetrahydropyrido[1,2-*a*]benzimidazoles quaternized at *N*(5) are susceptible to nucleophilic ring-opening under alkaline conditions.[65] Processes of this type are illustrated (Scheme 7.40) by the reaction of 1,2,3,4-tetrahydropyrido[1,2-*a*]benzimidazole (**7.209**) with benzoyl chloride in the presence of aqueous sodium hydroxide to afford 1-(2-benzamidophenyl)piperidin-2(1*H*)-one (**7.212**).[43] This transformation is rationalized[43] in terms of the intermediate

TABLE 7.20. HYDROLYTIC AND RELATED REACTIONS OF PYRIDO[1,2-*a*]BENZIMIDAZOLE DERIVATIVES

Reaction conditions[a]	Pyrido[1,2-*a*]benzimidazole Substrate	Product	Yield (%)	m.p. (°C)	Solvent of crystallization	Ref.
A	3-Me-4-CN-	3-Me-4-CO_2H-	75	240	Dimethylformamide–water	12
A	1,3-di-Me-4-CN-	1,3-di-Me-4-CO_2H-	95	253	Dimethylformamide–water	12
A	1,2,3-tri-Me-4-CN-	1,2,3-tri-Me-4-CO_2H-	66	242–243	Dimethylformamide–water	12
A	1,3-di-Me-2-HO_2CCH_2-4-CN-	1,3-di-Me-2-HO_2CCH_2-4-CO_2H-	95	268	Dimethylformamide–water	12
A	2-NO_2-4-CN-	2-NO_2-4-CO_2H-	86	176	Dimethylformamide–water	12
A	1,3,7(8)-tri-Me-4-CN-	1,3,7(8)-tri-Me-4-CO_2H-	80	240 (decomp.)	Dimethylformamide water	12
A	1,3-di-Me-4-CN-7(8)-MeO-	1,3-di-Me-4-CO_2H-7(8)-MeO-	56	244	Dimethylformamide–water	12
A	1,3-di-Me-4-CN-7(8)-Cl-	1,3-di-Me-4-CO_2H-7(8)-Cl-	91	305	Dimethylformamide water	12
A	1,3-di-Me-4-CN-7(8)-CO_2H-	1,3-di-Me-4,7(8)-di-CO_2H-	80	345 (decomp.)	Dimethylformamide water	12
A	1,3-di-Me-4-CN-7-Cl-9-NH_2-	1,3-di-Me-4-CO_2H-7-Cl-9-NH_2-	96	280 (decomp.)	Dimethylformamide water	12
A	1,3-di-Me-4-CN-7,8-di-Cl-	1,3-di-Me-4-CO_2H-7,8-di-Cl-	82	258 (decomp.)	Dimethylformamide water	12
B	3,5-di-Me-4-CO_2Et-, perchlorate	3,5-di-Me-, perchlorate	55	240–245	Ethanol	11a
C	3-MeO_2CCH_2-4-CN-1,5-dihydro-1(5*H*)-one	3-HO_2CCH_2-4-CN-1,5-dihydro-1(5*H*)-one[b]	—[c]	>310°	—	15
D	3-Me-4-CN-1,5-dihydro-1(5*H*)-one	3-Me-4-$CONH_2$-1,5-dihydro-1(5*H*)-one	93	270 (decomp.)	—[d]	14
E	3-CO_2Et-4-CN-1,5-dihydro-1(5*H*)-one	3-CO_2Et-4-$CONH_2$-1,5-dihydro-1(5*H*)-one	—[c]	>290	—[d]	15

TABLE 7.20 (*Continued*)

Reaction conditions[a]	Pyrido[1,2-*a*]benzimidazole Substrate	Product	Yield (%)	m.p. (°C)	Solvent of crystallization	Ref.
F	3-CO_2Et-4-CN-1,5-dihydro-1(5*H*)-one	3,4-di-CO_2Bu^n-1,5-dihydro-1(5*H*)-one	81	>170 (decomp.)	—[d]	15
G	3-CO_2Et-4-CN-1,5-dihydro-1(5*H*)-one	3-CONHBu^n-4-CN-1,5-dihydro-1(5*H*)-one	93	315 (decomp.)	—[d]	15
B	3,5-di-Me-4-CO_2Et-1,2-dihydro-	3,5-di-Me-1,2-dihydro-	52	206–210	Ethanol–acetonitrile	11a
H	4-CN-4-Ph-1,2,3,4-tetrahydro-	4-CO_2H-4-Ph-1,2,3,4-tetrahydro-	86	285–287	Ethanol–water	26
I	4-CO_2H-4-Ph-1,2,3,4-tetrahydro-	4-CO_2Et-4-Ph-1,2,3,4-tetrahydro-	83	146–148	Hexane	26
J	7-MeCONH-8-NO_2-1,2,3,4-tetrahydro-	7-NH_2-8-NO_2-1,2,3,4-tetrahydro-[e]	—[c]	266–267	Ethylene glycol monoethyl ether	29
K	7-MeCONH-8-NO_2-1,2,3,4-tetrahydro-	7-NH_2-8-NO_2-1,2,3,4-tetrahydro-	—[c]	253	Ethanol–water	49
K	7-NO_2-8-MeCONH-1,2,3,4-tetrahydro-	7-NO_2-8-NH_2-1,2,3,4-tetrahydro-	—[c]	310	Ethylene glycol monoethyl ether	49
L	4-PhCO_2NH-1,2,3,4-tetrahydro-1-one	4-NH_2-1,2,3,4-tetrahydro-1-one	83	147	Ether–methanol	23

[a] *A* = conc. H_2SO_4, AcOH, H_2O/(150°)(20 hr); *B* = conc. HCl, dimethylacetamide/(reflux)(40–60 min); *C* = 50% NaOH aq./(20–40°)(14 hr); *D* = 80% H_2SO_4 aq./(100°)(1 hr); *E* = 85% H_2SO_4 aq./(100–110°)(4 hr); *F* = Bu^nOH, 96% H_2SO_4 aq./(reflux)(30 hr); *B* = Bu^nNH_2, toluene-*p*-sulfonic acid/(reflux)(8 hr); *H* = conc. H_2SO_4/(145–150°)(2 hr); *I* = HCl gas, EtOH/(room temp.)(8 days); *J* = 2 M HCl/(reflux)(15 min); *K* = conc. HCl/(reflux)(1 hr); *L* = 25% HBr—AcOH/(room temp.)(16 hr).

[b] Sodium salt.

[c] Yield not quoted.

[d] Solvent of crystallization not specified.

[e] Forms a hydrochloride, m.p. 304° (decomp.).

formation and ring-opening of 5-benzoyl-1,2,3,4-tetrahydropyrido[1,2-*a*]-benzimidazolium chloride [Scheme 7.40; (**7.210**) → (**7.211**) → (**7.212**)].

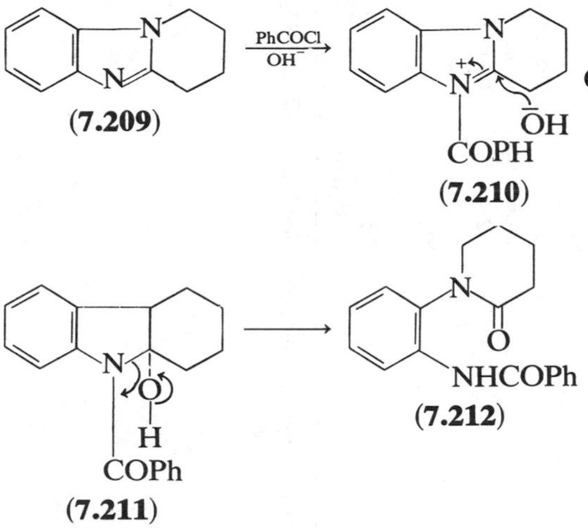

Scheme 7.40

1,2,3,4-Tetrahydropyrido[1,2-*a*]benzimidazole 5-*N*-oxide unlike its 2,3-dihydro-1*H*-pyrrolo[1,2-*a*]benzimidazole counterparts (cf. Chapter 6, section 6.1.3, "Reactions with Nucleophiles") gives only tars on attempted nucleophilic halogenation with acid halides.[74] On the other hand, nuclear diazonium salts derived from 1,2,3,4-tetrahydropyrido[1,2-*a*]benzimidazoles behave like the corresponding 2,3-dihydro-1*H*-pyrrolo[1,2-*a*]benzimidazole derivatives (cf. Chapter 6, section 6.1.3, "Reactions with Nucleophiles") in undergoing replacement by cyanide[60] and azide[48,49] ion affording methods (Table 7.21) for the synthesis of cyano- and azido-1,2,3,4-tetrahydropyrido[1,2-*a*]benzimidazoles. Azidopyrido[1,2-*a*]benzimidazoles are likewise accessible from the corresponding diazonium salts.[27]

Oxidation

The pyrido[1,2-*a*]benzimidazole ring system is relatively stable to oxidation under a variety of conditions. Thus, bromine effects the oxidative dimerisation (Scheme 7.41) of the *N*-aminopyrido[1,2-*a*]benzimidazolium salt (**7.213**) to the azo compound (**7.214**),[72] and also promotes the dehydrogenation (Table 7.22) of a 1,5-dihydropyrido[1,2-*a*]benzimidazole to the corresponding benzimidazolium cation.[2] Pyrido[1,2-*a*]benzimidazoles are also formed as the stable end-products of the palladium–charcoal catalyzed dehydrogenation (Table 7.22) of 1,2-dihydro-,[11a] 1,2,3,4-tetrahydro-,[29] and

TABLE 7.21. DISPLACEMENT REACTIONS OF PYRIDO[1,2-*a*]BENZIMIDAZOLE DIAZONIUM SALTS

Reaction conditions[a]	Pyrido[1,2-*a*]benzimidazole Substrate	Product	Yield (%)	m.p. (°C)	Solvent of crystallization	Ref.
A	8-NH_2-1,2,3,4-tetrahydro-	8-CN-1,2,3,4-tetrahydro-	—[b]	194	—[c]	60
A	7-Cl-8-NH_2-1,2,3,4-tetrahydro-	7-Cl-8-CN-1,2,3,4-tetrahydro-	—[b]	212	—[c]	60
A	7-Br-8-NH_2-1,2,3,4-tetrahydro-	7-Br-8-CN-1,2,3,4-tetrahydro-	—[b]	210	—[c]	60
A	7-F-8-NH_2-1,2,3,4-tetrahydro-	7-F-8-CN-1,2,3,4-tetrahydro-	—[b]	253	—[c]	60
B	6,8-di-NH_2-	6,8-di-N_3-	—[b]	167 (decomp.)	Light petroleum, b.p. 80–100°	27
B	6,8-di-NH_2-1,2,3,4-tetrahydro-	6,8-di-N_3-1,2,3,4-tetrahydro-	—[b]	132	Light petroleum b.p. 100–120°	27
C	7-NH_2-8-NO_2-1,2,3,4-tetrahydro-	7-N_3-8-NO_2-1,2,3,4-tetrahydro-	—[b]	194 (decomp.)	—[d]	49
C	7-NO_2-8-NH_2-1,2,3,4-tetrahydro-	7-NO_2-8-N_3-1,2,3,4-tetrahydro-	—[b]	195 (decomp.)	—[d]	49

[a] *A* = $NaNO_2$, HCl aq./(0°), then CuCN, KCN, Na_2CO_3/(room temp.)(30 min), and finally (50–60°)(15 min); *B* = $NaNO_2$, H_2SO_4, H_3PO_4/(10°), then treat with NaN_3; *C* = $NaNO_2$, HCl aq./(0°), then treat with NaN_3.

[b] Yield not quoted.

[c] Solvent of crystallization not specified.

[d] Decomposes on attempted crystallization.

TABLE 7.22. DEHYDROGENATION REACTIONS OF DIHYDRO- AND TETRAHYDROPYRIDO[1,2-*a*]BENZIMIDAZOLE DERIVATIVES

Reaction conditions[a]	Pyrido[1,2-*a*]benzimidazole		Yield (%)	m.p. (°C)	Solvent of crystallization	Ref.
	Substrate	Product				
A	3,5-di-Me-1,2-dihydro-, perchlorate	3,5-di-Me-, perchlorate	58	240–245	Ethanol	11a
A	3,5-di-Me-4-CO_2Et-1,2-dihydro-, perchlorate	3,5-di-Me-4-CO_2Et-, perchlorate	54	242–245 (decomp.)	Ethanol	11a
A	3-Me-4-CO_2Et-5-Ph-1,2-dihydro-, perchlorate	3-Me-4-CO_2Et-5-Ph- perchlorate	50	230–232	Ethanol	11a
B	1,2,3,4-tetra-CO_2Me-5-Me-1,5-dihydro-, perchlorate	1,2,3,4-tetra-CO_2Me-5-Me-, perchlorate	75	188–189	Methanol (trace of $HClO_4$)	2
C	1,3-di-Me-3,4-dihydro-	1,3-di-Me-	57	113–115	Light petroleum b.p. 30–70°	19
D	1,2,3,4-tetrahydro-	Unsubstituted	—[b]	178–179	Chlorobenzene	29
E	6,7,8,9-tetrahydro-	Unsubstituted	—[b]	176–177	Benzene	37
F	2-NO_2-6,7,8,9-tetrahydro-	2-NO_2-	4	240–241	Ethanol	32

[a] *A* = 10% PdC, dimethylacetamide/(reflux)(1 hr); *B* = Br_2, $HClO_4$, AcOH/(100°)(30 min); *C* = MnO_2, benzene/(reflux)(2 hr); *D* = PdC/(306°)(20 min); *E* = PdC/(300°)(6 hr); *F* = chloranil, xylene/(reflux)(44 hr).

[b] Yield not quoted.

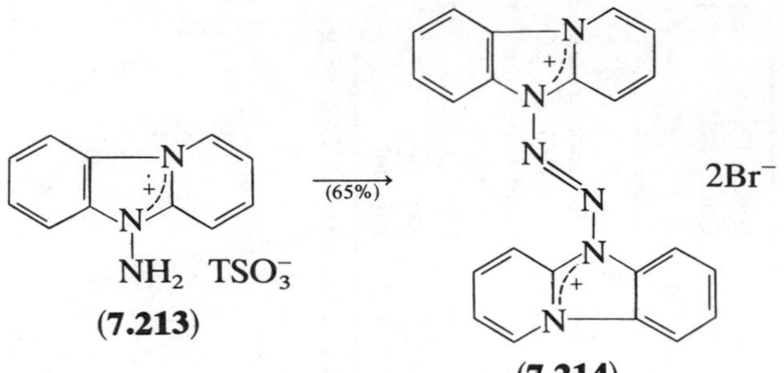

NH_2 TSO_3^-

(**7.213**)

(65%)

$2Br^-$

(**7.214**)

(i) Br_2/(room temp.)(few min)
[T = *p*-tolyl]

Scheme 7.41

6,7,8,9-tetrahydropyrido[1,2-*a*]benzimidazoles.[37] The dehydrogenation of molecules of the latter type can also be accomplished using chloranil as the oxidant.[32] The stable aromatic character of pyrido[1,2-*a*]benzimidazoles is further illustrated by their synthesis in good yield (Table 7.22) by the manganese dioxide oxidation of readily accessible 3,4-dihydropyrido[1,2-*a*]-benzimidazoles.[19] The 1,2,3,4-tetrahydropyrido[1,2-*a*]benzimidazole ring system, in contrast, is relatively stable to oxidants such as manganese dioxide, hydrogen peroxide, lead tetraacetate, and chloranil, as demonstrated by the utility of these reagents for the conversion of 1,2,3,4,4a,5-hexahydropyrido[1,2-*a*]benzimidazole derivatives into 1,2,3,4-tetrahydropyrido[1,2-*a*]benzimidazoles (Table 7.22).[58] The lead tetraacetate oxidation (Scheme 7.42) of the benzenesulfonate salt of the 1,2,3,4-tetrahydropyrido-[1,2-*a*]benzimidazole derivative (**7.215**) gives a pink solution, whose spectroscopic properties are consistent with the presence of the cation (**7.216**).[66]

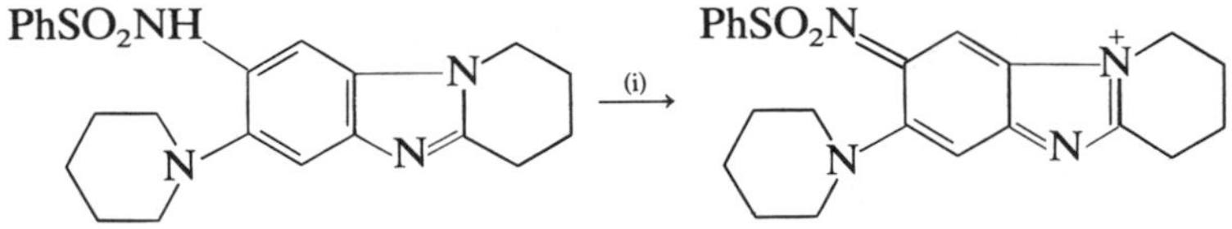

$PhSO_2NH$ $PhSO_2N$

(**7.215**) (**7.216**)

(i) $Pb(OAc)_4$, AcOH/(room temp.)(12 hr) (λ_{max} 540 nm)

Scheme 7.42

Reduction

The fully unsaturated and 1,5-dihydropyrido[1,2-*a*]benzimidazole ring systems are stable to reduction (Table 7.23) with iron and hydrochloric

TABLE 7.23. REDUCTION OF PYRIDO[1,2-*a*]BENZIMIDAZOLE DERIVATIVES

Reactions conditions[a]	Pyrido[1,2-*a*]benzimidazole Substrate	Product	Yield (%)	m.p. (°C)	Solvent of crystallization	Ref.
A	8-NH_2-	Unsubstituted	50	179	—[b]	28
B	8-$\overset{+}{N}_2$-, sulfate	Unsubstituted	71	—	—	27
C	6,8-di-NH_2-	Unsubstituted	—[c]	178–179	—[d]	27
B	6,8-di-$\overset{+}{N}_2$-	Unsubstituted	—[c]	—	—[d]	27
B	4-Me-8-$\overset{+}{N}_2$-	4-Me-	—[c]	162	Light petroleum (b.p. 40–60°)	27
B	4-Me-6,8-di-$\overset{+}{N}_2$-	4-Me-	—[c]	—	—	27
A	6-NH_2-8-NO_2-	8-NO_2-	89	260–262	Pyridine	27
A	4-Me-6-NH_2-8-NO_2-	4-Me-8-NO_2-	88	260–262	Xylene	27
D	2-NO_2-	2-NH_2-	60	220 (decomp.)	—	32
E	6-NO_2-	6-NH_2-	93	133–134	Benzene	29
F	8-NO_2-	8-NH_2-	81	229–230	Xylene	27, 28
F	4-Me-8-NO_2-	4-Me-8-NH_2-	80	185–187	Light petroleum (b.p. 100–120°)	27
G	6,8-di-NO_2-	6-NH_2-8-NO_2-	73	>280	Nitrobenzene	27
G	4-Me-6,8-di-NO_2-	4-Me-6-NH_2-8-NO_2-	—[c]	269–270	Xylene	27
F	6,8-di-NO_2-	6,8-di-NH_2-	—[c]	204–205	Ethanol	27
F	4-Me-6,8-di-NO_2-	4-Me-6,8-di-NH_2-	—[c]	130[e]	Benzene	27
H	1,2,3,4-tetra-CO_2Me-5-($MeO_2C\overset{\vert}{C}{=}CHCO_2Me$)-1,5-dihydro-	1,2,3,4-tetra-CO_2Me-5-($Me\underset{\vert}{C}HCH_2CO_2Me$)-1,5-dihydro-	66	173–174	Methanol	2
I	7-NH_2-8-NO_2-1,2,3,4-tetrahydro-	8-NO_2-1,2,3,4-tetrahydro-	85	217–219	Methanol	29
J	6-NH_2-	6-NH_2-1,2,3,4-tetrahydro-	—[c]	176–177	Ethyl acetate	29
K	7-NO_2-1,2,3,4-tetrahydro-	7-NH_2-1,2,3,4-tetrahydro-	90	218–220	Methanol	29
K	8-NO_2-1,2,3,4-tetrahydro-	8-NH_2-1,2,3,4-tetrahydro-	—[c]	198–200	Ethyl acetate–methanol	29
J	8-NH_2-	8-NH_2-1,2,3,4-tetrahydro-	—[c]	195–197	Ethyl acetate–methanol	29

TABLE 2.23 (*Continued*)

Reactions conditions[a]	Pyrido[1,2-*a*]benzimidazole		Yield (%)	m.p. (°C)	Solvent of crystallization	Ref.
	Substrate	Product				
L	9-NO_2-1,2,3,4-tetrahydro-	9-NH_2-1,2,3,4-tetrahydro-	—[c]	186–187	Benzene	29
M	7-Cl-8-NO_2-1,2,3,4-tetrahydro-	7-Cl-8-NH_2-1,2,3,4-tetrahydro-	—[c]	210	—[f]	60
M	7-Br-8-NO_2-1,2,3,4-tetrahydro-	7-Br-8-NH_2-1,2,3,4-tetrahydro-	—[c]	217	—[f]	60
M	7-F-8-NO_2-1,2,3,4-tetrahydro-	7-F-8-NH_2-1,2,3,4-tetrahydro-	—[c]	199	—[f]	60
K	7-NO_2-8-(1-piperidyl)-1,2,3,4-tetrahydro-	7-NH_2-8-(1-piperidyl)-1,2,3,4-tetrahydro-	—[c]	189–191	Ethyl acetate	29
N	1,2,3,4-tetrahydro-	1,2,3,4,4a,5-hexahydro-	quant.	119	Ethyl acetate–light petroleum	58
O	5-(2-Cl-5-$O_2NC_6H_3CH_2$)-1,2,3,4-tetrahydro-, chloride	5-(2-Cl-5-$O_2NC_6H_3CH_2$)-1,2,3,4,4a,5-hexahydro-	—[c]	96–97	Light petroleum	62

[a] *A* = $NaNO_2$, H_2SO_4, H_3PO_4/(−5°), then EtOH, H_2O/reflux (reaction time not specified); *B* = EtOH, H_2O/reflux (reaction time not specified); *C* = $NaNO_2$, HCl aq./(0°), then Al powder, EtOH/reflux (reaction time not specified); *D* = H_2, 5% PdC/room temp., atm press; *E* = Fe, AcOH, H_2O/(reflux)(3 hr); *F* = H_2, PtO_2, EtOH/70°, 3–5 atm; *G* = Na_2S, S, acetone, H_2O/(reflux)(2 hr); *H* = H_2, PdC, MeOH/(room temp., 3–5 atm)(17 hr); *I* = $NaNO_2$, HCl aq./(3–5°), then H_3PO_2/(39°)(30 min); *J* = H_2, PtO_2, MeOH/(90–100°, 60 atm)(3 hr); *K* = H_2, Raney-Ni, MeOH/(50–60°)(5 atm); *L* = Fe, NaOAc, HCl, AcOH, H_2O/(reflux)(3.5 hr); *M* = H_2, Raney-Ni, ethyleneglycol monomethyl ether/room temp., atm. press; *N* = $LiAlH_4$, ether/(reflux)(8 days); *O* = $NaBH_4$, H_2O/(room temp.)(5 min).

[b] Purified by crystallization from light petroleum (b.p. 40–60°), followed by sublimation.

[c] Yield not quoted.

[d] Purified by sublimation.

[e] Melting point indefinite.

[f] Solvent of crystallization not specified.

acid[29] or hydrogen over platinum[27,28] or palladium[2,32] catalysts at atmospheric pressure or somewhat above (1–5 atm.) under conditions which serve to convert nitro into amino substituents[27–29,32] or saturate unsaturated side chains.[2] However, the hydrogenation (Table 7.23)[29] of pyrido[1,2-*a*]benzimidazole derivatives over a platinum catalyst at elevated temperature (90–100°) and pressure (50–60 atm.) results in reduction of the pyridine ring giving the corresponding 1,2,3,4-tetrahydropyrido[1,2-*a*]benzimidazoles. The stability of the 1,2,3,4-tetrahydropyrido[1,2-*a*]benzimidazole ring system to catalytic reduction thereby demonstrated is further substantiated by the utility of hydrogenation over Raney nickel at atmospheric pressure[60] or above[29] as a means for the conversion of nitro-1,2,3,4-tetrahydropyrido[1,2-*a*]benzimidazoles into the corresponding amines (Table 7.23). On the other hand, 1,2,3,4-tetrahydropyrido[1,2-*a*]benzimidazole derivatives are readily reduced (Table 7.23) to 1,2,3,4,4a,5-hexahydropyrido[1,2-*a*]benzimidazoles using lithium aluminum hydride[58] or sodium borohydride.[62]

The reductive removal of diazonium substituents from pyrido[1,2-*a*]-benzimidazoles or their 1,2,3,4-tetrahydro derivatives is accomplished in an orthodox manner using hypophosphorus acid[29] or simply by heating with ethanol.[27,28] Selective reduction of the *C*(6) nitro substituent in a 6,8-dinitropyrido[1,2-*a*]benzimidazole can also be achieved in classical fashion using sodium polysulfide.[27]

Miscellaneous Reactions

The thermolysis of azido-1,2,3,4-tetrahydropyrido[1,2-*a*]benzimidazoles under different conditions provides the basis for the synthesis of tetracyclic structures containing a 1,2,3,4-tetrahydropyrido[1,2-*a*]benzimidazole nucleus. Annelation reactions of this type are represented by the oxazolopyrido[1,2-*a*]benzimidazole[48] and oxadiazolopyrido[1,2-*a*]benzimidazole[49]

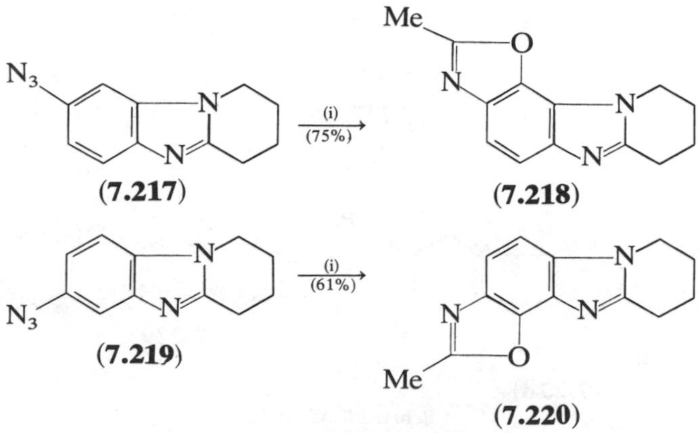

N_3 (7.217) —(i) (75%)→ Me, O, N (7.218)

N_3 (7.219) —(i) (61%)→ N, O, Me (7.220)

(i) polyphosphoric acid, AcOH/(reflux)(2 hr)

Scheme 7.43

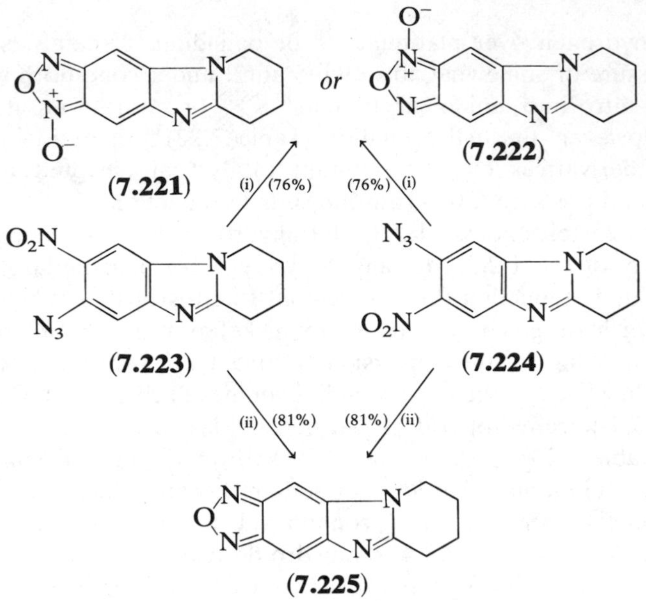

(7.221) or (7.222)

(i) (76%) (76%) (i)

(7.223) (7.224)

(ii) (81%) (81%) (ii)

(7.225)

(i) 2-ethoxyethanol/(reflux)(1 hr)
(ii) ethylene glycol/(reflux)(15 min)

Scheme 7.44

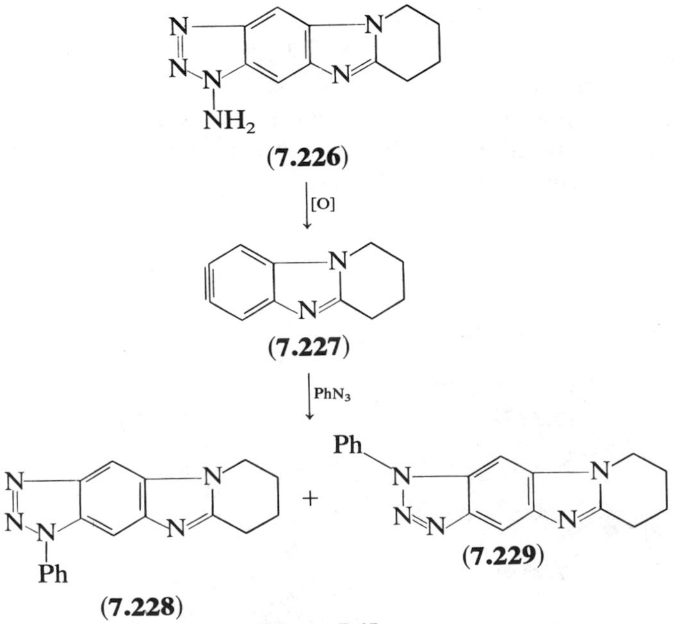

(7.226)

[O]

(7.227)

PhN_3

(7.228) + (7.229)

Scheme 7.45

syntheses outlined in Schemes 7.43 and 7.44. The oxidative transformation (Scheme 7.45)[75] of the fused *N*-amino-1,2,3-triazole (**7.226**) in the presence of phenyl azide into the isomeric 1,2,3-triazolopyrido[1,2-*a*]benzimidazoles (**7.228**) and (**7.229**) is indicative of the intermediacy of the benzyne derivative (**7.227**).

7.1.4. Practical Applications

Biological Properties

Analgetic activity[76] apart, the biological properties of pyrido[1,2-*a*]benzimidazoles appear to have attracted little attention.

Dyestuffs

Pyrido[1,2-*a*]benzimidazole derivatives have found widespread use as components of azomethine,[11b,77] azo,[15,71,78] and cyanine[60] dyestuffs, and as photographic sensitizing agents.[79]

7.2. Tricyclic 6-5-6 Fused Benzimidazoles with One Additional Heteroatom

The known oxygen-containing structures in the title category (cf. Scheme 7.46 and Table 7.24) include the 3,4-dihydro-2*H*-[1,3]oxazino[3,2-*a*]benzimidazole (**7.232**), the 3,4,4a,5-tetrahydro-1*H*-[1,3]oxazino[3,4-*a*]benzimidazole (**7.234**), and the 3,4-dihydro-1*H*-[1,4]oxazino[4,3-*a*]benzimidazole (**7.237**) ring systems. Derivatives of the corresponding, parent, unsaturated 2*H*- and 4*H*-[1,3]oxazino[3,2-*a*]benzimidazole ring systems (**7.230**) and (**7.231**), the 1*H*-[1,3]oxazino[3,4-*a*]benzimidazole ring system (**7.233**) and the 1*H*- and 10*H*-[1,4]oxazino[4,3-*a*]benzimidazole ring systems (**7.235**) and (**7.236**) do not appear to have been described to date.

Tricyclic 6-5-6 fused benzimidazole frameworks having sulfur as an additional heteroatom are represented (Scheme 7.47 and Table 7.24) in the literature by the 2*H*-[1,3]thiazino[3,2-*a*]benzimidazole ring system (**7.238**), its 4*H* isomer (**7.239**) and 3,4-dihydro derivative (**7.240**), and by the 1*H*-[1,4]thiazino[4,3-*a*]benzimidazole ring system (**7.242**), its 10*H* isomer (**7.243**), and 3,4-dihydro (**7.244**) and 10,10a-dihydro (**7.245**) derivatives. A search of the literature has failed to reveal any reference to the 1*H*-[1,3]-thiazino[3,4-*a*]benzimidazole ring system (**7.241**) or its derivatives.

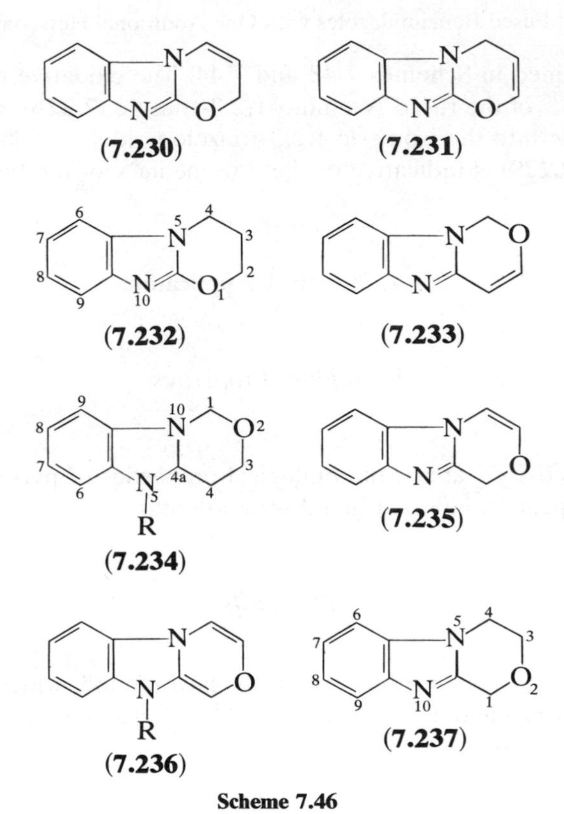

Scheme 7.46

TABLE 7.24. TRICYCLIC 6-5-6 FUSED BENZIMIDAZOLE RING SYSTEMS WITH ONE ADDITIONAL HETERO-ATOM (OXYGEN OR SULFUR)

Structure[a]	Name[b]
(**7.232**)	3,4-Dihydro-2*H*-[1,3]oxazino[3,2-*a*]benzimidazole
(**7.234**)	3,4,4*a*,5-Tetrahydro-1*H*-[1,3]oxazino[3,4-*a*]benzimidazole
(**7.237**)	3,4-Dihydro-1*H*-[1,4]oxazino[4,3-*a*]benzimidazole
(**7.238**)	2*H*-[1,3]Thiazino[3,2-*a*]benzimidazole
(**7.239**)	4*H*-[1,3]Thiazino[3,2-*a*]benzimidazole
(**7.240**)	3,4-Dihydro-2*H*-[1,3]thiazino[3,2-*a*]benzimidazole
(**7.242**)	1*H*-[1,4]Thiazino[4,3-*a*]benzimidazole
(**7.243**)	10*H*-[1,4]Thiazino[4,3-*a*]benzimidazole
(**7.244**)	3,4-Dihydro-1*H*-[1,4]thiazino[4,3-*a*]benzimidazole
(**7.245**)	10,10a-Dihydro-1*H*-[1,4]thiazino[4,3-*a*]benzimidazole

[a] Cf. Schemes 7.46 and 7.47.
[b] Based on the Ring Index.

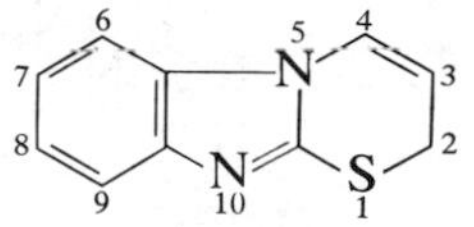

(7.238) 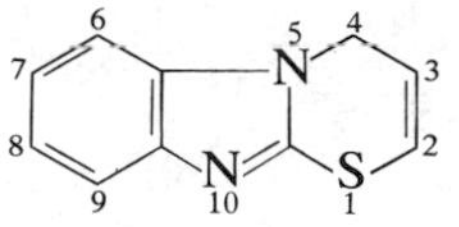 **(7.239)**

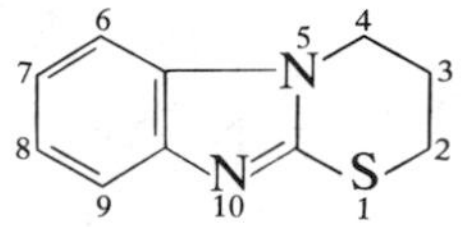

(7.240) 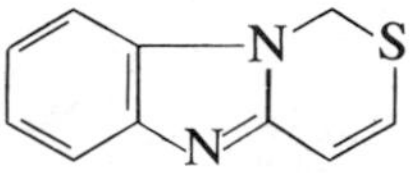 **(7.241)**

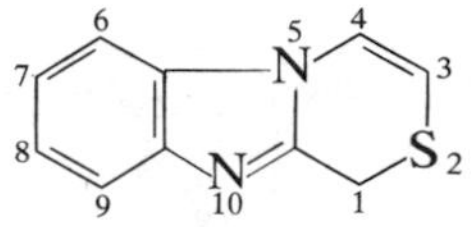

(7.242) 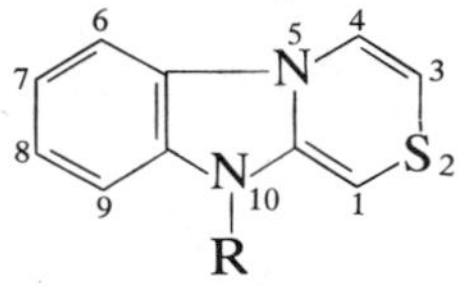 **(7.243)**

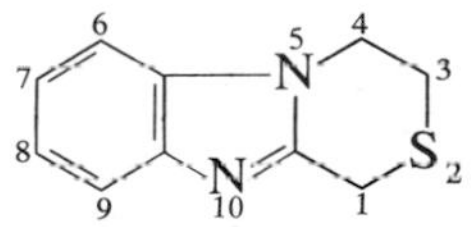

(7.244) 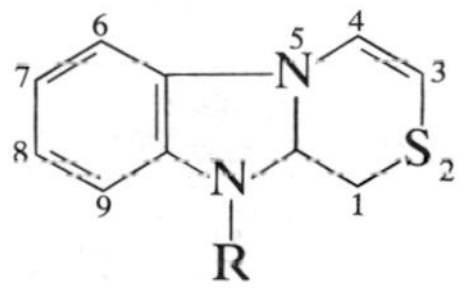 **(7.245)**

Schemc 7.47

Fully nitrogen-containing tricyclic 6-5-6 fused benzimidazolcs having one additional heteroatom comprise some 16 ring systems (Scheme 7.48 and Table 7.25). Of these, the pyrimido[1,2-*a*]benzimidazole ring system (**7.247**) and its various dihydro (**7.248**)–(**7.252**), tetrahydro (**7.253**), (**7.254**), and (**7.256**), and hexahydro (**7.257**) derivatives have been most extensively studied. The chemistry of pyrimido[1,2-*a*]benzimidazoles was briefly reviewed in 1961.[80]

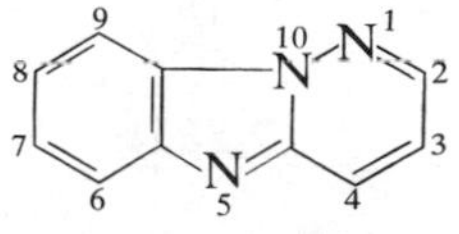

(7.246) 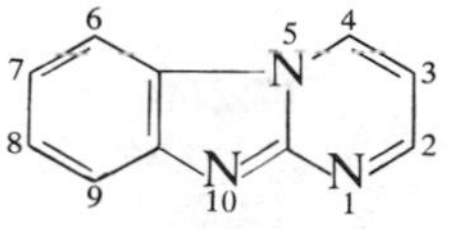 **(7.247)**

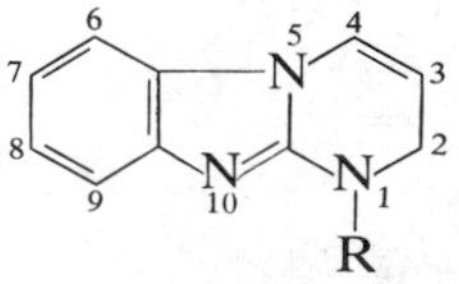

(7.248) 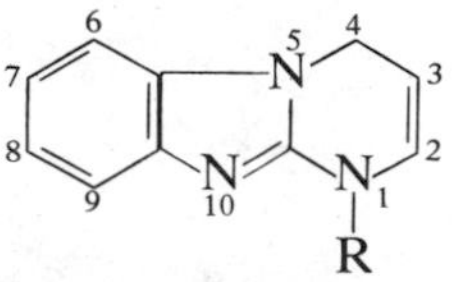 **(7.249)**

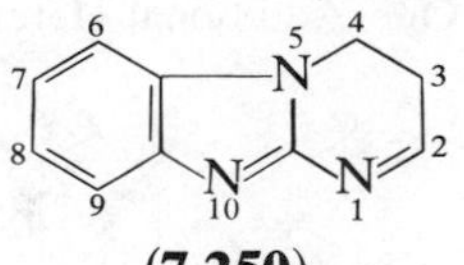

(**7.250**)

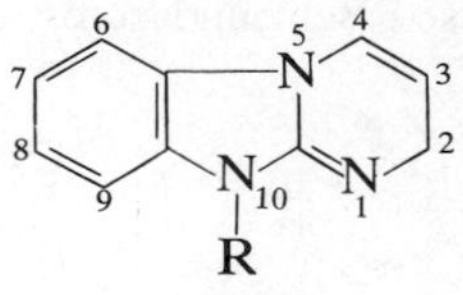

(**7.251**)

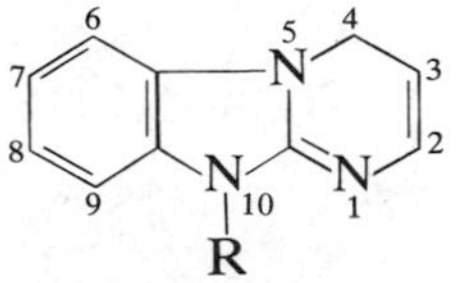

(**7.252**)

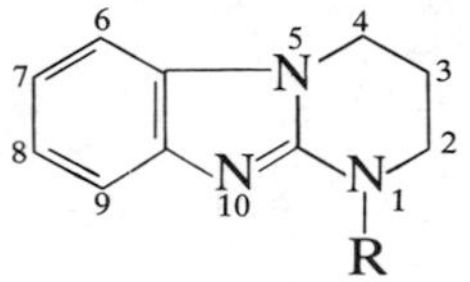

(**7.253**)

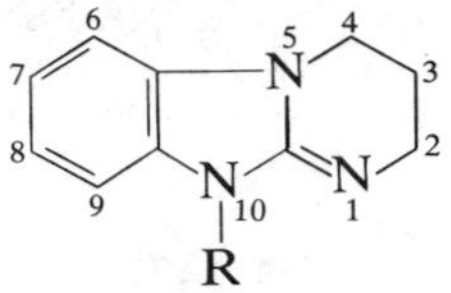

(**7.254**)

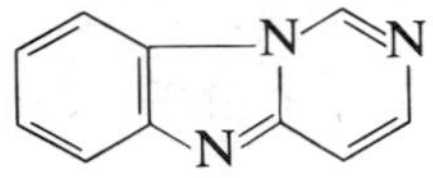

(**7.255**)

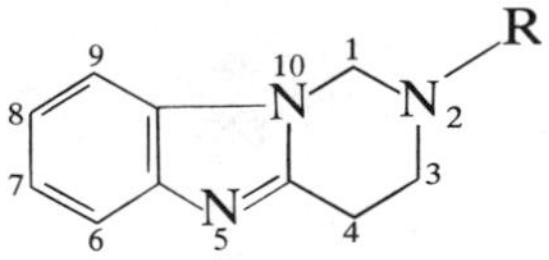

(**7.256**)

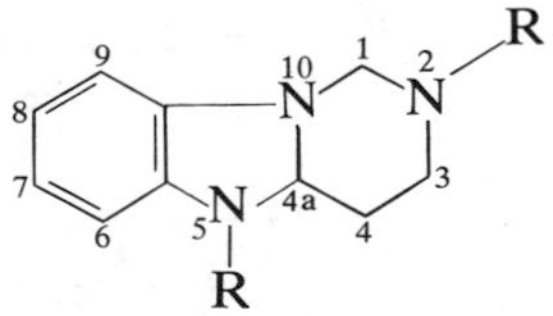

(**7.257**)

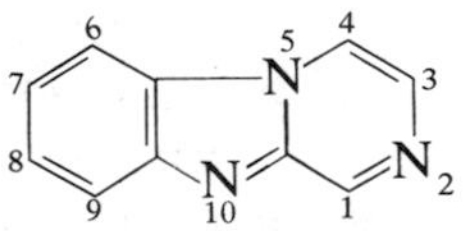

(**7.258**)

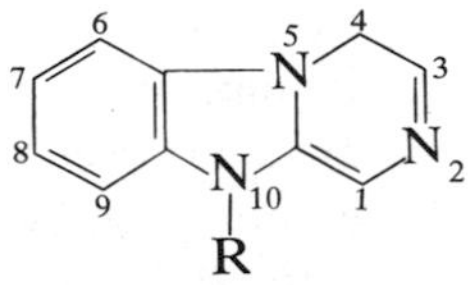

(**7.259**)

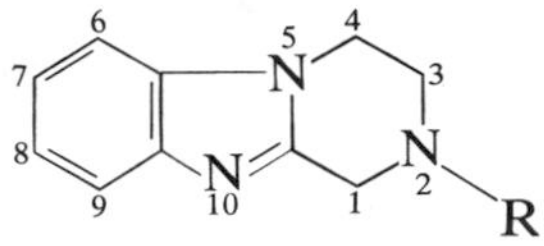

(**7.260**)

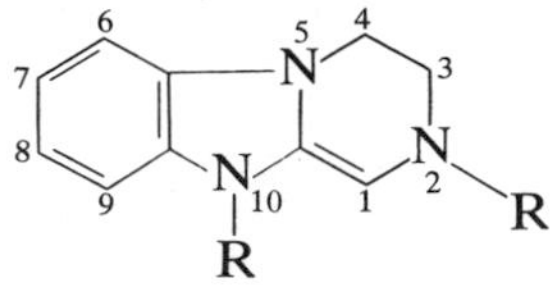

(**7.261**)

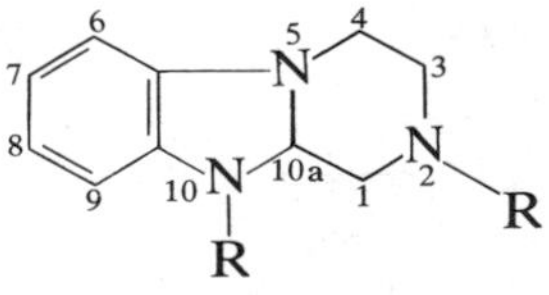

(**7.262**)

Scheme 7.48

Though the reduced frameworks (**7.256**) and (**7.257**) have been described in the literature, derivatives of the fully unsaturated pyrimido[3,4-*a*]benzimidazole ring system (**7.255**) have yet to be reported.

TABLE 7.25. TRICYCLIC 6-5-6 FUSED BENZIMIDAZOLE RING SYSTEMS WITH ONE ADDITIONAL HETEROATOM (NITROGEN)

Structure[a]	Name[b]
(**7.246**)	Pyridazino[2,3-*a*]benzimidazole
(**7.247**)	Pyrimido[1,2-*a*]benzimidazole
(**7.248**)	1,2-Dihydropyrimido[1,2-*a*]benzimidazole
(**7.249**)	1,4-Dihydropyrimido[1,2-*a*]benzimidazole
(**7.250**)	3,4-Dihydropyrimido[1,2-*a*]benzimidazole
(**7.251**)	2,10-Dihydropyrimido[1,2-*a*]benzimidazole
(**7.252**)	4,10-Dihydropyrimido[1,2-*a*]benzimidazole
(**7.253**)	1,2,3,4-Tetrahydropyrimido[1,2-*a*]benzimidazole
(**7.254**)	2,3,4,10-Tetrahydropyrimido[1,2-*a*]benzimidazole
(**7.256**)	1,2,3,4-Tetrahydropyrimido[3,4-*a*]benzimidazole
(**7.257**)	1,2,3,4,4a,5-Hexahydropyrimido[3,4-*a*]benzimidazole
(**7.258**)	Pyrazino[1,2-*a*]benzimidazole
(**7.259**)	4,10-Dihydropyrazino[1,2-*a*]benzimidazole
(**7.260**)	1,2,3,4-Tetrahydropyrazino[1,2-*a*]benzimidazole
(**7.261**)	2,3,4,10-Tetrahydropyrazino[1,2-*a*]benzimidazole
(**7.262**)	1,2,3,4,10,10a-Hexahydropyrazino[1,2-*a*]benzimidazole

[a] Cf. Scheme 7.48.
[b] Based on the Ring Index.

7.2.1. Synthesis

Ring-closure Reactions of Benzimidazole Derivatives

The 3,4-dihydro-2*H*-[1,3]oxazino[3,2-*a*]benzimidazole ring system is most readily constructed by the base-catalyzed ring-closure of 2-benzimidazolones having a suitably γ-functionalized propyl side chain at *N*(1). Ring formation of this type involves the intramolecular nucleophilic displacement of a leaving group at the γ-position of the propyl side chain by the *C*(2) oxo group of the imidazolone ring and is exemplified (Scheme 7.49) by the sodium ethoxide catalyzed cyclization of 1-(3-hydroxypropyl)-benzimidazol-2(3*H*)-one tosylate (**7.264**; $X = p\text{-}MeC_6H_4SO_2O$) to 3,4-dihydro-2*H*-[1,3]oxazino[3,2-*a*]benzimidazole (**7.265**) in good yield (Table 7.26).[81] The sodium hydride catalyzed condensation of 2-benzimidazolone (**7.263**) with 1,3-dibromopropane to afford 3,4-dihydro-2*H*-[1,3]oxazino-[3,2-*a*]benzimidazole (**7.265**) in unspecified yield (Table 7.26)[82,83] may likewise be rationalized in terms of the intermediate formation and base-catalyzed cyclization of 1-(3-bromopropyl)benzimidazol-2(3*H*)-one (**7.264**; X = Br). The products, formed in good yield (Table 7.26) by the reaction

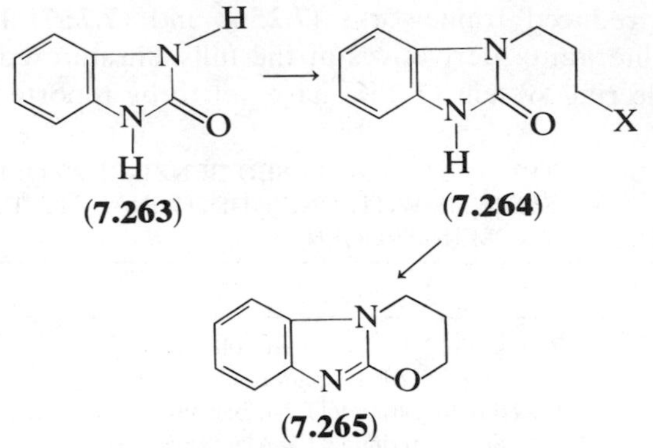

(7.263) (7.264) (7.265)

Scheme 7.49

(Scheme 7.50) of *C*(2)-unsubstituted benzimidazoles containing an electron-withdrawing group at *N*(1) (**7.266**; R = COMe, $COCHPh_2$, or SO_2Me) with diphenylketene, and originally formulated[84] as pyrido[1,2-*a*]benzimidazole derivatives, have been reassigned 3,4,4a,5-tetrahydro-1*H*-[1,3]oxazino[3,4-*a*]benzimidazole structures (**7.267**)[85] on the basis of their spectroscopic properties. On the other hand, the adducts derived by the analogous

TABLE 7.26. SYNTHESIS OF OXAZINOBENZIMIDAZOLES BY RING-CLOSURE REACTIONS OF BENZIMIDAZOLE DERIVATIVES

Starting materials	Reaction conditions[a]	Product	Yield (%)	m.p. (°C)	Solvent of crystallization	Ref.
(**7.263**)	*A*	(**7.265**)	—[b]	—[c]	—[d]	82, 83
(**7.264**; X = OSO_2Me)	*B*	(**7.265**)	74	116.5–118.5	Toluene	81
(**7.266**; R = H)	*C*	(**7.267**; R = $COCHPh_2$)	83	152–153	Benzene–light petroleum (b.p. 40–60°)	84, 86
(**7.266**; R = $COCHPh_2$)	*C*	(**7.267**; R = $COCHPh_2$)	86	—	—	84
(**7.266**; R = COMe)	*D*	(**7.267**; R = COMe)	73	135–138 (decomp.)	—[d]	85
(**7.266**; R = SO_2Me)	*D*	(**7.267**; R = SO_2Me)	70	130–135 (decomp.)	Ethyl acetate–light petroleum (b.p. 60–80°)	85
(**7.269**) + (**7.272**)	*E*	(**7.271**)	2	215–217	—[d]	87
(**7.272**)	*F*	(**7.273**)	35	161–162	Methanol–water	88

[a] *A* = $Br(CH_2)_3Br$, NaH, dimethylformamide/20° (reaction time not specified); *N* = NaOEt, EtOH/(reflux)(2 hr); *C* = $Ph_2C{=}C{=}O$ (no cosolvent)/(100°)(1 hr); *D* = $Ph_2C{=}C{=}O$, benzene/(room temp.)(several days); *E* = dimethylformamide/(110°)(2 hr); *F* = 47% HBr/(reflux)(2.5 hr).

[b] Yield not quoted.

[c] Melting point not quoted.

[d] Solvent of crystallization not specified.

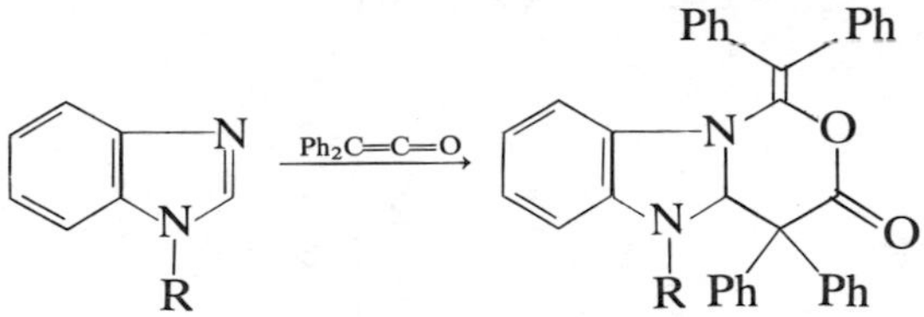

(7.266) $\xrightarrow{Ph_2C{=}C{=}O}$ (7.267)

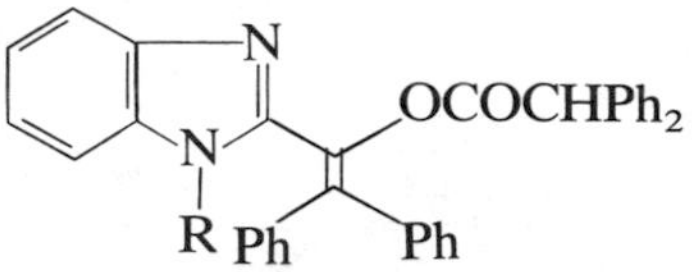

(7.268)

Scheme 7.50

reaction of simple N(1) alkyl and aryl benzimidazoles (**7.266**; R = Me, CH_2Ph, or Ph) with diphenylketene are variously assigned ring-closed oxazinobenzimidazole [e.g. (**7.267**; R = Me)][86] and open-chain benzimidazole (**7.268**)[85] structures. Derivatives of the 3,4-dihydro-1H-[1,4]-oxazino-[4,3-a]benzimidazole ring system are most efficiently and conveniently synthesized by ring-closure of 1- (2-substituted phenyl) morpholine derivatives (see later). However, because of the problem of isomer formation, this approach does not lend itself to the unambiguous synthesis of 3,4-dihydro-1H[1,4]oxazino[4,3-a]benzimidazoles substituted in the morpholine ring. Syntheses, albeit in low yield (Table 7.26) of molecules of the latter type are exemplified (Scheme 7.51) by the thermal reaction of ethyl 2-benzimidazolecarboxylate (**7.269**) with 2-phenyloxirane (**7.270**) to give the oxazinobenzimidazolone (**7.271**)[87] and the acid-catalyzed ring-closure of the benzimidazole derivative (**7.272**) to 1-phenyl-3,4-dihydro-1H-[1,4]oxazino[4,3-a]benzimidazole (**7.273**).[88]

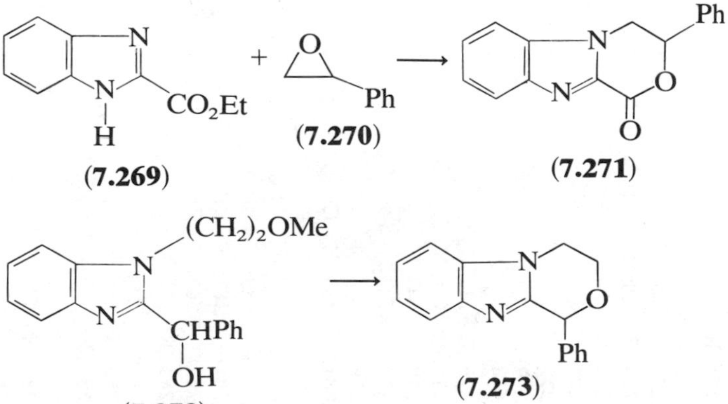

(7.269) + (7.270) → (7.271)

(7.272) → (7.273)

Scheme 7.51

TABLE 7.27. SYNTHESIS OF [1,3]THIAZINO[3,2-*a*]BENZIMIDAZOLES BY RING-CLOSURE REACTIONS OF 2-BENZIMIDAZOLETHIONE DERIVATIVES

Starting materials ($R \rightarrow R^3$ unspecified = H)	Reaction conditions[a]	Product ($R \rightarrow R^3$ unspecified = H)	Yield (%)	m.p. (°C)	Solvent of crystallization	Ref.
(**7.274**; $R^1 = Me$)	*A*	(**7.276**; $R^1 = Me$)	—[b]	160–161 (decomp.)	Ethanol	89
(**7.275**; $R^1 = Ph$)	*B*	(**7.276**; $R^1 = Ph$)	70	148–149	Chloroform	90
(**7.275**; $R^1 = Ph$, $R^2 = Cl$)	*B*	(**7.276**; $R^1 = Ph$, R^2 or $R^3 = Cl$)[c]	61	180–181	Chloroform	90
(**7.275**; $R^1 = Ph$, $R^2 = R^3 = Me$)	*B*	(**7.276**; $R^1 = Ph$, $R^2 = R^3 = Me$)	63	213–214	Chloroform	90
(**7.277**)	*C*	(**7.278**)	23	350 (decomp.)	Ethanol	91
(**7.279**)	*D*	(**7.280**)	48	207–208	Acetic acid	92
(**7.279**)	*E*	(**7.280**)	81	—	—	92
(**7.279**)	*F*	(**7.280**; $R^1 = Me$)	44	167–170	Methanol	93
(**7.279**; $R = Me$)	*F*	(**7.280**; $R = R^1 = Me$)[c]	32	169–172	Methanol	93
(**7.279**; $R = Cl$)	*F*	(**7.280**; $R = Cl$, $R^1 = Me$)[c]	24	156–158	Methanol	93
(**7.279**; $R = NO_2$)	*F*	(**7.280**; $R = NO_2$, $R^1 = Me$)[c]	15	201–203	Ether–methanol	93
(**7.279**; $R = CO_2H$)	*F*	(**7.280**; $R = CO_2H$, $R^1 = Me$)	25	297–300	Acetone	93
(**7.279**)	*G*	(**7.280**; $R^1 = OH$)	100	—[d]	—[e]	94
(**7.281**) + (**7.283**; $X = Cl$)	*H*	(**7.286**)[f]	—[b]	138–139	Benzene–light petroleum (b.p. 60–80°)	95
(**7.281**) + (**7.283**; $X = Br$)	*I*	(**7.286**)[g]	63	147–148	Ethanol	98
(**7.281**) + (**7.283**; $X = Br$)	*J*	(**7.286**)	98	—[d]	—[e]	99
(**7.281**) + (**7.283**; $X = Br$)	K	(**7.286**)	—[b]	—[d]	—[e]	83
(**7.281**; $R^2 = Cl$) + (**7.283**; $X = Cl$)	*I*	(**7.286**; R^1 or $R^2 = Cl$)[c,h]	—[b]	178–179	—[e]	98
(**7.281**; $R^1 = R^2 = Me$) + (**7.283**; $X = Cl$)	*I*	(**7.286**; $R^1 = R^2 = Me$)[i]	75	225–226	—[e]	98
(**7.281**; $R^1 = R^2 = Me$) + (**7.283**; $X = Br$)	*J*	(**7.286**; $R^1 = R^2 = Me$)	95	210–211	—[e]	99
(**7.281**) + (**7.283**; $R^3 = Me$, $X = Cl$)	*J*	(**7.286**; $R^3 = Me$)	99	119–120	—[e]	99

(**7.281**; $R^1 = R^2 = Me$) + (**7.283**; $R^3 = Me$, X = Cl)	*J*	(**7.286**; $R^1 = R^2 = R^3 = Me$)	92	250–251	—[e]	99
(**7.281**) + (**7.284**)	*I*	(**7.285**)[j]	—[b]	214–215	—[e]	98
(**7.281**) + (**7.283**; $R^3 = OH$, X = Cl)	*J*	(**7.285**)	53	214–215	—[e]	99
(**7.281**; $R^1 = R^2 = Me$) + (**7.284**)	*I*	(**7.285**; $R^1 = R^2 = Me$)	—[b]	266–267	Ethanol	98
(**7.281**; $R^1 = R^2 = Me$) + (**7.283**; $R^3 = OH$, X = Cl)	*J*	(**7.285**; $R^1 = R^2 = Me$)	45	268–269	—[e]	99
(**7.288**)	*L*	(**7.289**)	90	126–128	Ethanol–water	100
(**7.288**; R = Me)	*L*	(**7.289**; R = Me)	90	248–250	—[e]	100
(**7.282**)	*M*	(**7.287**)	76	151–152	Ethanol–water	101
(**7.282**)	*N*	(**7.287**)	47	168	—[e]	102
(**7.282**)	*O*	(**7.287**)	72	168	Ethanol	103
(**7.282**; $R^2 = Cl$)	*P*	(**7.287**; $R^2 = Cl$)	67	175	Ethanol	105
(**7.282**; $R^2 = NO_2$)	*P*	(**7.287**; $R^1 = NO_2$)	40	190	Ethanol	106
(**7.282**; $R^1 = R^2 = Me$)	*P*	(**7.287**; $R^1 = R^2 = Me$)	74	173–174	Ethanol	104
(**7.290**)	*Q*	(**7.292**)	92	189–191	—[e]	108

[a] *A* = polyphosphoric acid/(155–160°)(15 min); *B* = $Hg(OAc)_2$, conc. H_2SO_4, AcOH/(reflux)(5 hr); *C* = $MeO_2CC{\equiv}CCO_2Me$, EtOH, H_2/(reflux)(4 hr); *D* = *cis*-ClCH=CHCO$_2$H, EtOH, xylene/(80°)(10 hr); *E* = $HC{\equiv}CCO_2H$, EtOH, xylene/(reflux)(1 hr); *F* = ClC(Me)=CHCO$_2$Et, NaOEt, EtOH/(reflux)-(4 hr); *G* = C_3O_2, tetrahydrofurane, ether/−70° then 20° (reaction time not specified); *H* = 20% KOH, $NaHCO_3$, Pr^iOH/(reflux)(4 hr); *I* = KOH, EtOH/(reflux)(3 hr); *J* = KI, $NaHCO_3$, Pr^iOH (reaction conditions not specified); *K* = NaH, dimethylformamide/20° (reaction conditions not specified); *L* = NaOH, H_2O, EtOH/(reflux)(1 hr); *M* = Ac_2O, pyridine/(reflux)(1 hr); *N* = Ac_2O, pyridine (reaction conditions not specified); *O* = dicyclohexylcarbodiimide, pyridine/(5–10°)(10–12 hr); *P* = Ac_2O, pyridine/(100°)(10–30 min).

[b] Yield not quoted.

[c] *C*(7) or *C*(8) position for the substituent in the benzene ring not established.

[d] Melting point not quoted.

[e] Solvent of crystallization not specified.

[f] Forms a picrate, m.p. 242–4° (decomp.).

[g] Forms a hydrochloride, m.p. 198–199°.

[h] Forms a hydrochloride, m.p. 182–183°.

[i] Forms a hydrochloride, m.p. 218–220°.

[j] Forms a hydrochloride, m.p. 211–212°.

The 2*H*-[1,3]thiazino[3,2-*a*]benzimidazole ring system has been constructed in a single instance, in unspecified yield (Table 7.27), by the polyphosphoric acid catalyzed acylative ring-closure of a 2-(γ-oxoalkylthio)-benzimidazole derivative [Scheme 7.52; (**7.274**; $R^1 = Me$, $R^2 = R^3 = H$) $\rightarrow$ (**7.276**; $R^1 = Me$, $R^2 = R^3 = H$)].[89] Ring-closure of this type cannot be accomplished, as in related cyclizations (see later), by heating with acetic anhydride in the presence of pyridine.[89] More general synthetic access to the

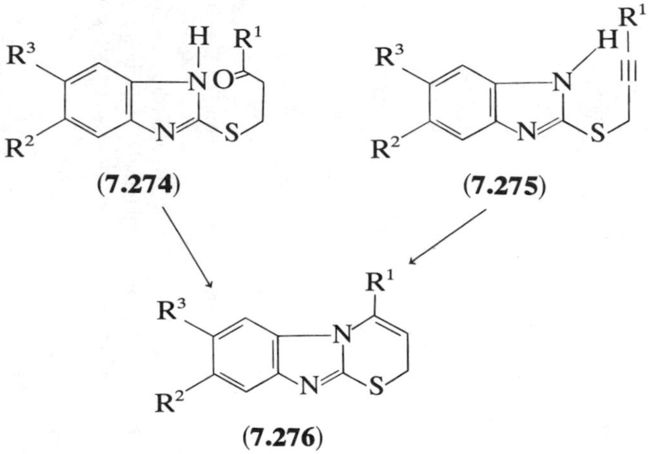

(**7.274**) (**7.275**)

(**7.276**)

Scheme 7.52

2*H*-[1,3]thiazino[3,2-*a*]benzimidazole ring system is provided by the mercuric acetate catalyzed cyclization[90] of 2-(3-phenyl-2-propynylthio)benzimidazoles (**7.275**), which affords 3-phenyl-2*H*-[1,3]thiazino[3,2-*a*]benzimidazoles (**7.276**; $R^1 = Ph$) in good yield (Table 7.27). The uncatalyzed reaction (Scheme 7.53) of 2-benzimidazolylthioacetonitrile (**7.277**) with dimethyl acetylenedicarboxylate affords a low yield (Table 7.27) of a product tentatively formulated[91] as the 4*H*-[1,3]thiazino[3,2-*a*]benzimidazole derivative (**7.278**). More orthodox ring-closure to the 4*H*-[1,3]thiazino[3,2-*a*]-benzimidazole ring system is exemplified (Scheme 7.53) by the uncatalyzed or base-catalyzed condensation of 2-benzimidazolethiones (**7.279**) with 3-chloroacrylic acid,[92] 3-chlorocrotonic acid,[93] or propiolic acid,[92] to give good yields (Table 7.27) of [1,3]thiazino[3,2-*a*]benzimidazol-4-ones (**7.280**). The use of unsymmetrically substituted 2-benzimidazolethiones in such cyclizations is reported[93] to give only one of the two possible thiazinobenzimidazolone products of as yet unestablished orientation. The unstable product obtained in quantitative yield (Table 7.27) by the reaction of 2-benzimidazolethione with carbon suboxide is formulated,[94] without structure proof, as 2-hydroxy-[1,3]thiazino[3,2-*a*]benzimidazol-4-one (**7.280**; $R = H$, $R^1 = OH$).

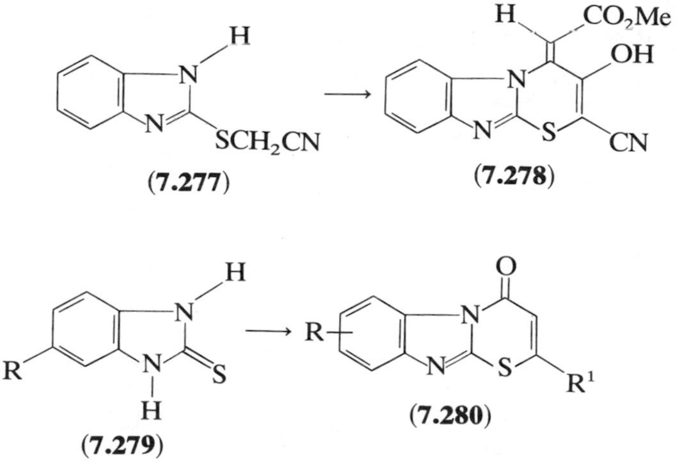

(7.277) (7.278)

(7.279) (7.280)

Scheme 7.53

3,4-Dihydro-2*H*-[1,3]thiazino[3,2-*a*]benzimidazoles are generally accessible in good yield (Table 7.27) by the base-catalyzed condensation of 2-benzimidazolethiones with 1,3-dihalogenopropanes [Scheme 7.54; (**7.281**) + (**7.283**; R^3 = H, Me, or OH, X = Cl or Br) → (**7.286**; R^3 = H, Me, or OH)].[83,95–99] The use of epichlorohydrin as the reagent in annelation reactions of this type provides an alternative method for the synthesis in good yield (Table 7.27) of 3-hydroxy-3,4-dihydro-2*H*-[1,3]thiazino[3,2-*a*]benzimidazoles [Scheme 7.54; (**7.281**) + (**7.284**) → (**7.285**)].[98] The base-catalyzed transformation (Scheme 7.55) of readily accessible 2-(γ-chloropropylthio)-benzimidazole derivatives (**7.288**) in high yield (Table 7.27) into 4-phenyl-3,4-dihydro-2*H*-[1,3]thiazino[3,2-*a*]benzimidazoles (**7.289**) exemplifies a further synthetic approach[100] to the 3,4-dihydro-2*H*-[1,3]thiazino[3,2-*a*] benzimidazole ring system. 2*H*-[1,3]Thiazino[3,2-*a*]benzimidazol-4(3*H*)-ones (**7.287**) are simply prepared (Scheme 7.54), usually in high yield (Table 7.27), by the cyclodehydration of 2-benzimidazolylthioacetic acid derivatives (**7.282**) available by the condensation of 2-benzimidazolethiones with chloroacetic acid.[101–106] Ring formation of this type is readily effected by heating with acetic anhydride[101,102,104–106] or dicyclohexylcarbodiimide[103] in each case in conjunction with pyridine. The cyclization of 2-benzimidazolylthioacetic acids unsymmetrically substituted in the benzene ring can lead to two possible, isomeric, 2*H*-[1,3]thiazino[3,2-*a*]benzimidazol-4(3*H*)-ones, depending on which of the nonequivalent benzimidazole nitrogen atoms is involved in the ring-closure step. In one of the few instances of this situation[106] reported to date, the orientation established for the single isomer obtained is as would be expected, consistent with cyclization via the more basic of the two available nitrogen centers. Other transformations leading to 2*H*-[1,3]thiazino[3,2-*a*]benzimidazol-4(3*H*)-ones include the

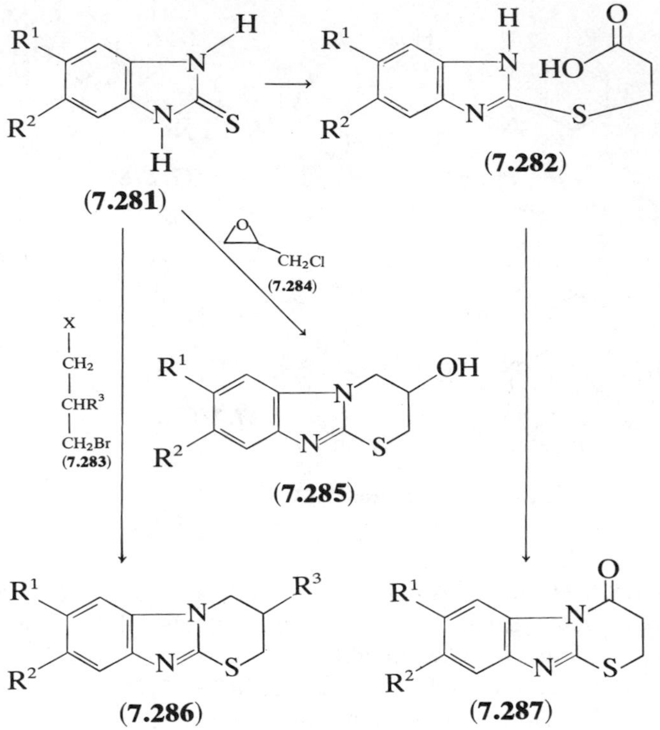

R1 H R1 H O
N N HO
→
R2 N S R2 N S
H (7.282)
(7.281)
O
CH2Cl
(7.284)
X
CH2 R1 OH
CHR3 N
CH2Br R2 N S
(7.283)
(7.285)

O
R1 R3 R1
N N
R2 N S R2 N S
(7.286) (7.287)

Scheme 7.54

reaction of 2-benzimidazolethione with acryloyl chloride to give 2*H*-[1,3]-thiazino[3,2-*a*]benzimidazol-4(3*H*)-one itself,[107] and the ring expansion of a 2,3-dihydrothiazolo[3,2-*a*]benzimidazole derivative rationalized[108] by the ring-opening/ring-closure sequence outlined in Scheme 7.56.

The thermal reaction (Scheme 7.57) of 1-methyl-2-benzimidazolylthioacetonitrile (**7.293**) with dimethyl acetylenedicarboxylate affords two products in low yield (Table 7.28) whose structures have been demonstrated[91] unambiguously by X-ray analysis to be the 10*H*-[1,4]thiazino[4,3-*a*]benzimidazole (**7.294**) and its dihydro derivative (**7.295**). The novel transposition of the sulfur atom and methylenecyano moiety required to account for these

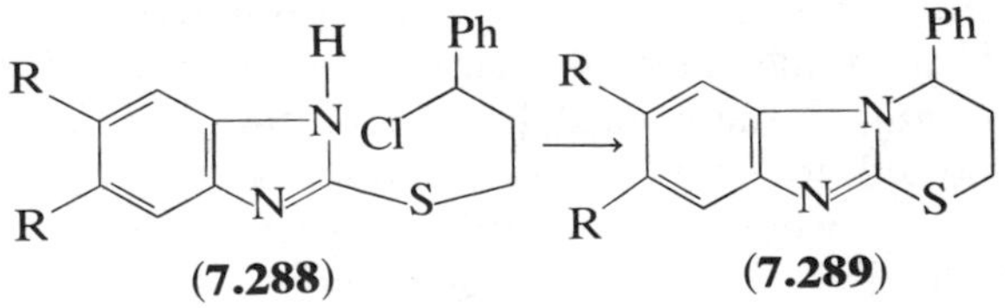

H Ph Ph
R R
N Cl N
→
R N S R N S
(7.288) (7.289)

Scheme 7.55

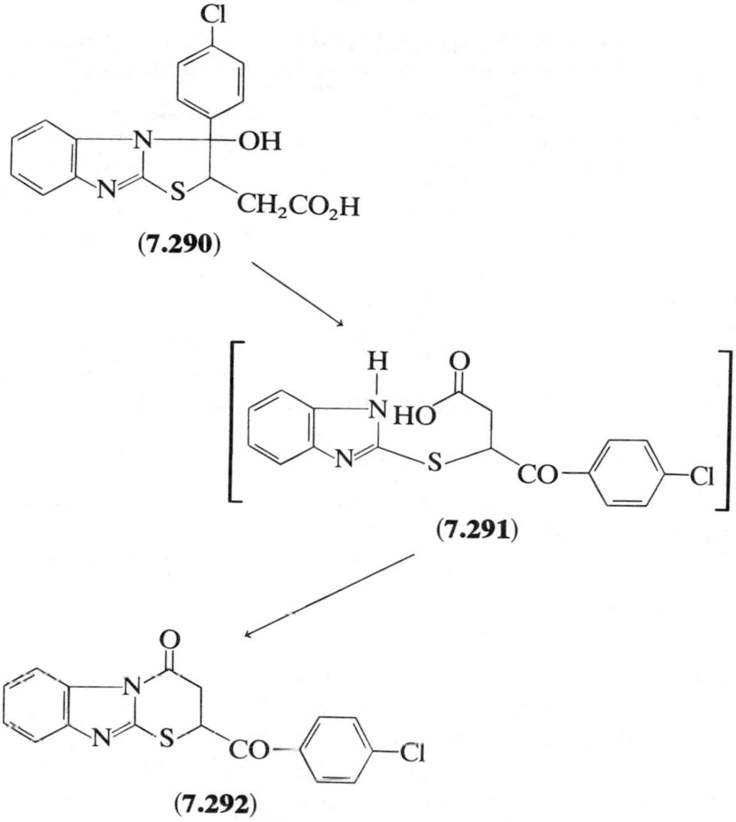

(**7.290**)

(**7.291**)

(**7.292**)

Scheme 7.56

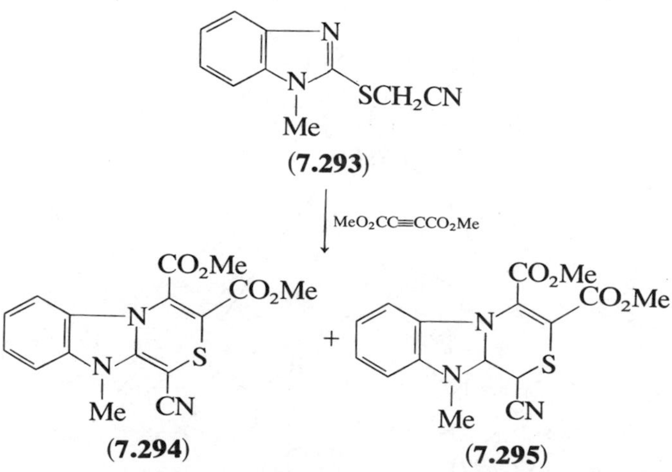

(**7.293**)

$MeO_2CC{\equiv}CCO_2Me$

(**7.294**) + (**7.295**)

Scheme 7.57

TABLE 7.28. SYNTHESIS OF [1,4]THIAZINO[4,3-*a*]BENZIMIDAZOLES BY RING-CLOSURE REACTIONS OF BENZIMIDAZOLE DERIVATIVES

Starting material	Reaction conditions[a]	Product	Yield (%)	m.p. (°C)	Solvent of crystallisation	Ref.
(**7.293**)	*A*	(**7.294**) + (**7.295**)	1 1	199–200 165–167	Ethyl acetate–light petroleum Ether	91
(**7.296**; R = H, R[1] = *p*-tolyl)	*B*	(**7.297**)	34	300	—[b]	109
(**7.296**; R = H, R[1] = OH)	*C*	(**7.298**; R = H)	50	104–105	Light petroleum (b.p. 40–60°)	101
(**7.296**; R = H, R[1] = OH)	*D*	(**7.298**; R = H)	54	104	Ethanol	103
(**7.296**; R = Cl, R[1] = OH)	*E*	(**7.298**; R = Cl)	75	213	—[b]	110

[a] *A* = $MeO_2CC{\equiv}CCO_2Me$, dimethylformamide/(100°)(3 hr); *B* = polyphosphoric acid/(160–170°)(2 hr); *C* = Ac_2O, pyridine/(reflux)(15 min); *D* = dicyclohexylcarbodiimide, pyridine/(5–10°)(10–12 hr).
[b] Solvent of crystallization not specified.

products is suggested[91] to involve a thiirane derivative as the crucial intermediate. The polyphosphoric acid catalyzed cyclization (Scheme 7.58) of the 2-(*β*-oxoalkylthiomethyl)benzimidazole (**7.296**; R = H, R[1] = *p*-tolyl) to the 1*H*-[1,4]thiazino[4,3-*a*]benzimidazole derivative (**7.297**)[109] exemplifies a more obvious, if still relatively inefficient (Table 7.28), method for the construction of a [1,4]thiazino[4,3-*a*]benzimidazole framework. 1*H*-[1,4]-thiazino[4,3-*a*]benzimidazol-4(3*H*)-ones are readily accessible in moderate

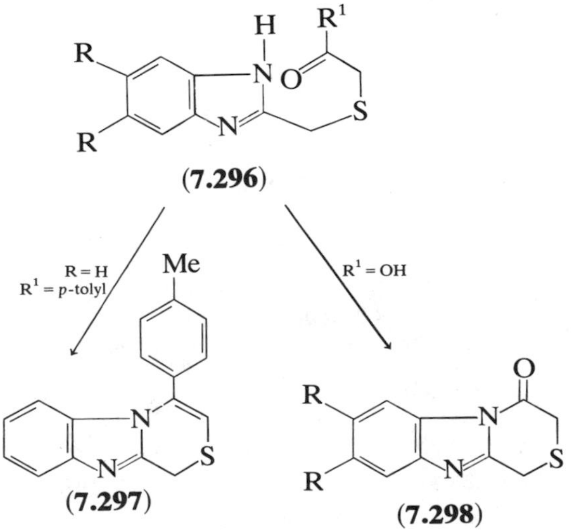

(**7.296**)

R = H
R[1] = *p*-tolyl

R[1] = OH

(**7.297**)

(**7.298**)

Scheme 7.58

to good yield (Table 7.28) by the orthodox cyclodehydration of 2-benzimidazolylmethylthioacetic acids catalyzed by acetic anhydride–pyridine[101,110] or dicyclohexylcarbodiimide–pyridine[103] [Scheme 7.58; (**7.296**; $R^1 = OH$) → (**7.298**)].

Syntheses of the various ring systems containing a pyrimido[1,2-*a*]-benzimidazole framework (cf. Scheme 7.48) are largely based on ring-closure reactions of 2-aminobenzimidazole derivatives. Thus, pyrimido[1,2-*a*]benzimidazole and its simple alkyl and aryl derivatives are directly accessible in moderate yield (Table 7.29) by the thermal condensation of *N*(1)-unsubstituted 2-aminobenzimidazoles with malondialdehyde diethyl acetal, β-keto aldehydes, and β-diketones [Scheme 7.59; (**7.299**; R = H) → (**7.300**; $R^3 = R^4 = H$, alkyl, or aryl)].[111–117] The use of unsymmetrical β-dicarbonyl compounds in such annelation reactions can in theory lead to two possible pyrimido[1,2-*a*]benzimidazole products, though in practice, one isomer is often formed largely in preference to the other. The condensation of 2-aminobenzimidazole with acetoacetaldehyde dimethyl acetal gives a single product[111–113] whose ^1H NMR spectrum [111] is consistent with its being 2-methylpyrimido[1,2-*a*]benzimidazole (**7.300**; $R^1 = R^2 = R^3 = H$, $R^4 = Me$) and not the 4-methyl isomer (**7.300**; $R^1 = R^2 = R^4 = H$, $R^3 = Me$). However, the former structure is open to question in the light of the report[114] that a methylpyrimido[1,2-*a*]benzimidazole having essentially the same m.p. as the acetoacetaldehyde dimethyl acetal derived product is formed in good yield (Table 7.29) in the perchloric acid catalyzed condensation of 2-aminobenzimidazole with methyl β-chlorovinyl ketone, as the perchlorate salt, whose ^1H NMR absorption requires its formulation as the 4-methyl isomer (**7.300**; $R^1 = R^2 = R^4 = H$, $R^3 = Me$). The 2-phenylpyrimido[1,2-*a*]-benzimidazole[115] and 2-methyl-4-phenylpyrimido[1,2-*a*]benzimidazole[117] structures (**7.300**; $R^1 = R^2 = R^3 = H$, $R^4 = Ph$) and (**7.300**; $R^3 = Ph$, $R^4 = Me$) assigned without structure proof to the products of the condensation of 2-aminobenzimidazoles with benzoylacetaldehyde and benzoylacetone, respectively, also require verification. *N*(1)-Alkyl-2-aminobenzimidazole hydrochlorides also condense thermally with β-diketones to give, after treatment with perchloric acid, generally good yields (Table 7.29) of the corresponding 10-alkylpyrimido[1,2-*a*]benzimidazolium perchlorates (**7.301**; R = alkyl).[114] The ^1H NMR absorption[114] of the products so-derived from benzoylacetone are consistent with their formulation[114] as 10-alkyl-2-methyl-4-phenylpyrimido[1,2-*a*]benzimidazolium perchlorates (**7.301**; R = alkyl, $R^3 = Ph$, $R^5 = Me$). This orientation implies that the reaction of an unsymmetrical β-diketone with an *N*(1)-alkyl-2-aminobenzimidazole hydrochloride involves preferential initial condensation of the amino substituent in the latter with the more electrophilic of the carbonyl centers in the diketone. Dibenzoylmethane tends to give only low yields (Table 7.29) in annelation reactions of this type, but can be effectively replaced by 2,4,6-triphenylpyrilium perchlorate to afford high yields (Table 7.29) of 10-alkyl-2,4-diphenylpyrimido[1,2-*a*]benzimidazolium perchlorates [Scheme 7.60;

TABLE 7.29. SYNTHESIS OF ALKYL AND ARYL PYRIMIDO[1,2-*a*]BENZIMIDAZOLES AND PYRIMIDO[1,2-*a*]BENZIMIDAZOLIUM SALTS BY RING-CLOSURE REACTIONS OF 2-AMINOBENZIMIDAZOLE DERIVATIVES

Starting materials ($R \rightarrow R^2$ unspecified = H)	Reaction conditions[a]	Product ($R \rightarrow R^5$ unspecified = H)	Yield (%)	m.p. (°C)	Solvent of crystallization	Ref.
(**7.299**)	A	(**7.300**)	11	197–200	2-Propanol	111
(**7.299**)	*B*	(**7.300**; $R^4 = Me$)	59	233–234	Ethanol	112
(**7.299**)	*C*	(**7.300**; $R^4 = Me$)	94	233	Ethanol	113
(**7.299**)	*D*	(**7.300**; $R^4 = Me$)	24	230–232	Ethanol	111
(**7.299**)[b]	*E*	(**7.300**; $R^3 = Me$)[c]	54	210–211	Ethanol or ethanol–water	114
(**7.299**)	*F*	(**7.300**; $R^4 = Ph$)	61	287–290	Dimethylformamide–water	115
(**7.299**)	*G*	(**7.300**; $R^4 = Ph$)	47	—	—	115
(**7.299**)[b]	*H*	(**7.300**; $R^4 = Ph$)[d]	32	268–269	Ethanol or ethanol–water	114
(**7.299**)	*I*	(**7.300**; $R^3 = R^4 = Me$)	—[d]	230–231	Water	116
(**7.299**)	*J*	(**7.300**; $R^3 = R^4 = Me$)	—[d]	234–235	—[f]	117
(**7.299**)	*K*	(**7.300**; $R^3 = R^4 = Me$)	43	239–241	—[f]	111
(**7.299**; $R^1 = Me$)	*J*	(**7.300**; R^1 or $R^2 = R^3$ or $R^4 = Me$)[g]	—[e]	230–231	—[f]	117
(**7.299**; $R^1 = OMe$)	*J*	(**7.300**; R^1 or $R^2 = OMe$, $R^3 = R^4 = Me$)[g]	—[e]	188–190	—[f]	117
(**7.299**; $R^1 = Cl$)	*J*	(**7.300**; R^1 or $R^2 = Cl$, $R^3 = R^4 = Me$)[g]	—[e]	210–214	—[f]	117
(**7.299**)	*L*	(**7.300**; R^3 or $R^4 = Me$ or Ph)[h]	—[e]	172–173	—[f]	117
(**7.299**; $R^1 = OMe$)	*L*	(**7.300**; R^1 or $R^2 = OMe$, R^3 or $R^4 = Me$ or Ph)[g,h]	—[e]	>245	—[f]	117
(**7.299**; $R^1 = Cl$)	*L*	(**7.300**; R^1 or $R^2 = OMe$, R^3 or $R^4 = Me$ or Ph)[g,h]	—[e]	195–196	—[f]	117
(**7.299**)	*M*	(**7.300**; $R^3 = R^4 = Ph$)	48	312–315	Dimethylformamide–water	115
(**7.299**)	*N*	(**7.300**; $R^3 = R^4 = Ph$)	16	311	—[f]	118
(**7.299**; $R = Me$)[b]	*O*	(**7.301**; $R = R^3 = Me$)	11	230–231	Ethanol or ethanol–water	114
(**7.299**; $R = Me$)[b]	*P*	(**7.301**; $R = R^3 = R^5 = Me$)	42	228–230	Ethanol or ethanol–water	114
(**7.299**; $R = Me$)	*Q*	(**7.301**; $R = R^3 = R^5 = Me$)	88	228–229	Acetic acid	118
(**7.299**; $R = Et$)	*Q*	(**7.301**; $R = Et$, $R^3 = R^5 = Me$)	63	222	Acetic acid	118
(**7.299**; $R = C_9H_{19}$)	*Q*	(**7.301**; $R = C_9H_{19}$, $R^3 = R^5 = Me$)	17	138–139	Methanol	118
(**7.299**; $R = CH_2Ph$)	*Q*	(**7.301**; $R = CH_2Ph$, $R^3 = R^5 = Me$)	48	245	Acetic acid	118
(**7.299**; $R = Ph$)[i]	*P*	(**7.301**; $R = Ph$, $R^3 = R^5 = Me$)[i]	20	164–166	Ethanol or ethanol–water	114

(**7.299**; R = Me)[b]	*R*	(**7.301**; R = R^3 = R^4 = R^5 = Me)	34	181–182	Ethanol or ethanol–water	114
(**7.299**; R = Me)[b]	*S*	(**7.301**; R = R^4 = R^5 = Me, R^3 = Et)	24	154–155	Ethanol or ethanol–water	114
(**7.299**; R = Me)[b]	*I*	(**7.301**; R = R^5 = Me, R^3 = Ph)	50	230–231	Ethanol or ethanol–water	114
(**7.299**; R = Me)	*N*	(**7.301**; R = Me, R^3 = R^5 = Ph)	46	279–281	Acetic acid	118
(**7.299**; R = R^1 = R^2 = Me)	*U*	(**7.301**; R = R^1 = R^2 = Me, R^3 = R^5 = Ph)	13	270	Acetic acid	118
(**7.299**; R = R^1 = R^2 = Me)	*N*	(**7.301**; R = R^1 = R^2 = Me, R^3 = R^5 = Ph)	92	—	—	118
(**7.299**; R = Et)	*U*	(**7.301**; R = Et, R^3 = R^5 = Ph)	6	282	Nitromethane	118
(**7.299**; R = Et)	*V*	(**7.301**; R = Et, R^3 = R^5 = Ph)	13	—	—	118
(**7.299**; R = Et)	*N*	(**7.301**; R = Et, R^3 = R^5 = Ph)	47	—	—	118
(**7.299**; R = Pr^n)	*N*	(**7.301**; R = Pr^n, R^3 = R^5 = Ph)	86	268–270	Acetic acid	118
(**7.299**; R = C_9H_{19})	*N*	(**7.301**; R = C_9H_{19}, R^3 = R^5 = Ph)	84	187	Acetic acid	118
(**7.299**; R = CH_2Ph)	*N*	(**7.301**; R = CH_2Ph, R^3 = R^5 = Ph)	96	294–295	Acetic acid	118
(**7.310**) + (**7.311**)	*W*	(**7.312**)	62	210–212	Ethanol	119

[a] *A* = $(MeO)_2CH_2CH(OMe)_2$, conc. HCl, AcOH/(reflux)(2 hr); *B* = $MeCOCH_2CH(OMe)_2$, xylene/heat (reaction time not specified); *C* = $MeCOCH_2CH(OMe)_2$, xylene/(reflux)/(5 hr); *D* = $MeCOCH_2CH(OMe)_2$ (no cosolvent)/(130–140°)/(3.5 hr); *E* = ClCH=CHCOMe, 42% $HClO_4$, EtOH/(room temp.)(3–4 days); *F* = $PhCOCH_2CHO$, tetrahydrofurane/(140°)(2 hr); *G* = $PhCOCH_2COCO_2H$, xylene/(140°)(3 hr); *H* = PhCOCH=CHCl, 42% $HClO_4$, EtOH/reflux till solution is obtained; *I* = $MeCOCH_2COMe$, H_2O/(reflux)(3 hr); *J* = $MeCOCH_2COMe$/heat (reaction conditions not specified); *K* = $MeCOCH_2COMe$/(100°)(2 hr); then Pr^iOH/(reflux)(18 hr); *L* = $PhCOCH_2COMe$/heat (reaction conditions not specified); *M* = $PhCOCH_2COPh$, xylene/(140°)(5 hr); *N* = [Ph, Ph, Ph, $\overset{+}{O}$] ClO_4^-/dimethylformamide/(reflux)(1 hr); *O* = $MeCOCH_2CHO$ (no cosolvent)/(150–170°)(3 hr), then treat with $HClO_4$; *P* = $MeCOCH_2COMe$ (no cosolvent)/(150–170°)(3 hr), then treat with $HClO_4$; *Q* = $MeCOCH_2COMe$, 70% $HClO_4$, AcOH/(reflux)(1 hr); *R* = MeCOCH(Me)COMe (no cosolvent)/(150–170°)(3 hr), then treat with $HClO_4$; *S* = EtCOCH(Me)COMe (no cosolvent)/(150–170°)(3 hr), then treat with $HClO_4$; *T* = $PhCOCH_2COMe$ (no cosolvent)/(150–170°)(3 hr), then treat with $HClO_4$; *U* = $PhCOCH_2COPh$, 70% $HClO_4$, dimethylformamide/(reflux)(2 hr); *V* = PhCH=CHCOPh, 70% $HClO_4$, dimethylformamide/(reflux)(3 hr); *W* = AcOH/reflux (reaction time not specified).

[b] Perchlorate.

[c] Perchlorate; free base has m.p. 235–237° (from ethanol–water).

[d] Perchlorate; free base has m.p. 174–176°.

[e] Yield not quoted.

[f] Solvent of crystallization not specified.

[g] *C*(7) or *C*(8) position for the benzene substituent not established.

[h] Respective positions of the pyrimidine ring substituents not established.

[i] Picrate.

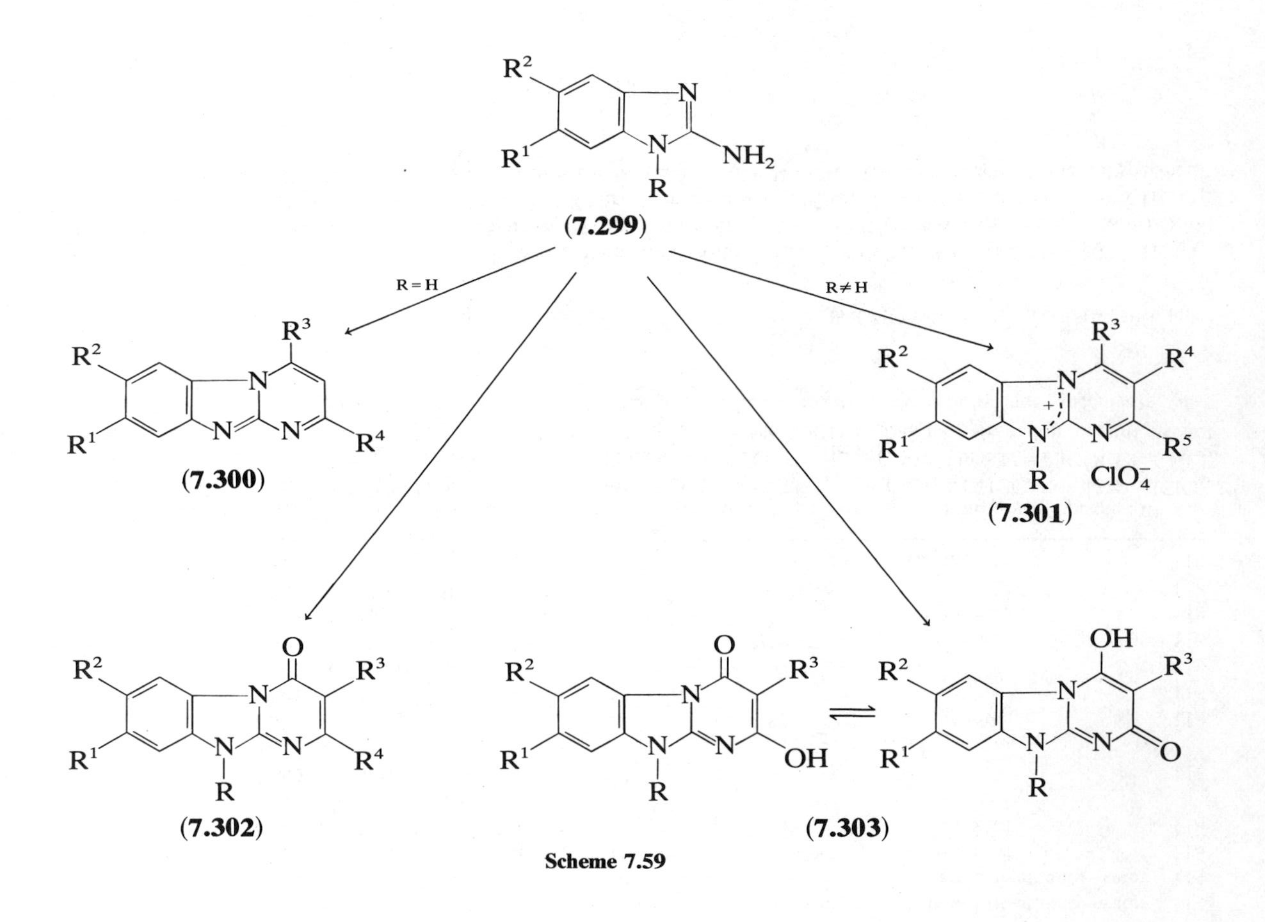

Scheme 7.59

[(**7.304**; R = alkyl) + (**7.305**) → → (**7.309**; R = alkyl)].[118] The mechanism suggested[118] to account for this variant is outlined in Scheme 7.60. Ethyl ethoxymethyleneacetoacetate behaves as a β-keto aldehyde rather than a β-keto ester component in its thermal condensation with 2-aminobenzimidazole in acetic acid solution which affords in good yield (Table 7.29) a product formulated[119] as ethyl 2-methylpyrimido[1,2-*a*]benzimidazole-3-carboxylate. However, this structure probably requires revision to that of ethyl 4-methylpyrimido[1,2-*a*]benzimidazole-3-carboxylate [Scheme 7.61; (**7.312**)] in the light of more recent studies[120] which demonstrate conclusively that the thermal reaction of a 2-aminobenzimidazole with an ethoxymethylene compound occurs by preferential initial condensation between the primary amino and ethoxymethylene substituents. In particular, the product obtained in moderate yield (Table 7.30) by heating 2-aminobenzimidazole with ethyl ethoxymethylenecyanoacetate has been shown[120] to be ethyl 4-aminopyrimido[1,2-*a*]benzimidazole-3-carboxylate (**7.315**; $R^1 = H$, $R^2 = CO_2Et$) and not, as previously reported[119] the 2-amino isomer. It follows that the 2-amino-3-cyanopyrimido[1,2-*a*]benzimidazole structure

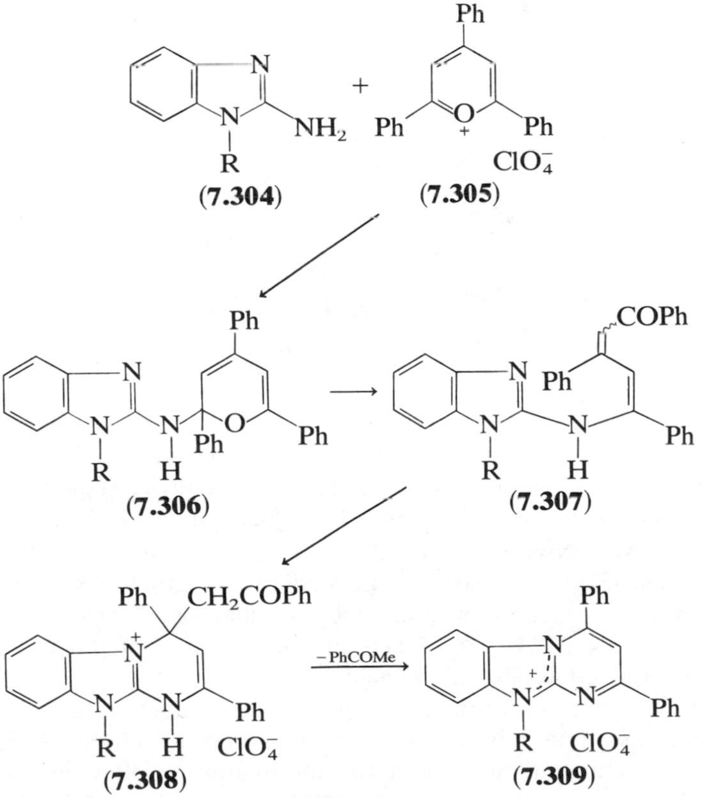

Scheme 7.60

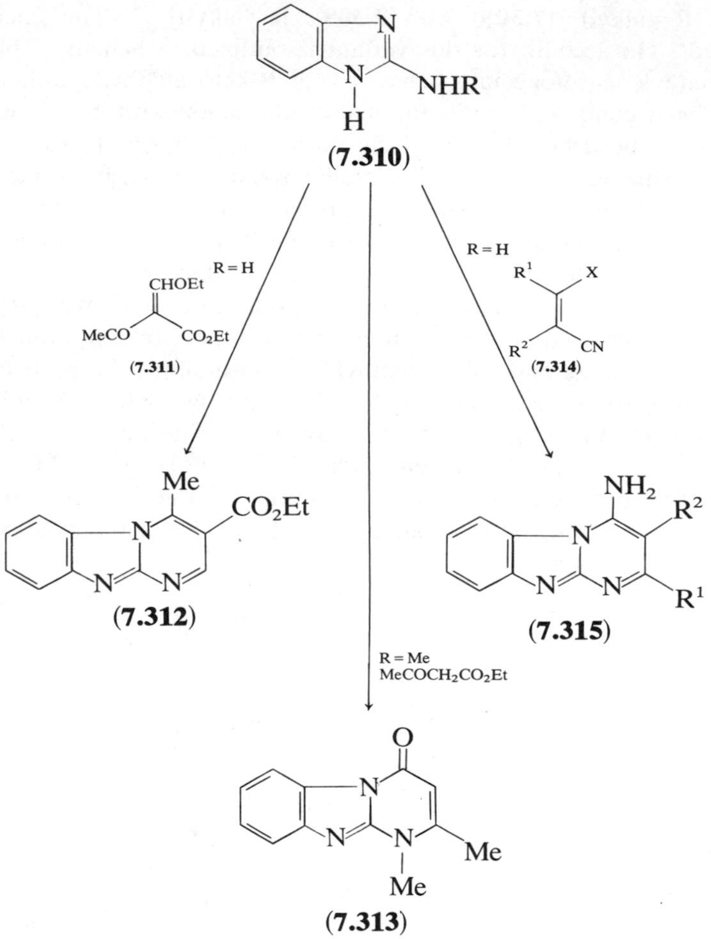

Scheme 7.61

assigned[119] to the product (Table 7.30) of the thermal condensation of 2-aminobenzimidazole with ethoxymethylenemalononitrile should be revised to that of 4-amino-3-cyanopyrimido[1,2-*a*]benzimidazole (**7.315**; $R^1 = H$, $R^2 = CN$). 4-Aminopyrimido[1,2-*a*]benzimidazoles (**7.315**) are also generally accessible (Scheme 7.61)[121,122] in moderate to good yield (Table 7.30) by the thermal reaction of 2-aminobenzimidazole (**7.310**; R = H) with β-cyanoenamines (**7.314**; $X = NH_2$) in the absence of solvent[122] or in solvents such as ethanol or acetic acid.[121]

In a simple extension of the β-diketone condensations already discussed, the reaction (Scheme 7.59) of 2-aminobenzimidazoles (**7.299**) with β-keto esters provides a general method for the synthesis, often in good yield (Table 7.31) of pyrimido[1,2-*a*]benzimidazol-4(10*H*)-ones (**7.301**).[111,120–127]

TABLE 7.30. SYNTHESIS OF 4-AMINOPYRIMIDO[1,2-*a*]BENZIMIDAZOLES BY RING-CLOSURE REACTIONS OF 2-AMINOBENZIMIDAZOLE

Starting materials	Reaction conditions[a]	Product	Yield (%)	m.p. (°C)	Solvent of crystallization	Ref.
(**7.310**) + (**7.314**; R^1 = Me, R^2 = H, X = NH_2)	*A*	(**7.315**; R^1 = Me, R^2 = H)	—[b]	>300	Ethanol–acetic acid	121
(**7.310**) + (**7.314**; R^1 = Ph, R^2 = H, X = NH_2)	*A*	(**7.315**; R^1 = Ph, R^2 = H)	—[b]	>300	Acetic acid–water	121
(**7.310**) + (**7.314**; R^1 = Ph, R^2 = H, X = NH_2)	*B*	(**7.315**; R^1 = Ph, R^2 = H)	50	>330	Ethanol	122
(**7.310**) + (**7.314**; R^1 = R^2 = Me, X = NH_2)	*B*	(**7.315**; R^1 = R^2 = Me)	25	325–330	Ethanol	122
(**7.310**) + (**7.314**; R^1 = H, R^2 = CO_2Et, X = OEt)	*C*	(**7.315**; R^1 = H, R^2 = CO_2Et)	30	248–250	Formic acid–water	119
(**7.310**) + (**7.314**; R^1 = H, R^2 = CO_2Et, X = OEt)	*D*	(**7.315**; R^1 = H, R^2 = CO_2Et)	36	270–272	Dimethylformamide	120
(**7.310**) + (**7.314**; R^1 = H, R^2 = CN, X = OEt)	*C*	(**7.315**; R^1 = H, R^2 = CN)	95	>350	Cyclohexanone	119

[a] *A* = no cosolvent/180° (reaction time not specified); *B* = AcOH or EtOH/reflux (reaction time not specified); *C* = no cosolvent/reflux (reaction time not specified); *D* = no cosolvent/(120°) 1.25 hr).
[b] Yield not quoted.

Pyrimido[1,2-*a*]benzimidazole synthesis of this type is readily accomplished by heating the reactants together in the absence of solvent[111,120,122–125,127] or in ethanol,[111,121,127] acetic acid,[121] or dimethylformamide.[111] In some instances (Table 7.31) condensation catalysts such as sodium ethoxide,[111] toluene-*p*-sulfonic acid,[111] and polyphosphoric acid[126] are employed with advantage. The β-keto ester components in condensation reactions of this type can be replaced (Table 7.31) by β-keto amides[116] or nitriles[128a] and in the specific case of ethyl acetoacetate by ethyl β-aminocrotonate[111,122] or

TABLE 7.31. SYNTHESIS OF PYRIMIDO[1,2-*a*]BENZIMIDAZOL-4(10*H*)-ONES BY RING-CLOSURE REACTIONS OF 2-AMINOBENZIMIDAZOLE DERIVATIVES

Starting materials ($R \rightarrow R^2$ unspecified = H)	Reaction conditions[a]	Product ($R \rightarrow R^4$ unspecified = H)	Yield (%)	m.p. (°C)	Solvent of crystallization	Ref.
(**7.299**)	*A*	(**7.302**; $R^4 = Me$)	—[b]	280	Ethanol	120, 123
(**7.299**)	*B*	(**7.302**; $R^4 = Me$)	—[b]	294 (decomp.)	Ethanol	122
(**7.299**)	*C*	(**7.302**; $R^4 = Me$)	—[b]	280	Ethanol	124
(**7.299**)	*D*	(**7.302**; $R^4 = Me$)	32	292–296	Ethanol	111
(**7.299**)	*E*	(**7.302**; $R^4 = Me$)	20	—	—	111
(**7.299**)	*F*	(**7.302**; $R^4 = Me$)	40	—	—	111
(**7.299**)	*G*	(**7.302**; $R^4 = Me$)	73	—	—	111
(**7.299**)	*H*	(**7.302**; $R^4 = Me$)	80	306–308	Ethanol–water	121
(**7.299**)	*I*	(**7.302**; $R^4 = Me$)	73	287–288	Methanol	127
(**7.299**)	*J*	(**7.302**; $R^4 = Me$)	38	292–296	Ethanol	111
(**7.299**)	*K*	(**7.302**; $R^4 = Me$)	60	—	—	111
(**7.299**)	*L*	(**7.302**; $R^4 = Me$)	—[b]	294 (decomp.)	Ethanol	122
(**7.299**)	*M*	(**7.302**; $R^4 = Me$)	14	292–296	Ethanol	111
(**7.299**)	*N*	(**7.302**; $R^4 = Me$)	60	287–288	Methanol	127
(**7.299**)	*N*	(**7.302**; $R^4 = Me$)	90	298–302 (decomp.)	Dimethylformamide–water	115
(**7.299**; $R^1 = Cl$)	*O*	{(**7.302**); $R^1 = Cl$, $R^4 = Me$) + (**7.302**; $R^2 = Cl$, $R^4 = Me$)}[c]	50	312–316	Ethanol	121
(**7.299**; $R^1 = Cl$)	*D*	{(**7.302**; $R^1 = Cl$, $R^4 = Me$) + (**7.302**; $R^1 = Cl$, $R^4 = Me$)}[c]	21	305–310	Dimethylformamide	111
(**7.299**; $R^1 = R^2 = Me$)	*P*	(**7.302**; $R^1 = R^2 = R^4 = Me$)	36	291–294	Ethanol	111
(**7.310**; $R = Me$)	*Q*	(**7.313**)	97	165–167	Ethyl acetate	120
{(**7.299**; $R = Me$) + (**7.299**; $R = Me$, $R^1 = Br$) + (**7.299**; $R = Me$, $R^2 = Br$)}[d]	*R*	(**7.302**; $R = R^4 = Me$)	21	231.5–232.5	Methanol	125
		(**7.302**; $R = R^4 = Me$, $R^1 = Br$)	7	224–225.5	Methanol	
		(**7.302**; $R = R^4 = Me$, $R^2 = Br$)	4	261.5–263.5	Methanol	

(**7.299**)	*S*	(**7.302**; $R^3 = R^4 = Me$)	60	330	Cyclohexanone	121
(**7.299**)	*T*	(**7.302**; $R^3 = Et$, $R^4 = Me$)	—[b]	284 (decomp.)	Benzene–ethanol	122
(**7.299**)	*T*	(**7.302**; $R^3 = Pr^n$, $R^4 = Me$)	—[b]	253	Benzene–ethanol	122
(**7.299**)	*O*	(**7.302**; $R^3 = CH_2COPh$, $R^4 = Me$)	31	286–288	1-Butanol	121
(**7.299**)	*O*	(**7.302**; $R^3 = CN$, $R^4 = Me$)	80	293–295	1-Butanol	121
(**7.299**)	*U*	(**7.302**; $R^3 = Cl$, $R^4 = Me$)	48	342	—[e]	128a
(**7.299**)	*H*	(**7.302**; $R^3 = Ph$)	87	295–297	Ethanol	121
(**7.299**)	*U*	(**7.302**; $R^4 = CH_2Cl$)	47	300	Ethanol–water	126
(**7.299**)	*D*	(**7.302**; $R^4 = CF_3$)	14	315–318	Ethanol	111
(**7.299**)	*S*	(**7.302**; $R^4 = CH_2CO_2Et$)	43	202–204	Isoamyl alcohol	121
(**7.299**)	*O*	(**7.302**; $R^3 = OEt$, $R^4 = CH_2OEt$)	81	223–225	Ethanol	121
(**7.299**)	*S*	(**7.302**; $R^4 = Ph$)	40	315–317	Ethanol	121
(**7.299**)	*C*	(**7.302**; $R^4 = Ph$)	—[b]	309–311	Ethanol	124
(**7.299**)	*I*	(**7.302**; $R^4 = Ph$)	46	296–297	Methanol	127
(**7.299**)	*V*	(**7.302**; $R^4 = Ph$)	—[b]	309–311	—[e]	116
(**7.299**; $R = Me$)	*R*	(**7.302**; $R = Me$, $R^4 = Ph$)	60	218–219	Chloroform	127
(**7.299**)	*H*	(**7.302**; $R^4 = 2$-furyl)	40	318–320	Cyclohexanone	121
(**7.299**; $N{=}CHNMe_2$ for NH_2)	*W*	(**7.302**; $R^3 = COMe$)	91	276–278	Ethanol	129
(**7.299**)	*H*	(**7.302**; $R^3 = CO_2Et$)	60	270–271	Acetic acid	119
(**7.299**)	*X*	(**7.302**; $R^3 = CO_2Et$)	37	292–294	Acetic acid	120
(**7.299**)	*E*	(**7.302**; $R^3 = CO_2Et$)	51	292–295	Dimethylformamide	127
(**7.299**)	*Y*	(**7.302**; $R^3 = CO_2Et$)	95	>280	Dimethylformamide	130
(**7.299**)	*F*	(**7.302**; $R^3 = CO_2Et$)	95	309–310	Acetic acid	131
(**7.299**; $R^1 = Cl$)	*B*	(**7.302**; R^1 or $R^2 = Cl$, $R^3 = CO_2Et$)[f]	80	>350	Formic acid	119
(**7.299**; $R = Me$)	*Y*	(**7.302**; $R = Me$, $R^3 = CO_2Et$)	97	190	—[e]	130
(**7.299**)	*Z*	(**7.302**; $R^3 = CN$)	14	358–362 (decomp.)	Acetic acid	120

(*Footnotes overleaf*)

(*Footnotes to Table 7.31*)

[a] $A = MeCOCH_2CO_2Et$ (no cosolvent)/(100°)(12 hr); $B = MeCOCH_2CO_2Et$ or $EtOCH{=}C(CO_2Et)_2$ (no cosolvent)/heat (reaction conditions not specified); $C = MeCOCH_2CO_2Et$ (no cosolvent)/(150–170°)(15–20 min); $D = R^4COCH(R^3)CO_2Et$/(130–135°)(1.25 hr); $E = MeCOCH_2CO_2Et$ or $EtOCH{=}C(CO_2Et)_2$, EtOH/(reflux)(5–45 hr); $F = MeCOCH_2CO_2Et$ or $EtOCH{=}C(CO_2Et)_2$, dimethylformamide/(80–100°)(3 hr); $G = MeCOCH_2CO_2Et$, NaOEt, EtOH/(reflux)(18 hr); $H = R^4COCH(R^3)CO_2Et$ or $EtOCH{=}C(CO_2Et)_2$, AcOH/reflux (reaction time not specified); $I = MeCOCH_2CO_2Et$/(reflux)(3 hr); $J = MeCOCH_2CO_2Bu^t$/(135°)(1.5 hr); $K = MeCOCH_2CO_2Bu^t$, toluene-*p*-sulfonic acid, benzene/(reflux)(20 hr); $L = MeCH{=}C(NH_2)CO_2Et$ (no cosolvent)/heat (reaction conditions not specified); $M = MeCH{=}C(NH_2)CO_2ET$ (no cosolvent)/(130°)(0.2 hr); $N =$ O, O, CH_2 , benzene/room temp. (reaction time not specified), or benzene/(reflux)(7 hr); $O = R^4COCH(R^3)CO_2Et$, EtOH/(reflux)(1 hr), then (140–150°)(few min); $P = MeCOCH_2CO_2Et$, dimethylformamide/(100°)(6 hr); $Q = MeCOCH_2CO_2Et$ (no cosolvent)/(160°)(21 hr); $R = R^4COCH_2CO_2Et$ (no cosolvent)/(140–160°)(5 hr); $S = R^4COCH(R^3)CO_2Et$, EtOH/reflux (reaction tine not specified); $T = MeCOCH(R^3)CO_2Et$ (no cosolvent)/130–140° (reaction time not specified); $U = R^4COCH(R^3)CO_2Et$, polyphosphoric acid/(140°)(30–40 min); $V = PhCOCH_2CONH_2$ (no cosolvent)/140° (reaction time not specified); $W =$ O, O, CH_2 , xylene/(room temp.)(1–2 hr), then (140°)(few min); $X = EtOCH{=}C(CO_2Et)_2$ (no cosolvent)/(110°)(0.5 hr); $Y = EtOCH{=}C(CO_2Et)_2$, 1,2,4-trichlorobenzene/reflux (reaction time not specified); $Z = EtOCH{=}C(CN)CO_2Et$ (no cosolvent)/(120°)(1.25 hr).

[b] Yield not quoted.

[c] Isomer mixture not separated.

[d] Mixture.

[e] Solvent of crystallization not specified.

[f] *C*(7) or *C*(8) position for the benzene substituent not established.

diketene.[127] Additionally, the use of α-alkylated[121,122] or arylated[121] or α-diazo[123] acetoacetic esters affords the corresponding *C*(3) functionalized pyrimido[1,2-*a*]benzimidazol-4(10*H*)-ones (Table 7.31). In this context it is noteworthy that the reaction[121] of 2-aminobenzimidazole with ethyl 2-cyanoacetoacetate affords 3-cyano-2-methylpyrimido[1,2-*a*]benzimidazol-4(10*H*)-one (**7.302**; $R^1 = R^2 = H$, $R^3 = CN$, $R^4 = Me$) in good yield (Table 7.31) thus demonstrating the apparent lack of involvement of the cyano substituent. 2-Amino-1-methylbenzimidazole (**7.299**; R = Me, $R^1 = R^2 = H$) and 2-methylaminobenzimidazole (**7.310**; R = Me) condense thermally with β-keto esters to afford high yields (Table 7.31) of the anticipated 1-methyl-pyrimido[1,2-*a*]benzimidazol-4(10*H*)-ones (**7.302**; R = Me)[125,127] and 1-methylpyrimido[1,2-*a*]benzimidazol-4(1*H*)-ones [e.g. (**7.313**)],[120] respectively. The reactions of β-keto esters with *N*(1)-unsubstituted 2-aminobenzimidazoles unsymmetrically substituted in the benzene ring is variously reported to lead to isomer mixtures[111] or single products of unestablished orientation.[121] The *C*(4)-oxo as opposed to alternative *C*(2)-oxo structures assigned in general to the products of the condensation reactions of 2-aminobenzimidazoles with β-keto esters and related substrates have been firmly established by ^{1}H NMR studies.[111,127] The product formed in high yield (Table 7.31) by the reaction[129] of the 2-formamidinobenzimidazole (**7.299**; $R = R^1 = R^2 = H$, N=CHNMe$_2$ for NH$_2$) with diketene has also been assigned a pyrimido[1,2-*a*]benzimidazol-4(10*H*)-one structure (**7.302**; $R = R^1 = R^2 = R^4 = H$, $R^3 = COMe$). 3-Acylpyrimido[1,2-*a*]benzimidazol-4(10*H*)-ones and related compounds of this type [e.g. (**7.302**; $R^3 = CO_2Et$ or CN)] are generally accessible in high yield (Table 7.31) by the more straightforward thermal condensation of 2-aminobenzimidazoles with substrates such as diethyl ethoxymethylenemalonate[119,120,127,130,131] and ethyl ethoxymethylenecyanoacetate[120] in the absence of solvent[120] or in ethanol,[127] acetic acid,[119] dimethylformamide,[131] or trichlorobenzene.[130] The *C*(4)-oxo as opposed to *C*(2)-oxo formulations for the products of these reactions are substantiated by the results of ^{1}H NMR studies.[120,127,130] 2-Aminobenzimidazole derivatives also condense readily with malonic esters on heating the reactants alone or in solvents such as ethanol or acetic acid giving moderate to high yields (Table 7.32) of tautomeric 2-hydroxypyrimido[1,2-*a*]benzimidazol-4(10*H*)-ones [Scheme 7.59; (**7.303**)].[116,121,132] Products, also assigned 2-hydroxypyrimido[1,2-*a*]benzimidazol-4(10*H*)-one structures (**7.303**) are formed in high yield (Table 7.32) by the reaction of 2-aminobenzimidazoles with carbon suboxide.[133]

In contrast to the formation of *C*(4)-oxo products observed with β-keto esters (see before), the thermal condensation (Scheme 7.62) of 2-aminobenzimidazole with ethyl cyanoacetate is reported[124] to afford the *C*(2)-oxo product (**7.321**), albeit in unspecified yield (Table 7.33). The possible intermediacy of the amide (**7.318**), formed by preferential initial condensation between the amino group of 2-aminobenzimidazole and the ester substituent of ethyl cyanoacetate, is indicated by its smooth base-catalyzed

TABLE 7.32. SYNTHESIS OF 2-HYDROXYPYRIMIDO[1,2-*a*]BENZIMIDAZOL-4(1*H*)-ONES BY RING-CLOSURE REACTIONS OF 2-AMINOBENZIMIDAZOLE DERIVATIVES

Starting material ($R \rightarrow R^2$ unspecified = H)	Reaction conditions[a]	Product ($R \rightarrow R^3$ unspecified = H)	Yield (%)	m.p. (°C)	Solvent of crystallization	Ref.
(**7.299**)	*A*	(**7.303**)	53	>340	Formic acid–water	121
(**7.299**)	*B*	(**7.303**)	100	330	Aniline	132
(**7.299**)	*C*	(**7.303**)	92	>310	Ethanol	133
(**7.299**; $R^1 = Me$)	*C*	(**7.303**; R^1 or $R^2 = Me$)[b]	85	283–285	Ethanol	133
(**7.299**; R = Me)	*C*	(**7.303**; R = Me)	88	262–264	Ethanol	133
(**7.299**; R = Et)	*C*	(**7.303**; R = Et)	90	281–283	Ethanol	133
(**7.299**; $R = Pr^n$)	*C*	(**7.303**; $R = Pr^n$)	85	253–255	Ethanol	133
(**7.299**; $R = Bu^n$)	*C*	(**7.303**; $R = Bu^n$)	86	245–246	Ethanol	133
(**7.299**)	*D*	(**7.303**; $R^3 = Et$)	42	284–285	Formic acid	121
(**7.299**)	*E*	(**7.303**; $R^3 = Bu^n$)	—[c]	278	Acetic acid–water	116
(**7.299**)	*F*	(**7.303**; $R^3 = CH_2Ph$)[d]	95	337	Nitrobenzene	132
(**7.299**)	*G*	(**7.303**; $R^3 = Ph$)	—[c]	285	Formic acid–water	116
(**7.299**)	*H*	(**7.303**; $R^3 = CO_2Et$)	80	>330	Ethylene glycol	121

[a] *A* = $CH_2(CO_2Et)_2$, EtOH/(reflux) 1 hr, then 140–150°)(few min); *B* = $(2,4\text{-}Cl_2C_6H_3O_2C)_2CH_2$ (no cosolvent)/(125°)(3 min); *C* = C_3O_2, acetone–benzene/60–70° (reaction time not specified); *D* = $EtO_2CCH(Et)CO_2Et$, EtOH/(reflux)(1 hr), then (140–150°)(few min); *E* = $EtO_2CCH(Bu^n)CO_2Et$ (no cosolvent)/(180°)(30 min); *F* = $(2,4\text{-}Cl_2C_6H_3O_2C)_2CHCH_2Ph$/[260° (melt)](5 min); *G* = $EtO_2CCH(Ph)CO_2Et$ (no cosolvent)/(180°)(30 min); *H* = $EtO_2CCH(CO_2Et)CO_2Et$, EtOH/(reflux)(1 hr), then (140–150°)(few min).

[b] *C*(7) or *C*(8) position for the benzene substituent not established.

[c] Yield not quoted.

[d] Sublimes at 300°.

cyclization[124] to the aminopyrimido[1,2-*a*]benzimidazole (**7.321**). General synthetic access (Scheme 7.62) to pyrimido[1,2-*a*]benzimidazol-2(1*H*)-ones (**7.319**) is provided by the condensation reactions of 2-aminobenzimidazoles with acetylenic esters (e.g., ethyl and methyl propiolate, ethyl phenylpropiolate, and dimethyl acetylenedicarboxylate).[120,127,130,134–136] Pyrimido[1,2-*a*]benzimidazol-2(1*H*)-one formation of this type is simply accomplished in moderate to high yield (Table 7.33) by heating the reactants together in solvents such as acetone,[120] ethanol,[127,130,134] dioxane,[135] or tetrahydrofurane.[136] The products of these reactions are assigned *C*(2)-oxo structures (**7.319**) as opposed to alternative *C*(4)-oxo structures on the basis of their ^{1}H NMR absorption.[120,127,130,135,136] Preferential cyclization to *C*(2)-oxo products is consistent with a mechanism[127] for pyrimido[1,2-*a*]benzimidazol-2(1*H*)-one formation involving Michael addition of the aminobenzimidazole to the acetylenic ester through a ring nitrogen atom (as opposed to via the amino group) followed by ring-closure of the adduct produced.

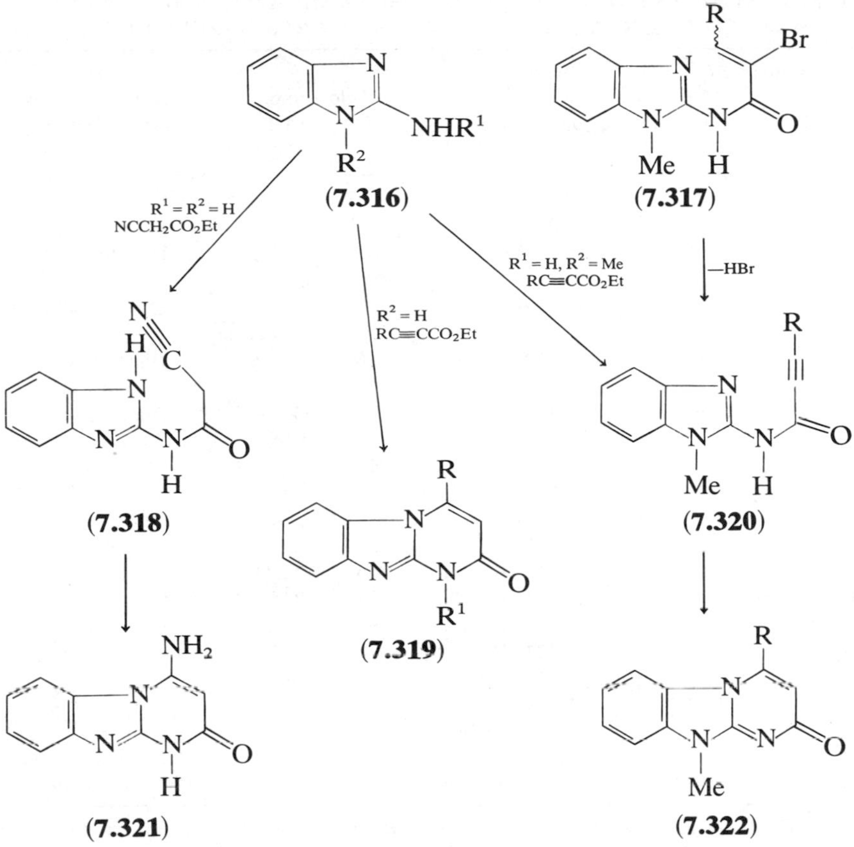

(7.316) (7.317) (7.318) (7.319) (7.320) (7.321) (7.322)

Scheme 7.62

However, the observed mode of ring-closure could also be the result of initial condensation between the amino substituent of the aminobenzimidazole and the ester group of the acetylenic ester, followed by cyclization of the resulting amide. The feasibility of the latter pathway is demonstrated by the reported[136] cyclization (Scheme (7.62) of the acetylenic amide (**7.320**) or its vinyl bromide precursor (**7.317**) to the pyrimido[1,2-*a*]benzimidazol-2(10*H*)-one (**7.322**) in high yield (Table 7.33).

The reaction (Scheme 7.63) of *N*(1)-unsubstituted 2-aminobenzimidazoles (**7.323**) with halogenoalkyloxiranes (**7.324**) to afford admittedly low yields (Table 7.34) of 3-hydroxy-1,2,3,4-tetrahydropyrimido[1,2-*a*]benzimidazoles (**7.325**)[137,138] nonetheless represents probably the most general method available for the synthesis of simple 1,2,3,4-tetrahydropyrimido[1,2-*a*]-benzimidazole derivatives. Condensation reactions of this type are readily accomplished by heating the reactants under reflux in 2-butanone[137] or with sodium hydroxide in ethanol.[138] The orientation established unambiguously

TABLE 7.33. SYNTHESIS OF PYRIMIDO[1,2-*a*]BENZIMIDAZOL-2(1*H*)-ONES BY RING-CLOSURE REACTIONS OF 2-AMINOBENZIMIDAZOLE DERIVATIVES.

Starting material ($R \rightarrow R^2$ unspecified = H)	Reaction conditions[a]	Product ($R \rightarrow R^1$ unspecified = H)	Yield (%)	m.p. (°C)	Solvent of crystallization	Ref.
(**7.316**)	*A*	(**7.321**)	—[b]	303–305	Ethanol	124
(**7.318**)	*B*	(**7.321**)	—[b]	—	—	124
(**7.316**)	*C*	(**7.319**)	42	336–339	Dimethylformamide	120
(**7.316**)	*D*	(**7.319**)	29	338–340 (decomp.)	Acetic acid	134
(**7.316**)	*E*	(**7.319**)	56	>300	Dimethylformamide–water	130
(**7.316**)	*F*	(**7.319**)	69	313–315	Dimethylformamide	127
(**7.316**; R^1 = Me)	*G*	(**7.319**; R^1 = Me)	18	178–180	Ethyl acetate	120
(**7.316**; R^1 = Me)	*E*	(**7.319**; R^1 = Me)	56	178	—[c]	130
(**7.316**; R^2 = Me)	*E*	(**7.322**)	31	274	—[c]	130
(**7.316**)	*H*	(**7.319**; R = Ph)	21	300	Methanol	127
(**7.316**)	*I*	(**7.319**; R = Ph)	46	>300	—[c]	130
(**7.316**; R^2 = Me)	*J*	(**7.322**; R = Ph)	31	235	Chloroform	127
(**7.316**; R^2 = Me)	*K*	(**7.322**; R = Ph)	94	222	—[c]	136
(**7.320**; R = Ph)	*L*	(**7.322**; R = Ph)	83	—	—	136
(**7.317**; R = Ph)	*L*	(**7.322**; R = Ph)	83	—	—	136
(**7.316**)	*M*	(**7.319**; R = CO_2Me)	73	221	Dimethylformamide	127
(**7.316**)	*N*	(**7.319**; R = CO_2Me)	52	229	Ether–dimethylformamide	135

[a] *A* = EtO_2CCH_2CN (no cosolvent)/(140°)(15 min); *B* = NaOH aq./room temp. (reaction time not specified); *C* = $HC{\equiv}CCO_2Et$, acetone/(reflux)(4.5 hr); *D* = $HC{\equiv}CCO_2Me$, EtOH/(reflux)(15 min); *E* = $HC{\equiv}CCO_2Et$, $Me_4\overset{+}{N}OH^-$, EtOH/(reflux)(6 hr); *F* = $HC{\equiv}CCO_2Et$, EtOH/(reflux)(0.5–3 hr); *G* = $HC{\equiv}CCO_2Et$, acetone/(reflux)(20 hr); *H* = $PhC{\equiv}CCO_2Et$, EtOH/(reflux)(0.5–3 hr); *I* = $PhC{\equiv}CCO_2Et$, $Me_4\overset{+}{N}OH^-$, EtOH/(reflux)(6 hr); *J* = $PhC{\equiv}CCO_2Et$, EtOH/(reflux)(3 hr); *K* = $PhC{\equiv}CCOCl$, pyridine, tetrahydrofurane/(−10°)(20 min), then (room temp.)(6 hr); *L* = NaOEt, EtOH/(reflux)(0.5 hr); *M* = $MeO_2CC{\equiv}CCO_2Me$, EtOH/(reflux)(0.5–3 hr); *N* = $MeO_2CC{\equiv}CCO_2Me$, dioxane/(32–35°)(6 hr).

[b] Yield not quoted.

[c] Solvent of crystallization not specified.

for the products of these reactions by spectroscopic means,[137] is consistent with a course for their formation involving initial ring *N*-alkylation followed by epoxide ring-opening in the *N*(1)-epoxyalkylbenzimidazole produced, and subsequent recyclization. Reaction also succeeds with *N*(1)-alkyl-2-aminobenzimidazoles, but here the eventual product (Table 7.34) is a 10-alkyl-2,3,4,10-tetrahydropyrimido[1,2-*a*]benzimidazole [e.g., Scheme 7.63; (**7.323**; $R^1 = Me$, $R^2 = CH_2Ph$) + (**7.324**; $R^2 = R^3 = R^4 = R^5 = H$) → (**7.326**)].[138] The base-catalyzed condensation[120] (Scheme 7.64) of 2-aminobenzimidazole (**7.327**; R = H) with $\alpha\beta$-unsaturated carboxylic acid chlorides

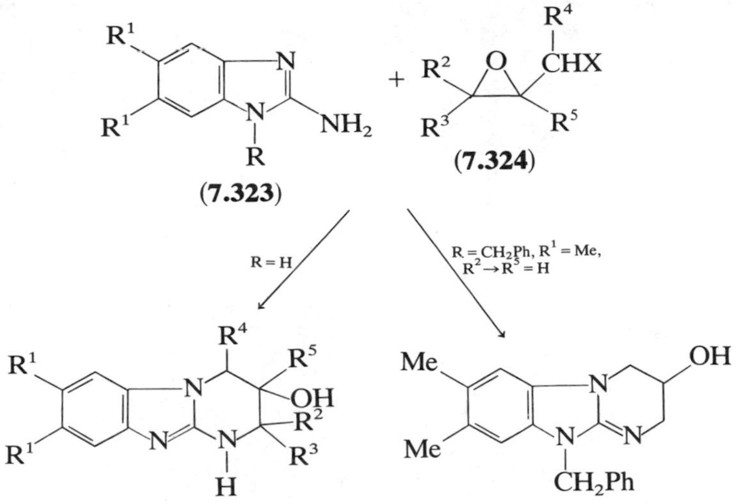

(7.323) (7.324) (7.325) (7.326)

Scheme 7.63

affords low yields (Table 7.34) of products whose ^{1}H NMR absorption is consistent with their formulation[120] as 3,4-dihydropyrimido[1,2-*a*]benzimidazol-2(1*H*)-ones (**7.328**; R = H). Formation of these compounds can be rationalized in terms of the initial acylation of the amino group in the aminobenzimidazole followed by Michael-type ring-closure in the $\alpha\beta$-unsaturated amide intermediate produced. This course is supported by the

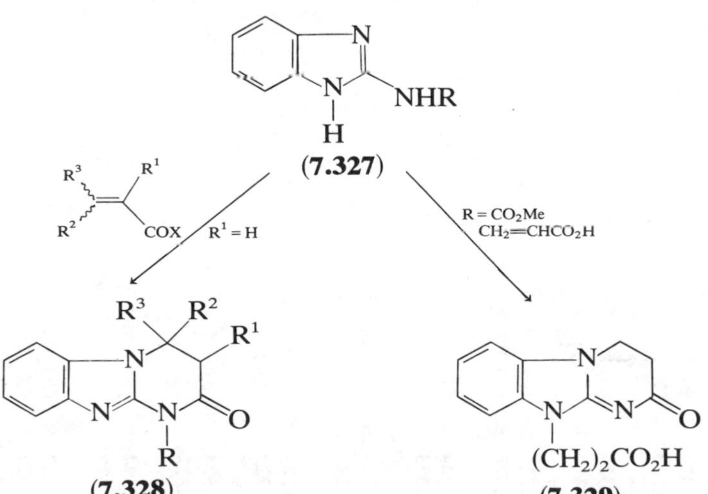

(7.327) (7.328) (7.329)

Scheme 7.64

TABLE 7.34. SYNTHESIS OF TETRAHYDROPYRIMIDO[1,2-*a*]BENZIMIDAZOLES BY RING-CLOSURE REACTIONS OF 2-AMINOBENZIMIDAZOLE DERIVATIVES

Starting materials ($R \rightarrow R^5$ unspecified = H)	Reaction conditions[a]	Product ($R \rightarrow R^5$ unspecified = H)	Yield (%)	m.p. (°C)	Solvent of crystallization	Ref.
(**7.323**) + (**7.324**; X = Br)	*A*	(**7.325**)	17	205 (decomp.)	—[b]	137
(**7.323**) + (**7.324**; X = Cl)	*B*	(**7.325**)[c]	80	158–160	—[d]	138
(**7.323**; R^1 = Me) + (**7.324**; X = Br)	*A*	(**7.325**; R^1 = Me)	31	175 (decomp.)	—[b]	137
(**7.323**; R = CH_2Ph, R^1 = Me) + (**7.324**; X = Cl)	*B*	(**7.326**)[e]	82	258–260	Ethanol–ether	138
(**7.323**) + (**7.324**; R^5 = Me, X = Br)	*A*	(**7.325**; R^5 = Me)	23	230–232	—[b]	137
(**7.323**) + (**7.324**; R^4 = Me, X = Br)	*A*	(**7.325**; R^4 = Me)	7	230 (decomp.)	—[b]	137
(**7.323**; R^1 = Me) + (**7.324**; R^4 = Me, X = Br)	*A*	(**7.325**; $R^1 = R^4$ = Me)	10	260 (decomp.)	—[b]	137
(**7.323**) + (**7.324**; $R^2 = R^3$ = Me, X = Br)	*A*	(**7.325**; $R^2 = R^3$ = Me)	24	240–242	—[b]	137
(**7.327**)	*C*	(**7.328**)	35	261–262	Dimethylformamide	120
(**7.327**; R = CO_2Me)	*D*	(**7.328**)	44	260–261	Ethanol	139
(**7.327**; R = CO_2Me)	*E*	(**7.328**)	18	—	—	139
(**7.327**; R = CO_2Me)	*F*	(**7.328**)	6	—	—	139
(**7.327**)	*G*	(**7.328**; R^1 = Me)	15	264–266	Ethanol–water	120
(**7.327**; R = CO_2Me)	*H*	(**7.328**; R^1 = Me)	82	260–262	Ethanol	139
(**7.327**)	*I*	(**7.328**; R^2 = Me)	29	260–262	Ethanol	120

(**7.327**; R = CO_2Me)	*J*	(**7.328**; R^2 = Me)	97	256–257	Ethanol	139
(**7.327**; R = CO_2Me)	*K*	(**7.328**; $R^2 = R^3$ = Me)	33	244–246	Ethanol	139
(**7.327**; R = CO_2Me)	*L*	(**7.328**; R^2 = Ph)	93	289–290	Ethanol	139
(**7.327**; R = CO_2Me)	*M*	{[**7.328**; R = $(CH_2)_2CO_2H$] + (**7.329**)}	55 25	245–246 216–217	Ethanol Water	139
(**7.330**)	*N*	(**7.331**)	86	170–171	Ethanol	136
(**7.332**)	*O*	(**7.333**)	50	206–208	—[f]	135
(**7.334**; Ar = *p*-$MeOC_6H_4$)	*P*	(**7.335**; R = $COCHPh_2$, $R^1 = R^2$ = Ph, Ar = *p*-$MeOC_6H_4$)	81	223–224	—[d]	141
(**7.334**; Ar = *p*-$O_2NC_6H_4$)	*P*	(**7.335**; R = $COCHPh_2$, $R^1 = R^2$ = Ph, Ar = *p*-$O_2NC_6H_4$)	75	217–218	—[d]	141
(**7.334**; Ar = *p*-$Me_2NC_6H_4$)	*P*	(**7.335**; R = $COCHPh_2$, $R^1 = R^2$ = Ph, Ar = *p*-$Me_2NC_6H_4$)	84	237–239	—[d]	141
(**7.334**; Ar = *p*-$Me_2NC_6H_4$)	*Q*	(**7.335**; R^1 = COMe, Ar = *p*-$Me_2NC_6H_4$)	66	233–235	—[d]	129
(**7.327**)	*R*	(**7.336**; R^1 = Et, $R^2 = Bu^n$)	—[g]	300	—[d]	116
(**7.327**)	*S*	(**7.336**; R^1 = Me, R^2 = Ph)	—[g]	300	—[d]	116
(**7.327**; R = CO_2Me)	*T*	(**7.336**; R = CO_2Me, $R^1 = R^2$ = Me)	46	164–166	Cyclohexane	142

[a] *A* = methyl ethyl ketone/(reflux)(1–4 days); *B* = NaOH, H_2O, EtOH/(reflux)(3 hr); *C* = $CH_2{=}CHCOCl$, Et_3N, MeCN/(room temp.)(1.5 hr); *D* = $CH_2{=}CHCO_2H$, dimethylformamide/(reflux)(6 hr); *E* = $CH_2{=}CHCO_2H$, pyridine/(reflux)(2 hr); *F* = $Cl(CH_2)_2CO_2H$, NaOMe, MeOH, dimethylformamide/(room temp.)(2 hr), then (100°)(2 hr); *G* = $CH_2{=}C(Me)COCl$, Et_3N, tetrahydrofurane/(room temp.)(5 hr); *H* = $CH_2{=}C(Me)CO_2H$ (no cosolvent)/(140–180°)(2 hr); *I* = $MeCH{=}CHCOCl$, Et_3N, acetone, tetrahydrofurane/(room temp.)(6 hr 15 min); *J* = $MeCH{=}CHCO_2H$ (no cosolvent)/(140–180°)(2 hr); *K* = $Me_2C{=}CHCO_2H$ (no cosolvent)/(150–200°)(2 hr); *L* = $PhCH{=}CHCO_2H$ (no cosolvent)/(150–200°)(2 hr); *M* = $CH_2{=}CHCO_2H$ (no cosolvent)/(140–190°)(2 hr); *N* = pyridine, EtOH/(reflux)(6–8 hr); *O* = dicyclohexylcarbodiimide, dioxane/(room temp.)(2 hr), then (50°)(2 hr); *P* = $Ph_2C{=}C{=}O$, xylene/(reflux)(10 hr); *Q* = O O CH_2, benzene/(reflux)(5 hr); *R* = $(EtO_2C)_2C(Bu^n)Et$ (no cosolvent)/(180°)(30 min); *S* = $(EtO_2C)_2C(Ph)Me$ (no cosolvent)/(180°)(30 min); *T* = $(ClCO)_2CMe_2$, K_2CO_3, $CHCl_3$/(room temp.)(72 hr).

[b] Not crystallized.

[c] Forms a hydrochloride, m.p. 214–216° (from ethanol–ether).

[d] Solvent of crystallization not specified.

[e] Hydrochloride.

[f] Purified by chromatography.

[g] Yield not quoted.

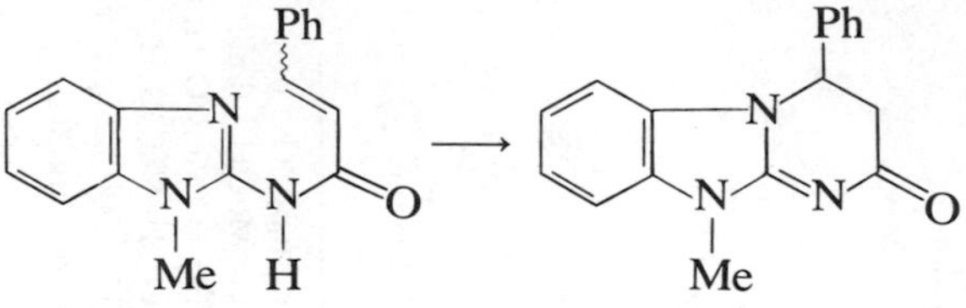

(7.330) → (7.331)

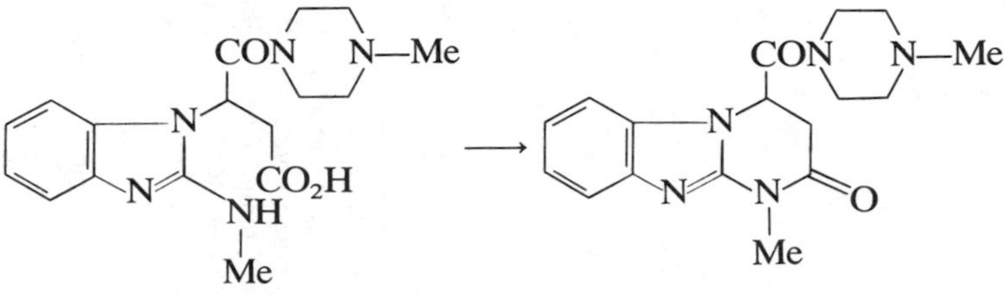

(7.332) → (7.333)

Scheme 7.65

demonstration[136] of the smooth base-catalyzed cyclization (Scheme 7.65) of the $\alpha\beta$-unsaturated amide (**7.330**) to give the 3,4-dihydropyrimido[1,2-*a*]-benzimidazol-2(10*H*)-one (**7.331**) in high yield (Table 7.34). 3,4-Dihydropyrimido[1,2-*a*]benzimidazol-2(1*H*)-ones are also formed in high yield (Table 7.34) by the thermal condensation of 2-methoxycarbonylaminobenzimidazole with $\alpha\beta$-unsaturated carboxylic acids [Scheme 7.64; (**7.327**; R = CO_2Me) → (**7.328**; R = H)].[139] 3,4-Dihydropyrimido[1,2-*a*]benzimidazol-2(1*H*)-one synthesis of this type is in some instances complicated by concomitant addition of the reagent to the NH group in the tautomeric product with consequent formation of a mixture of 3,4-dihydropyrimido-[1,2-*a*]benzimidazol-2(1*H*)-one and 3,4-dihydropyrimido[1,2-*a*]benzimidazol-2(10*H*)-one isomers [e.g., Scheme 7.64; (**7.327**; R = CO_2Me) + $CH_2=CHCO_2H$ → (**7.328**; R = $(CH_2)_2CO_2H$, $R^1 = R^2 = H$) + (**7.329**)].[139] The cyclization [Scheme 7.65; (**7.332**) → (**7.333**)] represents an alternative if synthetically limited (Table 7.34) approach to 3,4-dihydropyrimido[1,2-*a*]-benzimidazol-2(1*H*)-ones.[135] Products, formulated as 2,3-dihydropyrimido-[1,2-*a*]benzimidazol-4(10*H*)-one derivatives, are obtained in high yield (Table 7.34) by the cycloaddition of diketene[129] and diphenylketene[140,141] to 2-arylideneaminobenzimidazoles [Scheme 7.66; (**7.334**) → (**7.335**)]. Conversely, pyrimido[1,2-*a*]benzimidazole-2,4(1*H*,3*H*)-diones are formed in more orthodox fashion (Table 7.34) by the condensation of 2-aminobenzimidazole derivatives with α,α-disubstituted malonic esters[116] or malonyl chlorides[142] [Scheme 7.66; (**7.327**; R = H or CO_2Me) → (**7.336**; R = H or CO_2Me)].

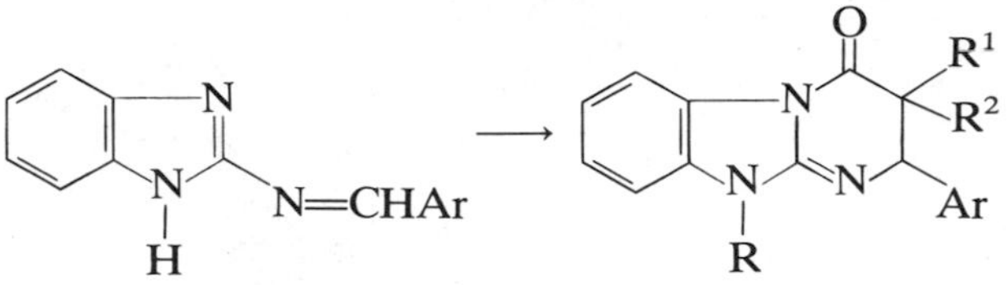

(**7.334**) (**7.335**)

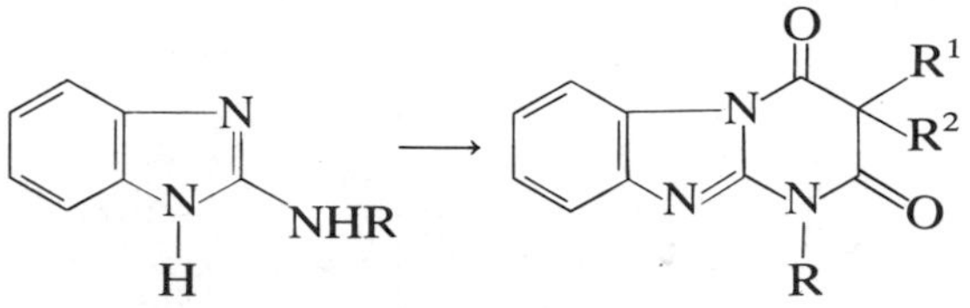

(**7.327**) (**7.336**)

Scheme 7.66

With only one exception (see later) methods for the construction of the elusive pyrimido[3,4-*a*]benzimidazole framework are based on ring-closure reactions of 2-(β-aminoethyl)benzimidazole derivatives. Typical of such methods is the general synthesis (Scheme 7.67) of 1-aryl-1,2,3,4-tetrahydropyrimido[3,4-*a*]benzimidazoles (**7.338**) in high yield (Table 7.35) by the sodium hydroxide catalyzed condensation of 2-(β-aminoethyl)benzimidazole (**7.337**; R = H) with aromatic aldehydes.[143] Open-chain benzylideneamino structures for the products of these reactions can be excluded on the basis of their ^{1}H NMR absorption.[143] Ring-closure of 2-(β-aminoethyl)benzimidazole can also be effected by base-catalyzed reaction with carbon disulfide, the product formed in good yield (Table 7.35) being 3,4-dihydropyrimido[3,4-*a*]benzimidazole-1(2*H*)-thione (**7.339**; X = S).[143] In related

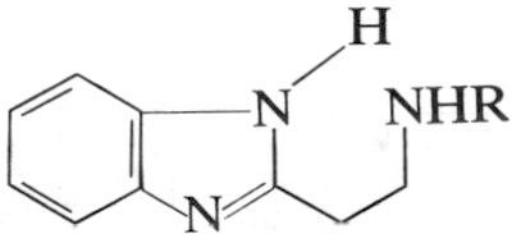

(**7.337**)

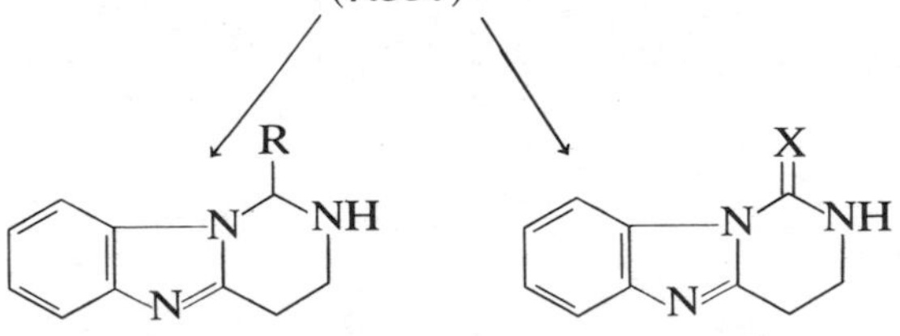

(**7.338**) (**7.339**)

Scheme 7.67

TABLE 7.35. SYNTHESIS OF 1,2,3,4-TETRAHYDROPYRIMIDO[3,4-*a*]BENZIMIDAZOLES BY RING-CLOSURE REACTIONS OF 2-(β-AMINOALKYL)BENZIMIDAZOLE DERIVATIVES

Starting materials	Reaction conditions[a]	Product	Yield (%)	m.p. (°C)	Solvent of crystallization	Ref.
(**7.337**; R = H)[b]	*A*	(**7.338**; R = H)	—[c]	197–199	—[d]	143
(**7.337**; R = H)[b]	*A*	(**7.338**; R = Ph)	92	156–157	Methanol–water	143
(**7.337**; R = H)[b]	*A*	(**7.338**; R = *p*-ClC_6H_4)	—[c]	190–191	—[d]	143
(**7.337**; R = H)[b]	*A*	(**7.338**; R = *p*-BrC_6H_4)	—[c]	221–222	—[d]	143
(**7.337**; R = H)[b]	*A*	(**7.338**; R = *m*-$O_2NC_6H_4$)	—[c]	189–190	—[d]	143
(**7.337**; R = H)[b]	*A*	(**7.338**; R = *p*-$Me_2NC_6H_4$)	—[c]	225–226	—[d]	143
(**7.337**; R = H)[b]	*A*	(**7.338**; R = 2-furyl)	—[c]	184 (decomp.)	—[d]	143
(**7.337**; R = H)[b]	*A*	(**7.338**; R = 2-thienyl)	—[c]	224–225	—[d]	143
(**7.337**; R = H)[b]	*A*	(**7.338**; R = 4-pyridyl)	—[c]	160–161	—[d]	143
(**7.337**; R = CO_2Et)	*A*	(**7.339**; X = O)	39	245–248	Ethanol	144
(**7.337**; R = CO_2Et)[e]	*C*	(**7.339**; X = O)	45	—	—	144
(**7.337**; R = H)[b]	*D*	(**7.339**; X = S)	67	216 (decomp.)	Dimethylformamide	143

[a] *A* = RCHO, 1 *M* NaOH/(80–90°)(15 min); *B* = no cosolvent/(170°)(75 min); *C* = Na_2CO_3, H_2O/(reflux)(few min); *D* = CS_2, NaOH, H_2O, EtOH/(reflux)(20 hr).

[b] Dihydrochloride.

[c] Yield not quoted.

[d] Solvent of crystallization not specified.

[e] Hydrochloride.

transformations (Scheme 7.67) the readily accessible 2-(β-ethoxycarbonylaminoethyl)benzimidazole (**7.337**; $R = CO_2Et$) undergoes[144] thermal or sodium carbonate catalyzed ring-closure to afford, albeit in only moderate yield (Table 7.35) 3,4-dihydropyrimido[3,4-*a*]benzimidazol-1(2*H*)-one. The 3,4,4a,5-tetrahydropyrimido[3,4-*a*]benzimidazole-1(2*H*)-thione [Scheme 7.68; (**7.342**)] is claimed[145] to be the somewhat unexpected end-product (obtained nevertheless in high yield—cf. Scheme 7.68) of the uncatalyzed reaction of *ortho*-phenylenediamine (**7.340**) with the keto isothiocyanate (**7.341**).

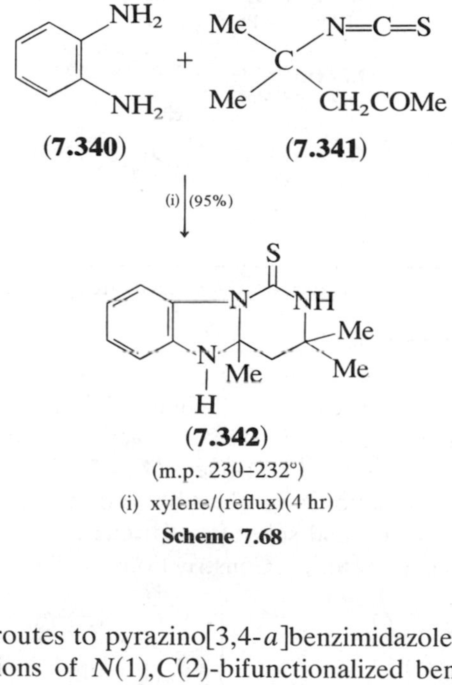

NH_2 NH_2 + Me Me C N=C=S CH_2COMe

(**7.340**) (**7.341**)

(i) (95%)

S N NH Me Me N Me H

(**7.342**)

(m.p. 230–232°)

(i) xylene/(reflux)(4 hr)

Scheme 7.68

Most synthetic routes to pyrazino[3,4-*a*]benzimidazoles are dependent on ring-closure reactions of *N*(1),*C*(2)-bifunctionalized benzimidazole derivatives. For example (Scheme 7.69), derivatives of the fully unsaturated pyrazino[4,3-*a*]benzimidazole ring system (**7.344**) are generally accessible in high yield (Table 7.36) by the ammonium acetate-mediated ring-closure of readily available 1-(2-oxoalkyl)-2-acylbenzimidazoles (**7.343**; $X_2 = Y_2 = O$) or the corresponding acetals (**7.343**; X_2 or $Y_2 = O$, Y or X = OR).[146,147] The

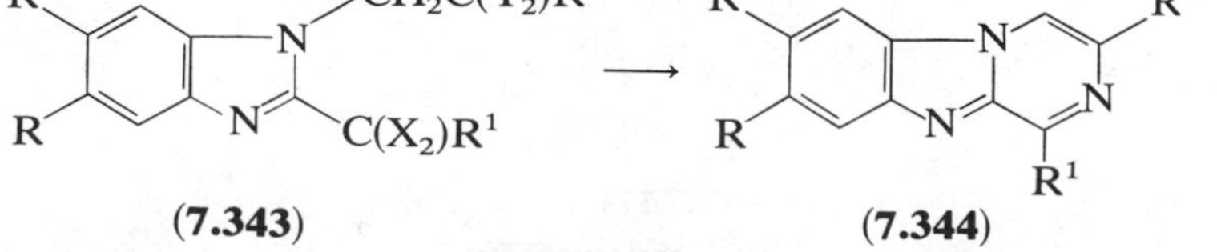

R R N N $CH_2C(Y_2)R^2$ $C(X_2)R^1$ → R R N N N R^2 R^1

(**7.343**) (**7.344**)

Scheme 7.69

TABLE 7.36. SYNTHESIS OF PYRAZINO[4,3-*a*]BENZIMIDAZOLES BY RING-CLOSURE REACTIONS OF BENZIMIDAZOLE DERIVATIVES

Starting material	Reaction conditions[a]	Product	Yield (%)	m.p. (°C)	Solvent of crystallization	Ref.
(**7.343**; $R=R^2=H$, $R^1=Ph$, $X_2=O$, $Y=OBu^n$)	*A*	(**7.344**; $R=R^2=H$, $R^1=Ph)^a$	79	218–219 (decomp.)	—[c]	146
(**7.343**; $R=R^1=H$, $R^2=Ph$, $X=OEt$, $Y_2=O$)	*B*	(**7.344**; $R=R^1=H$, $R^2=Ph$)	91	170–172	Ethanol	147
(**7.343**; $R=H$, $R^1=R^2=Ph$, $X_2=Y_2=O$)	*A*	(**7.344**; $R=H$, $R^1=R^2=Ph$)	97	223–224	Acetone	146
(**7.343**; $R=Me$, $R^1=R^2=Ph$, $X_2=Y_2=O$)	*A*	(**7.344**; $R=Me$, $R^1=R^2=Ph$)	91	227–228	Acetone	146
(**7.343**; $R=H$, $R^1=CO_2Et$, $R^2=Ph$, $X_2=Y_2=O$)	*B*	(**7.344**; $R=H$, $R^1=CO_2Et$, $R^2=Ph$)	91	170–172	Ethanol	147

[a] $A=NH_4OAc$, AcOH/(reflux)(1–2 hr); $B=NH_4OAc$, AcOH/(reflux)(15 min).
[b] Hydrochloride.
[c] Solvent of crystallization not specified.

less obvious thermolytic cyclization (Scheme 7.70) of 2-azidoalkylbenzimidazoles (**7.345**) having an $\alpha\beta$-unsaturated acyl side chain at *N*(1) affords pyrazino[4,3-*a*]benzimidazol-4(10*H*)-ones (**7.347**) in good yield (Table 7.37).[148] These transformations are plausibly rationalized (Scheme 7.70) of the intermediate formation and subsequent breakdown of tetracyclic 1,2,3-triazoline derivatives (**7.346**).[148] Construction of the 1,2,3,4-tetrahydro-

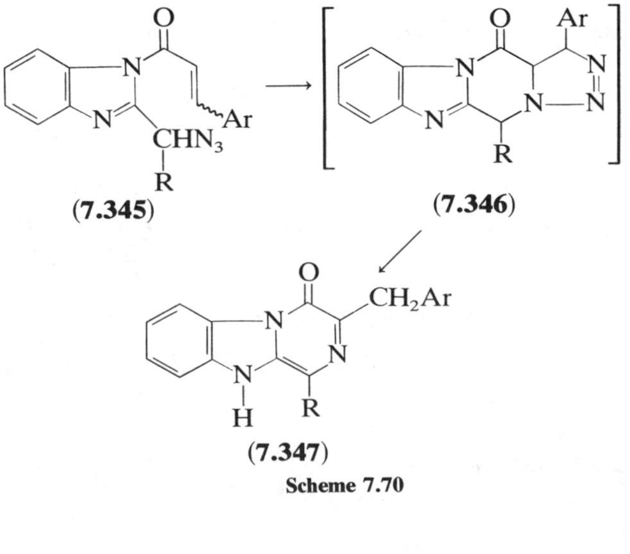

Scheme 7.70

TABLE 7.37. SYNTHESIS OF DIHYDROPYRAZINO[4,3-*a*]BENZIMIDAZOLES AND TETRAHYDROPYRAZINO[4,3-*a*]-BENZIMIDAZOLES BY RING-CLOSURE REACTIONS OF BENZIMIDAZOLE DERIVATIVES

Starting material	Reaction conditions[a]	Product	Yield (%)	(°C) m.p.	Solvent of crystallization	Ref.
(**7.345**; R = H, Ar = Ph)	*A*	(**7.347**; R = H, Ar = Ph)	50	207–209	Ethanol–dimethyl sulfoxide	148
(**7,345**; R = H, Ar = *o*-ClC_6H_4)	*A*	(**7.347**; R = H, Ar = *o*-ClC_6H_4)	49	217–219	Ethanol–dimethyl sulfoxide	148
(**7.345**; R = H, Ar = 1-naphthyl)	*A*	(**7.347**; R = H, Ar = 1-naphthyl)	30	200–202	Ethanol–dimethyl sulfoxide	148
(**7.345**; R = Me, Ar = Ph)	*A*	(**7.347**; R = Me, Ar = Ph)	67	217–218	Ethanol–dimethyl sulfoxide	148
(**7.345**; R = Me, Ar = *o*-ClC_6H_4)	*A*	(**7.347**; R = Me, Ar = *o*-ClC_6H_4)	50	215–216	Ethanol–dimethyl sulfoxide	148
(**7.345**; R = Me, Ar = 1-naphthyl)	*A*	(**7.347**; R = Me, Ar = 1-naphthyl)	65	226–228	Ethanol–dimethyl sulfoxide	148
(**7.348**; R = H)[b]	*B*	(**7.349**; R = H, $R^1 = Bu^n$)	38	92–94	Ether–light petroleum (b.p. 30–60°)	26
(**7.348**; R = Cl)	*B*	(**7.349**; R = Cl, $R^1 = Bu^n$)	27	127–129	Ether	26
(**7.348**; R = NO_2)	*B*	(**7.349**; R = NO_2, $R^1 = Bu^n$)	22	106–109	—[d]	26
(**7.348**; R = CO_2Et)	*B*	(**7.349**; R = CO_2Et, $R^1 = Bu^n$)	37	128–130	—[d]	26
(**7.348**; R = H)	*B*	(**7.349**; R = H, $R^1 = CH_2Ph$)	41	124–125	—[d]	26
(**7.348**; R = H)[b]	*B*	(**7.349**; R = H, $R^1 = (CH_2)_2Ph$)	40	108–111	—[d]	26
(**7.351**; R = H, $R^1 = (CH_2)_2Cl$)	*C*	(**7.349**; R = H, $R^1 = (CH_2)_2Cl$)	28	90–93	Ether–light petroleum	26
(**7.348**; R = H)	*D*	(**7.350**; R = Me, X = Cl)[c]	54	233–235	2-Propanol	149
(**7.348**; R = H)	*E*	(**7.350**; R = CH_2Ph, X = Cl)	70	193–194 (decomp.)	Ethanol–ether	149
(**7.352**; R = CH_2Ph)	*F*	(**7.350**; R = CH_2Ph, X = Cl)	—[d]	—	—	149
(**7.352**; R = CH_2Ph)	*G*	(**7.350**; R = CH_2Ph, X = Br)	65	187–189 (decomp.)	Ethanol–ether	149

TABLE 7.37 (*Continued*)

Starting material	Reaction conditions[a]	Product	Yield (%)	m.p. (°C)	Solvent of crystallization	Ref.
(**7.353**)	*H*	(**7.355**; R = H)	28	292–294	Water	150
(**7.354**; R = H)	*I*	(**7.355**; R = H)	65	—	—	150
(**7.354**; R = CH_2Ph)	*I*	(**7.355**; R = CH_2Ph)	79	200–201	Ethanol	150
(**7.356**; R = Bu^n)	*J*	(**7.357**; R = Bu^n, R^1 = H)	38	188–190	Ethanol–water	151
(**7.356**; R = cyclohexyl)	*J*	(**7.357**; R = cyclohexyl, R^1 = H)	52	265–267	Ethanol	151
(**7.356**; R = CH_2Ph)	*J*	(**7.357**; R = CH_2Ph, R^1 = H)	47	215–216	Ethanol	151
(**7.356**; R = *p*-MeC_6H_4)	*J*	(**7.357**; R = *p*-MeC_6H_4, R^1 = H)	25	278–280	Ethanol	151
(**7.356**; R = cyclohexyl)	*K*	(**7.357**; R = cyclohexyl, R^1 = Ph)	63	297–299	Dimethylformamide	151
(**7.356**; R = cyclohexyl)	*L*	(**7.358**; R = cyclohexyl)	79	168–170	Benzene–light petroleum	151
(**7.356**; R = CH_2Ph)	*L*	(**7.358**; R = CH_2Ph)	61	148–150	Benzene–light petroleum	151
(**7.356**; R = *p*-MeC_6H_4)	*L*	(**7.358**; R = *p*-MeC_6H_4)	85	185–187	Benzene–light petroleum	151
(**7.356**; R = *p*-$MeOC_6H_4$)	*L*	(**7.358**; R = *p*-$MeOC_6H_4$)	82	160	Benzene–light petroleum	151
(**7.359**)	*M*	(**7.360**)	75	305–306	Acetic acid–water	152
(**7.359**)	*N*	(**7.361**)	60	294–296	Dioxane	152

[a] *A* = toluene/(reflux)(1 hr); *B* = RNH_2, benzene/(room temp.)(44 hr), then (100°, pressure)(6 hr); *C* = Bu^nNH_2, benzene/(room temp.)(32 hr), then 100°, (pressure)(12 hr); *D* = Me_2NH, acetone/(reflux)(10 min); *E* = $PhCH_2NHMe$, acetone/(reflux)(10 min), then Pr^iOH/(reflux)(6 hr); *F* = $SOCl_2$, then Pr^iOH/(reflux)(6 hr); *G* = PBr_3, $CHCl_3$/(80–85°)(2 hr), then Pr^iOH/(reflux)(6 hr); *H* = aziridine, HCl, EtOH/(100°, pressure)(8 hr); *I* = $SOCl_2$, dimethylformamide/(0–5°), then (reflux)(2 hr); *J* = $ClCH_2COCl$, Et_3N, tetrahydrofurane/(room temp.)(1 hr), then (reflux)(3 hr); *K* = PhCH(Cl)COCl, Et_3N, tetrahydrofurane/(room temp.)(1 hr), then (reflux)(3 hr); *L* = $(COCl)_2$, Et_3N, tetrahydrofurane/(0°)(1 hr), then (room temp.)(1 hr); *M* = conc. H_2SO_4, AcOH/(65°)(1.5 hr); *N* = conc. NH_3 aq., EtOH/(30°)(2 weeks).

[b] Hydrochloride.

[c] Hygroscopic.

[d] Solvent of crystallization not specified.

[e] Yield not quoted.

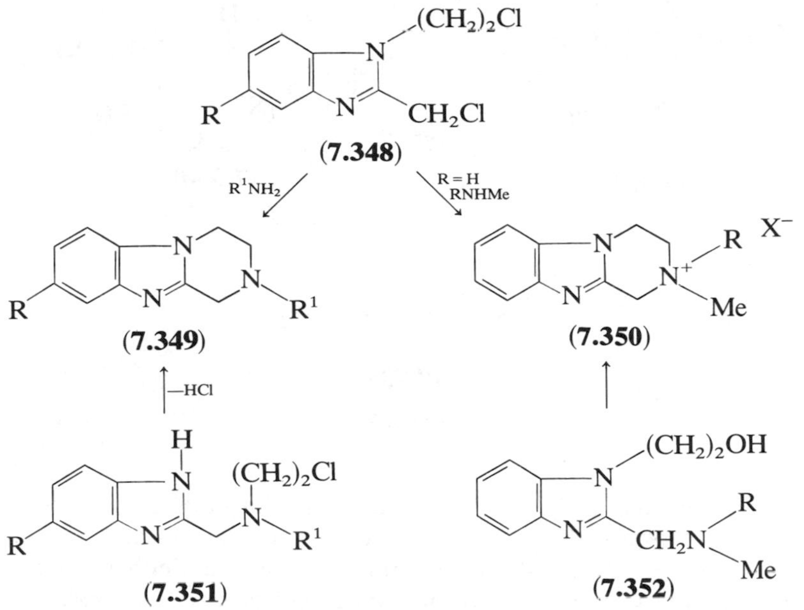

(**7.348**) (**7.349**) (**7.350**) (**7.351**) (**7.352**)

Scheme 7.71

pyrazino[4,3-*a*]benzimidazole ring system is accomplished in more orthodox fashion (Scheme 7.71) by the aminolytic ring-closure of conveniently synthesized 1-(2-chloroethyl)-2-chloromethylbenzimidazoles (**7.348**).[26,149] Reaction of the latter with aliphatic primary amines requires elevated temperature and pressure and affords only low to moderate yields (Table 7.37) of 2-alkyl-1,2,3,4-tetrahydropyrazino[4,3-*a*]benzimidazoles (**7.349**; R = alkyl).[26] Nor is the alternative approach[26] of dehydrohalogenative ring-closure [e.g., Scheme 7.71; (**7.351**; R = H, $R^1 = (CH_2)_2Cl$) → (**7.349**; R = H, $R^1 = (CH_2)_2Cl$)] any more efficient (Table 7.37). On the other hand, the ring-closure (Scheme 7.71) of 1-(2-chloroethyl)-2-chloromethylbenzimidazoles (**7.348**) with secondary aliphatic amines gives good yields (Table 7.37) of the corresponding 2,2-dialkyl-1,2,3,4-tetrahydropyrazino[4,3-*a*]-benzimidazolium chlorides (**7.350**).[149] Products of the latter type are also efficiently synthesized (Table 7.37) by the halogenative–dehydrohalogenative cyclization (Scheme 7.71) of readily accessible 1-(2-hydroxyethyl)-2-aminomethylbenzimidazoles (**7.352**), using reagents such as thionyl chloride or phosphorus tribromide.[149] Heating with thionyl chloride in dimethylformamide also effects the analogous ring-closure (Scheme 7.72) of benzimidazole-2-(*N*-hydroxyethyl)carboxamides (**7.354**) to 3,4-dihydropyrazino[4,3-*a*]benzimidazol-1(2*H*)-ones (**7.355**) in good yield (Table 7.37).[150] The parent 3,4-dihydropyrazino[4,3-*a*]benzimidazol-1(2*H*)-one (**7.355**; R = H) is also conveniently prepared, though only in low yield

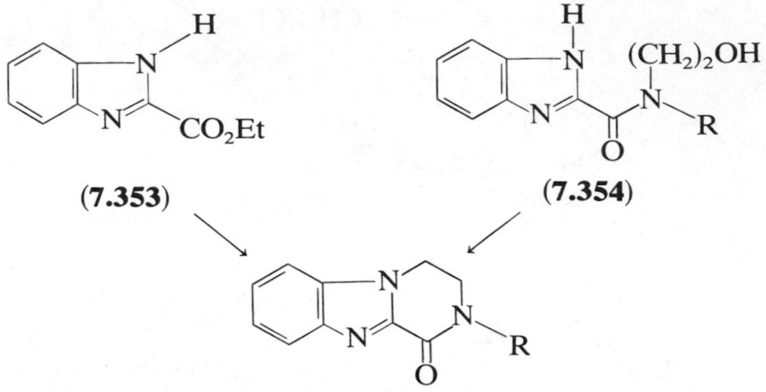

(**7.353**) (**7.354**)

(**7.355**)

Scheme 7.72

(Table 7.37) by the acid-catalyzed condensation of ethyl 2-benzimidazolecarboxylate (**7.353**) with aziridine at elevated temperature and pressure.[150] The triethylamine catalyzed condensation of 2-(alkylaminobenzyl)benzimidazole derivatives with α-chlorocarboxylic acid chlorides provides a convenient general method for the synthesis in moderate to good yield (Table 7.37) of 1,2-dihydropyrazino[4,3-*a*]benzimidazol-3(4*H*)-ones [Scheme 7.73; (**7.356**) → (**7.357**)].[151] The assignment of *C*(3)-oxo as opposed to alternative *C*(4)-oxo structures to these products is based on their IR carbonyl absorption and their relative stability to hydrolysis.[151] 2-(Alkylaminobenzyl)benzimidazoles (**7.356**) also react readily with oxalyl chloride in the presence of triethylamine to good yields (Table 7.37) of pyrazino[4,3-*a*]benzimidazole-3,4(1*H*,2*H*)-diones (**7.358**).[151] Pyrazino[4,3-*a*]benzimidazole-1,3(2*H*,4*H*)-dione, on the other hand, is accessible in high yield

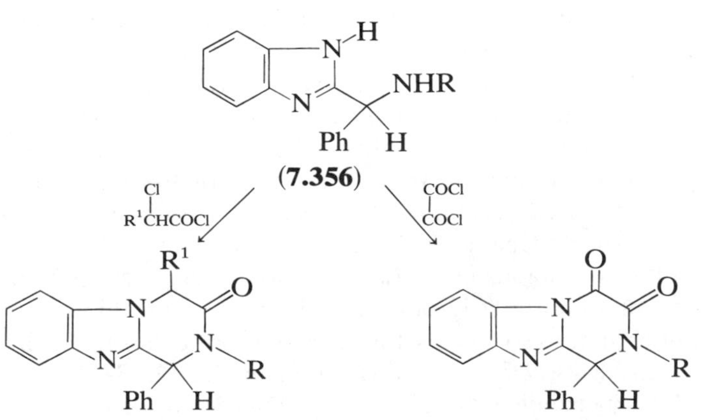

(**7.356**)

(**7.357**) (**7.358**)

Scheme 7.73

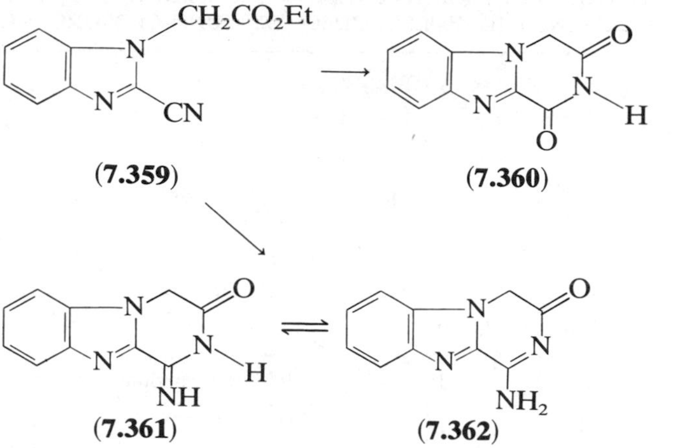

(7.359) (7.360)

(7.361) (7.362)

Scheme 7.74

(Table 7.37) by the simple sulfuric acid catalyzed hydrolytic ring-closure of 1-(ethoxycarbonylmethyl)-2-cyanobenzimidazole [Scheme 7.74; (**7.359**) → (**7.360**)].[152] The tautomeric *C*(1)-imino derivative [(**7.361**) ⇌ (**7.362**)] of the latter compound is likewise obtained in good yield (Table 7.37) by the aminolytic cyclization of the benzimidazole derivative (**7.359**).[152]

Ring-closure Reactions of Other Heterocycles

Ring-closure reactions of *N*-(2-substituted phenyl)morpholine derivatives provide the basis of a variety of methods (Table 7.38) for the general synthesis of 3,4-dihydro-1*H*-[1,4]oxazino[4,3-*a*]benzimidazoles. Transformations of this type are essentially analogous to those already discussed in detail for the synthesis of 1,2,3,4-tetrahydropyrido[1,2-*a*]benzimidazoles (cf. section 7.1.1, "Ring-closure Reactions of Other Heterocycles"), and consequently merit only brief comment under the present heading. Perhaps the simplest example of the ring-closure of an *N*-arylmorpholine derivative to a 3,4-dihydro-1*H*-[1,4]oxazino[4,3-*a*]benzimidazole is provided by the acid-catalyzed cyclodehydration of 1-(2-benzamidophenyl)morpholin-2(1*H*)-one to 3,4-dihydro-1*H*-[1,4]oxazino[4,3-*a*]benzimidazole [Scheme 7.75; (**7.363**) → (**7.364**)].[42] Of greater utility for the general synthesis in high yield (Table 7.38) of 3,4-dihydro-1*H*-[1,4]oxazino[4,3-*a*]benzimidazoles is the oxidative cyclization of 1-(2-aminophenyl)morpholines or their *N*-acyl derivatives, using reagents such as performic acid[46–48,153,154] pertrifluoroacetic acid,[45] or mercuric oxide in conjunction with EDTA.[52] In the sulfuryl chloride promoted cyclization of 1-(2-aminophenyl)morpholines, 3,4-dihydro-1*H*-[1,4]oxazino[4,3-*a*]benzimidazole formation (Table 7.38) is accompanied by perchlorination of the benzene nucleus.[67,154] The thermolysis

TABLE 7.38. SYNTHESIS OF 3,4-DIHYDRO-1*H*-[1,4]OXAZINO[4,3-*a*]BENZIMIDAZOLES [4,3-*a*]BENZIMIDAZOLES (**7.368**) BY RING-CLOSURE REACTIONS OF

(**7.365**) (**7.366**)

Substrate (**7.365**)							
R	R^1	R^2	R^3	R^4	Reaction conditions[a]	Product	R
NH_2	H	H	H	H	*A*	(**7.366**)	H
NHCHO	H	H	H	H	*B*	(**7.366**)	H
NHCOMe	H	H	H	H	*B*	(**7.366**)	H
NHCOPh	H	H	H	H	*B*	(**7.366**)	H
NH_2	H	H	H	H	*C*	(**7.366**)	H
N_3	H	H	H	H	*D*	(**7.366**)	H
NH_2	H	Me	H	H	*A*	(**7.366**)	H
NH_2	H	CF_3	H	H	*E*	(**7.366**)	H
NH_2	H	Cl	H	H	*A*	(**7.366**)	H
N_3	H	Cl	H	H	*D*	(**7.366**)	H
NH_2	H	NO_2	H	H	*A*	(**7.366**)	H
NHCOMe	H	NO_2	H	H	*B*	(**7.366**)	H
NHCOMe	H	H	NO_2	H	*B*	(**7.366**)	H
N_3	H	CN	H	H	*D*	(**7.366**)	H
NHCOMe	H	NHCOMe	H	H	*B*	(**7.366**)	H
NHCOMe	H	H	NHCOMe	H	*B*	(**7.366**)	H
NH_2	NH_2	H	OEt	H	*B*	(**7.366**)	H
N_3	H	Cl	Cl	H	*D*	(**7.366**)	H
NH_2	H	H	H	H	*F*	(**7.366**)	H
NH_2	H	CF_3	H	H	*F*	(**7.366**)	H
NO_2	H	NO_2	H	H	*G*	(**7.366**)	OAc
NO_2	H	H	H	H	*H*	(**7.367**)	H[e]
NO_2	H	H	H	H	*I*	(**7.367**)	H
NO_2	H	Bu^t	H	H	*J*	(**7.367**)	Bu^t
NO_2	H	CF_3	H	H	*K*	(**7.367**)	CF_3[f]
NO_2	H	NO_2	H	H	*H*	(**7.367**)	NO_2[e]
N_3	H	H	H	H	*L*	(**7.368**)	H
N_3	H	NO_2	H	H	*L*	(**7.368**)	NO_2

(**7.366**) AND (**7.367**), AND 1,2,3,4,10,10a-HEXAHYDRO[1,4]OXAZINO-
N-ARYLMORPHOLINE DERIVATIVES (**7.365**)

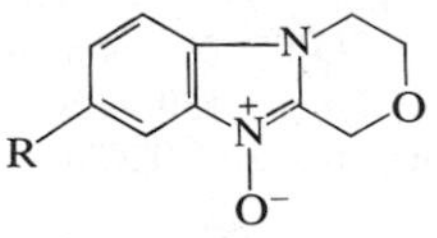

(**7.367**) 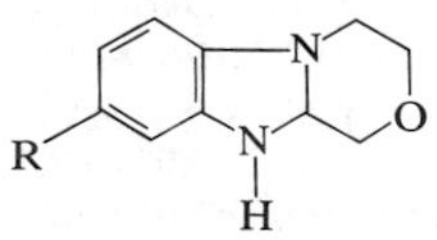 (**7.368**)

R^1	R^2	R^3	R^4	Yield (%)	m.p. (°C)	Solvent of crystallization	Ref.
H	H	H	H	73	129–130	Cyclohexane	45
H	H	H	H	85–95	—[b]	—[c]	46
H	H	H	H	85–95	—[b]	—[c]	46
H	H	H	H	85–95	—[b]	—[c]	46
H	H	H	H	47	129–130	—[c]	52
H	H	H	H	69	130	light petroleum (b.p. 40–60°)	59
H	Me	H	H	61	170–171	Benzene–cyclo hexane	45
H	CF_3	H	H	46	130–132	Chloroform-*n*-hexane	154
H	Cl	H	H	62	196	Ethyl acetate	45
H	Cl	H	H	50	200–200.5	Ethyl acetate	29, 60
H	NO_2	H	H	76	214–215	Methyl ethyl ketone	45
H	NO_2	H	H	80–90	217	—[c]	47
H	H	NO_2	H	80–90	205	—[c]	47
H	CN	H	H	—[d]	186	—[c]	60
H	NHCOMe	H	H	17	210	—[c]	48
H	H	NHCOMe	H	67	258	—[c]	48
NO_2	H	OEt	H	—[d]	—[b]	—[c]	153
H	Cl	Cl	H	—[d]	192	—[c]	60
Cl	Cl	Cl	Cl	50	143	—[c]	67
Cl	CF_3	Cl	Cl	30	156–158	—[c]	154
H	NO_2	H	H	15	196	Ethyl acetate–light petroleum (b.p. 60–80°)	56
—	—	—	—	26–30	201	Ethyl acetate	55
—	—	—	—	55	—	Ethyl acetate	55
—	—	—	—	3	136–138 (decomp.)	Chloroform–hexane	155
—	—	—	—	12	146–148 (decomp.)	Methanol–ether	155
—	—	—	—	13	215–218	Ethyl acetate	55
—	—	—	—	75	125	Ethyl acetate–light petroleum	58
—	—	—	—	75	155	Ethyl acetate–light petroleum	58

(Footnotes overleaf)

(*Footnotes to Table 7.38*)

[a] A = 30% H_2O_2, CF_3CO_2H, CH_2Cl_2/(reflux)(15–30 min); B = 30% H_2O_2, 98% HCO_2H/(100°)-(10–15 min); *C* = HgO, EDTA, 50% ethanol–water/(room temp.)(150 min); *D* = nitrobenzene/(165–175°)(0.5 hr); *E* = 30% H_2O_2, 88% HCO_2H/(60–75°)(15 min); *F* = SO_2Cl_2/0°, room temp. (reaction time not specified); *G* = $ZnCl_2$, Ac_2O/(reflux)(4 hr); *H* = conc. HCl/(110–150°)(12–20 hr); *I* = $h\nu$, conc. HCl, MeOH/(room temp.)(80 hr); *J* = 20% HCl/(110°)(28 hr); *K* = 20% HCl/(reflux)-(20 hr) *L* = diethylene glycol dimethyl ether/(reflux)(10 min).
[b] Melting point not quoted.
[c] Solvent of crystallization not specified.
[d] Yield not quoted.
[e] Hydrochloride.
[f] Monohydrate.

of 1-(2-azidophenyl)morpholines in nitrobenzene[29,59,60] provides an alternative method to oxidative cyclization for the synthesis in high yield (Table 7.38) of 3,4-dihydro-1*H*-[1,4]oxazino[4,3-*a*]benzimidazole derivatives. The thermal cyclization[58] of 1-(2-azidophenyl)morpholines in nonoxidizing solvents such as diethylene glycol dimethyl ether affords the corresponding 3,4,10,10a-tetrahydro-1*H*-[1,4]oxazino[4,3-*a*]benzimidazoles (Table 7.38)

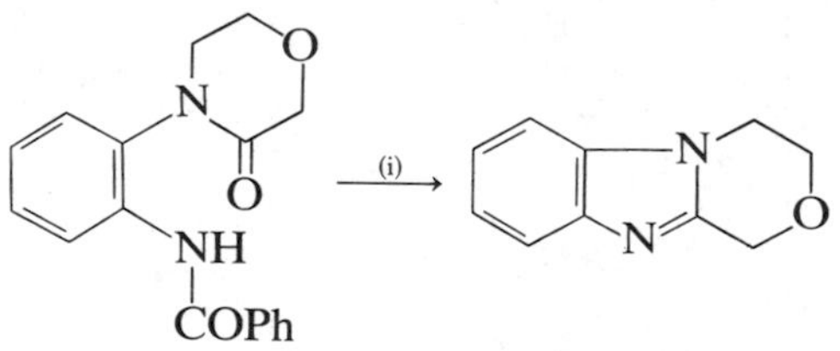

(**7.363**) (**7.364**)

(i) 2 *M* H_2SO_4/(reflux)(2 hr)

Scheme 7.75

thus demonstrating the probable intermediacy of the latter in the nitrobenzene-promoted thermolyses. The acid-catalyzed thermal[55,154,155] and photochemical[55] cyclizations of 1-(2-nitrophenyl)morpholines afford low yields (Table 7.38) of 3,4-dihydro-1*H*-[1,4]oxazino[4,3-*a*]benzimidazole 10-*N*-oxides as opposed to the parent 3,4-dihydro-1*H*-[1,4]oxazino[4,3-*a*]-benzimidazoles. 8-Nitro-3,4-dihydro-1*H*-[1,4]oxazino[4,3-*a*]benzimidazole 10-*N*-oxide is also a plausible intermediate in the zinc chloride–acetic anhydride-mediated cyclization of 1-(2,4-dinitrophenyl)morpholine to 1-acetoxy-8-nitro-3,4-dihydro-1*H*-[1,4]oxazino[4,3-*a*]benzimidazole (Table 7.38).[56] Other isolated examples (Scheme 7.76) of 1-arylmorpholine to 3,4-dihydro-1*H*-[1,4]oxazino[4,3-*a*]benzimidazole cyclizations include the acid-catalyzed conversion of the benzylidene derivative (**7.369**) into the salt

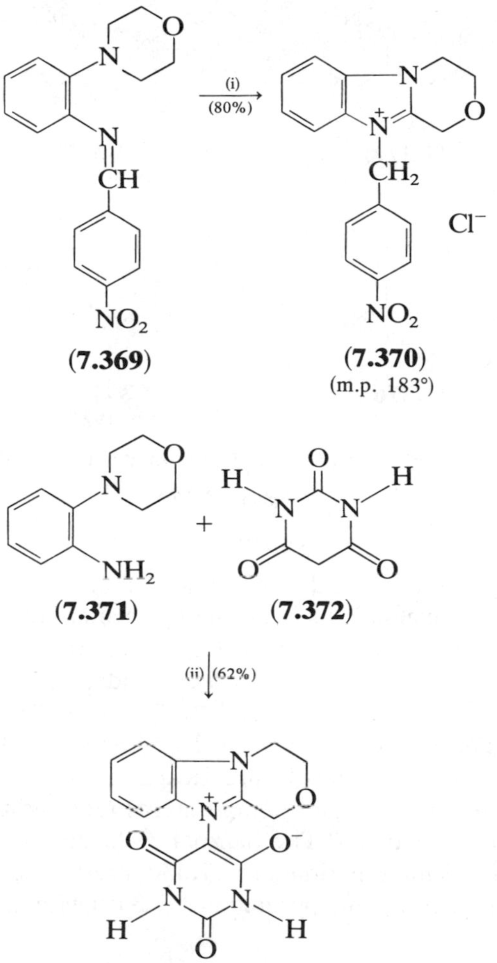

(i) (80%)

(**7.369**)

(**7.370**)
(m.p. 183°)

(**7.371**) + (**7.372**)

(ii) (62%)

(**7.373**)
(m.p. > 360°)

(i) conc HCl, EtOH/(room temp.)(14 hr)
(ii) conc. HCl, EtOH/(room temp.)(3 days)

Scheme 7.76

(**7.370**)[62] and the related acid-catalyzed condensation of 1-(2-aminophenyl)-morpholine (**7.371**) with alloxan (**7.372**) to afford the betaine (**7.373**).[63]

The polyphosphoric acid-catalyzed transformation (Scheme 7.77) of 1-(3-pyridazinyl)benzo-1,2,3-triazole (**7.374**) in good yield into 2-hydroxy-pyridazino[2,3-*a*]benzimidazole (**7.375**) represents what appears to be the only reported synthesis[156] of the pyridazino[2,3-*a*]benzimidazole ring system. The analogous thermolysis[156] of 1-(2-pyrimidyl)benzo-1,2,3-triazole

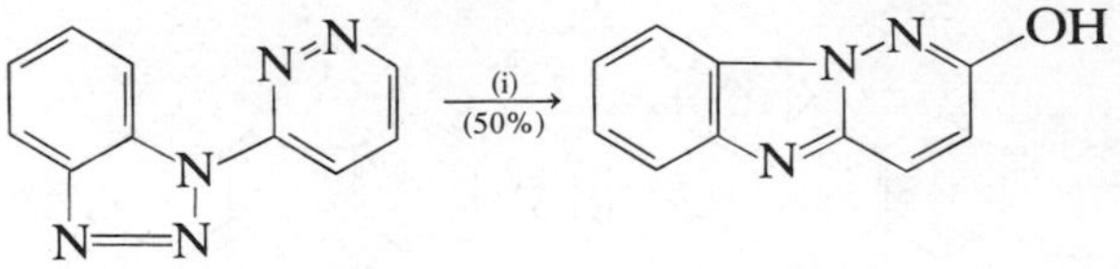

(**7.374**) →(i) (50%)→ (**7.375**) (m.p. 320°)

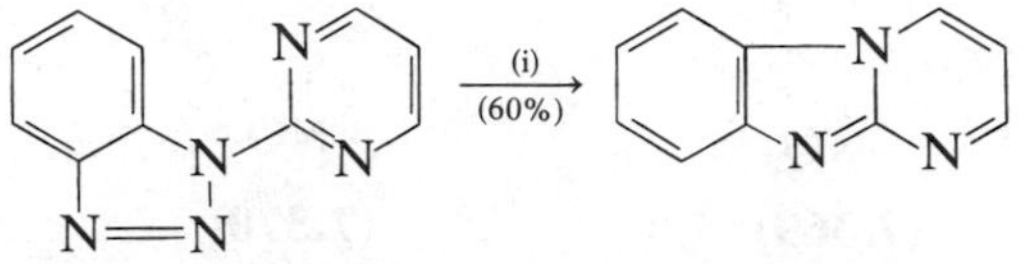

(**7.376**) →(i) (60%)→ (**7.377**) (m.p. 192°)

(i) polyphosphoric acid/150° (time not specified)

Scheme 7.77

(**7.376**) in polyphosphoric acid exemplifies an alternative method to 2-aminobenzimidazole/malondialdehyde diethyl acetal condensation (cf. section 7.2.1, "Ring-closure Reactions of Other Heterocycles") for the synthesis in high yield of pyrimido[1,2-*a*]benzimidazole (**7.377**). The acid-catalyzed nature of the cyclizations [(**7.374**) → (**7.375**)] and [(**7.376**) → (**7.377**)], which are akin to those of 1-(2-pyridyl)benzo-1,2,3-triazoles described in section 7.1.1 ("Ring-closure Reactions of Benzimidazole Derivatives") is indicated by the higher temperatures required to achieve product formation in aprotic media.[156] The thermal cyclization (Scheme 7.78) of 2-(2,4,6-trinitrophenylamino)pyrimidines (**7.380**) preformed or prepared *in situ* by the condensation of 2-aminopyrimidines (**7.379**) with 2,4,6-trinitrochloro-

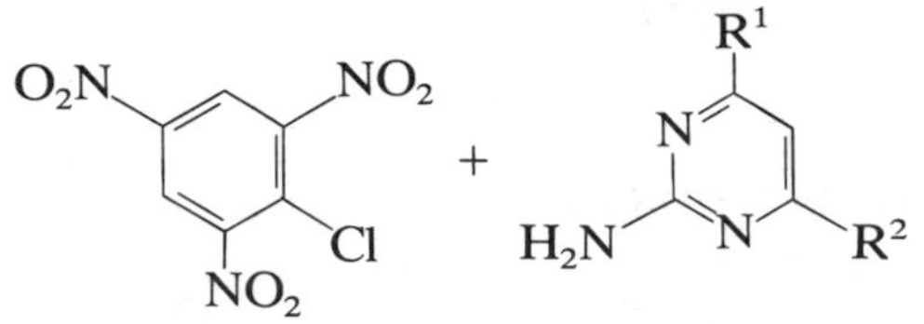

(**7.378**) + (**7.379**) 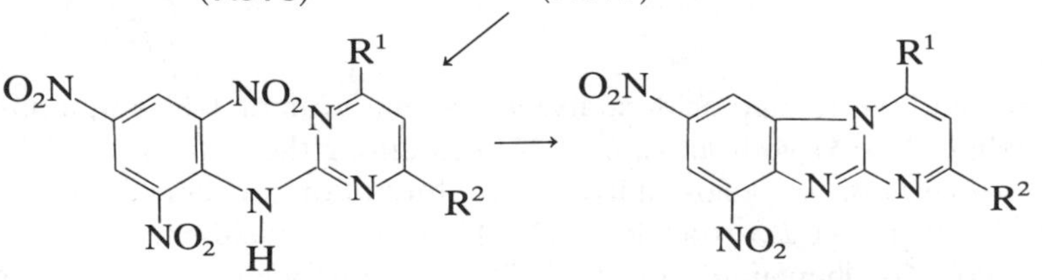 → (**7.380**) → (**7.381**)

Scheme 7.78

TABLE 7.39. SYNTHESIS OF NITROPYRIMIDO[1,2-*a*]BENZIMIDAZOLES BY RING-CLOSURE REACTIONS OF 2-(*N*-2,4,6-TRINITROPHENYLAMINO)PYRIMIDINE DERIVATIVES[a]

Starting material	Reaction conditions[b]	Product	Yield (%)	m.p. (°C)
(**7.374**) + (**7.375**; $R^1 = R^2 = H$)	*A*	(**7.377**; $R^1 = R^2 = H$)	15	196 (decomp.)[c]
(**7.376**; $R^1 = Me$, $R^2 = H$)	*B*	(**7.377**; $R^1 = Me$, $R^2 = H$)	24	>300[c]
(**7.376**; $R^1 = Me$, $R^2 = NH_2$)	*C*	(**7.377**; $R^1 = Me$, $R^2 = NH_2$)	—[d]	>330[c]
(**7.376**; $R^1 = Me$, $R^2 = NHCOMe$)	*C*	(**7.377**; $R^1 = Me$, $R^2 = NHCOMe$)	69	323–325

[a] From Refs. 157 and 158.
[b] *A* = benzene/(100°)(2 hr); *B* = nitrobenzene, phenol/heat (reaction temp. and time not specified); *C* = phenol, benzene/heat (reaction temp. and time not specified).
[c] Solvent of crystallization not specified.
[d] Yield not quoted.

benzene (**7.378**), is reported[157] to afford moderate yields (Table 7.39) of 7,9-dinitropyrimido[1,2-*a*]benzimidazole derivatives (**7.381**). However, the yields claimed for these transformations could not be substantiated.[158] The rearrangement [Scheme 7.79; (**7.382**) → (**7.383**)][159] exemplifies a mechanistically interesting if synthetically limited method for the construction of the 2,3-dihydropyrimido[1,2-*a*]benzimidazol-4(1*H*)-one ring system. The reductive cyclization (Scheme 7.79) of the pyrimidine derivative (**7.384**)

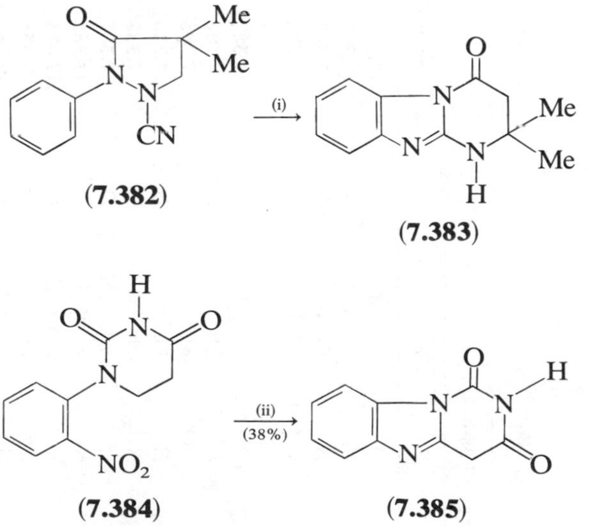

(**7.382**) —(i)→ (**7.383**)

(**7.384**) —(ii) (38%)→ (**7.385**)

(i) 270° (no co-solvent)/few min
(ii) H_2, 10% PdC, AcOH/(room temp.)(atm. press)

Scheme 7.79

TABLE 7.40. SYNTHESIS OF 1,2,3,4-TETRAHYDROPYRAZINO[4,3-*a*]BENZIMIDAZOLES (**7.387**) AND (**7.388**), AND 1,2,3,4,10,10a-HEXAHYDROPYRAZINO[4,3-*a*]BENZIMIDAZOLES (**7.389**) BY RING-CLOSURE REACTIONS OF *N*-ARYLPIPERAZINE DERIVATIVES (**7.386**)

(**7.386**) (**7.387**) (**7.388**) (**7.389**)

Substrate (**7.386**)												
R	R^1	R^2	Reaction conditions[a]	Product	R^1	R^2	Yield (%)	m.p. (°C)	Solvent of crystallization	m.p. (°C) (picrate)	Ref.
N_3	H	CO_2Et	*A*	(**7.387**)	H	CO_2Et	6	126–127	Ethyl acetate	—	29
N_3	Cl	CO_2Et	*A*	(**7.387**)	Cl	CO_2Et	29	129–131	Ethyl acetate	—	29
N_3	H	COMe	*B*	(**7.387**)	H	COMe	24	140–141	Acetone–ether	—	149
NO_2	CO_2Me	Me	*C*	(**7.387**)	CO_2Me	Me	20	158–160	Ethyl acetate–hexane	—	160
NO_2	CO_2Me	CH_2Ph	*C*	(**7.387**)	CO_2Me	CH_2Ph	—[b]	149–150	—[c]	—	160
NO_2	H	Me	*D*	(**7.388**)	H	Me	—[d]	90	—[e]	145	161
NO_2	NO_2	Me	*D*	(**7.388**)	NO_2	Me	—[d]	103	—[e]	205	161
NO_2	H	Ph	*D*	(**7.388**)	H	Ph	—[d]	70	—[e]	168	161
NO_2	NO_2	Ph	*D*	(**7.388**)	NO_2	Ph	—[d]	70	—[e]	95	161
NO_2	H	p-ClC_6H_4	*D*	(**7.388**)	H	p-ClC_6H_4	—[d]	105	—[e]	157	161
NO_2	NO_2	p-ClC_6H_4	*D*	(**7.388**)	NO_2	p-ClC_6H_4	—[d]	117	—[e]	110	161
NO_2	H	o-MeC_6H_4	*D*	(**7.388**)	H	o-MeC_6H_4	—[d]	88	—[e]	105	161
NO_2	NO_2	o-MeC_6H_4	*D*	(**7.388**)	NO_2	o-MeC_6H_4	—[d]	142	—[e]	153	161
NO_2	H	m-MeC_6H_4	*D*	(**7.388**)	H	m-MeC_6H_4	—[d]	130	—[e]	175	161
NO_2	NO_2	m-MeC_6H_4	*D*	(**7.388**)	NO_2	m-MeC_6H_4	—[d]	100	—[e]	140	161
N_3	NO_2	Me	*E*	(**7.389**)	—	—	75	175	Ethyl acetate–light petroleum	—	58

[a] *A* = nitrobenzene/170–180° (reaction time not specified); *B* = nitrobenzene/(180–183°)(10 min); *C* = $(EtO)_3P$/(180°)(2 hr); *D* = 20% HCl/(reflux)(25 hr); *E* = diethylene glycol dimethyl ether/(reflux)(12 min).

[b] Yield not quoted.

[c] Purified by chromatography.

[d] Yields 15–30%.

[e] Crystallized from ethyl acetate-chloroform.

provides a simple, though relatively inefficient means for the synthesis of pyrimido[3,4-*a*]benzimidazole-1,3(2*H*,4*H*)-dione (**7.385**).[144]

1,2,3,4-Tetrahydropyrazino[4,3-*a*]benzimidazoles, like their 3,4-dihydro-1*H*-[1,4]oxazino[4,3-*a*]benzimidazole and tetrahydropyrido[1,2-*a*]benzimidazole counterparts (cf. this section and section 7.1.1, "Ring-closure Reactions of Other Heterocycles"), are synthetically accessible albeit in low yield (Table 7.40) by the thermolytic cyclization of 1-(2-azidophenyl)piperazines in nitrobenzene.[29,149] In accord with their intermediacy in these transformations, 1,2,3,4,10,10a-hexahydropyrazino[4,3-*a*]benzimidazoles are the products (Table 7.40) of the thermolytic decomposition of 1-(2-azidophenyl)-piperazines in nonoxidizing solvents such as diethylene glycol dimethyl ether.[58] 1,2,3,4-Tetrahydropyrazino[4,3-*a*]benzimidazole derivatives are also formed in low yield (Table 7.40) by the triethyl phosphite catalyzed cyclization of 1-(2-nitrophenyl)piperazines.[160] Cyclization of the latter in hot concentrated hydrochloric acid, on the other hand, leads to the formation, again in low yield (Table 7.40) of 1,2,3,4-tetrahydropyrazino[4,3-*a*]benzimidazole 10-*N*-oxides.[161] The transformations outlined in Scheme 7.80 exemplify less orthodox routes to tetrahydropyrazino[4,3-*a*]benzimidazole derivatives.[162]

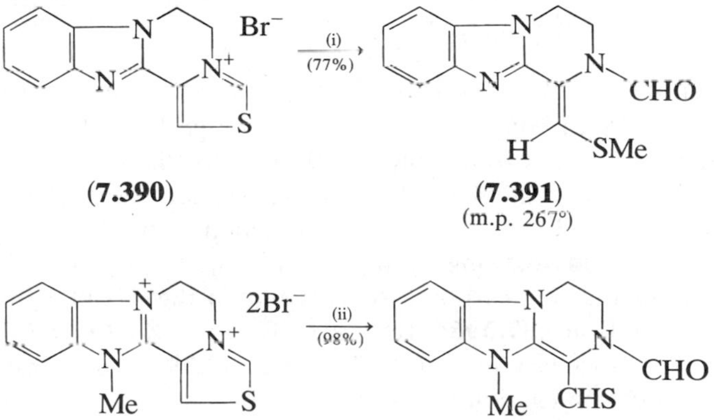

Br^- (i) (77%)

(**7.390**) (**7.391**) (m.p. 267°)

$2Br^-$ (ii) (98%)

(**7.392**) (**7.393**) (m.p. 267–269°)

(i) MeI, 2 *M* NaOH/(room temp.)(2 hr)
(ii) 2 *M* NaOH/(room temp)(few min)

Scheme 7.80

7.2.2. Physicochemical Properties

Spectroscopic Studies

INFRARED SPECTRA. In accord with their lactone-like structures, the 3,4-dihydro-1*H*-[1,4]oxazino[4,3-*a*]benzimidazole derivative (**7.394**) and the

TABLE 7.41. INFRARED SPECTRA OF OXAZINOBENZIMIDAZOLONE DERIVATIVES (**7.394**) AND (**7.395**)

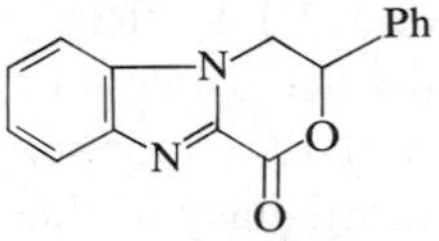

(**7.394**) 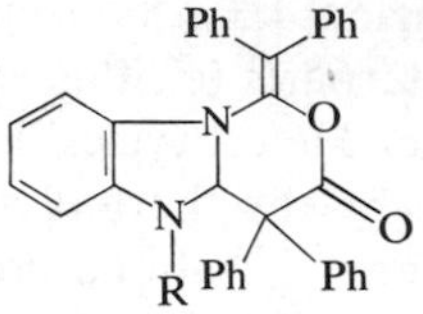 (**7.395**)

Compound	R	Medium	ν_{max} (cm^{-1}) C═O	C═C	Ref.
(**7.394**)	—	KBr	1740	—	87
(**7.395**)	COMe	Nujol	1768, 1686	1662	86
(**7.395**)	COMe	$CHCl_3$	1764, 1670	—	86
(**7.395**)	$COCHPh_2$	Nujol	1756, 1678	1640	86
(**7.395**)	$COCHPh_2$	$CHCl_3$	1763, 1668	—	86
(**7.395**)	SO_2Me	KBr	1758	1657	86
(**7.395**)	SO_2Me	$CHCl_3$	1764	1640	86

3,4,4a,5-tetrahydro-1*H*-[1,3]oxazino[3,4-*a*]benzimidazoles (**7.395**) exhibit high-frequency IR carbonyl absorption in the range 1770–1750 cm^{-1} (Table 7.41). The relatively low frequency (2180 cm^{-1}) for the IR cyano absorption of the 10*H*-[1,4]thiazino[4,3-*a*]benzimidazole derivative [Scheme 7.81; (**7.412**)] is attributed[91] to resonance interaction between the cyano substituent and the imidazole ring. Comparison of the IR stretching frequencies (Table 7.42) of the ring carbonyl substituents in the 4*H*-[1,3]thiazino[3,2-*a*]benzimidazol-4-one (**7.396**; $R^1 = Me$, $R^2 = H$, $X = O$) and the 2*H*-[1,3]thiazino[3,2-*a*]benzimidazol-4(3*H*)-ones (**7.397**) demonstrates the marked effect of conjugation in the former molecule. The carbonyl group in 1*H*-[1,4]thiazino[4,3-*a*]benzimidazol-4(3*H*)-one [Scheme 7.81; (**7.413**)] absorbs in the IR at 1728 cm^{-1}.[103]

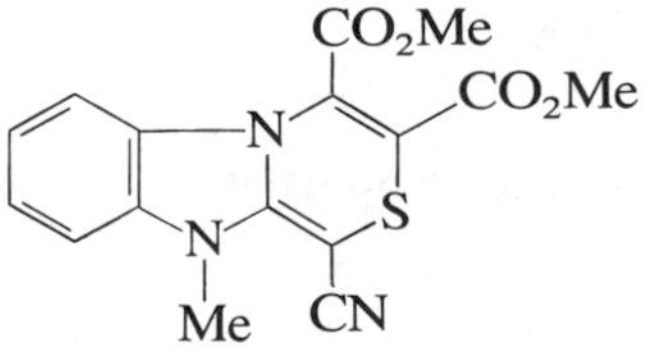

(**7.412**) 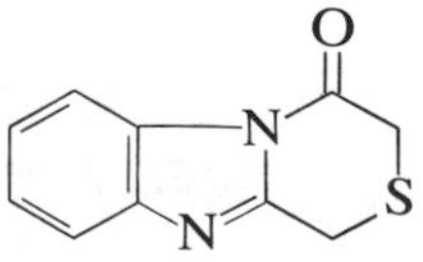 (**7.413**)

Scheme 7.81

TABLE 7.42. INFRARED SPECTRA OF [1,3]THIAZINO[3,2-*a*]BENZIMIDAZOLE DERIVATIVES (**7.396**) AND (**7.397**)

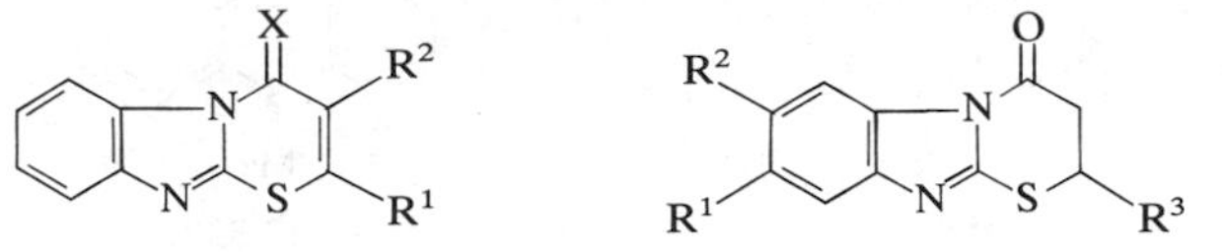

(**7.396**) (**7.397**)

Compound	R^1	R^2	R^3	X	Medium	ν_{max} (cm^{-1}) OH	C=O	C=C	Ref.
(**7.396**)	CN	OH	—	$CHCO_2Me$	KBr	3200–2200br	1740, 1720	—	91
(**7.396**)	Me	H	—	O	KBr	—	1690	1620	93
(**7.397**)	H	H	H	—	CS_2	—	1670	—	101
(**7.397**)	H	H	H	—	KBr	—	1731	—	103
(**7.397**)	H	H	H	—	KBr	—	1630	—	92
(**7.397**)	Cl	H	H	—	Nujol	—	1733	—	105
(**7.397**)	NO_2	H	H	—	KBr	—	1724	—	106
(**7.397**)	Me	Me	H	—	KBr	—	1680	—	104
(**7.397**)	H	H	*p*-ClC_6H_3CO	—	KBr	—	1725, 1680	—	108

Comparison of the IR carbonyl absorption (Table 7.43) of *N*-unsubstituted pyrimido[1,2-*a*]benzimidazol-2-ones with that of specifically *N*(1)- and *N*(10)-methylated derivatives demonstrates the *N*-unsubstituted molecules to exist in the solid state and in solution predominantly in the pyrimido[1,2-*a*]benzimidazol-2(1*H*)-one tautomeric form (**7.398**; R^1 = H).[130] A similar IR study[130] of pyrimido[1,2-*a*]benzimidazol-4-one derivatives (Table 7.43) demonstrates the pyrimido[1,2-*a*]benzimidazol-4(10*H*)-one tautomeric form (**7.401**; R = H) to predominate in the case of *N*-unsubstituted compounds. *N*(1)-substituted 3,4-dihydropyrimido[1,2-*a*]benzimidazol-2(1*H*)-ones [Table 7.44; (**7.403**)] are readily differentiated from their *N*(10)-substituted isomers [e.g. (**7.404**)] on the basis of the lower IR carbonyl stretching frequencies (Table 7.44) in molecules of the latter type as a result of increased conjugation. The close similarity of the IR frequencies (Table 7.44) of the ring carbonyl substituents in *N*-unsubstituted 3,4-dihydropyrimido[1,2-*a*]benzimidazol-2-ones and their *N*(1)-substituted derivatives indicates the former molecules to exist predominantly in the 3,4-dihydropyrimido[1,2-*a*]benzimidazol-2(1*H*)-one tautomeric form (**7.403**; R^1 = H). The IR carbonyl absorption (1680–1670 cm^{-1}) reported[133] for *N*(10)-substituted pyrimido[1,2-*a*]benzimidazole-2,4(3*H*,10*H*)-diones [Scheme 7.82; (**7.414**)] appears anomalous when compared with the IR carbonyl stretching frequencies (Table 7.44) of the 2,3-dihydropyrimido[1,2-*a*]benzimidazol-4(1*H*)-one (**7.405**; R = R^3 = R^4 = H, R^1 = R^2 = Me) [ν_{max} (C=O) = 1730 cm^{-1}], on the one hand, and the 3,4-dihydropyrimido-[1,2-*a*]benzimidazol-2(10*H*)-one (**7.404**) [ν_{max} (C=O) = 1625 cm^{-1}], on the

TABLE 7.43. INFRARED SPECTRA OF PYRIMIDO[1,2-*a*]BENZIMIDAZOLONES (**7.398**)–**7.401**)

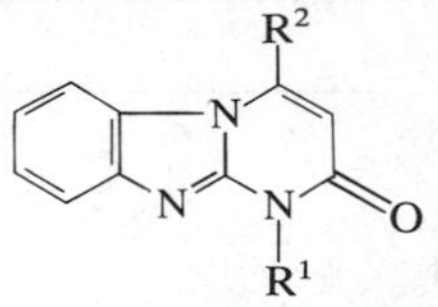

(**7.398**) 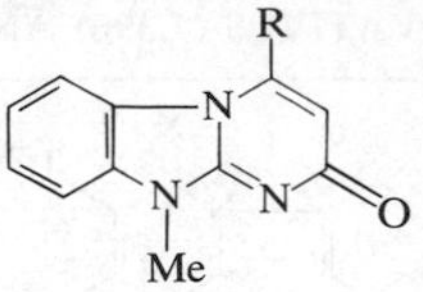 (**7.399**)

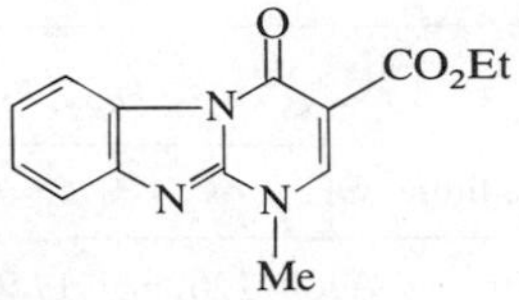

(**7.400**) 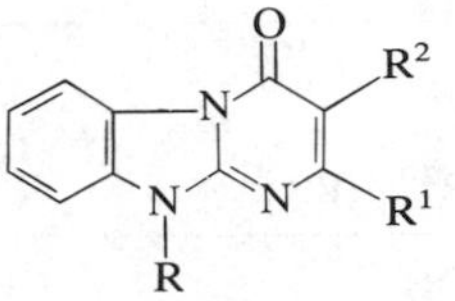 (**7.401**)

Compound	R	R^1	R^2	Medium	ν_{max} (cm^{-1}) NH	C≡N	C=O	C=N	Ref.
(**7.398**)	—	H	H	KBr	—	—	1675	—	127
(**7.398**)	—	H	H	KBr	—	—	1680	—	130
(**7.398**)	—	H	H	Me_2SO	—	—	1680	—	130
(**7.398**)	—	Me	H	KBr	—	—	1670	—	130
(**7.399**)	H	—	—	KBr	—	—	1630	—	130
(**7.398**)	—	H	Ph	KBr	3610, 3410	—	1650	—	127
(**7.398**)	—	H	Ph	KBr	—	—	1675	—	130
(**7.398**)	—	Me	Ph	KBr	—	—	1675	—	127
(**7.399**)	Ph	—	—	KBr	—	—	1685	—	127
(**7.399**)	Ph	—	—	$CHCl_3$	—	—	1615	—	136
(**7.398**)	—	H	CO_2Me	KBr	—	—	1740, 1690	—	127
(**7.398**)	—	H	CO_2Me	Nujol	—	—	1745, 1690, 1680	1660	135
(**7.398**)	—	Me	CO_2Me	CH_2Cl_2	—	—	1750, 1685	1640	135
(**7.400**)	—	—	—	KBr	—	—	1670	—	130
(**7.401**)	H	H	H	KBr	—	—	1678	—	130
(**7.401**)	Me	H	H	KBr	—	—	1690	—	130
(**7.401**)	H	Me	H	—[a]	—	—	1690	—	111
(**7.401**)	H	Me	H	KBr	—	—	1680	—	127
(**7.401**)	Me	Me	H	KBr	—	—	1685	—	125
(**7.401**)	Me	Me	H[b]	KBr	—	—	1680	—	125
(**7.401**)	Me	Me	H[c]	KBr	—	—	1680	—	125
(**7.401**)	H	CH_2Cl	H	KBr	3200–3000 br	—	1682	—	126
(**7.401**)	H	Ph	H	KBr	3650	—	1680	—	127
(**7.401**)	Me	Ph	H	KBr	—	—	1670	—	127
(**7.401**)	H	H	CO_2Et	KBr	—	—	1720, 1645	—	127
(**7.401**)	H	H	CO_2Et	KBr	—	—	1650	—	130
(**7.401**)	Me	H	CO_2Et	KBr	—	—	1653	—	120
(**7.401**)	H	H	CO_2H	Nujol	—	—	1725	—	120
(**7.401**)	H	H	CN	Nujol	—	2230	—[d]	—	120

[a] Medium not specified.
[b] *C*(7) bromo derivative.
[c] *C*(8) bromo derivative.
[d] ν_{max}(C=O) not specified.

TABLE 7.44. INFRARED SPECTRA OF TETRAHYDROPYRIMIDO[1,2-*a*]BENZIMIDAZOLE DERIVATIVES (**7.402**)–(**7.405**)

(**7.402**) (**7.403**) (**7.404**) (**7.405**)

Compound	R	R^1	R^2	R^3	R^4	Medium	ν_{max} (cm^{-1}) NH/OH	C=O	C=N	Ref.
(**7.402**)	—	H	H	H	H	KBr	3380–2500br	—	1620[a]	137
(**7.402**)	—	H	H	Me	H	KBr	3350–2550br	—	1627[a]	137
(**7.402**)	—	H	H	H	Me	KBr	3350–2600br	—	1628[a]	137
(**7.402**)	—	Me	Me	H	H	KBr	3400–2600br	—	1632[a]	137
(**7.403**)	—	H	H	—	—	Nujol	—	1680	—	120
(**7.403**)	—	H	H	—	—	KBr	—	1659–1640	—	139
(**7.403**)	—	H	Me	—	—	Nujol	—	1680		120
(**7.403**)	—	H	Me	—	—	KBr	—	1690	—	139
(**7.403**)	—	Me	CO_2Me	—	—	Nujol	—	1735, 1695	1620	135
(**7.403**)	—	Me	$CONH(CH_2)_2NMe_2$	—	—	CH_2Cl_2	—	1700, 1685	1620	135
(**7.403**)	—	Me	CON(piperazine)NMe	—	—	CH_2Cl_2	—	1695, 1660	1620	135
(**7.404**)	—	—	—	—	—	$CHCl_3$	—	1625	—	136
(**7.405**)	H	Me	Me	H	H	Nujol	3150	1730	1650	159
(**7.405**)	H	*p*-$Me_2NC_6H_4$	H	COMe	H	KBr	—	1729, 1715	1657	129
(**7.405**)	$COCHPh_2$	*p*-$MeOC_6H_4$	H	Ph	Ph	KBr	—	1733, 1681	1614	141
(**7.405**)	$COCHPh_2$	*p*-$Me_2NC_6H_4$	H	Ph	Ph	KBr	—	1733, 1678	1614	141
(**7.405**)	$COCHPh_2$	*p*-$O_2NC_6H_4$	H	Ph	Ph	KBr	—	1737, 1691	1618	141

[a] May also be assigned to NH def.[137]

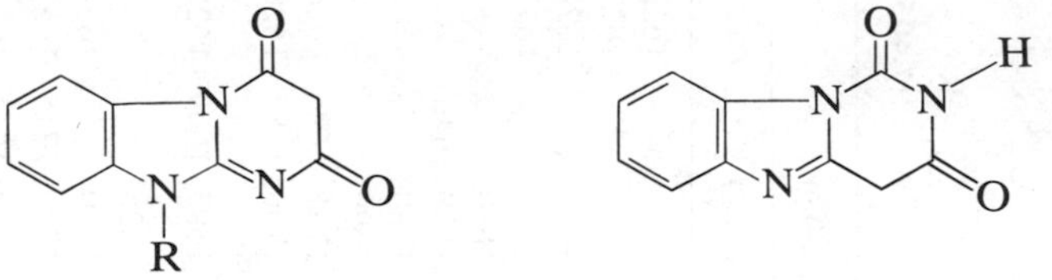

(**7.414**) (**7.415**)

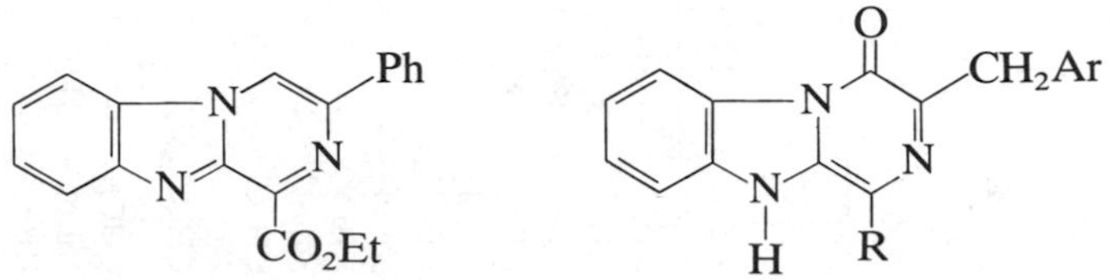

(**7.416**) (**7.417**)

Scheme 7.82

other. Pyrimido[3,4-*a*]benzimidazol-1,3(2*H*,4*H*)-dione [Scheme 7.82; (**7.415**)] exhibits IR carbonyl absorption (1750, 1722 cm^{-1}) typical of a diacyl amide structure.[144]

The ester substituent in the pyrazino[4,3-*a*]benzimidazole derivative [Scheme 7.82; (**7.416**)] gives rise to IR carbonyl absorption at 1720 cm^{-1}.[147] The IR carbonyl absorption (1630 cm^{-1}) of pyrazino[4,3-*a*]benzimidazol-4(10*H*)-ones (**7.417**)[148] is consistent with the highly conjugated nature of the carbonyl substituent in these molecules. 1,2-Dihydropyrazino[4,3-*a*]-benzimidazol-3(4*H*)-ones [Table 7.45; (**7.407**)] are readily distinguished from their 4(3*H*)-one counterparts [e.g. (**7.408**)] on the basis of their significantly lower IR carbonyl stretching frequencies (Table 7.45).

Ultraviolet Spectra. The similarity of the UV absorption[45] of simple 3,4-dihydro-1*H*-[1,4]oxazino[4,3-*a*]benzimidazoles to that of benzimidazole is consistent with the saturated nature of the fused oxazine ring in the former molecules. On the other hand, the conjugative unsaturation present in the thiazine ring of [1,3]thiazino[3,2-*a*]benzimidazol-4-ones [Table 7.46; (**7.418**)] accounts for the presence in their UV spectra (Table 7.46) of two intense maxima at 219–229 and 245–265 nm.

The UV spectra (Table 7.47) of fully unsaturated pyrimido[1,2-*a*]benzimidazoles (**7.420**) and pyrimido[1,2-*a*]benzimidazolium salts (**7.421**) are typified by the presence of a single intense maximum at ca. 240 nm accompanied by two lower intensity bands at longer wavelength. The UV spectrum (Table 7.48) of pyrimido[1,2-*a*]benzimidazol-2(1*H*)-one (**7.422**; $R^1 = R^2 =$ H), on the other hand, contains two intense absorption maxima at 236 and 250 nm and two lower intensity bands at 262 and 298 nm. Protonation of the molecule has the effect of suppressing the bands at 250 and 262 nm and inducing a hypsochromic shift in the long-wavelength band. Deprotonation,

TABLE 7.45. INFRARED SPECTRA OF 1,2,3,4-TETRAHYDROPYRAZINO[4,3-*a*]BENZIMIDAZOLE DERIVATIVES (**7.406**)–(**7.411**)

(**7.406**) (**7.407**) (**7.408**)

(**7.409**) (**7.410**) (**7.411**)

Compound	R	Medium	ν_{max} (cm^{-1}) NH	C═O	Ref.
(**7.406**)	H	KBr	—	1680	150
(**7.406**)	CH_2Ph	KBr	—	1660	150
(**7.407**)	Bu^n	—[a]	—	1660–1645	151
(**7.407**)	Cyclohexyl	—[a]	—	1660–1645	151
(**7.407**)	CH_2Ph	—[a]	—	1660–1645	151
(**7.408**)	—	—[a]	—	1710	151
(**7.409**)	—	Nujol	3220, 3115	1740, 1700	152
(**7.410**)	—	Nujol	3250	1650	152
(**7.411**)	Cyclohexyl	—[a]	—	1740–1730, 1685–1680	151
(**7.411**)	CH_2Ph	—[a]	—	1740–1730, 1685–1680	151

TABLE 7.46. ULTRAVIOLET SPECTRA OF [1,3]THIAZINO[3,2-*a*]-BENZIMIDAZOLE DERIVATIVES (**7.418**) AND (**7.419**)

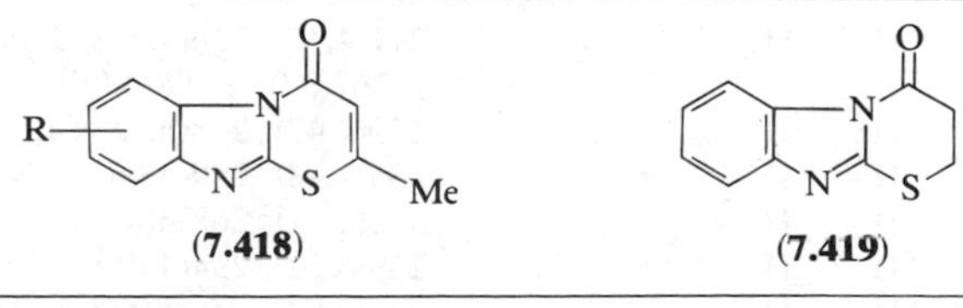

Compound	R	Solvent[a]	λ_{max} (nm) (log ε)	Ref.
(**7.418**)	H	*A*	223(4.17), 263(4.34)	93
(**7.418**)	Me	*A*	224(4.19), 266(4.42)	93
(**7.418**)	Cl	*A*	219(4.28), 265(4.38)	93
(**7.418**)	NO_2	*A*	221(4.25), 245(4.32)	93
(**7.418**)	CO_2H	*A*	229(4.46), 259(4.41)	93
(**7.419**)	—	*B*	295–297 (4.10–4.30)	92

[a] *A* = methanol; *B* = ethanol.

TABLE 7.47. ULTRAVIOLET SPECTRA[a] OF PYRIMIDO[1,2-*a*]BENZIMIDAZOLES (**7.420**) AND PYRIMIDO[1,2-*a*]BENZIMIDAZOLIUM SALTS (**7.421**)

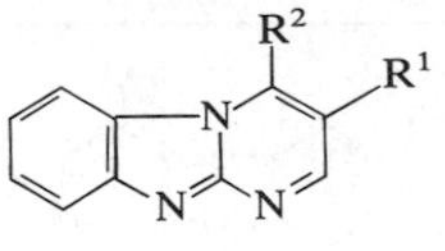

(**7.420**) 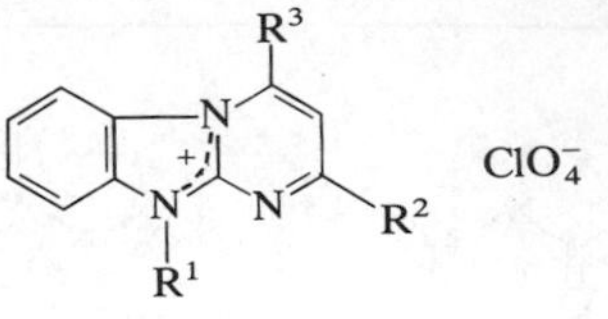 (**7.421**) ClO_4^-

Compound	R^1	R^2	R^3	λ_{max} (nm) (log ε)	Ref.
(**7.420**)	H	H	—	205(4.37), 245(4.62), 320(3.79), 365(3.38)	156
(**7.420**)	H	Me	—	247(4.14), 285(3.30),315(2.30)	114
(**7.420**)	CO_2Et	NH_2	—	236sh, 247(4.50), 256(4.48), 263sh, 310(4.11), 370sh	120
(**7.421**)	H	H	Ph	240(4.15), 300(3.66), 345(3.51)	114
(**7.421**)	Me	Me	Me	240(4.51), 294(3.96), 330(3.73)	114
(**7.421**)	Me	Me	Ph	240(4.37), 330(4.00), 365sh	114

[a] Measured for solutions in ethanol.

TABLE 7.48. ULTRAVIOLET SPECTRA OF PYRIMIDO[1,2-*a*]BENZIMIDAZOLONES (**7.422**)–(**7.425**)

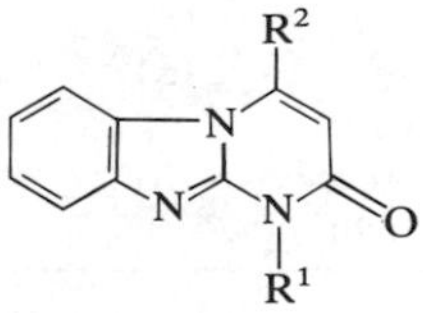

(**7.422**) 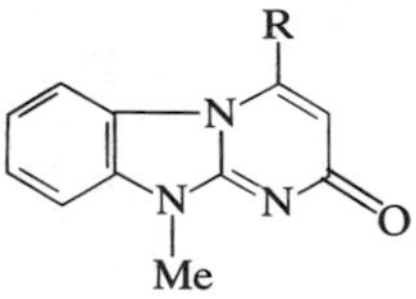 (**7.423**) 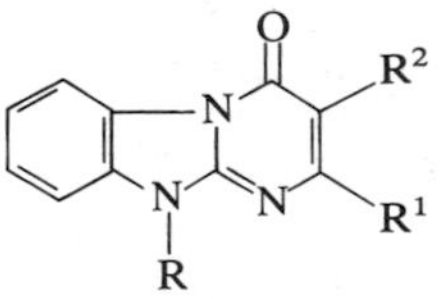 (**7.424**) 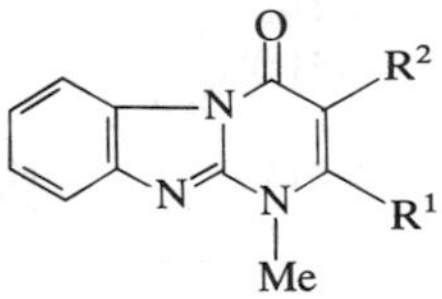 (**7.425**)

Compound	R	R^1	R^2	Solvent[a]	λ_{max} (nm) (log ε)	Ref.
(**7.422**)	—	H	H	*A*	211(4.41), 236(4.51), 250(4.26), 262(3.93), 298(4.06)	134
(**7.422**)	—	H	H	*B*	239(4.47), 248sh, 264sh, 298(4.03)	120
(**7.422**)	—	H	H	—[b]	236(4.48), 289(3.96)	120
(**7.422**)	—	H	H	—[c]	253(4.71), 268sh(4.06)	120
(**7.422**)	—	H	H	*B*	239(4.22), 294(4.05)	127
(**7.422**)	—	H	H	*A*	211(2.05), 237sh(4.24), 248(4.03), 263(4.66), 283(4.79)	130
(**7.422**)	—	Me	H	*B*	249(4.47), 263(4.20), 290(3.95)	120
(**7.422**)	—	Me	H	*A*	212(4.32), 227sh(4.10), 248(4.36), 263sh(4.11), 290(3.96)	130
(**7.422**)	—	H	Ph	*B*	242(4.46), 305(4.03)	127
(**7.422**)	—	H	Ph	*A*	212(4.76), 239(4.76), 297(4.30)	130
(**7.422**)	—	Me	Ph	*B*	252(4.00), 296(3.00)	127
(**7.422**)	—	Me	Ph	*C*	251(4.00), 298(4.00)	127

TABLE 7.48 (*Continued*)

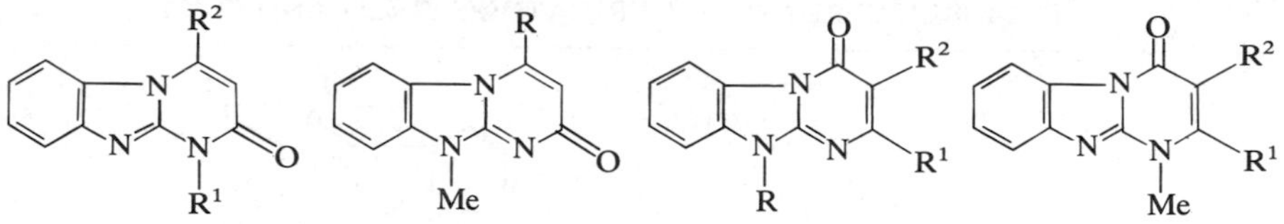

(7.422) (7.423) (7.424) (7.425)

Compound	R	R[1]	R[2]	Solvent[a]	λ_{max} (nm) (log ε)	Ref.
(**7.422**)	—	Me	Ph	*D*	253(4.00), 288(3.00)	127
(**7.422**)	—	H	CO_2Me	*B*	213(4.35), 239(4.42)	127
(**7.422**)	—	H	CO_2Me	*A*	207(4.52), 243(4.40), 262(4.07), 310(3.85)	135
(**7.422**)	—	Me	CO_2Me	*A*	212(4.51), 250(4.36), 264sh(4.08), 270sh(4.02), 306(3.80)	135
(**7.423**)	H	—	—	*A*	215(4.33), 237(4.52), 303(4.09)	130
(**7.423**)	Ph	—	—	*B*	241(4.00), 307(3.00)	127
(**7.423**)	Ph	—	—	*C*	240(4.00), 295(4.00)	127
(**7.423**)	Ph	—	—	*D*	242(4.00), 294(3.00)	127
(**7.424**)	H	H	H	*A*	228(4.33), 250sh(4.07), 321(4.02), 331(4.05)	130
(**7.424**)	Me	H	H	*A*	228(4.36), 250sh(4.10), 324(4.11), 337(4.15)	130
(**7.424**)	H	Ph	H	*B*	254(4.47), 288(4.22), 345(3.97)	127
(**7.424**)	Me	Ph	H	*B*	255(4.00), 287(4.00), 347(3.00)	127
(**7.424**)	Me	Ph	H	*C*	254(4.00), 298(4.00)	127
(**7.424**)	Me	Ph	H	*D*	253(4.00), 287(4.00), 348(3.00)	127
(**7.424**)	H	H	CO_2Et	*B*	231sh, 247sh, 276sh, 331(4.26), 341(4.26), 358sh	120
(**7.424**)	H	H	CO_2Et	*B*	248(4.31), 273(4.11), 291(4.01)	127
(**7.424**)	H	H	CO_2Et	*A*	220(4.38), 243sh(4.21), 264(3.87), 275sh(3.73), 331(4.27), 342(4.26)	130
(**7.424**)	Me	H	CO_2Et	*A*	217(4.45), 240(4.19), 267(3.68), 276(3.64), 326(4.33), 343(4.31)	130
(**7.424**)	COMe	H	CO_2Et	*A*	210(4.38), 231(4.25), 265sh(3.72), 273(3.56), 334(4.29), 342sh(4.28)	130
(**7.424**)	H	H	CO_2H	*B*	228(4.42), 248sh, 256sh, 330(4.23), 342(4.21)	120
(**7.424**)	Me	H	CO_2H	*A*	218(4.41), 241(4.17), 269(3.66), 334(4.23), 244sh(4.19)	130
(**7.424**)	H	H	CN	*B*	228sh(4.34), 244sh(4.28), 264sh, 276sh, 284sh, 296sh, 322(4.24), 345(4.23)	120
(**7.425**)	Me	Me	H	*B*	233(4.22), 248sh(3.99), 316(3.45)	120
(**7.425**)	Me	Me	H	—[b]	229(4.12), 244sh(3.94), 306(3.51)	120
(**7.425**)	Me	Me	H	—[c]	231sh(4.21), 250sh(3.96), 318(3.44)	120
(**7.425**)	Me	H	CO_2Et	*B*	228(4.41), 240(4.37), 329(4.00)	120

[a] *A* = MeOH; *B* = EtOH; *C* = 0.1 *M* NaOH, EtOH; *D* = 0.1 *M* HCl, EtOH.
[b] Measured at pH = 1.
[c] Measured at pH = 13.

TABLE 7.49. ULTRAVIOLET SPECTRA OF 1,2,3,4-TETRAHYDROPYRIMIDO-[1,2-*a*]BENZIMIDAZOLE DERIVATIVES (**7.426**) AND (**7.427**)

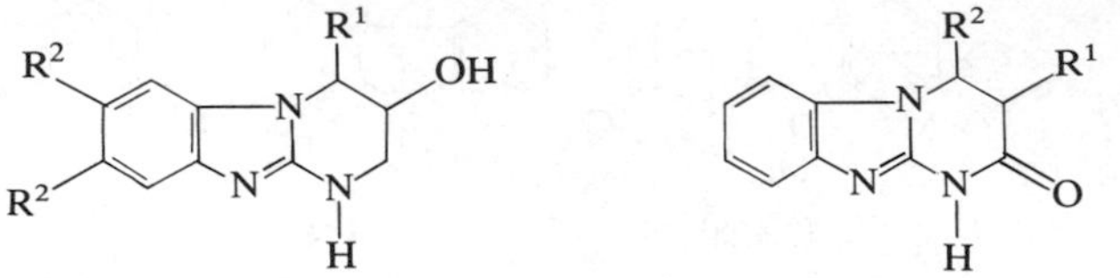

(**7.426**) (**7.427**)

Compound	R[1]	R[2]	Solvent[a]	λ_{max} (nm) (log ϵ)	Ref.
(**7.426**)	H	H	A	228(3.72), 253(3.69), 291(3.72)	137
(**7.426**)	H	Me	A	233(3.69), 250(3.62), 299(3.67)	137
(**7.426**)	Me	Me	A	231(3.61), 256(3.51), 302(3.57)	137
(**7.427**)	H	H	A	252(4.06), 260 *sh*, 284(4.12), 292(4.13)	120
(**7.427**)	H	H	—[b]	234(4.13), 279(4.15), 287(4.18)	120
(**7.427**)	H	H	—[c]	263(3.89), 272(3.87), 302(4.32)	120
(**7.427**)	Me	H	A	251(4.05), 258sh, 284(4.13), 291(4.13)	120
(**7.427**)	Me	H	—[b]	234(4.14), 279(4.16), 287(4.19)	120
(**7.427**)	Me	H	—[c]	263(4.00), 272(3.97), 302(4.35)	120
(**7.427**)	H	Me	A	252(4.06), 260sh, 285(4.13), 292(4.14)	120
(**7.427**)	H	Me	—[b]	234(4.14), 279(4.16), 287(4.19)	120
(**7.427**)	H	Me	—[c]	262(4.10), 272(4.07), 303(4.37)	120

[a] A = ethanol.
[b] Measured at pH = 1.
[c] Measured at pH = 13.

on the other hand, causes the disappearance of the bands at 236 and 298 nm and produces a small but significant bathochromic shift in the remaining maxima at 250 and 262 nm. Pyrimido[1,2-*a*]benzimidazol-4(10*H*)-ones [Table 7.48; (**7.424**)] can be differentiated from pyrimido[1,2-*a*]benzimidazol-2(1*H*)-ones (**7.422**) on the basis of their generally longer wavelength UV absorption (Table 7.48). UV studies[130] demonstrate the preferential existence of pyrimido[1,2-*a*]benzimidazol-4-ones in the 10*H*-tautomeric form (**7.424**; R = H). Not unexpectedly, the UV absorption (Table 7.49) of simple 1,2,3,4-tetrahydropyrimido[1,2-*a*]benzimidazole derivatives is similar to that of 3,4-dihydro-1*H*-[1,4]oxazino[4,3-*a*]benzimidazoles (see before) and benzimidazole. The UV spectra of 1,2,3,4-tetrahydropyrazino[4,3-*a*]benzimidazoles are likewise similar to those of 2-(dialkylaminomethyl)benzimidazoles.[149]

NUCLEAR MAGNETIC RESONANCE SPECTRA. The protons at the *C*(1) position in 3,4-dihydro-1*H*-[1,4]oxazino[4,3-*a*]benzimidazoles resonate at consistently lower field (Table 7.50) compared with the protons at the *C*(3) or *C*(4) positions due to the cumulative deshielding effect of the oxygen atom and the azomethine substituent. In 3,4-dihydro-2*H*-[1,3]thiazino[3,2-*a*]benzimidazole derivatives [Table 7.51; (**7.434**)], on the other hand, *H*(4)

TABLE 7.50. ^{1}H NMR SPECTRA[a,b] OF 3,4-DIHYDRO-1*H*-[1,4]OXAZINO[4,3-*a*]BENZIMIDAZOLE DERIVATIVES (**7.428**)–(**7.432**)

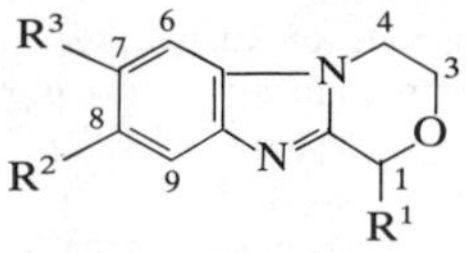

(**7.428**) 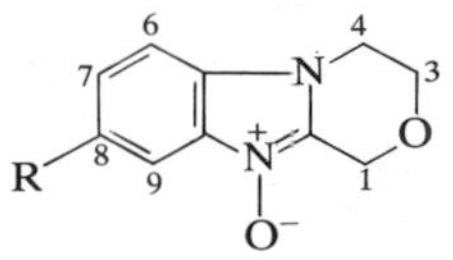 (**7.429**) 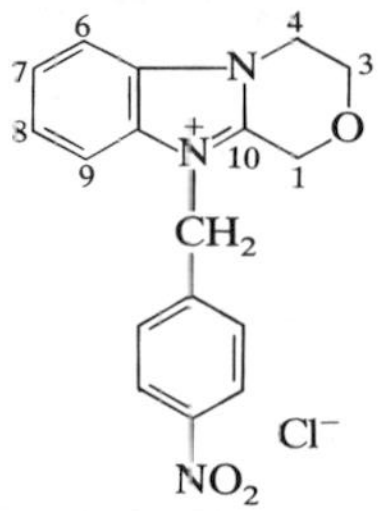 (**7.430**) 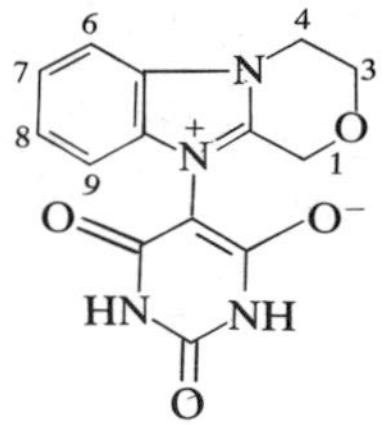 (**7.431**) 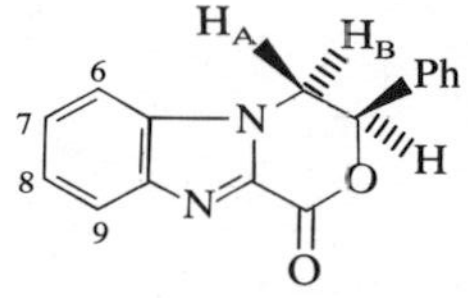 (**7.432**)

Compound ($R \rightarrow R^3$ unspecified = H)	Solvent[c]	H(1)	H(3)	H(4)	H(6)	H(7)	H(8)	H(9)	H(10)	Ref.
(**7.428**; $R^2 = NO_2$, $R^3 = NH_2$)	*A*	5.45br	← 4.42–4.65m →		7.88	—	—	8.93	—	49
(**7.428**; $R^2 = NO_2$, $R^3 = NHCOMe$)	*A*	5.47br	← 4.43–4.65m →		8.93	—	—	9.23	—	49
(**7.428**; $R^2 = NO_2$, $R^3 = N_3$)	*A*	5.35–5.55m	← 4.40–4.70m →		7.85	—	—	8.55	—	49
(**7.428**; $R^1 = OCOMe$, $R^2 = NO_2$)	*A*	7.28 2.19[d]	4.20	4.60	7.53[e]	8.28dd[e,f]	—	8.65[f]	—	56
(**7.429**)[g]	*B*	5.38	← 4.42 →		← 7.69–7.83m →				—	55
(**7.429**; $R = NO_2$)[g]	*C*	5.08	← 4.27 →		8.49d[f]	8.22dd[e,f]	—	7.62 d[f]	—	55
(**7.430**)	*B*	5.40	← 4.50 →		← 7.40–8.30m →				5.82[h]	62
(**7.431**)	*D*	5.31	← 4.55 →		← 7.50–8.00m →				—	63
(**7.432**)	*E*	—	6.25[i,j]	4.63[k,i,l] 5.05[m,j,l]	← —[n] →				—	87

[a] δ values in ppm measured from TMS.
[b] Signals are sharp singlets unless denoted as d = doublet; dd = doublet; m = multiplet.
[c] *A* = $CDCl_3$; *B* = D_2O; *C* = CF_3CO_2D; *D* = CF_3CO_2H; *E* = $(CD_3)_2SO$.
[d] COMe.
[e] $J = 9.1$–9.5 Hz.
[f] $J = 2.0$–2.1 Hz.
[g] Hydrochloride.
[h] CH_2Ph.
[i] $J_{H(3)-H_A} = 10.5$ Hz.
[j] $J_{H(3)-H_B} = 4.0$ Hz.
[k] H_A.
[l] $J_{H_A-H_B} = 13.5$ Hz.
[m] H_B.
[n] δ values not specified.

TABLE 7.51. ^{1}H NMR SPECTRA[a,b] OF [1,3]THIAZINO[3,2-*a*]BENZIMIDAZOLE DERIVATIVES (**7.433**)–(**7.435**)

(**7.433**) (**7.434**) (**7.435**)

Compound	Solvent[c]	H(2)	H(3)	H(4)	H(6)	H(7)	H(8)	H(9)	Ref.
(**7.433**)	*A*	2.04[d]	6.44	—	⟵		7.68–8.25m	⟶	93
(**7.434**; $R^1 = R^2 = H$)	*B*	3.70t[e]	2.75m	4.48t[e]	⟵		7.66	⟶	98
(**7.434**; $R^1 = H$, $R^2 = Me$)[f]	*B*	3.50t[e]	1.80–2.60m	4.15m	7.25[g]	—	1.28[d]	7.15[g]	98
(**7.434**; $R^1 = OH$, $R^2 = H$)	*C*	3.80m	5.25m	4.64m	⟵		7.17	⟶	98
(**7.434**; $R^1 = OH$, $R^2 = H$)[f]	*B*	3.78t[e]	5.05m	4.54t[e]	⟵		7.67	⟶	98
(**7.434**; $R^1 = OH$, $R^2 = Me$)	*C*	3.80m	5.15–5.35m	4.50–4.70m	7.50[g]	⟵ 2.52[d]	⟶	7.42[g]	98
(**7.435**; $R^1 = R^2 = H$)	*A*	⟵ 3.38m	⟶	—	8.12m	⟵ 7.29m	⟶	7.56 m	106
(**7.435**; $R^1 = Cl$, $R^2 = H$)	*D*	⟵ 3.29m	⟶	—	8.12d[h]	7.30dd[h,i]	—	7.60 d[i]	105
(**7.435**; $R^1 = H$, $R^2 = NO_2$)	*A*	⟵ [j]	⟶	—	8.79[k]	—	8.23[k]	7.68[k]	106

[a] δ values in ppm measured from TMS.
[b] Signals are sharp singlets unless denoted as d = doublet; dd = double doublet; t = triplet; m = multiplet.
[c] $A = (CD_3)_2SO$; $B = D_2O$; $C = CF_3CO_2H$; $D = CDCl_3$.
[d] C(Me).
[e] *J* value not quoted.
[f] Hydrochloride.
[g] These signal assignments may be reversed.
[h] $J_{H(6)-H(7)} = 9$ Hz.
[i] $J_{H(7)-H(9)} = 2$ Hz.
[j] δ values not quoted.
[k] Multiplicity not specified.

resonates at lowest field, the observed order of shielding (Table 7.51) for the thiazine ring protons being $H(3) > H(2) > H(4)$. The deshielding effect of the $C(4)$ carbonyl substituent in 2*H*-[1,3]thiazino[3,2-*a*]benzimidazol-4(3*H*)-ones [Table 7.51; (**7.435**)] results in similar chemical shifts for $H(2)$ and $H(3)$ in such molecules (Table 7.51). $H(6)$ in 2*H*-[1,3]thiazino[3,2-*a*]-benzimidazol-4(3*H*)-ones absorbs at lower field (Table 7.51) than the other aromatic protons as a result of the anisotropic deshielding effect of the $C(4)$ carbonyl substituent. The olefinic proton at the $C(3)$ position in 4*H*-[1,3]thiazino[3,2-*a*]benzimidazol-4-ones [e.g., Table 7.51; (**7.433**)] gives rise to singlet absorption in the range δ 2.45–2.55.[93]

$H(4)$ and $H(9)$ are the most deshielded (Table 7.52) of the pyrimidine ring and benzenoid protons, respectively, in fully unsaturated pyrimido[1,2-*a*]benzimidazole derivatives (**7.436**). The $C(2)$ and $C(4)$ methyl substituents in 2,4-dimethylpyrimido[1,2-*a*]benzimidazole (**7.436**; $R^1 = R^2 = Me$) can be differentiated[111] on the basis of the lower field resonance (Table 7.52) of the protons in the latter and their appearance as a doublet ($J = 0.9$ Hz) due to coupling with $H(3)$. Consequently, the lack of splitting (Table 7.52) associated with the protons of the methyl substituent in the pyrimido[1,2-*a*]-benzimidazole derivative obtained[111] by condensing 2-aminobenzimidazole with acetoacetaldehyde dimethyl acetal (cf. section 7.2.1, "Ring-closure Reactions of Benzimidazole Derivatives") supports its formulation[111] as 2-methylpyrimido[1,2-*a*]benzimidazole (**7.436**; $R^1 = Me$). Conversely, the deshielded nature of the methyl group in the major product derived[114] by the reaction of 2-aminobenzimidazole with methyl β-chlorovinyl ketone in the presence of perchloric acid (cf. section 7.2.1, "Ring-closure Reactions of Benzimidazole Derivatives"), requires its adjacency to the quaternary bridgehead nitrogen in the product, and hence indicates the latter to be 4-methylpyrimido[1,2-*a*]benzimidazolium perchlorate.[114] The finding[114] that the free base of this compound and the presumed 2-methylpyrimido[1,2-*a*]-benzimidazole from 2-aminobenzimidazole and acetoacetaldehyde dimethyl acetal (see before) are apparently identical, reveals a possible flaw in one or other of the ^{1}H NMR interpretations involved.

$H(3)$ and $H(4)$ in pyrimido[1,2-*a*]benzimidazol-2(1*H* or 10*H*)-ones [Table 7.53; (**7.437**) and (**7.438**)] resonate uniformly at δ 5.8–7.0 and δ 8.0–8.9, respectively, with $J_{H(3)-H(4)} = 7.5$–8.0 Hz. Correspondingly, $H(2)$ is considerably more deshielded (Table 7.54) than $H(3)$ in pyrimido[1,2-*a*]-benzimidazol-4(10*H*)-ones (**7.439**). A noteworthy feature of the ^{1}H NMR absorption (Table 7.54) of the latter molecules and their $N(1)$H isomers (**7.440**) is the enhanced deshielding of $H(6)$ as a consequence of the anisotropic effect of the proximate $C(4)$ carbonyl substituent. This feature is of course lacking in the ^{1}H NMR spectra (Table 7.53) of pyrimido[1,2-*a*]-benzimidazol-2-ones (**7.437**) and (**7.438**) and therefore permits the differentiation[111,120,130] of the two structural types. The normality of the NMR absorption (Table 7.55) of $H(6)$ in 3,4-dihydropyrimido[1,2-*a*]benzimidazol-2(1*H*)-ones (**7.441**) is consistent with the $C(2)$ position for the

TABLE 7.52. ^{1}H NMR SPECTRA[a,b] OF PYRIMIDO[1,2-*a*]BENZIMIDAZOLE DERIVATIVES (**7.436**)

(**7.436**)

R^1	R^2	Solvent[c]	H(2)	H(3)	H(4)	H(6)	H(7)	H(8)	H(9)	Ref.
H	H	*A*	8.89dd[d,e]	7.20dd[d,f]	9.49dd[e,f]	7.90m[g]	←7.50m→		8.30m[g]	111
Me	H	*A*	2.62[h]	7.00d[f]	9.33d[f]	7.90m[g]	←7.50m→		8.30m[g]	111
Me	Me	*A*	2.53[h]	6.76	2.97d[h,i]	7.80m[g]	←7.40m→		8.10m[g]	111
Me	Cl	*A*	2.32[h]	5.85	—	7.90m	←7.50m→			111
Me	$NH(CH_2)_2NMe_2$	*A*	2.42[h]	6.05	—[j]	8.20m	←7.50m→			111
CO_2Et	NH_2	*B*	1.55t[k,l] 4.66q[m,l]	9.39	—[j]	8.48m	←8.00m→			120

[a] δ values in ppm measured from TMS.

[b] Signals are sharp singlets unless denoted as d = doublet; dd = double doublet; t = triplet; q = quartet; m = multiplet.

[c] $A = (CD_3)_2SO$; $B = CF_3CO_2H$.

[d] $J_{H(2)-H(3)} = 4.0$ Hz.

[e] $J_{H(2)-H(4)} = 2.0$ Hz.

[f] $J_{H(3)-H(4)} = 6.5$ Hz.

[g] These signal assignments may be reversed.

[h] C(Me).

[i] $J_{H(3)-Me(4)} = 0.9$ Hz.

[j] δ values not quoted.

[k] Me of Et group.

[l] J value not specified.

[m] CH_2 of Et group.

TABLE 7.53. ^{1}H NMR SPECTRA[a,b] OF PYRIMIDO[1,2-*a*]BENZIMIDAZOL-2-ONES (**7.437**) AND (**7.438**)

(**7.437**) (**7.438**)

Compound	Solvent[c]	H(1)	H(3)	H(4)	H(6)	H(7)	H(8)	H(9)	H(10)	Ref.
(**7.437**; $R^1 = R^2 = H$)	*A*	—	7.00d[d]	8.86d[d]		← 7.86m →			—	120
(**7.437**; $R^1 = R^2 = H$)	*B*	—	6.05d[e]	8.70d[e]		← 7.10–8.00m →			—	127
(**7.437**; $R^1 = R^2 = H$)	*C*	—	6.13[f]	8.71[f]		← 7.20–8.10m →			—	130
(**7.437**; $R^1 = R^2 = H$)	*A*	—	6.94d[d]	8.83d[d]		← 7.97m →			—	134
(**7.437**; $R^1 = Me$, $R^2 = H$)	*C*	3.67[g]	6.12d[d]	8.00d[d]		← 7.10–7.80m →			—	120
(**7.437** $R^1 = Me$, $R^2 = H$)	*C*	—[h]	6.15[f]	8.02[f]		← 7.10–7.80m →			—	130
(**7.437**; $R^1 = H$, $R^2 = Ph$)	*B*	—	5.81	6.60–7.70m[i]		← 6.60–7.70m →			—	127
(**7.437**; $R^1 = H$, $R^2 = Ph$)	*C*	—	6.15	7.29–7.90m[i]	6.65m	← 7.29–7.90m →			—	130
(**7.437**; $R^1 = Me$, $R^2 = Ph$)	*C*	3.80[g]	6.26	7.25–7.70m[i]		← 7.25–7.70m →			—	127
(**7.437**; $R^1 = H$, $R^2 = CO_2Me$)	*B*	—	6.50	3.70[j]		← 7.05–7.75m →			—	127
(**7.437**; $R^1 = H$, $R^2 = CO_2Me$)	*B*	—	6.50	4.10[j]		← 7.10–7.65m →			—	135
(**7.437**; $R^1 = Me$, $R^2 = CO_2Me$)	*B*	3.60[g]	6.60	4.10[j]		← 7.15–7.70m →			—	135
(**7.437**; $R^1 = Me$, $R^2 = CH_2OH$)	*B*[k]	3.60[g]	6.20	4.95d[l,e] 5.80t[m,e]		← 7.15–7.85m →			—	135
(**7.437**; $R^1 = (CH_2)_2NMe_2$, $R^2 = Me$)	*C*	2.40[g] 2.75t[n,f] 4.50t[n,f]	6.00	2.75[o]		← 7.10–7.90m →			—	135
(**7.438**; $R^1 = Me$, $R^2 = H$)	*C*	—	6.37[f]	8.08[f]		← 7.30–7.70m →			—[h]	130
(**7.438**; $R^1 = Me$, $R^2 = Ph$)	*C*	—	6.39	7.30–7.80m[i]		← 7.30–7.80m →			7.90[g]	127

[a] δ values in ppm measured from TMS.
[b] Signals are sharp singlets unless denoted as d = doublet; t = triplet; m = multiplet.
[c] $A = CF_3CO_2H$; $B = (CD_3)_2SO$; $C = CDCl_3$.
[d] $J = 7.5–8.0$ Hz.
[e] $J = 5.0$ Hz.
[f] Multiplicity not specified.
[g] NMe.
[h] δ values not quoted.
[i] ArH.
[j] MeO.
[k] Spectrum taken at 100°.
[l] CH_2OH
[m] CH_2OH; exchangeable with D_2O.
[n] NCH_2.
[o] *C*(Me).

TABLE 7.54. ^{1}H NMR SPECTRA[a,b] OF PYRIMIDO[1,2-*a*]BENZIMIDAZOL-4-ONES (**7.439**) AND (**7.440**)

(**7.439**) (**7.440**)

Compound ($R^1 \rightarrow R^3$ unspecified = H)	Solvent[c]	H(1)	H(2)	H(3)	H(6)	H(7)	H(8)	H(9)	H(10)	Ref.
(**7.439**)	*A*	—	7.99[d]	6.03[d]	8.47m	←	7.30–7.75m	→	—	130
(**7.439**; R^1 = Me)	*B*	—	7.92[d]	6.08[d]	8.52m	←	7.10–7.50m	→	—	130
(**7.439**; R^2 = Me)	*A*	—	2.32[e,f]	5.81	8.40m	←	7.50m	→	—	111
(**7.439**; R^2 = Me)	*C*	—	2.70[e]	6.45	8.30m	←	7.70–8.00m	→	—	127
(**7.439**; R^1 = R^2 = Me)	*B*	—	2.36[e]	5.97	8.57dd[g,h]	←	7.17–7.44m	→	3.72[i]	125
(**7.439**; R^1 = R^2 = Me)[j]	*B*	—	2.34[e]	5.99	8.75[h]	—	7.54dd[g,h]	7.13d[g]	3.72[i]	125
(**7.439**; R^1 = R^2 = Me)[k]	*B*	—	2.36[e]	6.00	8.43d[g]	7.35–7.50 m	—	7.35–7.50m	3.70[i]	125
(**7.439**; R^2 = Ph)	*A*	—	7.45–8.21m[l]	6.70	8.35m	←	7.45–8.21m	→	—	127
(**7.439**; R^1 = Me, R^2 = Ph)	*B*	—	7.35–8.10m[l]	6.65	8.40	←	7.35–8.10m	→	3.90	127
(**7.439**; R^2 = CO_2Et)	*C*	—	9.21	— 1.54t[m,d] 4.63q[n,d]	8.72m	←	7.88m	→	—	120

(**7.439**; $R^3 = CO_2Et$)	*C*	—	8.70	1.30t[m,d] 4.30q[n,d]	8.20m	←— 6.90–7.60m —→	—	127
(**7.439**; $R^3 = CO_2Et$)	*A*	—	8.67	—[o]	8.50m	←— 7.30–7.60m —→	—	130
(**7.439**; $R^1 = Me$, $R^3 = CO_2Et$)	*D*	—	8.78	—[o]	8.63m	←— 7.20–7.70m —→	—	130
(**7.439**; $R^1 = COMe$, $R^3 = CO_2Et$)	*B*	—	8.75	—[o]	8.52m	←— 7.40–7.50m —→	—	130
(**7.439**; $R^3 = CO_2H$)	*E*	—	8.79	—	8.36m	←— 7.39m —→	—	120
(**7.439**; $R^1 = Me$, $R^3 = CO_2H$)	*C*	—	9.22	—	8.80m	←— 7.80–8.00m —→	—	130
(**7.439**; $R^3 = CN$)	*C*	—	8.89	—	8.72m	←— 7.92m —→	—	120
(**7.440**; $R^1 = Me$)	*C*	4.13[i]	2.73[e]	6.67	8.72m	←— 7.85m —→	—	120
(**7.440**; $R^2 = CO_2Et$)	*B*	—[o]	8.40	—[o]	8.57m	←— 7.30–7.90m —→	—	130

[a] δ values in ppm measured from TMS.
[b] Signals are sharp singlets unless denoted as d = doublet; dd = double doublet; t = triplet; q = quartet; m = multiplet.
[c] *A* = $(CD_3)_2SO$; *B* = $CDCl_3$; *C* = CF_3CO_2H; *D* = $CDCl_3$—$(CD_3)_2SO$; *E* = KOD—D_2O.
[d] Multiplicity not specified.
[e] *C*(Me).
[f] Singlet at 500 Hz, which resolves to a doublet ($J = 0.6$ Hz) at 50 Hz.
[g] $J = 8.5$–9.0 Hz.
[h] $J = 2.0$ Hz.
[i] NMe.
[j] *C*(7)bromo derivative.
[k] *C*(8) bromo derivative.
[l] Ar*H*.
[m] Me of Et.
[n] CH_2 of Et.
[o] δ values not quoted.

TABLE 7.55. ^{1}H NMR SPECTRA[a,b] OF 3,4-DIHYDROPYRIMIDO[1,2-*a*]BENZIMIDAZOL-2(1*H*)-ONES (**7.441**)

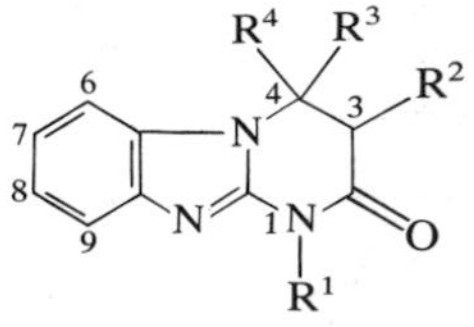

(**7.441**)

R^1	R^2	R^3	R^4	Solvent[c]	H(1)	H(3)	H(4)	H(6)	H(7)	H(8)	H(9)	Ref.
H	H	H	H	*A*	—	3.43t[d]	4.72t[d]	←	7.68		→	120
H	H	H	H	*A*	—	4.15t[e]	2.90t[e]	←	7.15		→	139
$(CH_2)_2CO_2H$	H	H	H	*A*	2.61–2.80m[f] 2.82–3.11m[g]	← 3.95–4.42m	→	←	—[h]		→	139
H	Me	H	H	*A*	—	1.58d[i,j] 3.30–3.70m	4.00–5.00oct[k]	←	7.67		→	120
H	H	Me	H	*A*	—	3.00–3.80oct[l]	1.72[i,j] 5.00–5.70m	←	7.73		→	120
H	H	Me	Me	*A*	—	2.90	1.56[i]	←	7.35		→	139
Me	H	CH_2OH	H	*B*	3.38[m]	2.84q[n,o] 3.31q[p,o]	4.78m 3.68m[q] 5.21t[r]	←	7.00–7.70m		→	135
H	H	Ph	H	*A*	—	3.10–3.40m	5.50–5.80m	←	7.00–7.40m		→	139
Me	H	CON NMe	H	*C*	3.55[m]	2.65–3.35m[s,t,u]	5.35dd[s,t] 2.35[m] 2.26–2.60m[v] 3.40–3.75m[v]	←	6.90–7.75m		→	135

Me	NHMe	CONHMe	H	D	3.55[m]	3.40–3.65m[w] 2.30m[x] 4.00–4.40m[y]	3.40–3.65m[w] 2.90m[x] 9.00m[y]	←——7.10–8.00m——→	135

[a] δ values in ppm measured from TMS.
[b] Signals are sharp singlets unless denoted as d = doublet; dd = double doublet; t = triplet; q = quartet; oct = octet; m = multiplet.
[c] A = CF_3CO_2H; B = $(CD_3)_2SO$; C = $CDCl_3$; D = C_5D_5N.
[d] $J = 7–8$ Hz.
[e] J values not specified.
[f] CH_2CO_2H.
[g] NCH_2.
[h] δ values not specified.
[i] C(Me).
[j] $J = 7$ Hz.
[k] $J_{H(3)-H(4)cis} \sim 8$ Hz; $J_{H(3)-H(4)trans} \sim 10$ Hz; $J_{gem} \sim 13$ Hz.
[l] $J_{H(3)cis-H(4)} \sim 2$ Hz; $J_{H(3)trans-H(4)} \sim 7$ Hz; $J_{gem} \sim 17$ Hz.
[m] NMe.
[n] $J_{H(3)cis-H(4)} = 1.5$ Hz.
[o] $J_{gem} = 16.5$ Hz.
[p] $J_{H(3)trans-H(4)} = 7.5$ Hz.
[q] CH_2OH.
[r] OH; $J_{OH-CH_2} = 5$ Hz.
[s] $J_{H(3)cis-H(4)} = 3$ Hz.
[t] $J_{H(3)trans-H(4)} = 7$ Hz.
[u] $J_{gem} = 16$ Hz.
[v] Piperazine H.
[w] H(3) and H(4).
[x] NHMe.
[y] NH.

carbonyl substituent in these molecules. Despite their adjacency to the carbonyl center, the protons at the *C*(3) position in 3,4-dihydropyrimido-[1,2-*a*]benzimidazol-2(1*H*)-ones (**7.441**) resonate at higher field (Table 7.55) than those at the *C*(4) position. The chemical shifts (δ 4.15 and 2.90) (cf. Table 7.55) reported[139] for *H*(3) and *H*(4) in 3,4-dihydropyrimido[1,2-*a*]benzimidazol-2(1*H*)-one (**7.441**; $R^1 = R^2 = R^3 = R^4 = H$) are not in accord with the values (δ 3.43 and 4.72) (cf. Table 7.55) obtained in other studies[120] and considered in the context of the ^1H NMR absorption (Table 7.55) of 3,4-dihydropyrimido[1,2-*a*]benzimidazol-2(1*H*)-ones as a whole, appear to be incorrect.

The deshielding effect (Table 7.56) of the *N*(10)-acetyl substituent in the pyrazino[4,3-*a*]benzimidazol-4(10*H*)-one (**7.442**; $R^1 = COMe$, $R^2 = H$,

TABLE 7.56 ^1H NMR SPECTRA[a,b] OF PYRAZINO[4,3-*a*]BENZIMIDAZOL-4(10*H*)-ONES

(**7.442**)

R^1	R^2	R^3	H(1)	H(3)	H(6)	H(7)	H(8)	H(9)	H(10)
H	H	Ph	7.80	4.22[d] 7.10–7.80m[e]	8.65d[f]	← 7.10–7.80m[e] →			—
H	Me	Ph	2.52[g]	4.22[c] 7.00–7.50m[e]	8.82d[f]	← 7.00–7.50m[e] →			—
H	H	o-ClC_6H_4	7.10–7.60m[f]	4.16[d] 7.10–7.60m[h]	8.62d[f]	← 7.10–7.60m[h] →			—
H	Me	o-ClC_6H_4	2.44[g]	4.18[d] 6.90–7.60m[e]	8.71d[f]	← 6.90–7.60m[e] →			—
H	H	1-Naphthyl	7.10–8.40m[h]	4.42[d] 7.10–8.40m[h]	8.60d[f]	← 7.10–8.40m[h] →			—
H	Me	1-Naphthyl	2.40[g]	4.50[d] 7.10–8.10m[e]	8.70[f]	← 7.10–8.10m[e] →			—
COMe	H	1-Naphthyl	8.14	4.20[d] 7.10–7.70m	8.70d[f]	← —[i] →			2.78[j]

[a] δ values in ppm measured from TMS in $(CD_3)_2SO$ as solvent.
[b] Signals are sharp singlets unless denoted as d = doublet; m = multiplet.
[c] From Ref. 148.
[d] CH_2.
[e] *H*(7), *H*(8), *H*(9), and Ar*H*.
[f] $J = 7.2$–8.4 Hz.
[g] *C*(Me).
[h] *H*(1), *H*(7), *H*(8), *H*(9), and Ar*H*.
[i] δ values not specified.
[j] COMe.

TABLE 7.57. ^{1}H NMR SPECTRA[a,b] OF TETRAHYDRO- AND HEXAHYDROPYRAZINO[4,3-*a*]BENZIMIDAZOLES (**7.443**)–(**7.447**)

(**7.443**) (**7.444**) (**7.445**) (**7.446**) (**7.447**)

Compound	Solvent[c]	H(1)	H(2)	H(3)	H(4)	H(6)	H(7)	H(8)	H(9)	H(10a)	Ref.
(**7.443**; $R^1 = R^2 = H$)	*A*	4.00	2.32[d]	←—3.03t[e]—	—→	←—	7.20m	—→	7.63m	—	150
(**7.443**; $R^1 = CO_2Me$, $R^2 = Me$)	*A*	3.33	2.50[f]	—[g]	2.92t[h]	7.25	7.92	3.90[i]	8.33	—	160
(**7.443**; $R^1 = H$, $R^2 = CH_2Ph$)	*A*	3.87	3.66[j] 7.28m[k]	2.86t[l]	3.92t[l]	←—	7.28m[k]	—→	7.70m	—	150
(**7.443**; $R^1 = CO_2Me$, $R^2 = CH_2Ph$)	*A*	3.75–4.50m[m]	3.75–4.50m[m]	3.75–4.50m[m]	2.83–3.17t[m]	←—[g]	—→	3.92[i]	8.35	—	160
(**7.444**)	*B*	—	4.76[j]	←—3.96m—	—→	←—	7.30–7.79m[o]		—→	—	150
(**7.445**)	*A*	2.53[p]	8.58[r]	←—4.20—	—→	←—	7.25m		—→	—	162
(**7.446**)	*C*	—	—	—	5.90	←—	8.00		—→	—	152
(**7.447**)	*A*	3.50m[s]	3.49[j] 7.29[t]	2.18m	2.73m	←—	6.54m		—→	4.79q[u]	150

[a] δ values in ppm measured from TMS.
[b] Signals are sharp singlets unless denoted as t = triplet; q = quartet; m = multiplet.
[c] $A = CDCl_3$; $B = (CD_3)_2SO$; $C = CF_3CO_2H$.
[d] N*H*; exchangeable with D_2O.
[e] $J = 5.7$ Hz.
[f] NMe.
[g] δ values not quoted.
[h] $J = 6$ Hz.
[i] OMe.
[j] *C*H_2Ph.
[k] *H*(6), *H*(7), *H*(8), and CH_2*Ph*.
[l] *J* values not quoted.
[m] *H*(1), *H*(3), and *C*H_2Ph.
[n] $J = 6$ Hz.
[o] *H*(6), *H*(7), *H*(8), *H*(9), and CH_2*Ph*.
[p] SMe.
[q] C=CH.
[r] CHO.
[s] *H*(1) and NH.
[t] CH_2*Ph*
[u] X part of ABX spin system $J_{AX} + J_{BX} = 6$ Hz.

R^3 = 1-naphthyl) is apparent in the low field position (δ 8.14) for *H*(1), which in the *N*(10)-unsubstituted derivative (**7.442**; $R^1 = R^2$ = H, R^3 = Ph) resonates at δ 7.80 (Table 7.56). *H*(1) in 1,2,3,4-tetrahydropyrazino[4,3-*a*]-benzimidazole (Table 7.57; (**7.443**; $R^1 = R^2$ = H)] absorbs at lower field (Table 7.57) than *H*(3) or *H*(4). The deshielding effect of the *C*(3) carbonyl substituent in pyrazino[4,3-*a*]benzimidazole-1,3(2*H*,4*H*)-dione (**7.446**) accounts for the low field position (Table 7.57) of the *C*(4) proton resonance in this molecule. The ^{1}H NMR signal (Table 7.57) of the bridgehead proton at the *C*(10a) position in 2-benzyl-1,2,3,4,10,10a-hexahydropyrazino[4,3-*a*]benzimidazole (**7.447**) appears as a quartet centered at δ 4.79 due to coupling with the protons of the *C*(1) methylene group.

Mass Spectra. Pyrimido[1,2-*a*]benzimidazol-2(1*H*)-ones and pyrimido-[1,2-*a*]benzimidazol-4(10*H*)-ones are readily distinguished on the basis of their mass spectral fragmentation.[127] The mass spectra of both structural

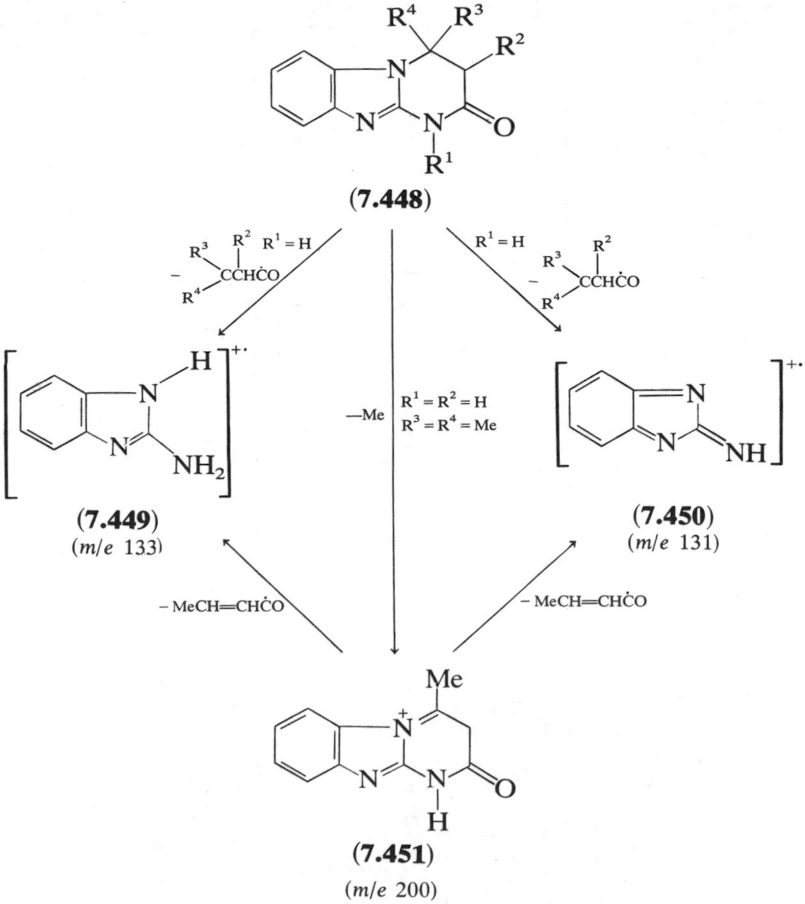

(**7.448**)

(**7.449**)
(*m/e* 133)

(**7.450**)
(*m/e* 131)

(**7.451**)
(*m/e* 200)

Scheme 7.83

types exhibit base peaks corresponding to the molecular ion, but whereas pyrimido[1,2-*a*]benzimidazol-4(10*H*)-ones undergo initial fragmentation by simple loss of the *C*(4) carbonyl moiety, pyrimido[1,2-*a*]benzimidazol-2(1*H*)-ones give rise to fragment ions resulting from the primary extrusion of the *N*(1)–*C*(2)–*C*(3) unit with its attached substituents.[127] Isomeric 3-hydroxy-1,2,3,4-tetrahydropyrimido[1,2-*a*]benzimidazoles can also be differentiated on the basis of their mass spectral fragmentation.[137] The electron-impact induced breakdown of 3,4-dihydropyrimido[1,2-*a*]benzimidazol-2(1*H*)-ones [Scheme 7.83; (**7.448**)][139] typically involves the primary excision of *C*(2)–*C*(3)–*C*(4) and attached substituents as a discrete unit giving fragment ions of the type (**7.449**) and (**7.450**). In the case of 4,4-disubstituted 3,4-dihydropyrimido[1,2-*a*]benzimidazol-2(1*H*)-ones [e.g. (**7.448**; $R^1 = R^2 = H$, $R^3 = R^4 = Me$) major fragmentation is preceded by loss of a *C*(4) substituent giving stable cations [e.g., Scheme 7.83; (**7.451**)].[139]

General Studies

Crystallography. The structures of the 10*H*-[1,4]thiazino[4,3-*a*]benzimidazole [Scheme 7.81; (**7.412**)] and its 1,10a-dihydro derivative, formed[91] in novel fashion from 2-benzimidazolylthioacetonitrile and dimethyl acetylenedicarboxylate, were established[91] by X-ray analysis. The determination of the crystal structure of a 3,4-dihydropyrimido[1,2-*a*]benzimidazol-2(1*H*)-one has revealed the slightly pyramidal configuration of *N*(5) in such molecules.[163]

Ionization Constants. Quantitative information on the basicity of tricyclic 6-5-6 fused benzimidazole structures with one additional heteroatom appears to be available only in the single instance of the 1,2,3,4-tetrahydropyrazino[4,3-*a*]benzimidazole ring system. The strongly basic character of the latter is indicated by the ionization constants ($pK_1 = 2.55 \pm 0.05$ and $pK_2 = 5.39 \pm 0.05$) reported[149] for 2-methyl-1,2,3,4-tetrahydropyrazino[4,3-*a*]benzimidazole.

7.2.3. Reactions

Reactions with Electrophiles

Protonation. The basicity of 3,4-dihydro-1*H*-[1,4]oxazino[4,3-*a*]benzimidazoles is indicated by their ready protonation in mineral acids at *N*(10) to give stable salts from which the parent bases are liberated unchanged on basification.[45] The stable perchlorate salts formed by fully unsaturated pyrimido[1,2-*a*]benzimidazoles are likewise derived by protonation at the *N*(10) position.[114]

TABLE 7.58. ALKYLATION REACTIONS OF 3,4-DIHYDRO-1*H*-[1,4]OXAZINO[4,3-*a*]BENZIMIDAZOLE DERIVATIVES

Reaction conditions[a]	1,2,3,4-Tetrahydro[1,4]oxazino[4,3-*a*] benzimidazole Substrate	Product[b]	m.p. (°C)[c]	Ref.
A	8-Cl-	8-Cl-10-Me-, iodide	242–244[d]	29
B	8-Cl-	8-Cl-10-Et-, iodide	186	60
B	8-CN-	8-CN-10-Et-, iodide	200–210	60
B	7,8-Cl_2-	7,8-Cl_2-10-Et-, iodide	202–205	60
C	7-CN-8-Cl-	7-CN-8-Cl-10-Me, iodide	170	60
B	7-CN-8-Cl-	7-CN-8-Cl-10-Et-, iodide	—[e]	60

[a] *A* = MeI/100° (reaction time not specified); *B* = EtI/(100°, sealed tube) (15–16 hr); *C* = MeI/(110°, sealed tube)(4 hr).
[b] Yields not quoted.
[c] Solvents of crystallization not specified.
[d] Crystallized from water.
[e] Melting point not quoted.

Alkylation. 3,4-Dihydro-1*H*-[1,4]oxazino[4,3-*a*]benzimidazoles react readily with alkyl iodides at elevated temperatures under sealed-tube conditions, giving the corresponding 10-alkyl-3,4-dihydro-1*H*-[1,4]oxazino[4,3-*a*]benzimidazolium iodides in unspecified yield (Table 7.58).[60] The similar ethylation of fully unsaturated pyrimido[1,2-*a*]benzimidazoles using ethyl iodide or ethyl sulfate affords quaternary salts of undetermined structure.[117,164] In accord with their *N*(1)*H* tautomeric structures, pyrimido[1,2-*a*]benzimidazol-2(1*H*)-ones are methylated by diazomethane[120] or methyl iodide in the presence of potassium hydroxide[127] or sodium hydride,[135] specifically at *N*(1) to afford the corresponding methyl derivatives in moderate to good yield (Table 7.59). The outcome of the base-catalyzed alkylation of tautomeric pyrimido[1,2-*a*]benzimidazol-4(10*H*)-ones, on the other hand, appears to be markedly dependent on the nature of the catalyst employed as base. Whereas reaction with methyl iodide–potassium hydroxide is reported[127] to give the *N*(10)-methyl derivative in low yield (Table 7.59), the use of alkyl halides in conjunction with potassium carbonate is variously claimed to yield (Table 7.59) the *N*(1)-alkyl derivative[130] or a mixture of the latter and its *N*(10)-alkyl isomer.[131] The attempted alkylation[144] of pyrimido[3,4-*a*]benzimidazol-1,3(2*H*,4*H*)-dione is reported to afford intractable gums. In contrast, 3,4-dihydropyrimido[3,4-*a*]benzimidazol-1(2*H*)-one is readily alkylated under basic conditions, though in low yield, at *N*(2) [Scheme 7.84; (**7.452**)→(**7.453**)].[144] 1,2,3,4-Tetrahydropyrazino-[4,3-*a*]benzimidazole reacts at the NH substituent with alkyl halides in the presence of triethylamine to give moderate to good yields (Table 7.60) of 2-alkyl-1,2,3,4-tetrahydropyrazino[4,3-*a*]benzimidazole derivatives.[149,150] The methylation of 2-methyl-1,2,3,4-tetrahydropyrazino[4,3-*a*]benzimidazole

TABLE 7.59. ALKYLATION REACTIONS OF PYRIMIDO[1,2-*a*]BENZIMIDAZOLE DERIVATIVES

Reaction conditions[a]	Pyrimido[1,2-*a*]benzimidazole: Substrate	Pyrimido[1,2-*a*]benzimidazole: Product	Yield (%)	m.p. (°C)	Solvent of crystallization	Ref.
A	2(1*H*)-one	1-Me-2(1*H*)-one	47	178–180	Ethyl acetate	120
B	4-Ph-2(1*H*)-one	1-Me-4-Ph-2(1*H*)-one	22	222–224	—[b]	127
C	4-CO_2Me-2(1*H*)-one	1-Me-4-CO_2Me-2(1*H*)-one	70	170–172	Methyl chloride–ether	135
B	2-Ph-4(10*H*)-one	2-Ph-10-Me-4(10*H*)-one	36	218–219	Chloroform	127
D	3-CO_2Et-4(10*H*)-one	3-CO_2Et-10-Me-4(10*H*)-one	47	244	—[c]	130
E	3-CO_2Et-4(10*H*)-one	1-Me-3-CO_2Et-4(1*H*)-one +	—[d]	233–235	Acetone	131
		10-Me-3-CO_2Et-4(10H)-one	—[d]	172–175	Acetone	
E	3-CO_2Et-4(10*H*)-one	1-Et-3-CO_2Et-4(1*H*)-one +	35	211–213	Acetone	131
		10-Et-3-CO_2Et-4(10*H*)-one	32	160–161	Acetone	
F	3-CO_2Et-4(10*H*)-one	1-Pr^n-3-CO_2Et-4(1*H*)-one +	—[d]	158–161	Acetone	131
		10-Pr^n-3-CO_2Et-4(10*H*)-one	—[d]	155–158	Acetone	
F	3-CO_2Et-4(10*H*)-one	1-$(CH_2)_2CHMe_2$-3-CO_2Et-4(1*H*)-one +	—[d]	134–136	Acetone	131
		10-$(CH_2)_2CHMe_2$-3-CO_2Et-4(10*H*)-one	—[d]	123–125	Acetone	
G	3-CO_2Et-4(10*H*)-one	1-$(CH_2)_3NMe_2$-3-CO_2Et-4(1*H*)-one +	—[d]	125–126	Acetone	131
		10-$(CH_2)_3NMe_2$-3-CO_2Et-4(10*H*)-one	—[d]	142–143	Acetone	
F	3-CO_2Et-4(10*H*)-one	1-CH_2Ph-3-CO_2Et-4(1*H*)-one +	—[d]	186–187	Acetone	131
		10-CH_2Ph-3-CO_2Et-4(10*H*)-one	—[d]	188–190	Acetone	
H	2,4-Me_2-	1(10)-Et-2,4-Me_2-, iodide[e]	—[d]	254–255	Ethanol	117
I	3,4-dihydro-2(1*H*)-one	10-$(CH_2)_2CO_2H$-3,4-dihydro-2(10*H*)-one	80	260–262	Water	139

[a] A = CH_2N_2, ether, MeOH, dimethylformamide/(0°)(4.5 hr); *B* = MeI, KOH, EtOH/(room temp.)(14 hr); *C* = NaH, toluene, dimethylformamide/(60°), then MeI/(room temp.)(1 hr) and (60°)(1 hr); *D* = MeI, K_2CO_3, dimethylformamide/(room temp.)(1 hr); *E* = RI, K_2CO_3, dimethylformamide/(80°)(72 hr). *F* = RBr, K_2CO_3, dimethylformamide/(80°)(72 hr); *G* = $Me_2N(CH_2)_3Cl$, K_2CO_3, dimethylformamide/(80°)(72 hr); *H* = EtI/(100°, sealed tube)(4 hr); *I* = $CH_2{=}CHCO_2H$/(100°)(few min., then room temp.)(4 hr).

[b] Solvent of crystallization not specified.

[c] Purified by chromatography.

[d] Yields not quoted.

[e] *N*(1) or *N*(10) position for the Et group not established.

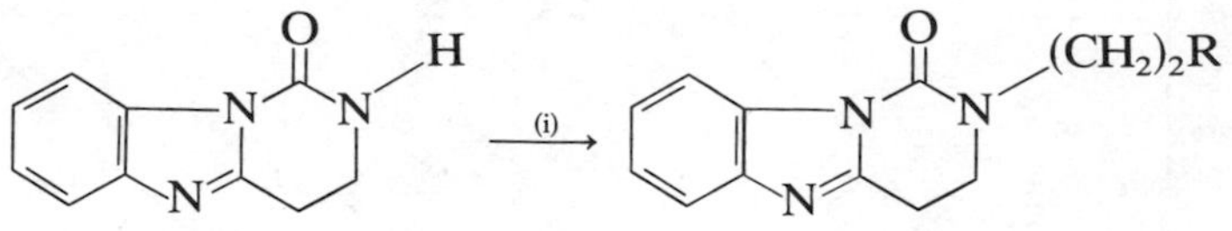

(7.452) —(i)→ (7.453)

R	Yield (%)	m.p. (°C)
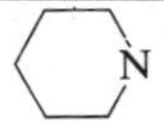 piperidino	21	81–84
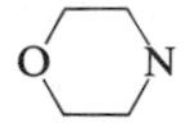 morpholino	34	116–119

(i) $R(CH_2)_2Cl$, $NaNH_2$, dioxane/reflux (time not specified)

Scheme 7.84

also takes place at the *N*(2) position affording the quaternary salt, 2,2-dimethyl-1,2,3,4-tetrahydropyrazino[4,3-*a*]benzimidazolium iodide, in unspecified yield.[149]

ACYLATION. As also observed with 1,2,3,4-tetrahydropyrido[1,2-*a*]-benzimidazole (cf. section 7.1.3, "Reactions with Electrophiles"), the benzoylation of 3,4-dihydro-1*H*-[1,4]oxazino[4,3-*a*]benzimidazole results in both ring-opening to 1-(2-benzamidophenyl)morpholin-2(1*H*)-one and acylation of the oxazine nucleus to give the enol benzoate [Scheme 7.85; (**7.454**)].[42] Analogy with the behavior of 7-amino-1,2,3,4-tetrahydropyrido-[1,2-*a*]benzimidazole (cf. section 7.1.3, "Reactions with Electrophiles") also

TABLE 7.60. ALKYLATION REACTIONS OF 1,2,3,4-TETRAHYDROPYRAZINO-[4,3-*a*]BENZIMIDAZOLE DERIVATIVES

Reaction conditions[a]	1,2,3,4-tetrahydropyrazino[4,3-*a*] benzimidazole: Substrate	Product	Yield (%)	m.p. (°C)	Solvent of crystallization	Ref.
A	Unsubstituted	2-Me	32	145–147	Acetone–ether	149
B	Unsubstituted	2-Et	36	107–108	Cyclohexane	150
C	Unsubstituted	2-Pr^n	26	75.5–77	Hexane	150
D	Unsubstituted	2-CH_2CN	70	176–176.5	Acetone	150
E	Unsubstituted	2-CH_2Ph	66	124–125	Cyclohexane	150
F	Unsubstituted	2-*p*-$MeOC_6H_4CH_2$	57	150–151	Cyclohexane	150
G	2-Me	2,2-Me_2, iodide	—[b]	212–215 (decomp.)	—[c]	149

[a] A = MeI, acetone/(reflux)(10 min); *B* = EtBr, Et_3N, acetone/(reflux)(138 hr); *C* = Pr^nBr, Et_3N, acetone/(reflux)(138 hr); *D* = $NCCH_2Cl$ (reflux)(48 hr); *E* = $PhCH_2Cl$/(reflux)(31 hr); *F* = *p*-$MeOC_6H_4CH_2Cl$/(reflux)(18 hr); *G* = MeI (reaction conditions not specified).
[b] Yield not quoted.
[c] Solvent of crystallization not specified.

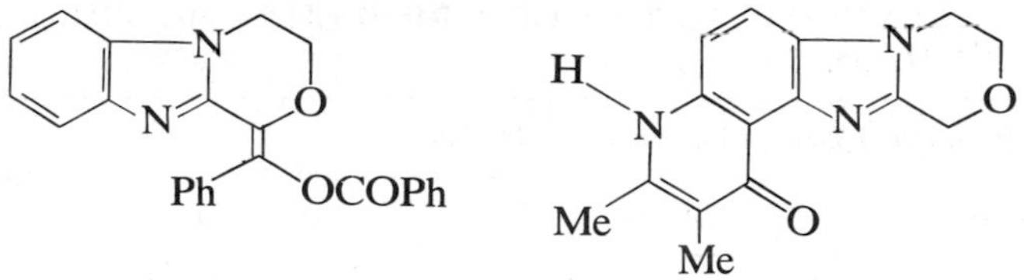

(**7.454**) (**7.455**)

Scheme 7.85

extends to the reaction of 8-amino-3,4-dihydro-1*H*-[1,4]oxazino[4,3-*a*]-benzimidazole with ethyl 2-methylacetoacetate in the presence of ethyl polyphosphate to give the fused pyridinone derivative [Scheme 7.85; (**7.455**)].[69] The methylene substituent adjacent to the carbonyl group in 4*H*-[1,3]thiazino[3,2-*a*]benzimidazol-4(3*H*)-one unlike that in 2,3-dihydrothiazolo[3,2-*a*]benzimidazol-3-one (cf. Chapter 6, section 6.2.3, "Reactions with Electrophiles") is insufficiently reactive to undergo acylative condensation with aromatic aldehydes even under forcing conditions.[102] In contrast, the *C*(3) methylene substituent in 1*H*-[1,4]thiazino[4,3-*a*]benzimidazol-4(3*H*)-one is reactive in this sense and condenses with aromatic aldehydes in hot glacial acetic acid to afford (albeit in low yield) arylidene derivatives of the type [Scheme 7.86; (**7.456**)].[103] The related acylative condensation of 1*H*-[1,4]thiazino[4,3-*a*]benzimidazol-4(3*H*)-one with heterocyclic enamines provides the basis for the synthesis of cyanine dyes containing a 1*H*-[1,4]thiazino[4,3-*a*]benzimidazol-4(3H)-one nucleus.[110]

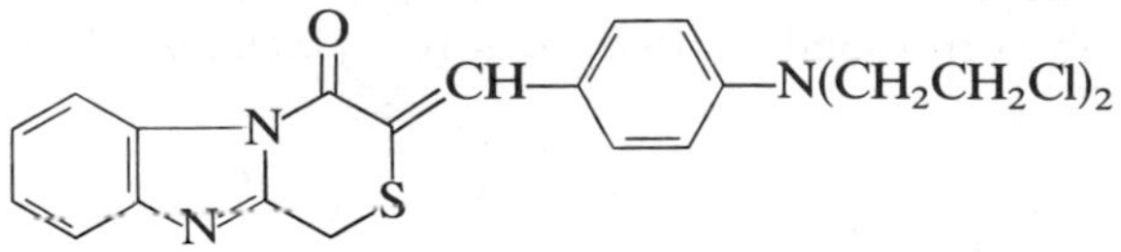

(**7.456**)

(m.p. 217–218°)

Scheme 7.86

The acylation of pyrimido[1,2-*a*]benzimidazolone derivatives can take place at nitrogen, oxygen, or carbon, depending on the nature of the substrate and the type of acylating agent used. Acetylation of 3-ethoxycarbonylpyrimido[1,2-*a*]benzimidazol-4(10*H*)-one in hot acetic anhydride takes place at nitrogen giving the *N*(10)-acetyl derivative in moderate yield (Table 7.61).[130] On the other hand, the sodium ethoxide catalyzed *p*-nitrobenzoylation of pyrimido[1,2-*a*]benzimidazol-2,4(3*H*,10*H*)-diones affords products in 60–70% yield (Table 7.61) formulated[133] as *C*(2) *p*-nitrobenzoyloxy derivatives on the basis of their IR carbonyl absorption and ready

TABLE 7.61. ACYLATION REACTIONS OF PYRIMIDO[1,2-*a*]BENZIMIDAZOL-4(10*H*)-ONES

Reaction conditions[a]	Pyrimido[1,2*a*]benzimidazol-4(10*H*)-one: Substrate	Pyrimido[1,2*a*]benzimidazol-4(10*H*)-one: Product	Yield (%)	m.p. (°C)	Solvent of crystallization	Ref.
A	3-CO_2Et-4(10*H*)-one	3-CO_2Et-10-COMe-4(10*H*)-one	52	181	Toluene–light petroleum	130
B	2-OH-10-Me-4(10*H*)-one	2-*p*-$O_2NC_6H_4CO_2$-10-Me-4(10*H*)-one	60–70	258–260	Methanol	133
B	2-OH-10-Et-4(10*H*)-one	2-*p*-$O_2NC_6H_4CO_2$-10-Et-4(10*H*)-one	60–70	195–196	Methanol	133
B	2-OH-10-Pr^n-4(10*H*)-one	2-*p*-$O_2NC_6H_4CO_2$-10-Pr^n-4(10*H*)-one	60–70	177–178	Methanol	133
B	2-OH-10-Bu^n-4(10*H*)-one	2-*p*-$O_2NC_6H_4CO_2$-10-Bu^n-4(10*H*)-one	60–70	168–169	Methanol	133
C	2-OH-4(10*H*)-one	3-(=CHNHPh)-2,4-(3*H*,10*H*)-dione	64	—[b]	—[c]	165

[a] *A* = Ac_2O/(reflux)(2 hr); *B* = NaOEt, EtOH, then *p*-$O_2NC_6H_4COCl$, benzene/(100°)(2 hr); *C* = PhN=CHOEt, *N*-methylpyrrolidone/(heat)(few min.).
[b] Melting point not quoted.
[c] Solvent of crystallization not specified.

hydrolysis to the starting materials. In further contrast, pyrimido[1,2-*a*]-benzimidazol-2,4(3*H*,10*H*)-dione is aminomethyleneated at the *C*(3) position in good yield (Table 7.61) by heating with ethyl isoformanilide in *N*-methylpyrrolidone.[165] The NH substituents in 1-(*p*-chlorophenyl)-1,2,3,4-tetrahydropyrimido[3,4-*a*]benzimidazole[143] and 1,2,3,4,10,10a-hexahydro-pyrazino[4,3-*a*]benzimidazole[150] react in orthodox fashion with isothiocyanates to give the corresponding thioureas [e.g., Scheme 7.87; (**7.457**) and (**7.458**)].

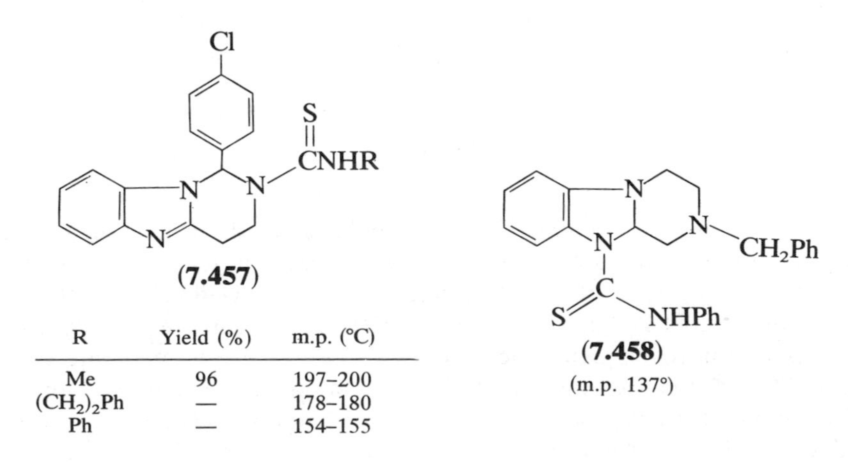

R	Yield (%)	m.p. (°C)
Me	96	197–200
$(CH_2)_2Ph$	—	178–180
Ph	—	154–155

Scheme 7.87

TABLE 7.62. CHLORINATION AND NITRATION REACTIONS OF 3,4-DIHYDRO-1*H*-[1,4]OXAZINO[4,3-*a*]BENZIMIDAZOLE DERIVATIVES

Reaction conditions[a]	1,2,3,4-tetrahydro[1,4]oxazino[4,3-*a*]-benzimidazole: Substrate	Product	Yield (%)	m.p. (°C)	Solvent of crystallization	Ref.
A	7-NH_2-8-CF_3-	6-Cl-7-NH_2-8-CF_3-	90	156–157	—[b]	154
B	8-CF_3-	7-NO_2-8-CF_3-	92	186–187	—[b]	154
C	8-Cl-	7-NO_2-8-Cl-	—[c]	220	—[b]	60
D	7-NHCOMe-	7-NHCOMe-8-NO_2-	65	262	Ethanol	49
D	8-NHCOMe-	7-NO_2-8-NHCOMe-	42	255	Ethanol	49
B	7-Cl-8-CF_3-	6-NO_2-7-Cl-8-CF_3-	95	196–197.5	Chloroform–*n*-hexane	154

[a] *A* = SO_2Cl_2, AcOH/(13°)(15 min); *B* = 90% HNO_3, conc. H_2SO_4/(0°) then (room temp.)(2 hr); *C* = conc. HNO_3, conc. H_2SO_4/0–5° (reaction time not specified); *D* = KNO_3, conc. H_2SO_4/(0–5°), then (room temp.)(1 hr).
[b] Solvent of crystallization not specified.
[c] Yield not quoted.

Electrophilic Substitution Reactions. With the exception of studies of the orthodox chlorination,[154] nitration,[29,49,60,154] and diazotization[48,49,60] reactions (Table 7.62) of 3,4-dihydro-1*H*-[1,4]oxazino[4,3-*a*]benzimidazoles, and the isolated report[123] of the bromination of 2-methylpyrimido[1,2-*a*]-benzimidazol-4(10*H*)-one to give the dibromo derivative [Scheme 7.88; (**7.459**)], the investigation of the susceptibility of tricyclic 6-5-6 fused benzimidazole ring systems with one additional heteroatom, to electrophilic substitution has been virtually neglected.

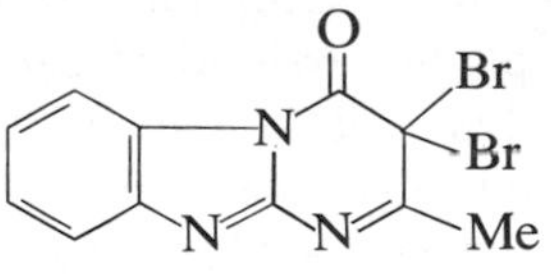

(7.459)
(m.p. 300°)

Scheme 7.88

Reactions with Nucleophiles

Hydroxylation and Related Reactions. The 3,4-dihydro-1*H*-[1,4]oxazino[4,3-*a*]benzimidazole ring system is stable to acidic hydrolysis under conditions applicable to the conversion of acetamido derivatives into the corresponding amines [e.g., Scheme 7.89; (**7.460**; $R^1 = NH_2$, $R^3 = NO_2$) and (**7.460**; $R^1 = NO_2$, $R^2 = NH_2$)].[48,49] However, as in the case of 1,2,3,4-tetrahydropyrido[1,2-*a*]benzimidazole (cf. section 7.1.3, "Reactions with Nucleophiles,"

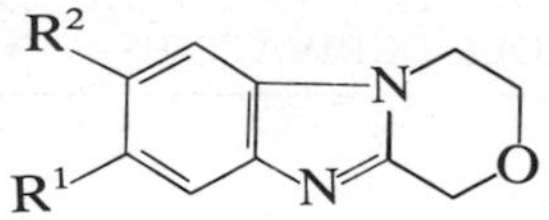

(**7.460**)

R^1	R^2	m.p. (°C)
NH_2	NO_2	262
NO_2	NH_2	300
Cl	CN	300
N_3	NO_2	163 (decomp.)
NO_2	N_3	145 (decomp.)

Scheme 7.89

Scheme 7.40) coordination at the *N*(10) position destabilizes 3,4-dihydro-1*H*-[1,4]oxazino[4,3-*a*]benzimidazole to hydrolytic attack, thus accounting for the ring-opening to 1-(2-benzamidophenyl)morpholin-2(1*H*)-one which occurs on treatment with benzoyl chloride in the presence of alkali.[42]

Acidic hydrolysis of the ester substituent in 4-amino-3-ethoxycarbonyl-pyrimido[1,2-*a*]benzimidazole is accompanied by hydrolytic removal of the amino group, the end-product obtained in good yield (Table 7.63) being pyrimido[1,2-*a*]benzimidazol-4(10*H*)-one-3-carboxylic acid.[120] The hydrolytic stability of the pyrimido[1,2-*a*]benzimidazol-4-one framework thereby demonstrated is further substantiated by its survival intact under both acidic[120,130] and alkaline[131] conditions suitable for the hydrolysis (Table 7.63) of *C*(3) cyano and ester derivatives of pyrimido[1,2-*a*]benzimidazol-4(1*H* and 10*H*)-ones to the corresponding carboxylic acids. However, the pyrimidinone ring in pyrimido[1,2-*a*]benzimidazol-4(1*H* and 10*H*)-ones is cleaved under more forcing alkaline conditions giving benzimidazole derivatives.[123,130] The dihydropyrimidinone ring in 3,4-dihydropyrimido[1,2-*a*]-benzimidazol-2(1*H*)-ones is also susceptible to hydrolytic scission under alkaline conditions.[135] The parent pyrimido[1,2-*a*]benzimidazol-2(1*H*)-one ring system, on the other hand, is stable enough to allow the alkaline hydrolysis of ester to carboxyl substituents (Table 7.63),[135] and the conversion of pyrimido[1,2-*a*]benzimidazol-2(1*H*)-one into pyrimido[1,2-*a*]benz-imidazol-2(1*H*)-thione by heating with phosphorus pentasulfide in pyridine (Table 7.63).[120] 1-Substituted 1,2,3,4-tetrahydropyrimido[3,4-*a*]benzimid-azole derivatives are decomposed by acids and bases to aldehydes and 2-(β-aminoethyl)benzimidazole in processes[143] corresponding to the reverse of their formation (cf. section 7.2.1, "Ring-closure Reactions of Benzimidazole Derivatives").

The pyrazino[4,3-*a*]benzimidazole and 1,2,3,4-tetrahydropyrazino[4,3-*a*]benzimidazole ring systems are stable under acidic conditions, which effect the hydrolytic removal of a *C*(3) ethoxycarbonyl substituent in the former[147] and an *N*(2) acetyl group in the latter.[149] The success of the sodium methoxide catalyzed Stevens rearrangement [Scheme 7.90; (**7.461**)→(**7.462**)][149] also implies the relative stability of the 1,2,3,4-tetrahydro-pyrazino[4,3-*a*]benzimidazole ring system to basic solvolysis.

TABLE 7.63. HYDROLYTIC AND RELATED REACTIONS OF PYRIMIDO[1,2-*a*]BENZIMIDAZOLE DERIVATIVES

Reaction conditions[a]	Pyrimido[1,2-*a*]benzimidazole Substrate	Product	Yield (%)	m.p. (°C)	Solvent of crystallization	Ref.
A	1-Me-3-CO_2Me-2(1*H*)-one	1-Me-3-CO_2H-2(1*H*)-one	—[b]	207	Water	135
B	3-CO_2Et-4-NH_2-	3-CO_2H-4(10*H*)-one	79	280–282	Dimethylformamide	120
C	3-CO_2Et-4(10*H*)-one	3-CO_2H-4(10*H*)-one	—[b]	—	—	130
B	3-CN-4(10*H*)-one	3-CO_2H-4(10*H*)-one	—[b]	—	—	120
C	3-CO_2Et-10-Me-4(10*H*)-one	3-CO_2H-10-Me-4(10*H*)-one	74	>300	Dimethylformamide	130
D	3-CO_2Et-10-Me-4(10*H*)-one	3-CO_2H-10-Me-4(10*H*)-one	59	213–216	—[c]	131
D	1-Et-3-CO_2Et-4(1*H*)-one	1-Et-3-CO_2H-4(1*H*)-one	81	252–253	—[c]	131
D	3-CO_2Et-10-Et-4(10*H*)-one	3-CO_2H-10-Et-4(10*H*)-one	79	228–230	Dimethylformamide	131
E	2(1*H*)-one	2(1*H*)-thione	—[b]	310–313 (decomp.)	Dimethylformamide	120

[a] *A* = 1 *M* NaOH/(room temp.)(1.5 hr); *B* = dil. HCl/(reflux)(0.5 hr); *C* = conc. HCl/(reflux)(2 hr); *D* = KOH, EtOH/(room temp.)(15 hr); *E* = P_2S_5, pyridine/(reflux)(1.5 hr).

[b] Yield not quoted.

[c] Not crystallized.

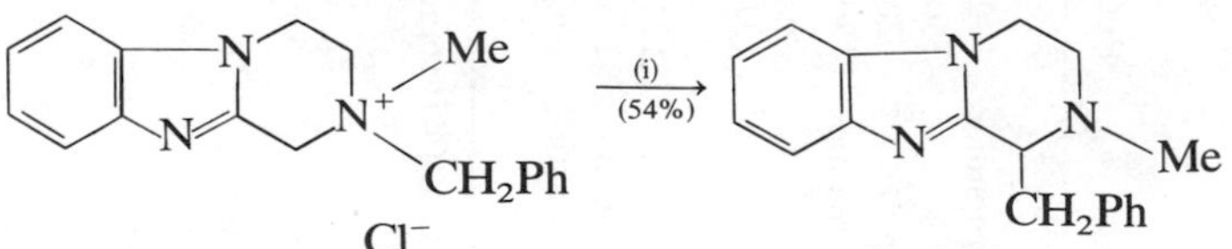

(**7.461**) Cl^- → (i) (54%) → (**7.462**) (m.p. 116–117°)

(i) NaOMe, MeOH/(20°)(40 min)

Scheme 7.90

Amination. 4-Chloropyrimido[1,2-*a*]benzimidazoles, derived by the orthodox chlorination of pyrimido[1,2-*a*]benzimidazol-4(10*H*)-ones with phosphorus oxychloride,[111,116] react with amines under mild conditions to afford moderate to good yields (Table 7.64) of 4-aminopyrimido[1,2-*a*]-benzimidazole derivatives.[111,116] Pyrimido[1,2-*a*]benzimidazol-2(1*H*)-thione is reported[120] to react with ammonia in diethylene glycol dimethyl ether to give 2-aminopyrimido[1,2-*a*]benzimidazole, though in unspecified yield. The aminolysis of ester substituents in pyrimido[1,2-*a*]benzimidazol-4(10*H*)-ones[131] and pyrimido[1,2-*a*]benzimidazol-2(1*H*)-ones[135] proceeds readily to give good yields (Table 7.64) of the corresponding amides. In the reactions of ester substituted pyrimido[1,2-*a*]benzimidazol-2(1*H*)-ones with amines at elevated temperatures, amide formation is accompanied by addition of the amine across the *C*(3)–*C*(4) double bond (Table 7.64).[135] 3,4-Dihydropyrimido[1,2-*a*]benzimidazol-2(1*H*)-one carboxylic esters can also be converted into amides on brief treatment with amines under mild conditions (Table 7.64).[135] However, on prolonged treatment aminolytic scission of the *N*(1)–*C*(2) bond tends to occur with ring-opening to benzimidazole derivatives.[135] 3,4-Dihydropyrimido[3,4-*a*]benzimidazol-1(2*H*)-thione is also ring-opened at the *C*(1)–*N*(2) bond by reaction with amines or hydrazine giving the corresponding *N*-[2-(2-benzimidazolyl)ethyl]thiourea and thiosemicarbazide derivatives, respectively.[143] Treatment of pyrazino[4,3-*a*]benzimidazol-3,4-(1*H*,2*H*)-diones with amines, even under mild conditions, results in cleavage of the pyrazinedione ring giving benzimidazole derivatives.[151]

Miscellaneous Reactions. The formally nucleophilic displacement of diazonium substituents in 3,4-dihydro-1*H*-oxazino[4,3-*a*]benzimidazoles under the conditions of the Sandmeyer reaction[60] and by reaction with azide ion[48,49] provides synthetic access to chloro- and azido-substituted 3,4-dihydro-1*H*-[1,4]oxazino[4,3-*a*]benzimidazole derivatives (cf. Scheme 7.89).

Oxidation

Information on the susceptibility of tricyclic 6-5-6 fused benzimidazole structures with one additional heteroatom to oxidation is in general terms

TABLE 7.64. AMINATION REACTIONS OF PYRIMIDO[1,2-*a*]BENZIMIDAZOLE DERIVATIVES

Reaction conditions[a]	Pyrimido[1,2-*a*]benzimidazole Substrate	Product	Yield (%)	m.p. (°C)	Solvent of crystallization	Ref.
A	2-Me-4-Cl-	2-Me-4-NH$(CH_2)_3NMe_2$-	60	198–200	Ethyl acetate	111
A	2-Me-4-Cl-	2-Me-4-NH$(CH_2)_3NEt_2$-	45	100–102	*n*-Heptane	111
A	2-Me-4-Cl-	2-Me-4-NH$(CH_2)_2NEt_2$-	45	180–181	Ethyl acetate	111
B	2-Me-4,7-Cl_2- +	2-Me-4-NH$(CH_2)_2NEt_2$-7-Cl-	10	158–160	Ethanol–water	111
	2-Me-4,8-Cl_2-	2-Me-4-NH$(CH_2)_2NEt_2$-8-Cl-	16	209–211	Ethyl acetate	
C	2-Me-4-Cl-	2-Me-4-NH$(CH_2)_3$N[ring: piperidine]	60	155	Benzene	166
C	2-Me-4-Cl-	2-Me-4-NH$(CH_2)_2$OH-	75	260–262	Ethanol	166
D	2-Me-4-Cl-	2-Me-4-NEt_2-	53	107–108	Light petroleum	166
C	2-Me-4-Cl-	2-Me-4-N[ring: piperidine]	50	188–189	—[b]	166
E	2-Me-4-Cl-	2-Me-4-N[ring: morpholine]	40	248	Amyl acetate	166
F	2-Me-4(10*H*)-one	2-Me-4-*p*-MeC_6H_4NH-	95	312–313	Dimethylformamide	166
G	2(1*H*)-thione	2-NH_2-	—[c]	—[d]	—[e]	120
H	1-Me-4-CO_2Me-2(1*H*)-one	1-Me-4-CONHMe-2(1*H*)-one	96	302–303	Dimethylformamide-ether	135
I	1-Me-4-CO_2Me-2(1*H*)-one	1-Me-3-NHMe-4-CONHMe-3,4-dihydro-2(1*H*)-one	52	148–149[f]	Methanol	135
J	1-Me-3-CO_2Et-4(1*H*)-one	1-Me-3-CONHMe-4(1*H*)-one	65	325–326	Dimethylformamide	131
J	1-Et-3-CO_2Et-4(1*H*)-one	1-Et-3-CONHMe-4(1*H*)-one	88	270–272	Dimethylformamide	131
J	1-CH_2Ph-3-CO_2Et-4(1*H*)-one	1-CH_2Ph-3-CONHMe-4(1*H*)-one	71	280–283	Dimethylformamide	131
K	1-Me-3-CO_2Et-4(1*H*)-one	1-Me-3-CONHBu^n-4(1*H*)-one	88	266–267	Dimethylformamide	131
K	1-Et-3-CO_2Et-4(1*H*)-one	1-Et-3-CONHBu^n-4(1*H*)-one	76	232–234	Dimethylformamide	131
K	1-Pr^n-3-CO_2Et-4(1*H*)-one	1-Pr^n-3-CONHBu^n-4(1*H*)-one	81	219–221	Dimethylformamide	131
K	1-CH_2Ph-3-CO_2Et-4(1*H*)-one	1-CH_2Ph-3-CONHBu^n-4(1*H*)-one	75	242–243	Dimethylformamide	131
K	1-Et-3-CO_2Et-4(1*H*)-one	1-Et-3-CONHCH(Me)Et-4(1*H*)-one	85	260–261	Dimethylformamide	131
L	3-CO_2Et-10-Me-4(10*H*)-one	3-CONHMe-10-Me-4(10*H*)-one	68	180–182	Dimethylformamide	131
L	3-CO_2Et-10-Et-4(10*H*)-one	3-CONHMe-10-Et-4(10*H*)-one	63	178–180	Acetone	131
L	3-CO_2Et-10-CH_2Ph-4(10*H*)-one	3-CONHMe-10-CH_2Ph-4(10*H*)-one	70	165–167	Dimethylformamide	131
K	3-CO_2Et-10-Me-4(10*H*)-one	3-CONHBu^n-10-Me-4(10*H*)-one	81	175–177	Ethyl acetate	131

TABLE 7.64 (*Continued*)

Reaction conditions[a]	Pyrimido[1,2-*a*]benzimidazole Substrate	Product	Yield (%)	m.p. (°C)	Solvent of crystallization	Ref.
K	3-CO_2Et-10-Et-4(10*H*)-one	3-CONHBu^n-10-Et-4(10*H*)-one	74	118–120	Ethyl acetate	131
K	3-CO_2Et-10-Pr^n-4(10*H*)-one	3-CONHBu^n-10-Pr^n-4(10*H*)-one	69	211–212	Ethyl acetate	131
K	3-CO_2Et-10-$(CH_2)_2NMe_2$-4(10*H*)-one	3-CONHBu^n-10-$(CH_2)_2NMe_2$-4(10*H*)-one	77	112–113	Ethyl acetate	131
K	3-CO_2Et-10-Et-4(10*H*)-one	3-CONHCH(Me)Et-10-Et-4(10*H*)-one	68	108–110	Ethyl acetate	131
K	3-CO_2Et-10-Et-4(10*H*)-one	3-CONH$(CH_2)_3NMe_2$-10-Et-4(10*H*)-one	71	80–82	Ethyl acetate	131
M	3-CO_2H-10-Me-4(10*H*)-one	3-CON O-10-Me-4(10*H*)-one	72	188–190	Dimethylformamide–water	131
M	3-CO_2H-10-Me-4(10*H*)-one	3-CON NMe-10-Me-4(10*H*)-one	63	288–290	Ethyl acetate	131
M	3-CO_2H-10-Et-4(10*H*)-one	3-CON -10-Et-4(10*H*)-one	70	145–147	Ethyl acetate	131
M	3-CO_2H-10-Et-4(10*H*)-one	3-CON O-10-Et-4(10*H*)-one	76	200–202	Ethyl acetate	131
M	3-CO_2H-10-Et-4(10*H*)-one	3-CON NMe-10-Et-4(10*H*)-one	73	180–182	Ethyl acetate	131
M	3-CO_2H-10-Et-4(10*H*)-one	3-CONHPh-10-Et-4(10*H*)-one	68	241–242	Ethyl acetate	131
M	3-CO_2H-10-CH_2Ph-4(10*H*)-one	3-CON O-10-CH_2Ph-4(10*H*)-one	65	203–205	Dimethylformamide	131
N	1-Me-3-CO_2Me-3,4-dihydro-2(1*H*)-one	1-Me-3-CONH$(CH_2)_2NMe_2$-3,4-dihydro-2(1*H*)-one	82	206–207	Methylene chloride–ether	135

[a] *A* = amine, $CHCl_3$/(room temp.)(21 hr); *B* = $Et_2N(CH_2)_2NH_2$, $CHCl_3$/(room temp.)(3 days); *C* = amine, EtOH/(reflux)(2–3 hr); *D* = Et_2NH, Pr^{n}OH/(60°)(4 hr); *E* = Morpholine, Bu^{n}OH/(100°)(2 hr); *F* = $POCl_3$/(reflux)(1 hr), then *p*-$MeC_6H_4NH_2$/(reflux)(1.5 hr); *G* = NH_3, diethylene glycol dimethyl ether/(120°)(3 hr); *H* = $MeNH_2$, CH_2Cl_2/(−20°)(1 hr); *I* = $MeNH_2$, CH_2Cl_2/(room temp.)(1 hr); *J* = $MeNH_2$, Bu^{n}OH/(120°, autoclave)(12 hr); *K* = amine/(reflux)(12 hr); *L* = $MeNH_2$, EtOH/(100°, autoclave)(12 hr); *M* = $SOCl_2$/(reflux) (reaction time not specified), then amine, toluene/(room temp.)(6 hr); *N* = $Me_2N(CH_2)_2NH_2$/(room temp.)(1.5 hr).

[b] Not crystallized.

[c] Yield not quoted.

[d] Melting point not quoted.

[e] Solvent of crystallization not specified.

[f] Remelts at 250°.

lacking. However, the relative stability of the 3,4-dihydro-1*H*-[1,4]oxazino-[4,3-*a*]benzimidazole ring system to oxidation may be inferred from the dehydrogenation of 3,4,4a,5-tetrahydro-1*H*-[1,4]oxazino[4,3-*a*]benzimidazole to 3,4-dihydro-1*H*-[1,4]oxazino[4,3-*a*]benzimidazole by a variety of oxidizing agents.[58] The relative stability of the 1,2,3,4-tetrahydropyrazino[4,3-*a*]-benzimidazole ring system to oxidizing conditions is also indicated by the successful formation of its derivatives by palladium–charcoal-mediated dehydrogenation of 1,2,3,4,10,10a-hexahydropyrazino[4,3-*a*]benzimidazoles.[150]

TABLE 7.65. CATALYTIC REDUCTION OF 3,4-DIHYDRO-1*H*-[1,4]OXAZINO [4,3-*a*]BENZIMIDAZOLE DERIVATIVES

Reaction conditions[a]	1,2,3,4-Tetrahydro[1,4]oxazino[4,3-*a*]benzimidazole		m.p.(°C)	Ref.
	Substrate	Product		
A	7-NO_2-8-CF_3-	7-NH_2-8-CF_3-[b]	204–205[c]	154
B	7-NO_2-8-Cl-	7-NH_2-8-Cl-	264	60
C	6-NO_2-7-Cl-8-CF_3-	6-NH_2-7-Cl-8-CF_3-[d]	212–212.5	154
D	8-Me-, *N*-oxide	8-Me-	165–168	154
D	8-Pr^i-, *N*-oxide	8-Pr^i-	95–97.5	154
D	1-Me-8-Pr^i-, *N*-oxide	1-Me-8-Pr^i-	91–93	154
D	3-Me-8-Pr^i-, *N*-oxide	3-Me-8-Pr^i-	148–150	154
D	8-Bu^t-, *N*-oxide	8-Bu^t-	127–130	154
D	8-CF_3-, *N*-oxide	8-CF_3-[e]	129–131	155
D	1-Me-8-CF_3-, *N*-oxide	1-Me-8-CF_3-	114–116	154
D	3-Me-8-CF_3-, *N*-oxide	3-Me-8-CF_3-	155–156	154
D	1,3-Me_2-8-CF_3-, *N*-oxide[f]	1,3-Me_2-8-CF_3-[f]	83–85	154
D	1,3-Me_2-8-CF_3-, *N*-oxide[g]	1,3-Me_2-8-CF_3-[g]	95–98	154
D	4,4-Me_2-8-CF_3-, *N*-oxide	4,4-Me_2-8-CF_3-	82–84	154
D	1-Et-8-CF_3-, *N*-oxide	1-Et-8-CF_3-	133–134	154
D	3-Et-8-CF_3-, *N*-oxide	3-Et-8-CF_3-	122–123	154
D	4-Et-8-CF_3-, *N*-oxide	4-Et-8-CF_3-	108–109	154
D	8-Cl-, *N*-oxide	8-Cl-	180–185	154
D	8-F-, *N*-oxide	8-F-	126–128	154
D	8-SO_2Me-, *N*-oxide	8-SO_2Me-	214–215	154
D	7,8-Me_2-, *N*-oxide	7,8-Me_2-	187.5–188.5	154
D	7-Cl-8-Me-, *N*-oxide	7-Cl-8-Me-	191–193	154
D	7-Cl-8-CF_3-, *N*-oxide	7-Cl-8-CF_3-	160–161	154
D	7-OMe-8-CF_3-, *N*-oxide	7-OMe-8-CF_3-	118–120	154
D	7-CN-8-CF_3-, *N*-oxide	7-CN-8-CF_3-	188–190	154
D	7,8-Cl_2-, *N*-oxide	7,8-Cl_2-	187–189.5	154
D	7-Cl-8-SO_2Me-, *N*-oxide	7-Cl-8-SO_2Me-	219–221	154
D	7-OMe-8-SO_2Me-, *N*-oxide	7-OMe-8-SO_2Me-	209–212	154

[a] *A* = H_2, 10% PdC, MeOH, ethylene glycol monomethyl ether/room temp., atm. press; *B* = H_2, Raney Ni, ethylene glycol monomethyl ether/room temp., atm. press; *C* = H_2, PtO_2, MeOH, dimethoxyethane/room temp., atm. press; *D* = H_2, PtO_2, MeOH/room temp., atm. press.

[b] Yield 97%.

[c] Crystallized from chloroform.

[d] Yield 92%.

[e] Yield 88%.

[f] *Cis* isomer.

[g] *Trans* isomer.

TABLE 7.66. REDUCTION OF PYRIMIDO[1,2-*a*]BENZIMIDAZOLE DERIVATIVES

Reaction conditions[a]	Pyrimido[1,2-*a*]benzimidazole: Substrate	Product	Yield (%)	m.p. (°C)	Solvent of crystallization	Ref.
A	2-Me-4-Cl-	2-Me-	49	230–232	Ethanol	111
B	1-Me-4-CO_2Me-2(1*H*)-one	1-Me-4-CO_2Me-3,4-dihydro-2(1*H*)-one	94	177–179	Methanol	135
C	1-$(CH_2)_2NMe_2$-4-CH_2OH-2(1*H*)-one	1-$(CH_2)_2NMe_2$-4-Me-2(1*H*)-one	—[b]	152–154	Methanol–ether	135
D	1-Me-4-CO_2Me-2(1*H*)-one	1-Me-4-CH_2OH-2(1*H*)-one	54	287–288	Methanol–ether	135
		+ 1-Me-4-CH_2OH-3,4-dihydro-2(1*H*)-one	28	194–197	Methanol–ether	
D	1-$(CH_2)_2NMe_2$-4-CO_2Me-2(1*H*)-one	1-$(CH_2)_2NMe_2$-4-CH_2OH-2(1*H*)-one	—[b]	197–198	Ethyl acetate	135
		+ 1-$(CH_2)_2NMe_2$-4-CH_2OH-3,4-dihydro-1(1*H*)-one	—[b]	152–154	Methylene chloride–ether	

[a] *A* = H_2, 5% PdC, NaOAc, EtOH/room temp., atm. press; *B* = H_2, PtO_2, MeOH/room temp., atm. press; *C* = H_2, PtO_2, tartaric acid, H_2O/(room temp., atm. press.)(7 hr); *D* = $NaBH_4$, MeOH/(room temp.)(30 min).

[b] Yield not quoted.

TABLE 7.67. REDUCTION OF PYRAZINO[4,3-*a*]BENZIMIDAZOLE DERIVATIVES

Reaction conditions[a]	Pyrazino[4,3-*a*]benzimidazole Substrate	Product	Yield (%)	m.p. (°C)	Solvent of crystallization	Ref.
A	1-CO_2Et-3-Ph-	1-CO_2Et-3-Ph-1,2,3,4-tetrahydro-	55	194–196	Ethyl acetate	147
B	2-CH_2Ph-1,2,3,4-tetrahydro-[b]	1,2,3,4-tetrahydro-	97	130–132	—[c]	150
C	2-Bu^n-8-NO_2-1,2,3,4-tetrahydro-	2-Bu^n-8-NH_2-1,2,3,4-tetrahydro-	77	122–124	Ether–light petroleum (b.p. 30–60°)	26
			—			
D	2-CH_2Ph-3,4-dihydro-1(2*H*)-one	2-CH_2Ph-1,2,3,4,10,10a-hexahydro-	62	82–88	—[d]	150
E	3,4-dihydro-1(2*H*)-one	1,2,3,4-tetrahydro-	60	130–132	Benzene	150
E	2-CH_2Ph-3,4-dihydro-1(2*H*)-one	2-CH_2Ph-1,2,3,4-tetrahydro-	76	124–125	Cyclohexane	150
F	2-CH_2Ph-3,4-dihydro-1(2*H*)-one	1,2,3,4-tetrahydro-	88	130–132	Benzene	150
G	2-CH_2CN-1,2,3,4-tetrahydro-	2-$(CH_2)_2NH_2$-1,2,3,4-tetrahydro-[e]	40	256 (decomp.)	—[c]	150
H	2-Me-2-CH_2Ph-1,2,3,4-tetrahydro-, bromide	2-Me-1,2,3,4-tetrahydro-[f]	92	243–245	—[c]	149

[a] $A = H_2$, Raney Ni, EtOH/(70°, 100 atm.)(10 hr); $B = H_2$, 10% PdC, EtOH/(50°, 50 psi)(7 hr); $C = H_2$, PtO_2, EtOH/room temp. atm. press; $D = LiAlH_4$, ether/(reflux)(60 hr); $E = LiAlH_4$, tetrahydrofurane/(reflux)(68–70 hr), then 10% PdC, EtOH/(room temp.)(17 hr); $F = LiAlH_4$, tetrahydrofurane/(reflux)(60 hr), then 10% PdC, EtOH/(12 hr), then H_2, 10% PdC, HCl/(55°, 58 psi)(7 hr); $G = LiAlH_4$, tetrahydrofurane/(reflux)(72 hr), then 10% PdC/(room temp.)(20 min), then HCl; $H = H_2$, 10% PdC, MeOH/room temp., atm. press.

[b] Hydrochloride.

[c] Solvent of crystallization not specified.

[d] Unstable.

[e] Trihydrochloride.

[f] Dihydrobromide; free base has m.p. 146–147° (from acetone–ether).

Reduction

Nitro,[60,154] and *N*-oxide[154,155] substituents in 3,4-dihydro-1*H*-[1,4]oxazino[4,3-*a*]benzimidazoles can be reduced catalytically (Table 7.65) without affecting the ring system. The fully unsaturated pyrimido[1,2-*a*]-benzimidazole ring system is likewise stable to catalytic reduction under conditions that accomplish the hydrogenolytic removal of a *C*(4)-chloro substituent (Table 7.66).[111] In contrast, the sodium borohydride reduction of *C*(4)-methoxycarbonyl substituents in pyrimido[1,2-*a*]benzimidazol-2(1*H*)-ones is accompanied by reduction of the *C*(3)–*C*(4) double bond, but not the lactam carbonyl substituent, giving good yields (Table 7.66) of 3,4-dihydropyrimido[1,2-*a*]benzimidazol-2(1*H*)-one derivatives.[135] The reduction of the *C*(3)–*C*(4) double bond in pyrimido[1,2-*a*]benzimidazol-2(1*H*)-ones can also be achieved catalytically (Table 7.66).[135]

The Raney nickel catalyzed hydrogenation[147] of 1-ethoxycarbonyl-2-phenylpyrazino[4,3-*a*]benzimidazole at elevated temperature and pressure to give the corresponding 1,2,3,4-tetrahydro derivative in moderate yield (Table 7.67) is indicative of the resistance of the 1,2,3,4-tetrahydropyrazino-[4,3-*a*]benzimidazole ring system to reduction of this type. This stability also extends to hydrogenation over platinum and palladium catalysts under conditions (Table 7.67) suitable for the reduction of nitro to amino groups[26] and for the hydrogenolysis of *N*(2)-benzyl substituents.[149,150] In the lithium aluminum hydride reduction of 3,4-dihydropyrazino[4,3-*a*]benzimidazol-1(2*H*)-ones (Table 7.67), on the other hand, reductive removal of the carbonyl substituent is accompanied by hydrogenation of the azomethine double bond giving unstable 1,2,3,4,10,10a-hexahydropyrazino[4,3-*a*]benzimidazoles.[150] These can be isolated or dehydrogenated *in situ* with palladium charcoal giving the corresponding 1,2,3,4-tetrahydropyrazino[4,3-*a*]benzimidazole derivatives (Table 7.67).[150]

7.2.4. Practical Applications

Biological Properties

1-Phenyl-3,4-dihydro-1*H*-[1,4]oxazino[4,3-*a*]benzimidazole has been reported[167] to afford protection against polio virus. Other 3,4-dihydro-1*H*-[1,4]oxazino[4,3-*a*]benzimidazole derivatives have been patented[155] as herbicides. 3-Hydroxy-3,4-dihydro-2*H*-[1,3]thiazino[3,2-*a*]benzimidazole is useful in the treatment of pulmonary asthma[98] and has also attracted attention because of its antiviral action.[96,98]

The antiviral properties of tetrahydropyrimido[1,2-*a*]benzimidazole derivatives have also been reported.[138,168] 1-Alkoxycarbonyl-1,2,3,4-tetrahydropyrimido[1,2-*a*]benzimidazoles[169] and pyrimido[1,2-*a*]benzimidazole-

2,4(1*H*,3*H*)-diones[142,166] are associated with pesticidal and fungicidal properties. Pyrimido[1,2-*a*]benzimidazol-4(1*H*)-ones have been patented[131] as central nervous system depressants and as psychotropic and antiinflammatory agents.

Dyestuffs

The 3,4-dihydro-1*H*-[1,4]oxazino[4,3-*a*]benzimidazole,[60] 3,4-dihydro-1*H*-[1,4]thiazino[4,3-*a*]benzimidazole,[110] and pyrimido[1,2-*a*]benzimidazole[117,165] ring systems have been employed as chromophoric units in cyanine dyes. Pyrimido[1,2-*a*]benzimidazole derivatives have also found use as azo dyestuffs[128b] and as photographic emulsion stabilizers.[170]

7.3. TRICYCLIC 6-5-6 FUSED BENZIMIDAZOLES WITH TWO ADDITIONAL HETEROATOMS

Only fully nitrogen-containing structures (Scheme 7.91 and Table 7.68) having a tricyclic 6-5-6 fused benzimidazole framework with two additional heteroatoms appear to be known. These include the fully unsaturated 1,2,4-triazino[2,3-*a*]benzimidazole (**7.463**), 1,2,4-triazino[4,5-*a*]benzimidazole (**7.467**), and 1,3,5-triazino[1,2-*a*]benzimidazole (**7.470A**) ring systems and derived di- and tetrahydro- structures (cf. Scheme 7.91 and Table 7.68). The

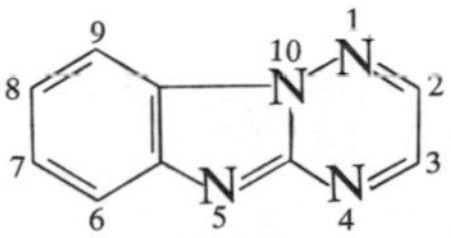

(**7.463**) 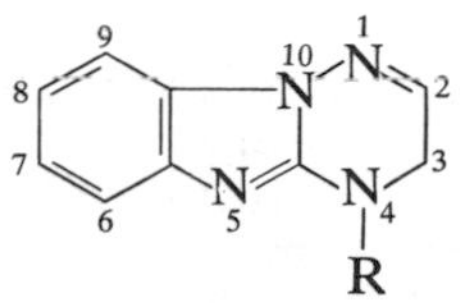 (**7.464**)

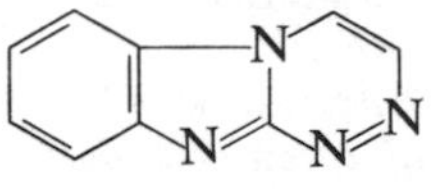

(**7.465**) 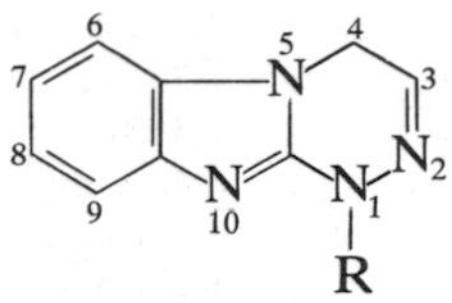 (**7.466A**)

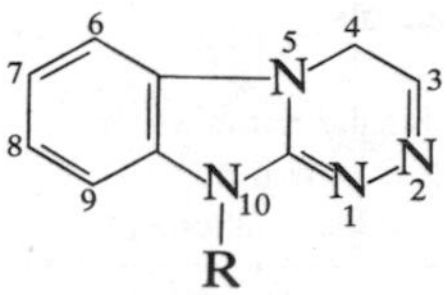

(**7.466B**) 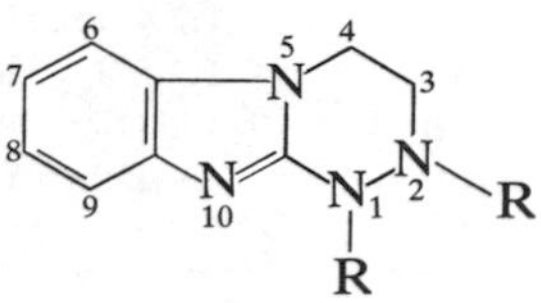 (**7.466C**)

Scheme 7.91

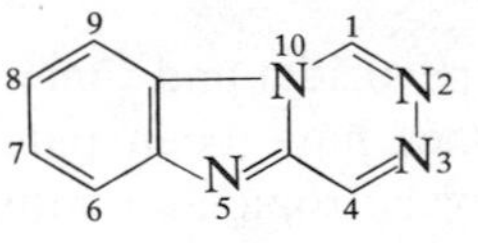

(**7.467**) 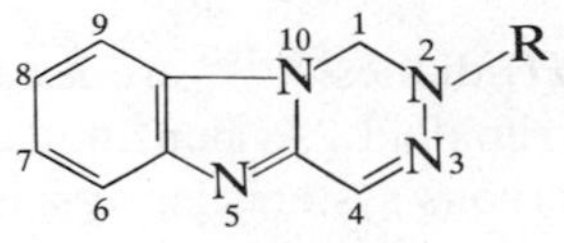 (**7.468A**)

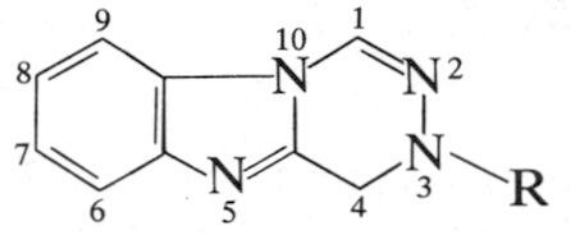

(**7.468B**) 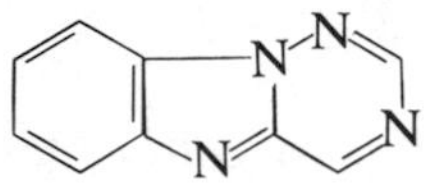 (**7.469**)

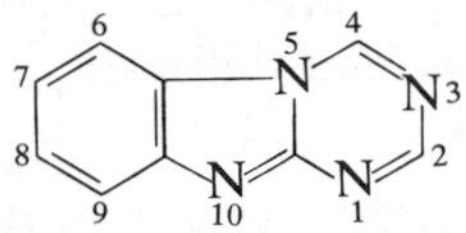

(**7.470A**) 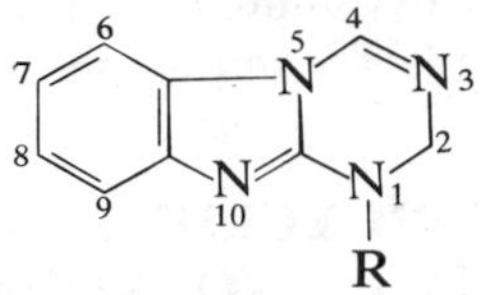 (**7.470B**)

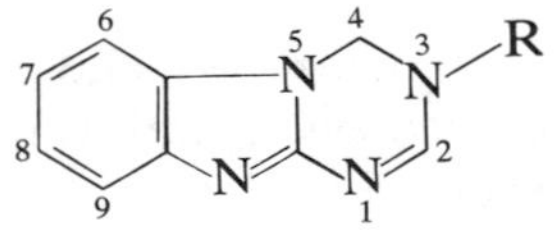

(**7.470C**) 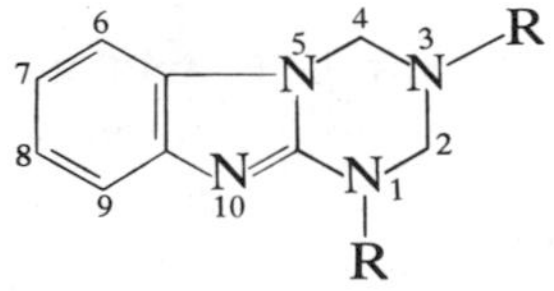 (**7.470D**)

Scheme 7.91 (*Contd.*)

TABLE 7.68 TRICYCLIC 6-5-6 FUSED BENZIMIDAZOLE RING SYSTEMS WITH TWO ADDITIONAL HETEROATOMS

Structure[a]	Name[b]
(**7.463**)	1,2,4-Triazino[2,3-*a*]benzimidazole
(**7.464**)	3,4-Dihydro-1,2,4-triazino[2,3-*a*]benzimidazole
(**7.465**)	1,2,4-Triazino[4,3-*a*]benzimidazole
(**7.466A**)	1,4-Dihydro-1,2,4-triazino[4,3-*a*]benzimidazole
(**7.466B**)	4,10-Dihydro-1,2,4-triazino[4,3-*a*]benzimidazole
(**7.466C**)	1,2,3,4-Tetrahydro-1,2,4-triazino[4,3-*a*]benzimidazole
(**7.467**)	1,2,4-Triazino[4,5-*a*]benzimidazole
(**7.468A**)	1,2-Dihydro-1,2,4-triaziono[4,5-*a*]benzimidazole
(**7.468B**)	3,4-Dihydro-1,2,4-triazino[4,5-*a*]benzimidazole
(**7.469**)	1,2,4-Triazino[1,6-*a*]benzimidazole
(**7.470A**)	1,3,5-Triazino[1,2-*a*]benzimidazole
(**7.470B**)	1,2-Dihydro-1,3,5-triazino[1,2-*a*]benzimidazole
(**7.470C**)	3,4-Dihydro-1,3,5-triazino[1,2-*a*]benzimidazole
(**7.470D**)	1,2,3,4-Tetrahydro-1,3,5-triazino[1,2-*a*]benzimidazole

[a] Cf. Scheme 7.91.

[b] Based on the Ring Index.

1,2,4-triazino[4,3-*a*]benzimidazole framework, as yet unknown in the fully unsaturated form (**7.465**) is represented by the 1,4-dihydro- and 4,10-dihydro-1,2,4-triazino[4,3-*a*]benzimidazole ring systems (**7.466A**) and (**7.466B**) and by the 1,2,3,4-tetrahydro-1,2,4-triazino[4,3-*a*]benzimidazole ring system (**7.466C**). A search of the literature has failed to reveal any compound having a structure based on the 1,2,4-triazino[1,6-*a*]benzimidazole skeleton (**7.469**).

Of the 12 known ring systems (Scheme 7.91 and Table 7.68) to be dealt with under the present heading those having a 1,3,5-triazino structure [Scheme 7.91; (**7.470A**)–(**7.470D**)] have been most extensively investigated. The chemistry of 1,3,5-triazino[1,2-*a*]benzimidazoles was briefly reviewed in 1961.[171]

7.3.1. Synthesis

Ring-closure Reactions of Benzimidazole Derivatives

The condensation of the readily available 1,2-diaminobenzimidazole with α-diketones provides a simple, and potentially general method for the synthesis of alkyl and aryl derivatives of the fully unsaturated 1,2,4-triazino-[2,3-*a*]benzimidazole ring system [Scheme 7.92; (**7.471**) + (**7.472**) → (**7.474**; R = alkyl or aryl)].[172] Ring-closure of this type proceeds in high yield (Table 7.69) and is usually effected by simply heating the reactants together in alcoholic solvents. However, condensation fails[172a] in the case of benzil unless potassium hydroxide is present as catalyst.[172b] The uncatalyzed condensation of 1,2-diaminobenzimidazole (**7.471**) with α-keto acids (**7.473**) is

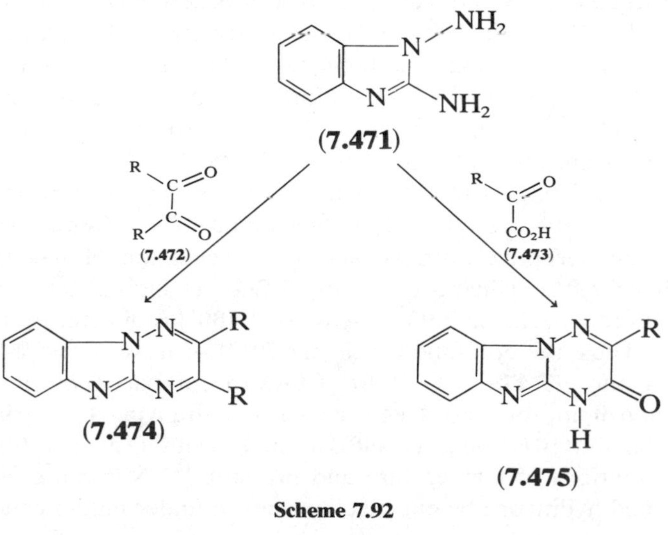

(**7.471**)

(**7.472**)

(**7.473**)

(**7.474**)

(**7.475**)

Scheme 7.92

TABLE 7.69. SYNTHESIS OF 1,2,4-TRIAZINO[2,3-*a*]BENZIMIDAZOLES (**7.474**) AND (**7.475**) BY RING-CLOSURE OF 1,2-DIAMINOBENZIMIDAZOLE (**7.471**) WITH α-DICARBONYL COMPOUNDS[a]

Starting materials	Reaction conditions[b]	Product	Yield (%)	m.p. (°C)	Solvent of crystallization	Ref.
(**7.471**) + (**7.472**; R = Me)	*A*	(**7.474**; R = Me)	58	236–239	Ethanol	172a
(**7.471**) + (**7.472**; R = Ph)	*B*	(**7.474**; R = Ph)	94	278–281	—[c]	172b
(**7.471**) + (**7.473**; R = Me)	*C*	(**7.475**; R = Me)	72	350–355	Dimethylformamide	172a
(**7.471**) + (**7.473**; R = Ph)	*C*	(**7.475**; R = Ph)	68	355–358	Dimethylformamide–water	172a

[a] From Ref. 172.
[b] *A* = MeOH(reflux)(2 hr); *B* = 2M KOH, EtOH/(reflux)(30 min); *C* = EtOH/(reflux)(2 hr).
[c] Not crystallized.

reported[172a] to afford the corresponding 1,2,4-triazino[2,3-*a*]benzimidazol-2(1*H*)-ones (**7.475**) in good yield (Table 7.69). However, the orientation of these products was not rigorously established, but if correct, is consistent with preferential initial condensation between the more nucleophilic *N*(1)-amino group in 1,2-diaminobenzimidazole and the more electrophilic keto center in the α-keto acid.

Syntheses of ring systems containing a 1,2,4-triazino[4,3-*a*]benzimidazole skeleton (Scheme 7.93 and Table 7.70) are largely based on ring-closure reactions of 2-hydrazinobenzimidazole derivatives. In particular, the triethylamine catalyzed reaction (Scheme 7.93) of 2-hydrazinobenzimidazoles (**7.477**) with α-bromoketones affords, via presumed hydrazone intermediates (**7.476**), moderate yields (Table 7.70) of 1,4-dihydro-1,2,4-triazino[4,3-*a*]benzimidazole derivatives (**7.479**).[173] Alternative synthetic access (Scheme 7.93) to molecules of the latter type is provided by the reaction of 1-(β-oxoalkyl)-2-chlorobenzimidazoles (**7.482**) with hydrazine.[174] The deceptively simple acylative cyclization of α-keto acid or ester 2-benzimidazolylhydrazones to 1,2,4-triazino[4,3-*a*]benzimidazol-4(10*H*)-ones [e.g., Scheme 7.93; (**7.478**)→(**7.480**)] is difficult to accomplish in practice. Thus, the cyclization (Scheme 7.93) of α-keto acid 2-benzimidazolylhydrazones (**7.478**; CO_2H for CO_2Me) (prepared *in situ* from 2-hydrazinobenzimidazole and α-keto acids) though giving 1,2,4-triazino[4,3-*a*]benzimidazol-4(10*H*)-ones (**7.480**) in high yield (Table 7.70) requires extreme conditions of temperature and pressure.[175] Nor can cyclization of the preformed hydrazone be effectively achieved under milder conditions, as

Scheme 7.93

TABLE 7.70. SYNTHESIS OF 1,2,4-TRIAZINO[4,3-*a*]BENZIMIDAZOLES BY RING-CLOSURE REACTIONS OF BENZIMIDAZOLE DERIVATIVES

Starting material	Reaction conditions[a]	Product	Yield (%)	m.p. (°C)	Solvent of crystallization	Ref.
(**7.477**; R = H)	*A*	(**7.479**; R = H, R^1 = Me)	43	293–295	Dimethylform-amide–water	173
(**7.477**; R = H)	*A*	(**7.479**; R = H, R^1 = *p*-$O_2NC_6H_4$)	51	298–299	Dimethylform-amide–water	173
(**7.477**; R = Me)	*A*	(**7.479**; R = Me, R^1 = *p*-ClC_6H_4)	45	309–310	Dimethylform-amide–water	173
(**7.477**; R = H)	*B*	(**7.480**; R = Me)	66–85	—[b]	—[c]	175
(**7.477**; R = H)	*B*	(**7.480**; R = Ph)	66–85	—[b]	—[c]	175
(**7.478**; R = CH_2CO_2Me)	*C*	(**7.480**; R = CH_2CO_2Me)	23	286	Dimethylform-amide–water	176
(**7.477**; R = H)	*D*	(**7.481**)	—[d]	287–288	Dimethylform-amide–water	177
(**7.482**; R = R^2 = H, R^1 = OEt)	*E*	(**7.481**)	84	—	—	177
(**7.483**)	*F*	(**7.481**)	84	—	—	177
(**7.484**)	*G*	(**7.485**)	76	223–225 (decomp.)	Methanol	179

[a] *A* = R^1COCH_2Br, Et_3N, EtOH or dimethylformamide/reflux (reaction time not specified); *B* = $RCOCO_2H$(150°, autoclave)(5 hr); *C* = Et_3N, MeOH/(reflux)(2.5 hr); *D* = $ClCH_2CO_2Et$ (reaction conditions not specified); *E* = N_2H_4, Et_3N or pyridine, dimethylformamide/150–160° (reaction time not specified); *F* = Et_3N or pyridine, dimethylformamide/heat (reaction conditions not specified); *G* = 85% N_2H_4/150–160°, sealed tube (reaction time not specified).

[b] Melting point not quoted.

[c] Solvent of crystallization not specified.

[d] Yield not quoted.

demonstrated by the low yield (23%) obtained[176] in the triethylamine catalyzed cyclization (Table 7.70) of the keto ester hydrazone (**7.478**; R = CH_2CO_2Me) to the 1,2,4-triazino[4,3-*a*]benzimidazol-4(10*H*)-one derivative (**7.480**; R = CH_2CO_2Me). The condensation (Scheme 7.93) of 2-hydrazinobenzimidazole (**7.477**; R = H) with ethyl chloroacetate affords a route[177] to 1,4-dihydro-1,2,4-triazino[4,3-*a*]benzimidazol-3(2*H*)-one (**7.481**). This product is also formed (Scheme 7.93) in high yield (Table 7.70), via the presumed intermediacy of the hydrazide (**7.483**), by the reaction of 1-(ethoxycarbonylmethyl)-2-chlorobenzimidazole (**7.482**; R = R^2 = H, R^1 = OEt) with hydrazine.[177] The related reaction (Scheme 7.94) of 1-(2-chloroethyl)-2-chlorobenzimidazole (**7.484**) with hydrazine at high temperature and pressure affords 1,2,3,4-tetrahydro-1,2,4-triazino[4,3-*a*]-benzimidazole (**7.485**) in good yield (Table 7.70).[178]

Alkyl and aryl derivatives of the fully unsaturated 1,2,4-triazino[4,5-*a*]-benzimidazole ring system are generally accessible in good yield (Table 7.71) by the thermal ring-closure of readily available 2-benzimidazolyl-ketone hydrazones with orthoesters [Scheme 7.95; (**7.488**; R = R^2 = H) → →

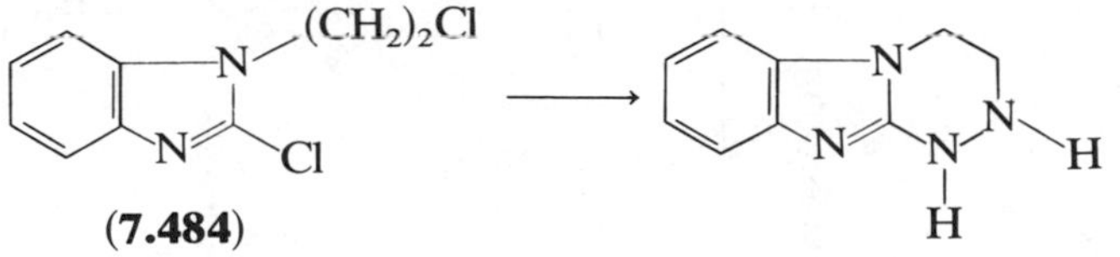

(7.484) (7.485)

Scheme 7.94

(**7.489**)].[179] The imidic ester intermediates in these reactions can be isolated in some instances and cyclize as expected on heating (Table 7.71) to give the corresponding 1,2,4-triazino[4,5-*a*]benzimidazole derivatives [Scheme 7.95; (**7.487**) → (**7.489**)].[179] Arylhydrazones (**7.488**; R = H, R^1 = CN, R^2 = aryl), derived by the coupling of arenediazonium salts with 2-cyanomethylbenzimidazole (**7.486**; R = H, R^1 = CN), undergo acylative cyclization on heating with ethyl chloroformate in pyridine to afford 1,2,4-triazino[4,3-*a*]-benzimidazol-1(2*H*)-ones (**7.490**; R^1 = CN, R^2 = aryl) in excellent yield

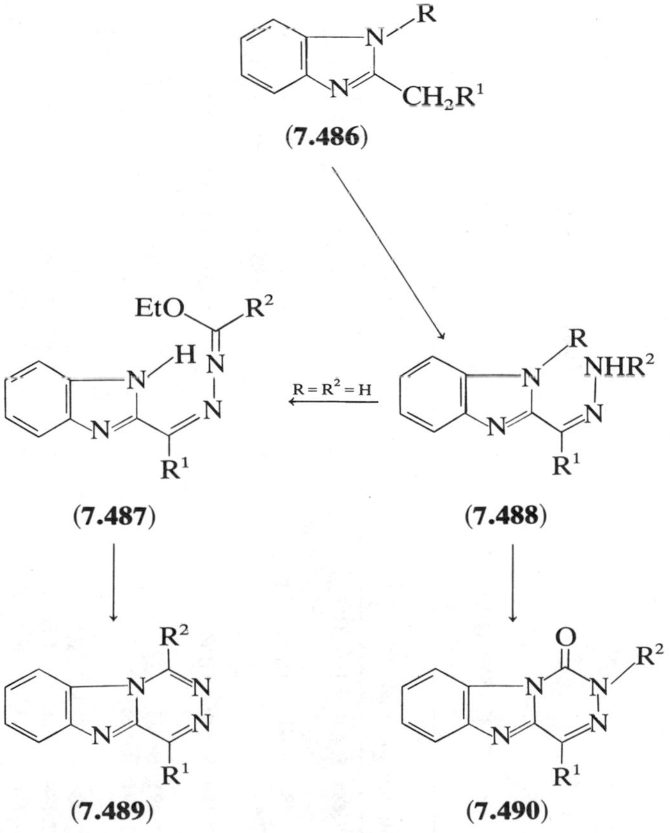

(7.486)

R = R^2 = H

(7.487) (7.488)

(7.489) (7.490)

Scheme 7.95

TABLE 7.71. SYNTHESIS OF 1,2,4-TRIAZINO[4,5-*a*]BENZIMIDAZOLES BY RING-CLOSURE REACTIONS OF BENZIMIDAZOLE DERIVATIVES

Starting material	Reaction conditions[a]	Product	Yield (%)	m.p. (°C)	Solvent of crystallization	Ref.
(**7.488**; $R = R^1 = R^2 = H$)	*A*	(**7.489**; $R^1 = R^2 = H$)	65	235–236 (decomp.)	Ethanol	179
(**7.488**; $R = R^2 = H$, $R^1 = Ph$)	*A*	(**7.489**; $R^1 = Ph$, $R^2 = H$)[b]	80	205–206	Benzene	179
(**7.487**; $R^1 = H$, $R^2 = Me$)	*B*	(**7.489**; $R^1 = H$, $R^2 = Me$)	—[c]	162 (decomp.)	Ethanol	179
(**7.488**; $R = R^2 = H$, $R^1 = Ph$)	*C*	(**7.489**; $R^1 = Ph$, $R^2 = Me$)	42	170–172 (decomp.)	Ethanol	179
(**7.491**; $R = H$)	*D*	(**7.492**)	50	154–155	Ethanol	7
(**7.491**; $R = Et$)	*D*	(**7.493**)	3	224–225	Dimethyl sulfoxide–ethanol	7
(**7.486**; $R = CO_2Et$, $R^1 = CN$)	*E*	(**7.490**; $R^1 = CN$, $R^2 = Ph$)	95	200–201	Ethanol	180
(**7.488**; $R = H$, $R^1 = CN$, $R^2 = Ph$)	*F*	(**7.490**; $R^1 = CN$, $R^2 = Ph$)	96	—	—	180
(**7.486**; $R = CO_2Et$, $R^1 = CN$)	*E*	(**7.490**; $R^1 = CN$, $R^2 = p\text{-}MeOC_6H_4$)	97	220–221	Ethanol	180
(**7.488**; $R = H$, $R^1 = CN$, $R^2 = p\text{-}MeOC_6H_4$)	*F*	(**7.490**; $R^1 = CN$, $R^2 = p\text{-}MeOC_6H_4$)	96	—	—	180

(**7.486**; R = CO_2Et, R^1 = CN)	*E*	(**7.490**; R^1 = CN, R^2 = *p*-$EtOC_6H_4$)	98	298–299	Ethanol	180
(**7.488**; R = H, R^1 = CN, R^2 = *p*-$EtOC_6H_4$)	*F*	(**7.490**; R^1 = CN, R^2 = *p*-$EtOC_6H_4$)	97	—	—	180
(**7.486**; R = CO_2Et, R^1 = CN)	*E*	(**7.490**; R^1 = CN, R^2 = *p*-BrC_6H_4)	99	265–267	Ethanol	180
(**7.488**; R = H, R^1 = CN, R^2 = *p*-BrC_6H_4)	*F*	(**7.490**; R^1 = CN, R^2 = *p*-BrC_6H_4)	100	—	—	180
(**7.488**; R = H, R^1 = R^2 = CON(piperidino))	*G*	(**7.490**; R^1 = CON(piperidino), R^2 = H)	quant.	266–268	Ethanol	7
(**7.494**; R = R^1 = H)	*A*	(**7.495**; R^1 = R^2 = H)	80	336	Dimethylformamide–water	181
(**7.494**; R = R^1 = H)	*C*	(**7.495**; R^1 = H, R^2 = Me)	47	345	Dimethylformamide–water	181
(**7.494**; R = COMe, R^1 = H)	*H*	(**7.495**; R^1 = H, R^2 = Me)	70	—	—	181
(**7.494**; R = H, R^1 = NO_2)	*I*	(**7.495**; R^1 = NO_2, R^2 = H)	15	305	Dimethylformamide–benzene	182

[a] *A* = $HC(OEt)_3$/(160–210°)(1.25–2 hr); *B* = 200–210° (no cosolvent)/(1 hr); *C* = $MeC(OEt)_3$/(150–210°)(1.25–3 hr); *D* = Br_2, diisopropylamine, CH_2Cl_2/(room temp.)(10–15 min); *E* = $Ar\overset{+}{N}_2$, pyridine/(0°)(3–12 hr); *F* = $ClCO_2Et$, pyridine/(0°)(3 hr), then (room temp.)(15–20 hr), and finally (reflux)(3 hr); *G* = melt or heat in ethanol; *H* = $POCl_3$, benzene/(reflux)(12 hr); *I* = $HC(OEt)_3$/(reflux)(6 hr).

[b] Forms a hydrochloride, m.p. 203–203.5° (from ethanol).

[c] Yield not quoted.

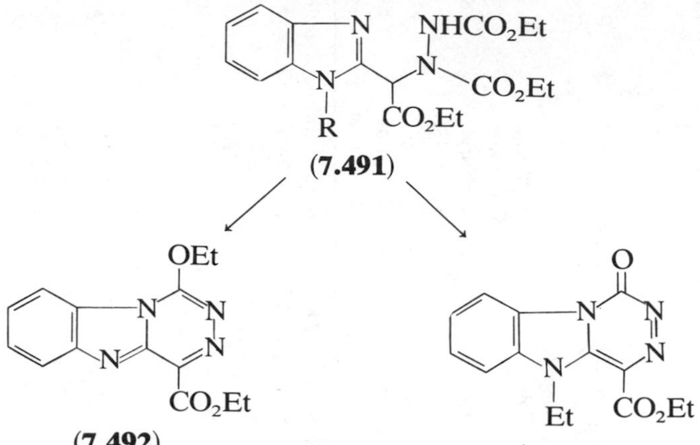

(7.491)

(7.492)

(7.493)

Scheme 7.96

(Table 7.71).[180] The oxidative ring-closure (Scheme 7.96) of the adduct (**7.491**; R = H) (obtained by the hetero-ene reaction of ethyl 2-benzimidazolylacetate with diethyl azodicarboxylate) using bromine in the presence of diisopropylamine results in the formation of the 1,2,4-triazino[4,5-*a*]benzimidazole derivative (**7.492**) in 50% yield (Table 7.71).[7] The analogous oxidative cyclization of the adduct (**7.491**; R = Et) affords the 1,2,4-triazino[4,5-*a*]benzimidazol-4(10*H*)-one (**7.493**), though only in 3% yield (Table 7.71).[7] Readily accessible 2-benzimidazolylhydrazides undergo cyclization on heating with orthoesters affording a simple high yield (Table 7.71) route to 1,2,4-triazino[4,5-*a*]benzimidazol-4(3*H*)-ones [Scheme 7.97; (**7.494**; R = H) → (**7.495**)].[181,183] The phosphorus oxychloride promoted cyclodehydration (Scheme 7.97) of the *N*-acetyl hydrazide (**7.494**; R = COMe, R^1 = H) to give the 1,2,4-triazino[4,5-*a*]benzimidazol-4(3*H*)-one (**7.495**; R^1 = H, R^2 = Me) exemplifies an alternative approach to the synthesis of molecules of this type.[181]

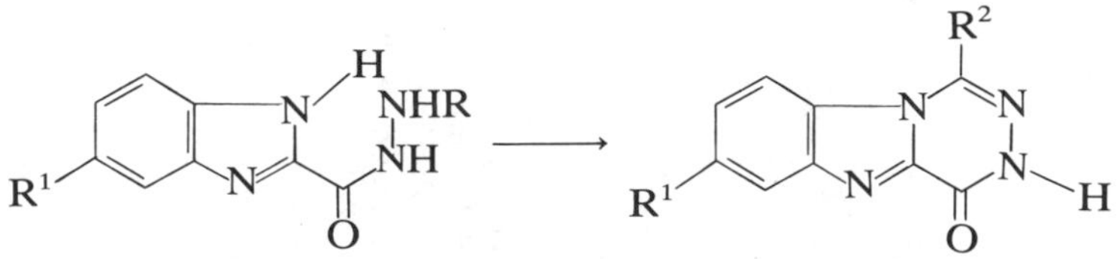

(7.494)

(7.495)

Scheme 7.97

Methods for the construction of 1,3,5-triazino[1,2-*a*]benzimidazole derivatives (Scheme 7.98 and Table 7.72) rely heavily on cyclization reactions of 2-aminobenzimidazoles (**7.496**) or simple derivatives such as *N*-(2-benzimidazolyl)guanidines (**7.497**; X = NR), ureas (**7.497**; X = O), or thioureas

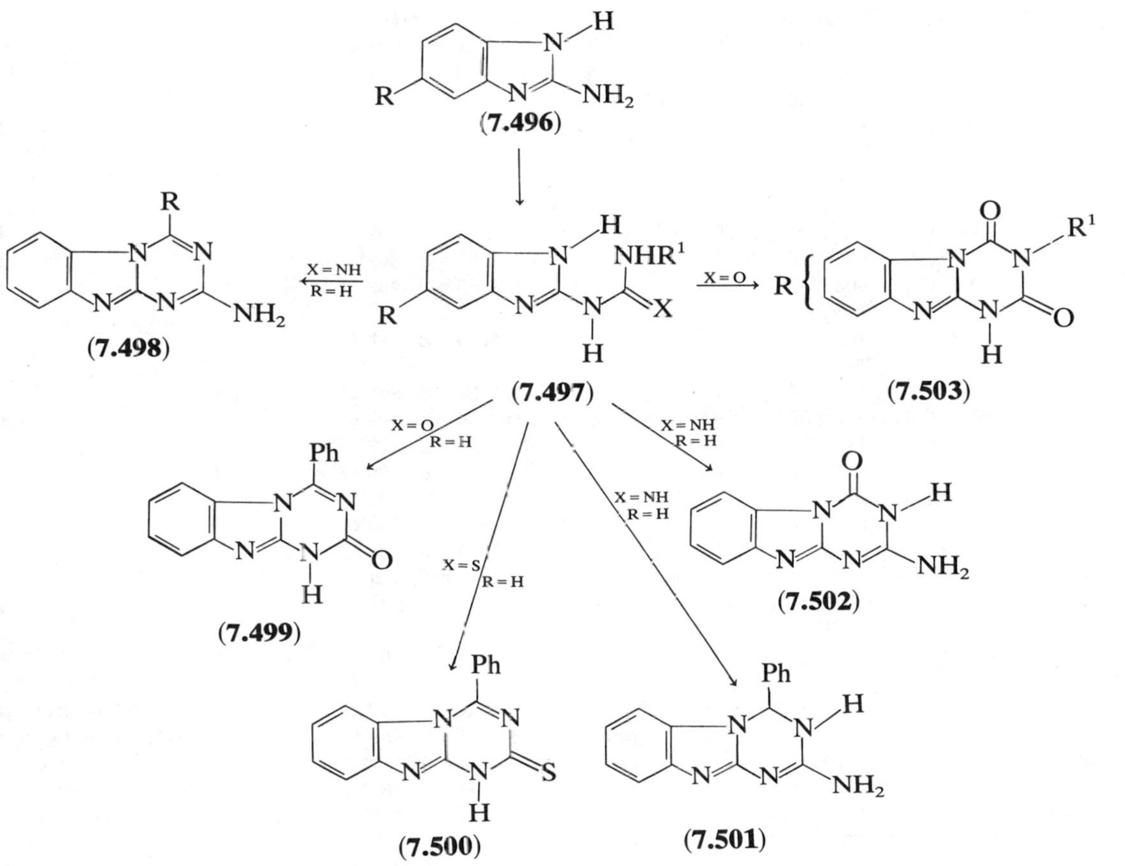

(**7.496**)

(**7.498**)

(**7.497**)

(**7.503**)

(**7.499**)

(**7.502**)

(**7.500**)

(**7.501**)

Scheme 7.98

TABLE 7.72. SYNTHESIS OF 1,3,5-TRIAZINO[1,2-*a*]BENZIMIDAZOLES BY RING-CLOSURE REACTIONS OF 2-AMINOBENZIMIDAZOLE DERIVATIVES

Starting material	Reaction conditions[a]	Product	Yield (%)	m.p. (°C)	Solvent of crystallization	Ref.
(**7.497**; $R = R^1 = H$, X = NH)	*A*	(**7.498**; R = H)	42	305–306	Dimethylformamide	185
(**7.497**; $R = R^1 = H$, X = NH)	*B*	(**7.498**; R = H)	73	—	—	185
(**7.497**; $R = R^1 = H$, X = NH)	*C*	(**7.498**; R = Ph)	78	>360	Dimethylformamide	186
(**7.497**; R = H, R^1 = COPh, X = NH)	*D*	(**7.498**; R = Ph)	—[b]	—	—	186
(**7.497**; R = H, R^1 = CSNHCOPh, X = NH)	*D*	(**7.498**; R = Ph)	38	—	—	186
(**7.504**; R = OPh)	*E*	(**7.505**; R = OPh, R^1 = Ph)	72	250–253 (decomp.)	Toluene	187
(**7.504**; R = OPh)	*E*	(**7.505**; R = OPh, $R^1 = p\text{-}n\text{-}C_5H_{11}C_6H_4$	66	245–246	Toluene	187
(**7.504**; R = OPh)	*E*	(**7.505**; R = OPh, $R^1 = p\text{-}ClC_6H_4$)	77	271–273	Toluene	187
(**7.504**; R = OPh)	*E*	(**7.505**; R = OPh, $R^1 = p\text{-}BrC_6H_4$)	79	268–270	Toluene	187
(**7.504**; R = OPh)	*E*	(**7.505**; R = OPh, $R^1 = p\text{-}O_2NC_6H_4$)	87	302–305	Toluene	187
(**7.504**; $R = NH_2$)	*E*	(**7.505**; $R = NH_2$, R^1 = Ph)	49	276–278	Toluene	187
(**7.504**; $R = NH_2$)	*E*	(**7.505**; $R = NH_2$, $R^1 = p\text{-}ClC_6H_4$)	55	281–282	Toluene	187
(**7.504**; $R = NH_2$)	*E*	(**7.505**; $R = NH_2$, $R^1 = p\text{-}O_2NC_6H_4$)	63	305–306	Toluene	187
(**7.496**; R = H)	*F*	(**7.498**; $R = NH_2$)[c]	42	277 (decomp.)	—[d]	183
(**7.496**; R = H)	*F*	(**7.498**; $R = NH_2$)[e]	70	304–306 (decomp.)	Water	184
(**7.504**; R = OPh)	*G*	(**7.506**; R = OPh, R^1 = Ph)	69	205–206	Toluene	187
(**7.504**; R = OPh)	*G*	(**7.506**; R = OPh, R^1 = diphenylyl)	73	236–238	Toluene	187
(**7.504**; R = OPh)	*G*	(**7.506**; R = OPh, $R^1 = p\text{-}Me_2NC_6H_4$)	53	>320 (decomp.)	Toluene	187
(**7.504**; R = OPh)	*G*	(**7.506**; R = OPh, $R^1 = p\text{-}O_2NC_6H_4$)	76	226–227 (decomp.)	Toluene	187
(**7.504**; R = OPh)	*G*	(**7.506**; R = OPh, R^1 = 2-furyl)	85	224–225	Toluene	187
(**7.504**; $R = NH_2$)	*G*	(**7.506**; $R = NH_2$, $R^1 = p\text{-}O_2NC_6H_4$)	87	203–205	Toluene	187
(**7.504**; $R = NMe_2$)	*G*	(**7.506**; $R = NMe_2$, R^1 = Ph)	63	225–226	Toluene	187
(**7.505**; $R = NEt_2$)	*G*	(**7.506**; $R = NEt_2$, $R^1 = p\text{-}O_2NC_6H_4$)	83	203–204	Toluene	187
(**7.496**; R = H)	*H*	(**7.499**)	69	>360	Dioxane–water	186

(**7.497**; R = H, R^1 = COPh, X = O)	*I*	(**7.499**)	28	—	—	186
(**7.496**; R = H)	*J*	(**7.500**)[e]	—[b]	323	Pyridine	188
(**7.496**; R = H)	*K*	(**7.500**)	46	311	Dimethylformamide	189
(**7.497**; R = R^1 = H, X = NH)	*L*	(**7.501**)	94	294–295	Dimethylformamide	143
(**7.497**; R = R^1 = H, X = NH)	*M*	(**7.502**)	quant.	>360	Dimethylformamide	186
(**7.497**; R = H, R^1 = CONHCOPh, X = NH)	*N*	(**7.502**)	78	—	—	186
(**7.497**; R = H, R^1 = $CONHCO_2Et$, X = NH)	*N*	(**7.502**)	78	—	—	186
(**7.497**; R = H, R^1 = $CSNHCO_2Et$, X = NH)	*N*	(**7.502**)	78	—	—	186
(**7.512**; R^1 = R^2 = H)	*O*	(**7.513**; R = Me)	—[b]	138	—[f]	190
(**7.512**; R^1 = R^2 = H)	*O*	(**7.513**; R = Et)	—[b]	167–168	—[f]	190
(**7.512**; R^1 = R^2 = H)	*O*	(**7.513**; R = Pr^n)	—[b]	130–132	—[f]	190
(**7.512**; R^1 = R^2 = H)	*O*	(**7.513**; R = Pr^i)	—[b]	112–114	—[f]	190
(**7.512**; R^1 = R^2 = H)	*O*	(**7.513**; R = Bu^n)	71	109	*n*-Hexane	190
(**7.512**; R^1 = R^2 = H)	*P*	(**7.513**; R = Bu^n)	74	109–110	*m*-Xylene	195
(**7.512**; R^1 = R^2 = H)	*O*	(**7.513**; R = Bu^i)	—[b]	138	—[f]	190
(**7.512**; R^1 = R^2 = H)	*O*	(**7.513**; R = Bu^t)	—[b]	143	—[f]	190
(**7.512**; R^1 = R^2 = H)	*O*	(**7.513**; R = CH_2Ph)	—[b]	145–146	—[f]	190
(**7.512**; R^1 = R^2 = H)	*O*	(**7.513**; R = Ph)	—[b]	196–197	—[f]	190
(**7.512**; R^1 = R^2 = H)	*P*	(**7.513**; R = Ph)	81	152–153	—[f]	195
(**7.508**; R = R^1 = H)	*Q*	(**7.509**; R^1 = *p*-$O_2NC_6H_4$, R^2 = Ph)	34	224–226	Dimethylformamide–water	187
(**7.508**; R = R^1 = H)	*Q*	(**7.509**; R^1 = *p*-$MeOC_6H_4$, R^2 = *p*-BrC_6H_4)	50	268–269	Dimethylformamide–water	187
(**7.508**; R = R^1 = H)	*Q*	(**7.509**; R^1 = *p*-MeC_6H_4, R^2 = *p*-$O_2NC_6H_4$)	58	>360	Dimethylformamide–water	187
(**7.508**; R = R^1 = H)	*Q*	(**7.509**; R^1 = *p*-$MeOC_6H_4$, R^2 = *p*-$O_2NC_6H_4$)	56	>360	Dimethylformamide–water	187
(**7.508**; R = R^1 = H)	*Q*	(**7.509**; R^1 = 2-naphthyl, R^2 = *p*-$O_2NC_6H_4$)	43	238–240	Dimethylformamide–water	187
(**7.508**; R = R^1 = H)	*R*	(**7.510**; R = Ph)	41	308–309	Dimethylformamide–water	187
(**7.508**; R = R^1 = H)	*R*	(**7.510**; R = *p*-MeC_6H_4)	46	>360	Dimethylformamide–water	187

TABLE 7.72 (*Continued*)

Starting material	Reaction conditions[a]	Product	Yield (%)	m.p. (°C)	Solvent of crystallization	Ref.
(**7.508**; $R = R^1 = H$)	*R*	(**7.510**; $R = p\text{-}ClC_6H_4$)	35	>350	Dimethylformamide–water	187
(**7.508**; $R = R^1 = H$)	*R*	(**7.510**; $R = m\text{-}ClC_6H_4$)	40	>350	Dimethylformamide–water	187
(**7.507**; R = H)	*S*	(**7.511**)	—[b]	>320	Acetic acid	197
(**7.507**; R = H)	*T*	(**7.511**)	quant.	348	Dioxane–water	186
(**7.507**; R = Me)	*S*	(**7.511**; R = Me)[g]	—[b]	>320	Acetic acid	197
(**7.507**; R = Cl)	*S*	(**7.511**; R = Cl)[g]	—[b]	>320	—[h]	197
(**7.496**; R = H)	*U*	(**7.503**; $R = H, R^1 = Me$)	70	>350	Acetic acid	200
(**7.512**; $R^1 = R^2 = H$)	*V*	(**7.515**; $R = Me, R^1 = R^2 = H$)	93	>350	—[d]	199
(**7.514**; $R = Me, R^1 = R^2 = H$)	*W*	(**7.515**; $R = Me, R^1 = R^2 = H$)	44	>360	Dimethylformamide	199
(**7.512**; $R^1 = H, R^2 = Me$)	*V*	(**7.515**; $R = Me$, R^1 or $R^2 =$ H or Me)[g]	56	320	—[f]	199
(**7.512**; $R^1 = R^2 = H$)	*V*	(**7.515**; $R = Et, R^1 = R^2 = H$)	84	325–326	Dioxane or dimethylformamide	199
(**7.512**; $R^1 = R^2 = H$)	*V*	(**7.515**; $R = Pr^n, R^1 = R^2 = H$)	53	285–287	Dioxane or dimethylformamide	199
(**7.512**; $R^1 = R^2 = H$)	*V*	(**7.515**; $R = Pr^i, R^1 = R^2 = H$)	41	300	Dioxane or dimethylformamide	199
(**7.512**; $R^1 = H, R^2 = Me$)	*V*	(**7.515**; $R = Pr^i$, R^1 or $R^2 =$ H or Me)[g]	50	325	Dimethylformamide	199
(**7.496**; R = H)	*X*	(**7.503**; $R = H, R^1 = Bu^n$)	84	282	Dioxane or dimethylformamide	199
(**7.496**; R = H)	*U*	(**7.503**; $R = H, R^1 = Bu^n$)	87	285–287	Acetic acid	200
(**7.512**; $R^1 = R^2 = H$)	*V*	(**7.515**; $R = Bu^n, R^1 = R^2 = H$)	78	280	Dioxane or dimethylformamide	199
(**7.497**; $R = H, R^1 = Bu^n, X = O$)	*Y*	(**7.503**; $R = H, R^1 = Bu^n$)	54	280	Dioxane	199
(**7.512**; $R^1 = H, R^2 = Me$)	*V*	(**7.515**; $R = Bu^n, R^1 = Me, R^2 = H$) +	17	310	Dimethylformamide–water	199
		(**7.515**; $R = Bu^n, R^1 = H, R^2 = Me$)	18	255	Dioxane	

(**7.512**; $R^1 = H$, $R^2 = CF_3$)	*V*	(**7.515**; $R = Bu^n$, R^1 or $R^2 = H$ or CF_3)[g]	—[b]	300	Dioxane	199
(**7.512**; $R^1 = H$, $R^2 = Et$)	*V*	(**7.515**; $R = Bu^n$, R^1 or $R^2 = H$ or Et)[g]	35	315	Dioxane	199
(**7.512**; $R^1 = H$, $R^2 = Bu^n$)	*V*	(**7.515**; $R = Bu^n$, R^1 or $R^2 = H$ or Bu^n)[g]	43	315	—[d]	199
(**7.512**; $R^1 = H$, $R^2 = OMe$)	*V*	{(**7.515**; $R = Bu^n$, $R^1 = OMe$, $R^2 = H$) + (**7.515**; $R = Bu^n$, $R^1 = H$, $R^2 = OMe$)}	26 18	270 205	Dioxane Dioxane	199
(**7.512**; $R^1 = H$, $R^2 = Cl$)	*V*	(**7.515**; $R = Bu^n$, R^1 or $R^2 = H$ or Cl)[g]	—[b]	320	Dioxane	199
(**7.512**; $R^1 = R^2 = Me$)	*V*	(**7.515**; $R = Bu^n$, $R^1 = R^2 = Me$)	67	305	Dioxane	199
(**7.512**; $R^1 = R^2 = Cl$)	*V*	(**7.515**; $R = Bu^n$, $R^1 = R^2 = Cl$)	15	250	Dioxane–water	199
(**7.496**; R = H)	*V*	(**7.503**; R = H, $R^1 = Ph$)	93	340	Dioxane or dimethylformamide	199
(**7.512**; $R^1 = R^2 = H$)	*V*	(**7.515**; R = Ph, $R^1 = R^2 = H$)	87	340	Dioxane or dimethylformamide	199
(**7.497**; R = H, $R^1 = NMe_2$, X = R)	*Z*	(**7.503**; R = H, $R^1 = NMe_2$)	79	>330	—[h]	201, 202
(**7.497**; R = H, $R^1 = CH_2CO_2Me$, X = O)	*Z*	(**7.503**; R = H, $R^1 = CH_2CO_2Me$)	—[b]	252	—[f]	201

[a] *A* = HCO_2Et, MeOH/(reflux)(48 hr); *B* = 1,3,5-triazine, piperidine, EtOH/(reflux)(10 hr); *C* = PhCON═C═S, pyridine/(reflux)(24 hr); *D* = pyridine/(reflux)(15 hr); *E* = R^1CO_2H, toluene/(reflux)(3 hr); *F* = dicyandiamide/(125–200°)(6 hr); *G* = R^1CHO, toluene/(reflux)(1–3 hr); *H* = PhCON═C═O, pyridine/(reflux)(48 hr); *I* = pyridine/(reflux)(24 hr); *J* = PhCON═C═S, pyridine, benzene/(room temp.)(16 hr); *K* = PhCON═C═S, pyridine, benzene/(room temp.)(24 hr); *L* = PhCHO, KOH, EtOH/(reflux)(5 hr); *M* = PhCON═C═O, EtO_2CN═C═O, or EtO_2CN═C═S, Et_3N, dioxane/(80°)(3 hr), *N* = dimethylformamide/(reflux)(few min); *O* = RNH_2 35% HCHO aq., CH_2Cl_2/(38°)(1 hr); *P* = RNH_2 30–37% HCHO aq., dioxane/(70–90°)(1–5 hr); *Q* = R^1N═CHR^2, dimethylformamide/(130°)(1 hr); *R* = RN═C═O, dimethylformamide/(130°)(1–2 hr); *S* = BrCN, $KHCO_3$ EtOH/(50–60°)(few min); *T* = EtO_2CN═C═O, Et_3N/(80°)(4 hr); *U* = MeNHCON(Me)COCl or PhNHCON(Bu^n)COCl, Et_3N, acetone/(reflux)(2–3 hr); *V* = RN═C═O, Et_3N, toluene/(reflux)(8–16 hr); *W* = Et_3N, toluene/(reflux)(12 hr); *X* = Bu^nN═C═O, $PhNMe_2$, methyl isobutyl ketone/(reflux)(2 hr); *Y* = $COCl_2$, Et_3N, toluene/(room temp.)(2 hr), then (reflux)(8 hr); *Z* = $(PhO)_2C$═O, PhOH, PhCN/(160°)(24 hr).

[b] Yield not quoted.

[c] Nitrate.

[d] Not crystallized.

[e] Pyridinium salt; free base has m.p. 310° (from dioxane).

[f] Solvent of crystallization not specified.

[g] *C*(7) or *C*(8) position for the substituent in the benzene ring not established.

[h] Purified by precipitation from alkaline solution with dilute acid.

(**7.497**; X = S). The direct conversion of a 2-aminobenzimidazole derivative into a 1,3,5-triazino[1,2-*a*]benzimidazole is exemplified by the thermal reaction of 2-aminobenzimidazole with dicyandiamide to give, depending on the temperature used, moderate[183] to good[184] yields (Table 7.72) of 2,4-diamino-1,3,5-triazino[1,2-*a*]benzimidazole (**7.498**; R = NH_2). 2-Amino-1,3,5-triazino[1,2-*a*]benzimidazole (**7.498**; R = H) is likewise formed in good yield (Table 7.72) by the acylative ring-closure of *N*-(2-benzimidazolyl)-guanidine (**7.497**; R = R^1 = H, X = NH) with ethyl formate or less orthodoxly with 1,3,5-triazine in the presence of piperidine.[185] 2-Amino-4-phenyl-1,3,5-triazino[1,2-*a*]benzimidazole (**7.498**; R = Ph) is the somewhat unexpected product obtained in high yield (Table 7.72) by the pyridine catalyzed condensation of *N*-(2-benzimidazolyl)guanidine (**7.497**; R = R^1 = H, X = NH) with benzoylisothiocyanate.[186] The first-formed intermediate in this transformation is the *N*-benzoylthiourea derivative (**7.497**; R = H, R^1 = CSNHCOPh, X = NH) which can be isolated and shown[186] to be convertible in the presence of pyridine into the 1,3,5-triazino[1,2-*a*]benzimidazole (**7.498**; R = Ph) (Table 7.72), or thermally, in the absence of catalyst, into the *N*-benzoylquanidine derivative (**7.497**; R = H, R^1 = COPh, X = NH). The demonstrable pyridine catalyzed cyclization of the latter (Table 7.72) then accounts satisfactorily for the original formation of the 1,3,5-triazino[1,2-*a*]-benzimidazole (**7.498**; R = Ph) from *N*-(2-benzimidazolyl)guanidine (**7.497**; R = R^1 = H, X = NH). Ring-closure reactions (Scheme 7.99) of 2-amino-1-(iminoacyl)benzimidazoles (**7.504**) constitute an alternative general synthetic approach to derivatives of the fully unsaturated 1,3,5-triazino[1,2-*a*]benzimidazole ring system.[187] Processes of this type are exemplified by the thermal reactions of the imidate (**7.504**; R = OPh) and the amidines (**7.504**; R = NR_2) with aromatic carboxylic acids to give good yields (Table 7.72) of 2-aryl-4-phenoxy- and 4-amino-2-aryl-1,3,5-triazino[1,2-*a*]benzimidazoles (**7.505**; R = OPh or NR_2, R^1 = aryl), respectively.[187] The related thermal ring-closure reactions of the imidate (**7.504**; R = OPh) and the amidines (**7.504**; R = NR_2) with aromatic aldehydes[187] afford efficient (Table 7.72) general methods for the synthesis of 1,2-dihydro-1,3,5-triazino[1,2-*a*]-benzimidazoles of the type (**7.506**; R = OPh or NR_2, R^1 = aryl). In ring formation (Scheme 7.99) akin to that affording 2-amino-4-phenyl-1,3,5-triazino[1,2-*a*]benzimidazole (**7.498**; R = Ph) (see before), 4-phenyl-1,3,5-triazino[1,2-*a*]benzimidazol-2(1*H*)-one (**7.499**) is obtained[186] in moderate yield (Table 7.72) by the pyridine catalyzed cyclization of the *N*-benzoylurea derivative (**7.497**; R = H, R^1 = COPh, X = O), which can be preformed or prepared *in situ* by the reaction of 2-aminobenzimidazole with benzoylisocyanate. 4-Phenyl-1,3,5-triazino[1,2-*a*]benzimidazole-2(1*H*)-thione (**7.500**) is likewise formed in moderate yield (Table 7.72) by the ring-closure of 2-aminobenzimidazole with benzoylisothiocyanate or ethoxycarbonylisothiocyanate.[188,189] The potassium hydroxide catalyzed condensation (Scheme 7.98) of *N*-(2-benzimidazolyl)guanidine (**7.497**; R = R^1 = H, X = NH) with benzaldehyde to give the 1,3,5-triazino[1,2-*a*]benzimidazole de-

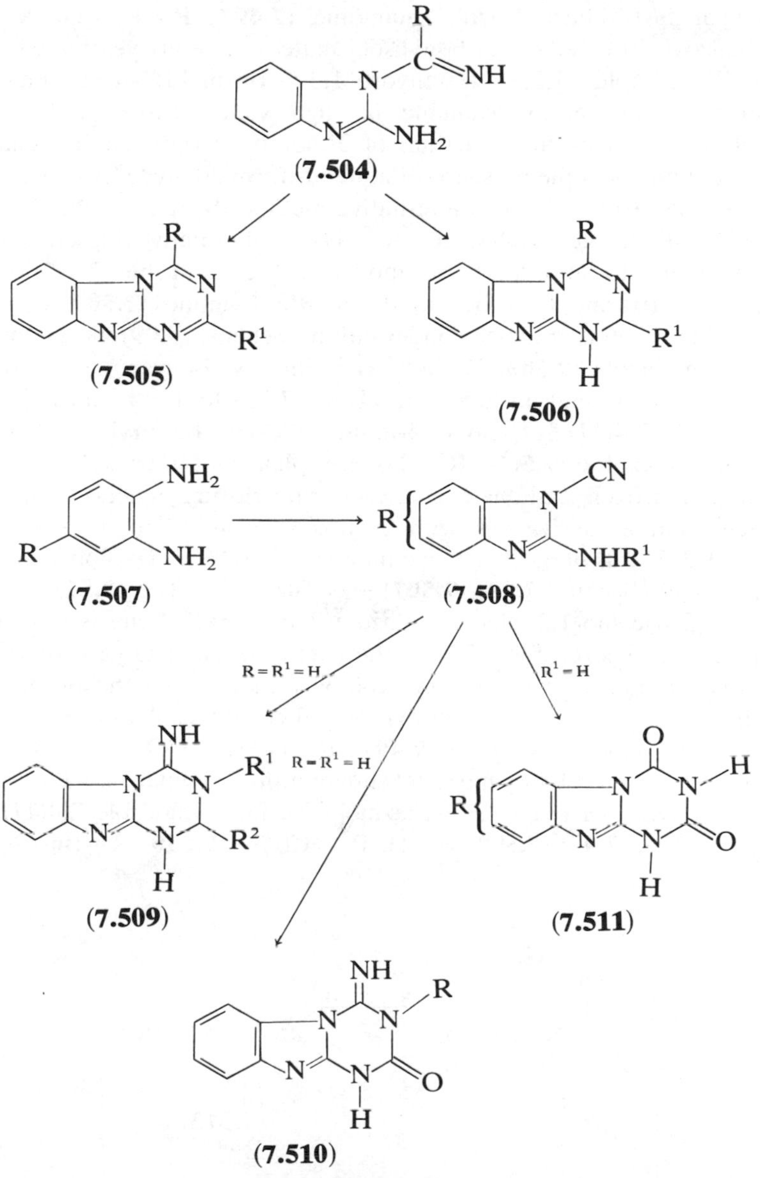

(7.504)

(7.505)

(7.506)

(7.507)

(7.508)

(7.509)

(7.511)

(7.510)

Scheme 7.99

rivative (**7.501**)[143] exemplifies an efficient (Table 7.72) and potentially general method for the synthesis of 3,4-dihydro-1,3,5-triazino[1,2-*a*]benzimidazoles. 2-Amino-1,3,5-triazino[1,2-*a*]benzimidazol-4(3*H*)-one (**7.502**) is the end-product (Table 7.72) of a series of thermal ring-closure reactions originating in *N*-acyl derivatives (**7.497**; R = H, R^1 = $CONHCO_2Et$, CONHCOPh, or $CSNHCO_2Et$) readily accessible by the reaction of *N*-(2-benzimidazolyl)guanidine (**7.497**; $R = R^1 = H$, X = NH) with benzoyl- or ethoxycarbonylisocyanate or ethoxycarbonylisothiocyanate.[186] Simple 1,2,3,4-tetrahydro-1,3,5-triazino[1,2-*a*]benzimidazole derivatives are generally available in high yield (Table 7.72) by the Mannich-like condensation reactions of 2-methoxycarbonylaminobenzimidazoles with amines in the presence of aqueous formaldehyde [Scheme 7.100; (**7.512**) → (**7.513**)].[190–196] In an alternative method (Scheme 7.99), 2-amino-1-cyanobenzimidazole (**7.508**; $R = R^1 = H$) is thermally ring-closed with arylideneimines to afford low to moderate yields (Table 7.72) of 1,2-dihydro-1,3,5-triazino[1,2-*a*]benzimidazole-4(3*H*)-imines (**7.509**; $R^1 = R^2 =$ aryl).[187] The related thermal condensation (Scheme 7.99) of 2-amino-1-cyanobenzimidazole (**7.508**; $R = R^1 = H$) with aryl isocyanates[187] provides synthetic access, though in low yield (Table 7.72) to 1,3,5-triazino[1,2-*a*]-benzimidazole-2,4(1*H*,3*H*)-dione 4-imines (**7.510**; R = aryl). 2-Amino-1-cyanobenzimidazoles (**7.508**; $R^1 = H$) are plausible intermediates in the potassium hydrogen carbonate catalyzed ring-closure reactions of *ortho*-phenylenediamine derivatives with cyanogen bromide to give *N*-unsubstituted 1,3,5-triazino[1,2-*a*]benzimidazole-2,4(1*H*,3*H*)-diones in unspecified yield [Scheme 7.99; (**7.507**) → (**7.508**; $R^1 = H$) → (**7.511**)],[197] The parent 1,3,5-triazino[1,2-*a*]benzimidazole-2,4(1*H*,3*H*)-dione is obtained[186] in quantitative yield (Table 7.72) by the triethylamine catalyzed reaction of 2-aminobenzimidazole with ethoxycarbonylisocyanate via the probable intermediacy of an *N*-ethoxycarbonyl *N*-(2-benzimidazolyl)urea derivative [Scheme 7.98; (**7.496**; R = H) → (**7.497**; R = H, $R^1 = CO_2Et$, X = O)]. The thermal ring-closure of 1-cyano-2-(cyanoamino)benzimidazole with *n*-butylamine to give 3-*n*-butyl-1,3,5-triazino[1,2-*a*]benzimidazole-2,4(1*H*,3*H*)-diimine [Scheme 7.99; (**7.508**; R = H, $R^1 = CN$) → (**7.510**; $R = Bu^n$, NH for

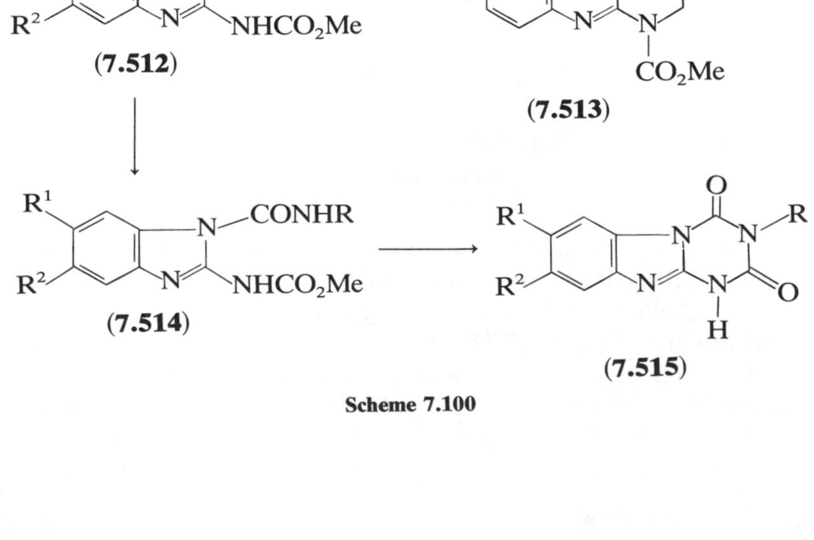

(**7.512**) (**7.513**) (**7.514**) (**7.515**)

Scheme 7.100

O)][198] represents a potentially useful method for the synthesis of molecules of this type. 3-Substituted 1,3,5-triazino[1,2-*a*]benzimidazole-2,4(1*H*,3*H*)-diones are generally accessible, often in high yield (Table 7.72) by the triethylamine catalyzed ring-closure of 2-aminobenzimidazole derivatives with isocyanates[199] or allophanoyl chlorides,[200] via the possible intermediacy of urea derivatives, which can likewise be cyclized by thermal treatment with diaryl carbonates[201,202] or by triethylamine catalyzed condensation with phosgene[199] [Scheme 7.98; (**7.496**)→(**7.497**; X=O)→(**7.503**)]. Alternatively the ring-closure of 2-aminobenzimidazoles with isocyanates and allophanoyl chlorides involves *N*-carboxamide intermediates, which can be preformed and cyclized under basic conditions to afford 3-substituted 1,3,5-triazino[1,2-*a*]benzimidazole-2,4(1*H*,3*H*)-diones [e.g., Scheme 7.100; (**7.512**)→(**7.514**)→(**7.515**)],[203] though only in low yield (Table 7.72).

Ring-closure Reactions of Other Heterocycles

The dehydrative ring-closure (Scheme 7.101) of 2-(2-aminophenyl)-1,2,4-triazine-3,5(2*H*,3*H*)-diones (**7.516**) which gives 1,2,4-triazino[2,3-*a*]benzimidazol-2(1*H*)-ones (**7.517**) in good yield, represents the sole reported example of a reaction in the title category.

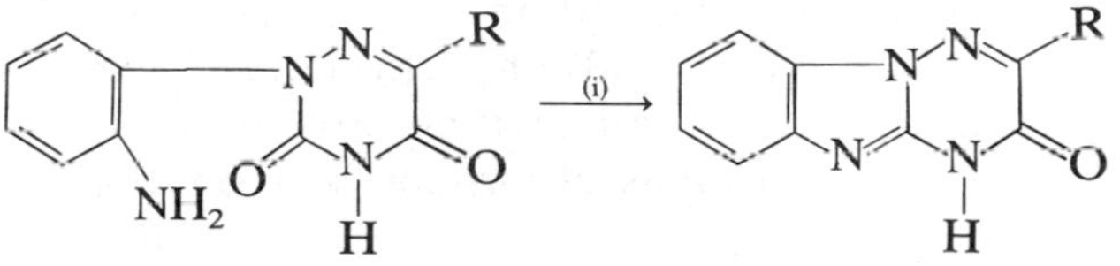

(**7.516**) —(i)→ (**7.517**)

R	Yield (%)	m.p. (°C)
CO_2H	76	260–265 (decomp.)
CN	71	>360

(i) AcOH/(reflux)(10 hr)

Scheme 7.101

7.3.2. Physicochemical Properties

Spectroscopic Studies

INFRARED SPECTRA. The NH substituents in the tetrahydro-1,2,4-triazino-[4,3-*a*]benzimidazole derivative [Scheme 7.102; (**7.536**)] give rise to well-defined IR NH absorption at 3230 cm^{-1}. Compared with the IR carbonyl stretching frequencies (ca. 1700 cm^{-1}) of simple 1,2,4-triazino[2,3-*a*]benzimidazol-2(1*H*)-ones [e.g., Table 7.73; (**7.518**; R=Me or Ph)] that

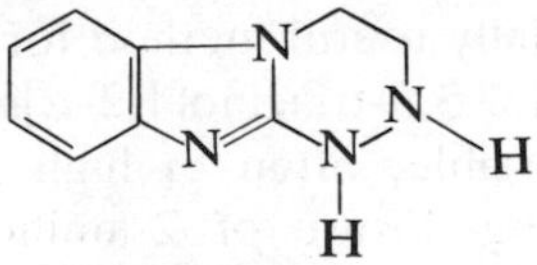

(7.536)

Scheme 7.102

(1670 cm^{-1}) reported[206] for the cyano derivative (**7.518**; R = CN) is abnormally low. The IR stretching frequency (1693 cm^{-1}) of the ring carbonyl group in an *N*-unsubstituted 1,2,4-triazino[4,3-*a*]benzimidazol-4-one[176] is consistent with the existence of such molecules in the *N*(10)H as opposed to the *N*(1)H tautomeric form. The more conjugated ring carbonyl substituent in 1,2,4-triazino[4,5-*a*]benzimidazol-1(5*H*)-ones [e.g., Table 7.74; (**7.520**)] absorbs in the IR at lower frequencies than that in 1,2,4-triazino[4,5-*a*]-benzimidazol-1(2*H*)-ones [Table 7.74; (**7.521**)]. The similar IR frequencies (Table 7.74) of the ring carbonyl substituents in *N*-unsubstituted 1,2,4-triazino[4,5-*a*]benzimidazol-1-ones [Table 7.74; (**7.521**; R^1 = H)] and specifically *N*(2)-substituted derivatives [e.g., Table 7.74; (**7.521**; R^1 = Et)] are indicative of the preferential existence of the former molecules in the *N*(2)H as opposed to *N*(5)H tautomeric form. The ring carbonyl substituent in 1,2,4-triazolo[4,5-*a*]benzimidazol-4(3*H*)-ones [Table 7.74; (**7.522**)] absorbs uniformly in the IR at 1680 cm^{-1}. 1,3,5-Triazino[1,2-*a*]benzimidazole-2,4(1*H*,3*H*)-dione 4-imines [Table 7.75; (**7.526**)] exhibit IR NH absorption at 3350–3340 cm^{-1}, carbonyl absorption in the range 1745–1735 cm^{-1} and well-defined C═N absorption at 1655–1645 cm^{-1}. The IR

TABLE 7.73. INFRARED SPECTRA OF 1,2,4-TRIAZINO[2,3-*a*]BENZIMIDAZOL-3(4*H*)-ONES (**7.518**)

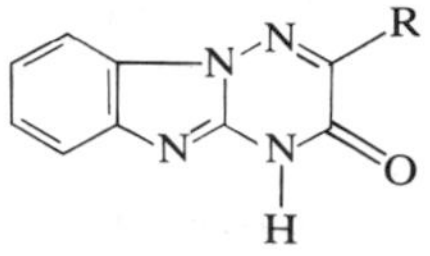

(**7.518**)

		ν_{max} (cm^{-1})				
R	Medium	OH, NH	C≡N	C═O	C═N	Ref.
Me	—[a]	—	—	1700	—	172a
Ph	—[a]	—	—	1700	—	172a
CO_2H	Nujol	—	—	1732, 1714	1640	206
CN	Nujol	—	2242	1670	1628	206
$C(NH_2)$═NOH	Nujol	3465, 3365	—	1705	1638	206

[a] Medium not specified.

TABLE 7.74. INFRARED SPECTRA OF 1,2,4-TRIAZINO[4,5-*a*]BENZIMIDAZOLE DERIVATIVES (**7.519**)–(**7.522**)

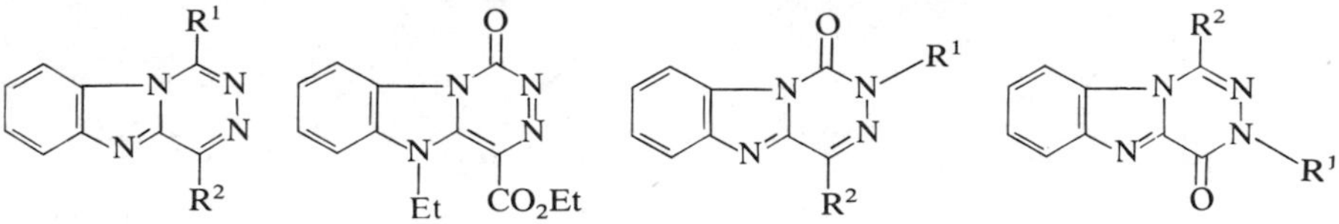

(**7.519**) (**7.520**) (**7.521**) (**7.522**)

Compound	R^1	R^2	Medium	ν_{max} (cm^{-1}) NH, OH	C=O	Ref.
(**7.519**)	OEt	CO_2Et	Nujol	—	1735	7
(**7.519**)	OCOCMe	CO_2Et	Nujol	—	1800, 1740	7
(**7.519**)	$OCOCH_2Cl$	CO_2Et	Nujol	—	1800, 1730	7
(**7.520**)	—	—	Nujol	—	1700, 1676	7
(**7.521**)	H	CO_2Et	Nujol	3180	1735, 1720	7
(**7.521**)	H	CO_2Et	MeCN	—	1730	7
(**7.521**)	H	$CONH_2$	Nujol	—	1735, 1635	7
(**7.521**)	Et	CO_2Et	Nujol	—	1734, 1710	7
(**7.521**)	Et	CO_2Et	MeCN	—	1734, 1710	7
(**7.522**)	H	H	Nujol	3200	1680	181, 182
(**7.522**)	H	Me	Nujol	3200	1680	181, 182
(**7.522**)	Me	H	Nujol	—	1680	182
(**7.522**)	CH_2OH	H	Nujol	—	1680	207
(**7.522**)	$(CH_2)_2CN$	H	Nujol	—	1680	207

spectra (Table 7.75) of 3-substituted 1,3,5-triazino[1,2-*a*]benzimidazole-2,4-(1*H*,3*H*)-diones (**7.527**) are typified by the presence of two bands in the carbonyl region in the ranges 1745–1715 and 1705–1680 cm^{-1}. Bands in the ranges 1696–1670 and 1650–1630 cm^{-1} in the IR spectra of 1,2-dihydro-1,3,5-triazino[1,2-*a*]benzimidazoles [Table 7.75; (**7.524**)] are attributed[187] to C=N absorption.

Ultraviolet Spectra. The UV spectra (Table 7.76) of derivatives of the fully unsaturated 1,2,4-triazino[4,5-*a*]benzimidazole ring system [Table 7.76; (**7.530**)] are characterized by the presence of three intense absorption maxima in the range 223–335 nm. Comparison of the UV absorption (Table 7.76) of specifically *N*(2) and *N*(5) alkylated 1,2,4-triazino[4,5-*a*]benzimidazol-1-ones [e.g., Table 7.76; (**7.531**) and (**7.532**; R^1 = Et, R^2 = CO_2Et)] with that of *N*-unsubstituted 1,2,4-triazino[4,5-*a*]benzimidazol-1-ones [e.g., Table 7.76 (**7.532**; R^1 = H)] clearly demonstrates molecules of the latter type to exist preferentially in the *N*(2)H tautomeric form. Correspondingly, the essential coincidence of the UV spectra (Table 7.76) of *N*-unsubstituted and *N*(3)-methyl-1,2,4-triazino[4,5-*a*]benzimidazol-4(3*H*)-ones [Table 7.76; (**7.533**; R^1 = H or Me)] and their dissimilarity from that of

TABLE 7.75. INFRARED SPECTRA OF 1,3,5-TRIAZINO[1,2-*a*]BENZIMIDAZOLE DERIVATIVES (**7.523**)–(**7.527**)

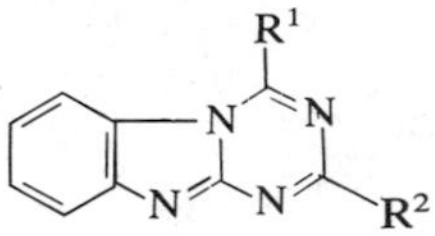

(**7.523**) 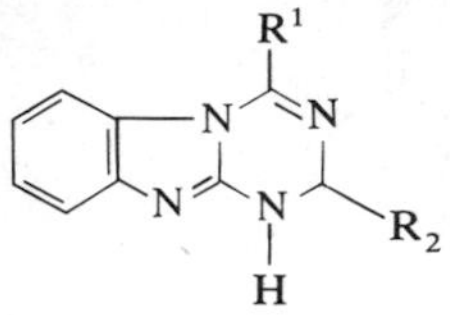 (**7.524**) 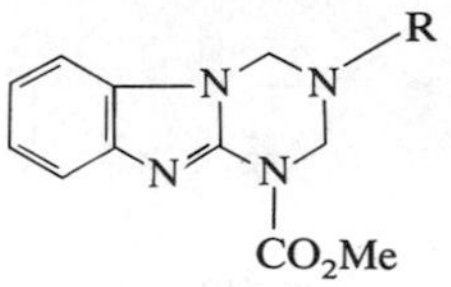 (**7.525**)

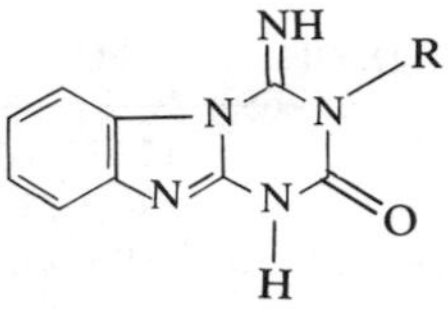

(**7.526**) 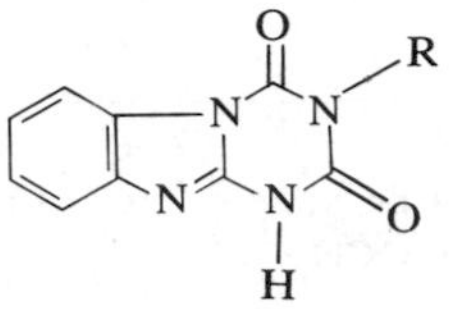 (**7.527**)

Compound	R	R^1	R^2	Medium	ν_{max} (cm^{-1}) OH, NH	C=O	C=N	Ref.
(**7.523**)	—	OPh	Aryl	—[a]	—	—	1680–1640	187
(**7.523**)	—	H	NH_2	KBr	3335, 3250	1690	1650	185
(**7.523**)	—	NH_2	Aryl	—[a]	3400-3300	—	1675–1635	187
(**7.523**)	—	NH_2	NH_2	KBr	3190	—	1670, 1620	184
(**7.524**)	—	OPh	Aryl	—[a]	3200–3100	—	1696–1670, 1645–1630	187
(**7.524**)	—	NR_2	Aryl	—[a]	3160–3140	—	1695–1680 1650–1635	187
(**7.525**)	Bu^n	—	—	—[a]	—	1745	1620	190
(**7.526**)	Aryl	—	—	—[a]	3350–3340	1745–1735	1655–1645	187
(**7.527**)	CH_2CO_2Me	—	—	KBr	—	1755, 1740, 1695	—	201
(**7.527**)	$CH_2CO_2Bu^n$	—	—	KBr	—	1740, 1715, 1680	—	201
(**7.527**)	CH_2CO_2Ph	—	—	KBr	—	1765, 1738, 1690	—	201
(**7.527**)	NMe_2	—	—	KBr	—	1745, 1705–1695	1640–1620	201, 202

[a] Medium not specified

the *O*-methyl derivative [Table 7.76; (**7.530**; $R^1 = H$, $R^2 = OMe$)] demonstrates the existence of the *N*-unsubstituted compounds in the keto as opposed to hydroxy form.

Nuclear Magnetic Resonance Spectra. The protons of the *C*(3) and *C*(4) methylene substituents in 1,2,3,4-tetrahydro-1,2,4-triazino[4,3-*a*]-benzimidazole [Scheme 7.102; (**7.536**)] resonate as triplets centered at

TABLE 7.76. ULTRAVIOLET SPECTRA OF 1,2,4-TRIAZINOBENZIMIDAZOLE DERIVATIVES (**7.528**)–(**7.533**)

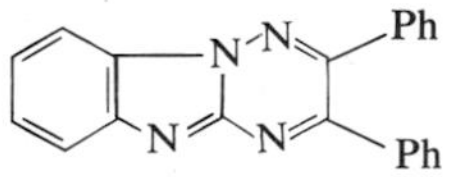

(**7.528**) 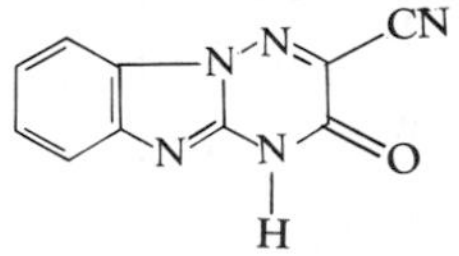 (**7.529**) 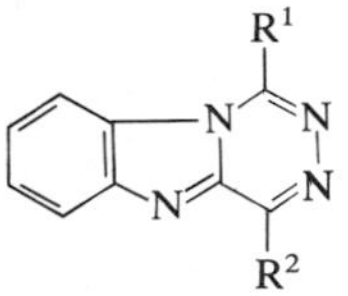 (**7.530**)

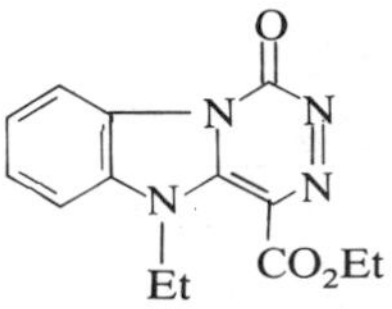

(**7.531**) 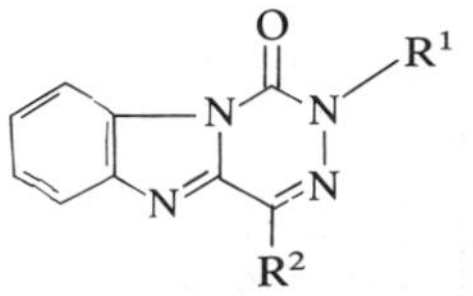 (**7.532**) 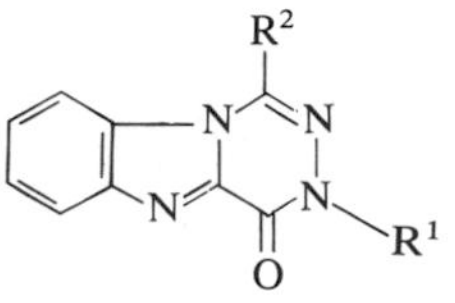 (**7.533**)

Compound	R[1]	R[2]	Solvent[a]	λ_{max} (nm) (log ϵ)	Ref.
(**7.528**)	—	—	*A*	202(4.54), 270(4.38), 375(4.01)	172b
(**7.529**)	—	—	*A*	214(4.51), 265(4.54), 279(4.58), 286(4.60) 343(3.96)	206
(**7.530**)	H	H	*A*	246(4.40), 316(3.78)	179
(**7.530**)	Me	H	*A*	226(4.18), 245(4.40), 320(3.98), 335(3.90)	179
(**7.530**)	H	Ph	*A*	255(4.36), 268(4.34), 315(4.07)	179
(**7.530**)	H	OMe	*A*	223(4.27), 258(4.28), 334(3.84)	182
(**7.530**)	Me	Ph	*A*	251(4.43), 268(4.40), 315(4.02)	195
(**7.530**)	OEt	CO_2Et	*B*	227(4.21), 250(4.49), 320(3.77)	7
(**7.530**)	OCOMe	CO_2Et	*B*	243(4.34), 324(3.92)	7
(**7.530**)	$OCOCH_2Cl$	CO_2Et	*B*	243(4.34), 326(3.92)	7
(**7.531**)	—	—	*B*	245(4.24), 265(4.26), 359(4.07)	7
(**7.532**)	H	CO_2Et	*B*	218(4.37), 244(4.37), 326(3.93)	7
(**7.532**)	H	CON  (piperidino)	*B*	230(4.31), 246(4.22), 320(4.05)	7
(**7.532**)	Et	CO_2Et	*B*	247(4.39), 332(3.97)	7
(**7.533**)	H	H	*C*	250(4.49), 295(3.94), 305(3.92)	182
(**7.533**)	Me	H	*A*	251(4.46), 295(3.97), 305(3.96)	182
(**7.533**)	CH_2OH	H	*A*	247(4.43), 295(3.99), 395(3.99)	207
(**7.533**)	H	Me	*A*	245(4.44), 297(3.97)	182

[a] *A* = ethanol; *B* = methanol; *C* = dioxane.

δ 3.90 and 4.40, respectively.[178] The signal due to *H*(1) which appears at δ 10.2 (Table 7.77) in fully unsaturated 1,2,4-triazino[4,5-*a*]benzimidazole derivatives [Table 7.77; (**7.534**)] is shifted upfield to ca. δ 9.3 in 1,2,4-triazino[4,5-*a*]benzimidazol-4(3*H*)-ones [Table 7.77; (**7.535**)]. *H*(9) in the latter molecules resonates at lower field than the other benzenoid protons (Table 7.77).

TABLE 7.77. ^1H NMR SPECTRA[a,b] OF 1,2,4-TRIAZINO[4,5-*a*]BENZIMIDAZOLE DERIVATIVES (**7.534**) AND (**7.535**)

(**7.534**) (**7.535**)

Compound	Solvent[c]	H(1)	H(3)	H(4)	H(6)	H(7)	H(8)	H(9)	Ref.
(**7.534**; $R^1 = R^2 = H$)	*A*	10.20d[d]	—	9.60d[d]	← —[e] →				179
(**7.534**; $R^1 = H$, $R^2 = Ph$)	*B*	10.20	—	7.58m[f]	← 8.03–8.69m →				179
(**7.534**; $R^1 = Me$, $R^2 = Ph$)	*B*	3.15[g]	—	7.58m[f]	← 8.07m, 8.62m, 8.82m →				195
(**7.535**; $R = R^1 = R^2 = H$)	*C*	9.30	—	—	← 8.16–8.43m →				207
(**7.535**; $R = R^1 = R^2 = H$)	*B*	—[e]	—	—	7.97m	7.60m	7.60m	8.30m	182
(**7.535**; $R = R^2 = H$, $R^1 = NO_2$)	*D*	—[e]	—	—	8.17d[f]	8.48dd[h,i]	—	9.43d[i]	182
(**7.535**; $R = R^1 = H$, $R^2 = NO_2$)	*D*	—[e]	—	—	8.78d[i]	—	8.42dd[h,i]	9.43d[h]	182
(**7.535**; $R = CH_2OH$, $R^1 = R^2 = H$)	*C*	9.30	3.82[j]	—	← 8.15–8.43m →				207
(**7.535**; $R = (CH_2)_2CN$, $R^1 = R^2 = H$)	*C*	9.34	3.23t[k,l] 4.75t[m,l]	—	← 8.16–8.44m →				207

[a] δ values in ppm measured from TMS.
[b] Signals are sharp singlets unless denoted as d = doublet; dd = double doublet; t = triplet; m = multiplet.
[c] *A* = $Me_2SO—CCl_4$; *B* = Me_2SO; *C* = CF_3CO_2H; *D* = Me_2NCHO.
[d] $J = 2.0$ Hz.
[e] δ values not quoted.
[f] Ph*H*.
[g] C(Me).
[h] $J = 8.8–10.0$ Hz.
[i] $J = 2.0–2.3$ Hz.
[j] CH_2OH.
[k] NCH_2.
[l] *J* value not quoted.
[m] CH_2CN.

General Studies

IONIZATION CONSTANTS. The pK_a value (5.32±0.02) measured spectroscopically in aqueous ethanol for 2-cyano-1,2,4-triazino[2,3-*a*]benzimidazol-3(4*H*)-one indicates this molecule, as expected, to be a weak *N*-acid.[206]

7.3.3. Reactions

Reactions with Electrophiles

ALKYLATION. In accord with their *N*(2)H as opposed to *N*(5)H tautomeric structures, 1,2,4-triazino[4,5-*a*]benzimidazol-1(2*H*)-ones are alkylated in high yield (Table 7.78) at *N*(2) under basic conditions.[7] Correspondingly, the base-catalyzed alkylation of 1,2,4-triazino[4,5-*a*]benzimidazol-4(3*H*)-ones takes place at *N*(3) also in high yield (Table 7.78).[182,207] In contrast, the uncatalyzed thermal methylation of 1,2,4-triazino[4,5-*a*]benzimidazol-4(3*H*)-one takes place at oxygen giving 4-methoxy-1,2,4-triazino[4,5-*a*]-benzimidazole in good yield (Table 7.78).[182] 1,2,4-Triazino[4,5-*a*]benzimidazol-4(3*H*)-one is cyanoethylated in good yield at *N*(3) under basic conditions [Scheme 7.103; (**7.537**)→(**7.538**)].[207] The reaction of 1,3,5-triazino-[1,2-*a*]benzimidazole-2(1*H*)-thione with ethyl iodide in the presence of

TABLE 7.78. ALKYLATION REACTIONS OF 1,2,4-TRIAZINO[4,5-*a*]BENZIMIDAZOLONES

Reaction conditions[a]	1,2,4-Triazino[4,5-*a*]benzimidazolone		Yield (%)	m.p. (°C)	Solvent of crystallization	Ref.
	Substrate	Product				
A	4-CO_2Et-1(2*H*)-one	2-Et-4-CO_2Et-1(2*H*)-one	66	161–162	Ethanol	7
B	4(3*H*)-one	4-OMe	64	207	Ethanol	181, 182
C	4(3*H*)-one	3-Me-4(3*H*)-one	83	308–310	Dimethylformamide	181, 182
C	1-Me-4(3*H*)-one	1,3-Me_2-4(3*H*)-one	84	241–243	Ethanol	181, 182
C	4(3*H*)-one	3-Et-4(3*H*)-one	42	232	Ethanol	181, 182
D	4(3*H*)-one	3-CH_2Ph-4(3*H*)-one	57	289–290 (decomp.)	Dimethylformamide	207

[a] *A* = EtI, NaH, dimethylformamide/(100°)(18 hr); *B* = Me_2SO_4, $PhNO_2$/(130–140°)(45 min.); *C* = MeI or EtI, NaOEt, EtOH/(reflux)(4.5 hr); *D* = $PhCH_2Cl$, KI, NaOEt, EtOH dimethyl sulfoxide/(reflux)(7 hr).

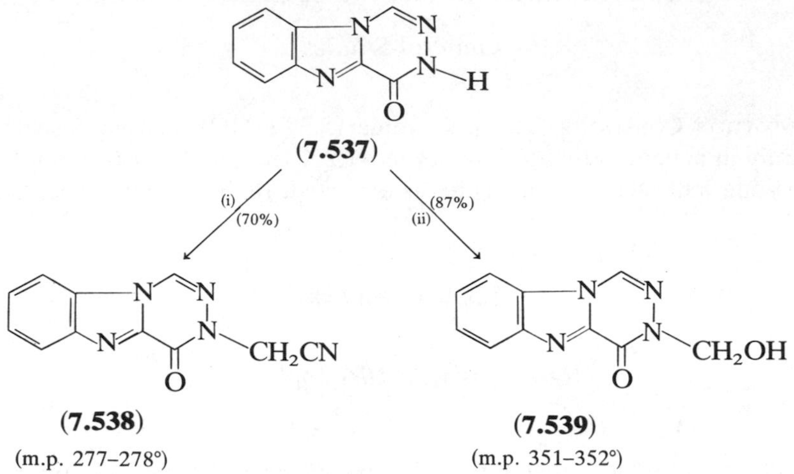

(7.537)

(i) (70%)   (ii) (87%)

(7.538)
(m.p. 277–278°)

(7.539)
(m.p. 351–352°)

(i) $CH_2{=}CHCN$, Triton W/(reflux)(3 hr)
(ii) HCHO aq., EtOH/(reflux)(3.5 hr)

Scheme 7.103

TABLE 7.79. ALKYLATION REACTIONS OF 1,3,5-TRIAZINO[1,2-*a*]BENZIMIDAZOLE DERIVATIVES

Reaction conditions[a]	1,3,5-Triazino[1,2-*a*]benzimidazole Substrate	Product	m.p. (°C)	Solvent of crystallization	Ref.
A	2,4(1*H*,3*H*)-dione	1,3-Me_2-2,4(1*H*,3*H*)-dione	232	—[b]	199
A	2,4(1*H*,3*H*)-dione	1,3-Et_2-2,4(1*H*,3*H*)-dione	156	—[b]	199
B	2,4(1*H*,3*H*)-dione	1,3-Bu^n_2-2,4(1*H*,3*H*)-dione	116–117	—[b]	199
C	2,4(1*H*,3*H*)-dione	1,3($CH_2Ph)_2$-2,4(1*H*,3*H*)-dione	200-202	—[b]	199
A	3-Bu^n-2,4(1*H*,3*H*)-dione	1-Me-3-Bu^n-2,4(1*H*,3*H*)-dione	150–151	—[b]	199
B	3-Me-2,4(1*H*,3*H*)-dione	1-Bu^n-3-Me-2,4(1*H*,3*H*)-dione	180	Acetonitrile	199
C	3-Ph-2,4(1*H*,3*H*)-dione	1-Me-3-Ph-2,4(1*H*,3*H*)-dione[b] +	222–224	—[e]	199
		2-OMe-3-Ph-4(3*H*)-one[d]	280–281	—[e]	
A	3-Et-2,4(1*H*,3*H*)-dione	1-Pr^n-3-Et-2,4(1*H*,3*H*)-dione	170–175	—[b]	199
A	3-Ph-2,4(1*H*,3*H*)-dione	1-Pr^n-3-Ph-2,4(1*H*,3*H*)-dione	185–192	—[b]	199
D	4-Ph-2(1*H*)-thione	2-SEt-4-Ph-	190	Ethanol	188

[a] *A* = NaOMe, MeOH, benzene/(room temp.)(30 min, then RI, dimethylformamide)/(80°)(1 hr); *B* = NaH, dimethylformamide/(room temp.)(10 min), then Bu^nBr)/(120°)(1 hr); *C* = MeI, K_2CO_3, dimethylformamide/(100°)(1 hr); *D* = EtI, NaOH, MeOH(room temp.)(16 hr).

[b] Solvent of crystallization not specified.

[c] Yield 26%.

[d] Yield 5%.

[e] Purified by chromatography.

sodium hydroxide is reported[188] to occur at sulfur giving a thioethyl derivative (Table 7.79) in unspecified yield. *N*(3)-Substituted 1,3,5-triazino[1,2-*a*]benzimidazole-2,4(1*H*,3*H*)-diones undergo orthodox alkylation at *N*(1) under basic conditions.[199]

Acylation. 1,2,4-Triazino[4,5-*a*]benzimidazol-1(2*H*)-ones are acylated at oxygen rather than nitrogen by acid chlorides under basic conditions [e.g., Scheme 7.104; (**7.540**)→(**7.541**)].[7] On the other hand, the reaction of 1,2,4-triazino[4,5-*a*]benzimidazol-4(3*H*)-one with formaldehyde results in hydroxymethylation at *N*(3) [Scheme 7.103; (**7.537**)→(**7.539**)].[207] Also, the acylation of *N*(3)-substituted 1,3,5-triazino[1,2-*a*]benzimidazole-2,4-(1*H*,3*H*)-diones by acid chlorides and acid anhydrides occurs at *N*(1) rather than at oxygen (Table 7.80).[199]

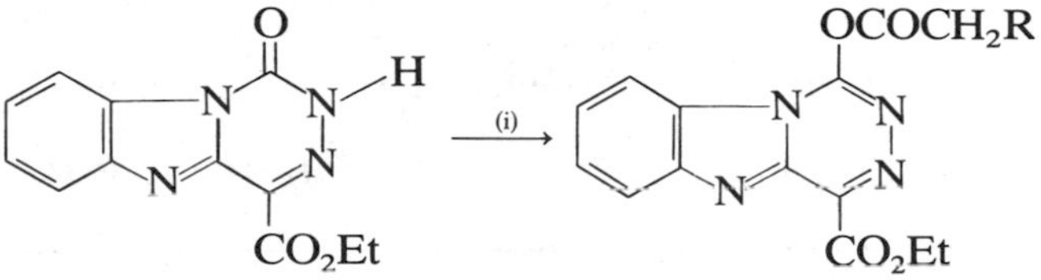

(**7.540**) →(i) (**7.541**)

R	Yield (%)	m.p. (°C)
H	76	150–151
Cl	—	188–190

(i) MeCOCl, $(Pr^i)_2NH$, CH_2Cl_2/(room temp.)(2 hr)

Scheme 7.104

Electrophilic Substitution Reactions. The nitration of 1,2,4-triazino-[4,5-*a*]benzimidazol-4(3*H*)-one by mixed acid affords the 8-nitro derivative [m.p. 338° (decomp.)] in 70% yield.[182] This appears to be the only reported example of electrophilic substitution in a tricyclic 6-5-6 fused benzimidazole ring system of the type under consideration.

Reactions with Nucleophiles

The stability of the 1,2,4-triazino[2,3-*a*]benzimidazol-4(3*H*)-one ring system to acidic hydrolysis is demonstrated by the conversion of 2-cyano-1,2,4-triazino[2,3-*a*]benzimidazol-3(4*H*)-one in hot concentrated hydrochloric acid into the corresponding carboxylic acid [m.p. 260–265° (decomp.)] in 71% yield.[206] In contrast, the fused triazine ring in fully unsaturated 1,2,4-triazino[4,5-*a*]benzimidazoles,[7,179] and in 1,2,4-triazino[4,5-*a*]-benzimidazol-1(2*H*)-ones[208] and 1,2,4-triazino[4,5-*a*]benzimidazol-4(3*H*)]-ones[182] is readily opened under both acidic and basic conditions giving

TABLE 7.80. ACYLATION REACTIONS OF 1,3,5-TRIAZINO[1,2-*a*]BENZIMIDAZOL-2,4(1*H*,3*H*)-DIONES[a]

Reaction conditions[b]	1,3,5-Triazino[1,2-*a*]benzimidazole Substrate	Product	m.p. (°C)	Solvent of crystallization
A	3-Et-2,4(1*H*,3*H*)-dione	1-COMe-3-Et-2,4(1*H*,3*H*)-dione	225	—[c]
A	3-Pr^n-2,4(1*H*,3*H*)-dione	1-COMe-3-Pr^n-2,4(1*H*,3*H*)-dione	203	—[c]
A	3-Pr^i-2,4(1*H*,3*H*)-dione	1-COMe-3-Pr^i-2,4(1*H*,3*H*)-dione	208	—[c]
B	3-Bu^n-2,4(1*H*,3*H*)-dione	1-CO_2Et-3-Bu^n-2,4(1*H*,3*H*)-dione	148–152	Acetonitrile
B	3-Bu^n-2,4(1*H*,3*H*)-dione	1-CO_2Ph-3-Bu^n-2,4(1*H*,3*H*)-dione	170–172	Benzene
C	3-Ph-2,4(1*H*,3*H*)-dione	1-CONHMe-3-Ph-2,4(1*H*,3*H*)-dione	350	Dimethylformamide
C	3-Bu^n-2,4(1*H*,3*H*)-dione	1-CONHBu^n-3-Bu^n-2,4(1*H*,3*H*)-dione	280	—[c]
C	3-Ph-2,4(1*H*,3*H*)-dione	1-CONHPh-3-Ph-2,4(1*H*,3*H*)-dione	330–345	—[c]

[a] From Ref. 199.
[b] *A* = Ac_2O/(reflux)(2–7 hr); *B* = $ClCO_2Et$ or $ClCO_2Ph$, Et_3N, tetrahydrofurane/(65°)(2 hr); *C* = RN=C=O, Et_3N, toluene/(reflux)(5 hr).
[c] Solvent of crystallization not specified.

benzimidazole derivatives. The 1,3,5-triazine ring in 1,3,5-triazino[1,2-*a*]-benzimidazoles is also susceptible to base-catalyzed ring-opening.[205]

Oxidation

Investigations of the behavior of the various 6-5-6 fused benzimidazole ring systems with two additional heteroatoms toward oxidation have been limited to the study[174] of the electrochemical oxidation of 1,4-dihydro-1,2,4-triazino[4,3-*a*]benzimidazole derivatives and the demonstration of the dehydrogenation[182] of 1,2-dihydro-1,3,5-triazino[1,2-*a*]benzimidazoles to fully unsaturated 1,3,5-triazino[1,2-*a*]benzimidazole derivatives.

7.3.4. Practical Applications

Biological Properties

1,3,5-Triazino[1,2-*a*]benzimidazole derivatives of various types have found widespread application as antiparasitic agents[190–194,199] and as fungicides,[190–193,196,199,201–203,209] pesticides,[199,202] and herbicides.[199]

Other Applications

1,3,5-Triazino[1,2-*a*]benzimidazole-2,4(1*H*,3*H*)-dione has been patented[197] as a photographic emulsion stabilizer.

REFERENCES

1. (a) W. L. Mosby in *Systems with Bridgehead Nitrogen Part* 1, *The Chemistry of Heterocyclic Compounds*, A. Weissberger, (Ed), Interscience, New York, **1961,** Chap. 5, pp. 507–517; (b) R. M. Acheson and N. F. Elmore, *Adv. Heterocycl. Chem.*, **23,** 263 (1978).
2. R. M. Acheson, M. W. Foxton, P. J. Abbott, and K. R. Mills, *J. Chem. Soc.* (*C*), **1967,** 882.
3. P. J. Abbott, R. M. Acheson, U. Eisner, J. D. Watkin, and J. R. Carruthers, *J. Chem. Soc. Perkin Trans. I,* **1976,** 1269.
4. R. M. Acheson and M. S. Verlander, *J. Chem. Soc. Perkin Trans. I,* **1974,** 430.
5. R. M. Acheson and W. R. Tully, *J. Chem. Soc.* (*C*), **1968,** 1623.
6. R. M. Acheson and M. S. Verlander, *J. Chem. Soc. Perkin Trans. I,* **1972,** 1577.
7. N. Finch and C. W. Gemenden, *J. Org. Chem.*, **35,** 3114 (1970).
8. H. Ogura and K. Kikuchi, *J. Org. Chem.*, **37,** 2679 (1972).
9. I. Zugravescu, J. Herdan, and I. Druta, *Rev. Roum. Chim.*, **19,** 649 (1974); *Chem. Abstr.*, **81,** 25603 (1974).
10. O. Meth-Cohn, *Tetrahedron Lett.*, **1975,** 413.
11. (a) D. D. Chapman, J. K. Elwood, D. W. Heseltine, H. M. Hess, and D. W. Kurtz, *J. Org. Chem.*, **42,** 2474 (1977); (b) D. W. Heseltine, D. W. Kurtz, D. D. Chapman, and J. K. Elwood, *Fr. Patent*, 2,228,090; *Chem. Abstr.*, **83,** 81245 (1975).
12. B. Mencke and K. Schmitt, *Arch. Pharm.*, **300,** 481 (1967); *Chem. Abstr.*, **67,** 116846 (1967).
13. H. Schaefer and K. Gewald, *Z. Chem.*, **16,** 272 (1976); *Chem. Abstr.*, **86,** 55339 (1977).
14. G. Lamm, *Ger. Patent,* 2,025,427; *Chem. Abstr.*, **76,** 72519 (1972).
15. J. Dehnert and G. Lamm, *Ger. Patent,* 2,022,817; *Chem. Abstr.*, **76,** 142405 (1972).
16. V. A. Chuiguk and Y. M. Volovenko, *Chem. Heterocycl. Compd.*, **1975,** 467, *Khim. Geterotsikl. Soedin.*, **1975,** 530; *Chem. Abstr.*, **83,** 116911 (1975).
17. K. Kurata, H. Awaya, Y. Tominaga, Y. Matsuda, and G. Kobayashi, *Bunseki Kiki*, **15,** 413 (1977); through *Chem. Abstr.*, **88,** 121047 (1978).
18. A. S. Afridi, A. R. Katritzky, and C. A. Ramsden, *J. Chem. Soc. Perkin Trans. I,* **1977,** 1436.
19. S. Miyano, N. Abe, K. Takeda, and K. Sumoto, *Synthesis,* **1978,** 451; *Chem. Abstr.*, **89,** 109245 (1978).
20. S. Linke and C. Wunsche, *J. Heterocycl. Chem.*, **10,** 333 (1973).
21. T. Kato and M. Daneshtalab, *Chem. Pharm. Bull.* (*Tokyo*), **24,** 1640 (1976); *Chem. Abstr.*, **85,** 177317 (1976).
22. N. A. Malichenko and L. M. Yagupolskii, *Khim. Geterotsikl. Soedin.*, **1978,** 388; through *Chem. Abstr.*, **89,** 43238 (1978).
23. K. Maekawa and J. Ohtani, *Agric. Biol. Chem.*, **42,** 482 (1978); *Chem. Abstr.*, **88,** 190682 (1978).
24. W. L. Mosby, *J. Org. Chem.*, **24,** 419 (1959).
25. R. C. De Selms, *J. Org. Chem.*, **27,** 2165 (1962).
26. H. Matrick and A. R. Day, *J. Org. Chem.*, **26,** 1511 (1961).
27. G. Morgan and J. Stewart, *J. Chem. Soc.*, **1938,** 1292.
28. G. Morgan and J. Stewart, *J. Chem. Soc.*, **1939,** 1057.
29. K. H. Saunders, *J. Chem. Soc.*, **1955,** 3275.

30. A. J. Hubert, *J. Chem. Soc.* (*C*), **1969,** 1334.
31. P. Nantka-Namirski and J. Kalinowski, *Acta Pol. Pharm.*, **27,** 525 (1970); *Chem. Abstr.*, **75,** 5805 (1971).
32. L. Stephenson and W. K. Warburton, *J. Chem. Soc.* (*C*), **1970,** 1355.
33. P. Nantka-Namirski and J. Kalinowski, *Acta Pol. Pharm.*, **28,** 221 (1971); *Chem. Abstr.*, **75,** 76687 (1971).
34. P. Nantka-Namirski and J. Kalinowski, *Acta Pol. Pharm.*, **30,** 1 (1973); *Chem. Abstr.*, **79,** 91934 (1973).
35. P. Nantka-Namirski and J. Kalinowski, *Acta Pol. Pharm.*, **31,** 137 (1974); *Chem. Abstr.*, **81,** 152112 (1974); *Polish Patent*, 68,831; through *Chem. Abstr.*, **80,** 133436 (1974).
36. R. A. Abramovitch, D. H. Hey, and R. D. Mulley, *J. Chem. Soc.*, **1954,** 4263.
37. N. Campbell and E. B. McCall, *J. Chem. Soc.*, **1951,** 2411.
38. S. Kajihara, *Nippon Kagaku Zasshi*, **86,** 839 (1965); through *Chem. Abstr.*, **65,** 16935c (1966).
39. L. Schmid and H. Czerny, *Monatsh.*, **83,** 31 (1952).
40. G. Cauquis and J. L. Cros, *Bull. Soc. Chim. France*, **1971,** 3760.
41. G. Cauquis, J. L. Cros, and M. Genies, *Bull. Soc. Chim. France*, **1971,** 3765; J. J. Basselier, G. Cauquis, and J. L. Cros, *Chem. Commun.*, **1969,** 1171.
42. O. Meth-Cohn and H. Suschitzky, *Adv. Heterocycl. Chem.*, **14,** 211 (1972).
43. O. Meth-Cohn, *J. Chem. Soc.*, **1964,** 5245.
44. (a) L. Spiegel and H. Kaufmann, *Chem. Ber.*, **41,** 679 (1908); (b) H. Suschitzky and M. E. Sutton, *Tetrahedron*, **24,** 4581 (1968).
45. M. D. Nair and R. Adams, *J. Am. Chem. Soc.*, **83,** 3518 (1961).
46. O. Meth-Cohn and H. Suschitzky, *J. Chem. Soc.*, **1963,** 4666.
47. D. P. Ainsworth and H. Suschitzky, *J. Chem. Soc.* (*C*), **1966,** 111.
48. R. Garner and H. Suschitzky, *J. Chem. Soc.* (*C*), **1967,** 74.
49. R. C. Perera, R. K. Smalley, and L. G. Rogerson, *J. Chem. Soc.* (*C*), **1971,** 1348.
50. R. Huisgen and H. Rist, *Ann. Chem.*, **594,** 159 (1955).
51. H. Takahashi, S. Sato, and H. Otomasu, *Chem. Pharm. Bull.* (*Tokyo*), **22,** 1921 (1974); *Chem. Abstr.*, **82,** 43248 (1975).
52. H. Mohrle and J. Gerloff, *Arch. Pharm.*, **311,** 381 (1978); *Chem. Abstr.*, **89,** 109231 (1978).
53. H. Mohrle and H. J. Hemmerling, *Arch. Pharm.*, **311,** 586 (1978); *Chem. Abstr.*, **89,** 163531 (1978).
54. H. Suschitzky and M. E. Sutton, *Tetrahedron Lett.*, **1967,** 3933.
55. R. Fielden, O. Meth-Cohn, and H. Suschitzky, *J. Chem. Soc. Perkin Trans. I*, **1973,** 696.
56. R. K. Grantham and O. Meth-Cohn, *J. Chem. Soc.* (*C*), **1969,** 70.
57. A. Darchen and D. Peltier, *Bull. Soc. Chim. France*, **1974,** 673.
58. R. Garner, G. V. Garner, and H. Suschitzky, *J. Chem. Soc.* (*C*), **1970,** 825.
59. O. Meth-Cohn, R. K. Smalley, and H. Suschitzky, *J. Chem. Soc.*, **1963,** 1666.
60. M. J. Libeer, H. Depoorter, G. G. Van Mierlo, and R. G. Lemahieu, *U.S. Patent*, 3,931,156; *Chem. Abstr.*, **85,** 64804 (1976); *Belgian Patent*, 618,235; *Chem. Abstr.*, **58,** 14164f (1963).
61. R. K. Grantham, O. Meth-Cohn, and M. A. Naqui, *J. Chem. Soc.* (*C*), **1969,** 1438.
62. J. W. Clark-Lewis, K. Moody, and M. J. Thompson, *Aust. J. Chem.*, **23,** 1249 (1970).
63. R. K. Grantham and O. Meth-Cohn, *J. Chem. Soc.* (*C*), **1969,** 1444.
64. D. Price, H. Suschitzky, and J. I. Hollies, *J. Chem. Soc.* (*C*), **1969,** 1967.
65. D. P. Ainsworth, O. Meth-Cohn, and H. Suschitzky, *J. Chem. Soc.* (*C*), **1968,** 923.
66. I. Baxter and D. W. Cameron, *J. Chem. Soc.* (*C*), **1968,** 1747.
67. J. Martin, O. Meth-Cohn, and H. Suschitzky, *Tetrahedron Lett.*, **1973,** 4495.
68. E. Schefczik, *Ger. Patent*, 2,611,665; *Chem. Abstr.*, **87,** 203077 (1977).
69. E. B. Mullock, R. Searby, and H. Suschitzky, *J. Chem. Soc.* (*C*), **1970,** 829.
70. F. A. L'Eplattenier, L. Vuitel, H. Junek, and O. S. Wolfbeis, *Synthesis*, **1976,** 543.
71. G. Lamm and J. Dehnert, *Ger. Patent*, 2,023,295; *Chem. Abstr.*, **76,** 101201 (1972); *Ger.*

Patent, 2,004,488; *Chem. Abstr.*, **75,** 130767 (1971); A. Roueche and F. L'Eplattenier, *Ger. Patent*, 2,510,373; *Chem. Abstr.*, **84,** 6479 (1976).
72. E. E. Glover and K. T. Rowbottom, *J. Chem. Soc. Perkin Trans. I*, **1976,** 367.
73. B. Rudner, *U.S. Patent*, 2,891,060; *Chem. Abstr.*, **54,** 4652b (1960); *Brit. Patent*, 823,332; *Chem. Abstr.*, **55,** 3641e (1961).
74. R. Fielden, O. Meth-Cohn, and H. Suschitzky, *J. Chem. Soc. Perkin Trans. I*, **1973,** 705.
75. R. C. Perera and R. K. Smalley, *Chem. Commun.*, **1970,** 1458.
76. H. G. Alpermann, *Arzneim.-Forsch.*, **16,** 1641 (1966); *Chem. Abstr.*, **66,** 64072 (1967).
77. F. L'Eplattenier, A. Pugin, and L. Vuitel, *Ger. Patent*, 2,442,315; *Chem. Abstr.*, **83,** 61667 (1975).
78. W. Mueller, *Ger. Patent*, 2,542,408; *Chem. Abstr.*, **85,** 22749 (1976).
79. O. Sus, M. P. Schmidt, and M. Glos, *U.S. Patent*, 2,773,765; *Chem. Abstr.*, **51,** 11139h (1957); W. Neugebauer, *U.S. Patent*, 2,958,599; *Chem. Abstr.*, **55,** 6222e (1961); D. W. Heseltine, D. W. Kurtz, D. D. Chapman, and J. K. Elwood, *U.S. Patent*, 4,003,750; *Chem. Abstr.*, **87,** 60746 (1977).
80. W. L. Mosby in *Systems with Bridgehead Nitrogen Part* 2, *The Chemistry of Heterocyclic Compounds*, A. Weissberger, (Ed.), Interscience, New York, **1961,** Chap. 6, pp. 808–814.
81. J. B. Van Den Berk, L. E. J. Kennis, M. J. M. C. Van Der Au, and A. A. M. T. Van Heertum, *Ger. Patent*, 2,632,870; *Chem. Abstr.*, **87,** 23274 (1977).
82. M. M. Htay and O. Meth-Cohn, *Tetrahedron Lett.*, **1976,** 79.
83. R. J. Hayward, M. M. Htay, and O. Meth-Cohn, *Chem. Ind.* (*London*), **1977,** 373.
84. R. D. Kimbrough, *J. Org. Chem.*, **29,** 1242 (1964).
85. G. A. Taylor, *J. Chem. Soc. Perkin Trans. I*, **1975,** 1001.
86. M. J. Haddadin and A. Hassner, *J. Org. Chem.*, **38,** 2650 (1973).
87. J. Rokach, Y. Girard, and J. G. Atkinson, *Can. J. Chem.*, **51,** 3765 (1973).
88. D. G. O'Sullivan and A. K. Wallis, *J. Chem. Soc.*, **1965,** 2331.
89. H. Singh and S. Singh, *Indian J. Chem.*, **9,** 918 (1971); *Chem. Abstr.*, **76,** 3761 (1972).
90. K. K. Balasubramanian and R. Nagarajan, *Synthesis*, **1976,** 189.
91. A. Davidson, I. E. P. Murray, P. N. Preston, and T. J. King, *J. Chem. Soc. Perkin Trans. I*, **1979,** 1239.
92. L. V. Zavyalova, N. K. Rozhkova, and K. L. Seitanidi, *Chem. Heterocycl. Compd.*, **1975,** 38; *Khim. Geterotsikl. Soedin.*, **1975,** 47; *Chem. Abstr.*, **83,** 9941 (1975).
93. K. C. Liu, J. Y. Tuan, B. J. Shih, and L. C. Lee, *Arch. Pharm.*, **310,** 522 (1977); *Chem. Abstr.*, **87,** 167956 (1977).
94. E. Ziegler and R. Wolf, *Monatsh. Chem.*, **93,** 1441 (1962); *Chem. Abstr.*, **59,** 2819c (1963).
95. S. L. Mukherjee, G. Bagavant, V. S. Dighe, and S. Somasekhara, *Current Sci.* (*India*), **32,** 454 (1963); *Chem. Abstr.*, **59,** 15275h (1963).
96. D. V. Kovpak and A. K. Bagrii, *Khim. Issled. Farm.*, **1970,** 33; through *Chem. Abstr.*, **75,** 151744 (1971).
97. P. M. Kochergin, A. K. Bagrii, D. V. Kovpak, and G. F. Galenko, *USSR Patent*, 390,094; through *Chem. Abstr.*, **79,** 126515 (1973).
98. K. Hideg. O. Hankovszky, E. Palosi, G. Hajos, and L. Szporny, *Ger. Patent*, 2,429,290; *Chem. Abstr.*, **82,** 156307 (1975); *Hung. Teljes*, 11,219; through *Chem. Abstr.*, **85,** 94382 (1976).
99. A. K. Bagrii, G. F. Galenko, and P. M. Kochergin, *Dopov. Akad. Nauk Ukr. RSR Ser. B*, **1975,** 801; through *Chem. Abstr.*, **84,** 43959 (1976).
100. H. O. Hankovszky and K. Hideg, *Acta Chim.* (*Budapest*), **63,** 447 (1970); *Chem. Abstr.*, **72,** 100597 (1970).
101. A. L. Misra, *J. Org. Chem.*, **23,** 897 (1958).
102. I. I. Chizhevskaya, L. I. Gapanovich, and L. V. Poznyak, *J. Gen. Chem. USSR*, **33,** 931 (1963); *Zh. Obshch. Khim.*, **33,** 945 (1963); *Chem. Abstr.*, **59,** 8725a (1963).
103. I. I. Chizhevskaya, N. N. Khovratovich, and Z. M. Grabovskaya, *Chem. Heterocycl.*

Compd., **1968,** 329; *Khim. Geterotsikl. Soedin.*, **1968,** 443; *Chem. Abstr.*, **69,** 86906 (1968).
104. H. S. Chaudhary, C. S. Panda, and H. K. Pujari, *Indian J. Chem.*, **8,** 10 (1970); *Chem. Abstr.*, **72,** 121432 (1970).
105. J. Mohan, V. K. Chadha, and H. K. Pujari, *Indian J. Chem.*, **11,** 1119 (1973); *Chem. Abstr.*, **80,** 108448 (1974).
106. J. Mohan, V. K. Chadha, K. S. Sharma, and H. K. Pujari, *Indian J. Chem.*, *Sect. B*, **14B,** 723 (1976); *Chem. Abstr.*, **86,** 106508 (1977).
107. K. V. Ananeva and N. K. Rozhkova, *Uzb. Khim. Zh.*, **17,** 56 (1973); *Chem. Abstr.*, **80,** 47905 (1974).
108. S. C. Bell and P. H. L. Wei, *J. Med. Chem.*, **19,** 524 (1976); *Chem. Abstr.*, **84,** 130238 (1976).
109. S. Singh, H. Singh, M. Singh, and K. S. Narang, *Indian J. Chem.*, **8,** 230 (1970); *Chem. Abstr.*, **72,** 132624 (1970).
110. *Netherlands Patent*, 6,515,398; *Chem. Abstr.*, **65,** 2401g (1966).
111. L. M. Werbel, A. Curry, E. F. Elslager, C. A. Hess, M. P. Hutt, and C. Youngstrom, *J. Heterocycl. Chem.*, **6,** 787 (1969).
112. C. F. H. Allen, H. R. Beilfuss, D. M. Burness, G. A. Reynolds, J. F. Tinker, and J. A. Van Allan, *J. Org. Chem.*, **24,** 796 (1959).
113. D. M. Burness, *U.S. Patent*, 2,837,521; *Chem. Abstr.*, **53,** 2262e (1959).
114. G. M. Golubushina and V. A. Chuiguk, *Ukr. J. Chem.*, **37,** 45 (1971); *Ukr. Khim. Zh.*, **37,** 1132 (1971); *Chem. Abstr.*, **76,** 46158 (1972).
115. W. Ried and W. Muller, *J. Prakt. Chem.*, **8,** 132 (1959).
116. M. Ridi. S. Checchi, and P. Papini, *Ann. Chim. (Rome)*, **44,** 769 (1954); *Chem. Abstr.*, **52,** 17285d (1958).
117. L. Basaglia and B. Mariani, *Ann. Chim. (Rome)*, **53,** 755 (1963); *Chem. Abstr.*, **59,** 15411d (1963).
118. E. A. Zvezdina, M. P. Zhdanova, A. M. Simonov, and G. N. Dorofeenko, *Chem. Heterocycl. Compd.*, **1975,** 1025; *Khim. Geterotsikl. Soedin.*, **1975,** 1180; *Chem. Abstr.*, **84,** 30994 (1976); *USSR Patent*, 490,801; *Chem. Abstr.*, **84,** 74304 (1976).
119. A. De Cat and A. Van Dormael, *Bull. Soc. Chim. Belges*, **60,** 69 (1951); *Chem. Abstr.*, **46,** 5020e (1952).
120. A. W. Chow, D. R. Jakas, B. P. Trotter, N. M. Hall, and J. R. E. Hoover, *J. Heterocycl. Chem.*, **10,** 71 (1973); A. W. Chow, *U.S. Patent*, 3,468,888; *Chem. Abstr.*, **72,** 3489 (1970).
121. H. Antaki and V. Petrow, *J. Chem. Soc.*, **1951,** 551.
122. A. De Cat and A. Van Dormael, *Bull. Soc. Chim. Belges*, **59,** 573 (1950); *Chem. Abstr.*, **45,** 10247b (1951).
123. G. B. Crippa and G. Perroncito, *Gazz. Chim. Ital.*, **65,** 1067 (1935).
124. M. Ridi and S. Checchi, *Ann. Chim. (Rome)*, **44,** 28 (1954); *Chem. Abstr.*, **49,** 4658g (1955).
125. Y. Shiokawa and S. Ohki, *Chem. Pharm. Bull (Tokyo)*, **19,** 401 (1971); *Chem. Abstr.*, **74,** 124471 (1971).
126. H. Bohme and K. H. Weisel, *Arch. Pharm*, **310,** 26 (1977); *Chem. Abstr.*, **86,** 189837 (1977).
127. H. Ogura, M. Kawano, and T. Itoh, *Chem. Pharm. Bull. (Tokyo)*, **21,** 2019 (1973); *Chem. Abstr.*, **79,** 137078 (1973).
128. (a) H. Bohme and K. H. Weisel, *Chem. Ber.*, **109,** 2908 (1976); (b) N. Heimbach, *U.S. Patent*, 2,432,419; *Chem. Abstr.*, **42,** 2193c (1948).
129. M. Sakamoto, K. Miyazawa, and Y. Tomimatsu, *Chem. Pharm. Bull. (Tokyo)*, **25,** 3360 (1977); *Chem. Abstr.*, **88,** 89621 (1978).
130. D. W. Dunwell and D. Evans, *J. Chem. Soc. Perkin Trans. I*, **1973,** 1588.
131. T. Denzel and H. Hoehn, *U.S. Patent*, 4,072,679; *Chem. Abstr.*, **89,** 109553 (1978).
132. E. Ziegler and E. Nolken, *Monatsh. Chem.*, **92,** 1184 (1961).

133. L. B. Dashkevich and E. S. Korbelainen, *Chem. Heterocycl. Compd.*, **1966,** 457; *Khim. Geterotsikl. Soedin.*, **1966,** 602; *Chem. Abstr.*, **66,** 37836 (1967).
134. H. Reimlinger, M. A. Peiren, and R. Merenyi, *Chem. Ber.*, **105,** 794 (1972).
135. F. Troxler and H. P. Weber, *Helv. Chim. Acta*, **57,** 2356 (1974).
136. P. V. Tkachenko, A. M. Simonov, and I. I. Popov, *Chem. Heterocycl. Compd.*, **1978,** 73; *Khim. Geterotsikl. Soedin.*, **1978,** 90; *Chem. Abstr.*, **88,** 190727 (1978).
137. H. Alper and L. Pepper, *Can. J. Chem.*, **53,** 894 (1975).
138. H. O. Hankovszky, K. Hideg, and S. Pacsa, *Ger. Patent*, 2,420,108; *Chem. Abstr.*, **82,** 140203 (1975).
139. A. A. Shazhenov and C. S. Kadyrov, *Chem. Heterocycl. Compd.*, **1977,** 1114; *Khim. Geterotsikl. Soedin.*, **1977,** 1389; *Chem. Abstr.*, **88,** 50780 (1978).
140. M. Sakamoto, K. Miyazawa, K. Yamamoto, and Y. Tomimatsu, *Chem. Pharm. Bull. (Tokyo)*, **22,** 2201 (1974); *Chem. Abstr.*, **82,** 16767 (1975).
141. M. Sakamoto, K. Miyazawa, K. Yamamoto, and Y. Tomimatsu, *Chem. Pharm. Bull. (Tokyo)*, **24,** 2532 (1976); *Chem. Abstr.*, **86,** 89744 (1977).
142. W. P. Langsdorf, *Ger. Patent*, 2,043,811; *Chem. Abstr.*, **76,** 140910 (1972); *French Patent*, 2,102,857; *Chem. Abstr.*, **77,** 164772 (1972); *Brit. Patent*, 1,291,312; *Chem. Abstr.*, **78,** 43481 (1973); *U.S. Patent*, 3,804,830; *Chem. Abstr.*, **81,** 3942 (1974).
143. K. Nagarajan, V. R. Rao, and A. Venkateswarlu, *Indian J. Chem.*, **8,** 126 (1970); *Chem. Abstr.*, **73,** 3879 (1970).
144. H. J. Davies and C. H. Dickerson, *J. Chem. Soc.*, **1965,** 5125.
145. G. Zigeuner, W. B. Lintschinger, A. Fuchsgruber, and K. Kollmann, *Monatsh. Chem.*, **107,** 171 (1976); *Chem. Abstr.*, **85,** 5579 (1976).
146. V. I. Shvedov, L. B. Altukhova, L. A. Chernyshkova, and A. N. Grinev, *Chem. Pharm. J.*, **3,** 566 (1969); *Khim. Farm. Zh.*, **3,** 15 (1969); *Chem. Abstr.*, **72,** 66899 (1970).
147. A. N. Grinev, A. A. Druzhinina, and I. K. Sorokina, *Chem. Heterocycl. Compd.*, **1976,** 1048; *Khim. Geterotsikl. Soedin.*, **1976,** 1266; *Chem. Abstr.*, **86,** 29761 (1977).
148. K. Hideg and H. O. Hankovszky, *Synthesis*, **1978,** 313.
149. J. Schmutz and F. Kunzle, *Helv. Chim. Acta*, **39,** 1144 (1956).
150. W. B. Edwards and A. R. Day, *J. Org. Chem.*, **39,** 1519 (1974).
151. H. Schubert, H. Lettau, and J. Fischer, *Tetrahedron*, **30,** 1231 (1974).
152. B. Serafin and L. Konopski, *Pol. J. Chem.*, **52,** 51 (1978); *Chem. Abstr.*, **89,** 24218 (1978).
153. K. G. Barnett, J. P. Dickens, and D. E. West, *Chem. Commun.*, **1976,** 849.
154. K. K. W. Shen and W. S. Belles, *U.S. Patent*, 4,049,422; *Chem. Abstr.*, **88,** 22937 (1978).
155. K. K. W. Shen and W. S. Belles, *U.S. Patent*, 4,094,662; *Chem. Abstr.*, **89,** 163584 (1978).
156. A. J. Hubert and H. Reimlinger, *Chem. Ber.*, **103,** 2828 (1970).
157. E. Ochiai and M. Yanai, *J. Pharm. Soc. Japan*, **60,** 493 (1940); *Chem. Abstr.*, **35,** 743 (1941).
158. S. S. Berg and V. Petrow, *J. Chem. Soc.*, **1952,** 784.
159. C. W. Bird, *Tetrahedron*, **21,** 2179 (1965).
160. R. Prasad, G. Kumar, and A. P. Bhaduri, *Indian J. Chem.*, **15B,** 652 (1977); *Chem. Abstr.*, **88,** 89622 (1978).
161. S. S. Tiwari and S. B. Misra, *Indian J. Chem.*, **14B,** 725 (1976); *Chem. Abstr.*, **86,** 121291 (1977).
162. J. A. Maynard, I. D. Rae, D. Rash, and J. M. Swan, *Aust. J. Chem.*, **24,** 1873 (1971).
163. F. Troxler and H. P. Weber, *Helv. Chim. Acta*, **57,** 2364 (1974).
164. *Italian Patent*, 658,238; through *Chem. Abstr.*, **63,** 15026g (1965).
165. F. G. Webster, *Fr. Patent*, 1,577,440; *Chem. Abstr.*, **73,** 20441 (1970).
166. E. Tenor, T. Eckhard, and F. Fueller, *East Ger. Patent*, 76,515; *Chem. Abstr.*, **75,** 63822 (1971).
167. D. G. O'Sullivan, D. Pantic, and A. K. Wallis, *Nature*, **205,** 262 (1965).

168. H. O. Hankovszky, K. Hideg, and S. Pacsa, *Hung, Teljes*, 10,396; through *Chem. Abstr.*, **84,** 135623 (1976).
169. W. P. Langsdorf, *S. African Patent*, 7,005,123; through *Chem. Abstr.*, **76,** 21955 (1972).
170. K. Murobushi, Y. Kuwabara, S. Baba, and K. Aoki, *J. Chem. Soc.* (*Tokyo*), **58,** 440 (1955); through *Chem. Abstr.*, **49,** 14544i (1955); Y. Kuwabara and K. Aoki, *Konishiroku Rev.*, **6,** 1 (1955) through *Chem. Abstr.*, **49,** 11473f (1955); Y. Kuwabara and K. Aoki, *Yakugaku Zasshi*, **77,** 906 (1957); through *Chem. Abstr.*, **51,** 16159f (1957); Y. S. Moshkovskii and M. V. Deichmeister, *Trudy Vsesoyuz. Nauch, Issledovatel Kinofotoinst.*, **1960,** 74; through *Chem. Abstr.*, **56,** 8208h (1962).
171. W. L. Mosby in *Systems with Bridgehead Nitrogen, Part* 2, *The Chemistry of Heterocyclic Compounds*, A. Weissberger, (Ed.), Wiley-Interscience, New York, 1961, Chap. 7, pp. 907–910.
172. (a) R. I. Fu Ho and A. R. Day, *J. Org. Chem.*, **38,** 3084 (1973); (b) A. V. Zeiger and M. M. Joullie, *J. Org. Chem.*, **42,** 542 (1977).
173. M. V. Povstyanoi, E. V. Logachev, and P. M. Kochergin, *Chem. Heterocycl. Compd.*, **1976,** 603; *Khim. Geterotsikl. Soedin.*, **1976,** 715; *Chem. Abstr.*, **85,** 94322 (1976).
174. Y. I. Beilis, M. V. Povstyanoi, E. V. Logachev, and A. V. Shikarev, *J. Gen. Chem. USSR*, **46,** 420 (1976); *Zh. Obshch. Khim.*, **46,** 426 (1976); *Chem. Abstr.*, **84,** 142548 (1976); P. M. Kochergin and M. V. Povstyanoi, *USSR Patent*, 384,821; *Chem. Abstr.*, **79,** 105299 (1973).
175. M. V. Povstyanoi, E. V. Logachev, and P. M. Kochergin, *Ukr. Khim. Zh.*, **43,** 746 (1977); through *Chem. Abstr.*, **87,** 167984 (1977).
176. D. J. Le Count and A. T. Greer, *J. Chem. Soc. Perkin Trans. I*, **1974,** 297.
177. M. V. Povstyanoi, P. M. Kochergin, E. V. Logachev, E. A. Yakubovskii, A. V. Akimov, and V. P. Kruglenko, *Chem. Heterocycl. Compd.* **1976,** 1180; *Khim. Geterotsikl. Soedin*, **1976,** 1424; *Chem. Abstr.*, **86,** 55350 (1977).
178. M. V. Povstyanoi, P. M. Kochergin, E. V. Logachev, and E. A. Yakubovskii, *Chem. Heterocycl. Compd.* **1975,** 371; *Khim. Geterotsikl. Soedin.*, **1975,** 422; *Chem. Abstr.*, **83,** 28197 (1975).
179. Z. A. Pankina and M. N. Shchukina, *Chem. Pharm. J.*, **6,** 633 (1972); *Khim. Farm. Zh.*, **6,** 8 (1972); *Chem. Abstr.*, **78,** 92396 (1973).
180. J. Slouka, *Monatsh. Chem.*, **100,** 91 (1969); *Chem. Abstr.*, **70,** 96766 (1969).
181. Z. A. Pankina and M. N. Shchukina, *Chem. Heterocycl. Compd.*, **1970,** 228; *Khim. Geterotsikl. Soedin.*, **1970,** 245; *Chem. Abstr.*, **72,** 111426 (1970); *Chem. Heterocycl. Compd.*, **1968,** 281; *Khim. Geterotsikl. Soedin.*, **1968,** 380; *Chem. Abstr.*, **69,** 96680 (1968).
182. Z. A. Pankina, M. N. Shchukina, N. P. Kostyuchenko, and Y. N. Sheinker, *Chem. Pharm. J.*, **4,** 314 (1970); *Khim. Farm. Zh.*, **4,** 12 (1970); *Chem. Abstr.*, **73,** 56068 (1970).
183. H. Schlaepfer and J. Bindler, *U.S. Patent*, 3,309,366; *Chem. Abstr.*, **67,** 21939 (1967).
184. A. Kreutzberger and A. Tantawy, *Chem. Ber.*, **111,** 3007 (1978).
185. A. Kreutzberger, *Arch. Pharm.*, **309,** 794 (1976); *Chem. Abstr.*, **86,** 72588 (1977).
186. L. Capuano, H. J. Schrepfer, M. A. Jaeschke, and H. Porschen, *Chem. Ber.*, **107,** 62 (1974).
187. M. Augustin and K. R. Kuppe, *Tetrahedron*, **30,** 3533 (1974).
188. R. Sgarbi, *Chim. Ind.* (*Milan*), **48,** 18 (1966); *Chem. Abstr.*, **64,** 9727e (1966).
189. L. Capuano and H. J. Schrepfer, *Chem. Ber.*, **104,** 3039 (1971).
190. A. Roechling and K. Haertel, *Ger. Patent*, 2,224,244; *Chem. Abstr.*, **80,** 59970 (1974).
191. H. Roechling and K. Haertel, *Ger. Patent*, 2,308,067; *Chem. Abstr.*, **81,** 152284 (1974).
192. H. Roechling and K. Haertel, *Ger. Patent*, 2,349,911; *Chem. Abstr.*, **83,** 97387 (1975).
193. H. Roechling, and K. Haertel, K. Kirsch, and D. Duewel, *Ger. Patent*, 2,356,258; *Chem. Abstr.*, **83,** 97395 (1975).
194. H. Roechling, R. Kirsch, and D. Duewel, *Ger. Patent*, 2,452,365; *Chem. Abstr.*, **85,** 63096 (1976).

195. G. Kempter, W. Ehrlichmann, and R. Thomann, *Z. Chem.*, **17,** 262 (1977); *Chem. Abstr.*, **87,** 167987 (1977); *E. Ger. Patent* 118,881; *Chem. Abstr.*, **86,** 106666 (1977).
196. C. C. Beard, J. A. Edwards, and J. H. Fried, *South African Patent*, 7,603,751; through *Chem. Abstr.*, **89,** 43516 (1978).
197. N. Heimbach and R. H. Clark, *U.S. Patent* 2,444,609; *Chem. Abstr.*, **42,** 7180i (1948).
198. E. R. White, E. A. Bose, J. M. Ogawa, B. T. Manji, and W. W. Kilgore, *Agr. Food Chem.*, **21,** 616 (1973).
199. L. Schroeder, W. Ost, and K. Thomas, *Ger. Patent*, 2,144,505; *Chem. Abstr.*, **78,** 159686 (1973).
200. H. G. Werchan, G. Dittrich, and P. Held, *E. Ger. Patent*, 127,636; *Chem. Abstr.*, **88,** 136680 (1978).
201. W. Daum and P. E. Frohberger, *Ger. Patent*, 2,527,677; *Chem. Abstr.*, **86,** 155704 (1977).
202. W. Daum, W. Behrenz, I. Hammann, H. Scheinpflug, and W. Brandes, *Ger. Patent*, 2,528,623; *Chem. Abstr.*, **86,** 155796 (1977).
203. E. A. Bose and E. R. White, *U.S. Patent*, 3,725,406; *Chem. Abstr.*, **79,** 62595 (1973).
204. A. Ambrus and E. Hargitai, *Environ. Qual. Saf. Suppl.*, **3,** 113 (1975); *Chem. Abstr.*, **85,** 117793 (1976).
205. J. P. Calmon and D. R. Sayag, *J. Agric. Food Chem.*, **24,** 314 (1976). *Chem. Abstr.*, **84,** 131467 (1976).
206. J. Slouka, *Collect. Czech. Chem. Commun.*, **42,** 894 (1977); *Chem. Abstr.*, **87,** 102271 (1977).
207. Z. A. Pankina and M. N. Shchukina, *Chem. Pharm. J.*, **3,** 440 (1969); *Khim. Farm. Zh.*, **3,** 15 (1969); *Chem. Abstr.*, **72,** 55412 (1970).
208. J. Slouka and M. Budikova, *Acta Univ. Palacki Olomuc., Fac. Rerum Nat.*, **45,** 113 (1974); *Chem. Abstr.*, **82,** 125360 (1975).
209. B. Sachse, F. Schwerdtle, and H. Roechling, *Meded. Fac. Landbouwwet., Rijksuniv. Gent.*, **40,** 723 (1975); through *Chem. Abstr.*, **84,** 131292 (1976); K. Haertel, B. Sachse, and F. Schwerdtle, *Ger. Patent*, 2,519,520; *Chem. Abstr.*, **86,** 26911 (1977).

CHAPTER 8

Condensed Benzimidazoles of Type 6-5-7 and Higher Homologs

M. F. G. STEVENS

Tricyclic benzimidazoles of the 6-5-7 arrangement, and larger ring homologs, have not been extensively studied and most efforts in this area have been directed toward the development of synthetic methods with little systematic examination of the physical and chemical properties of the system.

Three main synthetic routes can be recognized:

1. Interaction of an ortho substituent (generally amino, substituted amino, azido, or nitro) with the α-methylene group of a saturated heterocyclic

(heteroalicyclic) ring. Yields are, in general, high and the mechanistic details and scope of this reaction in benzimidazole chemistry—"the *t*-amino effect"—have been reviewed by Meth-Cohn and Suschitzky.[1]

2. Cyclization of 1- or 2-substituted benzimidazoles. Included in this type are the controversial reactions of 2-substituted benzimidazoles with acetylenic esters, which afford azepino[1,2-*a*]benzimidazoles in meager yields, and more efficient syntheses of tricyclic systems bearing additional heteroatoms in the 7(8) ring.
3. Photolysis of phenazine *N*-oxides. Although of mechanistic interest, this approach is limited to the formation of benzimidazoles fused to an azepinone fragment.

8.1. TRICYCLIC BENZIMIDAZOLES WITH NO ADDITIONAL HETEROATOM

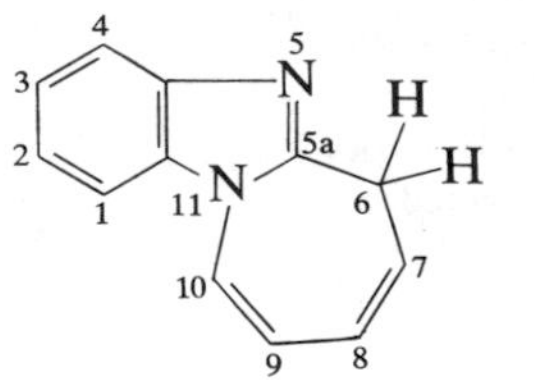

6*H*-Azepino[1,2-*a*]benzimidazole

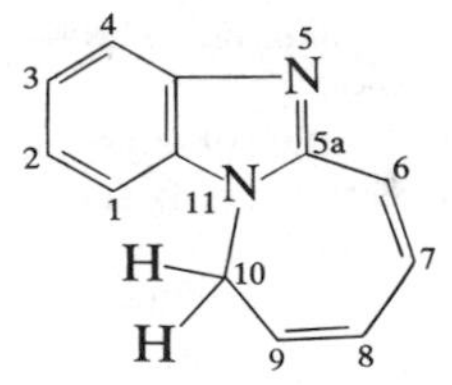

10*H*-Azepino[1,2-*a*]benzimidazole

8.1.1. Synthesis

Cyclization of ortho-Substituted N-Aryl Heterocycles

Oxidation of *N*-(*o*-aminophenyl)hexahydroazepine (**8.1**) with peroxytrifluoroacetic acid yields the azepinobenzimidazole (**8.2**) in high yield.[2] De-

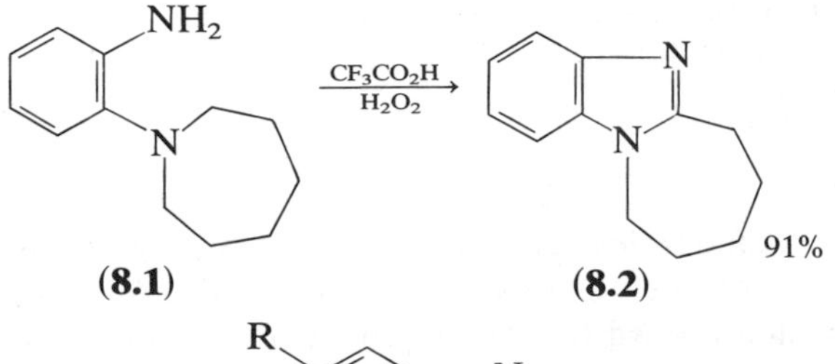

CF_3CO_2H, H_2O_2

(**8.1**) (**8.2**) 91%

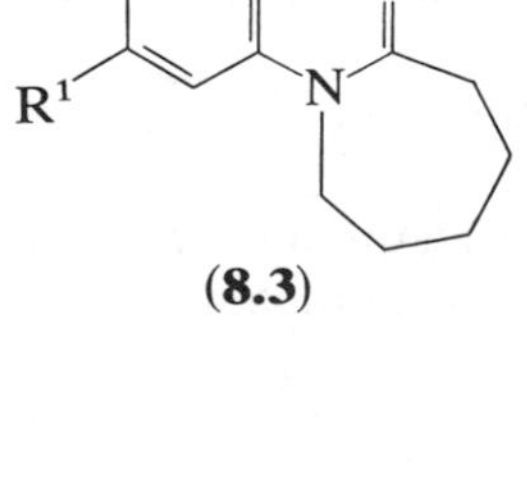

(**8.3**)

TABLE 8.1. AZEPINO[1,2-*a*]BENZIMIDAZOLES (**8.3**) FORMED BY PERACID OXIDATION OF *N*-(*o*-AMINOPHENYL)HEXAHYDROAZEPINES OR THEIR *N*-ACETYL DERIVATIVES

Substituents					
R	R^1	Oxidant	Yield (%)	m.p. (°C)	Ref.
H	H	CF_3CO_2H/H_2O_2	91	124–125	2
CO_2Et	H	HCO_2H/H_2O_2	58	108	3
NHAc	H	HCO_2H/H_2O_2	68	252	4
H	NHAc	HCO_2H/H_2O_2	66	222	4
NO_2[a]	H	HCO_2H/H_2O_2	—[c]	177	5
H[b]	NO_2	HCO_2H/H_2O_2	—[c]	196	5

[a] Starting material: *N*-(2-acetylamino-4-nitrophenyl)hexahydroazepine.
[b] Starting material: *N*-(2-acetylamino-5-nitrophenyl)hexahydroazepine.
[c] Yield not recorded.

rivatives of (**8.1**) substituted in the benzene ring[3] or bearing an acetylated *o*-aminophenyl group[4,5] similarly form substituted azepinobenzimidazoles (**8.3**) with performic acid being the oxidant of choice (Table 8.1). Cyclization of the unsubstituted amine (**8.1**) can also be accomplished by sulfuryl chloride.[6] The first step is probably formation of the tetrachloro derivative (**8.4**), followed by conversion to sulfonylamine (**8.5**). Cyclization to the tricycle (**8.6**) may occur either by way of a nitrene (following loss of sulfur dioxide) or by intramolecular *H*-abstraction and cyclization.[7] Diazotization of (**8.1**) in hydrochloric acid followed by treatment with sodium sulfite—sulfur dioxide yields a diazosulfonate (**8.7**), which cyclizes to a benzimidazolium sulfamate (**8.8**) with excess sulfur dioxide at 70°.[8]

(**8.1**) $\xrightarrow{SO_2Cl_2}$ (**8.4**) ⟶ (**8.5**) ⟶ (**8.6**) 85%

Cl Cl NH2 Cl Cl N

(**8.4**)

Cl Cl N=SO2 Cl Cl N

(**8.5**)

Cl Cl N Cl Cl N

(**8.6**)

85%

(**8.1**) (i) HNO_2 (ii) Na_2SO_3—SO_2 → (**8.7**) SO_2 at 70° → (**8.8**)

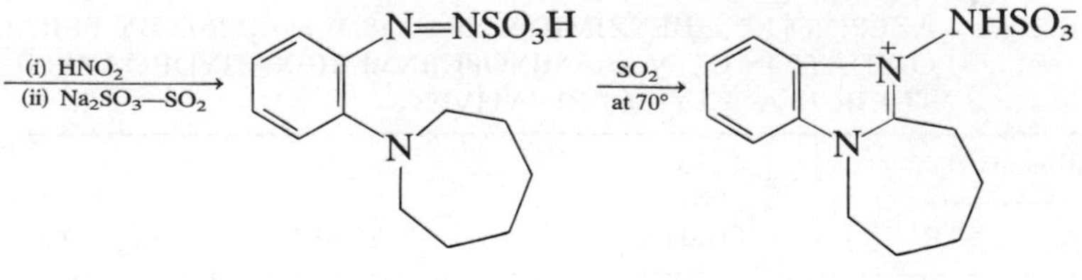

A series of anils (**8.9**), prepared from amine (**8.1**) and aromatic aldehydes, cyclize in ethanolic hydrochloric acid to benzimidazolium salts (**8.10**) (Table 8.2) together with an equal proportion of the benzylamines (**8.11**).[9] In one case—the anil from 2-chloro-5-nitrobenzaldehyde—the product is a red insoluble dihydrobenzimidazole (**8.12**), which is transformed to a benzimidazolium salt in boiling ethanolic hydrochloric acid. A similar anil cyclization is probably involved in the synthesis of betaine (**8.13**) when (**8.1**) is reacted with alloxan in acid.[10] Labeling experiments with the deuterated piperidine analog of (**8.9**; $R=C_6H_4Cl$-*o*) show that deuterium is incorporated into the benzylic positions of both the benzimidazolium salt and benzylamine by-product.[9] The anils probably react by way of their mesomeric forms (**8.9a**); subsequent cyclization of the immonium ions (**8.14**)

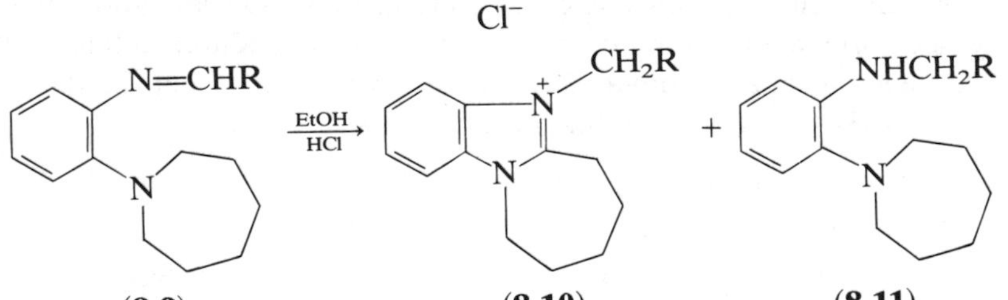

(**8.9**) EtOH/HCl → (**8.10**) + (**8.11**)

$R = Ph$, C_6H_4Cl-*o*, C_6H_4Cl-*p*

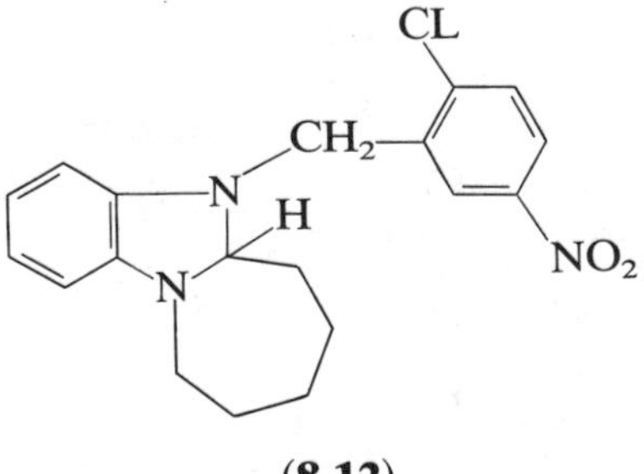

(**8.12**) 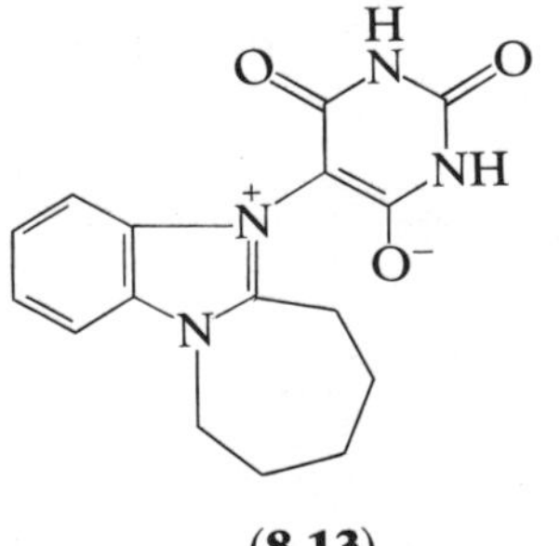 (**8.13**)

TABLE 8.2. AZEPINO[1,2-*a*]BENZIMIDAZOLES FORMED BY CYCLIZATION OF ANILS (**8.9**) IN ETHANOLIC HYDROCHLORIC ACID[a]

Compound	R	Yield (%)	m.p. (°C)
8.10	Ph	84	288 (decomp.)
8.10	2-ClC_6H_4	86	244–246
8.10	4-ClC_6H_4	90	252 (hemihydrate)
8.10	2-Cl,5-$NO_2C_6H_3$[b]	80	243 (decomp.)
8.12	—	94	144
8.13	—	—[c]	>300

[a] From Ref. 9.
[b] From compound (**8.12**) and ethanolic hydrochloric acid.
[c] Yield not recorded.

gives the protonated dihydrobenzimidazoles (**8.15**). The flexible 7-membered ring of the latter allows for a bimolecular transition state between protonated and free dihydrobenzimidazole (**8.16**) in such a way that hydride transfer can occur (Scheme 8.1).

N-(*o*-Azidophenyl)hexahydroazepines (**8.17**) thermolyze in nitrobenzene to afford the tricyclic benzimidazoles (**8.19**; R = H, Cl). The role of the

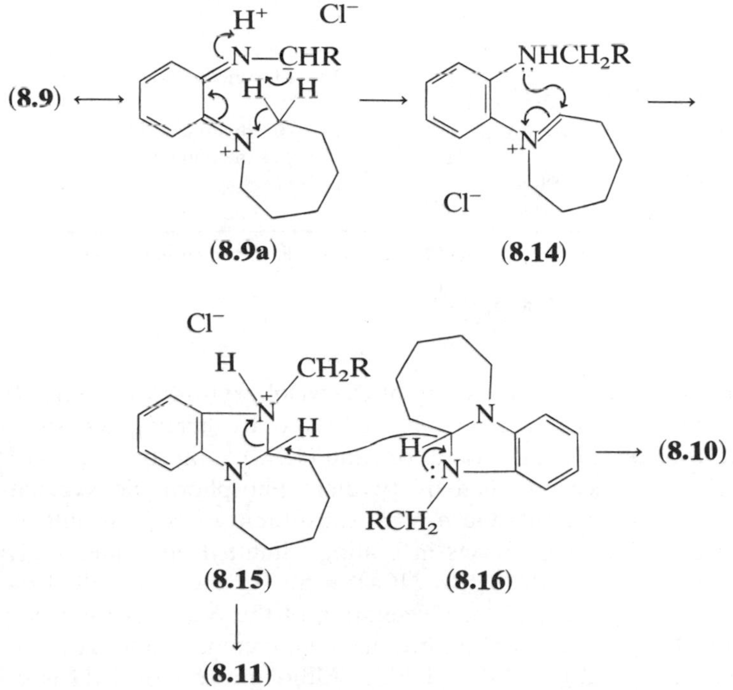

(**8.9**) ⟷ (**8.9a**) ⟶ (**8.14**) ⟶

(**8.15**) (**8.16**) ⟶ (**8.10**)

(**8.15**) ⟶ (**8.11**)

Scheme 8.1

(**8.17**) (**8.18**) (**8.19**)

R = H,Cl,NO_2

nitrobenzene is twofold: it initiates the thermal decomposition and oxidizes the intermediate dihydrobenzimidazoles (**8.18**).[11] A dihydro intermediate (**8.18**; R = NO_2) can be isolated when the solvent is hot diglyme and further transformed to the aromatic product (**8.19**; R = NO_2) by a range of oxidizing agents.[12] Physical characteristics of azepino[1,2-*a*]benzimidazoles prepared by this route are recorded in Table 8.3.

TABLE 8.3. AZEPINO[1,2-*a*]BENZIMIDAZOLES PREPARED BY THERMOLYSIS OF *N*-(*o*-AZIDOPHENYL)-HEXAHYDROAZEPINES (**8.17**)

Compound	R	Yield (%)	m.p. (°C)	Ref.
8.18	NO_2	75	119	12
8.19	H	85	125–126	11
			232–234 (methiodide)	11
8.19	Cl	72	107–109	11
			259–261 (hydrochloride)	11
			273–273 (methiodide)	11
8.19	NO_2	—[a]	174–175	11
8.19	NO_2	100[b]	—	12

[a] By decomposition of (**8.17**; R = NO_2) in nitrobenzene; yield not recorded.
[b] By oxidation of (**8.18**; R = NO_2).

Although reductive cyclization of *N*-cyclohexyl-*o*-nitroaniline (**8.20**) to azepinobenzimidazole (**8.2**) can be achieved by heating the substrate in either ferrous oxalate (21% yield) or sand (15%),[13] this type of cyclization is more effectively accomplished by trivalent phosphorus deoxygenating agents.[14] Conveniently, only the *o*-nitro substituent of *N*-(2,4-dinitrophenyl)-hexahydroazepine (**8.21**) reacts in boiling trimethyl phosphite,[12] giving the dihydrobenzimidazole (**8.18**; R = NO_2) in 50–60% yield; triethyl phosphite is a less satisfactory agent. Deoxygenation of the *N*-acetyl-*o*-nitrodiphenylamine (**8.22**) with triethyl phosphite in boiling cumene affords the azepinobenzimidazole (**8.23**) in 11% yield.[15] Although 6*H*- or 10*H*-isomers are compatible with the spectral data the 6*H*-arrangement is preferred, since it

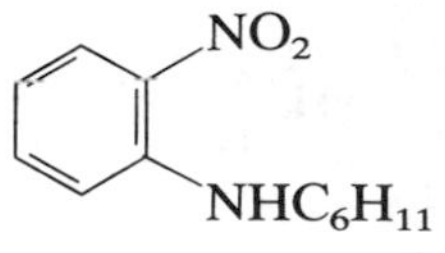

NO_2 NHC_6H_{11}

(8.20)

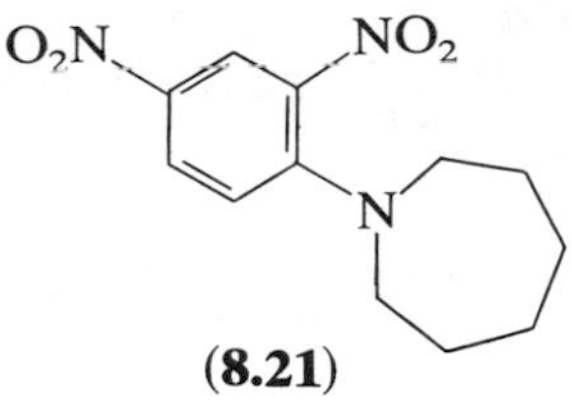

O_2N NO_2 N

(8.21)

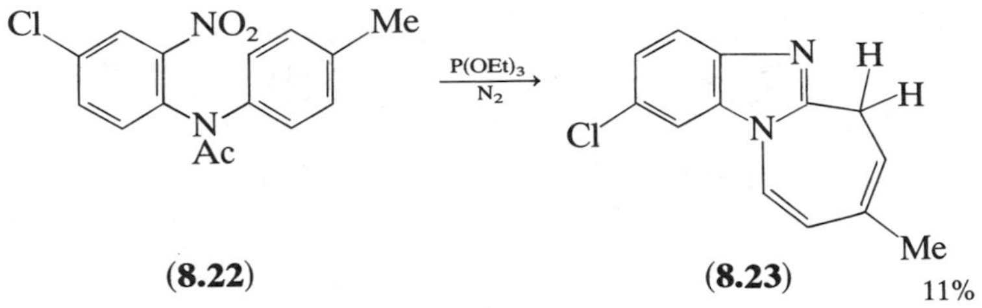

Cl NO_2 Me N Ac

(8.22)

$\xrightarrow[N_2]{P(OEt)_3}$

N H H Cl N Me

(8.23) 11%

accords with its (presumed) formation via an intermediate *o*-nitrene → azanorcaradiene → azepine transformation.

Conversion of benzimidazoles to their *N*-oxides cannot normally be accomplished by direct oxidation (see Chapter 2, section 2.2.1). However, when *N*-(*o*-nitrophenyl)hexahydroazepines or their higher ring homologs (**8.24**) are refluxed in 6*N*-hydrochloric acid,[16,17] photolyzed in methanolic hydrochloric acid,[17,18] or reduced with titanous chloride in hydrochloric acid[19] either tricyclic *N*-oxides (**8.27**) or their deoxygenated counterparts (**8.28**) are formed. In the acid cyclization the reaction is believed to proceed via nitronic acid species (**8.25**), which abstract a hydrogen atom from the α-methylene group (Scheme 8.2); the products are hydrochloride salts of *N*-oxides (**8.26**). The reaction can be exploited for the attachment of larger heteroalicyclic rings (Table 8.4): however, in the case of the 13-membered *N*-aryl heterocycles (**8.24**; R = H or Cl, $n = 11$) only 2-(ω-chlorododecylamino)nitrobenzenes (**8.29**) are formed in 90% yields.[17] Surprisingly, in the photocyclization route both *N*-oxides (**8.27**) and benzimidazoles (**8.28**) appear to be formed by different pathways[17,18] and the product distribution is exquisitely sensitive to the basicity of the tertiary nitrogen and the electronic influence of the substituent in the phenyl ring.

In the titanous chloride–hydrochloric acid reduction of *N*-(*o*-nitrophenyl)-hexahydroazepine (**8.24**; R = H, $n = 5$) the reaction probably involves an intermediate *N*-oxide, which is deoxygenated to the benzimidazole (**8.28**; R = H, $n = 5$) by the second equivalent of reducing agent.[19] With zinc chloride–acetic anhydride the reaction takes a different course and the zinc salt of the 6-hydroxyazepinobenzimidazole (**8.30**) is formed.[20] Similar treatment of *N*-(2,4-dinitrophenyl)hexahydroazepine yields the acetoxyazepine (**8.31**). Comparable rearrangements have been noted in 1,2-dimethylbenzimidazole-3-oxide.[21]

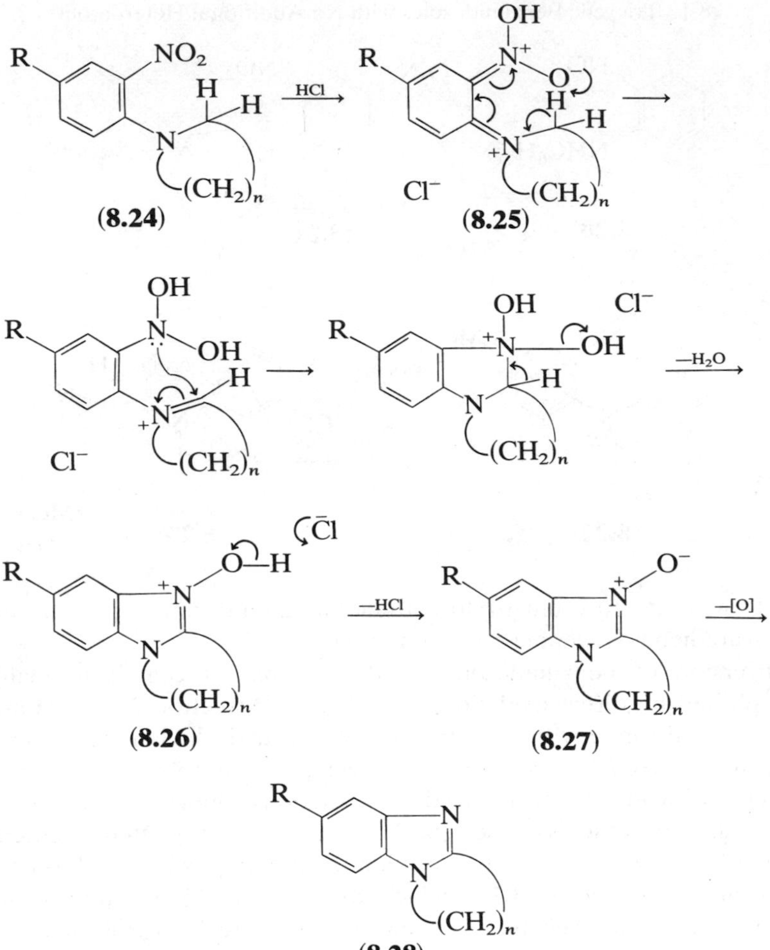

R NO_2 H H N $(CH_2)_n$ HCl

(**8.24**)

R OH N^+ O^- H H ^+N Cl^- $(CH_2)_n$

(**8.25**)

R OH N OH H ^+N Cl^- $(CH_2)_n$

R OH Cl^- ^+N OH H N $(CH_2)_n$ $-H_2O$

R ^+N O H $\bar{C}l$ N $(CH_2)_n$ $-HCl$

(**8.26**)

R O^- $\overset{+}{N}$ N $(CH_2)_n$ $-[O]$

(**8.27**)

R N N $(CH_2)_n$

(**8.28**)

R = H, Cl, NO_2

Scheme 8.2

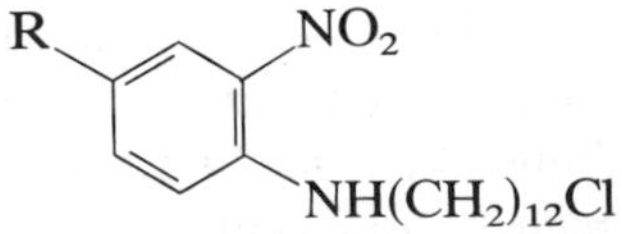

R NO_2 $NH(CH_2)_{12}Cl$

(**8.29**)

R = H,Cl

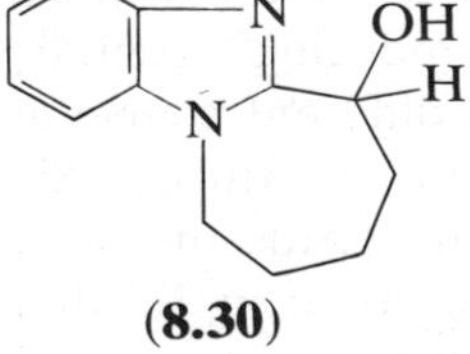

N OH H N

(**8.30**)

O_2N N OAc H N

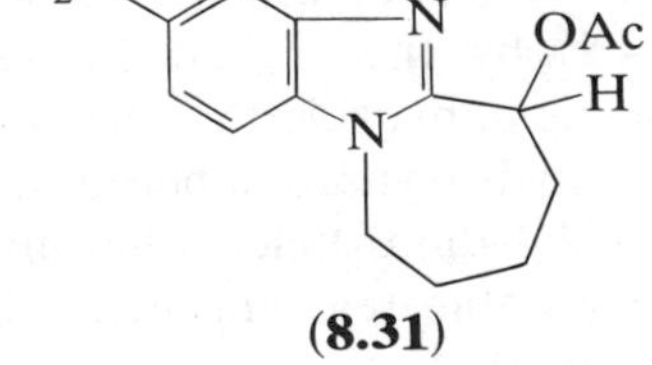

(**8.31**)

TABLE 8.4. AZEPINO[1,2-*a*]BENZIMIDAZOLE-5-OXIDES (**8.27**), AZEPINO[1,2-*a*]-BENZIMIDAZOLES (**8.28**), (**8.30**), AND (**8.31**), AND HIGHER HETEROALICYCLIC HOMOLOGS FORMED FROM *N*-(*o*-NITROPHENYL)HETEROCYCLES (**8.24**)

Compound	R	*n*	Synthetic method[a]	Conditions No. of hr/ temp. in °C	Yield (%)	m.p. (°C)	Ref.
8.27	H	5	*A*	40/110	74	212 (hydrochloride)	16, 17
8.27	Cl	5	*A*	48/110	62	129 (dihydrate)	17
			B	24/~20	79	—	17, 18
8.27	NO_2	5	*A*	12/150	76	188	17
						206 (hydrochloride)	16
8.27	Cl	6	*A*	72/110	61	125 (hydrate)	17
8.27	Cl	7	*A*	72/110	52	110 (hydrate)	17
8.28	H	5	*B*	24/~20	81	126	17, 18
			C	1/80	100	124	19
8.28	H	11	*B*	24/~20	86	111	17
8.28	Cl	11	*B*	24/~20	23	129	17
8.30	—	—	*D*	3/140	70	182	20
8.31	—	—	*E*	2.5/140	87	205	20

[a] *A* = Nitro compound in 6*N*-hydrochloric acid; *B* = Nitro compound photolyzed by a 200 W Hanovia medium-pressure lamp (Pyrex cooling jacket) in aqueous AnalaR methanol and hydrochloric acid deaerated with nitrogen; *C* = Nitro compound and titanous chloride (2 mol. equiv.) in 10*N*-hydrochloric acid; *D* = Nitro compound and zinc chloride (1 mol. equiv.) in acetic anhydride; *E* = Nitro compound and zinc chloride (2 mol. equiv.) in acetic anhydride.

Cyclization of 1- *and* 2-*Substituted Benzimidazoles*

The simplest cyclization of this type involves the ring-closure of ω-substituted 2-alkylbenzimidazoles. 2-(5-Bromopentyl)benzimidazole (**8.32a**) cyclizes in sodium ethoxide to the azepinobenzimidazole (**8.2**) in 94% yield;[22] heating the corresponding amine (**8.32b**) at 300° is less efficient.[23] Trimer (**8.33**) prepared from *o*-phenylenediamine and adipoyl chloride[24] cyclizes to an azepinone (**8.34**) in moderate yield on sublimation. The action

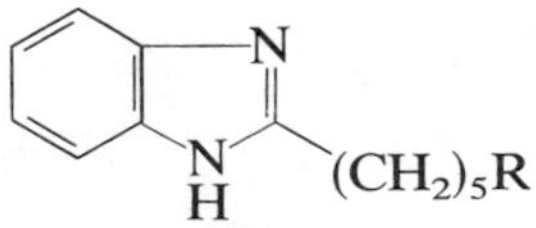

N
N H (CH$_2$)$_5$R

(**8.32**)

a, R = Br
b, R = NH_2

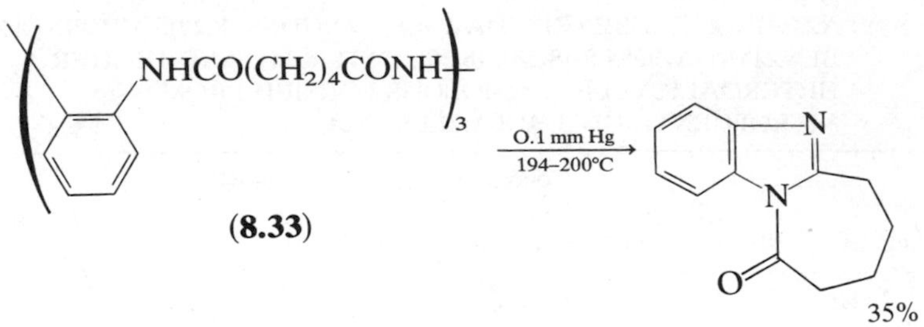

NHCO$(CH_2)_4$CONH

(8.33)

0.1 mm Hg, 194–200°C

35%

(8.34)

POCl$_3$; H_2O

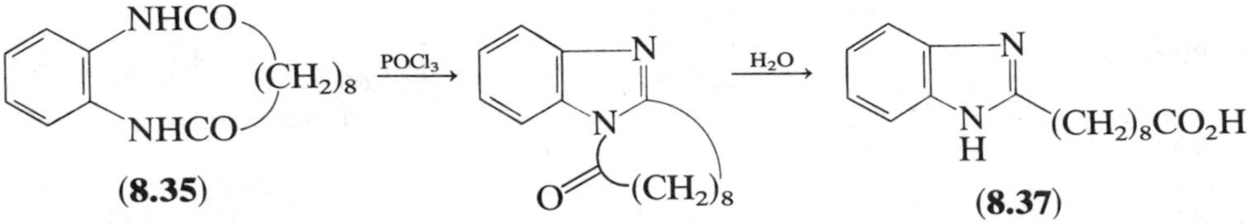

(8.35) (8.36) (8.37)

of phosphoryl chloride on the diamide (**8.35**) gives 9-(benzimidazol-2-yl)nonanoic acid (**8.37**) possibly by way of the 11-membered heterocycle (**8.36**).[24]

A methyl or activated methylene group in the 2-position of benzimidazoles can be elaborated into a 7-membered ring by reaction with acetylenic esters. 1,2-Dimethylbenzimidazole (**8.38**; R = Me) and the 1-ethyl analog react with excess dimethyl acetylenedicarboxylate (DMAD) at room temperature to afford complex mixtures: minor components of the mixtures are the azepino[1,2-*a*]benzimidazoles (**8.39**). Tetrahydrofuran is the solvent of choice.[25] Significantly the ethyl group in 2-ethyl-1-methylbenzimidazole will not participate in formation of a 7-membered ring. The first step in the reaction is the attack of DMAD at the vacant 3-position of the benzimidazole to give a zwitterion (**8.40**), which can react further by proton transfer from the reactive methyl—but not ethyl—group. The carbanionic intermediate (**8.41**) thus generated traps a further mole of DMAD (Scheme

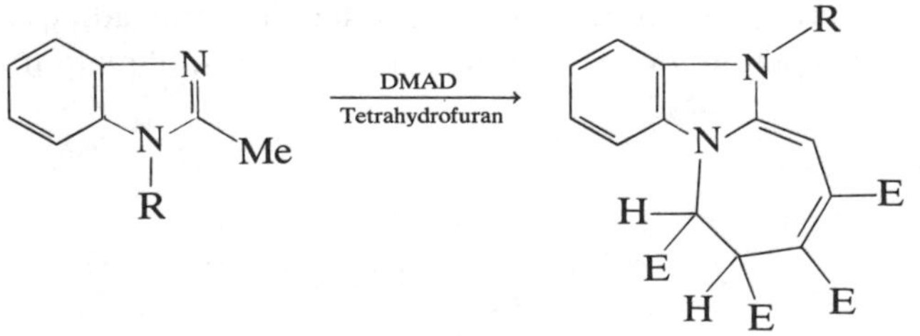

DMAD, Tetrahydrofuran

(8.38) (8.39)

R = Me,Et; E = CO_2Me

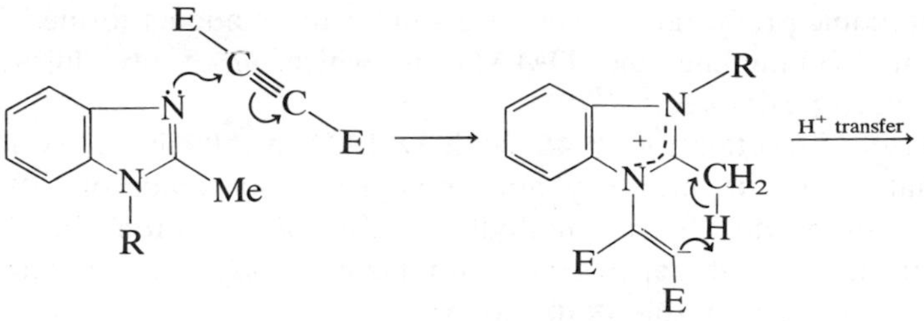

(**8.40**)

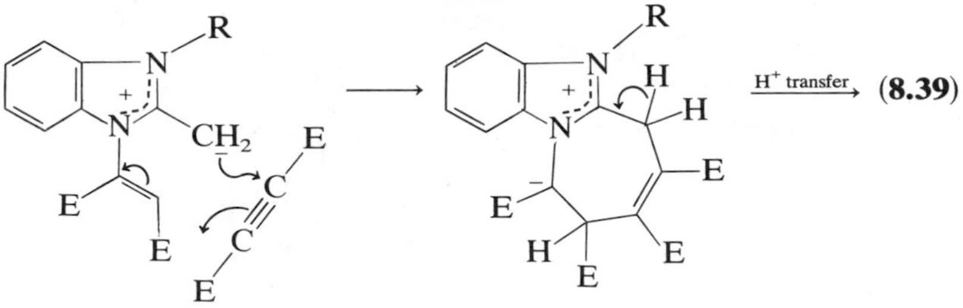

(**8.41**)

R = Me,Et; E = CO_2Me

Scheme 8.3

8.3). A similar mechanism has been advanced to explain the formation of azepines in the reactions of 2-alkylthiazoles and acetylenic esters.[26]

When 2-methyl- and 2-benzylbenzimidazole substituted with a *trans*-acrylate substituent in the 1-position (**8.42**) react with DMAD products are formed, which were originally claimed to be azepino[1,2-*a*]benzimidazoles;[27] this assignment has been corrected[28] and the overall position clarified.[29] For example, structure (**8.43**) originally proposed[27] as the product from (**8.42**; R = Me) has been reassigned the cyclobuta[4,5]pyrrolo[1,2-*a*]benzimidazole structure (**8.44**) on the basis of its ^{13}C NMR spectrum and

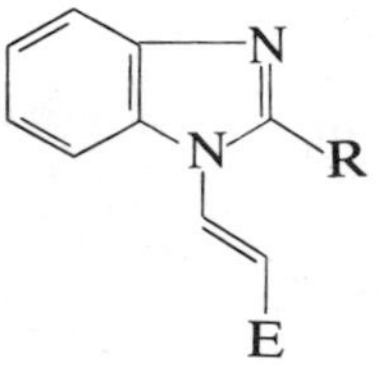

(**8.42**)
R = Me; Benzyl

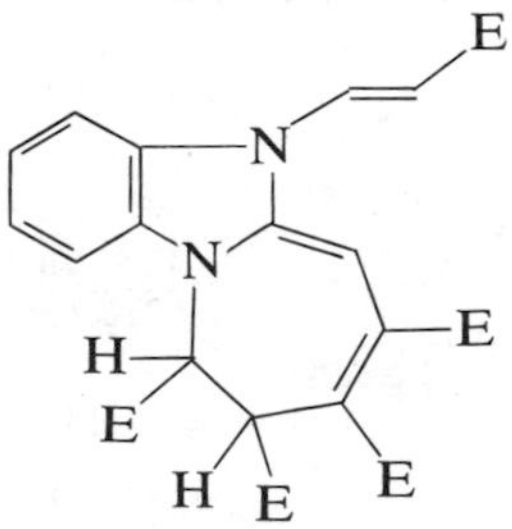

(**8.43**)

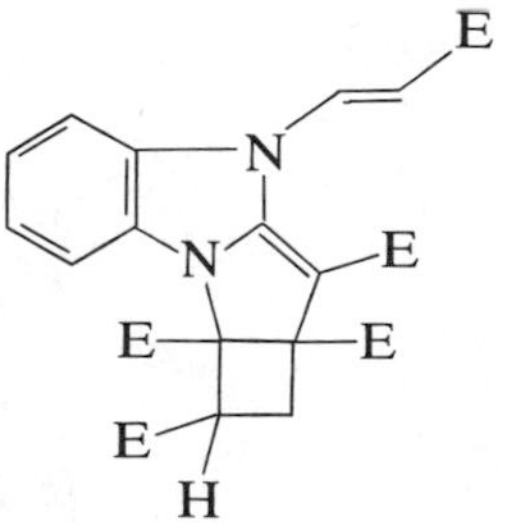

(**8.44**)
E = CO_2Me

other spectroscopic properties:[28] these are similar to an adduct formed from 6-bromo-2-methylquinoline and DMAD on which an X-ray diffraction analysis has been performed.[30,31]

Benzimidazole-2-acetonitrile (**8.45**) with DMAD in refluxing acetonitrile gives a small yield of the 6-cyanoazepino[1,2-*a*]benzimidazole (**8.46**), whereas the main product from 1-methylbenzimidazole-2-acetonitrile is now considered to be a cyclobutapyrrolobenzimidazole (**8.47**)[28] rather than the 10-cyanoazepinobenzimidazole (**8.48**) as originally proposed.[32]

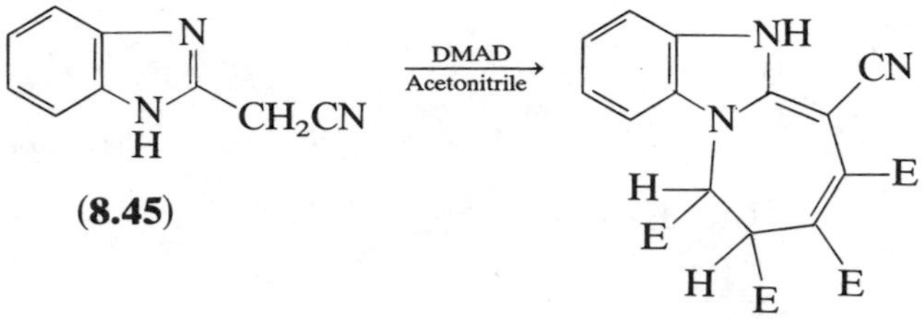

N N H CH_2CN (**8.45**) —DMAD, Acetonitrile→ NH CN N E H E H E E (**8.46**) <5%

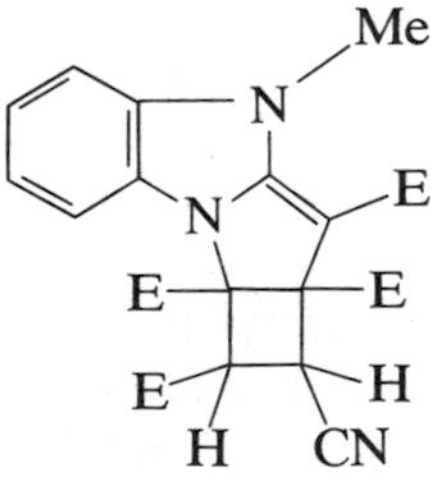

Me N N E E E E H H CN (**8.47**)

Me N E N E H CN H E E (**8.48**)

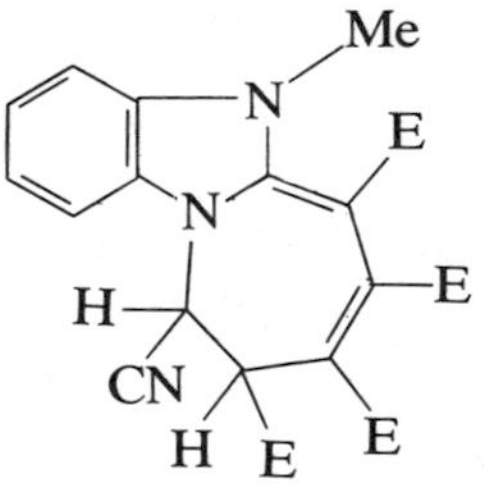

$E = CO_2Me$

Azepinobenzimidazoles of two types are formed from 2-benzyl-1-methylbenzimidazole (**8.49**) and acetylenic esters.[33] The "normal" products of this type of cyclization (**8.50**) are accompanied by more conjugated

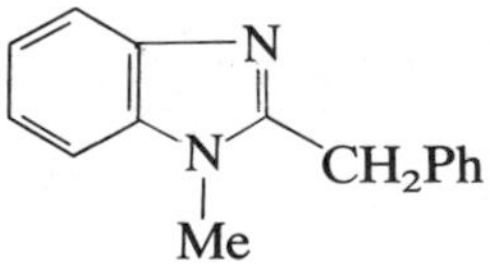

N N CH_2Ph Me (**8.49**)

Me N Ph N E H E H E E (**8.50**)

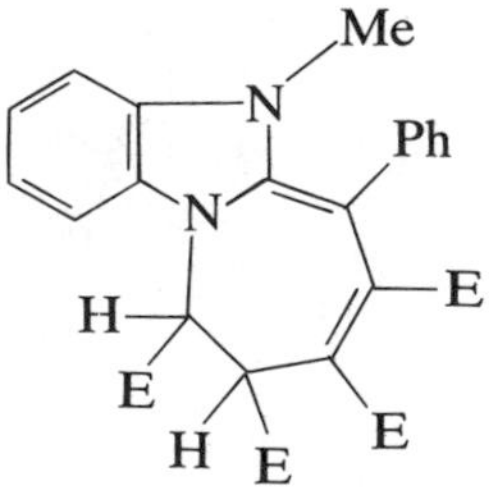

$\xrightarrow{-HE}$

(**8.51**) (**8.52**)

$E = CO_2Me, CO_2Et$

triesters (**8.52**): the latter compounds may be formed from starting benzimidazole by successive additions of ester and proton transfers, which generate dipolar species (**8.51**), followed by loss of the elements of methyl or ethyl formate. Physical characteristics of azepinobenzimidazoles prepared from benzimidazoles and acetylenic esters are compiled in Table 8.5.

TABLE 8.5. AZEPINO[1,2-*a*]BENZIMIDAZOLES FROM 2-SUBSTITUTED BENZIMIDAZOLES AND ACETYLENIC ESTERS

Starting benzimidazole		Product							
Compound	R	Compound	R	E	Reagent[a]	Reaction solvent	Yield (%)	m.p. (°C)	Ref.
8.38	Me	**8.39**	Me	CO_2Me	DMAD	THF	13	206–207	25
8.38	Et	**8.39**	Et	CO_2Me	DMAD	THF	17	204–205	25
8.45	—	**8.46**	—	CO_2Me	DMAD	CH_3CN	<5	253–255 (decomp.)	32
8.49	—	**8.50**	—	CO_2Me	DMAD	CH_3CN	8	210	33
8.49	—	**8.50**	—	CO_2Et	DEAD	CH_3CN	9	105–110 (decomp.)	33
8.49	—	**8.52**	—	CO_2Me	DMAD	CH_3CN	6	157	33
8.49	—	**8.52**	—	CO_2Et	DEAD	CH_3CN	<5	120	33

[a] DMAD = dimethyl acetylenedicarboxylate; DEAD = diethyl acetylenedicarboxylate.

Photolysis of Phenazine N-Oxides

Irradiation of phenazine-5-oxide (**8.53**; R = H) gives the unsaturated lactam (**8.54**) together with the deoxygenated product phenazine.[34] The yield of lactam is sensitive to solvent changes being a maximum (42%) in oxygen-flushed benzene, and a minimum (3%) in nitrogen-flushed methanol in which phenazine formation (68%) predominates. Irradiation of a series of 2-substituted phenazine-5-oxides (**8.53**) and their 10-oxide isomers (**8.55**) gives an insight into the mechanism of the rearrangement.[35,36] It is claimed that annulenes (**8.56**) or (**8.57**) cannot be intermediates, since each oxide

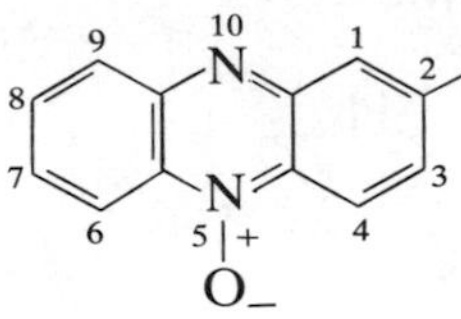

(**8.53**) 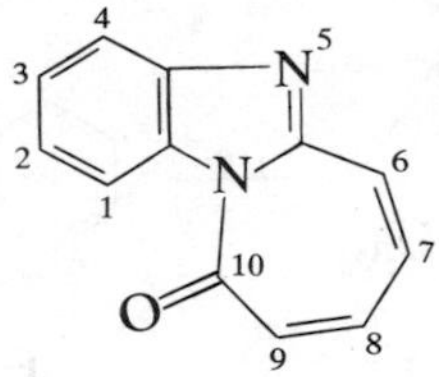 (**8.54**) 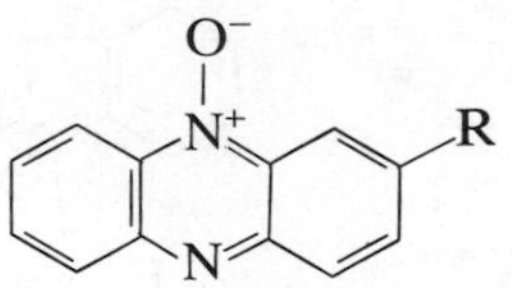 (**8.55**)

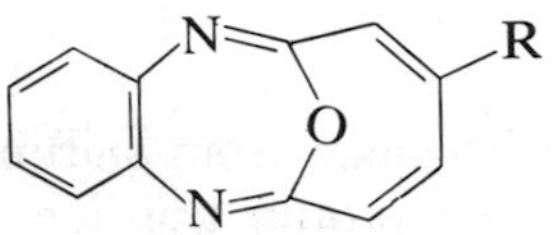

(**8.56**) 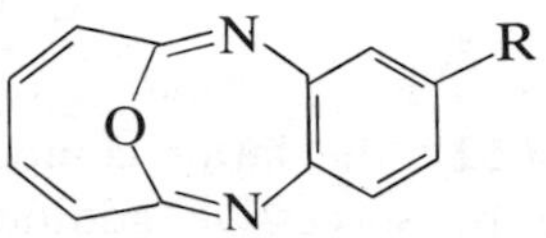 (**8.57**)

TABLE 8.6. AZEPINO[1,2-*a*]BENZIMIDZOLES (**8.54**) FORMED BY IRRADIATION[a] OF PHENAZINE OXIDES IN ACETONITRILE

Phenazine oxide		Product(s)			
Compound	R	[substituent in (**8.54**)]	Maximum yield[b] (%)	m.p. (°C)	Ref.
8.53	H[c]	—	42	176–177	34
8.53	Cl	2-Cl	9.5	176–177	35
		3-Cl	12	159–161	35
		7-Cl	25	138–139	35
8.53	OMe	2-OMe	1	153–154	35
		3-OMe	1	147–148	35
		7-OMe	2	—[d]	35
8.55	Cl	2-Cl	18.5		35
		3-Cl	20		35
		8-Cl	12.5	129–130	35
8.55	OMe	2-OMe	12		35
		3-OMe	24		35
		8-OMe	48	146–147	35
8.55	Me	2-Me	34.5[e]	—[d]	36
		3-Me	34.5[e]	113–116	36
		8-Me	34.5[e]	110–111	36
8.55	CN	2-CN	26.5	238–239	36
		3-CN	27.5	221–223	36
		8-CN	22.5	191–192	36

[a] Medium-pressure mercury lamp with Pyrex filter.
[b] Based on reacted phenazine oxide.
[c] In oxygen-flushed benzene.
[d] Melting point not recorded.
[e] Combined yield.

should then yield *four* different azepinobenzimidazoles—namely, the lactams (**8.54**) substituted in the 2-, 3-, 7-, and 8-positions. In practice only three of the four possible isomers are detected in each case (Table 8.6). 2-Nitrophenazine-10-oxide (**8.55**; R = NO_2) is the exception undergoing deoxygenation only.[36] The product distribution is compatible with a mechanism involving the sequence phenazine oxide (**8.53**)→oxaziridine (**8.58**)→ oxadiazepine (**8.59**)→spiro-benzimidazole (**8.60**)→azepinobenzimidazole (**8.54**) (Scheme 8.4).

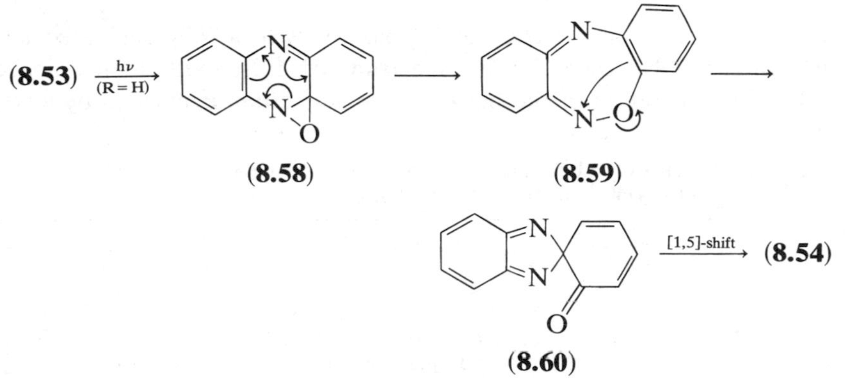

(**8.53**) $\xrightarrow[(R=H)]{h\nu}$ (**8.58**) ⟶ (**8.59**) ⟶ (**8.60**) $\xrightarrow{[1,5]\text{-shift}}$ (**8.54**)

Scheme 8.4

8.1.2. Physicochemical Studies

Infrared Spectra

No systematic study of the IR spectra of azepino[1,2-*a*]benzimidazoles or their larger ring homologs has been conducted: data recorded in the literature have been mainly used to corroborate the presence or absence of a particular functional group. A low-frequency nitrile absorption (2200 cm^{-1}) indicates that resonance involving the imidazole ring and the 6-cyano group is important in the 6-cyanoazepinobenzimidazole (**8.46**) (Scheme 8.5): that

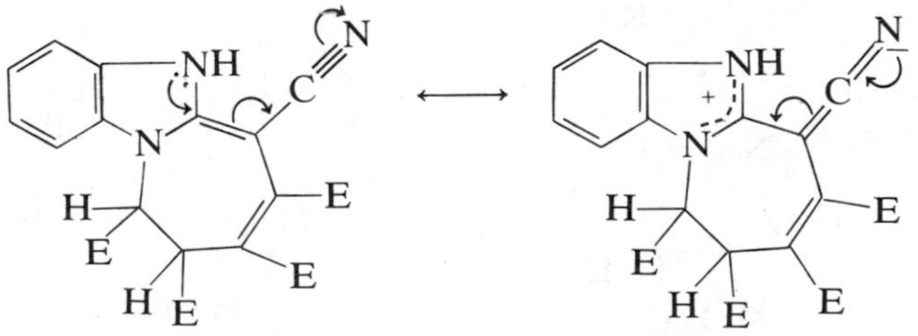

(**8.46**) ⟷ (**8.46a**)

E = CO_2Me

Scheme 8.5

the NH tautomer predominates in the solid state is inferred from a broad band at 3300–3000 cm^{-1}.[32]

The conjugated lactam (**8.54**) and its chloro, methoxy, methyl, and cyano derivatives show carbonyl absorptions in the range 1650–1682 cm^{-1};[34–36] in the saturated lactam (**8.34**) carbonyl absorbs at higher frequency (1715–1710 cm^{-1}).[24,34]

Ultraviolet Spectra

The UV spectra of tricyclic benzimidazoles with a fully saturated additional ring closely resemble the spectrum of 1,2-dimethylbenzimidazole (Table 8.7); generation of the benzimidazolium cation is marked by a weak

TABLE 8.7. ULTRAVIOLET SPECTRA OF 7,8,9,10-TETRAHYDRO-6*H*-AZEPINO[1,2-*a*]BENZIMIDAZOLES

Compound	Solvent	λ_{max} in nm (log ϵ)	Ref.
8.2	—[a]	250(3.45) 284(3.54) 292(3.53)	2
	—[b]	252(3.84) 276(3.79) 283(3.85)	22
	—[c]	241(3.65) 270(3.93) 277(4.00)	22
8.30	—[a]	216(3.94) 254(3.98) 269(3.86) 276(3.94) 283(3.89)	20
8.31	—[a]	240(4.58) 307(4.14)	20
1,2-Dimethyl-benzimidazole	—[d]	250(3.79) 274(3.78) 280(3.80)	37
	—[c]	240(3.69) 269(3.90) 275(3.97)	37

[a] In ethanol.
[b] In methanol.
[c] In 0.01*N*-HCl.
[d] In 0.01*N*-KOH.

hypsochromic shift. Unsaturated linkages in conjugation with the benzimidazole chromophore shift the absorption into the visible range (Table 8.8). The spectra of cations obtained from azepinobenzimidazoles (**8.39**) are more conjugated than that of the benzimidazolium chromophore,[25] and the proton probably adds to C-8 (**8.61**). In contrast, the spectra of derivatives

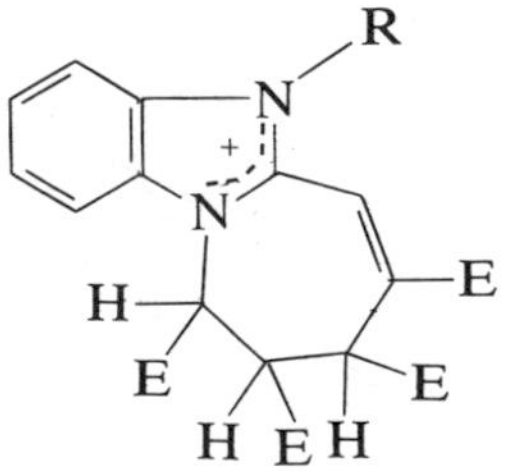

(**8.61**)

R = Me,Et
E = CO_2Me

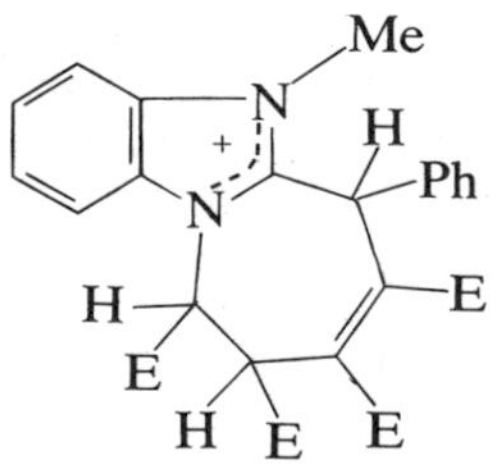

(**8.62**)

E = CO_2Me,CO_2Et

TABLE 8.8. ULTRAVIOLET AND VISIBLE SPECTRA OF AZEPINO[1,2-*a*]-BENZIMIDAZOLES WITH CONJUGATED GROUPS IN THE AZEPINE RING

	Substituents				
Compound	E	Others	Solvent	λ_{max} in nm (log ϵ)	Ref.
8.39	CO_2Me	R = Me	—[a]	217(4.27) 253.5(3.95) 296sh(3.62) 431(4.82)	25
			—[b]	240(4.38) 243.5(4.38) 249(4.36) 326(4.55)	25
8.39	CO_2Me	R = Et	—[a]	216.5(4.23) 253.5(3.91) 295sh(3.58) 431(4.76)	25
			—[b]	244(4.06) 249(4.05) 326(4.35)	25
8.46	CO_2Me	—	—[a]	208(4.23) 224(4.25) 250(4.10) 288sh(3.93) 298(4.01) 305sh(3.99) 308(4.59)	32
			—[b]	222(4.34) 248sh(4.05) 295(3.73) 370(4.02)	32
8.50	CO_2Me	—	—[a]	233(4.18) 344(4.44)	33
			—[c]	216sh(4.25) 274(4.11) 282(4.03)	33
8.50	CO_2Et	—	—[a]	233(4.35) 345(4.56)	33
			—[c]	220sh(4.34) 274(4.22) 282(4.14)	33
8.52	CO_2Me	—	—[a]	253(4.43) 279(4.29) 352(4.25)	33
			—[c]	252(4.43) 278(4.29) 351(4.25)	33
8.52	CO_2Et	—	—[a]	252(4.56) 279(4.40) 352(4.40)	33
			—[c]	252(4.59) 278(4.44) 350(4.44)	33
8.54	—	—	—[d]	236(4.23) 243(4.24) 272(4.34) 399(3.76)	34
8.54	—	2-Cl	—[d]	238(4.26) 245(4.32) 269(4.41) 278(4.43) 378(3.76) 401(3.76) 430(3.49)	35
8.54	—	3-Cl	—[e]	245.5(4.40) 263(4.43) 383(3.76)	35
8.54	—	7-Cl	—[d]	238(4.27) 244(4.31) 275(4.49) 397(3.73) 413(3.70) 444(3.38)	35
8.54	—	8-Cl	—[e]	228(4.25) 244(4.34) 272.5(4.56) 386(3.71)	35
8.54	—	2-OMe	—[d]	253(4.09) 273.5(4.09) 284.5(4.09) 405(3.61) 452(3.50)	35
8.54	—	3-OMe	—[d]	242(4.24) 276(4.19) 288(4.20) 410(3.78) 435(3.85)	35
8.54	—	7-OMe	—[d]	237.5(4.28) 245(4.32) 275(4.30) 281.5(4.30) 395(3.38) 418(3.55)	35
8.54	—	8-OMe	—[d]	226(4.18) 242(4.27) 255.5(4.41) 264(4.41) 350.5(3.66) 371.5(3.71) 393(3.42)	35
8.54	—	3-Me	—[d]	239(4.13) 246(4.20) 277(4.25) 402(3.73)	36
8.54	—	8-Me	—[d]	237(4.15) 243(4.18) 265(4.22) 274(4.20) 389(3.55)	36
8.54	—	2-CN	—[e]	264(4.49) 274(4.48) 370(3.82)	36
8.54	—	3-CN	—[e]	245(4.47) 259(4.36) 368(3.73)	36
8.54	—	8-CN	—[e]	223(4.32) 239(4.31) 278(4.42) 406(3.61)	36

[a] MeOH.
[b] MeOH–72% $HClO_4$, 2:1 v/v.
[c] MeOH acidified with 3 drops of 72% perchloric acid.
[d] Cyclohexane.
[e] Acetonitrile.

(**8.50**) are changed on acidification to the benzimidazolium cation type[33] suggesting protonation at C-6 (**8.62**). In the case of the 6-cyanoazepino-benzimidazole (**8.46**) the long-wavelength band is substantially retained in acid,[32] implying that protonation may not involve the azepine ring. Finally, the spectra of the fully conjugated azepines (**8.52**) are unmodified on acidification.[33]

Nuclear Magnetic Resonance Spectra

The bridgehead hydrogen at C-5a in the stable dihydrobenzimidazole (**8.12**) absorbs at δ 5.12.[9] Otherwise the ^{1}H NMR spectra of benzimidazoles fused to a fully saturated ring (**8.28**), or their *N*-oxide precursors (**8.27**), are as expected.[17] The chemical shifts of the proton(s) attached to C-6 in the azepine (**8.23**) and the hydroxy- (**8.30**) and acetoxyazepines (**8.31**) are at δ 3.47, 5.08, and 6.33, respectively (Table 8.9).

TABLE 8.9. CHEMICAL SHIFTS (δ) AND COUPLING CONSTANTS (*J*) OF

Compound	Substituent(s)	Solvent	H-1	H-2	H-3	H-4	H-6
8.23	—	$CDCl_3$	7.38d	—	7.21q	7.59d	3.47d
8.30	—	$CDCl_3$	←	7.4–6.9m	→	7.67m	5.08br,t
8.31	—	$CDCl_3$	7.46d	8.22dd	—	8.61d	6.33br
8.39	R = Me E = CO_2Me	$CDCl_3$	←	7.2–7.05m		→	4.45s
8.39	R = Et E = CO_2Me	$CDCl_3$	←	7.2–7.05m		→	4.55s
8.46	E = CO_2Me	$(CD_3)_2CO$	←	7.8–7.2m		→	—
8.50	E = CO_2Me	$CDCl_3$	←	7.35–6.90m		→	—
8.50	E = CO_2Et	$CDCl_3$	←	7.2–6.85m		→	—
8.52	E = CO_2Me	$CDCl_3$	8.73d	←	7.36–7.10m	→	—
8.52	E = CO_2Me	CF_3CO_2H	←		7.86s[b] and 7.50s[b]	→	—
8.52	E = CO_2Et	$CDCl_3$	8.71d	←	7.28–7.05m	→	—

[a] s = singlet; d = doublet; dd = double doublet; m = multiplet; br = broad absorption; t = triplet;
[b] Apparent singlet.
[c] Protonation possibly at C-10.
[d] Possibly interchangeable with ester CH_3.

Products formed from 2-substituted benzimidazoles and acetylenic esters, which were originally considered to be azepinobenzimidazoles on the basis of their ^{1}H NMR spectra, are now thought to be cyclobutapyrrolobenzimidazoles[28,29] (see section 8.1.1, "Cyclization of 1- and 2-Substituted Benzimidazoles"). The "normal" adducts with an azepine ring [e.g., structures (**8.39**), (**8.46**), and (**8.50**)] show a relatively low field AB system for the protons attached at saturated carbons C-9 and C-10.[28] The unusual azepinobenzimidazoles (**8.52**; E = CO_2Me, CO_2Et) exhibit an exceptionally low-field aromatic proton at C-1, which is deshielded by the nearly coplanar 10-ester groups:[33] in trifluoroacetic acid the deshielding effect disappears possibly because of protonation at C-10. The hydrogens in the azepine ring in the conjugated lactams (**8.54**) absorb in the olefinic region as expected.[34,35]

A ^{13}C NMR spectrum of azepinobenzimidazole (**8.39**; R = Me, E = CO_2Me) distinguishes between C-6, C-9, and C-10; the shifts (in ppm to low field of internal TMS) are 72.6, 46.0, and 58.0, respectively.[28]

AZEPINO[1,2-*a*]BENZIMIDAZOLES[a]

H-7	H-8	H-9	H-10	Others	*J* values (Hz)	Ref.
5.66br, t	—	6.03d	7.04d	1.86d (CH_3)	1,3(1.8); 3.4(8.3); 6,7(6.4); 7,CH_3(1.1); 9,10(9.0)	15
←	1.9m	→	4.6–3.5m	5.32br (OH)	—	20
←	2.04br	→	4.41br,t	2.19s (CH_3)	2,4(2.2); 1,2(9.1)	20
—	—	5.47d	5.95d	3.41s (NCH_3); 3.85, 3.74, 3.69, 3.51 (ester CH_3)	9,10(6)	25
—	—	5.45d	5.93d	1.32t (NCH_2CH_3); 3.83, 3.72, 3.68, 3.51 (ester CH_3 and NCH_2CH_3)	9,10(6)	25
—	—	5.25d	6.50d	3.76, 3.67, 3.61, 3.45 (ester CH_3)	9,10(5.8)	32
—	—	4.50d	5.64d	3.66s (NCH_3)[d]; 3.64, 3.64, 3.46, 3.16 (ester CH_3)	9,10(3.5)	33
—	—	4.43d	5.54d	3.59s, (NCH_3); 1.16, 1.16, 0.85, 0.77 (ester CH_3); 4.30–3.28m (ester CH_2)	9,10(3.5); CH_2,CH_3(7)	33
7.05s	—	—	—	3.69s (NCH_3); 3.89, 3.86, 3.69 (ester CH_3)	1,2(8)	33
7.69s	—	—	6.48br[c]	3.91s (NCH_3); 4.22, 4.07, 4.07 (ester CH_3)	—	33
6.98s	—	—	—	3.68s (NCH_3); 1.35, 1.30, 1.11 (ester CH_3); 4.49–3.92m (ester CH_2)	1,2(8); CH_2, CH_3(7)	33

q = quartet

Mass Spectra

The azepino[1,2-*a*]benzimidazole adducts from 2-substituted benzimidazoles and acetylenic esters can be distinguished from their cyclobutapyrrolo[1,2-*a*]benzimidazole isomers by their mass spectra. The adducts with a 7-membered ring primarily lose the ester groups, whereas the latter give an intense peak corresponding to loss of an acrylate residue.[28] For example, the mass spectra of azepines (**8.39**) display base peaks resulting from loss of one ester group (M-59) followed by losses of MeO·, MeOH, and MeOCO·;[38] the 6-cyanoazepine (**8.46**) gives a strong M–MeOH peak probably by sequential loss of MeO· (metastable-confirmed) and an H atom.[32]

The initial events in the fragmentation of the molecular ion of the 6*H*-azepinobenzimidazole (**8.23**) are loss of an H-atom or methyl radical,[15] whereas the conjugated lactam (**8.54**) loses CO to form the stable pyridino-[1,2-*a*]benzimidazole radical ion (**8.63**) as base peak.[34]

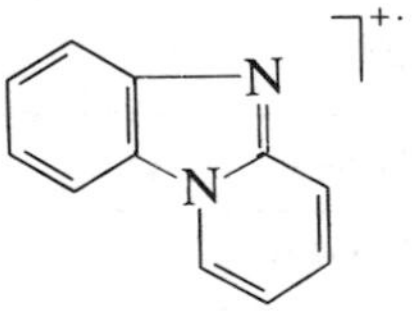

(**8.63**)

General Properties

7,8,9,10-Tetrahydro-6*H*-azepino[1,2-*a*]benzimidazole (**8.2**) and its derivatives forms salts with hydrochloric[11] and picric acids[11,13,22] and quaternary salts with methyl iodide.[11,13] Some *N*-oxides of tricyclic benzimidazoles also form hydrochlorides.[16,17]

The pK_a values of the aforementioned have not been determined and even the site of protonation has not been explicitly defined: moreover, although the quaternary salts are nominally classified as 11-methiodides (**8.64**), methylation of the related bisbenzimidazodiazocine (**8.65**) with methyl toluene-*p*-sulfonate yields a quaternary salt (**8.66**) methylated at the 5,13-positions.[39]

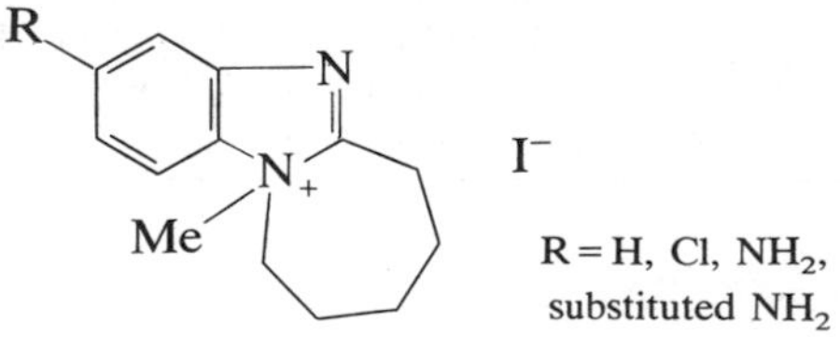

I^-

$R = H, Cl, NH_2,$ substituted NH_2

(**8.64**)

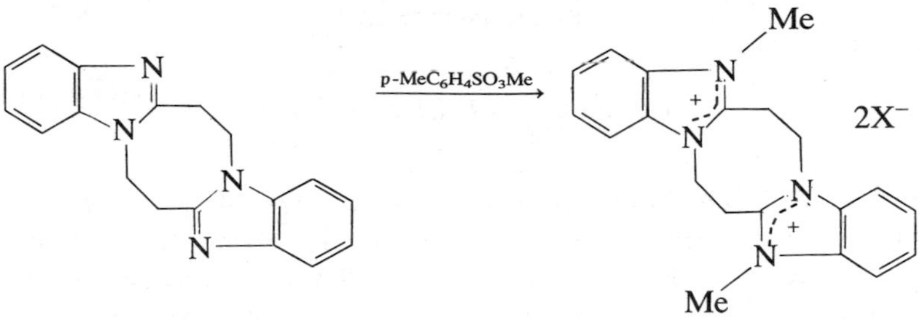

p-$MeC_6H_4SO_3Me$ → $2X^-$

(**8.65**) (**8.66**)

X = toluene-*p*-sulfonate

Reduction of *N*-(*o*-nitrophenyl)hexahydroazepine with titanous chloride–hydrochloric acid to the azepinobenzimidazole (**8.2**) (see section 8.1.1, "Cyclization of *ortho*-Substituted *N*-Aryl Heterocycles") is facilitated by complex formation between the intermediates and titanous salts (Scheme 8.6). Formation of an initial complex (**8.67**) is followed by proton transfer and ring-closure (**8.68**) onto the favorably aligned α-methylene group. The *N*-oxide (**8.70**) is envisaged as being formed via a π-complexed intermediate (**8.69**) formed by overlap of the metal atom with the π-orbital of the benzimidazole N and O atoms. Failure of morpholino or *N*-methylpiperazino derivatives to cyclize may be attributed to the titanous chloride being prevented from forming a complex because of preferential chelation with the other heteroatom (O or N) in the heteroalicyclic ring.[19]

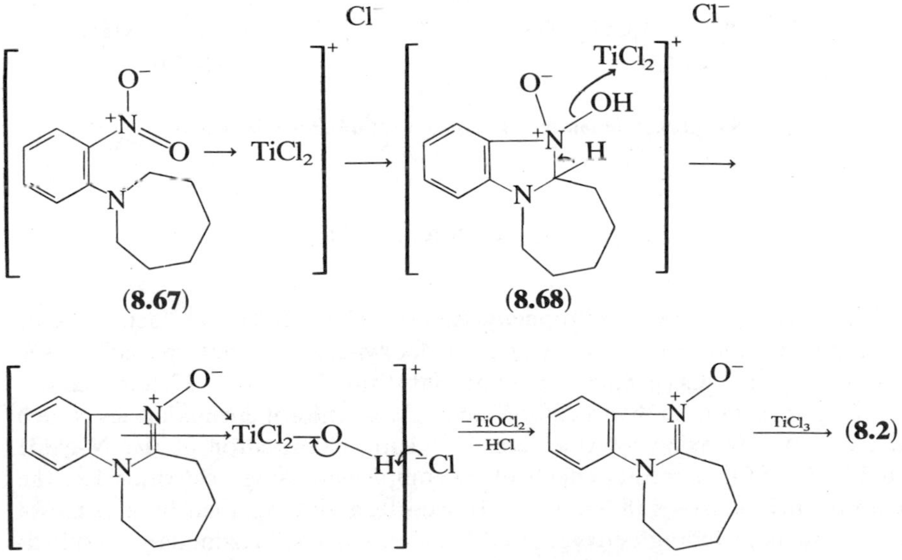

(**8.67**) (**8.68**)

$-TiOCl_2$, $-HCl$; $TiCl_3$ → (**8.2**)

(**8.69**) (**8.70**)

Scheme 8.6

8.1.3. Reactions

Reactions with Electrophiles

Nitration of 2-substituted azepinobenzimidazoles (**8.3**; R = H, R^1 = Cl, NHAc) yields the corresponding 3-nitro derivatives[40] and the 3-substituted analogs (**8.3**; R = Cl, NHAc, CO_2Et, R^1 = H) conversely undergo nitration in the 2-position.[3,11,40] Reduction of 2- or 3-nitroazepinobenzimidazoles or their derivatives yield the corresponding amines, which can be further transformed by conventional processes into a range of benzimidazoles substituted in the 2- or 3-positions (Table 8.10). 3-Amino-7,8,9,10-tetrahydro-6*H*-azepino[1,2-*a*]benzimidazole (**8.3**; R = NH_2, R^1 = H) condenses with 2-amino-4-chloro-6-methylpyrimidine-1-methiodide to yield the salt (**8.71**), with 1-hydroxynaphthalene-2-carboxylic acid and adipoyl chloride to yield amides (**8.72**) and (**8.73**), respectively, and with phosgene to form the urea (**8.74**).[11]

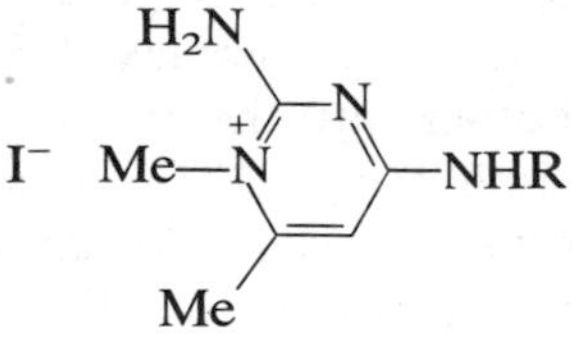

(**8.71**)

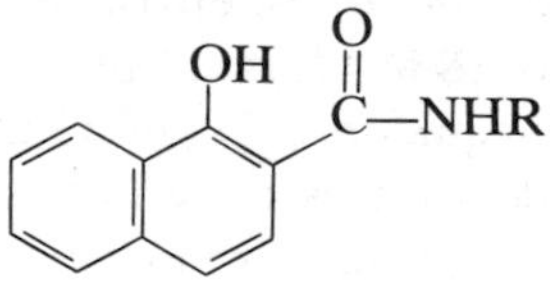

(**8.72**)

RHN—OC(CH_2)$_4$CO—NHR
(**8.73**)

RHN—CO—NHR
(**8.74**)

R = 7,8,9,10-tetrahydro-6H-azepino[1,2-*a*]benzimidazol-3-yl

Reactions with Nucleophiles

Cyclization of *N*-(*o*-nitrophenyl)heterocycles (**8.24**) in acids yields azepinobenzimidazole-5-oxides or their deoxygenated counterparts[16–18] (see section 8.1.1, "Cyclization of *ortho*-Substituted *N*-Aryl Heterocycles"). Prolonged action in hydrochloric acid gives chlorobenzimidazoles which derive from *N*-oxide intermediates.[41,42] Ring chlorination of the *N*-oxide (**8.27**; R = NO_2; $n = 5$) is effectively accomplished using acid chlorides: the chlorobenzimidazoles (**8.75**; R^1 = Cl) are often accompanied by a range of by-products including deoxygenated benzimidazoles, benzimidazoles substituted in the heteroalicyclic ring (**8.76**), and unusual 1,3-bridged benzimidazolones (**8.77**). Varying proportions of 1- and 4-chloro derivatives are

TABLE 8.10. PHYSICAL CHARACTERISTICS OF 2- AND 3-SUBSTITUTED AND 2,3-DISUBSTITUTED AZEPINOBENZIMIDAZOLES

Compound	R	R[1]	m.p. (°C)	Ref.
8.3	NH_2	H	180.5–181	11
8.3	N_3	H	108	4
8.3	NHAc	H	254–255	11
8.3	$NHCO_2Et$	H	238–240	11
	(methiodide)		270–272	11
			(decomp.)	11
8.3	OH	H	295–296	11
8.3	H	NO_2	196–197	11
8.3	H	NH_2	198–199	11
8.3	H	N_3	98	4
8.3	NHAc	NO_2	190–191	11
			183	40
8.3	NH_2	NO_2	295–296	11
			273	40
8.3	CO_2Et	NO_2	—[a]	3
8.3	CO_2H	NO_2	—[a]	3
8.3	CO_2H	NH_2	165	3
8.3	CO_2^-	N_2^+	—[a]	3
8.3	NO_2	NHAc	183	40
8.3	NO_2	NH_2	228	40
8.3	NO_2	N_3	178 (decomp.)	40
8.3	N_3	NO_2	140 (decomp.)	40
8.3	Cl	NO_2	171	40
8.3	NO_2	Cl	180	40
8.71	—	—	220	11
	(Methiodide, hydrate)		335	11
8.72	—	—	310–311	11
8.73	—	—	345 (decomp.)	11
	(Dimethiodide)		345 (decomp.)	11
8.74	—	—	335	11
	(Dimethiodide)		316–318 (decomp.)	11

[a] Melting point not recorded.

formed (Table 8.11). Although C-4 is the position most favored for nucleophilic substitution the large leaving group attached at N-5 (**8.78**) hinders approach of the nucleophile and substitution predominates at the electronically less-favored C-1 (Scheme 8.7).

When a reagent with a good potential leaving group (e.g., tosyl chloride) is combined with a nucleophile stronger than chloride ion (e.g., cyanide,

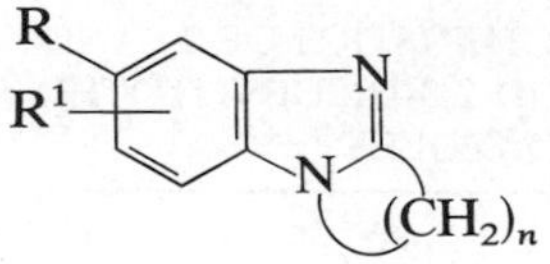

(**8.75**)

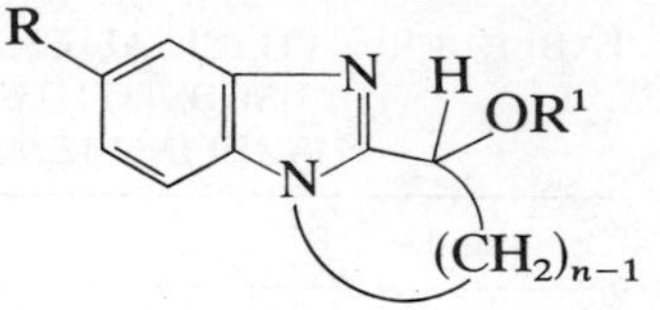

(**8.76**)

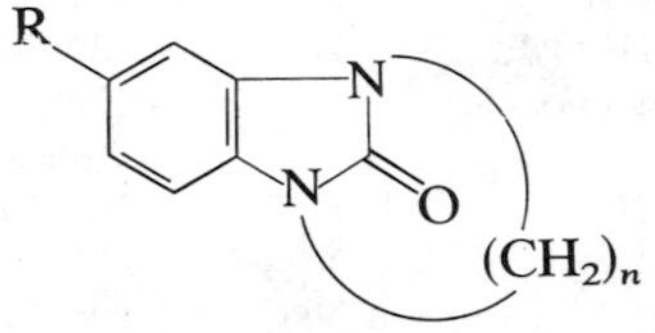

(**8.77**)

azide, or thiocyanate ions, thiols, and amines) the added nucleophile substitutes in the benzene ring.[42] With azide ion–tosyl chloride the *N*-oxide (**8.27**; $R = NO_2$, $n = 5$) gives two products in a 2:1 ratio. The major product, the 1-azido-3-nitroazepinobenzimidazole (**8.75**; $R = NO_2$, $R^1 = 1\text{-}N_3$, $n = 5$) is accompanied by the furoxan (**8.79**). The same furoxan is formed from the chloronitroazepine (**8.75**); $R = NO_2$, $R^1 = 4\text{-Cl}$, $n = 5$) and sodium azide in

TABLE 8.11. RING CHLORINATION OF 7,8,9,10-TETRAHYDRO-3-NITRO-6*H*-AZEPINO[1,2-*a*]BENZIMIDAZOLE-5-OXIDE (**8.27**: $R = NO_2$, $n = 5$)[a]

Reagent	Product(s) ($R = NO_2$, $n = 5$ in all cases)	R^1	Yield (%)	m.p. (°C)
$POCl_3$	8.75	1-Cl	65	136
		4-Cl	35	162
SO_2Cl_2	8.75	1-Cl	55	—
		4-Cl	27	—
$SOCl_2$	8.75	1-Cl	65	—
		4-Cl	35	—
AcCl	8.76	Ac	100	204
BzCl	8.75	4-Cl	50	—
	8.76	Bz	50	179
$4\text{-MeO}\cdot C_6H_4COCl$	8.75	4-Cl	10	—
	8.76	$4\text{-MeO}\cdot C_6H_4CO$	84	155
TsCl	8.75	1-Cl	27	—
		4-Cl	19	—
	8.77	—	7	131
$POBr_3$	8.75	H	—[b]	—[c]

[a] From Ref. 42.
[b] Yield not recorded.
[c] See Table 8.3: compound (**8.19**; $R = NO_2$).

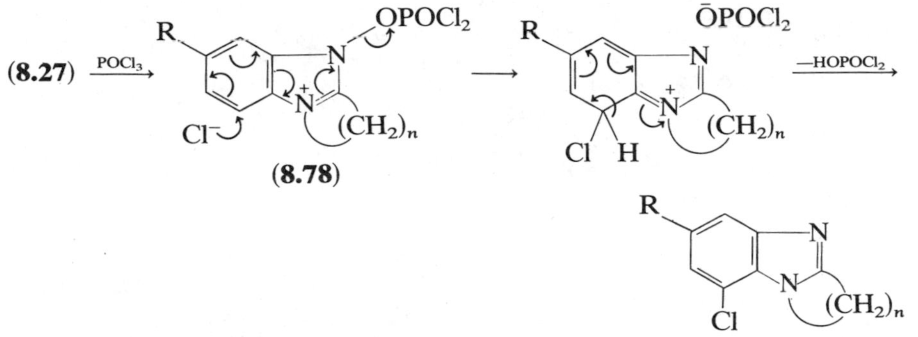

(**8.27**) $\xrightarrow{POCl_3}$ (**8.78**) $\longrightarrow$ $\xrightarrow{-HOPOCl_2}$

Scheme 8.7

dimethyl sulfoxide at 55°. In contrast the chloro groups in the isomeric chloronitro compounds (**8.3**; R = Cl, $R^1 = NO_2$) or (**8.3**; $R = NO_2$, R^1 = Cl) are insensitive to nucleophilic substitution, despite being activated by the adjacent nitro groups.[40] With thiocyanate anion as nucleophile the major product from (**8.27**; $R = NO_2$, $n = 5$) is the expected thiocyanate (**8.75**; $R = NO_2$, R^1 = 4-SCN, $n = 5$) together with a yellow dimeric substance tentatively identified as the disulfide (**8.80**). With oxidizable anions (e.g., bromide, iodide, and butan-1-thiolate) and tosyloxy as leaving group deoxygenation rather than ring substitution occurs.[42]

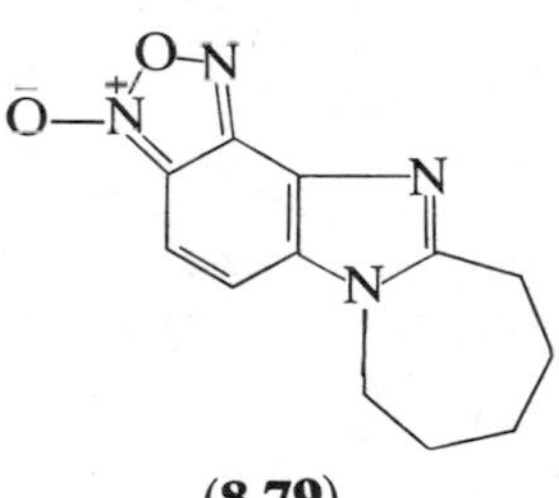

(**8.79**) 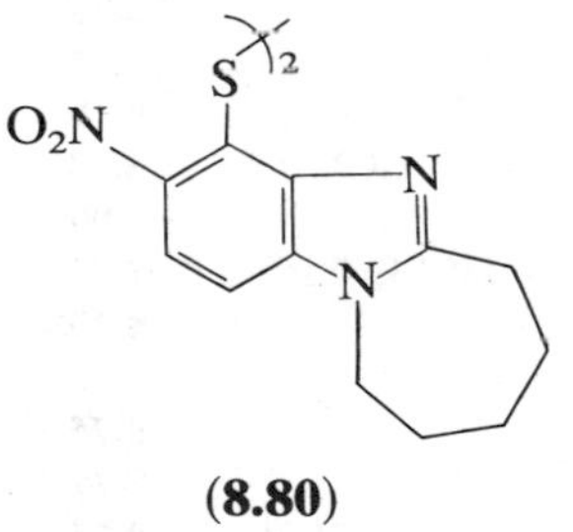 (**8.80**)

Major products of the coaction of an acid chloride and sodium hydroxide on *N*-oxides (**8.27**) are the polymethylene-bridged benzimidazolones (**8.77**; $n = 5$–7).[41,42] The mechanism of this reaction is best explained in terms of attack by hydroxide ion at C-5a of the tosylated *N*-oxide (**8.81**) followed by concerted loss of the tosyloxy group and rearrangement (Scheme 8.8). A similar nucleophilic attack at C-5a is probably involved in the alkaline degradation of the benzimidazolium sulfamate (**8.8**) to *N*-phenylcaprolactam (40%).[8]

Azepinobenzimidazoles formed by the coaction of an acid chloride and a range of nucleophiles are tabulated in Table 8.12.

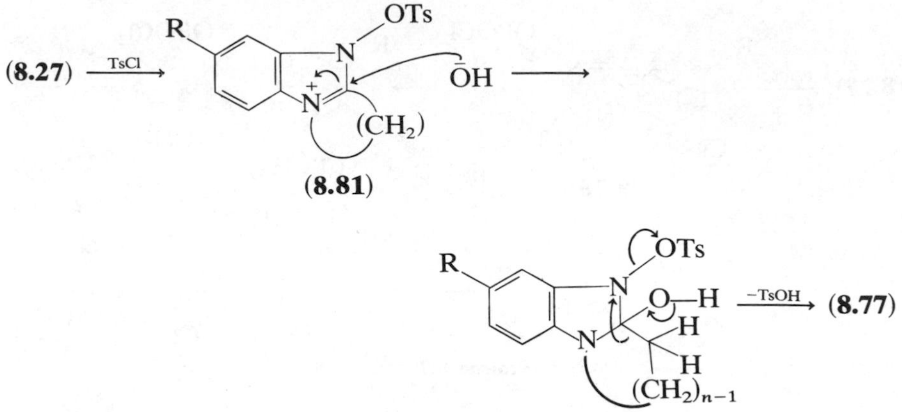

Scheme 8.8

TABLE 8.12. PRODUCTS OF THE COACTION OF AN ACID CHLORIDE AND NUCLEOPHILE ON THE *N*-OXIDES (**8.27**)[a]

Starting *N*-oxide (**8.27**)			Products				Yield	m.p.
R	*n*	Reagents	Compound	R	*n*	R^1	(%)	(°C)
NO_2	5	BzCl/CN^-	**8.75**	NO_2	5	4-CN	20	205
			8.76	NO_2	5	Bz	43	—[b]
NO_2	5	TsCl/CN^-	**8.75**	NO_2	5	4-CN	20	—
NO_2	5	TsCl/N_3^-	**8.75**	NO_2	5	1-N_3	66	154
			8.79	—	—	—	33	218
NO_2	5	TsCl/SCN^-	**8.75**	NO_2	5	4-SCN	60	142
			8.80	—	—	—	20	300
NO_2	5	TsCl/Bu^nSH	**8.75**	NO_2	5	4-Cl	27	—[b]
			8.75	NO_2	5	4-SBu^n	21	71
			8.75	NO_2	5	H	6	—[c]
			8.75	NO_2	5	1-Cl	14	—[b]
			8.77	NO_2	5	—	10	131
NO_2	5	TsCl/Bu^nNH_2	**8.75**	NO_2	5	4-$NHBu^n$	60	99
			8.77	NO_2	5	—	20	—
NO_2	5	TsCl/$PhNH_2$	**8.75**	NO_2	5	4-NHPh	90	150
			8.75	NO_2	5	2-NHPh	10	198
H	5	TsCl/OH^-	**8.77**	H	5	—	23	81
Cl	5	TsCl/OH^-	**8.77**	Cl	5	—	35	112
Cl	5	BzCl/OH^-	**8.76**	Cl	5	Bz	36	171
			8.77	Cl	5	—	27	—
NO_2	5	TsCl/OH^-	**8.77**	NO_2	5	—	55	—
Cl	6	TsCl/OH^-	**8.77**	Cl	6	—	85	105
Cl	7	TsCl/OH^-	**8.77**	Cl	7	—	63	143

[a] From Ref. 42.
[b] See Table 8.11.
[c] See Table 8.3: compound (**8.19**; R = NO_2).

Photochemical and Thermal Reactions

When *N*-oxides (**8.27**; R = H, Cl) with fewer than five methylene groups are photolyzed in methanol they are generally recovered unchanged.[43] *N*-Oxides with larger rings either undergo deoxygenation or rearrange to yield benzimidazolones (**8.77**) (Table 8.13). In one case (**8.27**; R = Cl, $n = 5$) ring substitution by the solvent to yield the 2-methoxyazepinobenzimidazole (**8.82**) was also observed.

TABLE 8.13. PRODUCTS FORMED BY PHOTOLYSIS[a] OF *N*-OXIDES (**8.27**)[b]

Starting *N*-oxide (**8.27**)		Product(s)				
R	*n*	Compound	R	*n*	Yield (%)	m.p. (°C)
H	5	**8.28**	H	5	32	124
		8.77	H	5	25	81
Cl	5	**8.28**	Cl	5	16	—[c]
		8.77	Cl	5	44	112
		8.82	—	—	15	164
Cl	6	**8.28**	Cl	6	22	117
		8.77	Cl	6	66	105
Cl	7	**8.77**	Cl	7	"low"	143

[a] Hanovia 200 W medium-pressure lamp with a Pyrex cooling jacket; AnalaR methanol under nitrogen as solvent.
[b] From Ref. 43.
[c] See Table 8.3: compound (**8.19**; R = Cl).

The benzofuroxan (**8.83**) (32%) together with traces of nitroamines are formed when the azidonitro compound (**8.3**; R = NO_2, $R^1 = N_3$) or its isomer (**8.3**; R = N_3, $R^1 = NO_2$) are photolyzed in acetic acid.[40] The method is inferior to a thermolysis route (57%) employing hot 2-ethoxyethanol (133–135°) as solvent. At higher temperatures (boiling ethylene glycol or propionic acid) the benzofurazan (**8.84**) is formed in 60% yield.[40] Benzoxazoles (**8.85**) and (**8.86**) result when 2-azido- or 3-azido-7,8,9,10-tetrahydro-6*H*-azepino[1,2-*a*]benzimidazoles, respectively, are thermolyzed in an acetic acid–polyphosphoric acid mixture.[4]

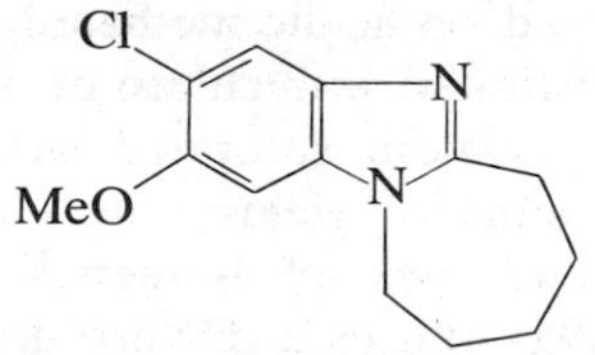

Cl MeO N N

(**8.82**)

O^- N^+ O N N N

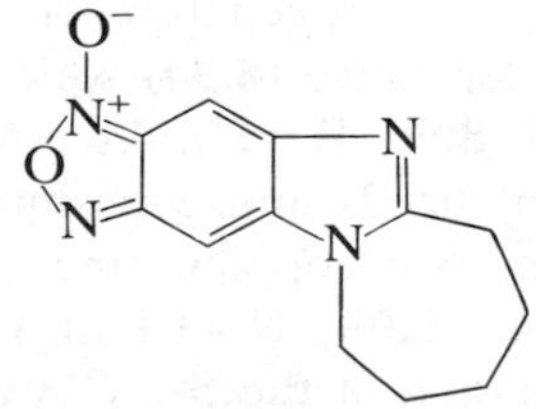

(**8.83**)

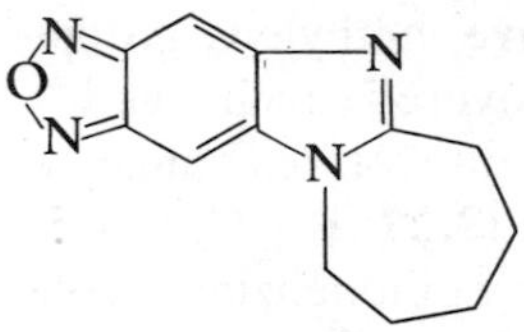

(8.84) 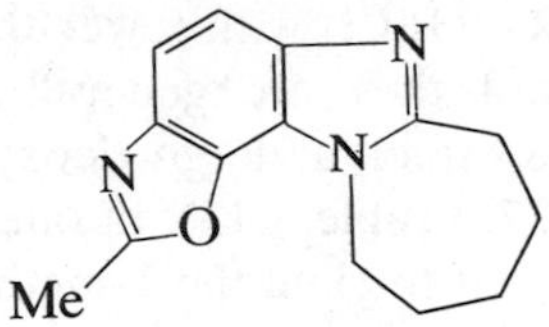 (8.85) 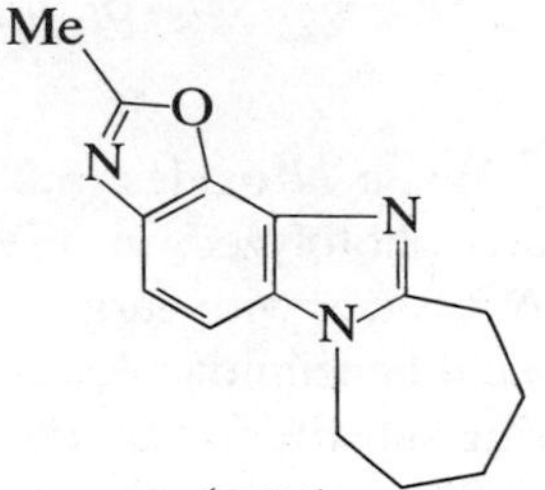 (8.86)

Fragmentation of the diazonium betaine (**8.87**) at high temperature proceeds via the aryne (**8.88**), which can be trapped by tetracyclone to give a low yield of the tetracycle (**8.89**).[3]

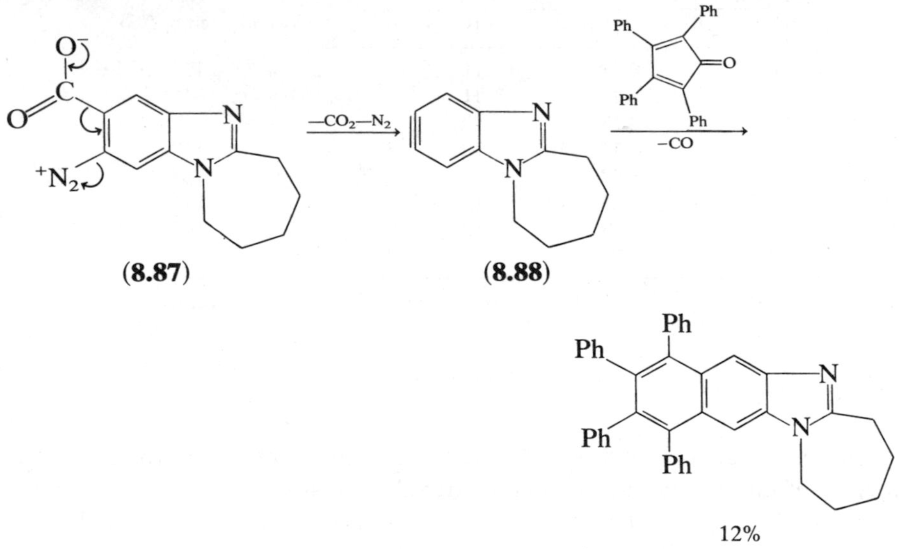

$-CO_2-N_2$ (8.87) (8.88) $-CO$ 12% (8.89)

Scheme 8.9

Reduction

Catalytic hydrogenation of the conjugated lactam (**8.54**) affords a tetrahydro derivative (**8.34**) which is ring-opened on acidic methanolysis[34] to give ester (**8.90**; R = H). Azepines (**8.54**) substituted with chloro or methoxy groups in the 2- and 3-positions form two isomeric saturated lactams on reduction. Nucleophilic ring-opening by methanol yields only one benzimidazole (**8.90**; R = Cl or OMe) from each pair of isomers.[35] Partial hydrogenation of the 8-methoxyazepine (**8.91**) affords a dihydro derivative (**8.92**), which cleaves in acidic methanol to form the ketone (**8.93**).

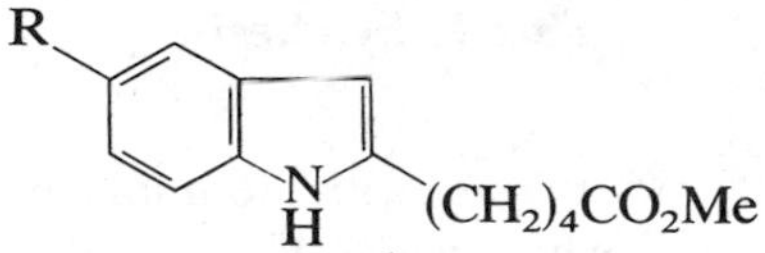

R

N H (CH$_2$)$_4$CO$_2$Me

(**8.90**)

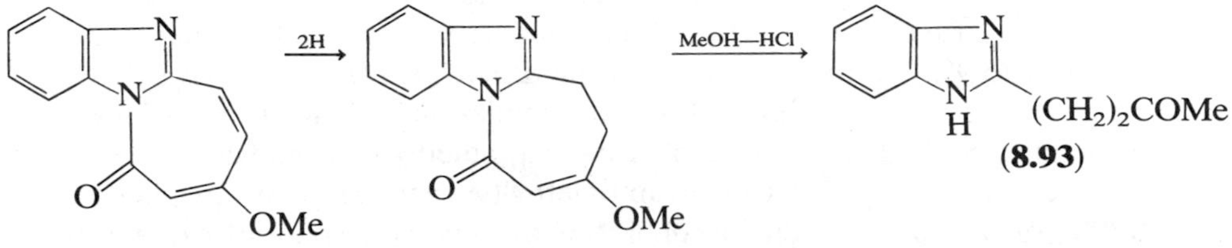

N N 2H → N N MeOH—HCl → N N H (CH$_2$)$_2$COMe

O OMe O OMe (**8.93**)

(**8.91**) (**8.92**)

8.2. TRICYCLIC BENZIMIDAZOLES WITH ONE ADDITIONAL HETEROATOM

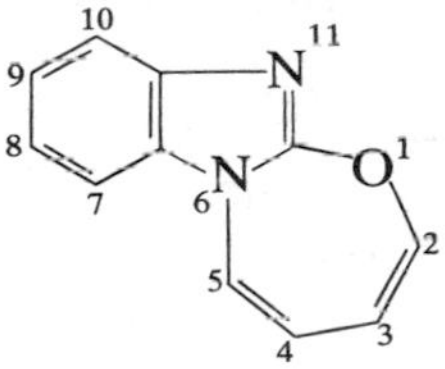

[1,3]Oxazepino[3,2-*a*]benzimidazole

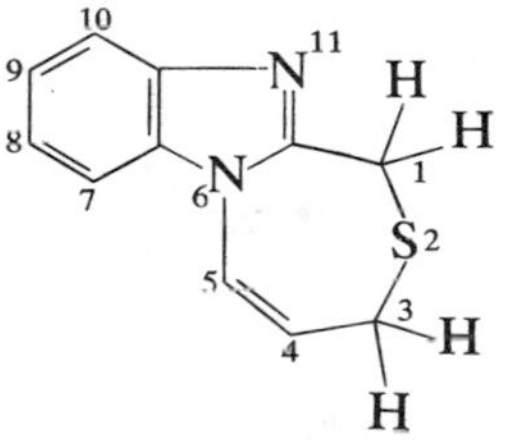

1*H*,3*H*-[1,4]Thiazepino[4,3-*a*]benzimidazole

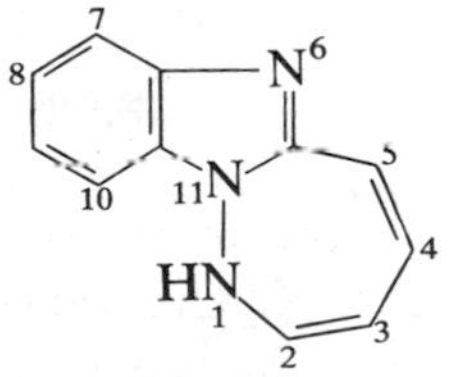

1*H*-[1,2]Diazepino[1,7-*a*]benzimidazole

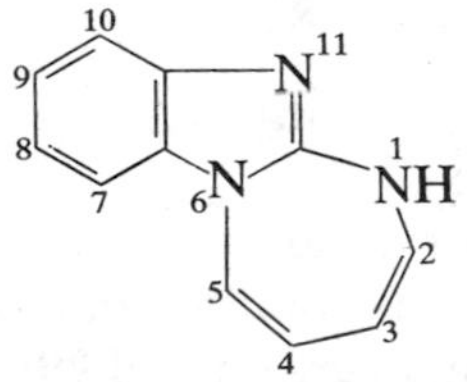

1*H*-[1,3]Diazepino[1,2-*a*]benzimidazole

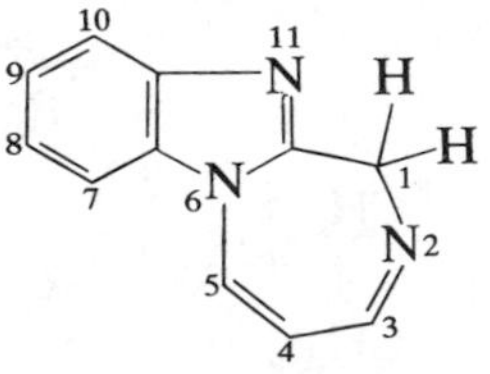

1*H*-[1,4]Diazepino[1,2]*a*]benzimidazole

8.2.1. Synthesis

Interaction of benzimidazolone with sodium hydride in dimethylformamide generates an anion which reacts with α,ω-dibromoalkanes. Products are either 1,3-polymethylene-bridged benzimidazolones (**8.77**; $n = 10$, 12) or the corresponding dimers and trimers when the methylene chain is shortened.[44] Reaction with 1,4-dibromobutane yields mainly trimer and a low yield (5.9%) of the oxazepinobenzimidazole (**8.94**).

Entry to the [1,4]thiazepino[4,3-*a*]benzimidazole system can be achieved by heating *S*-alkyl derivatives of 2-mercaptomethylbenzimidazole.[45] Cyclodehydration of 4-(2-mercaptomethylbenzimidazolyl)propiophenone (**8.95a**) or the corresponding alcohol (**8.95b**) in polyphosphoric acid at 100° furnishes thiazepinobenzimidazoles (**8.96**) and (**8.97**), respectively: the propionic acid (**8.95c**) yields thiazepinone (**8.98**) in refluxing xylene.

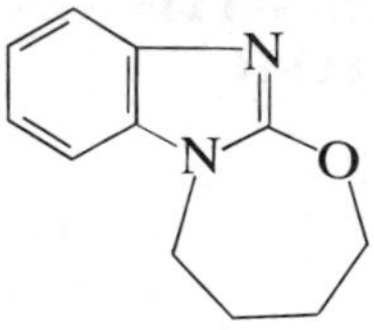

(**8.94**)

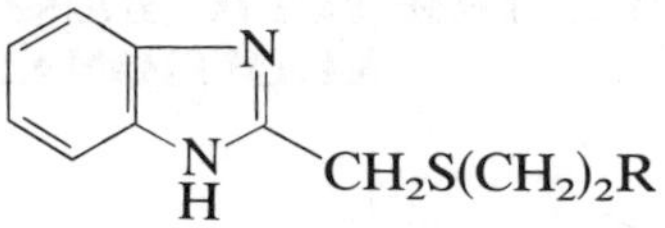

$CH_2S(CH_2)_2R$

(**8.95**)

a, R = Bz
b, R = CH(OH)Ph
c, R = CO_2H

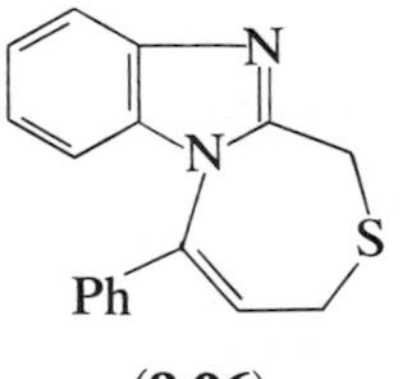

(**8.96**) 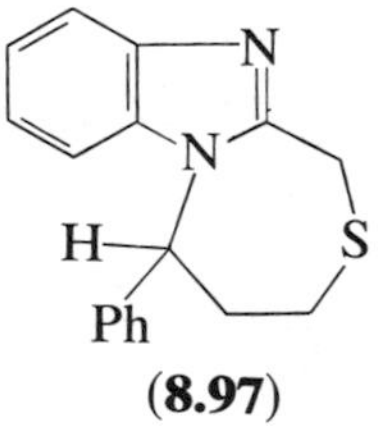 (**8.97**) 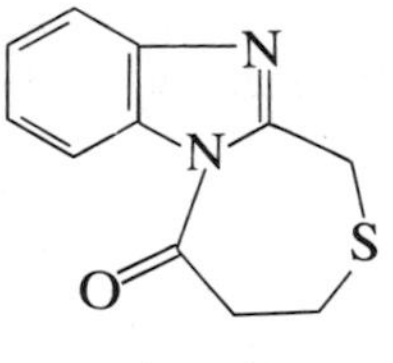 (**8.98**)

Acid catalyzed cyclization of *NN*-dialkyl-*N'*-(*o*-nitrophenyl)hydrazines represents an intriguing variant of the *t*-amino effect.[1,46,47] Cyclization of the parent hydrazine (**8.99**; $R = R^1 = H$, $n = 4$) is best achieved in refluxing trifluoroacetic acid or 48% hydrobromic acid, whereas substituted analogs cyclize efficiently in boiling 6*N*-hydrochloric acid to yield a series of [1,2]diazepino[1,7-*a*]benzimidazoles and higher ring homologs (**8.100**; $n = 4$, 5).

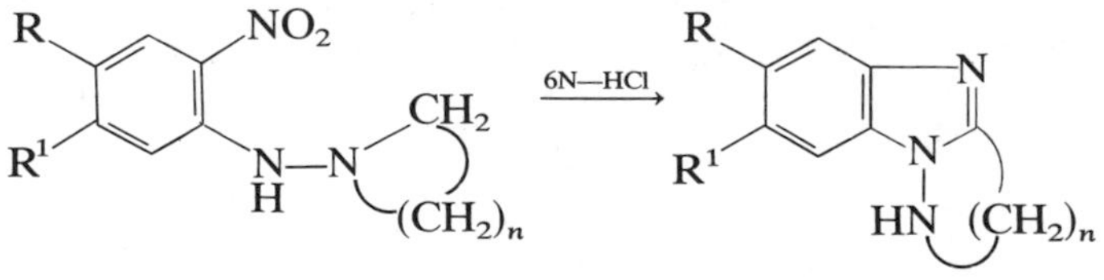

6N—HCl →

(**8.99**) (**8.100**)

TABLE 8.14. FORMATION OF 1*H*-[1,2]DIAZEPINO[1,7-*a*]BENZIMIDAZOLES AND HIGHER HETEROALICYCLIC HOMOLOGS (**8.100**) FROM THE ACID-CATALYZED CYCLIZATION OF HYDRAZINES (**8.99**)[a]

Starting hydrazine (**8.99**)			Acid (time of	Product (**8.100**)				
R	R^1	n	heating in hr)	R	R^1	n	Yield (%)	m.p. (°C)
H	H	4	TFA	H	H	4	50	226
H	H	4	HCl (3)	H	Cl	4	63	198–199
H	Cl	4	HCl (4)	H	Cl	4	23	—
CF_3	H	4	HCl (5)	CF_3	H	4	34	189
NO_2	H	4	HCl (8)	NO_2	H	4	6	224–226
NO_2	H	4	PPA (4)	NO_2	H	4	50	—
H	Cl	5	HCl (2)	H	Cl	5	46	203–205
CF_3	H	5	HCl (5)	CF_3	H	5	22	148
NO_2	H	5	HCl (5)	NO_2	H	5	8	222
NO_2	H	5	PPA (6)	NO_2	H	5	10	—

[a] From Ref. 47.

Exceptions are the dinitrophenylhydrazines, which give poor yields in hydrochloric acid but improved results in polyphosphoric acid (Table 8.14). The mechanism is believed to be initiated by attack of the *t*-amino nitrogen at the electron-deficient nitronic acid group (**8.101**) which ring-closes to a spiro-benzotriazole *N*-oxide (**8.102**). Subsequent rearrangement and cyclization generates *N*-oxides (**8.103**) which are deoxygenated under the reaction conditions (Scheme 8.10). The inability of pyrollidinohydrazines (**8.99**; $n=3$) to undergo cyclization may stem from steric constraints in the formation of the spiro-benzotriazole. Evidence for a benzimidazole *N*-oxide intermediate (**8.103**) can be adduced from the observation that the parent hydrazine (**8.99**; $R=R^1=H$, $n=4$) and the corresponding chloro analog (**8.99**; $R=H$, $R^1=Cl$) both give the same diazepinobenzimidazole (see section 8.1.3, "Reactions with Nucleophiles").

Methyl 2-benzimidazolylcarbamate (**8.104**) condenses with succinyl or dimethylsuccinyl chloride in alkali to yield representatives of the [1,3]diazepino[1,2-*a*]benzimidazole system (**8.105a,b**), which are claimed to have miticidal and fungicidal activities.[48–50]

2-Aminomethylbenzimidazole (**8.106**) is a convenient substrate for the preparation of [1,4]diazepino[1,2-*a*]benzimidazoles: it reacts with 1,3-diketones, α,β-unsaturated ketones and difunctional acids to yield diazepinobenzimidazoles (**8.107**), (**8.108**), and (**8.109**), respectively,[51] in reasonable yields (Table 8.15). Epichlorohydrin can serve as the difunctional cyclizing agent reacting with amine (**8.106**) to form the diazepinobenzimidazole (**8.110**). Virucidal[52] and virustatic activities[53] are claimed for this compound.

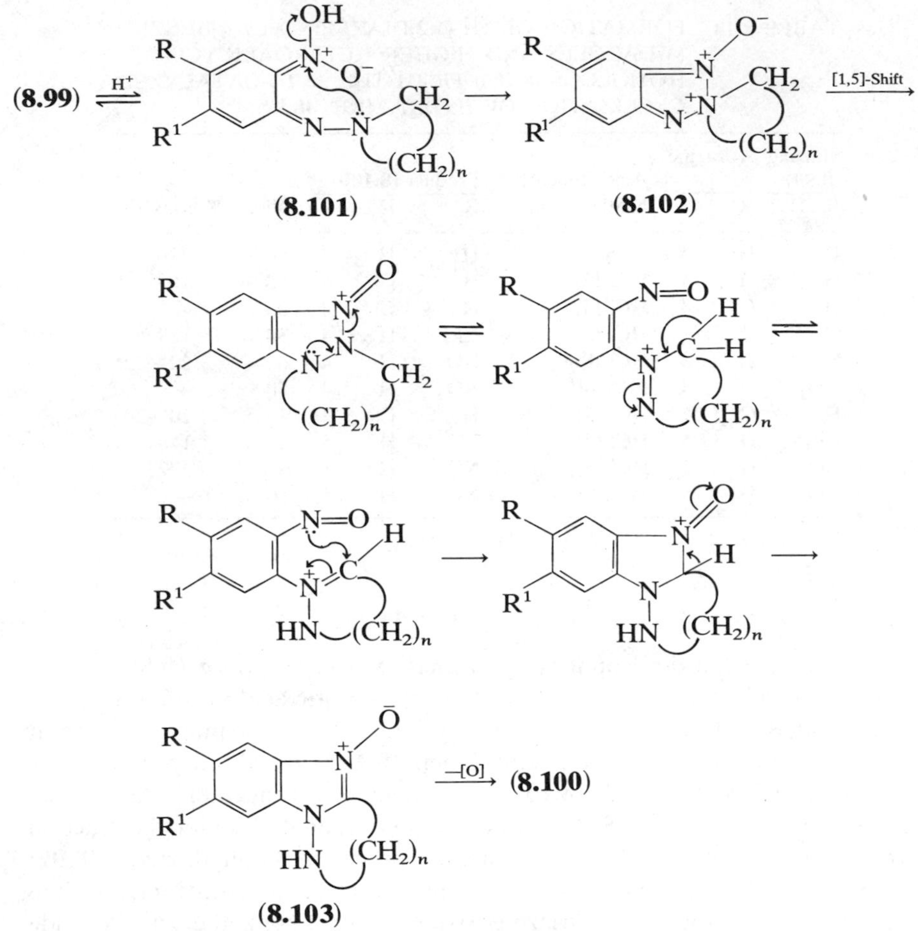

(8.99) $\overset{H^+}{\rightleftharpoons}$ (8.101) ⟶ (8.102) $\xrightarrow{[1,5]\text{-Shift}}$

⇌ ⇌

⟶ ⟶

(8.103) $\xrightarrow{-[O]}$ (8.100)

Scheme 8.10

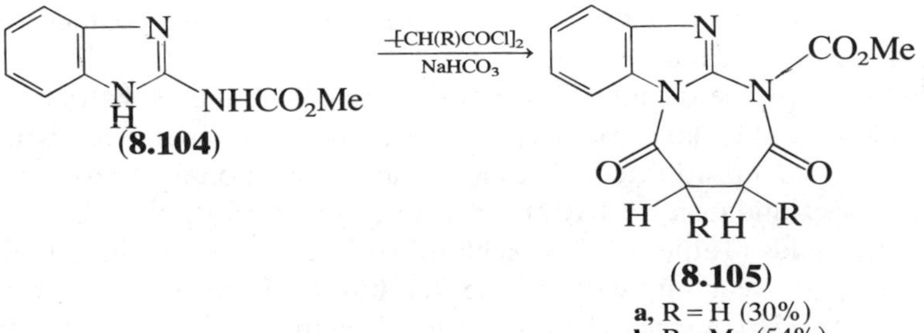

(8.104) $\xrightarrow[NaHCO_3]{-[CH(R)COCl]_2}$ (8.105)

a, R = H (30%)
b, R = Me (54%)

(**8.106**)

$RCOCH_2Ac$

$MeCH{=}CHCO_2H$ or $MeCH(Cl)CH_2CO_2H$

$RCH{=}CHBz$

(**8.107**)
R = Me,Ph

(**8.109**)

(**8.108**)
R = Ph,2-Thienyl

(**8.110**)
R = H,Me

TABLE 8.15. [1,4]DIAZEPINO[1,2-*a*]BENZIMIDAZOLES FROM 2-AMINOMETHYL-BENZIMIDAZOLE (**8.106**) AND CARBONYL COMPOUNDS[a]

Product	R	Reagent	Reaction conditions	Yield (%)	m.p. (°C)
8.107	Me (Picrate)	Acetylacetone	4*N*-HCl; reflux	81	222 258 (decomp.)
8.107	Ph	Benzoylacetone	160°; 4 hr	—[b]	176 (dihydrate)
8.108	Ph	Benzylidene acetophenone	DMF; reflux; 1–2 hr	64	243
8.108	2-Thienyl	Thien-2-ylidene acetophenone	DMF; reflux; 1–2 hr	58	256 (decomp.)
8.109	—	Crotonic acid	120–130°; 4 hr	67	194 (hydrate)
8.109	—	3-Chlorobutyric acid	DMF; reflux; 1 hr	56	234 (decomp.) (hydrochloride)

[a] From Ref. 51.
[b] Yield not recorded.

An unusual synthesis of [1,4]diazepino[1,2-*a*]benzimidazoles starts with the bromopropylbenzimidazole (**8.111**). Above 100° this compound cyclizes to the thiazolium salt (**8.112**): alternatively, the salt can be prepared by interaction of the anthelmintic thiabendazole (**8.113**) and 1,3-dibromopropane.[54] The unstable thiazolium ring can be cleaved in aqueous alkali to form a thiol which probably has the *Z*-configuration (**8.114a**). The thiol is not isolable but is readily transformed to a crystalline *S*-methyl derivative (**8.114b**) with methyl iodide. The *N*-methylbenzimidazolium dibromide (**8.115**) in aqueous alkali forms the stable vinylogous amide (**8.116**).

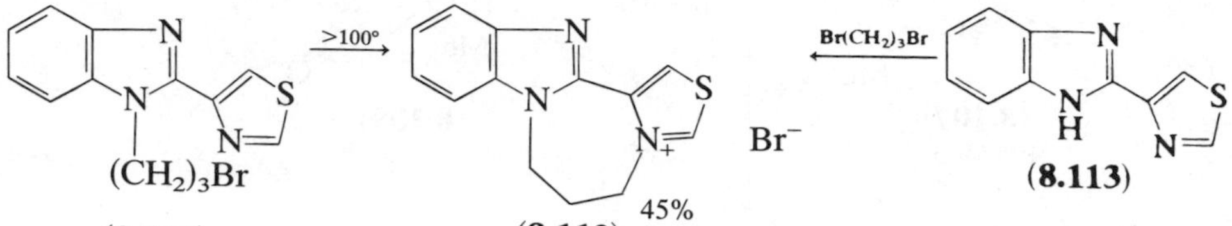

(**8.111**) $\xrightarrow{>100°}$ (**8.112**) 45% Br^- $\xleftarrow{Br(CH_2)_3Br}$ (**8.113**)

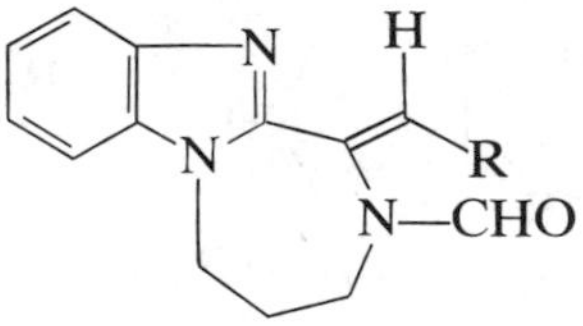

(**8.114**)
a, R = SH
b, R = SMe
c, R = $NHCH_2Ph$

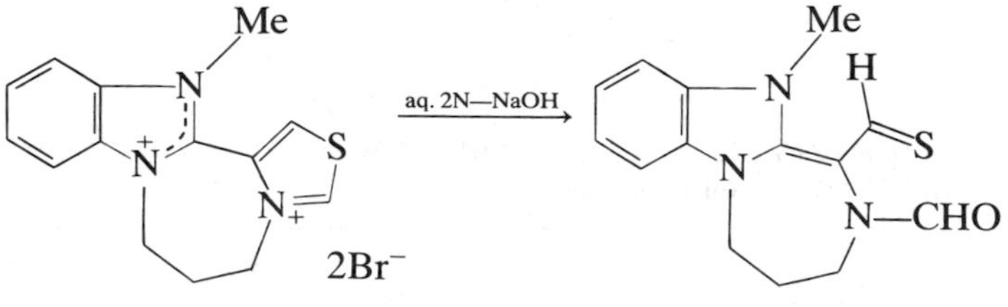

(**8.115**) 2Br⁻ $\xrightarrow{\text{aq. 2N—NaOH}}$ (**8.116**) 91%

8.2.2. Physicochemical Studies and Reactions

The UV spectrum of the 1*H*-[1,2]diazepinobenzimidazole (**8.100**; R = R^1 = H, $n = 4$) shows characteristic benzimidazole peaks at 282 and 288 nm and the IR spectrum an NH absorption at 3200 cm^{-1}. ^1H NMR spectra of compounds (**8.100**; $n = 4$, 5) exhibit a characteristic feature in that the NH proton appears as a triplet (in some cases), or broadened, because of coupling to the adjacent methylene group.[47]

Ring-opening of the [1,4]diazepine ring in the lactam (**8.109**) is effected in boiling 6*N*-hydrochloric acid;[51] the product is the 1,2-disubstituted benzimidazole (**8.117**).

The thiol (**8.114a**) can be converted to a series of products with physical characteristics recorded in Table 8.16. Oxidation of the anion of (**8.114a**)

TABLE 8.16. PHYSICAL CHARACTERISTICS OF [1,4]DIAZEPINO[1,2-*a*]-BENZIMIDAZOLES[a]

Compound	R	Yield (%)	m.p. (°C)	Chemical shifts (δ)[b]; *J* (Hz)
8.114	SMe	92	150–151	2.3br quintet (2*H*, CH_2, $J = 7$); 2.48s (3*H*, SCH_3); 3.90t (2H, CH_2NCHO, $J = 7$); 4.20m (2H, CH_2N); 7.2–7.4m (3*H*, arom.); 7.7m (1*H*, arom.); 7.70s (1*H*, $=CHSCH_3$); 8.38s (1*H*, NCHO)
8.114	$NHCH_2Ph$	52	121–122	2.1–2.5m (2*H*, CH_2); 3.87t (2*H*, CH_2NHCHO, $J = 7$); 4.0–4.2m (2*H*, CH_2N); 4.48d (2*H*, PhCH_2, $J = 7$); 6.66d[c] (1*H*, =CHN, $J = 12.5$); 7.1–7.25m (3*H*, arom.); 7.33s (5*H*, arom.); 7.6m (1*H*, arom.); 8.27s (1*H*, NCHO); 9.00br s[d] (1*H*, NH)
8.116	—	91	259–260 (decomp.)	—[e]
8.118	—	90	268 (decomp.)	2.35br quintet (2*H*, CH_2, $J = 7$); 3.91t (2*H*, CH_2NCHO, $J = 7$); 4.2m (2*H*, CH_2N); 7.2–7.4m (3*H*, arom.); 7.7m (1*H*, arom.); 7.7m (1*H*, =CHS); 8.40s (1*H*, NCHO)
8.119	CHO	65[f]	127–128	1.69[g] and 1.83[h]dd (3*H*, CH_3CH, $J = 7$); 5.23[h] and 6.05[g]qq (1*H*, CHCH_3, $J = 7$); 8.15[g] and 8.33[h]ss (1*H*, NCHO)
8.119	Me	35	146	1.70d (3*H*, CH_3CH, $J = 7$); 2.15s (3*H*, CH_3N); 1.82m (2*H*, CH_2); 3.2–3.45m (2*H*, CH_2NCH_3); 4.0–4.5m (3*H*, CH_2N and CHNCH_3); 7.2–7.3m (3*H*, arom.); 7.75m (1*H*, arom.)
8.119	H	81	152	1.73d (3*H*, CH_3CH, $J = 7$); 1.8m (2*H*, CH_2); 3.85–4.65m (3*H*, CH_2N and CHCH_3); 4.10t (2*H*, CH_2NH, $J = 7$); 7.15–7.3m (3*H*, arom.); 7.73m (1*H*, arom.)

[a] From Ref. 54.
[b] In $CDCl_3$. s = singlet; t = triplet; m = multiplet; dd = double doublet; qq = double quartet; br = broad absorption.
[c] Collapses to a singlet with D_2O.
[d] Removed by D_2O.
[e] Insoluble in $CDCl_3$ and DMSO.
[f] Mixture (2:1) of *E*- and *Z*-geometrical isomers.
[g] *Z*-isomer.
[h] *E*-isomer.

with hydrogen peroxide affords the sulfide (**8.118**): the sulfide reacts with benzylamine at 120° under nitrogen to form amine (**8.114c**). Two geometrical isomers of the *N*-formyldiazepinobenzimidazole (**8.119a**) are formed in 2:1 ratio when the anion of (**8.114a**) is treated with Raney nickel: a minor product of desulfuration is the *t*-amine (**8.119b**). The corresponding *sec*-amine (**8.119c**) is liberated when the *N*-formyldiazepinobenzimidazole is refluxed in 2*N*-hydrochloric acid.[54]

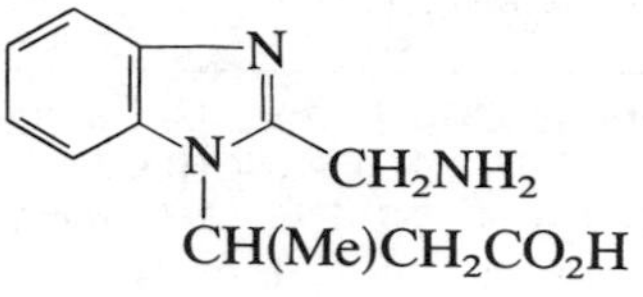

(**8.117**)

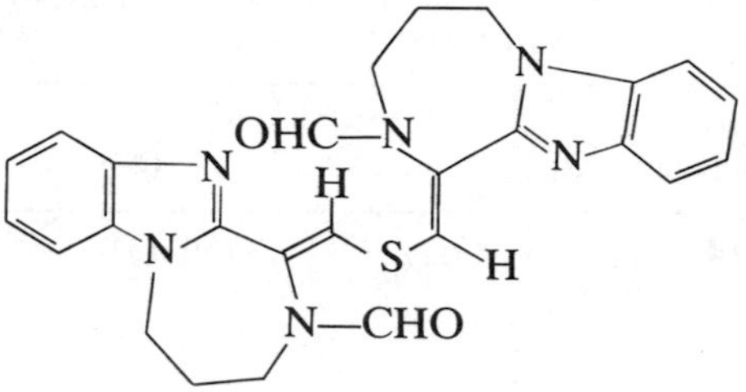

(**8.118**)

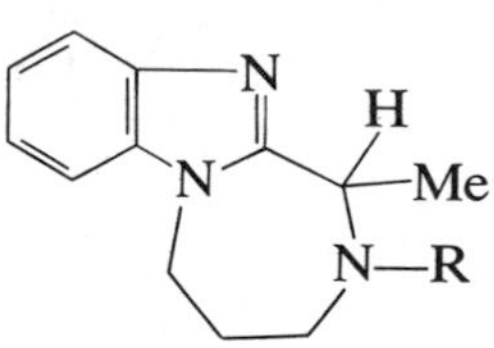

(**8.119**)

a, R = CHO
b, R = Me
c, R = H

8.3. TRICYCLIC BENZIMIDAZOLES WITH TWO ADDITIONAL HETEROATOMS

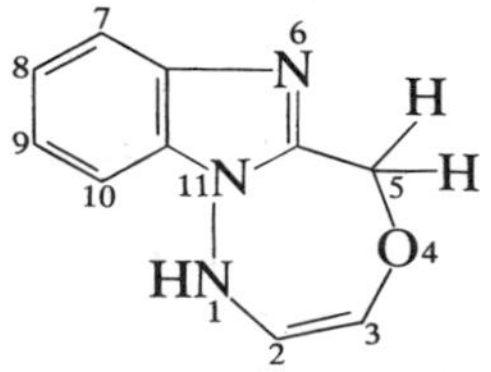

1*H*,5*H*-[1,4,5]Oxadiazepino[4,3-*a*]benzimidazole

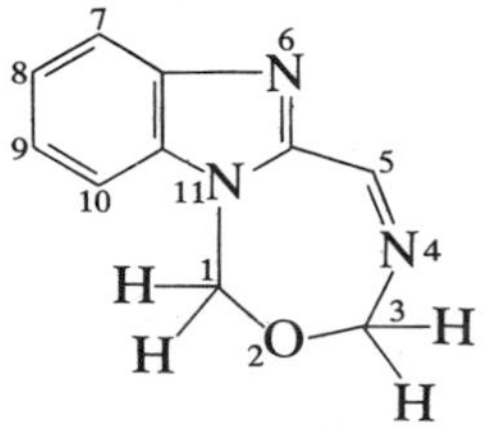

1*H*,3*H*-[1,3,6]Oxadiazepino[3,4-*a*]benzimidazole

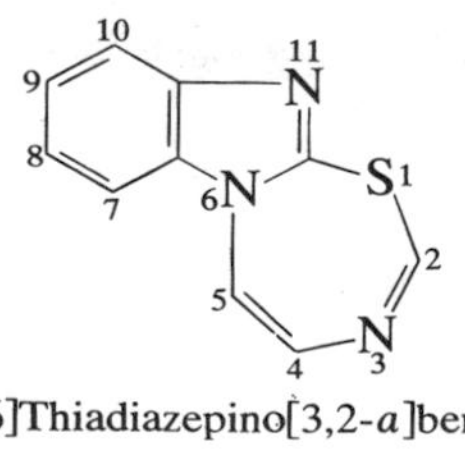

[1,3,6]Thiadiazepino[3,2-*a*]benzimidazole

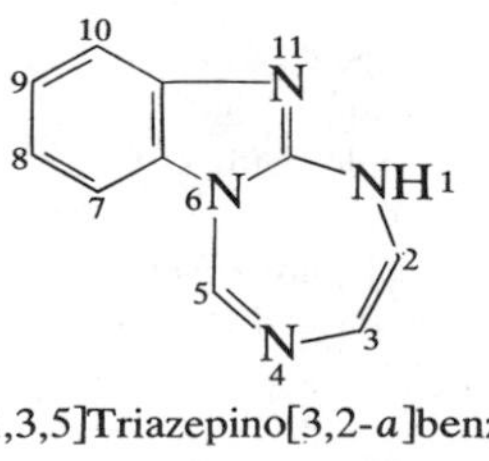

1*H*-[1,3,5]Triazepino[3,2-*a*]benzimidazole

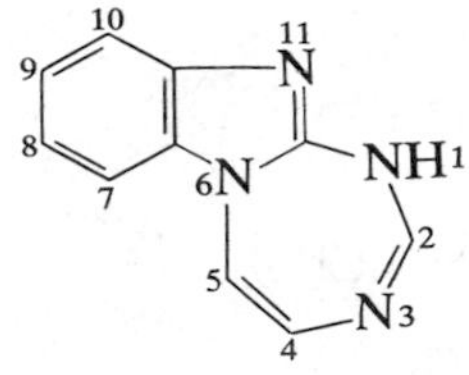

1H-[1,3,5]Triazepino[1,2-a]benzimidazole

8.3.1. Synthesis and Reactions

Unlike cyclizations of the corresponding piperido- or hexahydroazepino-hydrazines (see section 8.2.1) the morpholinohydrazine (**8.120**) forms a tricycle, the [1,4,5]oxadiazepino[4,3-a]benzimidazole (**8.121**), only with difficulty.[46,47] Unstable [1,3,6]oxadiazepino[3,4-a]benzimidazoles (**8.123**) can be prepared from substituted 2-aminomethylbenzimidazoles (**8.122**): the products revert to starting materials in 4N-hydrochloric acid at room temperature.[55] The ^{1}H NMR spectrum of derivative (**8.123**; R = Me) shows singlets for the methylene groups at δ 5.59, 4.79, and 4.31.

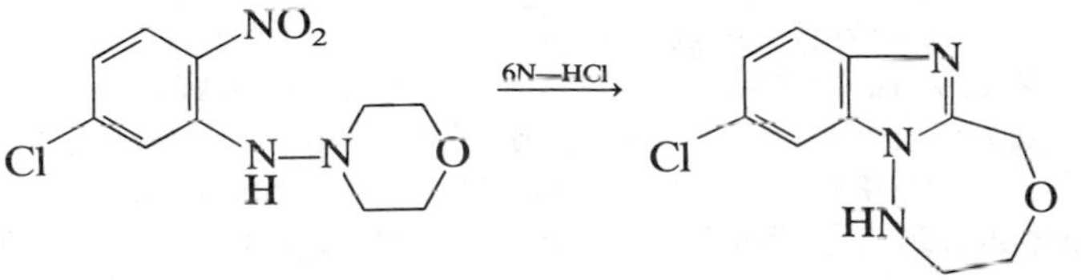

6N—HCl

(**8.120**) (**8.121**)

37% HCHO / 4N—HCl

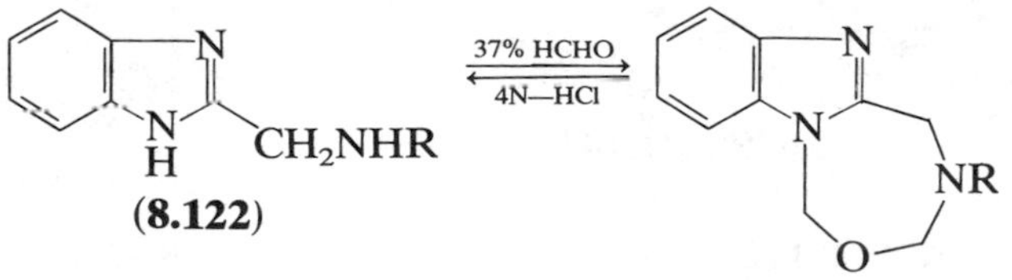

(**8.122**) (**8.123**)

R = Me,Et,Pr^n,Bu^n

Bis-(β-chloroethyl)carbamoyl chloride (BCC) reacts with the anion of 2-mercaptobenzimidazole, at −5 to −10° to yield the S-carbamoylated product (**8.124**). If the reaction mixture is warmed to 5–10° the thiadiazepino-benzimidazole (**8.125**) is formed.[56] 2-Amino- or 2-anilinobenzimidazoles (**8.126**), on the other hand, must initially undergo carbamoylation at N-1, since the products are [1,3,5]triazepino[3,2-a]benzimidazoles (**8.127**). These derivatives are worthy of examination as potential antitumor agents.

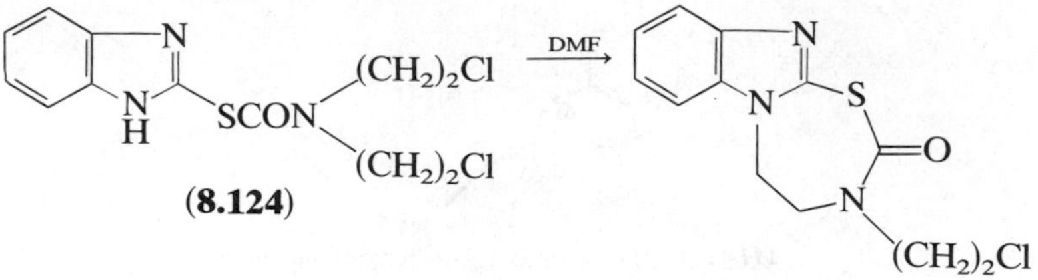

DMF

(**8.124**)

(**8.125**)

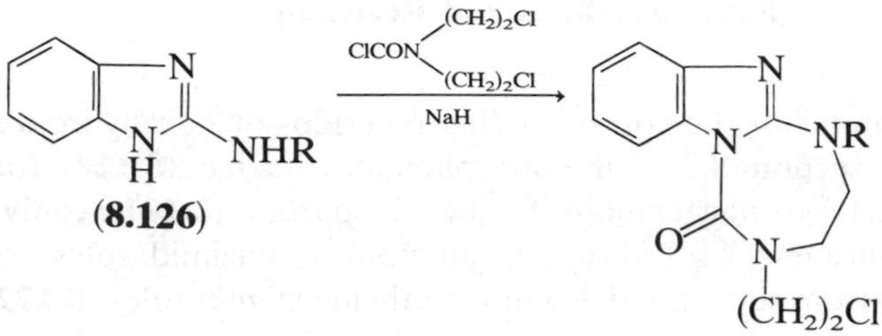

NaH

(**8.126**)

(**8.127**)

R = H,Ph

2-Amino-1-aminoethylbenzimidazoles (**8.128**) cyclize with a range of conventional "1-carbon" reagents[57] to form [1,3,5]triazepino[1,2-*a*]benzimidazoles (**8.129**) or (**8.130**). Reaction of diamines (**8.128**; R = H, R^1 = H, Bu^n) with carbon disulfide are the exceptions stopping at the inner salt stage (**8.131**). The salt (**8.131**; R = R^1 = H) loses hydrogen sulfide in boiling dimethylformamide to afford a thione (**8.129**; R = R^1 = H, X = S), which undergoes *S*-methylation in methyl iodide.

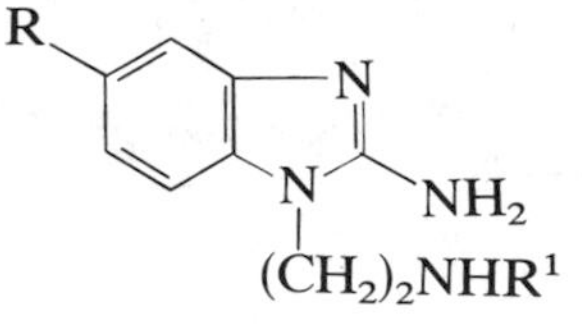

(**8.128**)

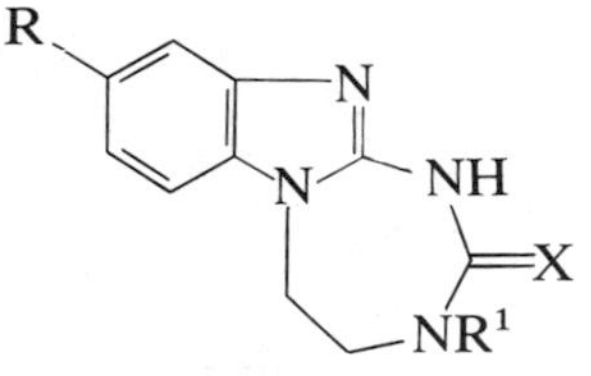

(**8.129**)
X = O,S,NTs

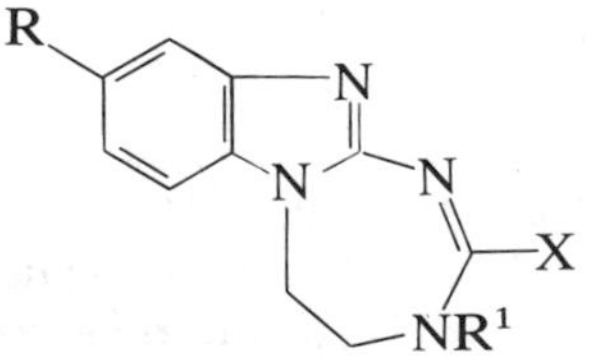

(**8.130**)
X = H,NH_2,SMe

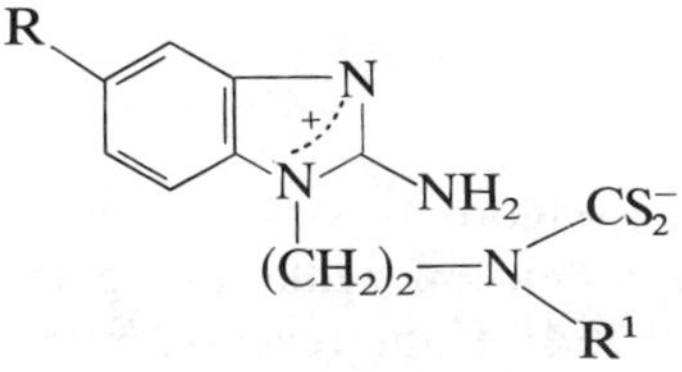

(**8.131**)

R = H,Cl
R^1 = H,Bu^n

TABLE 8.17. SYNTHESIS AND PHYSICAL CHARACTERISTICS OF OXADIAZEPINO-, THIADIAZEPINO-, AND TRIAZEPINOBENZIMIDAZOLES

Starting materials				Products							
Compound	R	R^1	X	Compound	R	R^1	X	Cyclizing reagent	Yield (%)	m.p. (°C)	Ref.
8.120	—	—	—	8.121	—	—	—	6*N*-HCl	36	215–217	47
8.122	Me	—	—	8.123	Me	—	—	37% HCHO	77	227.5–228.5	55
8.122	Et	—	—	8.123	Et	—	—	37% HCHO	71	202–203	55
8.122	Pr^n	—	—	8.123	Pr^n	—	—	37% HCHO	75	151.5–152.5	55
8.122	Bu^n	—	—	8.123	Bu^n	—	—	37% HCHO	77	141.5–142.5	55
8.124	—	—	—	8.125	—	—	—	NaH in DMF	30	171–173	56
8.126	H	—	—	8.127	H	—	—	BCC[a]–NaH	30	152–154	56
8.126	Ph	—	—	8.127	Ph	—	—	BCC–NaH	60	171–172	56
8.128	H	H	—	8.129	H	H	O	1,1′-carbonyl diimidazole	35	298–300	57
8.128	H	H	—	8.129	H	H	NTs	$(MeS)_2C{=}NTs$	51	334–335	57
8.128	Cl	Bu^n	—	8.129	Cl	Bu^n	O	1,1′-Carbonyl-diimidazole	53	291–294	57
8.128	Cl	Bu^n	—	8.129	Cl	Bu^n	S	1,1′-Thiocarbonyl-diimidazole	46	264–266	57
8.128	Cl	Bu^n	—	8.130	Cl	Bu^n	H	$HC(OEt)_3$	67	260–261	57
8.128	Cl	Bu^n	—	8.130	Cl	Bu^n	NH_2	BrCN	52	267–268	57
8.131	H	H	—	8.129	H	H	S	DMF	73	258–260	57
8.129	H	H	S	8.130	H	H	SMe	—[b]	90	308–9 275 (hydroiodide)	57

[a] Bis-(β-chloroethyl)carbamoyl chloride.
[b] Methyl iodide in methanol

Cleavage of the triazepine ring of the hydroiodide salt of the *S*-methyl derivative (**8.130**; $R = R^1 = H$, $X = SMe$) can be effected with aqueous 48% hydrobromic acid: the starting diamine (**8.128**; $R = R^1 = H$) is regenerated in 88% yield.[57] The close similarity in the UV spectra of the triazepinobenzimidazoles (**8.130**; $X = H$, NH_2, SMe) implies that they exist in their 3-*H*-tautomeric forms. Data on the synthesis and physical constants of tricyclic compounds with two additional hetero atoms is collected in Table 8.17.

REFERENCES

1. O. Meth-Cohn and H. Suschitzky, *Adv. Heterocycl. Chem.*, **14,** 211 (1972).
2. M. D. Nair and R. Adams, *J. Am. Chem. Soc.*, **83,** 3518 (1961).
3. R. C. Perera and R. K. Smalley, *Chem. Commun.*, **1970,** 1458.
4. R. Garner and H. Suschitzky, *J. Chem. Soc.* (*C*), **1967,** 74.
5. D. P. Ainsworth and H. Suschitzky, *J. Chem. Soc.* (*C*), **1966,** 111.
6. J. Martin, O. Meth-Cohn, and H. Suschitzky, *Tetrahedron Lett.* **1973,** 4495.
7. O. Meth-Cohn, R. K. Smalley, and H. Suschitzky, *J. Chem. Soc.*, **1963,** 1666.
8. D. P. Ainsworth, O. Meth-Cohn, and H. Suschitzky, *J. Chem. Soc.* (*C*), **1968,** 923.
9. R. K. Grantham and O. Meth-Cohn, *J. Chem. Soc.* (*C*), **1969,** 1444.
10. R. K. Grantham, O. Meth-Cohn, and M. A. Nagui, *J. Chem. Soc.* (*C*), **1969,** 1438.
11. K. H. Saunders, *J. Chem. Soc.*, **1955,** 3275.
12. R. Garner, G. V. Garner, and H. Suschitzky, *J. Chem. Soc.* (*C*), **1970,** 825.
13. R. H. Smith and H. Suschitzky, *Tetrahedron*, **16,** 80 (1961).
14. J. I. G. Cadogan, *Quart. Rev.*, **22,** 222 (1968).
15. R. G. R. Bacon and S. D. Hamilton, *J. Chem. Soc. Perkin Trans. I*, **1974,** 1975.
16. R. Fielden, O. Meth-Cohn, D. Price, and H. Suschitzky, *Chem. Commun.*, **1969,** 772.
17. R. Fielden, O. Meth-Cohn, and H. Suschitzky, *J. Chem. Soc. Perkin Trans. I*, **1973,** 696.
18. R. Fielden, O. Meth-Cohn, and H. Suschitzky, *Tetrahedron Lett.*, **1970,** 1229.
19. H. Suschitzky and M. E. Sutton, *Tetrahedron*, **24,** 4581 (1968).
20. R. K. Grantham and O. Meth-Cohn, *J. Chem. Soc.* (*C*), **1969,** 70.
21. S. Takahashi and H. Kano, *Chem. Pharm. Bull.* (*Tokyo*), **14,** 1219 (1966).
22. R. C. DeSelms, *J. Org. Chem.*, **27,** 2165 (1962).
23. *Ger. Offen.*, 2,435,406 (1976); *Chem. Abstr.*, **84,** 164782s (1976).
24. R. J. Hayward and O. Meth-Cohn, *J. Chem. Soc. Perkin Trans. I*, **1975,** 212.
25. R. M. Acheson, M. W. Foxton, P. J. Abbot, and K. R. Mills, *J. Chem. Soc.* (*C*), **1967,** 882.
26. R. M. Acheson, M. W. Foxton, and G. R. Miller, *J. Chem. Soc.*, **1965,** 3200.
27. R. M. Achéson and M. S. Verlander, *J. Chem. Soc. Perkin Trans. I*, **1974,** 430.
28. R. M. Acheson and G. Procter, *J. Chem. Soc. Perkin Trans. I*, **1977,** 1924.
29. R. M. Acheson and N. F. Elmore, *Adv. Heterocycl. Chem.*, **23,** 265 (1978).
30. R. M. Acheson, G. Procter, and S. R. Critchley, *J. Chem. Soc. Chem. Commun.*, **1976,** 692.
31. R. M. Acheson, G. Procter, and S. R. Critchley, *Acta. Cryst.*, **B33,** 916 (1977).
32. R. M. Acheson and M. S. Verlander, *J. Chem. Soc. Perkin Trans. I*, **1972,** 1577.
33. R. M. Acheson and W. R. Tully, *J. Chem. Soc.* (*C*), **1968,** 1623.
34. A. Albini, G. F. Bettinetti, and S. Pietra, *Tetrahedron Lett.*, **1972,** 3657.
35. A. Albini, G. F. Bettinetti, and S. Pietra, *Gazz. Chim. Ital.*, **105,** 15 (1975).
36. A. Albini, A Barinotti, G. F. Bettinetti, and S. Pietra, *J. Chem. Soc. Perkin Trans. II*, **1977,** 238.
37. G. H. Beaven, E. R. Holiday, and E. A. Johnson, *Spectrochim. Acta*, **4,** 338 (1951).

38. R. M. Acheson, R. T. Aplin, and D. R. Harrison, *J. Chem. Soc.* (*C*), **1968,** 383.
39. J. Elguero, A. R. Katritzky, B. S. El-Osta, R. L. Harlow, and S. H. Simonsen, *J. Chem. Soc. Perkin Trans. I,* **1976,** 312.
40. R. C. Perera, R. K. Smalley, and L. G. Rogerson, *J. Chem. Soc.* (*C*), **1971,** 1348.
41. R. Fielden, O. Meth-Cohn, and H. Suschitzky, *Chem. Commun.*, **1970,** 1658.
42. R. Fielden, O. Meth-Cohn, and H. Suschitzky, *J. Chem. Soc. Perkin Trans. I,* **1973,** 705.
43. R. Fielden, O. Meth-Cohn, and H. Suschitzky, *J. Chem. Soc. Perkin Trans. I,* **1973,** 702.
44. M. M. Htay and O. Meth-Cohn, *Tetrahedron Lett.*, **1976,** 79.
45. H. Singh and S. Singh, *Ind. J. Chem.*, **13,** 323 (1975); *Chem. Abstr.*, **83,** 164142n (1975).
46. D. W. S. Latham, O. Meth-Cohn, and H. Suschitzky, *J. Chem. Soc. Chem. Commun.*, **1973,** 41.
47. D. W. S. Latham, O. Meth-Cohn, and H. Suschitzky, *J. Chem. Soc. Perkin Trans. I,* **1977,** 478.
48. *Brit. Patent,* 1,291,312 (1972); *Chem. Abstr.*, **78,** 43481k (1973).
49. *FR. Patent,* 2,102,857 (1972); through *Chem. Abstr.*, **77,** 164772 (1972).
50. *U.S. Patent,* 3,804,830 (1974); *Chem. Abstr.*, **81,** 3942s (1974).
51. W. Reid and F. Grüll, *Chem. Ber.*, **96,** 130 (1963).
52. *Hung. Teljes,* 10,396 (1975); through *Chem. Abstr.*, **84,** 135623k (1976).
53. *Ger. Offen.*, 2,420,108 (1974); through *Chem. Abstr.*, **82,** 140203m (1975).
54. J. A. Maynard, I. D. Rae, D. Rash, and J. M. Swan, *Austr. J. Chem.*, **24,**1873 (1971).
55. S. Ishiwata and Y. Shiokawa, *Chem. Pharm. Bull.* (*Tokyo*), **18,** 1245 (1970); *Chem. Abstr.*, **73,** 56076d (1970).
56. W. Schulze and G. Letsch, *Pharmazie,* **28,** 367 (1973); *Chem. Abstr.*, **79,** 78753y (1973).
57. B. Ágai, G. Doleschall, Gy. Hornyák, K. Lempert, and Gy. Simig, *Tetrahedron,* **32,** 839 (1976).

CHAPTER 9

Condensed Benzimidazoles Bridged Between N-1 and C-7

M. F. G. STEVENS

9.1. TRICYCLIC BENZIMIDAZOLES WITH NO ADDITIONAL HETEROATOM

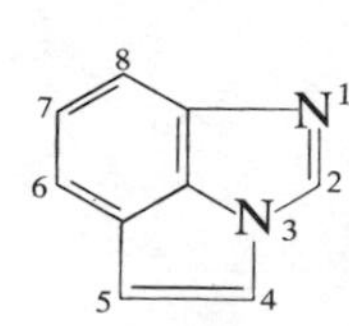
Pyrrolo[1,2,3-*cd*]benzimidazole

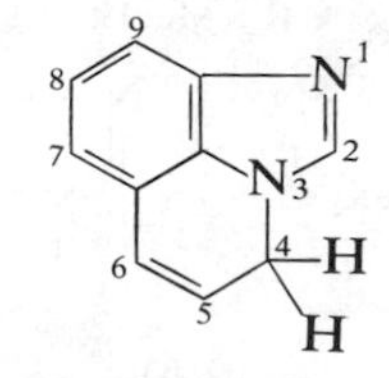
4H-Imidazo[4,5,1-*ij*]quinoline

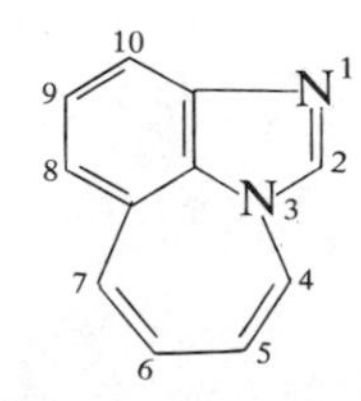
Imidazo[4,5,1-*jk*][1]benzazepine

9.1.1. Synthesis

Cyclisation of 8-aminoquinolines

Ring-closure of 8-amino-1,2,3,4-tetrahydroquinolines (**9.1**) with carboxylic acids, acid amides, anhydrides, or chlorides, continues to be the most general route to imidazo[4,5,1-*ij*]quinolines (**9.3**), and earlier work on this versatile synthesis has been summarized.[1] When a carboxylic acid is the reactant cyclization may often be achieved simply by heating the acid with the aminoquinoline alone, or in 4*N*-hydrochloric acid: alternatively (dihydrochloride) salts of the amine may be combined with the acid.[1–6] Decarboxylation of the acid is sometimes an interfering side reaction and the method, in general, fails with aromatic, heterocyclic, or unsaturated aliphatic acids; with the aromatic or heterocyclic derivatives the appropriate acid chloride is an effective substitute for the acid.[1] Occasionally the intermediate amides

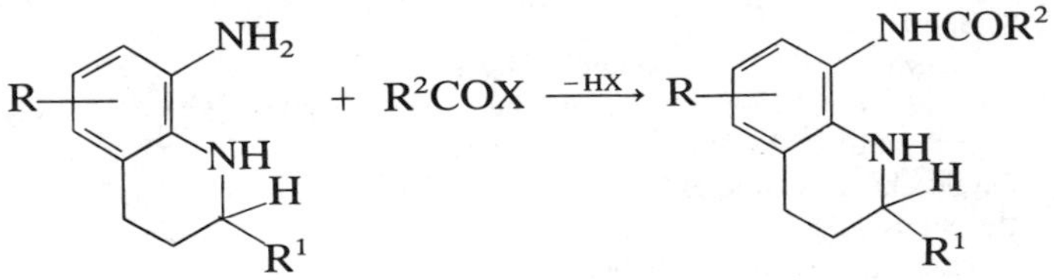

(**9.1**) + R^2COX $\xrightarrow{-HX}$ (**9.2**) $\xrightarrow{-H_2O}$ (**9.3**)

$X = OH, Cl, NH_2, OCOR^2$

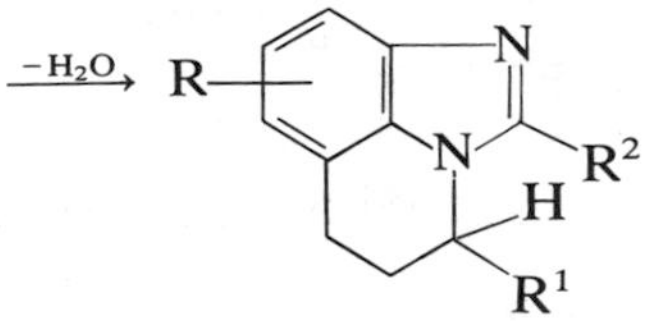

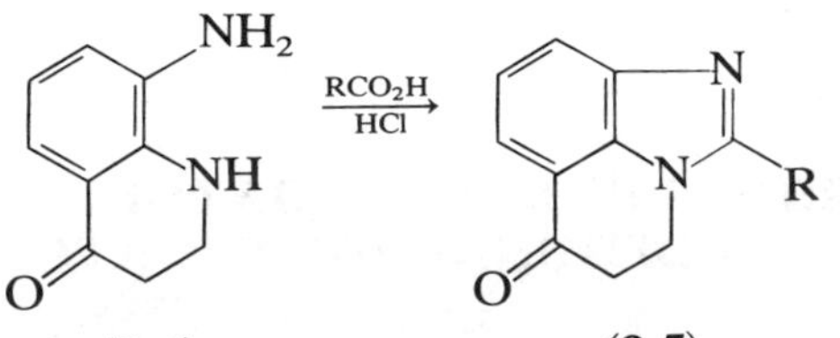

(**9.4**) $\xrightarrow[HCl]{RCO_2H}$ (**9.5**)

R = Me, Et

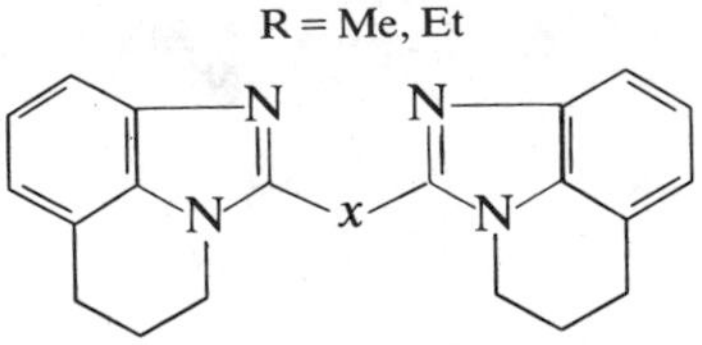

(**9.6**)

(**9.2**) can be isolated and separately cyclized with a range of dehydrating agents.[1] Cyclization of 8-amino-1,2,3,4-tetrahydroquinolin-4-ones (**9.4**) to imidazoquinolines (**9.5**) with acetic or propionic acids proceeds in only moderate yields, but reaction between 8-amino-1,2,3,4-tetrahydroquinoline and dibasic acids or their derivatives occasionally gives satisfactory yields of bisbenzimidazoles (**9.6**). Derivatives of imidazo[4,5,1-*ij*]quinoline prepared by the aforementioned methods are listed in Table 9.1.

A series of amides of 8-amino-6-methoxyquinoline (**9.7**) can be transformed to reduced tricyclic benzimidazoles (**9.8**) by catalytic hydrogenation at high pressure in acetic acid:[4,10] the formamide group surprisingly decreases the propensity of the pyridine ring toward reduction.[4] The bridged derivative (**9.9**) can similarly be prepared by the reduction–cyclization approach (Table 9.2). Representative 8-methoxybenzimidazoles show no antimalarial activity against mice infected with *Plasmodium berghei.*[10]

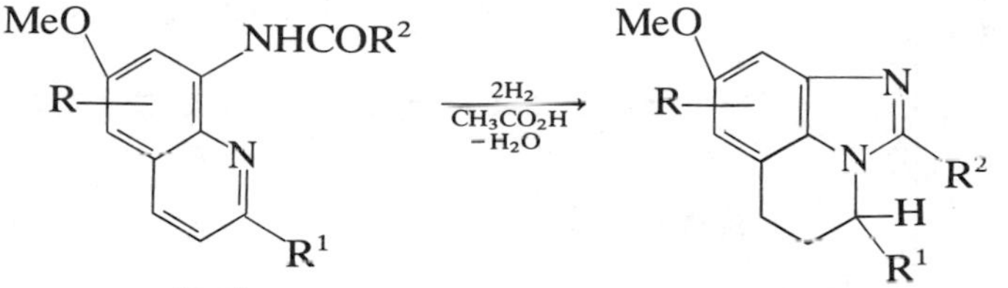

MeO NHCOR^2 R N R^1 $\xrightarrow[CH_3CO_2H \; -H_2O]{2H_2}$ MeO R N N R^2 H R^1

(**9.7**) (**9.8**)

MeO N N $(CH_2)_2$ N N OMe

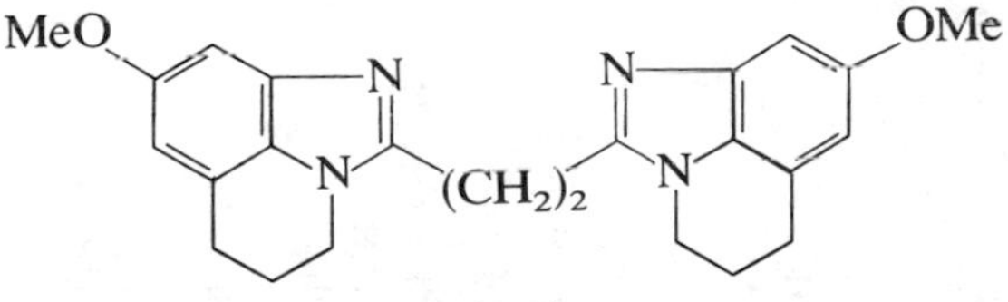

(**9.9**)

8-Amino-1,2,3,4-tetrahydroquinolines cyclize with cyanogen bromide to afford 2-aminobenzimidazoles (**9.10**).[1,2,4,11] Either phosgene[1,2,4] or urea[4] can act as the 1-carbon donor in the formation of benzimidazolin-2-one (**9.11**; X = O). Alternatively, synthesis of the 8-methoxybenzimidazolin-2-one (**9.11**; R = 8-OMe, X = O) can be effected by catalytic reduction of the urethan (**9.12**) in acetic acid, or hydrolytic cleavage of the urea (**9.13**) in 4*N*-hydrochloric acid.[4] Cyclization of the quinolone (**9.14**) to tricycle (**9.15**) proceeds smoothly at 190°.[13] Benzimidazolin-2-thiones (**9.11**; X = S) are formed from 8-amino-1,2,3,4-tetrahydroquinolines and carbon disulfide.[1,2,12] Cyclization of 5,7-diamino-2,3-dihydro-2-methylindole (**9.16**) with carbon disulfide furnishes a representative of the pyrrolo[1,2,3-*cd*]-benzimidazole ring system (**9.17**).[12]

TABLE 9.1. 5,6-DIHYDRO-4*H*-IMIDAZO[4,5,1-*ij*]QUINOLINES (**9.3**), (**9.5**), AND (**9.6**) FROM 8-AMINO-1,2,3,4-TETRAHYDROQUINOLINES AND CARBOXYLIC ACIDS OR THEIR DERIVATIVES

Compound	R	R^1	R^2	*x*	Cyclization method[a]	Yield (%)	m.p. of base (°C)	m.p. of picrate	Ref.
9.3	H	H	H	—	*A*	50	58–60	225	1
9.3	H	H	Me	—	*A*	35	128	237–238	1
							(methiodide: 249–250)		6
9.3	H	H	CH_2OH	—	*B*	74	185	216	1
9.3	H	H	CH_2SH	—	*B*	39	160–161	215–216 (decomp.)	1, 2
9.3	H	H	CF_3	—	*B*	83	103–104	—	2
9.3	H	H	$(CH_2)_2SH$	—	*B*	63	Oil	190	1, 2
9.3	H	H	Cyclohexyl	—	*C*	67	93–95 (hydrate)	183–184	1
9.3	H	H	Ph	—	*D*	100	80–82	196–197	1, 2
9.3	H	H	$4-NO_2C_6H_4$	—	*E*	43	179–180	214–215	1, 2
9.3	H	H	$3,4,5-Tri-OMeC_6H_2$	—	*D*	100	181–182	208–210	1, 2
9.3	H	H	$4-NH_2C_6H_4$	—	*B*	53	92–94	169	1, 2
9.3	H	H	4-Pyridyl	—	*F*	62	142.5	183–185	1, 2
							(hydrochloride: 256–258)		
9.3	H	H	2-Naphthyl	—	*F*	51	206–207	221–222	1, 2
							(hydrochloride: 250)		
9.3	H	H	3-Indolylmethyl	—	*B*	10	232–233	211–213	1, 2
9.3	H	H	$2-HO_2C\cdot C_6H_4$	—	*G*	30	314–315	—	2
9.3	7-OMe	H	H	—	*H*	80	127–128	—	3
9.3	8-OMe	H	H	—	*A*	48	89–91.5	—	4
9.3	9-Me	H	H	—	*H*	96	102–103	—	5
9.3	9-OPh	H	H	—	*H*	89	110–112	—	5
9.3	$9-NEt_2$	H	H	—	*H*	98	88–89	—	5
9.3	8-OMe	Me	H	—	*H*	86	115–116	—	3
9.3	8-Me	Me	H	—	*H*	92	106–107	—	3
9.3	8-Me	Me	CH_2OH	—	*H*	98	167–168	—	6

9.3	8-OMe	Me	CH_2OH	—	*H*	80	179–180	—	6
9.3	9-Me	H	CH_2OH	—	*H*	99	183–184	—	6
9.3	8-Me	Me	Me	—	*H*	80	154–155	—	6
9.3	9-Me	H	H	—	*H*	76	150–151	—	6
9.3	9-Piperidino	H	Me	—	*I*	100	157–158	—	7
9.5	Me	—	—	—	*J*	45	154–155	—	8
9.5	Et	—	—	—	*J*	—[b]	125–127	—	9
9.6	—	—	—	No bridge	*A*[c]	11	261–262	210	1
9.6	—	—	—	CH_2	*A*[d]	11	262–263	245–246	1
9.6	—	—	—	$(CH_2)_2$	*B*	73	256–258	>350	1
9.6	—	—	—	$(CH_2)_3$	*B*	61	198 (dihydrate)	268 (decomp.)	1
9.6	—	—	—	$(CH_2)_4$	*B*	70	217–218	>300	1
9.6	—	—	—	CH═CH	*B*	11	318–320 (dihydrate)	>360	1
9.6	—	—	—	CH_2OCH_2	*B*	61	171–172	254 (decomp.)	1
9.6	—	—	—	CH_2SCH_2	*B*	47	194 (hydrate)	247 (decomp.)	1
9.6	—	—	—	$(CH_2)_2S(CH_2)_2$	*K*	65	139.5–140	228 (decomp.)	1
9.6	—	—	—	$CH_2CH(OH)$	*B*	14	225–225.5 (hydrate)	>360	1

[a] *A* = amine (**9.1**) and carboxylic acid; reflux until effervescence ceases; *B* = amine (**9.1**) and carboxylic acid; reflux in 4*N*-hydrochloric acid; *C* = amide (**9.2**) with $POCl_3/P_2O_5$; reflux in xylene; *D* = amide (**9.2**) with P_2O_5; reflux in benzene; *E* = amide (**9.2**) with $POCl_3/PCl_5$; reflux; *F* = amine (**9.1**) and acid chloride; reflux in benzene–pyridine; *G* = amine (**9.1**) and phthalic anhydride; heat until effervescence ceases; *H* = dihydrochloride salt of amine (**9.1**) and carboxylic acid; reflux; *I* = amine (**9.1**) and acetic acid–acetic anhydride; *J* = amine (**9.4**) and carboxylic acid; reflux in hydrochloric acid; *K* = vacuum sublimation of the β-mercaptoethyl compound [**9.3**; $R = R^1 = H$, $R^2 = (CH_2)_2SH$].

[b] Yield not recorded.

[c] Ammonium oxalate not oxalic acid.

[d] Malonamide not malonic acid.

TABLE 9.2. SUBSTITUTED 8-METHOXY-5,6-DIHYDRO-4*H*-IMIDAZO[4,5,1-*ij*]-QUINOLINES (**9.8**) AND (**9.9**) FROM AMIDE DERIVATIVES OF 8-AMINO-6-METHOXYQUINOLINE (**9.7**)

Compound	R	R^1	R^2	Yield (%)	m.p. (°C)	Ref.
9.8	H	H	CH_2OH	21	190–192	4
9.8	H	Me	Me	48	140–142	4
9.8	H	H	CH_2NEt_2	53	78–81	4
9.8	H	H	$(CH_2)_2CO_2H$	31	221.5–223.5	4
9.8	H	H	$(CH_2)_2Me$	98	83–84	10
9.8	H	H	$(CH_2)_2CONEt_2$	64	73–74	10
9.8	H	H	$(CH_2)_{10}Me$	52	67.5–72.5	4
9.8	H	H	Ph	45	237–238	4
9.8	H	H	CH_2Ph	48	94.5–97	4
9.8	H	H	CH(OH)Ph	44	230.5–236	4
9.8	7-NH_2	H	Me	17.5	169–170	4
9.8	7-OMe	H	Me	13	161–163	4
9.9	—	—	—	43.5	239–240	4

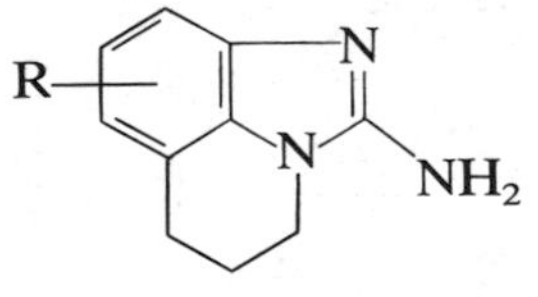

(**9.10**)

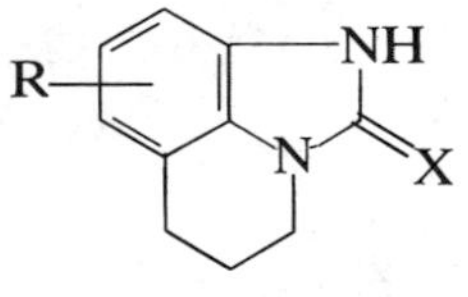

(**9.11**)
X = O,S

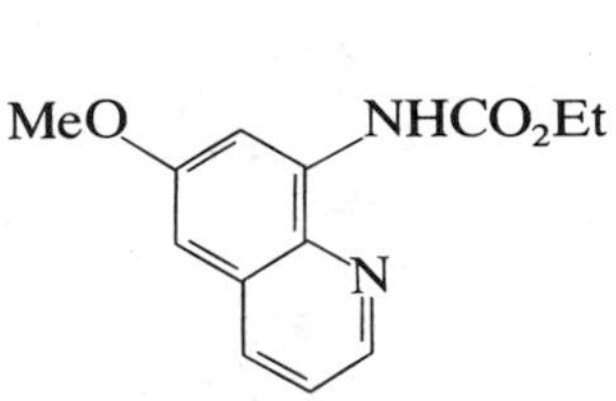

(**9.12**)

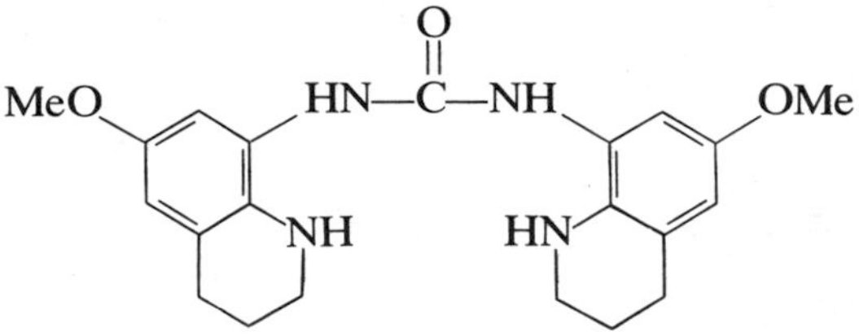

(**9.13**)

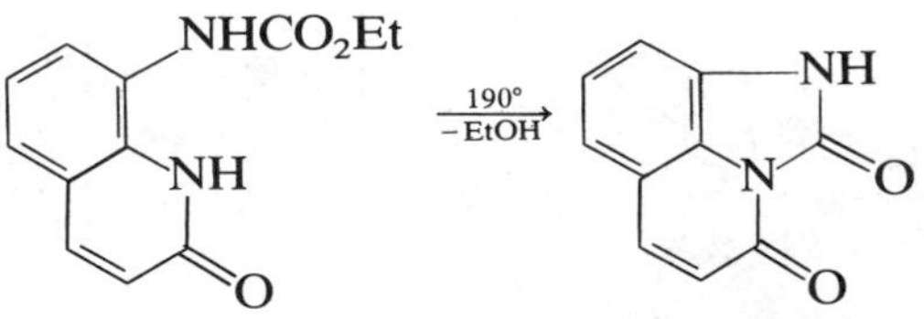

(**9.14**)

$\xrightarrow[-\text{EtOH}]{190°}$

(**9.15**)
93%

$\xrightarrow[\text{Pyridine}]{CS_2}$

(9.16) **(9.17)** 67%

Physical characteristics of 2-aminobenzimidazoles, benzimidazolin-2-ones, and 2-thiones prepared as above are recorded in Table 9.3 (see also section 9.1.3. "Reactions with Nucleophiles").

TABLE 9.3. IMIDAZO[4,5,1-*ij*]QUINOLINES AND A PYRROLO[1,2,3-*cd*]BENZIMIDAZOLE FROM 8-AMINO-1,2,3,4-TETRAHYDROQUINOLINES AND SUBSTITUTED 6-METHOXYQUINOLINES

Compound	R	X	Cyclization method[a]	Yield (%)	m.p. (°C)	Ref.
9.10	H	—	CNBr under N_2	40	201–202 (picrate: >320)	1, 2, 11
9.10	8-OMe	—	CNBr under N_2	39	229–230 (hydrobromide: 266–268)	4
9.11	H	O	ClCOCl in AcOH	75	213–214	1, 2
			H_2NCONH_2 at 140°[b]	97	210–211	3
9.11	8-OMe	O	ClCOCl in AcOH	49	234–235	4
			—[c]	76	234–235	4
			—[d]	5	234–235	4
9.11	H	S	CS_2 in EtOH	32	214.5–215.5	1, 2
9.11	8-OMe	S	CS_2 in pyridine	97	>280	12
9.15	—	—	—[e]	93	280–315 (decomp.)	13
9.17	—	—	CS_2 in pyridine[f]	67	210–212	12

[a] From 8-amino-1,2,3,4-tetrahydroquinoline(s) unless otherwise stated.
[b] From the dihydrochloride salt of 8-amino-1,2,3,4-tetrahydroquinoline.
[c] From urea (**9.13**) in 4*N*-hydrochloric acid.
[d] From urethan (**9.12**) and hydrogen over 20% platinum on carbon; in acetic acid.
[e] From urethan (**9.14**) at 190°.
[f] From (**9.16**).

Cyclization of N-(o-Acylamino)phenyl Heterocycles

Decomposition of the pyrrolidine (**9.18**) in polyphosphoric acid at 145–150° affords a product identical to the tricycle (**9.19**), independently prepared by cyclization of the diacetyltetrahydroquinoline (**9.20**), and different

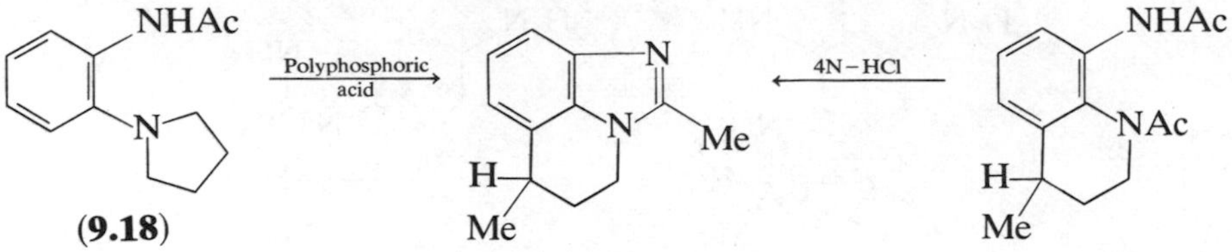

(**9.18**) (**9.19**) (**9.20**)

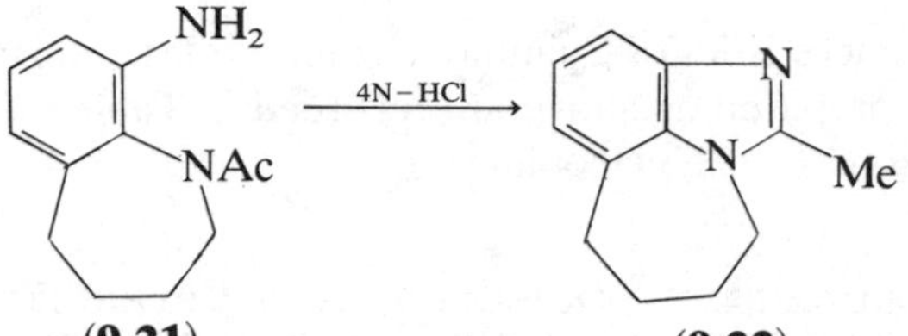

(**9.21**) (**9.22**) 50%

from the anticipated imidazobenzazepine (**9.22**), itself readily prepared from (**9.21**). This decomposition embodies features of the "*t*-amino effect"[15] (see Chapter 8, section 8.1.1 "Cyclization of *ortho*-Substituted *N*-Aryl Heterocycles") and occurs with a range of anilides substituted in the ortho position with heteroalicyclic groups[14,16]: the mechanism is summarized in Scheme 9.1.

The first step is thought to be conversion of the anilide (**9.23**) to iminopolyphosphate (**9.24**), which cyclizes to the spiro-benzimidazolium salt (**9.25**). Fission of the salt yields the polyphosphate (**9.27**) via the (incipient) carbonium ion (**9.26**). Elimination of polyphosphoric acid then generates an olefin (**9.28**), which under the prevailing Friedel–Crafts conditions cyclizes at the *penultimate* carbon atom of the chain to yield the observed benzimidazoles (**9.29**).

Yields of cyclic products are highest (Table 9.4) when R = H, Me or Ph; $n = 5$, but no cyclization occurs with electron-withdrawing R substituents (CF_3 or $ClCH_2$) or bulky groups (CMe_3). Similarly, cyclization is completely inhibited when the heteroalicyclic ring is morpholine, piperazine, or an *N*-substituted piperazine, and the rate of disappearance of starting material (**9.23**) is suppressed as the ring size increases ($n = >5$).

Cyclization of the hepta- or octamethyleneimine homologs (**9.23**); R = Me, $n = 7$ or 8) does not involve the penultimate carbon of the polymethylene chain: rather, cyclization only takes place after the primary carbonium ion has rearranged to contain a tertiary carbonium ion.[16] Thus the heptamethyleneimine gave product (**9.30**) in 35% yield and the octamethyleneimine analog gave a mixture of at least three components.

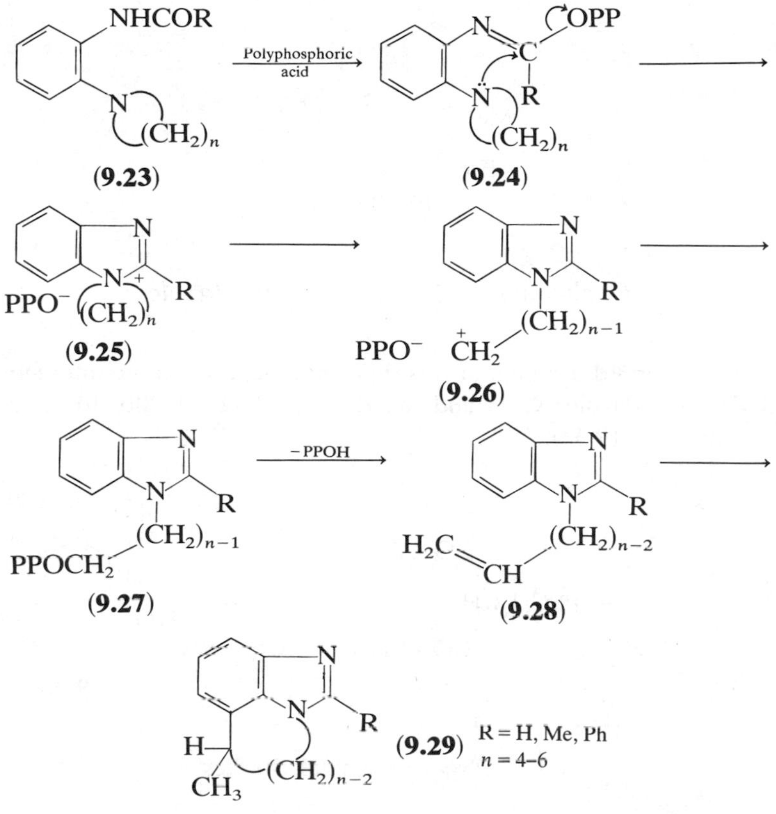

(9.23) → Polyphosphoric acid → (9.24) → (9.25) → (9.26) → (9.27) → −PPOH → (9.28) → (9.29)

(9.29) R = H, Me, Ph
n = 4–6

Scheme 9.1

TABLE 9.4. CONDENSED BENZIMIDAZOLES (**9.29**) FORMED FROM *N*-(*o*-ACYLAMINO)PHENYL HETEROCYCLES (**9.23**) IN POLYPHOSPHORIC ACID AT 145–150°

Starting material (**9.23**) R	Starting material (**9.23**) *n*	Heating time (hr)	Product (**9.29**) R	Product (**9.29**) *n*	Yield (%)	m.p. or b.p. of base [(°C)/mm]	m.p. of' derivative (°C)	Ref.
H	4	0.5	H	4	27	(125/0.5)	175[a]	14
Me	4	6	Me	4	21	140	219[a]	14
Ph	4	0.5	Ph	4	10	—	180[a]	14
H	5	1	H	5	58	(157/2)	220[a], 150[b]	14
Me	5	1	Me	5	90	117	195[a], 163[b]	14
Ph	5	1	Ph	5	55	130	224[b]	14
H	6	1	H	6	10	—	214[a]	14
Me	6	1	Me	6	35	120	165[a]	14
Ph	6	1	Ph	6	14	102	—	14
Me	7	1.5	—[c]		35	96	—	16

[a] Picrate.
[b] Methiodide.
[c] Compound (**9.30**).

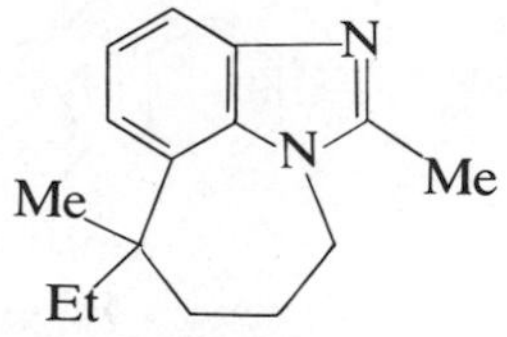

(**9.30**)

Cyclization of 2-Methylbenzimidazole

The only recorded example of this type of annelation is the interaction of 2-methylbenzimidazole (**9.31**) and the diester (**9.32**) at 200° to afford the imidazoquinolone (**9.33**).[17]

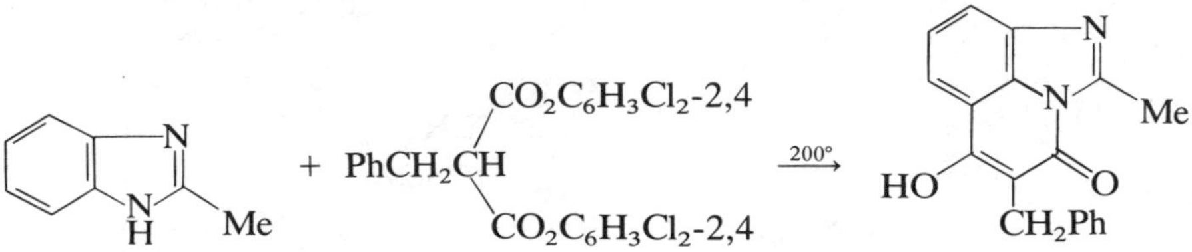

(**9.31**) + $PhCH_2CH(CO_2C_6H_3Cl_2\text{-}2,4)_2$ (**9.32**) $\xrightarrow{200°}$ (**9.33**) 34%

9.1.2. Physicochemical Studies

Absorptions at 1690 and 1215 cm^{-1} in the IR spectra of the products from 8-amino-1,2,3,4-tetrahydroquinoline and phosgene or carbon disulfide[1,2] confirm that the cyclic compounds exist in the 2-one or 2-thione tautomeric forms (**9.34**; X = O or S), respectively.[11] The related derivative cyclized with cyanogen bromide, although originally considered[11] to be the imine (**9.34**; X = NH) on the basis of misinterpreted IR and UV spectral comparisons with model compounds, is now known to exist predominantly as the 2-amino tautomer (**9.35**; X = NH).[3,18] In chloroform solution the IR spectrum shows bands at 3504 and 3412 cm^{-1} for the asymmetrical and symmetrical NH stretching modes: after partial deuteration a new band emerges at 3458 cm^{-1}. This new band is an uncoupled NH stretching

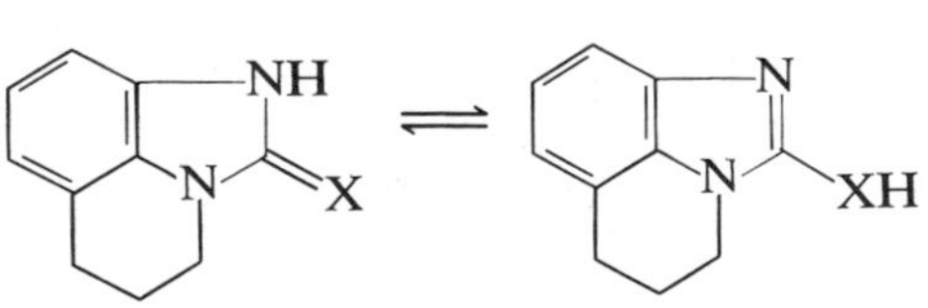

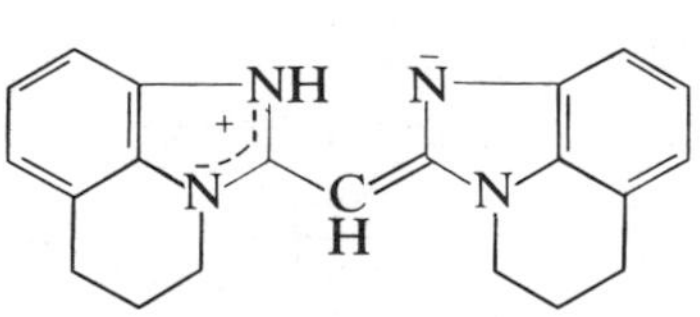

(**9.34**) ⇌ (**9.35**) (**9.36**)

X = O, S, NH

absorption arising from the NHD group, and such a triplet cannot be explained by invoking a partially deuterated imine contribution.[18]

The UV spectra of 2-substituted 5,6-dihydro-4*H*-imidazo[4,5,1-*ij*]-quinolines closely resemble the spectra of 1,2-dialkylbenzimidazoles. The only notable exception is the spectrum of the methylene-bridged benzimidazole (**9.6**; X=CH_2), which has an additional absorption in the visible region indicative of conjugation through the methylene group (**9.36**).[1]

The UV spectra of representative benzimidazoles bridged between N-1 and C-7 are recorded in Table 9.5.

Reports of ^1H NMR spectra of bridged benzimidazoles are rare,[8,10] and only in those compounds (**9.29**) formed by cyclization of *N*-(*o*-acylamino)phenyl heterocycles (**9.23**), which involve a contraction of the polymethylene chain, has NMR been important in structure determination.[14,16]

TABLE 9.5. ULTRAVIOLET SPECTRA OF BENZIMIDAZOLES BRIDGED BETWEEN N-1 AND C-7

Compound	R	R^1	R^2	X	*x* (or *n*)	Solvent[a]	λ_{max} (log ε)	Ref.
9.3	H	H	H	—	—	*A*	257(3.78) 274(3.65) 283.1(3.59)	1
9.3	H	H	Me	—	—	*A*	255.2(3.73) 273.6(3.63) 282.4(3.62)	1
9.3	H	H	CH_2OH	—	—	*A*	259(3.83) 276(3.77) 285(3.63)	1
9.3	H	H	Ph	—	—	*A*	240(4.19) 290(4.22)	1
9.3	H	H	4-Pyridyl	—	—	*A*	250.7(3.95) 306(4.21)	1
9.5	Me	—	—	—	—	*A*	219(4.01) 294(3.84) 322(3.81)	8
9.6	—	—	—	—	CH_2	*A*	258(4.14) 276(4.11) 284.8(4.09) 360(3.12)	1
9.10	H	—	—	—	—	*B*	257[b] 288	3
9.11	H	—	—	O	—	*A*	225–234(3.82) 284(3.76)	1
9.11	H	—	—	S	—	*A*	225(4.27) 247.6(4.26) 304.8(4.47)	1
9.11	8-NH_2	—	—	S	—	*A*	226(3.90) 293(3.91) 322(3.86)	12
9.17	—	—	—	—	—	*A*	223[b] 262 335	12
9.29	H	—	—	—	(4)	*C*[c]	212(4.13) 254(3.79) 275(3.70) 282(3.70)	14
9.29	Me	—	—	—	(4)	*C*	214(4.77) 257(3.80) 275(3.70) 283(3.69)	14
9.29	Me	—	—	—	(5)	*C*	213(4.59) 257(3.87) 276(3.68) 284(3.62)	14
9.29	Ph	—	—	—	(5)	*C*	215(4.76) 244(4.12) 291(4.12)	14
9.29	Me	—	—	—	(6)	*C*	215(4.51) 259(3.79) 276(3.67) 284(3.63)	14
1,2-Dimethyl benzimidazole			—	—	—	*D*	250.5(3.79) 274(3.78) 280(3.80)	19

[a] A = 95% EtOH; *B* = Dioxan; *C* = MeOH; *D* = 0.01 *N*–KOH.
[b] Log ε not recorded.
[c] Measured as picrate salt against a blank of picric acid of the same molar concentration.

The pK_a values for 9-substituted imidazo[4,5,1-*ij*]quinolines (**9.3**; R = OPh, Me, NEt_2, $R^1 = R^2 = H$) measured in 5% aqueous ethanol[5] are 5.0, 6.03, and 6.90, respectively. These values lie in the range where the Chichibabin reaction is pòssible (see section 9.1.3 "Reactions with Nucleophiles").

9.1.3. Reactions

Reactions with Electrophiles

Nitration of unsubstituted 5,6-dihydro-4*H*-imidazo[4,5,1-*ij*]quinoline with 50% nitric acid at −15° affords a mixture (81%) of three mononitro derivatives.[5] Controlled nitration of 9-methylimidazoquinolines (**9.37**); R = H, Me) in nitric–sulfuric acid at −10% affords 8-nitro compounds (**9.38**; R = H, Me, $R^1 = 8\text{-}NO_2$), whereas more vigorous treatment at 80° yields the 7,8-dinitro derivatives (see Table 9.6). The *o*-diamine (**9.39**) prepared by stannous chloride reduction of the corresponding dinitroimidazoquinoline can be transformed to the tetracyclic imidazole (**9.40**; X = CH) or triazole (**9.40**; X = N) with formic or nitrous acids, respectively.[5]

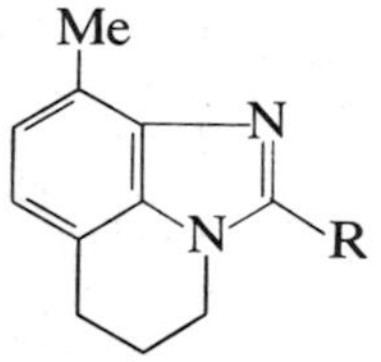

(**9.37**)

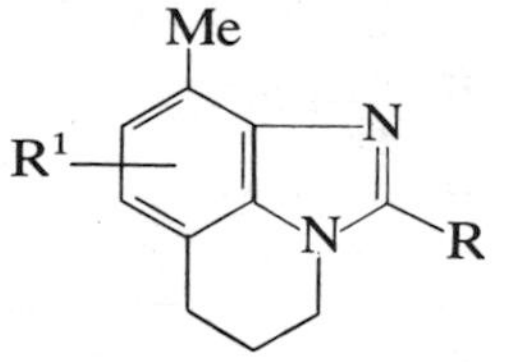

(**9.38**)

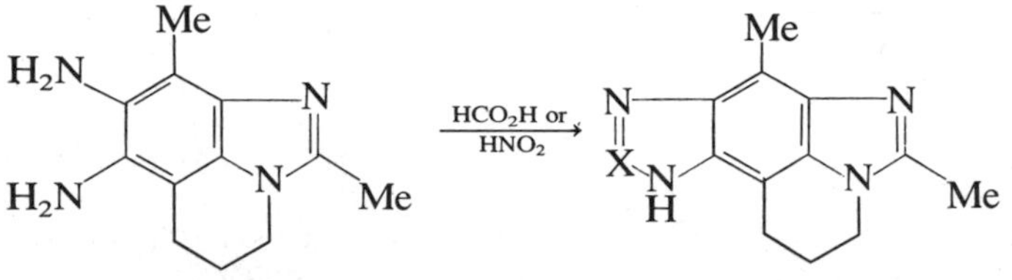

(**9.39**)

HCO_2H or HNO_2

(**9.40**)
X = CH (100%)
X = N (51%)

Quaternization of imidazo[4,5,1-*ij*]quinolines[3,6,14] or benzimidazoles bearing a larger methylene bridge,[14] with methyl iodide, affords the N-1 methiodides.

TABLE 9.6. NITRO- AND AMINO-SUBSTITUTED 9-METHYL-5,6-DIHYDRO-4*H*-IMIDAZO[4,5,1-*ij*]-QUINOLINES (**9.38**)[a]

R	R^1	Yield (%)	m.p. (°C)
H	8-NO_2	80	232–233
H	7,8-Di-NO_2	88.5	202–204
Me	8–NO_2	83	164–165
Me	8-NH_2	97	142–143
Me	7,8-Di-NO_2	67	209–210
Me	7,8-Di-NH_2	90	151–152

[a] From Ref. 5.

Reactions with Nucleophiles

Tricyclic benzimidazolin-2-ones (**9.41**); R = H, OMe) are transformed to reactive chloro compounds (**9.42**) with phosphorus oxychloride[1–4]: the chloro group can be displaced by sodium methoxide to yield ether (**9.43**; R = H)[1,2] or by a range of secondary and tertiary amines[3,4] to provide amines (**9.44**) (Table 9.7). The chloro group of the quaternary iodide (**9.45**) is particularly reactive toward nucleophiles:[3] it is displaced with (aqueous) ammonia or methylamine in absolute ethanol to yield amines (**9.46**; R = H, Me), which are isolated as hydroiodide salts; with aqueous methylamine the 1-methylbenzimidazolin-2-one (**9.47**) is formed. The UV spectra of imines (**9.46**) differ qualitatively from the spectra of representative 2-aminobenzimidazoles (**9.44**).[3]

R NH O POCl$_3$ → R N Cl NaOMe (R = H) → R N OMe

(**9.41**) (**9.42**) (**9.43**)

↓ R^1R^2NH

R N NR^1R^2

(**9.44**)

R = H, OMe
R^1R^2 = H,Me; Me,Me; C_5H_{10}

Me, $\overset{+}{N}$, I^-, N, Cl $\xrightarrow{RNH_2}$ Me, N, N, NR

(**9.45**) (**9.46**)

R = H, Me

$\xrightarrow{H_2O}$ Me, N, N, O

(**9.47**)

TABLE 9.7. PRODUCTS OF NUCLEOPHILIC SUBSTITUTION REACTIONS OF SUBSTITUTED 5,6-DIHYDRO-4*H*-IMIDAZO[4,5,1-*ij*]QUINOLINES

Compound	R	R^1	R^2	Yield (%)	m.p. (°C)	m.p. of derivative (°C)	Ref.
9.42	H	—	—	60	75–76	175 (decomp.)[a]	1, 2
				82	75–76		3
9.42	OMe	—	—	68	132–134	—	4
9.43	H	—	—	100	Oil	154–155[a], 212–214[b]	1, 2
9.44	H	H	Me	92	222–223	—	3
9.44	H	Me	Me	94	63–64[c]	—	3
9.44	OMe	Me	Me	22	189–191	—	4
9.44	OMe	$—(CH_2)_5—$		57	71–73	—	4
9.46	H	—	—	83	91–92	>305[d]	3
9.46	Me	—	—	61	93–95	284–285[d]	3
9.47	—	—	—	55	120–121	—	3
9.49	H	H	—	47	199–200	151–153[e]	20
9.49	8-OMe	H	—	75	229–230	210–211[e]	20
9.49	7-OMe	H	—	75	234–235	—	3
9.49	8-OMe	H	—	75	229–230	210–211[e]	20
9.49	8-OMe	Me	—	50	227–228	—	3
9.49	8-OMe	Me	—	40	234–235	—	3
9.49	9-Me	H	—	70	258–259 (decomp.)	—	5
9.49	9-OPh	H	—	60	249–250	—	5
9.49	$9-NEt_2$	H	—	45	124–125	—	5
9.50	9-Me	H	—	58	213–214	—	5
9.50	$9-NEt_2$	H	—	33	153–154	—	5
9.51	H	H	—	85	214–215	—	5
9.51	7-OMe	H	—	91	223–224	—	5
9.51	8-Me	Me	—	90	210–211	—	5

[a] Picrate.
[b] Hydrochloride.
[c] Hydrate.
[d] Hydroiodide.
[e] Nitrobenzylidene derivative.

Despite the generally accepted view that only π-deficient heterocycles undergo the Chichibabin reaction, a range of imidazoquinolines (**9.48**) react with sodamide in dimethylaniline at 120°.[3,5,20] Nucleophilic substitution in the 2-position can also be achieved by fusing imidazoquinolines with potassium hydroxide or sulfur.[5] Yields of 2-aminobenzimidazoles (**9.49**), benzimidazolin-2-ones (**9.50**) and 2-thiones (**9.51**) formed by the latter methods (Table 9.7) compare favorably with those from 8-amino-1,2,3,4-tetrahydroquinolines and 1-carbon donors (see section 9.1.1).

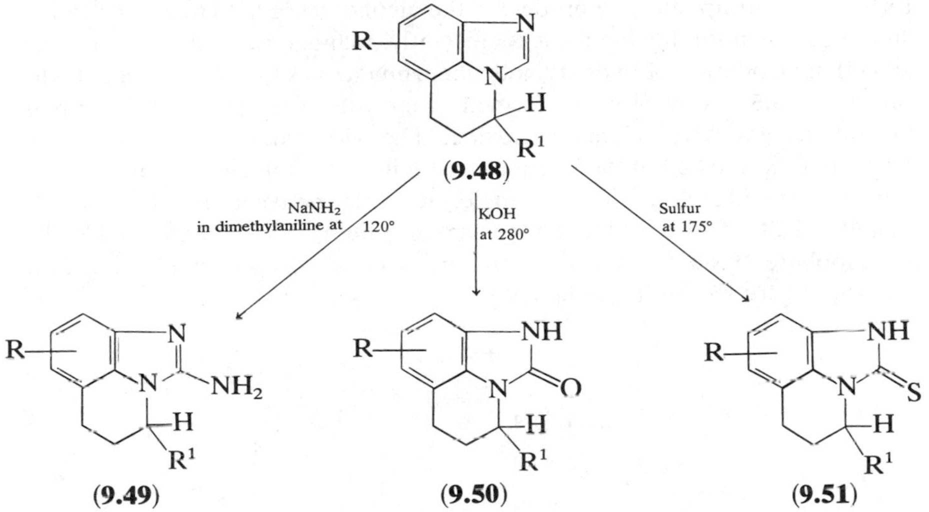

(**9.48**)

$NaNH_2$ in dimethylaniline at 120°; KOH at 280°; Sulfur at 175°

(**9.49**) (**9.50**) (**9.51**)

Hydrolysis of the 1,2,7-trimethylimidazobenzazepinium iodide (**9.52**) in aqueous alkali affords a quantitative yield of the ring-opened product (**9.53**).[14]

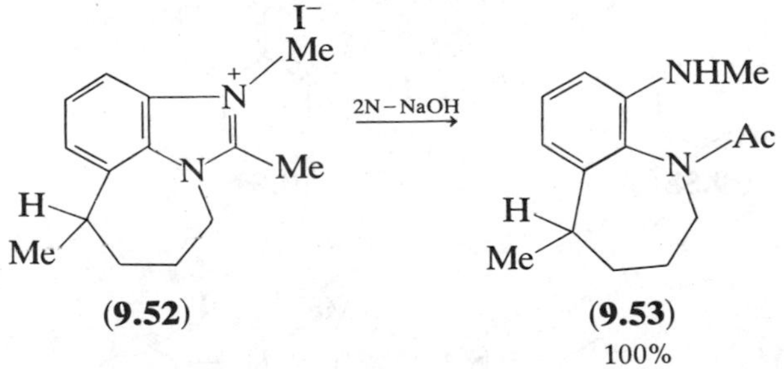

2N−NaOH

(**9.52**) (**9.53**) 100%

Oxidation

Although the imidazobenzazepine (**9.54**) has a potentially vulnerable 2-methyl group, oxidation with potassium permanganate only involves the

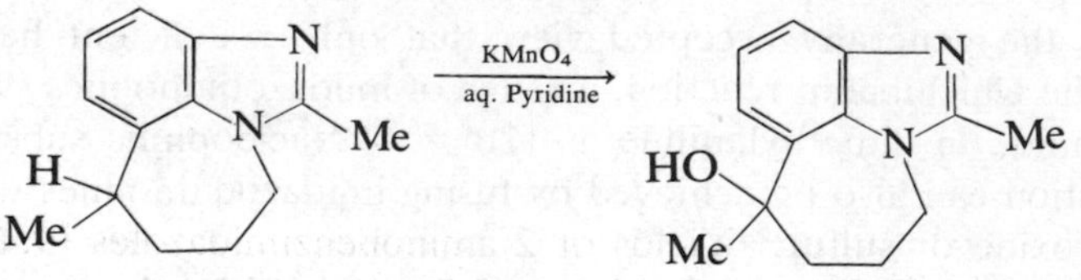

KMnO$_4$, aq. Pyridine

(9.54) → (9.55) 25%

tertiary CH group, and the product is the alcohol (**9.55**).[14] However, freshly sublimed selenium dioxide or, less efficiently, manganese dioxide effects the side-chain oxidation of hydroxymethylbenzimidazoles (**9.56**).[6] The aldehydic products (**9.57**) can also be formed from the corresponding 2-methylbenzimidazoles with selenium dioxide. The aldehydes form oximes and participate in crossed aldol reactions with either acetophenone[6] or other 2-methylbenzimidazoles.[21] For example, reaction between the tricyclic aldehyde (**9.58**) and the dimethylbenzimidazolium iodide (**9.59**) yields the luminophore (**9.60**).[6,21] Physical properties of aldehydes (**9.57**) and their derivatives are shown in Table 9.8.

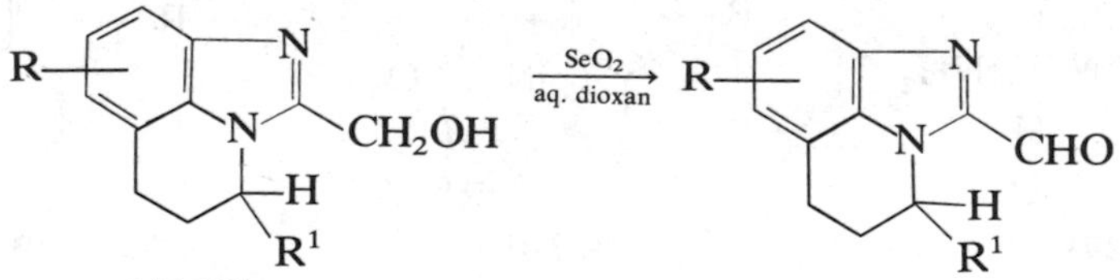

SeO_2, aq. dioxan

(9.56) → (9.57)

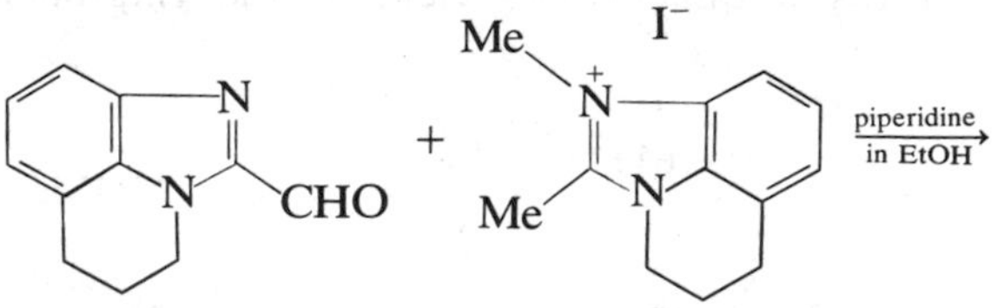

(9.58) + (9.59) —piperidine in EtOH→

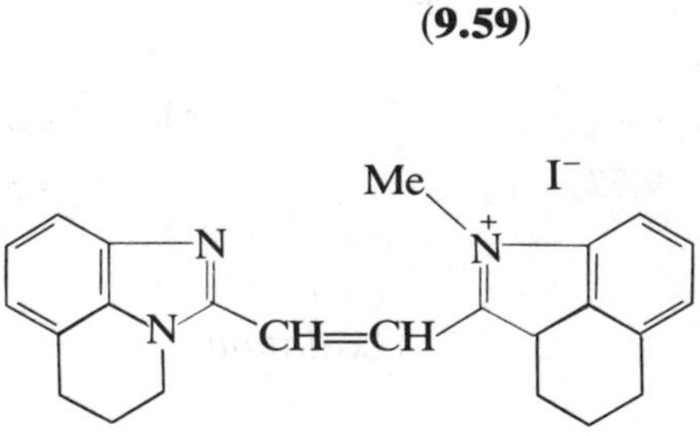

(9.60) 66%

TABLE 9.8. ALDEHYDES (**9.57**) OF THE IMIDAZO[4,5,1-*ij*]-QUINOLINE SERIES PREPARED BY SELENIUM DIOXIDE OXIDATION OF 2-HYDROXYMETHYL-BENZIMIDAZOLES (**9.56**)[a]

R	R¹	Yield (%)	m.p. (°C)	m.p. of derivatives (°C)
H	H	81	179–180	255 (decomp.)[b], 208–209,[c] 285–287 (decomp.)[d]
8-Me	Me	66	102–103	163–164 (decomp.)[b]
8-OMe	Me	58	136–138	140–141 (decomp.)[b]
9-Me	H	62	111–112	156–157 (decomp.)[b], 179–180[c]

[a] From Ref. 6.
[b] Oxime.
[c] Derivative with acetophenone.
[d] Compound (**9.60**).

Reduction

1-Methyl-5,6-dihydro-4*H*-imidazo[4,5,1-*ij*]quinolinium iodide (**9.61**; X = I) yields a benzimidazoline (**9.62**) on treatment with sodium borohydride:[22] reoxidation with moist silver nitrate regenerates the quaternary salt [isolated as the perchlorate (**9.61**; X = ClO_4)]. The benzimidazoline (**9.62**) is a powerful reducing agent and the main product from its reaction with benzoyl chloride is the quaternary chloride (**9.61**; X = Cl) (65%) together with the tetrahydroquinoline (**9.63**; R = Bz): in the process the benzoyl chloride is reduced to benzaldehyde. Oxidation of the benzimidazoline with sulfur or selenium affords high yields of the 2-thione or 2-selenone (**9.64**; X = S, Se). Interaction of benzimidazoline (**9.62**) and its 1-methiodide salt leads to disproportionation and the formation of the benzimidazolium iodide (**9.61**;

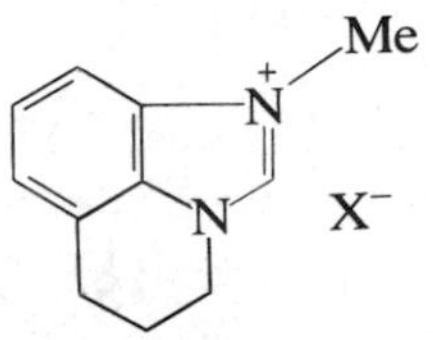

Me N⁺ N X⁻

(**9.61**)

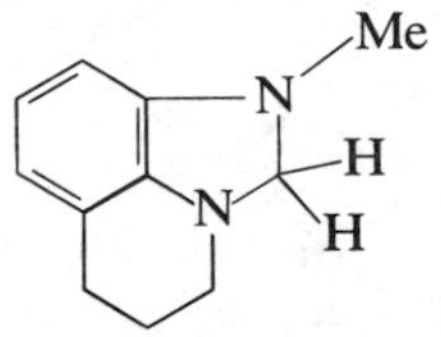

Me N H N H

(**9.62**)

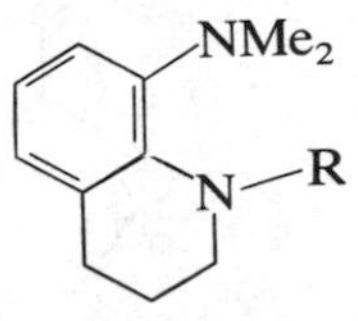

NMe_2 N R

(**9.63**)

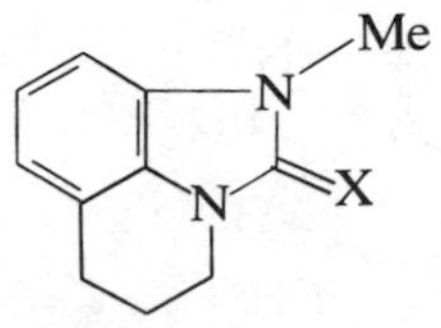

Me N N X

(**9.64**)

TABLE 9.9. 1-METHYL-5,6-DIHYDRO-4*H*-IMIDAZO-[4,5,1-*ij*]QUINOLINIUM SALTS (**9.61**), BENZIMIDAZOLINE (**9.62**), AND ITS DERIVATIVES[a]

Compound	X	Yield (%)	m.p. or b.p. [(°C)/mm]
9.61	I	95[b]	195.5–197
9.61	ClO_4	70[c]	170.5–171.5
9.61	Cl	65[d]	—[e]
9.62	—	75	(128/3) (methiodide: 122–126) (hydrochloride: 79–86)
9.64	S	78	157–159.5
9.64	Se	91	185–189

[a] From Ref. 22.
[b] Prepared from 5,6-dihydro-4*H*-imidazo[4,5,1-*ij*]quinoline and methyl iodide.
[c] From the iodide (**9.61**; X = I) and sodium perchlorate.
[d] From benzimidazoline (**9.62**) and benzoyl chloride.
[e] Melting point not recorded.

X = I) (92%) and a trimethyltetrahydroquinoline (**9.63**; R = Me).[22] Physical properties of the derivatives **9.61**, **9.62**, and **9.64** are listed in Table 9.9.

Surprisingly the UV spectrum of the benzimidazoline (**9.62**) shows more conjugation [λ_{max} 223, 268, and 308–310 nm (log ε 4.44, 3.67, and 3.60, respectively] than the aromatic benzimidazoles (Table 9.5).

Diborane has been employed for the reduction of the side-chain *NN*-diethyl carboxamide function of (**9.8**; $R = R^1 = H$, $R^2 = (CH_2)_2CONEt_2$) without concomitant reduction of the imidazole ring.[10]

9.2. TRICYCLIC BENZIMIDAZOLES WITH ONE ADDITIONAL HETEROATOM

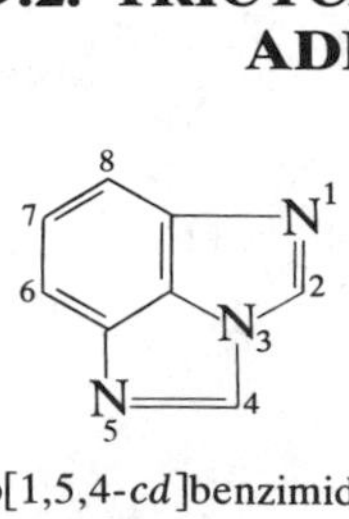

Imidazo[1,5,4-*cd*]benzimidazole

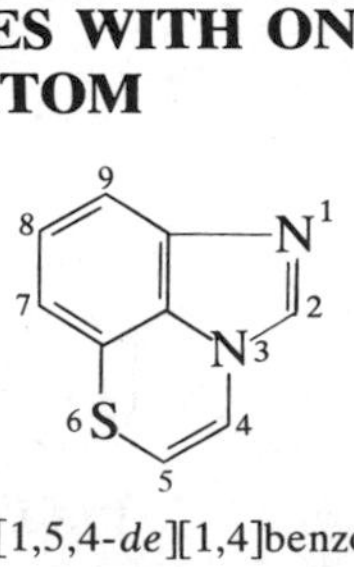

Imidazo[1,5,4-*de*][1,4]benzothiazine

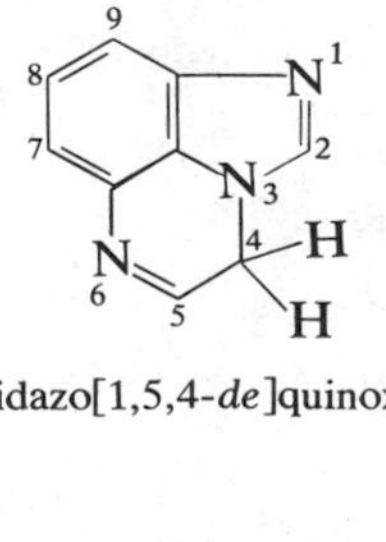

4*H*-Imidazo[1,5,4-*de*]quinoxaline

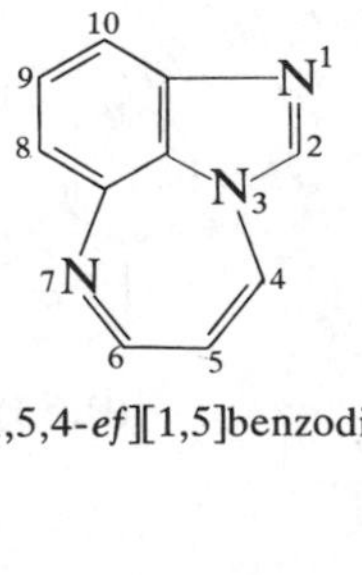

Imidazo[1,5,4-*ef*][1,5]benzodiazepine

9.2.1. Synthesis

Cyclization of 5-Aminoquinoxalines and Other Nonbenzimidazole Heterocycles

In its simplest form this type of cyclization is exemplified by the reaction between the dihydrobromide salts of triamines (**9.65**) and formic acid: the product is either an imidazo[1,5,4-*de*]quinoxaline (**9.66**; $n = 2$) or imidazo[1,5,4-*ef*][1,5]benzodiazepine (**9.66**; $n = 3$).[23] More ambitious syntheses can be achieved by reacting quinoxalinones (**9.67**) with carboxylic acids or urea to afford imidazoquinoxalines (**9.68**) and (**9.69**) in good yields.[24] A general alternative synthesis of type (**9.68**; $R^2 = H$) employs the dinitroanilines (**9.70**) as starting materials. These cyclize directly to imidazoquinoxalines on catalytic hydrogenation in formic acid. Di- and tricyclic hydroxamic acids (**9.71**) and (**9.72**) are intermediates in the latter reaction: one such intermediate (**9.71**; R = Cl, $R^1 = H$) cyclizes with loss of the chloro group.[24]

HCO_2H

(**9.65**) (**9.66**)

$n = 2$ or 3

R^2CO_2H

(**9.67**) (**9.68**)

urea

(**9.69**)

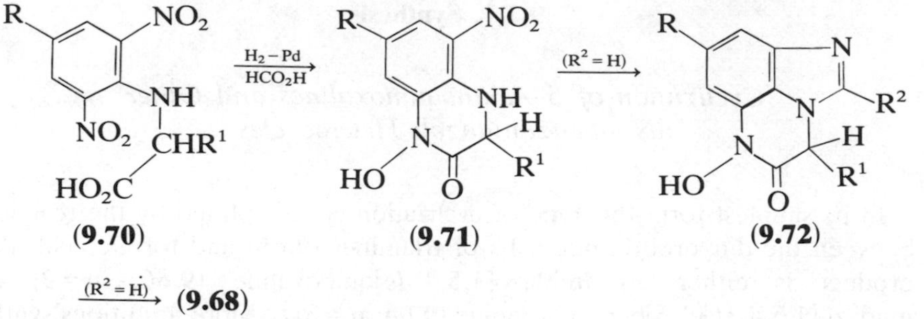

(9.70) $\xrightarrow[HCO_2H]{H_2-Pd}$ (9.71) $\xrightarrow{(R^2 = H)}$ (9.72)

$\xrightarrow{(R^2 = H)}$ (9.68)

Both reduction and cyclization can be achieved in one step when the aromatic quinoxalines (**9.73**) are submitted to high-pressure catalytic hydrogenation in acetic acid.[25] The nature of the catalytic reducing surface implies a cis arrangement of substituents in the reduced pyrazine ring, and this is confirmed by NMR spectroscopy (see section 9.2.2). When hydrogenation of quinoxaline (**9.73**; $R = R^2 = H$, $R^1 = Me$) is conducted in formic acid the product is the 6-formylimidazoquinoxaline (**9.75**). Physical constants of derivatives of the type (**9.66**, **9.68**, **9.69**, **9.74**, and **9.75**) are collected in Table 9.10).

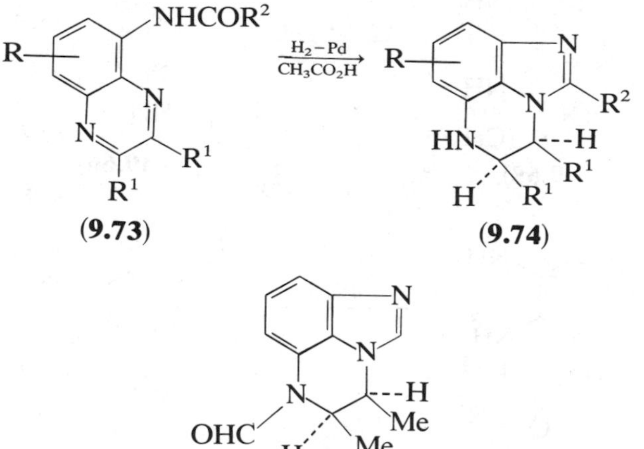

(9.73) $\xrightarrow[CH_3CO_2H]{H_2-Pd}$ (9.74)

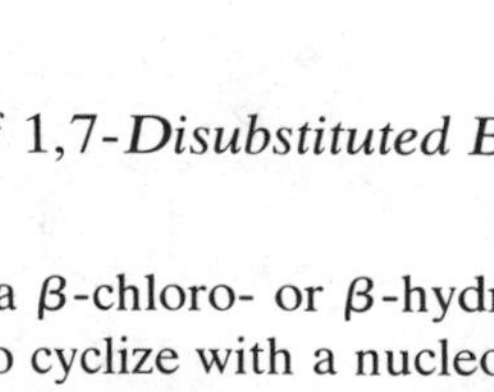

(9.75)

Cyclization of 1,7-*Disubstituted Benzimidazoles*

Benzimidazoles bearing a β-chloro- or β-hydroxy-ethyl substituent in the l-position can be induced to cyclize with a nucleophilic group at C-7. Hence, 1-(β-chloroethyl)-7-mercaptobenzimidazole (**9.76**) cyclizes in the presence of potassium hydroxide to yield the sole example of the imidazo[1,5,4-*de*]-[1,4]benzothiazine ring system (**9.77**).[23] This reaction can be extended to the

TABLE 9.10. IMIDAZO[1,5,4-*de*]QUINOXALINES AND AN IMIDAZO-[1,5,4-*ef*][1,5]BENZODIAZEPINE PREPARED FROM 5-AMINO-QUINOXALINES AND OTHER AMINO HETEROCYCLES

Compound	R	R^1	R^2	*n*	Yield (%)	m.p. (°C)	Ref.
9.66	—	—	—	2	70	162.5–166	23
9.66	—	—	—	3	59	129–132	23
9.68	H	H	H	—	83	288	24
9.68	H	H	Me	—	41	255	24
9.68	H	Me	H	—	69	300	24
9.68	Cl	Me	H	—	84	323	24
9.68	H	H	Ph	—	51	277	24
9.68	H	Me	Me	—	49	288	24
9.68	H	Me	Ph	—	16	216	24
9.69	H	H	—	—	80	250	24
9.69	H	Me	—	—	65	270–272	24
9.74	H	Me	H	—	93	150	25
9.74	H	Me	Me	—	70	178	25
9.74	9-OMe	Me	Me	—	72	138	25
9.74	7-OMe	Me	Me	—	75	165	25
9.74	H	Ph	Me	—	62	195	25
9.75	—	—	—	—	75	175	25

7-aminobenzimidazole substrates (**9.78**; X = OH or Cl) to achieve a general synthesis of imidazo[1,5,4-*de*]quinoxalines (**9.79**) (Table 9.11). In a minor modification of the method the tri-(β-hydroxyethyl)benzimidazoles (**9.80**; R = Me or SO_2NEt_2), formed from substituted 7-aminobenzimidazoles and excess ethylene oxide, can be cyclized in thionyl chloride to yield salts (**9.81**).[30]

SH $(CH_2)_2Cl$ —aq. ethanolic KOH→ 52%

(**9.76**) (**9.77**)

NH_2 $CH_2CH(R^1)X$ —−HX→ HN, H, R^1, R^2

(**9.78**) X = Cl, OH (**9.79**)

(9.80) $\xrightarrow{SOCl_2}$ (9.81)

a, R = Me (51%)
b, R = SO_2NEt_2 (60%)

The sodio derivative of 7-oxobenzimidazole (**9.82**) when treated with bromoacetophenone followed by ammonium acetate in acetic acid furnishes the tricycle (**9.84**) probably by way of the intermediate benzimidazole (**9.83**).[31] An acid-stable product formed by reacting 1,2,3-triaminobenzene with two mole equivalents of benzoic acid at 180° is claimed to be 2,4-diphenylimidazo[1,5,4-*cd*]benzimidazole (**9.86**) on the flimsy evidence that it shows no free amino group and can be recovered unchanged from boiling

TABLE 9.11. IMIDAZO[1,5,4-*de*]QUINOXALINES (**9.79**) FORMED BY CYCLIZATION OF 7-AMINO-1-(SUBSTITUTED)-BENZIMIDAZOLES (**9.78**)

Starting benzimidazole (**9.78**)				Cyclization	Product (**9.79**)			Yield	m.p.	
R	R¹	R²	X	method[a]	R	R¹	R²	(%)	(°C)	Ref.
H	H	H	Cl	*A*	H	H	H	63	162.5–166	23
H	H	H	OH	*B*	H	H	H	76	158	26
H	H	Me	OH	*B*	H	H	Me	—[b]	79–80[c]	26
									175–176[d]	26
									175–176.5[e]	26
									257–259[f]	26
H	CH_2NEt_2	Me	OH	*B*	H	CH_2NEt_2	Me	80	89–91	27
									95–98[c]	27
Cl	H	H	OH	*B*	Cl	H	H	—[b]	208–210	26
Cl	H	Me	OH	*B*	Cl	H	Me	—[b]	199–200	26
Me	H	Me	Cl	*A*	Me	H	Me	77	203–204	28
									78–79[d]	28
									188–189[g]	28
									114–115[h]	28
SO_2NEt_2	H	Me	Cl	—	SO_2NEt_2	H	Me	80	223–224	29
									181[d]	29
									197[g]	29
									225[i]	29

[a] *A* = In refluxing ethanol; *B* = Polyphosphoric acid at 185°.
[b] Yield not recorded.
[c] 6-Nitroso derivative.
[d] 6-Acetyl derivative.
[e] 6-Allylthiocarbamoyl derivative.
[f] Methiodide.
[g] 6-(β-Hydroxyethyl) derivative.
[h] 6-(β-Chloroethyl) derivative.
[i] Hydrochloride of 6-(β-chloroethyl) derivative.

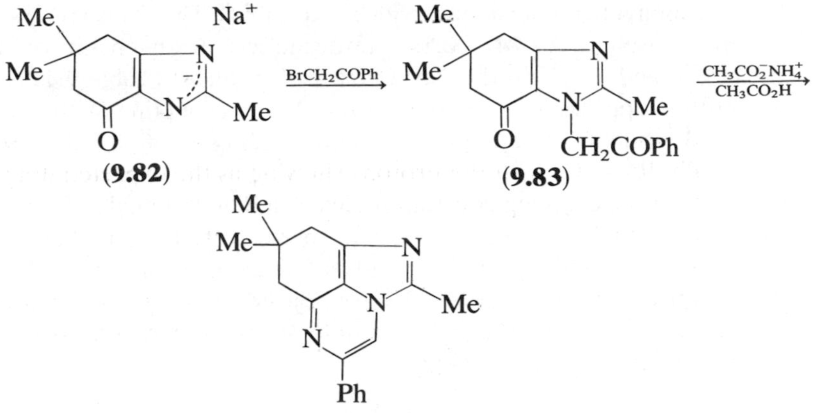

(9.82) $\xrightarrow{BrCH_2COPh}$ (9.83) $\xrightarrow[CH_3CO_2H]{CH_3CO_2^-NH_4^+}$ (9.84)

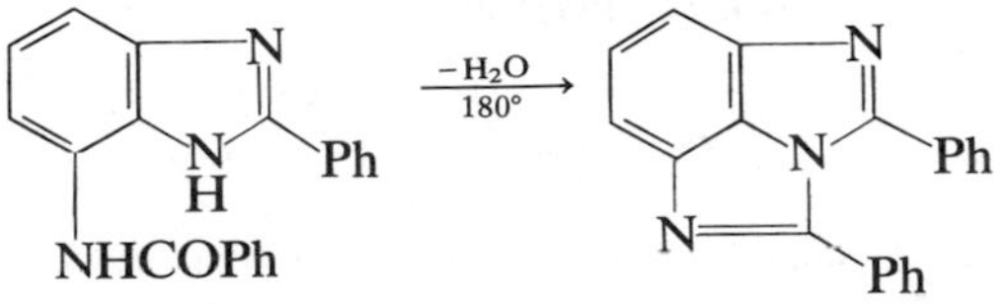

(9.85) $\xrightarrow[180°]{-H_2O}$ (9.86)

15% hydrochloric acid.[32] A plausible intermediate could be benzimidazole (9.85).

9.2.2. Physicochemical Studies and Reactions

The carbonyl absorptions in 2-substituted imidazoquinoxalin-5(6*H*)-ones (**9.68**) occur in the range 1680–1696 cm^{-1}, whereas the diones (**9.69**) show a

TABLE 9.12. ^{1}H NMR SPECTRA (δ VALUES) OF IMIDAZO[1,5,4-*de*]QUINOXALINES (*J* IN Hz)[a]

Compound	R	R¹	R²	H-2 (1*H*, s)[b]	H-4	H-5	H-6 (1*H*, s)[b]	2-Me (3*H*, s)[b]	4-Me (3*H*, d, *J* = 7)[c]	5-Me (3*H*, d, *J* = 7)[c]
9.74	H	Me	H	8.08	4.53[d]	3.59[d]	5.98	—	1.18	1.23
9.74	H	Me	Me	—	4.50[d]	3.51[d]	5.86	2.45	1.08	1.26
9.74	H	Ph	Me	—	5.77[e]	4.99[e]	6.49	2.21	—	—
9.74	7-OMe	Me	Me	—	4.48[d]	3.45[d]	5.23	2.42	1.09	1.28
9.74	9-OMe	Me	Me	—	4.42[d]	3.45[d]	5.49	2.42	1.06	1.20
9.75	—	—	—	8.30	4.98[d]	4.48[d]	—	—	0.86	1.64

[a] From Ref. 25.
[b] s = singlet.
[c] d = doublet.
[d] Doublet (*J* = 3 Hz) of quartet (*J* = 7 Hz).
[e] Doublet (*J* = 4 Hz).

broadened carbonyl band between 1650–1700 cm^{-1}.[24] The ^1H NMR spectra of imidazoquinoxalines (**9.74**) corroborate the cis arrangement of substituents at C-4 and C-5.[25] In the 4,5-dimethyl derivatives (Table 9.12) each methine proton appears as double quartet due to coupling with methyl ($J = 7$ Hz) and vicinal ($J = 3$ Hz) protons. The coupling is simplified in (**9.74**; $R = H$, $R^1 = Ph$, $R^2 = Me$) with the protons showing as the expected doublet: the magnitude of the coupling constant indicates a dihedral angle of ~60°.

No systematic study of the chemical properties of these tricyclic benzimidazoles has been attempted. The imidazoquinoxalines of type (**9.79**) are readily derivatized with a range of electrophilic reactants at the secondary amine function (N-6).[26–29] The 6-formamido substituent of (**9.75**) can be removed in 10% hydrochloric acid.[25]

REFERENCES

1. A. Richardson and E. D. Amstutz, *J. Org. Chem.*, **25,** 1138 (1960).
2. U.S. *Patent*, 3,200 123 (1965).
3. V. G. Poludnenko and A. M. Simonov, *Khim. Geterotsikl. Soedin.*, **1970,** 1410; *Chem. Abstr.*, **74,** 53657d (1971).
4. L. M. Werbel, J. Battaglia, and M. L. Zamora, *J. Heterocycl. Chem.*, **5,** 371 (1968).
5. A. M. Simonov and V. G. Poludnenko, *Khim. Geterotsikl. Soedin.*, **1972,** 242; *Chem. Abstr.*, **76,** 140645h (1972).
6. V. G. Poludnenko, A. M. Simonov, and L. G. Kogutnitskaya, *Khim. Geterotsikl. Soedin.*, **1971,** 967; *Chem. Abstr.*, **76,** 34175v (1972).
7. H. Suschitzky and M. E. Sutton, *J. Chem. Soc.* (*C*), **1968,** 3058.
8. J.-M. Kamenka and M. N. Alam, *J. Heterocycl. Chem.* **10,** 459 (1973).
9. D. S. Chothia, S. Y. Dike, A. B. Engineer, and J. R. Merchant, *Ind, J. Chem.* **14B,** 323 (1976).
10. F. I. Carroll, J. T. Blackwell, A. Philip, and C. E. Twine, *J. Medicin. Chem.*, **19,** 1111 (1976).
11. A. Richardson, *J. Org. Chem.*, **28,** 2581 (1963).
12. I. G. Il'ina, N. B. Kazennova, V. G. Bakhmutskaya, and A. P. Terent'ev, *Khim. Geterotsikl. Soedin.*, **1973,** 1112; *Chem. Abstr.*, **79,** 126396h (1973).
13. M. Hamana and S. Kumadaki, *Chem. Pharm. Bull.* (*Tokyo*), **22,** 1506 (1974).
14. O. Meth-Cohn and H. Suschitzky, *J. Chem. Soc.*, **1964,** 2609.
15. O. Meth-Cohn and H. Suschitzky, *Adv. Heterocycl. Chem.*, **14,** 211 (1972).
16. R. Garner and H. Suschitzky, *J. Chem. Soc.* (*C*), **1966,** 1572.
17. E. Ziegler, H. Junek, E. Noelken, K. Gelfert, and R. Salvador, *Monatsh. Chem.*, **92,** 814 (1961).
18. R. T. C. Brownlee, A. R. Katritzky, and R. D. Topsom, *J. Chem. Soc.* (*B*), **1966,** 726.
19. G. H. Beaven, E. R. Holiday, and E. A. Johnson, *Spectrochim. Acta*, **4,** 338 (1951).
20. A. M. Simonov and V. G. Poludnenko, *Khim. Geterotsikl. Soedin.*, **1968,** 567; *Chem. Abstr.*, **71,** 124323t (1969).
21. USSR. *Patent*, 384,854 (1973); through *Chem. Abstr.*, **80,** P21317a (1974).
22. A. V. El'tsov and V. N. Khokhlov, *Zh. Org. Khim.*. **6,** 2618 (1970); *Chem. Abstr.*, **74,** 64223k (1971).
23. W. Knoblock and G. Lietz, *J. Prakt. Chem.*, **308,** 113 (1967).
24. H. Otomasu, S. Ohmiya, H. Takahashi, K. Yoshida, and S. Sato, *Chem. Pharm. Bull.* (*Tokyo*), **21,** 353 (1973).

25. H. Otomasu, H. Takahashi, and K. Yoshida, *Chem. Pharm. Bull.* (*Toyko*), **21,** 492 (1973).
26. I. Molnár, *Chimia* (*Switz.*), **14,** 364 (1960); *Chem. Abstr.*, **55,** 9400g (1961).
27. I. Molnár, *Pharm. Acta Helv.*, **39,** 288 (1964); *Chem. Abstr.*, **62,** 1662d (1965).
28. A. Dikciuviene, V. Bieksa, and J. Degutis, *Liet. TSR Mokslu Akad. Darb.*, *Ser. B*, **1974** (No. 3), 81; through *Chem. Abstr.*, **83,** 43243n (1975).
29. A. Dikciuviene, V. Bieksa, and J. Degutis, *Liet. TSR Mokslu Akad. Darb.*, *Ser. B*, **1974** (No. 4), 83; through *Chem. Abstr.*, **85,** 21292s (1976).
30. A. Dikciuviene, V. Bieksa, and J. Degutis, *Liet. TSR Mokslu Akad. Darb.*, *Ser. B*, **1973** (No. 2), 105; through *Chem. Abstr.*, **79,** 115497r (1973).
31. V. I. Shvedov, L. B. Altukhova, and A. N. Grinev, *Khim. Geterotsikl. Soedin.*, **1972,** 131; *Chem. Abstr.*, **76,** 153710a (1972).
32. L. S. Efros, *Zh. Obshcheĭ Khim.* **23,** 957 (1953); *Chem. Abstr.*, **48,** 8223c (1954).

CHAPTER 10

Commercial Applications of Benzimidazoles

P. N. PRESTON

10.1. INTRODUCTION

In planning the layout of this book it was decided at the outset to include as a final chapter a compilation of commercially marketed benzimidazoles and congeneric tricyclic compounds. Derivatives of the latter type are extensively described in the patent literature, but there are no commercially marketed products in this category. The compounds described in this chapter thus relate to Chapters 1 and 3 only.

Benzimidazoles have been marketed extensively as pharmaceuticals, insecticides, and fungicides (section 10.2); the commercial outlook for polybenzimidazoles is considered (section 10.3), and other areas of potential commercial interest are summarized (section 10.4).

10.2. PHARMACEUTICALS, VETERINARY ANTHELMINTIC AGENTS, AND FUNGICIDES

The most widely used compounds in this group have been 2-(4-thiazolyl) derivatives and methyl benzimidazol-2-yl carbamates (see Tables 10.1 and 10.3). The successful commercial utilization of these compounds has stimulated great interest in the synthesis of closely related derivatives, and in Table 10.2 is compiled a list of patented products; it should be noted, however, that derivatives of the latter type are not yet in industrial production.

Benzimidazoles or their precursors have been used extensively as fungicides, and, although three such compounds are no longer commercially available [viz. Cypendazole[1] (**10.1**), Fenzaflor[2] (**10.2**), and Fungilon[3] (**10.3**)], a number remain in industrial production (see Table 10.3). The most important commercial product is methyl 1-(butylcarbamoyl)benzimidazol-2-ylcarbamate (Benomyl),[4] which was introduced in 1967 by E. I. du Pont de Nemours and Co. This material is a protective and eradicant fungicide with systemic activity, effective against a wide range of fungi affecting fruits, nuts, vegetables, and field crops; it is effective against mites, primarily as an ovicide. It is also used as pre- or post-harvest sprays or dips for the control of storage rots of fruits and vegetables.

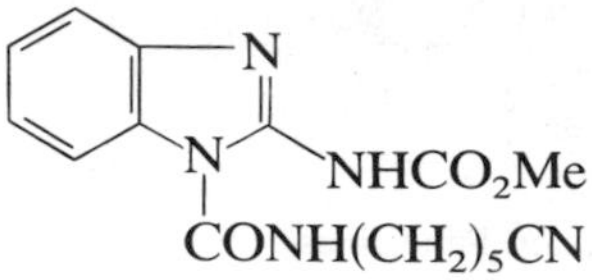

(**10.1**)

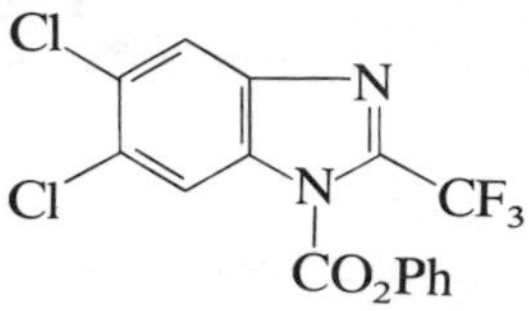

(**10.2**)

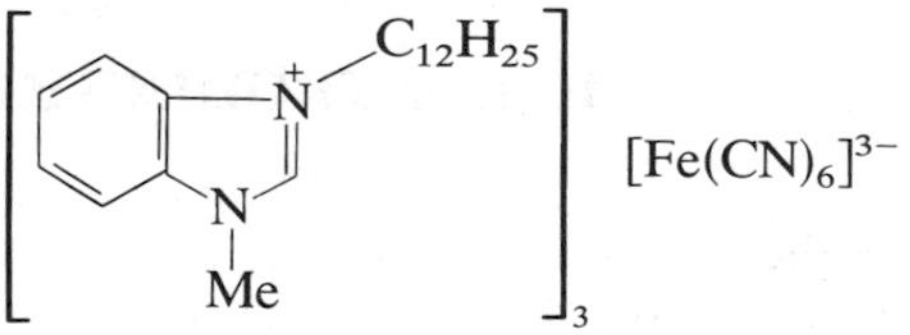

(**10.3**)

There has recently been a decline in activity in the use of benzimidazoles as pesticides, primarily because there have been many reports of pathogenic fungi acquiring resistance to benzimidazole fungicides.

10.3. POLYBENZIMIDAZOLES—OUTLOOK

The synthesis of polybenzimidazoles was pioneered by C. S. Marvel's group[5] and a number of reviews have appeared on this interesting class of heterocyclic polymers.[6,7]

Early studies in the 1920s were focused on the synthesis of polymers in which the benzimidazole moiety was attached as a side chain to the backbone and involved the use of 1-, 2-, and 5-vinylbenzimidazoles. Of more current interest are polymers in which the heterocyclic system forms part of the backbone, and these (cf. **10.4**) are readily prepared in the benzimidazole series by allowing tetramine monomers to react with dicarboxylic acids or their derivatives. Polymers of the latter type exhibit a high

TABLE 10.1. COMMERCIALLY AVAILABLE PHARMACEUTICALS AND VETERINARY ANTHELMINTICS

Approved name	Proprietary name	By whom marketed	Where marketed	Use	Structure
Thiabendazole	Mintezol	Merck Sharpe and Dohme		Human anthelmintic	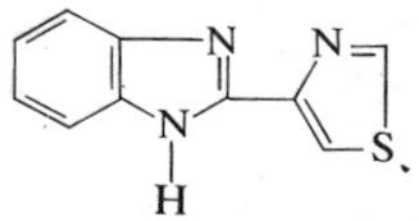 N, N, N, S, H
	Thibenzole	Merck Sharpe and Dohme		Veterinary anthelmintic	
Mebendazole	Telmin	Crown Chemicals	U.K., Ireland	Veterinary and human anthelmintic	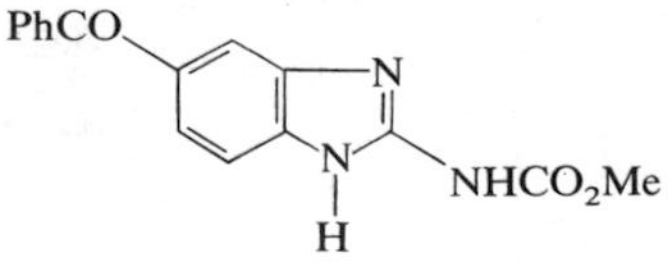 PhCO, N, N, H, $NHCO_2Me$
	Equiverm plus	Crown Chemicals	U.K.		
	Vermox	Janssen	Ireland[a] U.K.[b]		
		Ortho Pharmaceutical Corp.	West Germany[c] U.S.A.[d]		
	Pantelmin	Johnson & Johnson	Brazil[e]		
	Mebenvet	Janssen	West Germany[f]		
	Mebutor	Andromaco	Argentina[g]		
	Sirben	Andromaco	Brazil[h]		
	Ovitelmin	Crown Chemicals	U.K.[i]		
	Vermirax	Biosintetica	Brazil[j]		
	Nemasole	Janssen	Argentina[k]		
Cambendazole		Merck Sharpe and Dohme		Veterinary anthelmintic	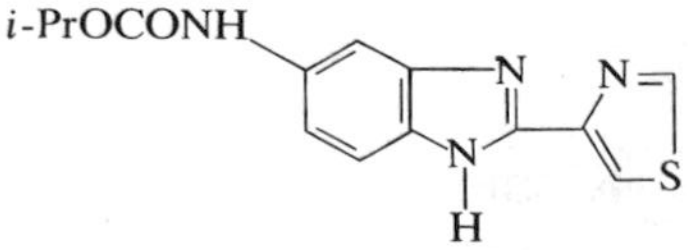 *i*-PrOCONH, N, N, N, S, H
Parbendazole	Helmatac	Smith Kline and French Laboratories		Veterinary anthelmintic	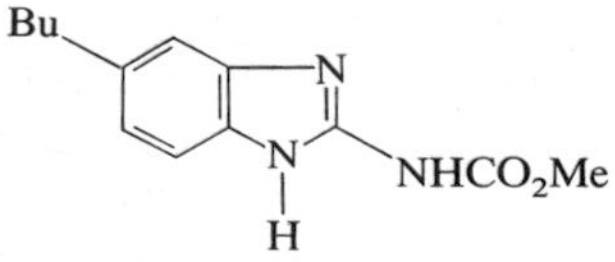 Bu, N, N, H, $NHCO_2Me$
	Neminil	Hydro-Chemie	West Germany[l]		

TABLE 10.1 (*Continued*)

Approved name	Proprietary name	By whom marketed	Where marketed	Use	Structure
Fenbendazole	Panacur	Hoechst[m]		Veterinary anthelmintic	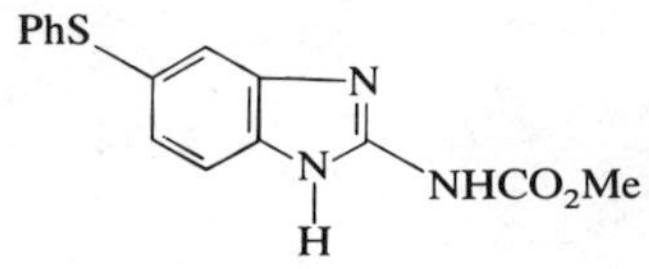 PhS, N, N–H, $NHCO_2Me$
Oxibendazole	Loditac	Silke Pharmaceuticals Ltd. (for Smith, Kline and French)	Ireland[n]	Veterinary anthelmintic	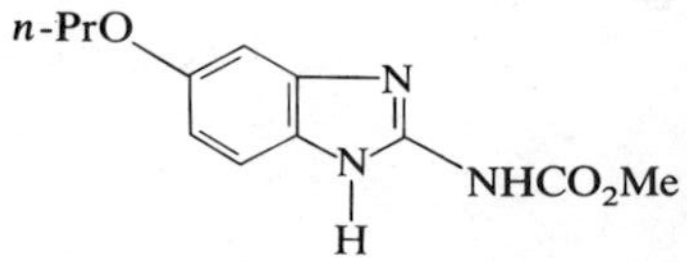 *n*-PrO, N, N–H, $NHCO_2Me$
Albendazole		Smith Kline and French Laboratories		Veterinary anthelmintic	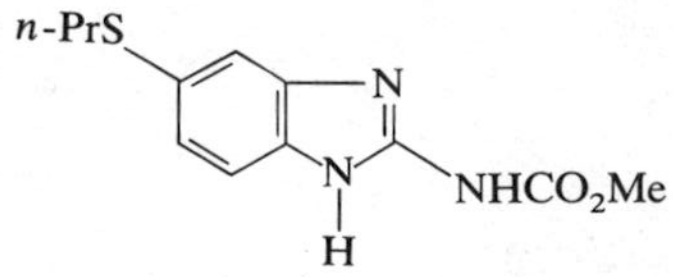 *n*-PrS, N, N–H, $NHCO_2Me$
Oxfendazole		Syntex		Veterinary anthelmintic	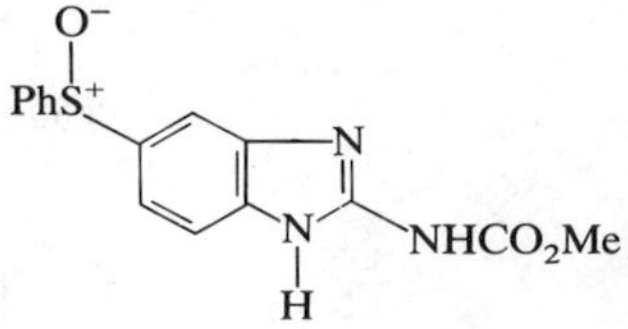 O^-, PhS^+, N, N–H, $NHCO_2Me$
Nocodazole		Janssen		Antineoplastic agent Microtubule inhibitor	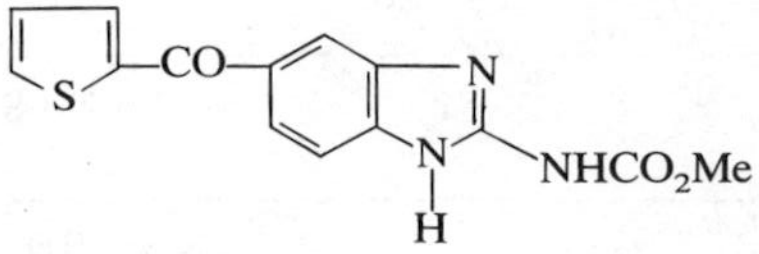 S, CO, N, N–H, $NHCO_2Me$

Flubendazole		Janssen		Antineoplastic agent Microtubule inhibitor	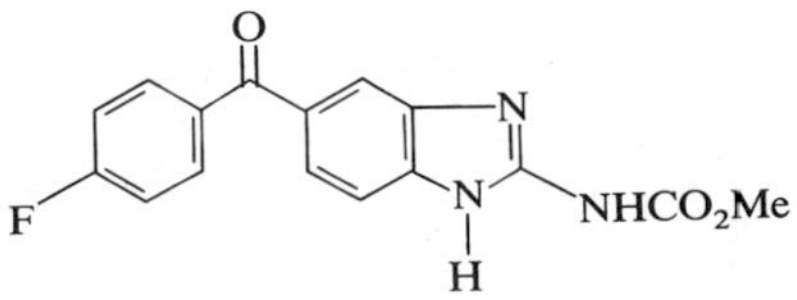 F–C_6H_4–CO–(benzimidazole)–$NHCO_2Me$, N–H
Bezitramide	Burgodin	Janssen		Analgesic	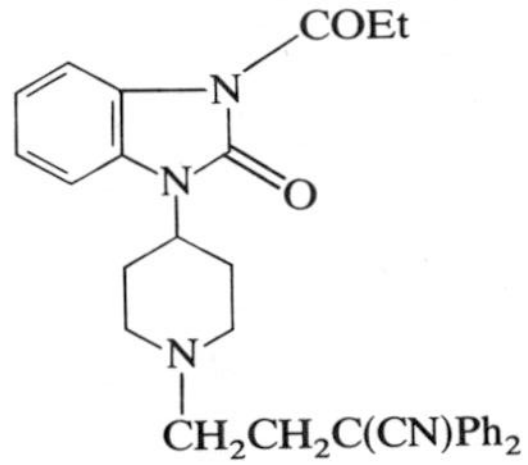 N-COEt benzimidazolone, N-piperidinyl–N–$CH_2CH_2C(CN)Ph_2$
Imet 3393	Cytostasan		Germany	Anticancer	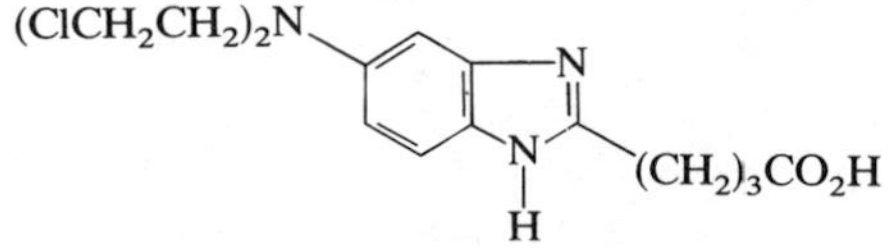 $(ClCH_2CH_2)_2N$–benzimidazole–$(CH_2)_3CO_2H$, N–H
Clemizole penicillin	Bellocillin	Laboratories Roger Bellon	France	Bactericide	Combination of benzylpenicillin and 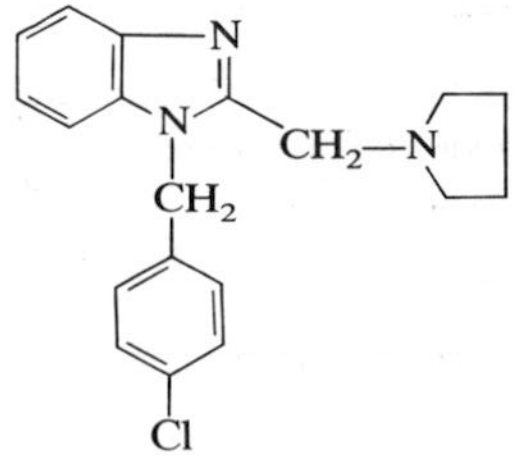
	Megacillin	Grünenthal	Germany		benzimidazole–CH_2–N(pyrrolidine), N–CH_2–C_6H_4–Cl

TABLE 10.1 (*Continued*)

Approved name	Proprietary name	By whom marketed	Where marketed	Use	Structure
					Combination of benzylpenicillin and
Clemizole	Allercur Hérol Histacur Histacuran Reactrol	Schering Chemicals		Antihistamine	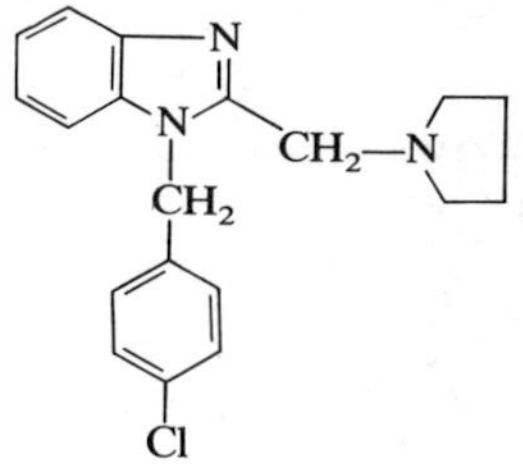 N, N, CH_2—N, CH_2, Cl
Diabazole Diabazol	Tromasédan	Laboratoires Millot	France	Vasodilator spasmolytic hypotensive	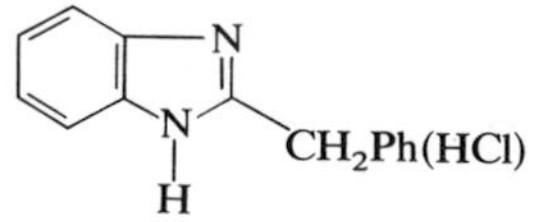 N, N, H, $CH_2Ph(HCl)$
Benperidol	Frénactil Glianimon	Laboratoires Clin-Comar-Byla Troponwerke Dinklage	France Germany	Pyschopharma-cological agent	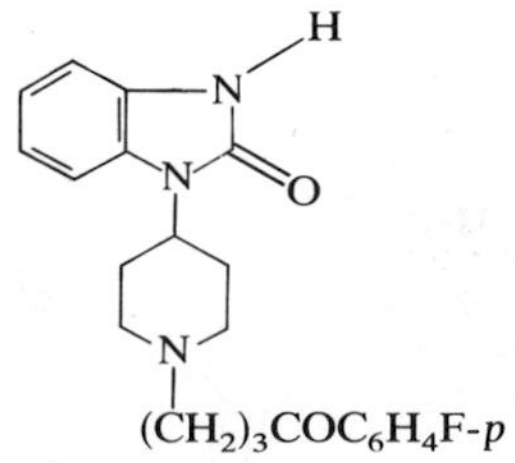 H, N, N, O, N, $(CH_2)_3COC_6H_4F$-*p*

Pimozide	Orap	Janssen	U.K.	Pyschopharma-cological agent	benzimidazolone–piperidine–N–$(CH_2)_3CH(C_6H_4F\text{-}p)_2$
	Opiran	Laboratoires Cassenne	France		
Droperidol	Droleptan	Janssen		Pyschopharma-cological agent	benzimidazolone–tetrahydropyridine–N–$(CH_2)_3COC_6H_4F\text{-}p$
	Inapsine	McNeil	U.S.		

[a] *Irish Pharmacy Journal*, **52,** 307 (1974).
[b] *Chemist and Druggist*, 28th August, 248 (1976).
[c] *Inpharma*, 5th March, 14 (1977).
[d] *Scrip*, 22nd February, (**144**), 15 (1975).
[e] *Unlisted Drugs*, **25,** 46i (1973).
[f] *Tieraerztliche Umschau*, **29,** 603 (1974) [Advert].
[g] *Unlisted Drugs*, **27,** 8e (1975).
[h] *Unlisted Drugs*, **27,** 75j, (1975).
[i] *British Farmer and Stockbreeder*, 20th March, 35 (1976).
[j] *Unlisted Drugs*, **28,** 77m (1976).
[k] *Inpharma*, 2nd October, 18 (1976).
[l] *Tieraerztliche Umschau*, **28,** 242 (1973).
[m] *Deutsche Tieraerztliche Wochenschrift*, **81,** 177 (1974).
[n] *Irish Pharmacy Journal*, **55,** 295 (1977) [Advert].

TABLE 10.2. PATENTED BENZIMIDAZOLES OF INTEREST AS POTENTIAL PHARMACEUTICALS AND VETERINARY ANTHELMINTICS (NOT COMMERCIALLY AVAILABLE)

Potential use	By whom patented	Structure[a]	Patent number or *Derwent Abstract* reference
Anthelmintic	Hoechst	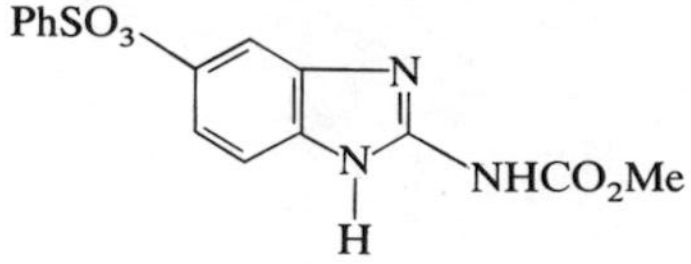 $PhSO_3$, N, N–H, $NHCO_2Me$	DT 2541–752
Anthelmintic	Meji Seika Kiosha Ltd.	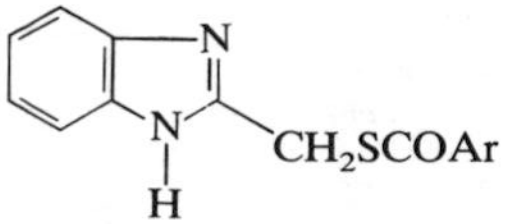 N, N–H, CH_2SCOAr	JA 4028511
Anthelmintic	Fisons	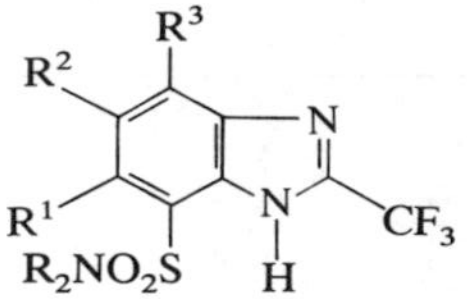 R^3, R^2, R^1, R_2NO_2S, N, N–H, CF_3 R^1, R^2, R^3 = halogeno	NL 74 02438
Anthelmintic	Squibb	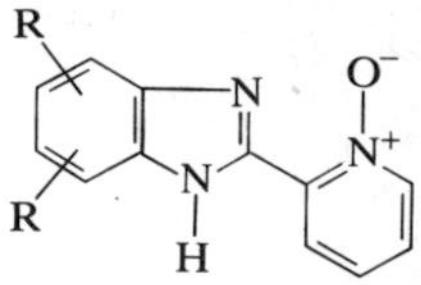 R, R, N, N–H, O^-, N^+	US 3,864,350
Anthelmintic	Squibb	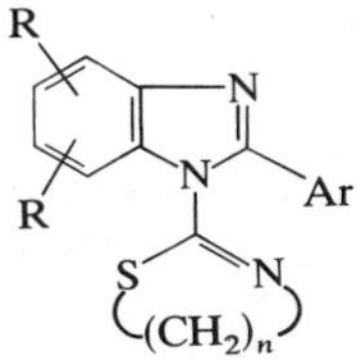 R, R, N, N, Ar, S, N, $(CH_2)_n$	DT 2446–259

Anthelmintic	Smith Kline and French	5-$R^1(CH_2)_nX$-benzimidazole-2-$NHCO_2Me$ (1-H) R^1 = furyl, thienyl X = O, S n = 0–4	US 3,969,526
Anthelmintic	Roche	5-R^1CO-benzimidazole-2-$NHCO_2Me$ (1-H) R^1 = 2-pyridyl	BE 844949
Antiviral	Lilly	5-R(C=Z)-benzimidazole-2-NHR (1-SO_2R) Z = O or NOR	BE 845641
Active vs poultry blackhead	Roche	5-ArX-6-Cl-benzimidazole-2-$NHCO_2R$ (1-H) X = O, S	DT 2606 531

[a] R and Ar denote alkyl and aryl, respectively.

TABLE 10.3. COMMERCIALLY AVAILABLE BENZIMIDAZOLE FUNGICIDES

Approved name	Proprietary name	By whom marketed	Structure
Thiophanate methyl[a]	Cercobin methyl or Tospsin methyl	Nippon Soda	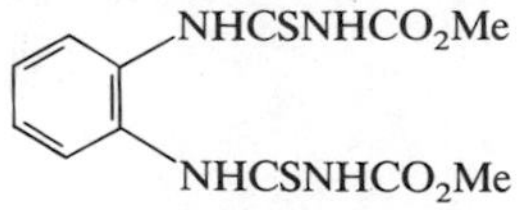 $C_6H_4(NHCSNHCO_2Me)_2$
Thiophanate[a]	Cercobin or Tospsin	Nippon Soda	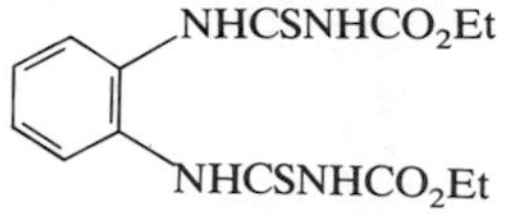 $C_6H_4(NHCSNHCO_2Et)_2$
Benomyl	Benlate	Dupont	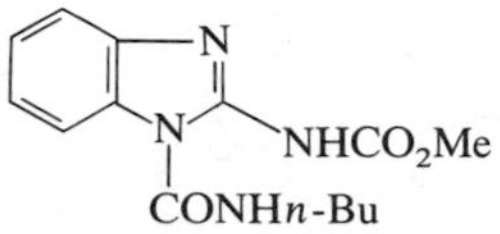 N; N; $NHCO_2Me$; CONH*n*-Bu
Fuberidazole	Voronit	Bayer	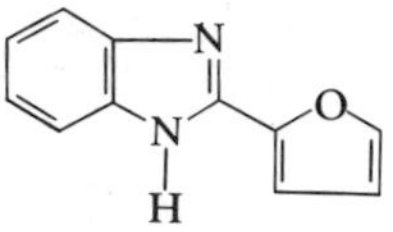 N; N; H; O
Thiabendazole	Mertect (TBZ)	Merck Sharpe and Dohme	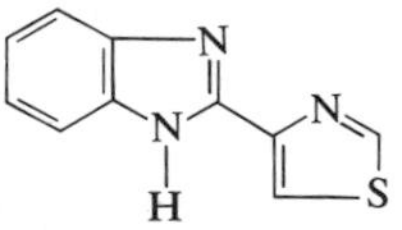 N; N; H; N; S

[a] These compounds are converted *in vivo* (and also under some *in vitro* conditions) to benzimidazole-2-carbamates (cf. H. Buchenauer, L. V. Edgington, and F. Grossman, *Pestic. Sci.*, **4**, 343 (1973).

degree of thermal and chemical stability: if the X group is aliphatic, the polymer degrades above 350°C, but all-aromatic polymers are stable beyond ca. 500°C in an atmosphere of nitrogen. The material of greatest potential in the series has been prepared by condensation of 3,3′,4,4′-tetraminobiphenyl with diphenylisophthalate.[5] This polymer (PBI) has a variety of valuable characteristics, including the following: high thermal stability, good adhesive properties, useful cryogenic properties, and high moisture regain. PBI has been used by Celanese Corporation[8] to prepare textile fibers and these have been used in the space program by U.S. Air Force Materials Laboratory, NASA. Among the valuable properties of this fiber are high moisture regain and nonflammability.

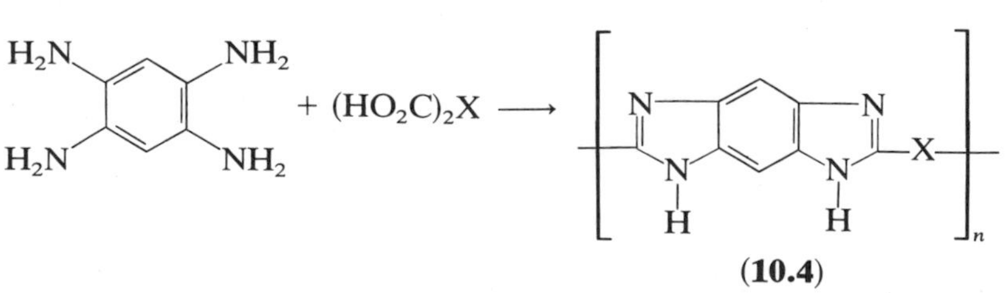

H_2N NH_2 H_2N NH_2 $+ (HO_2C)_2X \longrightarrow$ [N N N N X H H]$_n$

(10.4)

In summary, it appears that PBI has commercial potential, the most likely outlet being in the field of synthetic fiber production.

10.4. MISCELLANEOUS AREAS OF POTENTIAL COMMERCIAL INTEREST

Benzimidazole derivatives have been evaluated for use in the following areas, but to the knowledge of the author there are no materials on the market: corrosion inhibitors for copper,[9–12] brass[10,13,14] and aluminum;[15] photosensitive compounds for use in photothermography[16,17] and for the preparation of silver-free photographic films;[18,19] polymer additives as stabilizing[20] and shrinking[21] agents; and as fluorescent brighteners.[22] One benzimidazole-substituted azo compound (**10.5**) is cited in the Colour

(di- and tri-sulfonated)

(**10.5**)

Colour Index 14150 [Chrome Fast Yellow G(A)]

Index,[23] but this chrome mordant dye is now obsolete. The most likely commercial outlets for benzimidazoles in this field are as merocyanine derivatives for use as sensitizing dyes.[24]

REFERENCES

1. Marketed until recently by Bayer A. G. Leverkusen as Folcidin (cf. *U.S. Patent*, 3,673,210).
2. Marketed until recently by Fisons Pest Control as Lovazal (cf. D. T. Saggers and M. L. Clark, *Nature*, **215,** 275 (1967).
3. Marketed until recently by Farbenfabriken Bayer A.G. under code number Bayer 32394 as Fungilon (cf. *Fr. Patent*, 1,248,412).
4. See C. J. Delp and H. L. Klopping, *Pl. Dis. Reptr.*, **52,** 95 (1968); also *Netherlands Patent*, 6,706,331 and *U.S. Patent*, 3,631,176.
5. C. S. Marvel, *S.P.E. J*, **20,** 220 (March 1964).
6. V. V. Korshak and M. M. Teplyakov, *J. Macromol. Sci. Rev. Macromol. Chem.*, **5,** 409 (1971).
7. J. R. Leal, *Modern Plastics*, (August), 60 (1975) and review articles cited therein.
8. Celanese Research Company, Summit, N.J., U.S.A. (A division of Celanese Corporation).

9. A. B. Patel, N. K. Patel, and J. C. Vora, *Labdev*, Part A, **12A,** 86 (1974); *Chem. Abstr.*, **85,** 37062c.
10. N. K. Patel, J. Franco, and S. H. Mehta, *Chem. Era*, **11,** 9 (1975); *Chem. Abstr.*, **85,** 111793k.
11. G. Trabanelli, F. Zucchi, G. Brunoro, and V. Carassiti, *Proc. Int. Congr. Met. Corros. 5th*, 565 (1974); *Chem. Abstr.*, **84,** 81514e.
12. H. Tobe, N. Morito, K. Monma, and W. Suetaka, *Nippon Kinzoku Gakkaishi*, **38,** 770 (1974); *Chem. Abstr.*, **82,** 20777e.
13. N. K. Patel, *J. Inst. Chem. Calcutta*, **46,** 137 (1974); *Chem. Abstr.*, **82,** 128373j.
14. N. K. Patel, S. C. Makwana, and M. M. Patel, *Corros. Sci.*, **14,** 91 (1974); *Chem. Abstr.*, **81,** 15923z.
15. R. K. Shah, B. B. Patel, and N. K. Patel, *J. Electrochem. Soc. India*, **24,** 139 (1975); *Chem. Abstr.*, **84,** 109924q.
16. M. Sasaki, T. Kazami, A. Noguchi, Y. Tsujimoto, T. Yamamuro, and T. Saito, *Jap. Patent*, 76 01,114 (1976); *Chem. Abstr.*, **85,** 151783p.
17. M. Sasaki, T. Kazami, A. Nosuchi, Y. Tsujimoto, T. Yamamuro, and T. Saito, *Jap. Patent*, 76 01,113 (1976); *Chem. Abstr.*, **85,** 70696k.
18. M. Sasaki, T. Kazami, A. Noguchi, Y. Tsujimoto, T. Yamamuro, and T. Saito, *Jap. Patent*, 76 01,112 (1976); *Chem. Abstr.*, **85,** 151784q.
19. G. L. Eian, *Ger. Patent*, 2,433,831 (1975); *Chem. Abstr.*, **84,** 67887x.
20. S. I. Sadykh-Zade, R. M. Mamedov, P. K. Mamedova, A. I. Elchueva, A. P. Dzhafarov, and M. F. Ganieva, *USSR Patent*, 515,765 (1976); *Chem. Abstr.*, **85,** 64149t.
21. L. B. Palmer and R. P. Conger, *U.S. Patent*, 3,849,158 (1974); *Chem. Abstr.*, **82,** 99930f.
22. D. Guenther, G. Hitschfel, H. J. Nestler, and G. Riesch, *Ger. Patent*, 2,320,528 (1974); *Chem. Abstr.*, **83,** 61739h.
23. Colour Index (3rd ed.), Vol. 4. Society of Dyers and Colourists, Bradford, U.K.
24 (a) D. M. Sturmer and D. W. Heseltine in *The Theory of the Photographic Process*, 4th ed. (T. H. James, Ed.), Macmillan, New York, 1977, Chap. 8. (b) H. Meier, *Spectral Sensitization*, Focal Press, New York, p. 61. (c) G. F. Duffin, in *Dye Senzitization* (W. F. Berg, V. Mazzucato, H. Meier, and G. Semerano, Eds.), Focal Press, New York, 1970, p. 282.

Author Index

Numbers in *italics* indicate pages where reference appear.

Subject Index